Lecture Notes in Artificial Intelligence   827

Subseries of Lecture Notes in Computer Science
Edited by J. G. Carbonell and J. Siekmann

Lecture Notes in Computer Science
Edited by G. Goos and J. Hartmanis

Dov M. Gabbay   Hans Jürgen Ohlbach (Eds.)

Temporal Logic

First International Conference, ICTL '94
Bonn, Germany, July 11-14, 1994
Proceedings

Springer-Verlag
Berlin Heidelberg New York
London Paris Tokyo
Hong Kong Barcelona
Budapest

Series Editors

Jaime G. Carbonell
School of Computer Science, Carnegie Mellon University
Schenley Park, Pittsburgh, PA 15213-3890, USA

Jörg Siekmann
University of Saarland
German Research Center for Artificial Intelligence (DFKI)
Stuhlsatzenhausweg 3, D-66123 Saarbrücken, Germany

Volume Editors

Dov M. Gabbay
Department of Computing, Imperial College of Science, Technology and Medicine
180 Queen's Gate, London SW7 2AZ, United Kingdom

Hans Jürgen Ohlbach
Max-Planck-Institut für Informatik
Im Stadtwald, D-66123 Saarbrücken, Germany

CR Subject Classification (1991): I.2.3, I.2.8, F.4.1, F.3.1, H.2

ISBN 3-540-58241-X Springer-Verlag Berlin Heidelberg New York
ISBN 0-387-58241-X Springer-Verlag New York Berlin Heidelberg

CIP data applied for

This work is subject to copyright. All rights are reserved, whether the whole or part of the material is concerned, specifically the rights of translation, reprinting, re-use of illustrations, recitation, broadcasting, reproduction on microfilms or in any other way, and storage in data banks. Duplication of this publication or parts thereof is permitted only under the provisions of the German Copyright Law of September 9, 1965, in its current version, and permission for use must always be obtained from Springer-Verlag. Violations are liable for prosecution under the German Copyright Law.

© Springer-Verlag Berlin Heidelberg 1994
Printed in Germany

Typesetting: Camera ready by author
SPIN: 10472607   45/3140-543210 - Printed on acid-free paper

Preface

Time is a very exciting subject. It is one of the few subjects on which everyone is an expert. We move through time continuously and in order to survive and manage ourselves sensibly we constantly have to make temporal decisions. Philosophy, since the days of Aristotle, has been trying to analyse the way we make these decisions. With the rise of computer science, where ideally one wants the machine to do the job for or of the human, there is a new urgency in the precise logical analysis of human temporal activity.

Human (and hence computer) time related activity can be divided into several main areas, all very familiar to us. One of the simplest, and the most important area, is our handling of time dependent data. In computing this is the area of databases. To us ordinary people, it is just time dependent information, involving questions like when to go to the dentist, when to pick up the child from school, until when can one delay in not filing one's tax return and so on. There is another temporal dimension involved in the area of time dependent data besides direct dependency on time. This is the dimension of when a data item is presented to us. For example, if we get a bill to pay our tax on January 1st 1990, it is important when the bill was sent or received, e.g. received September 1989 for January 1990. In database terms there are two times involved: the time dependency of the data and the time when it was introduced into the database. Surprisingly, computing is only now beginning to cope with such things.

Another important area in both human activity and computing is planning. If I have to do the shopping and take my child to visit a friend and cook supper, I have to organise the sequence properly. Going shopping is a simple planning problem but to organise an airport is a more complex planning problem. To be able to let the computer solve it for us we need to develop a logical theory and correctly analyse the steps involved.

Everybody knows the term "time sharing"; what it means in practice is that if neither of us can afford something (e.g. a car or a flat in Spain) then we buy it together and time share. (Computers are more humble, they share things like memory or a printer in order to be more efficient.) We can formulate some intuitive principles on how to share (in computing this is called the *specification*) but there is always the question of exactly how we are going to manage it (what dates am I going to be in the flat and what dates are you, who is going to do the garden and collect the garbage etc.). This is the *implementation* of the principles. Given such an implementation, we have the problem of how to show that it is fair and square and satisfies the specification. One way of doing it is to formulate the procedures in "temporal" logic and then formally prove that it satisfies the specification. In computing the official name is program *specification and verification*.

The main present day research areas of temporal logic are:

1. Philosophical applications. Temporal logic is used in philosophy to clarify various concepts which have been studied since the time of Aristotle. Some examples are causality, historical necessity, identity through time, the notions of events and actions, etc.
2. Applications in computer science as described above.
3. Natural language. Logical analysis of the use of tense and aspect in natural languages. Logical time models for natural language.
4. Pure logical study of temporal logic within the framework of logic itself. Special topics here include:
   (a) Axiom systems, theorem proving and proof theory. Decidability. Model theory.
   (b) Expressive power of temporal languages.
   (c) Applications of temporal logic to the pure theory of other logics (e.g. the notion of provability as a modal logic etc.)
   (d) Deductive reasoning involving time.

   To computer science all the above four aspects of the pure logical theory are of great importance.

Temporal logics can be presented in several different ways:

1. Use predicate logic with an additional parameter for time.
2. Use special temporal logics to express temporal phenomena. There are two methods of presentation here.
   (a) Semantic presentation.
   (b) Presentation using axiomatic or other deductive systems for the connectives.
3. The final method is via direct reference to events.

Temporal logic has changed and developed incredibly since its conception as a discipline by Arthur Prior thirty years ago. It is studied by many researchers of numerous and different backgrounds. Different research groups have different conceptions of what temporal logic is and of what it is exactly they themselves do. On many occasions we have heard comments like "that is not logic" referring to a system presented by a colleague. The subject is certainly in a state of accelerated dynamic growth and a new orientation and point of view is currently needed as well as a good coverage of its mathematical and computational aspects. A good understanding, communication and cooperation will enable the subject and the community of researchers to face the challenges of the future.

The conference is the first *international conference* particularly dedicated to temporal logic. It started with four tutorials: *Programming with Temporal Logics* by Michael Fisher, Manchester Metropolitan University, England, *Incorporating Time in Databases* by Vram Kouramajian, Huston, Texas, USA, *Verification of Finite-State Systems* by Orna Grumberg, The Technion, Haifa, Israel and *Reasoning about Action and Change – Temporal Reasoning in AI* by Erik Sandewall, Linköping University, Sweden. The four invited lectures were given by some of the leading researchers in temporal logic, Johan van Benthem, Hans Kamp, James F. Allen and Amir Pnueli. The presentation of the technical papers included in this volume was accompanied by a workshop with more informal and spontaneous contributions.

We are indebted to the program committee for their effort and thought in organizing the program, to the invited speakers and to the presenters of the tutorials. Our special thanks go to the colleagues and secretaries for their support in organizing this conference: Mark Reynolds, Ruy de Queiroz, Lydia Rivlin, Janice Londsdale, Ellen Fries and Christine Kiesel, and in particular to Christine Harms, who has been an invaluable help ensuring that the event ran smoothly.

April 1994   Dov Gabbay and Hans Jürgen Ohlbach

Program Committee

James Allen	(University of Rochester, USA)
Howard Barringer	(University of Manchester, UK)
Johan van Benthem	(University of Amsterdam, The Netherlands)
Gerd Brewka	(GMD, Bonn, Germany)
Edmund Clarke	(Carnegie-Mellon Univ, USA)
Nissim Francez	(Technion, Israel)
Dov M. Gabbay	(Imperial College, London)
Michael Georgeff	(AAII, Australia)
Hans Kamp	(IMS, Stuttgart)
Istvan Nemeti	(Math Institute, Hungary)
Hans Jürgen Ohlbach	(Max-Planck-Institut, Saarbrücken)
Amir Pnueli	(Weizmann Institute, Israel)
Antonio Porto	(Univ Nova de Lisboa, Portugal)
Willem de Roever	(Kiel University, Germany)
Eric Sandewall	(Linköping University, Sweden)
Amílcar Sernadas	(INESC, Portugal)
Yoav Shoham	(Stanford, USA)
Andrzej Szalas	(University of Warsaw, Poland)

Additional Reviewers

B. Bani-Eqbal
O. Bernholtz
C. Brzoska
C. Caleiro
L. Fariñas del Cerro
G. Ferguson
M. Fisher
N. Foo
M. Grabowski
G. Gough
E. Hajnicz
J. Harland
D. Hutter
S. Katz
J. Lang

G. Lakemeyer
R. Li
B. Monahan
D. Niwinski
A. Nonnengart
A. Rao
P. Resende
C. Ribeiro
R. Ronnquist
M. Ryan
C. Sernadas
G. Tidhar
I. Walukiewicz
A. Williams

Contents

Combining Temporal Specification Techniques — 1
M.U. Sørensen, O.E. Hansen and H.H. Løvengreen

Global Equivalence Proofs for ISTL — 17
S. Katz

A Real Time Process Logic — 30
J.C.M. Baeten, J.A. Bergstra and R.N. Bol

Sometimes "Tomorrow" is "Sometime" – — 48
Action Refinement in a Temporal Logic of Objects
J.L. Fiadeiro and T. Maibaum

Applications of Transaction Logic to Knowledge Representation — 67
A.J. Bonner and M. Kifer

Circumscribing Features and Fluents — 82
P. Doherty and W. Lukaszewicz

Dealing with Time Granularity in a Temporal Planning System — 101
S. Badaloni and M. Berati

Axiomatizing U and S over Integer Time — 117
M. Reynolds

Temporal Logic with Reference Pointers — 133
V. Goranko

Completeness through Flatness in Two–Dimensional Temporal Logic — 149
Y. Venema

Efficient Computation of Nested Fix-Points, with Applications to Model Checking — 165
B. Vergauwen, J. Lewi, I. Avau and A. Poté

How Linear Can Branching-Time Be? — 180
O. Grumberg and R.P. Kurshan

First-Order Future Interval Logic — 195
G. Kutty, L.E. Moser, P.M. Melliar-Smith, L.K. Dillon and Y.S. Ramakrishna

Buy One, Get One Free !!! — 210
O. Bernholtz and O. Grumberg

Back and Forth Through Time and Events 225
P. Blackburn, C. Gardent and M. de Rijke

Interpreting Tense, Aspect and Time Adverbials: 238
A Compositional, Unified Approach
C.H. Hwang and L.K. Schubert

Synchronized Histories in Prior-Thomason Representation of 265
Branching Time
M.C. Di Maio and A. Zanardo

On the Completeness of Temporal Database Query Languages 283
M. Böhlen and R. Marti

The Abductive Event Calculus as a General Framework for 301
Temporal Databases
K. Van Belleghem, M. Denecker and D. De Schreye

A Decision Procedure for a Temporal Belief Logic 317
M. Wooldridge and M. Fisher

Decidability of Deliberative *Stit* Theories with Multiple Agents 332
M. Xu

Abduction in Temporal Reasoning 349
C. Ribeiro and A. Porto

A Temporal Logic Approach to Implementation and Refinement 365
in Timed Petri Nets
M. Felder and A. Morzenti

A Stuttering Closed Temporal Logic for Modular Reasoning about 382
Concurrent Programs
A. Mokkedem and D. Méry

A Hierarchy of Partial Order Temporal Properties 398
M. Kwiatkowska, D. Peled and W. Penczek

A Graph-Based Approach to Resolution in Temporal Logic 415
C. Dixon, M. Fisher and H. Barringer

Annotation-Based Deduction in Temporal Logic 430
H. McGuire, Z. Manna and R. Waldinger

Survey Papers

An Overview of Temporal and Modal Logic Programming   445
 M.A. Orgun and W. Ma

A Survey of Concurrent METATEM – The Language and its Applications   480
 M. Fisher

Temporal Query Languages: A Survey   506
 J. Chomicki

Position Papers and System Descriptions

Improving Temporal Logic Tableaux Using Integer Constraints   535
 R. Hähnle and O. Ibens

A System for Automated Deduction in Graphical Interval Logic   540
 P.M. Melliar-Smith, L.E. Moser, Y.S. Ramakrishna, G. Kutty
 and L.K. Dillon

$SCDBR$: A Reasoner for Specifications in the Situation Calculus of 543
Database Updates
 L.E. Bertossi and J.C. Ferretti

Author Index   546

Combining Temporal Specification Techniques

Morten Ulrik Sørensen   Odd Erik Hansen   Hans Henrik Løvengreen

Department of Computer Science, Technical University of Denmark
Building 244, DK-2800 Lyngby, Denmark
E-mail: mus@id.dtu.dk, odd@id.dtu.dk, hhl@id.dtu.dk

Abstract. This article presents a combination of different temporal specification and verification techniques for real time systems. We develop a semantic model that can be used as a model for both transition based formalisms like TLA and the temporal logics by Manna and Pnueli as well as interval based formalisms like the Duration Calculus. On this model a new temporal logic, TLD, is defined, which combines the strengths of both kinds of formalisms by including all the essential operators of all the considered logics, the meaning of the operators being preserved. Finally, a general framework for relating logics is established, and we prove that TLD is indeed an extension of both DC and TLA.

1 Introduction

In recent years, a number of logics for specifying and verifying real-time systems have been proposed. Some of these can be seen as encodings of real-time within existing untimed logics. Of these, we shall focus on classic linear-time temporal logic [9, 8] extended with a clock-variable as in [3, 10, 1]. Other logics introduce new, radically different models to capture real-time behaviour. Here, we shall concentrate on the logic of Duration Calculus (DC) [12].

Different formalisms have different strengths. For classic temporal logic, techniques have been established for proving safety and liveness properties of reactive systems and for proving refinement. These techniques are often *transition based*, in the sense that the proof can be reduced to a number of simple first-order properties of the individual state transitions of the system.

Duration Calculus, on the other hand, has proven very versatile for specifying real-time properties [12]. Typically, such specifications constrain the duration of certain patterns of state. Furthermore, pattern-based techniques for verifying real-time properties are emerging.

In our work, we would like to combine the ability to reason about functional properties found in the transition based logics with the power of real-time specification found in DC. One approach would be to define a new semantic model and a new logic, borrowing ideas from both worlds. Such an approach, however, does not guarantee that results and tools already developed can be reused in a meaningful manner. Instead, we aim for a combined logic which can be proven to be a *conservative extension* of each of the constituent logics—syntactically as well as semantically.

Such a conservative combination of classic linear-time temporal logic, represented by Lamport's Temporal Logic of Actions (TLA), and DC has been carried out in full detail in [2]. That report presents the various logics and accompanying models, shows their relationships and provides a number of applications. In the present paper, we concentrate on the logics and the framework for comparing them, leaving a presentation of the applications for another paper.

Following the introduction, Sect. 2 gives a brief introduction to the Duration Calculus (Sect. 2.1) and the Temporal Logic of Actions (Sect. 2.2), and a more elaborate introduction to the common model (Sect. 2.3) and the combined logic (Sect. 2.4). For the purpose of relating the logics it has been necessary to redefine the semantics of the logics in a uniform manner. In Sect. 3 we consider the issue of establishing a formal relation between temporal logics. First, in Sect. 3.1, we develop a framework for relating logics, and following, in Sections 3.2 and 3.3, we relate TLD to DC and TLA in this framework. Finally, in Sect. 4 we sum up the results and point out subjects for further work.

2 Logics

Later we will establish a formal relation between the existing formalisms, Duration Calculus and TLA, and the logic we are about to define, TLD, but in order to make this relation simpler we shall start by presenting the semantics of DC and TLA in a uniform and careful manner. For introductions to either formalism, refer to [12] or [8] respectively.

All the considered logics are based on an underlying first-order logic which we assume is defined in a suitable specification language like Z or VDM, [14, 5]. The domain of values for this first order logic is Val, and it contains at least the natural numbers and the usual arithmetic. Constants are considered 0-ary functions. We also assume a set Rigid of rigid variable names, and a set Flex of flexible variable names, and to each variable name v we assume a related type T_v. Common to the models for the logics is also that a *valuation* is a mapping from the rigid variable names to values, $\mathcal{V}$: Rigid $\to$ Val, and a *state* is a mapping from the flexible variable names to values, s : Flex $\to$ Val. The mappings are assumed to respect the types of the variables. We omit the formal semantics of state expressions in the logics, merely noting that the interpretation of the rigid variables is given by the valuation, the interpretation of the flexible variables is given by a state (the first state if unprimed, the second if primed), and functions are interpreted by evaluating the function with respect to the interpretation of the arguments.

2.1 Duration Calculus

Formulas $\mathcal{D}$ of the Duration Calculus are built from atomic duration formulas, duration terms r, and state expressions e in the following way:

$$\mathcal{D} ::= \text{true} \mid A^n(r_1,\ldots,r_n) \mid \mathcal{D}_1 \vee \mathcal{D}_2 \mid \neg \mathcal{D} \mid \mathcal{D}_1 \,;\, \mathcal{D}_2 \mid \forall u \in \text{T}_u \cdot \mathcal{D}$$
$$r ::= u \mid \int p \mid f^n(r_1,\ldots,r_n)$$
$$p ::= e$$
$$e ::= u \mid x \mid g^n(e_1,\ldots,e_n)$$

where A^n is an n-ary predicate symbol over reals, $r_1,\ldots,r_n$ are duration terms, u is a rigid variable, T is a type, p is a state formula, f^n and g^n are n-ary function symbols, and x is a flexible variable. Duration terms are real valued, state formulas are Boolean valued.

Formulas are interpreted with respect to a so-called *interpretation* $\mathcal{I}$ and an *observation interval* $[b,e]$. An interpretation in DC is a (piecewise continuous) function that associates a state to each point in time, modelled by the non-negative reals. Often this is illustrated by a timing diagram:

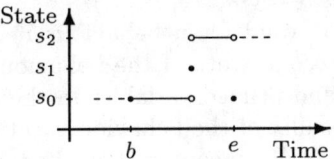

Intuitively, $\int p$ (the duration of p) is the amount of time in the interval where p is true, and $\mathcal{D}_1\,;\,\mathcal{D}_2$ ($\mathcal{D}_1$ chop $\mathcal{D}_2$) holds if the interval can be split into two subintervals such that $\mathcal{D}_1$ holds for the first and $\mathcal{D}_2$ holds for the second.

For the formal semantics in Table 1 we let $(\mathcal{I},\mathcal{V},[b,e]) \models_{\overline{\mathcal{D}}} \mathcal{D}$ denote that $\mathcal{D}$ holds for the interpretation $\mathcal{I}$, the valuation $\mathcal{V}$, and the observation interval $[b,e]$. For non-Boolean valued expressions we shall use *meaning functions*, such that $[\![e]\!]^{\mathcal{I},[b,e]}_{\mathcal{V}}$ denotes the value taken by expression e when evaluated with respect to the interpretation $\mathcal{I}$ and the valuation $\mathcal{V}$ in the interval $[b,e]$, and likewise $[\![e]\!]^{\mathcal{I},t}_{\mathcal{V}}$ denotes the value of e when evaluated with respect to $\mathcal{I}$ and $\mathcal{V}$ at point in time t.

Table 1. Semantics of DC

$[\![\int p]\!]^{\mathcal{I},[b,e]}_{\mathcal{V}} \triangleq \int_b^e \chi([\![p]\!]^{\mathcal{I},t}_{\mathcal{V}})\,dt$

$(\mathcal{I},\mathcal{V},[b,e]) \models_{\overline{\mathcal{D}}} \text{true} \triangleq \text{tt}$

$(\mathcal{I},\mathcal{V},[b,e]) \models_{\overline{\mathcal{D}}} A^n(r_1,\ldots,r_n) \triangleq \underline{A}^n([\![r_1]\!]^{\mathcal{I},[b,e]}_{\mathcal{V}},\ldots,[\![r_n]\!]^{\mathcal{I},[b,e]}_{\mathcal{V}})$

$(\mathcal{I},\mathcal{V},[b,e]) \models_{\overline{\mathcal{D}}} \mathcal{D}_1 \vee \mathcal{D}_2 \triangleq (\mathcal{I},\mathcal{V},[b,e]) \models_{\overline{\mathcal{D}}} \mathcal{D}_1 \vee (\mathcal{I},\mathcal{V},[b,e]) \models_{\overline{\mathcal{D}}} \mathcal{D}_2$

$(\mathcal{I},\mathcal{V},[b,e]) \models_{\overline{\mathcal{D}}} \neg \mathcal{D} \triangleq \neg (\mathcal{I},\mathcal{V},[b,e]) \models_{\overline{\mathcal{D}}} \mathcal{D}$

$(\mathcal{I},\mathcal{V},[b,e]) \models_{\overline{\mathcal{D}}} \mathcal{D}_1\,;\,\mathcal{D}_2 \triangleq \exists m \in [b,e] \cdot (\mathcal{I},\mathcal{V},[b,m]) \models_{\overline{\mathcal{D}}} \mathcal{D}_1 \wedge (\mathcal{I},\mathcal{V},[m,e]) \models_{\overline{\mathcal{D}}} \mathcal{D}_2$

$(\mathcal{I},\mathcal{V},[b,e]) \models_{\overline{\mathcal{D}}} \forall u \in \text{T}_u \cdot \mathcal{D} \triangleq \forall \mathcal{V}' \cdot \mathcal{V}' =_u \mathcal{V} \Rightarrow (\mathcal{I},\mathcal{V}',[b,e]) \models_{\overline{\mathcal{D}}} \mathcal{D}$

where $\mathcal{V}' =_u \mathcal{V} \triangleq \mathcal{V}'\backslash\{u\} = \mathcal{V}\backslash\{u\}$

$\chi(b) \triangleq \textbf{if } b \textbf{ then } 1 \textbf{ else } 0$

valid/tautology: $\models_{\overline{\mathcal{D}}} \mathcal{D} \triangleq \forall \mathcal{I},b,e,\mathcal{V} \cdot (\mathcal{I},[b,e],\mathcal{V}) \models_{\overline{\mathcal{D}}} \mathcal{D}$

meaning of formula: $[\![\phi]\!]_{\mathcal{D}} \triangleq \{(\mathcal{I},\mathcal{V},[b,e]) \mid (\mathcal{I},\mathcal{V},[b,e]) \models_{\overline{\mathcal{D}}} \phi\}$

2.2 The Temporal Logic of Actions

Formulas F, G of TLA are built from state and action expressions e and a:

$$F ::= p \mid \Box[\mathcal{A}]_e \mid \neg F \mid F \wedge G \mid \Box F \mid \exists u : F \mid \exists x : F$$
$$p ::= e \mid \text{enabled } \mathcal{A}$$
$$\mathcal{A} ::= a$$
$$e ::= u \mid x \mid f^n(e_1, \ldots, e_n)$$
$$a ::= u \mid x \mid x' \mid f^n(a_1, \ldots, a_n)$$

where p is a predicate, $\mathcal{A}$ is an action, u is a rigid variable, x is a flexible variable, and f^n is an n-ary function. Predicates and actions are Boolean valued state expressions and action expressions respectively. $[\mathcal{A}]_e$ (square $\mathcal{A}$ sub e) is an abbreviation for $\mathcal{A} \vee e' = e$.

In TLA, formulas are evaluated with respect to so-called *behaviours*, which are infinite sequences of states $\sigma = \langle s_0, \ldots, s_i, \ldots \rangle$.

Intuitively, the action $\mathcal{A}$ is valid for a behaviour if $\mathcal{A}$ describes the state change from the first to the second state of the behaviour, unprimed variables interpreted in the first state and primed variables in the second. Likewise, $\Box F$ is true if F is valid for every suffix of the behaviour. Actions can only occur in TLA formulas in the context $\Box[\mathcal{A}]_e$, meaning that all steps (state changes) are either $\mathcal{A}$ steps or does not alter the value of the state expression e, in order to ensure that TLA formulas are *stuttering insensitive*, that is, duplication of states in a behaviour does not alter the truth value of any formula.

In the formal semantics of TLA, presented in Table 2, we let $(\sigma, \mathcal{V}) \models_{\overline{A}} F$ denote that the formula F is *valid* for the behaviour σ and the valuation $\mathcal{V}$. We introduce valuations in TLA to handle the valuation of rigid variables explicitly. For expressions we use meaning functions, e.g., $s_1 [\![e]\!]^{\mathcal{V}} s_2$ is the value taken by e when evaluated with respect to the states s_1 and s_2 and the valuation $\mathcal{V}$.

Table 2. Semantics of TLA

$$s[\![\text{enabled } \mathcal{A}]\!]^{\mathcal{V}} \triangleq \exists s_1 \cdot s[\![\mathcal{A}]\!]^{\mathcal{V}} s_1$$
$$(\sigma, \mathcal{V}) \models_{\overline{A}} p \triangleq \sigma(0) [\![p]\!]^{\mathcal{V}}$$
$$(\sigma, \mathcal{V}) \models_{\overline{A}} \mathcal{A} \triangleq \sigma(0) [\![\mathcal{A}]\!]^{\mathcal{V}} \sigma(1)$$
$$(\sigma, \mathcal{V}) \models_{\overline{A}} \neg F \triangleq \neg (\sigma, \mathcal{V}) \models_{\overline{A}} F$$
$$(\sigma, \mathcal{V}) \models_{\overline{A}} F \wedge G \triangleq (\sigma, \mathcal{V}) \models_{\overline{A}} F \wedge (\sigma, \mathcal{V}) \models_{\overline{A}} G$$
$$(\sigma, \mathcal{V}) \models_{\overline{A}} \Box F \triangleq \forall \sigma' \cdot \sigma' \text{ suffix of } \sigma \Rightarrow (\sigma', \mathcal{V}) \models_{\overline{A}} F$$
$$(\sigma, \mathcal{V}) \models_{\overline{A}} \exists u : F \triangleq \exists \mathcal{V}' \cdot \mathcal{V}' =_u \mathcal{V} \wedge (\sigma, \mathcal{V}') \models_{\overline{A}} F$$
$$(\sigma, \mathcal{V}) \models_{\overline{A}} \exists x : F \triangleq \exists \sigma', \sigma'' \cdot \natural \sigma' = \natural \sigma \wedge \sigma'' =_x \sigma' \wedge (\sigma'', \mathcal{V}) \models_{\overline{A}} F$$

where $\natural \langle s_0, s_1, s_2, \ldots \rangle \triangleq$ **if** $\forall i \cdot s_i = s_0$ **then** $\langle s_0, s_1, s_2, \ldots \rangle$
  else if $s_1 = s_0$ **then** $\natural \langle s_1, s_2, s_3, \ldots \rangle$
  else $\langle s_0 \rangle \circ \natural \langle s_1, s_2, s_3, \ldots \rangle$

valid/tautology: $\models_{\overline{A}} F \triangleq \forall \sigma, \mathcal{V} \cdot (\sigma, \mathcal{V}) \models_{\overline{A}} F$
meaning of formula: $[\![F]\!]_{\mathcal{A}} \triangleq \{(\sigma, \mathcal{V}) \mid (\sigma, \mathcal{V}) \models_{\overline{A}} F\}$

2.3 The Common Model

A common model for transition-based logics like TLA and interval-based logics like DC must be convenient for modelling both *state changes* and *time intervals*. The real-time model that we suggest models a system run by an execution, which is a *finite or infinite* sequence of situations, a situation being a pair consisting of a state (an assignment of values to the variables) and a point in time. This means that time is more explicitly modelled than in TLA, where it is encoded as a variable, whereas state changes are more directly expressed than in the interpretations of DC. A finite execution corresponds to an interpretation and an observation interval, whereas infinite executions, like DC interpretations or TLA behaviours, model entire system runs.

An alternative to this model would be to model system runs by functions from time to sequences of states (the sequences being of length one at points in time with no state changes) as in [15], closer to the model for DC. We have chosen, for the sake of inductive reasoning, the former of these alternatives, but believe them to be isomorphic.

Thus, the *time (clock) variable t*, ranging over the non-negative reals $\mathbb{R}^+$, has a special status and is separated from the other variables in the model. The information about all non-rigid variables at an instant is gathered in a *situation*, which is a pair consisting of a state s and a point in time t, thus $\mathsf{s} : (\mathsf{Flex} \to \mathsf{Val}) \times \mathbb{R}^+$.

An *execution* $\sigma = \langle (s_1, t_1), (s_2, t_2), \ldots \rangle$ is a nonempty sequence, finite or infinite, of situations. Letting index σ and len σ denote the indices, respectively the number of situations, in σ (len $\sigma = \infty$ if σ is infinite), and letting $\inf(\sigma)$ and $\mathrm{fin}(\sigma)$ denote that σ is infinite, respectively finite, we can formulate the following well-formedness criteria on executions:

Monotonicity: $\quad \forall i \in (\mathsf{index}\,\sigma) \setminus \{1\} \cdot t_{i-1} \leq t_i \wedge (s_{i-1} = s_i \vee t_{i-1} = t_i)$

Non-Zenoness: $\quad \inf(\sigma) \Rightarrow \forall c \in \mathbb{R}^+ \cdot \exists i \in \mathsf{index}\,\sigma \cdot t_i > c$

Monotonicity states that time is a non-decreasing function (of the index), and that the state is unchanged in time steps (either the state or the time is the same in any two neighbour situations). Non-Zenoness states that time does not converge.

An execution can be considered to represent a *sampling* of a (continuous) system run, as illustrated by the following *timing diagram* representing an execution $\langle \ldots, (s_0, t_0), (s_1, t_0), (s_1, t_1), (s_1, t_2), (s_0, t_2), (s_2, t_2), (s_2, t_3), (s_1, t_3), (s_2, t_3), \ldots \rangle$

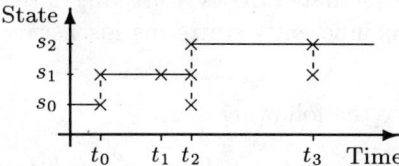

Note that an execution can have more than one state at each point in time.

We let $\sigma_{i \to j}$ denote the *subexecution* from the i'th to the j'th situation of σ, such that $\langle \mathsf{s}_1, \ldots, \mathsf{s}_i, \ldots, \mathsf{s}_j, \ldots \rangle_{i \to j}$ is $\langle \mathsf{s}_i, \ldots, \mathsf{s}_j \rangle$. As a special case we shall denote by $\sigma_{i\to}$ the suffix of σ from and including the i'th element. Two subexecutions, the first of which must be finite, can be *concatenated* by use of the concatenating operator $\circ$, such that $\langle s_1, \ldots, s_n \rangle \circ \langle s'_1, \ldots \rangle$ equals $\langle s_1, \ldots, s_n, s'_1, \ldots \rangle$. Also, we shall say that σ_1 is an *x-variant* of σ_2, denoted $\sigma_1 =_x \sigma_2$, if the two executions differ only in the value of the flexible variable x.

2.4 The Temporal Logic with Durations

The Temporal Logic with Durations, TLD, attempts to combine the strengths of transition based logics (TLA and the like) and interval based formalisms (DC). All operators are borrowed from either DC, TLA, or the logics proposed by Manna and Pnueli, [12, 8, 9], the intuitive meaning being preserved. However, the (semantic) meaning of the operators *has* changed, in that TLD is carefully designed to be *stuttering insensitive*.

Stuttering. Temporal logics are used for the specification of systems, and are used to show that one (the concrete) specification *implements* another (the abstract) specification, in the sense that every execution allowed by the former is also allowed by the latter. However, since the concrete specification is usually a refinement of the abstract specification, i.e., of greater detail, the concrete specification may need to allow state changes that do not change the abstract state. Such "invisible steps" are called *stuttering steps*. Therefore, it is considered a desirable property for a specification to be *stuttering insensitive*, such that every formula evaluates to the same truth-value when evaluated with respect to stuttering equivalent executions, that is, executions that differ only by a number of stuttering steps.

Compared to the model of TLA, our model is enriched by an explicit time for each situation. Close to the ideas of [4], we therefore consider two different kinds of stuttering steps: *situation stuttering steps* and *time stuttering steps*. When a situation in an execution is repeated we call it a situation stuttering step, and when a time-step is broken into smaller time-steps we say that a time stuttering step is inserted.

Then two executions σ_1 and σ_2 are stuttering equivalent, $\sigma_1 \simeq \sigma_2$, iff they differ only by (possibly infinitely many) stuttering steps. In the introduction to TLA in Sect. 2.2 we saw that TLA was *syntactically restricted* to be stuttering insensitive, by allowing actions to occur in the context $\Box[\mathcal{A}]_e$ only. We have chosen to ensure *semantically* that TLD is stuttering insensitive by defining every operator such that it is inherently stuttering insensitive.

Syntax. TLD formulas have the following form:

$$\phi ::= \inf \mid \mathcal{F} \mid \mathcal{D} \mid \phi_1 \vee \phi_2 \mid \neg \phi \mid \phi_1 \mathbin{;} \phi_2 \mid \forall u \in T_u \cdot \phi \mid \forall x \in T_x \cdot \phi$$

where $\mathcal{F}$ is a temporal formula, $\mathcal{D}$ is an interval formula, u is a rigid variable in Rigid, and x is a flexible variable in Flex. Temporal formulas and interval formulas have the forms:

$$\mathcal{F} ::= p \mid \mathcal{A} \mid \Box \mathcal{F}$$
$$\mathcal{D} ::= R^n(r_1, \ldots, r_n)$$

where p is a predicate, $\mathcal{A}$ is an action, and R^n is an n-ary relation over reals (duration terms). Predicates and actions are Boolean valued state expressions and action expression respectively, and duration terms are real valued.

$$p ::= e \mid \text{enabled } \mathcal{A} \qquad e ::= u \mid x \mid t \mid f^n(e_1, \ldots, e_n)$$
$$\mathcal{A} ::= a \qquad\qquad\qquad\quad a ::= u \mid x \mid t \mid x' \mid t' \mid f^n(a_1, \ldots, a_n)$$
$$r ::= u \mid \int \mathcal{F} \mid f^n(r_1, \ldots, r_n)$$

where u is a rigid variable in Rigid, x is a flexible variable in Flex, t is the time variable, f^n is a function symbol, and $\mathcal{F}$ is a temporal formula *without references to the time variable*. This constraint is needed to ensure stuttering insensitiveness of interval formulas.

Intuitively, the operators have the same meaning as in the formalism from which they are taken. Especially, $\int \phi$ (read "the duration of ϕ") is the amount of time in the considered execution where ϕ holds, and $\phi \,;\, \psi$ ("ϕ chop ψ") holds if the execution can be split into two subexecutions such that ϕ holds for the first subexecution and ψ for the second, and $\Box \phi$ ("henceforth ϕ") holds if ϕ holds at every point in the execution. Unlike the models for DC and TLA, the model for TLD has both finite and infinite executions, and *inf* is introduced to distinguish these.

Unprimed variables are evaluated in the first situation of the execution, and the time variable t refers to the point in time. Primed variables are evaluated in the first situation that differs from the absolute first situation of the execution (intuitively the second situation, but not when the execution starts with stuttering steps).

The version of TLD presented in [2] is a little richer than the above, in that also the operators $\boxminus$ ("always in the past"), $\mathcal{U}$ ("until"), $\mathcal{S}$ ("since") and their derived are defined. We omit these for the sake of simplicity.

Semantics. The semantics of TLD is defined in Table 3, $(\sigma, \mathcal{V}) \models \phi$ denoting that ϕ is *valid* for the execution σ and the valuation $\mathcal{V}$, and the meaning functions $[\![e]\!]^{\mathbf{s}}_\mathcal{V}$, $[\![e]\!]^{\mathbf{s},\mathbf{s}'}_\mathcal{V}$, and $[\![e]\!]^{\sigma}_\mathcal{V}$ denoting the value of expression e when evaluated with respect to the valuation $\mathcal{V}$ and the situation $\mathbf{s}$, the situations $\mathbf{s}$ and $\mathbf{s}'$, or the execution σ respectively.

With the mentioned syntactic restriction that the duration $\int \phi$ can only be taken of temporal formulas ϕ with no references to the time variable, it can be proven that TLD is stuttering insensitive. The proof, which is based on structural induction, is presented in [2].

From the primitive operators the usual abbreviations can be defined: $\Diamond \phi$ (eventually ϕ), ℓ (length in time of execution), $\lceil \phi \rceil$ (almost everywhere ϕ), $\Diamond\!\!\!\!\diamond\, \phi$ (in some subexecution ϕ), $\boxdot \phi$ (in every subexecution ϕ) etc.

Table 3. Semantics of TLD

$[\![enabled\ \mathcal{A}]\!]_{\mathcal{V}}^{\mathbf{s}} \triangleq \exists \mathbf{s}_2 \cdot [\![\mathcal{A}]\!]_{\mathcal{V}}^{\mathbf{s},\mathbf{s}_2}$

$[\![t]\!]_{\mathcal{V}}^{\mathbf{s}} \triangleq t(\mathbf{s})$

$[\![\int \phi]\!]_{\mathcal{V}}^{\sigma} \triangleq \sum_{j \in \text{index}\ \sigma \setminus \{\text{len}\ \sigma\}} \chi((\sigma_{j\to}, \mathcal{V}) \models \phi)(t_{j+1}(\sigma) - t_j(\sigma))$

$(\sigma, \mathcal{V}) \models \text{inf} \triangleq \text{inf}(\sigma)$

$(\sigma, \mathcal{V}) \models p \triangleq [\![p]\!]_{\mathcal{V}}^{\sigma(1)}$

$(\sigma, \mathcal{V}) \models \mathcal{A} \triangleq [\![enabled\ \mathcal{A}]\!]_{\mathcal{V}}^{\sigma(1)} \wedge$
$\quad \forall \sigma' \cdot (\sigma' \simeq \sigma \wedge \text{len}\ \sigma' > 1) \Rightarrow ([\![\mathcal{A}]\!]_{\mathcal{V}}^{\sigma'(1),\sigma'(2)} \vee \sigma'(1) = \sigma'(2))$

$(\sigma, \mathcal{V}) \models \Box \phi \triangleq \forall \sigma' \cdot \sigma' \simeq \sigma \Rightarrow \forall j \in \text{index}\ \sigma' \cdot (\sigma'_{j\to}, \mathcal{V}) \models \phi$

$(\sigma, \mathcal{V}) \models R^n(r_1, \ldots, r_n) \triangleq \underline{R}^n([\![r_1]\!]_{\mathcal{V}}^{\sigma}, \ldots, [\![r_n]\!]_{\mathcal{V}}^{\sigma})$

$(\sigma, \mathcal{V}) \models \phi \vee \psi \triangleq (\sigma, \mathcal{V}) \models \phi \vee (\sigma, \mathcal{V}) \models \psi$

$(\sigma, \mathcal{V}) \models \neg \phi \triangleq \neg (\sigma, \mathcal{V}) \models \phi$

$(\sigma, \mathcal{V}) \models \phi\ ;\ \psi \triangleq \exists \sigma_1, \sigma_2 \cdot \text{len}\ \sigma_1 > 0 \wedge \text{len}\ \sigma_2 > 0 \wedge$
$\quad \sigma \simeq \sigma_1 \circ \sigma_2 \wedge t_{\text{len}\ \sigma_1}(\sigma_1) = t_1(\sigma_2) \wedge$
$\quad (\sigma_1, \mathcal{V}) \models \phi \wedge (\sigma_2, \mathcal{V}) \models \psi$

$(\sigma, \mathcal{V}) \models \forall u \in T_u \cdot \phi \triangleq \forall \mathcal{V}' \cdot \mathcal{V}' =_u \mathcal{V} \Rightarrow (\sigma, \mathcal{V}') \models \phi$

$(\sigma, \mathcal{V}) \models \forall x \in T_x \cdot \phi \triangleq \forall \sigma', \sigma'' \cdot (\sigma' \simeq \sigma \wedge \sigma'' =_x \sigma') \Rightarrow (\sigma'', \mathcal{V}) \models \phi$

valid/tautology: $\models \phi \triangleq \forall \sigma, \mathcal{V} \cdot (\sigma, \mathcal{V}) \models \phi$

meaning of formula: $[\![\phi]\!] \triangleq \{(\sigma, \mathcal{V}) \mid (\sigma, \mathcal{V}) \models \phi\}$

Ensuring stuttering insensitiveness complicates the semantics somewhat. For instance, $\Box \phi$ is defined "for all suffixes of all stuttering equivalent" such that, e.g., $(\text{inf} \wedge t = 0) \Rightarrow \Diamond t = 7$ holds for any execution. This is desirable from the point of view that we are modelling time that changes continuously.

The semantics of chop, $\phi\ ;\ \psi$, is of interest. From this definition, $\phi\ ;\ \psi$ is valid if a stuttering equivalent execution can be split in a step that is not a time step. The fact that it is a stuttering equivalent execution being split implies that the execution can also be split in a situation (the situation is duplicated) or at any point in a time step (a time step stuttering is added and the execution is split in the new situation). Thus an execution can be split in the following three ways:

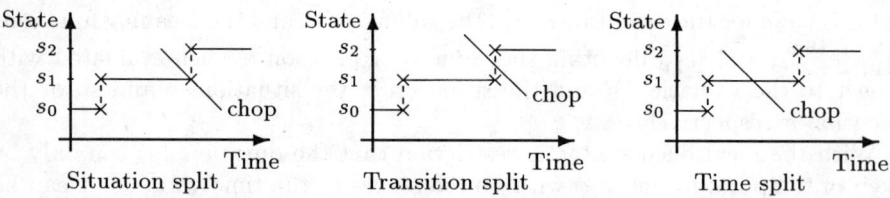

Situation split   Transition split   Time split

A chop in DC always splits an interval such that a point in time is part of both sub-intervals. However, DC is insensitive to values in single points, and could equally well be interpreted on open intervals. Allowing chops in transitions in TLD makes it meaningful to write useful formulas like $\Box p\ ;\ \Box \neg p$.

Also the semantics of an action deserves an explanation. The intention is that $\mathcal{A}$ is valid for an execution if $\mathcal{A}$ describes the first situation change that is not a (situation) stuttering step. Thus, the first step of all stuttering equivalent executions must be either an $\mathcal{A}$ step or a situation stuttering step. The first conjunct, that $\mathcal{A}$ is enabled, is added to prevent actions that do not depend on the post situation (that is, actions that could also be considered predicates) from being trivially true in the last situation of an execution.

3 Relating Logics

In the design of TLD and its model it has been a primary concern to create a common model for (among others) Duration Calculus and TLA, and a logic that would be an extension of these. In order to combine knowledge about systems given in different logics we will establish a framework for relating logics. In designing the framework, we have been much inspired by the similar work for hardware models in [16].

3.1 A Framework

The models of reactive systems that we consider agree in their linear-time view, i.e., an element of the model M represents an observation of a system run and a system is modelled by the set of all possible observations of all possible runs, i.e., by an element of $\mathcal{P}(M)$, the powerset of M. Thus, e.g., a DC observation (an interpretation and an observation interval) may be related to a TLD observation (an execution) in the sense that they model the same (partial) system run.

Therefore, to relate two logics and their respective models we will find a *formula translator* $*$ and a relation between sets of observations, such that a formula ϕ and its translation $*\phi$ satisfy related sets of observations. Naming one logic the concrete logic (subscript C) and the other logic the abstract one (subscript A), we let Φ_A and Φ_C denote the sets of formulas in the logics, while $[\![\ ]\!]_A$ and $[\![\ ]\!]_C$ are meaning functions assigning to every formula the set of observations satisfied by the formula. Representing the relation between observations via a function abs_* from sets of concrete observations to sets of abstract observations, we wish to obtain that the following diagram commutes:

$$\begin{array}{ccc} \Phi_C & \xleftarrow{\ *\ } & \Phi_A \\ {\scriptstyle [\![\ ]\!]_C} \downarrow & & \downarrow {\scriptstyle [\![\ ]\!]_A} \\ \mathcal{P}(M_C) & \xrightarrow[abs_*]{} & \mathcal{P}(M_A) \end{array}$$

That is, for any formula ϕ_A in the abstract logic we will establish the relation:

$$abs_*([\![*\phi_A]\!]_C) = [\![\phi_A]\!]_A$$

Given that abs_* intuitively is a correct translation between the models and that $*\phi_A$ is almost identical to ϕ_A, this equation may be interpreted as stating that any ϕ_A has the same meaning (delimits the same systems) in both models.

The Abstraction Function. Possibly, the most obvious (first) attempt to relate observations is to define a function from one concrete observation to one abstract observation—a standard abstraction function. Experience from our examples tells us, however, that observations do not correspond sufficiently well for this approach to work, as some concrete observations relate to a number of abstract observations, while other concrete observations do not relate to any abstract observations at all. Therefore, a function $abs : M_C \to \mathcal{P}(M_A)$ relating a concrete observation to a (possibly empty) set of abstract observations should be defined. From this function two functions relating sets of observations can be defined:

Definition 1
For sets A in $\mathcal{P}(M_A)$ of M_A and sets C in $\mathcal{P}(M_C)$, the set abstraction function abs^* from $\mathcal{P}(M_A)$ to $\mathcal{P}(M_C)$ and its opposite abs_* from $\mathcal{P}(M_C)$ to $\mathcal{P}(M_A)$ are defined from abs in the following way:

$$abs^*(A) \triangleq \{c \mid abs(c) \subseteq A\} \quad \text{and} \quad abs_*(C) \triangleq \{a \mid \exists c \in C \cdot a \in abs(c)\} \qquad \blacksquare$$

We are particularly interested in sets of abstract observations that respect the abstraction function, such that they contain either every abstraction of a concrete observation, or none. That is, sets of abstract observations with the property:

Definition 2 (Disjoint/subset)
A set $A \in \mathcal{P}(M_A)$ of abstract observations has the disjoint/subset property with respect to abs iff

$$\forall c \in M_C \cdot abs(c) \cap A = \emptyset \vee abs(c) \subseteq A \qquad \blacksquare$$

Sets of abstract observations delimited by formulas in the abstract logic, $[\![\phi_A]\!]_A$, should have the disjoint/subset property with respect to abs (this must be proven in each case—if not possible, probably an inappropriate abstraction function has been chosen).

Also, the abstraction function should be defined such that it is *onto*, i.e., any abstract observation should be related by abs to at least one concrete observation or, equivalently, $abs_*(M_C) = M_A$.

From Definition 1 some interesting properties of abs_* and abs^* can be proven, the most important being that the pair forms an adjunction. It is also a Galois connection [11], but we shall not use the additional properties that arises from this fact. This and the derived results are inspired by [16].

Proposition 1 (Adjunction)
The pair abs_*, abs^* forms an adjunction from $(\mathcal{P}(M_C), \subseteq)$ to $(\mathcal{P}(M_A), \subseteq)$ with left adjoint abs_* and right adjoint abs^*, i.e.,

$$\text{For any } C \in \mathcal{P}(M_C) \text{ and } A \in \mathcal{P}(M_A), \quad C \subseteq abs^*(A) \Leftrightarrow abs_*(C) \subseteq A \qquad \blacksquare$$

The proof is straightforward and can be found in [2].

If *abs* can be proven to be onto some useful properties can be derived from the adjunction property. For all sets of concrete observations, $C \in \mathcal{P}(M_C)$, and for all sets $A \in \mathcal{P}(M_A)$ with the disjoint/subset property:

$$C \subseteq abs^*(abs_*(C)) \quad \text{and} \quad A = abs_*(abs^*(A))$$

If A does not have the disjoint/subset property, $abs_*(abs^*(A))$ is a subset of A.

For sets with the disjoint/subset property, the equalities in Table 4 concerning *abs* and operators on sets hold. The proofs are straightforward and can be found in [2].

Table 4. abs^* and set operations

$$
\begin{aligned}
abs^*(M_A) &= M_C \\
abs^*(\emptyset) &= \{c \mid abs(c) = \emptyset\} \\
abs^*(A_1 \cap A_2) &= abs^*(A_1) \cap abs^*(A_2) \\
abs^*(A_1 \cup A_2) &= abs^*(A_1) \cup abs^*(A_2) \\
abs^*(M_A \backslash A) &= \{c \mid abs(c) = \emptyset\} \cup (M_C \backslash abs^*(A))
\end{aligned}
$$

The Relation. With proper definitions of *abs* and *, two logics and their models can be related. In [2] the following theorem is proven.

Theorem 1
If *abs* is onto, and for any formula ϕ in the abstract logic the set $[\![\phi]\!]_A$ has the disjoint/subset property with respect to *abs*, and $[\![{*\phi}]\!]_C = abs^*([\![\phi]\!]_A)$, then

- $c \models_{\overline{C}} {*\phi}$ iff $\forall a \in abs(c) \cdot a \models_{\overline{A}} \phi$
- $\models_{\overline{C}} {*\phi}$ iff $\models_{\overline{A}} \phi$
- $abs_*([\![{*\phi}]\!]_C) = [\![\phi]\!]_A$
- soundness in the abstract logic of the rule $\psi_1, \ldots, \psi_n \vdash_A \phi$ implies soundness in the concrete logic of the rule ${*\psi_1}, \ldots, {*\psi_n} \vdash_C {*\phi}$. ∎

This concludes the description in general terms of our framework for relating logics and their models. In the following we will use the framework to show that TLD can be seen as an extension of both DC and TLA.

3.2 Relating TLD and Duration Calculus

We first show that TLD is actually related to DC, taking TLD to be the concrete logic and DC to be the abstract one.

Abstraction Function. The models for DC and TLD do not take the same view of the relation between change of state and progress of time.

The model for TLD is point based, i.e., it is based on executions which are sequences of situations. In this view a system is always in a well-defined state.

State changes do not take time, and more state changes at the same point in time are possible. Between the state changes, the state of the system is stable except that time may progress (time steps). Contrary to this, the view of the Duration Calculus is that the system is in exactly one state at any point in time.

This fundamental difference may appear to make a proper relation of the models infeasible. However, since no formula of DC can distinguish interpretations that differ only in isolated points, and no formula of TLD can distinguish stuttering equivalent executions, *classes* of DC observations can be unambiguously related to *classes* of TLD observations. E.g., we take the view that all the DC observations sketched here:

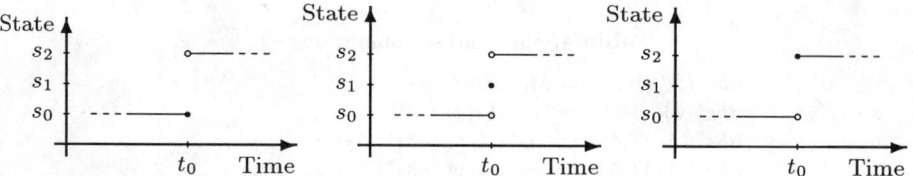

are related to the TLD execution fragment $\langle \ldots, (s_1, t_0), (s_2, t_0), (s_3, t_0), \ldots \rangle$ and all executions stuttering equivalent to this one.

To define the abstraction function *abs* we start by the auxiliary predicate *agree* that holds if the interpretation $\mathcal{I}$ agrees with the execution σ. By agree we mean that the value of all flexible variables are the same at time t. If t is a discontinuity point we say that $\mathcal{I}$ and σ agree if the state of one of the situations in σ with time t is equal to the state specified by $\mathcal{I}$.

$$\mathrm{agree}(\mathcal{I}, \sigma, t) = \left(\begin{array}{c} \exists s, t_i, t_j, \sigma_1, \sigma_2 \cdot \quad t_i < t < t_j \wedge \mathcal{I}(t) = s \\ \wedge \ \sigma = \sigma_1 \circ \langle (s, t_i), (s, t_j) \rangle \circ \sigma_2 \\ \vee \ \exists s, \sigma_1, \sigma_2 \cdot \sigma = \sigma_1 \circ \langle (s, t) \rangle \circ \sigma_2 \wedge \mathcal{I}(t) = s \end{array} \right)$$

The abstraction function that defines the conversion from an observation in M_{TLD} to a subset of M_{DC} is now defined using the auxiliary predicate:

$abs(\sigma, \mathcal{V}) =$
$\{ (\mathcal{I}, \mathcal{V}, [t_1(\sigma), t_{\mathsf{len}\, \sigma}(\sigma)]) \mid \mathrm{fin}(\sigma) \wedge \forall t' \cdot t_1(\sigma) \leq t' \leq t_{\mathsf{len}\, \sigma}(\sigma) \Rightarrow \mathrm{agree}(\mathcal{I}, \sigma, t') \}$

The infinite observations in TLD are related to the empty set. From *abs* the functions abs^* and abs_* are derived following Definition 1.

Since it can be proven, [2], that sets of the form $[\![\mathcal{D}]\!]_D$ have the disjoint/subset property with respect to *abs*, $\mathcal{D}$ being any duration formula, and that the abstraction function is onto, the set relations in Table 4 and the adjunction properties can be used.

Conversion of Formulas. TLD is carefully designed to be an extension of the Duration Calculus, hence the formula translator * is very simple. The only consideration is that *holds* is only defined for finite observation intervals, to which *abs* relates finite TLD executions, thus the proper definition of the formula translator is:

$$*\mathcal{D} \triangleq \mathit{fin} \Rightarrow \mathcal{D}$$

With these definitions of *abs* and * it is a relatively straightforward but tedious task to prove that $[\![*\mathcal{D}]\!] = \mathit{abs}^*([\![\mathcal{D}]\!]_D)$.

The proof is based on structural induction over formulas of the Duration Calculus. Bringing the entire proof is out of scope, but in order to give an impression we present the following simple case:

Proof of $\mathit{abs}^*([\![\mathcal{D}_1 \vee \mathcal{D}_2]\!]_D) = [\![\mathit{fin} \Rightarrow (\mathcal{D}_1 \vee \mathcal{D}_2)]\!]$:
The semantics of the Duration Calculus formula $\mathcal{D}_1 \vee \mathcal{D}_2$ can be rewritten as follows:

$$[\![\mathcal{D}_1 \vee \mathcal{D}_2]\!]_D$$
$$= \{(\mathcal{I}, \mathcal{V}, [b, e]) \mid (\mathcal{I}, \mathcal{V}, [b, e]) \models \mathcal{D}_1 \vee \mathcal{D}_2\}$$
$$= \{(\mathcal{I}, \mathcal{V}, [b, e]) \mid (\mathcal{I}, \mathcal{V}, [b, e]) \models \mathcal{D}_1 \vee (\mathcal{I}, \mathcal{V}, [b, e]) \models \mathcal{D}_2\}$$
$$= \{(\mathcal{I}, \mathcal{V}, [b, e]) \mid (\mathcal{I}, \mathcal{V}, [b, e]) \models \mathcal{D}_1\} \cup \{(\mathcal{I}, \mathcal{V}, [b, e]) \mid (\mathcal{I}, \mathcal{V}, [b, e]) \models \mathcal{D}_2\}$$
$$= [\![\mathcal{D}_1]\!]_D \cup [\![\mathcal{D}_2]\!]_D$$

Applying an equality from Table 4 yields:

$$\mathit{abs}^*([\![\mathcal{D}_1]\!]_D \cup [\![\mathcal{D}_2]\!]_D) = \mathit{abs}^*([\![\mathcal{D}_1]\!]_D) \cup \mathit{abs}^*([\![\mathcal{D}_2]\!]_D)$$

From the induction hypothesis it follows that $[\![\mathit{fin} \Rightarrow \mathcal{D}_i]\!] = \mathit{abs}^*([\![\mathcal{D}_i]\!]_D)$ for both $i = 1$ and $i = 2$. We insert this in the equality and rewrite:

$$\mathit{abs}^*([\![\mathcal{D}_1 \vee \mathcal{D}_2]\!]_D)$$
$$= \mathit{abs}^*([\![\mathcal{D}_1]\!]_D \cup [\![\mathcal{D}_2]\!]_D)$$
$$= \mathit{abs}^*([\![\mathcal{D}_1]\!]_D) \cup \mathit{abs}^*([\![\mathcal{D}_2]\!]_D)$$
$$= [\![\mathit{fin} \Rightarrow \mathcal{D}_1]\!] \cup [\![\mathit{fin} \Rightarrow \mathcal{D}_2]\!]$$
$$= \{(\sigma, \mathcal{V}) \mid (\sigma, \mathcal{V}) \models \mathit{fin} \Rightarrow \mathcal{D}_1\} \cup \{(\sigma, \mathcal{V}) \mid (\sigma, \mathcal{V}) \models \mathit{fin} \Rightarrow \mathcal{D}_2\}$$
$$= \{(\sigma, \mathcal{V}) \mid (\sigma, \mathcal{V}) \models \mathit{fin} \Rightarrow \mathcal{D}_1 \vee (\sigma, \mathcal{V}) \models \mathit{fin} \Rightarrow \mathcal{D}_2\}$$
$$= \{(\sigma, \mathcal{V}) \mid (\sigma, \mathcal{V}) \models \mathit{fin} \Rightarrow \mathcal{D}_1 \vee \mathit{fin} \Rightarrow \mathcal{D}_2\}$$
$$= \{(\sigma, \mathcal{V}) \mid (\sigma, \mathcal{V}) \models \mathit{fin} \Rightarrow (\mathcal{D}_1 \vee \mathcal{D}_2)\}$$
$$= [\![\mathit{fin} \Rightarrow (\mathcal{D}_1 \vee \mathcal{D}_2)]\!] \qquad \blacksquare$$

The rest of the induction follows along the same line. The exception is the base case, proving that the semantics of atomic duration formulas are equivalent in the logics. This proof includes a proof for equivalence of duration terms and state expressions. It is not entirely trivial, but feasible, and again the fully detailed proof can be found in [2].

The tautology concept of the Duration Calculus can be related to the tautology concept of TLD in the obvious way:

$$\models \mathit{fin} \Rightarrow \mathcal{D} \quad \mathrm{iff} \quad \models_{\overline{D}} \mathcal{D}$$

From the relation between tautologies it follows that the entire proof system of DC can be translated into a sound proof system for TLD, such that if $\mathcal{D}$ is a theorem of the Duration Calculus, the corresponding formula in TLD, $\mathit{fin} \Rightarrow \mathcal{D}$, is a theorem of TLD with the imported proof system.

3.3 Relating TLD and TLA

The models of TLA and TLD being so closely connected, it is relatively straightforward to relate TLA to TLD. The only point that demands for special consideration is the issue of *types*. TLD is a typed logic and every variable has an associated type, whereas TLA is an untyped logic.

We choose to solve this problem by assuming a *typed* TLA in which variables are typed and quantifications are of the form $\forall v \in T_v : F$, or of the form $\exists v \in T_v : F$. Such a typed version is equivalent to *type-restricted* TLA, which is the syntactic subset of TLA where all quantifications are of one of the forms $\forall v : v \in T_v \Rightarrow F$ and $\exists v : v \in T_v \wedge F$, in the sense that the validity of (closed) formulas is the same in both logics. Browsing practical examples of TLA reveals that most specifications are written in the type-restricted subset, such that types are actually used in TLA, but proven as invariants rather than being imposed as type constraints.

In the of this paper we thus assume TLA to be typed.

Abstraction Function. The abstraction function relating a TLD execution $\langle \ldots, (s_i, t_i), \ldots \rangle$ to the set of corresponding TLA behaviours is easily found, simply ignoring the time of the situations[1]:

$$abs(\sigma) \triangleq \{\sigma_A \mid \exists \sigma'_A \cdot \natural \sigma_A = \natural \sigma'_A \wedge \forall i \in \mathbb{N}_0 \cdot \sigma'_A(i) = s_{i+1}(\sigma)\}$$

remembering that TLA behaviours are numbered from 0, whereas TLD executions are numbered from 1. From *abs* the adjunction abs_*, abs^* is defined using Definition 1. Since every TLA formula is stuttering insensitive, and $abs(\sigma)$ is a set consisting of all the behaviours that are stuttering equivalent with one that immediately corresponds to σ, all sets of the form $[\![F]\!]_A$ have the disjoint/subset property with respect to *abs*. Furthermore, *abs* can easily be proven to be onto. What remains to be done is finding a formula translator * such that $[\![*F]\!] = abs^*([\![F]\!]_A)$ for every TLA formula F.

Conversion of Formulas. All TLA behaviours are infinite, whereas there are both finite and infinite TLD executions. Therefore, the formula translator * we are looking for is

$$*F \triangleq \inf \Rightarrow F$$

In [2] it is proven using structural induction over TLA formulas that for every TLA formula F, $[\![*F]\!] = abs^*([\![F]\!]_A)$. As a result it follows that for all TLA formulas

$$(\sigma, \mathcal{V}) \models \inf \Rightarrow F \quad \text{iff} \quad \forall \sigma_A \in abs(\sigma) \cdot (\sigma_A, \mathcal{V}) \models_{\overline{A}} F$$
$$\text{and} \quad \models \inf \Rightarrow F \quad \text{iff} \quad \models_{\overline{A}} F$$

---

[1] In [2] a different relation between TLA and TLD is also established, relating the time of the situations to the value of the special time variable *now* as used in [1].

Also, if $\psi_1, \ldots, \psi_n \vdash_A \phi$ is a sound rule in TLA, $(\mathit{inf} \Rightarrow \psi_1), \ldots, (\mathit{inf} \Rightarrow \psi_1) \vdash (\mathit{inf} \Rightarrow \phi)$ is a sound rule in TLD. Furthermore, importing every rule of TLA in this way ensures that if F is a theorem of TLA, $\mathit{inf} \Rightarrow F$ is a theorem of TLD.

4 Conclusions

In other approaches, temporal logics have been extended with durations [6], and the Duration Calculus has been extended to handle liveness [13]. Also, there have been other proposals for a unifying model [15]. None of these approaches, however, at the same time preserve the semantics of both DC and the temporal logic. Hence, although it may not seem surprising that it is in principle feasible to conservatively combine different real-time logics, we are not aware of other concrete results of doing so. Thus, we consider our work to exemplify how such a combination can be accomplished at the same time illustrating that great care must be exercised in order to ensure conservativeness.

In [2] TLD is used for the detailed specification and verification of a gas burner system in the style of [12] and an algorithm for (fast) mutual exclusion, suggested by Lamport in [7]. These two case studies, one which originates from DC and one which originates from TLA, together cover a wide variety of proof techniques: interval based, transition based, and combined. Especially the combined techniques are interesting, combining the intuitive reasoning based on invariants etc. with the expressive power concerning real-time properties from DC. These case studies lead us to believe that TLD could be a useful combination, although further applications are needed to substantiate this.

We are convinced that for various reasons, a number of formalisms are going to co-exist in the future, each accompanied by a set of useful tools. Practitioners, however, will probably prefer to "integrate" these tools into various development environments with the risk that the dependability of the end product is jeopardized. Our work illustrates a way of establishing formal relationships between the theories underlying the tools such that the exact ramifications of their integration can be analyzed.

For further work, a number of subjects arise. For the present combination, we would like to establish a relatively complete proof system for TLD and develop techniques for proving combined properties. To encompass communicating systems, the model should be extended with events. Finally, one could try to include yet other logics, e.g., logics dealing with continuous variables (hybrid systems) or probabilities.

References

1. Martin Abadi and Leslie Lamport. An old-fashioned recipe for real time. In J.W. Bakker et al., editors, *Real-Time: Theory in Practice. Lecture Notes in Computer Science, vol. 600*, pages 1–27. Springer-Verlag, 1992.
2. Odd Erik Hansen and Morten Ulrik Sørensen. Combining temporal specification techniques. Master's thesis, Technical University of Denmark, Department of Computer Science, July 1993.

3. Thomas A. Henzinger, Zohar Manna, and Amir Pnueli. Temporal proof methodologies for real-time systems. In *Proceedings of the 18th Annual Symposium on Principles of Programming Languages*, pages 353–384. ACM Press, 1991.
4. Thomas A. Henzinger, Zohar Manna, and Amir Pnueli. Timed transition systems. In J.W. de Bakker et al., editors, *Real-Time: Theory in Practice. Lecture Notes in Computer Science, vol. 600*, pages 226–251. Springer-Verlag, 1992.
5. Cliff B. Jones. *Systematic Software Development using VDM*. Prentice Hall, second edition, 1990.
6. Yassine Lakneche and Josef Hooman. Metric temporal logic with durations. To appear in a TCS special issue on Hybrid Systems (editors J. Sifakis and A. Pnueli), 1994.
7. Leslie Lamport. A fast mutual exclusion algorithm. Technical Report 7, Digital Equipment Corporation, Systems Research Center, November 1985.
8. Leslie Lamport. The Temporal Logic of Actions. Technical Report 79, Digital Equipment Corporation, Systems Research Center, 1991.
9. Zohar Manna and Amir Pnueli. *The Temporal Logic of Reactive and Concurrent Systems*, volume 1, Specification. Springer-Verlag, New York, 1992.
10. Zohar Manna and Amir Pnueli. Verifying hybrid systems. In *Workshop on Theory for Hybrid Systems*, 1992.
11. A. Melton, D.A. Scmidt, and G.E. Strecker. Galois connections and computer science applications. In D.H. Pitt et al., editors, *Category Theory and Computer Programming. Lecture Notes in Computer Science, vol. 240*, pages 299–312. Springer-Verlag, 1986.
12. Anders P. Ravn, Hans Rischel, and Kirsten M. Hansen. Specifying and verifying requirements of real-time systems. *IEEE Trans. Software Engineering*, 19(1):41–55, Jan. 1993.
13. Jens Ulrik Skakkebæk. Liveness and Fairness in a Duration Calculus. In *Proceedings of CONCUR '94*, Lecture Notes in Computer Science. Springer-Verlag, 1994. To appear.
14. J.M. Spivey. *The Z Notation*. Prentice Hall, 1989.
15. Burghardt von Karger. A simple model for the ProCoS languages. Technical Report OU BvK 10/2, ProCoSII, ESPRIT BRA 7071, Oxford University, September 1993.
16. Glynn Winskel. Relating two models of hardware. In D.H. Pitt, A. Poigné, and D.E. Rydeheard, editors, *Category Theory and Computer Science. Lecture Notes in Computer Science, vol. 283*, pages 98–113. Springer-Verlag, 1987.

Global Equivalence Proofs for ISTL

Shmuel Katz*

Computer Science Department
The Technion, Haifa, Israel
email: katz@cs.technion.ac.il

Abstract. A version of the temporal logic *ISTL* is used to demonstrate a two-stage approach to refinement of distributed programs. In each refinement, first convenient lower-level computations are shown to implement upper-level operations, and then in the second stage, all other computations are shown to be equivalent to one of the convenient ones. The equivalence maintains the ordering of all causally dependent events, but allows independent events to occur in different orders. The advantage of this separation is that different kinds of reasoning and induction can be used for the two aspects. The approach is demonstrated for a refinement that is part of the development of a cache consistency algorithm where synchrony is gradually loosened among the operations of different processes.

1 Introduction

Temporal logics have been defined that exploit information on partial order among events in a distributed system. The temporal logic we consider is based on the idea of a *partial order computation* (also called a *run*) which is simply a maximal set of occurrences of operations (called events) of a distributed system that have some partial ordering among them. The ordering includes any causality required among events, and may have additional restrictions. Events which are ordered are called *dependent*, and the others are *independent*. A program or system defines a collection of such runs. In this approach, presented previously in [KP90, KP92b, KP92a], the collection of all linearizations of the events that are consistent with the partial order are considered, in a temporal logic framework. Each such linearization is viewed as generating a sequence of alternating events and global states, that represents an execution sequence. All such execution sequences generated from a given run are called an *interleaving set* and are considered equivalent. Here 'equivalence' is used in the sense that the only difference between the execution sequences in an interleaving set is that strictly independent operations are executed in a different order.

In the temporal logic ISTL, a branching time assertion is interpreted as being *true* for a distributed system, if it is true for every interleaving set of the system. (This is analogous to the standard interpretation of a linear temporal

---

* This research was supported by the Fund for the Promotion of Research in the Technion, 120-845.

logic assertion being true of a system if it holds for every execution sequence.) Then it is easy to express that each equivalence class has some execution sequence satisfying a property p, simply as Ep. Such properties are often natural for distributed systems and allow expressing specifications for problems such as database serializability, distributed snapshots, and, as will be shown below, sequential consistency of cache-based shared memory systems.

In addition, for many properties it is true that $Ep \Rightarrow Ap$, i.e., if p is true of one execution in an interleaving set, then it is true for all the others in that set. For such properties, verification can be made more efficient by showing generically that p is a property for which $Ep \Rightarrow Ap$, then explicitly showing Ep, and using modus ponens to conclude Ap.

Thus properties of the form Ep can arise in a variety of contexts, and proof rules have been presented that allow concluding Ep. In such rules there are actually two tasks that are mixed together. One task is to show that p is true for the executions that are identified as the ones to be explicitly considered, and the other is to show that sufficient executions have been chosen to 'cover' all of the equivalence classes with at least one representative. The motivation for showing both properties at once is to allow a classic iterative proof on the computation, maintaining compositionality and modularity in the proof. At each step we can assume both that p is true for (some extension of) the parts of the computations considered so far, and that sufficient computations are being considered. This allows compositional proofs and proof rules to be used, but has the price of complicated proof rules[KP92b]. In the inductive step, it is necessary to show that the states reached so far all have a possible next state that will both maintain p and extend the existing computations to sufficient representatives.

Here a complete separation is suggested between showing that each of a chosen set of computations (called the *convenient computations*) fulfills the needed properties, and showing that every computation is equivalent to one of the convenient ones. The proof of the first aspect uses the usual iterative approaches, while the proof of the second aspect is global, and uses temporal logic assertions about the entire computation, along with formulas that encode which operations are independent of each other. The advantage of this separation is that different kinds of reasoning can be used for the two aspects, each most natural for the problem at hand.

This approach is demonstrated in the context of refinements of distributed systems, gradually replacing high level atomic operations by a collection of lower level operations that loosen the synchrony among distributed processes, but still maintain some key properties. Each refinement is divided into two independent proof stages. The first stage is showing that convenient executions of operations from the next lower level are a simple refinement of executions from the upper level, and can be demonstrated correct using standard refinement mappings.

Then we show that every additional execution sequence at the lower level is equivalent to one of the convenient ones. This stage could be considered as a 'loosening' of the ordering imposed by the convenient executions. The two-step reasoning at each level saves having to directly relate each lower-level sequence through a mapping to an upper level one. Although such a mapping exists, it may require the use of history and prophecy variables, and be extremely difficult to express and justify. This is because the collection of lower level operations that can be considered the 'implementation' of an upper level one are interleaved with an arbitrary number of operations that implement other higher level operations. Thus it is difficult to obtain an iterative proof that is uniform for all the computations when a direct mapping is required.

As an example of this approach, we treat the replacement of an abstract sequential global memory by a less synchronized version with queues between the processes and the global memory. In the abstract version, each process can execute atomic read and write operations directly from the memory. In the lower level version, a process can only write to a local queue, while later the head of the queue is written to the memory internally. This is one basic step in a series of refinements that can be used to justify that a cache-consistency protocol maintains what is known as *sequential consistency*. Intuitively, this means that the projection of local events of each process is consistent with use of the serial memory, even if a version with queues and local caches is being used instead.

The cache consistency protocol we treat is presented in [ABM93]. It has served as the basis for a variety of attempts to prove its correctness, in the framework of the Esprit REACT project [Ger93]. An algebraic approach related to the one here, that uses partial order information directly, can be found in that report, as well as in [JZ93, PZ93]. Sequential consistency seems, by its very definition, to favor the interleaving set view that considers the set of all total orders of events that are consistent with a partial order, as the semantic object to be considered.

Once we introduce queues, it is easy to define convenient executions for them and reason in terms of such sequences. The fact that all other sequences are equivalent to the convenient sequences is of course a crucial aspect of the correctness proof. It will be necessary to restrict the use of the queues on the implementation level, in order to guarantee this property. This will be expressed as another term in a temporal logic formula that is true iff all lower-level executions are equivalent to a convenient one. As we shall see below, care must be taken in defining which events are dependent, in order to obtain the appropriate equivalence relation and/or partial ordering.

The rest of this paper is structured as follows. We first explain in more detail the idea of (convenient) interleaving sequences and the dependency relation. The implications for independence of queue operations are also examined. The version of temporal logic used is then briefly described. Section 4 explains the conjuncts that define the independence relation and other temporal formulas that describe the lower-level execution sequences. In Section 5 these are summarized and used in a semantic version of the proof, basically a description of the temporal reasoning necessary to show that other executions are equivalent to the convenient ones. Some of the further steps in deriving a cache-consistency algorithm are also briefly described.

2 Defining dependencies and convenient executions

The convenient executions at the lower level are precisely those where the lower-level operations that implement a higher level one are all done sequentially, with no other lower-level operations interspersed. These are legal lower-level executions, even if they are unlikely to occur in practice because the operations are distributed in a collection of asynchronously executing processors. A mapping function from each convenient execution to some abstract computation is generally simple and iterative. After this first stage, we have only shown that every convenient execution sequence is a refinement of some higher-level abstract execution. The loosening stage requires precise reasoning about which operations are independent in which states. Each operation is viewed as a guard c (i.e., a condition for applicability on the state s) followed by a command f that is simply a function of s (with the operation written $c \rightarrow f$), as in [ABM93]. Note that such an interpretation of an event is reasonable only when a state is assumed as a semantic object, as part of the definition of an execution sequence. Then two operations, say $op1$ and $op2$, are *independent* in a state s, denoted $s \Rightarrow I(op1, op2)$, if beginning in state s neither affects the truth of the other's guard, and the result of executing them in either order is the same, i.e.,

$$c1(s) \Rightarrow (c2(f1(s)) \Leftrightarrow c2(s))$$

$$c2(s) \Rightarrow (c1(f2(s)) \Leftrightarrow c1(s))$$

$$(c1(s) \wedge c2(s)) \Rightarrow (f1(f2(s)) = f2(f1(s)))$$

The definition above is known as *conditional independence*[KP92a] because a pair of operations may be dependent in some states, and independent in others. The states in which two operations are independent are defined by a state predicate. Two execution sequences are considered equivalent if they differ only in that independent operations were done in a different order, but all dependent operations are done in the same order. The reasoning used to show the equivalence of two computations is quite different from that used to show the mapping from a higher to a lower level. If we are given a collection of independent operations in various states, then two sequences are equivalent if they differ only by interchanging two adjacent operations beginning at a state where they are independent. The equivalence class we consider is the transitive closure of this 'exchange' relation.

When more complex data structures are assumed, the dependencies become more complicated, and the extra freedom is exploited by the lower level implementation.

As a particularly relevant example, we consider the dependencies for a queue q with operations $empty(q)$, $put(q, e)$, and $get(q, e)$, where e is a data element. When the queue is non-empty, then $put(q, e)$ is independent of $get(q, f)$:

$$(\neg empty(q)) \Rightarrow I(put, get) \tag{1}$$

When the queue is empty, a *put* and a *get* operation will be dependent:

$$empty(q) \Rightarrow \neg I(put, get) \qquad (2)$$

All adjacent pairs of *put*'s are dependent:

$$\neg(I(put, put)) \qquad (3)$$

All adjacent pairs of *get*'s are dependent:

$$\neg(I(get, get)) \qquad (4)$$

The first rule is intuitively true because a *put* and a *get* by different processes on a nonempty queue are done at opposite ends of the queue, and never involve the same item, while this is not so when the queue is initially empty. In that case the *get* operation must follow a *put*.

The other rules follow from the fact that the contents of the queue differs according to the order of *put*'s, while the states of the rest of the system differ if *get*'s are done in a different order. A formal proof of these dependencies could be based, for example, on an algebraic specification of the queue axioms.

3 The logic

The version of temporal logic used in this paper will be briefly summarized. This is an adaptation of the logic $ISTL$ introduced in [KP90], with additions to facilitate showing equivalence of execution sequences. Most of the operators are those of CTL^* [EH86], but interpreted as true for a system if they hold for each interleaving set. An interleaving set is defined as an equivalence class of computations under exchanges of operations that can be done when the independence relation I holds. The syntax is thus standard, and the semantics (implicitly) universally quantifies over the interleaving sets:

Ap – for every computation in each interleaving set, p is true
Ep – for some computation in each interleaving set, p is true
Fp – eventually for some state, p is true
Gp – for every state from the present, p is true
Xp – for the next state, p is true
pUq – p is true until q becomes true (and q does become true)

In order to facilitate reasoning about sequences of operations, we add some conventions. First, an operation name also serves as a state predicate that is true precisely when that operation was executed in the transition from the previous state. (An alternative temporal logic that treats operations more directly can be seen in Lamport's TLA [Lam]). Then sequences of operations (or other predicates) can be denoted as

"$s;t$" – defined as $EX(s \wedge Xt)$ (in the next state s can hold, followed by a state with t).

4 Temporal Expression of Independence and Allowed Computations

The definition of sequential consistency used in this paper can be stated as follows.

A memory M is sequentially consistent with respect to a serial memory M_{serial}, iff

$$\forall \sigma \in Beh(M) \exists \tau \in Beh(M_{serial}) \forall i = 1 \ldots n \ \ \sigma|i = \tau|i$$

$Beh(M)$ is the set of execution sequences associated with a system M, and $Beh(M_{serial})$ is the set where read and write operations are atomically done on the global memory. The above asserts that the projections on each process are the same as those in some execution using a serial memory, even though the general behavior may have extra internal steps associated with the memory, so that a write operation may not affect the memory directly. This statement suggests the interleaving set approach, since it closely relates to the idea of convenient sequences: the behavior of the serial memory will be viewed as consisting of lower-level convenient sequences, where all lower-level executions are *equivalent* to such a convenient execution. That is, if we now view M_{serial} as a temporal logic predicate true of the lower-level serial computations, we require $E\ M_{serial}$.

We must define an independence relation so that the system is sequentially consistent if every execution is equivalent to a convenient serial one. That is, we require formulas in $ISTL^*$ that express the independence of adjacent operations (i.e., when I is true), that characterize the convenient serial computations, and that characterize every computation (including restrictions on when values can be read). Once these have been defined, we need to show that assuming the formula that defines the independence of operations, and the formula that defines all computations, $E\ M_{serial}$ is true.

In defining the independence relation so that it reflects sequential consistency, the local operations of each processor must be unchanged in the equivalent convenient version. Thus we assume a total order among local operations of a single processor. Since this order must be maintained for all equivalent execution sequences, we obtain the identity of local projections for every two equivalent execution sequences, as required in the definition of sequential consistency. For any two operations a_i and b_i, executed by process i, we therefore require

$$\neg I(a_i, b_i) \tag{5}$$

Of course, local operations a_i and b_j of *different* processes are independent:

$$i \neq j \Rightarrow I(a_i, b_j) \tag{6}$$

We consider how to refine abstract *read* and *write* actions. An abstract *write* action can be implemented by adding to the end of a queue the pair consisting of the value to be written and the memory address, later removing that pair from the head of the queue, and then writing it in the memory. If we denote the action of putting the value-address pair in the queue by $W(d,v)$, and the action of removing the pair from the head of the queue and writing to the memory by $MW(d,v)$ (standing for *Memory Write*) such a pair is the implementation of the abstract *write*. Thus W is associated with a *put* operation, and MW combines a *get* with a memory write.

Similarly, an abstract *read* could be implemented by reading from the memory, adding the value-location pair to another queue, and later reading the value-address pair from the head of that queue into the local process. However, the treatment of reads will be postponed to a second level of refinement, so for the present we assume a direct atomic read action denoted $R(d,v)$, meaning that value d is read from address (or variable) v.

In order to capture the intuition of reading and writing into memory, we express that the value returned for a variable or memory location x in an action $R(c,x)$ is the last value written into it by a $MW(d,x)$ action, in the assertion:

$$(MW(d,v) \wedge (\neg MW(b,v)) U R(c,v))) \Rightarrow c = d \qquad (7)$$

This is known as *read/write consistency* and is a fundamental assumption when truly atomic reads and writes are being used. However, when reads and writes occur at different processes, and are not atomic, we can weaken the requirement.

If we now replace the abstract read and write actions of the serial memory by the lower-level actions above, we arrive at a situation that can be viewed as the addition of abstract write queues to the serial memory. Since we have a collection of such write queues, the "lower" level involves operations on an Out_i queue between the processor i and the central memory, for each processor. Since there now is a queue for each processor, we denote a write to the end of the ith queue by W_i, and removing from the head of that queue plus writing to the memory by MW_i. Reading by process i is denoted by R_i. All of these have the same parameters as previously, namely the value and the address (or variable name). The events that are considered local to a process i are not independent, and these include all occurrences of W_i and R_i, but not MW_i. The MW_i and R_i operations can be considered local to the memory and must satisfy read/write consistency.

In the convenient executions, items are inserted by the process i using W_i operations into the corresponding Out_i queue and immediately removed and copied to the central memory by the MW_i action. In these very particular computations, every W_i is immediately followed by writing into the memory using MW_i, with no intervening operations anywhere in the system. The queues are thus always empty except when a single item has just been put in and has not yet been written to the memory in the next step. In temporal logic we can state the requirement for a convenient computation as simply

$$G(W_i(c,x) \Rightarrow X MW_i(c,x)) \qquad (8)$$

That is, throughout the computation, if a W_i has occurred, it is immediately followed by the corresponding MW_i. Every adjacent $W_i; MW_i$ pair is clearly a trivial implementation of the direct write on the abstract level. Since the read events R_i are still atomic, all convenient execution sequences can be easily shown to implement the abstract sequences, by a trivial induction on the sequence.

Then we need to claim that every execution of the lower level sastisfying the queue axioms and the memory consistency assumptions is equivalent under the independence relation I to one of the convenient executions. This is almost true, but we need to restrict the read operations of the lower level to maintain the total order among local actions of a single process. Consider a situation where a process has written a pair (d, x) to its Out queue, then executes a read operation (implemented as an R) on x, and only then does a MW execute on that queue, changing the memory. The value read is clearly whatever was in the memory before the last MW. This implies that there is a linearization consisting of

$$W_i(d, x); R_i(c, x); MW_i(d, x)$$

with $d \neq c$. But such a computation is not consistent with the dependency requirements, because we claim that it is not equivalent to any convenient computation. If we wish to find a convenient execution to which this one is equivalent, we must show that the R operation can be exchanged, either with the following MW or the preceding W. The former exchange would lead to

$$W_i(d, x); MW_i(d, x); R_i(c, x)$$

This is not a convenient execution, since it violates the restrictions on the value read being the last one written in the memory location (read/write consistency). Exchanging the R_i and W_i operations would lead to

$$R_i(c, x); W_i(d, x); MW_i(d, x)$$

This is a convenient sequence, but is not equivalent to the original one, because it does not have the same total order of the local operations in process i.

This difficulty is solved by simply requiring that the lower level operations be restricted so that any read operation by a process i, R_i, is 'delayed' until the Out_i queue is empty, i.e., until all of the 'pending' MW_i operations have been done. In that case the problematic computation described above is simply declared impossible. Of course, there is no such restriction for reads and writes from *different* processes. The restriction on the implementation is again a temporal logic formula and can be expressed in several ways. One is simply to state that

$$AG(R_i \Rightarrow empty(Out_i)).$$

Another treats the actions directly, using a # symbol to denote the number of times an operation has occurred:

$$AG(R_i \Rightarrow (\#W_i = \#MW_i)) \qquad (9)$$

That is, no R_i is between a W_i and an MW_i, because every W_i has a corresponding MW_i when R_i occurs.

The independence relations define what exchanges of operations can be made, and thus which computations are equivalent. This needs to be introduced into the logic explicitly, through the formula

$$AG(\ I(a,b)\ \Rightarrow\ ((\text{``}a;b\text{''}) \Leftrightarrow (\text{``}b;a\text{''}))) \tag{10}$$

In words, if $I(a,b)$ holds in a state, then the sequences that begin in that state and then have "$a;b$" are equivalent to those with "$b;a$". at that point.

5 Proving refinements

The proof requirements of showing a refinement that satisfies sequential consistency are obtained by using the relations from the previous section. The independence relations for queues (1–4) will have W_i corresponding to *put* and MW_i to *get* for each queue Out_i. We also have the independence and dependence relations on all local actions in each process (5–6). To these we add the read/write consistency rules for simple memory locations (7), the delay condition on reads defined above (9), and the formula connecting I and equivalence. We then claim that an execution sequence satisfying these dependencies must be equivalent (under the relations I) to one where all $W - MW$ pairs from the same queue are adjacent (8), i.e., to one of the convenient sequences. Note that the convenient sequences are assumed to have already been shown to correspond to abstract atomic read/write consistency. In terms of $ISTL^*$, the restrictions on the possible lower level computations must imply $EConvenient$, where $Convenient$ is the temporal logic definition of the convenient sequences.

The conjuncts in the correctness formula are summarized in Figure 1. With the restrictions we have added, this implication is not difficult to prove. Consider any sequence satisfying read/write consistency and read delays. Assuming the other formulas in Figure 1 (that define independence), we want to show

$$EG(W_i(c,x) \Rightarrow XMW_i(c,x)).$$

We prove by induction on the number of states between a $W_i(d,x) - MW_i(d,x)$ pair that correspond to putting a value in the Out_i queue and later removing it. If the two are adjacent, this pair is part of a convenient execution. If there is one state between them, and in that state $MW_j(c,y)$ for any j, c, and y, the independence relations show that there is an equivalent computation with the MW_j before the $W_i(d,x)$. The same is true of any R_j or W_j where $j \neq i$. If in that state there is another W_i it can be exchanged with the following $MW_i(d,x)$ (and recall that there cannot be an R_i). In general, note that there cannot be a 'matching' pair between another such pair from the same process, because that would violate the queue axioms. Assume that for all pairs with n states between them, we can find equivalent computations where the pairs are adjacent. For a pair with $n + 1$ states between them, if the first state is anything except W_i, it can be exchanged with the previous one, and the inductive hypothesis can be used. Otherwise, the n remaining actions can be exchanged either after the

queues, for process i:
$$(\neg empty(Out_i)) \Rightarrow I(W_i, MW_i)$$
$$empty(Out_i) \Rightarrow \neg I(W_i, MW_i)$$
locality, for a, b operations W or R in processes i, j:
$$\neg I(a_i, b_i)$$
$$i \neq j \Rightarrow I(a_i, b_j)$$
read/write memory consistency, for all processes i, j, and k:
$$AG(\ (MW_i(d, v) \wedge (\neg MW_j(b, v)) U R_k(c, v))) \Rightarrow c = d\ )$$
delay of reads, for process i:
$$AG(R_i \Rightarrow (\ \#W_i = \#MW_i))$$
independence and equivalence, for operations a and b:
$$AG(\ I(a, b)\ \Rightarrow\ ((\text{``}a; b\text{''}) \Leftrightarrow (\text{``}b; a\text{''}))\ )$$

Fig. 1. Conjuncts in the correctness formula of a refinement

$MW_i(d, x)$, or before the actions $W_i(d, x); W_i(c, y)$ because any action (except R_i's, which are excluded by assumption) that can be exchanged with $W_i(d, x)$ can also be exchanged with $W_i(c, y)$. Finally, the $W_i(c, y)$ can be exchanged with the $MW_i(d, x)$.

The proof here is simply a systematic analysis of which operations are independent of which others, in order to show that any computation is equivalent to a convenient one. We show exchanges that bring a general computation 'closer' according to some measure to a convenient one.

Just as for the abstract write actions actions, we could refine read actions into a pair of actions $MR_i(x, v); R_i(x, v)$, where in this case the memory read should precede the process read action. For these computations we could again show that we may precede the MR action with an empty queue test to obtain the desired behavior. The convenient sequences consist of sequences where after executing the MR action, immediately an R action is executed. In fact, the reading is handled in another way in the cache consistency algorithm.

6 Further refinements

Although we will not treat the further levels in detail, the convenient executions and the type of reasoning necessary is described in this section. The idea is that a local cache memory of bounded size is associated with each process, and updates to the global memory are also inserted in a queue for each process, from which they are transferred to the local memory. On this level, In queues are used. A lower-level MW_i operation, in addition to removing an element from the head of the Out_i queue and writing it to the memory, now also adds the update requests to the In queue of each process. Alternatively, we could view this strengthened MW_i operation as simply the previous MW_i that only wrote to the main memory, followed immediately by an automatic MR operation for each process, that adds that same value to the end of the In queue of the process. A CU_i event removes an update request from the head of the In_i queue, and writes in the cache according to the update. A read request R_i is now from the local cache rather than from central memory or from an abstract queue. The convenient sequences for this level are simply those computations for which every (strengthened) MW_i event is immediately followed by a subsequence with a single CU_j event for every process j. That is, again the queues will be empty, have one item inserted, and immediately use that item to write in the local caches. We view a strengthened "MW-subsequence of CU's" as the implementation of a simple MW which only wrote to the central memory, and a subsequent MR. In this case, for these convenient execution sequences, each cache is the same as the central memory when outside such subsequences. A lower level read operation is done now on the local cache, instead of as previously, and thus is denoted CR in order to distinguish it from the previous read. By the considerations above it will give the same result as the higher level read from the central memory, for the convenient sequences.

The CR_i read actions of a process i will again be restricted in order to allow showing equivalence to convenient sequences. Recall that this is intended to prevent local inconsistency where a process could sense that a W_i action has not yet 'taken effect' when a subsequent local read is done. As before, read actions of process i cannot occur while Out_i is nonempty. Similarly, those items in the In_i queue that represent updates that were originally initiated by process i itself, must be removed from the In_i queue and written to the cache before a read of that variable by process i. This is needed, just like the flushing of the Out queue required before a memory read operation by that process, in order to guarantee local total order. In terms of operations, CR_i occurs in an execution only after all CU_i events that correspond to previous local W_i actions have occurred.

In a real cache consistency algorithm, the possibility of cache misses must also be treated. The idea behind cache misses is that the cache has limited capacity and can therefore not mirror the whole of the central memory. Sometimes variables are removed from the cache so that values for other variables can be put into the cache. This is modelled by adding internal cache invalidate actions CI_i that remove value-address pairs from the cache. A cache read $CR_i(d, x)$ can only occur if the pair (d,x) is in the cache. If it is not, the read must be delayed until the needed pair is retrieved from the central memory. This is done by means of repeating MR operations to read a value for a variable from the central memory and putting it into the In queue of only the process that had the CI event, so that they eventually are put into the cache by a cache update. As previously, this description can be captured by a temporal logic assertion further restricting when a CR can occur:

$$AG(CR_i(d, x) \Rightarrow (d, x) \in Cache_i).$$

Convenient sequences at this level now consist of sequences where all MR and corresponding CU actions immediately follow one another, and each CI is followed immediately by an MR. The dependence relation is now such that all MW and MR actions are dependent. All we need to show is that the $CI; MR; CU$ sequence that will occur after a cache miss in these convenient executions can be mapped to not having done anything on the abstract level, and that all other executions are equivalent to those where all CI's are in such subsequences.

7 Concluding remarks

In this paper we gave a sketch of the derivation of the lazy caching algorithm by means of a series of refinements, starting from the definition of serial and sequentially consistent memory. Reasoning in terms of convenient sequences and their equivalence classes seems to be well-suited for this purpose. The independence relations and restrictions on possible implementations are easily expressed using $ISTL$.

The steps in such proofs of equivalence are uniform. First, predicates are needed that make the independence of adjacent operations explicit. These can be justified from the underlying semantics of the model, or by properties of the data structures used. In the case of sequential consistency, the independence is further restricted by the problem specification, namely that there is a total ordering among local process writes and reads. These properties can often be shown once for a large collection of related problems.

Second, the properties of the general computations are described as global temporal logic predicates. These follow from a description of the implementation level, including restrictions on when a read action is possible.

Next, the convenient computations are described, also using the temporal logic.

The claim to be proven is that under the equivalence defined by I, with the assumptions on the possible computations, $E\ Convenient$ is true. The proof of this fact is done by induction showing that each computation is equivalent to one that is 'closer' to a convenient one. A systematic examination of which operations can be exchanged is done using the independence information. This aspect seems amenable to automation, since it involves a large number of very simple assertions.

In the example given, the main concern is on showing equivalence, and the convenient sequences are chosen so that the refinement proof is particularly easy. This does not always have to be the optimal division, and sometimes more effort will have to be devoted to showing that the convenient executions indeed satisfy the needed property.

Acknowledgement: Job Zwiers and Wil Janssen suggested the gradual refinement stages and showed connections to algebraic partial orders, and Rob Gerth greatly helped by clarifying the cache consistency protocol.

References

[ABM93] Y. Afek, G. Brown, and M. Merritt. Lazy caching. *Transactions of Programming Languages*, 15(1):182–206, 1993.

[EH86] E.A. Emerson and J.Y. Halpern. Sometimes and not never revisited: on branching versus linear time temporal logic. *Journal of the ACM*, 33:151–178, 1986.

[Ger93] R. Gerth(editor). Verifying sequentially consistent memory. Technical report, Esprit React report, 1993.

[JZ93] W. Janssen and J. Zwiers. Specifying and proving communication closedness in protocols. In *Proceedings 13th IFIP symp. on Protocol Specification, Testing and Verification*, 1993.

[KP90] S. Katz and D. Peled. Interleaving set temporal logic. *Theoretical Computer Science*, 75:263–287, 1990.

[KP92a] S. Katz and D. Peled. Defining conditional independence using collapses. *Theoretical Computer Science*, 101:337–359, 1992.

[KP92b] S. Katz and D. Peled. Verification of distributed programs using representative interleaving sequences. *Distributed Computing*, 6:107–120, 1992.

[Lam] L. Lamport. The temporal logic of actions. *ACM Transactions on Programming Languages and Systems*. to appear.

[PZ93] M. Poel and J. Zwiers. Modular completeness for communication closed layers. In *Proceedings CONCUR '93*, 1993.

A Real Time Process Logic

J.C.M. Baeten[1]
Department of Computing Science, Eindhoven University of Technology,
P.O.Box 513, 5600 MB Eindhoven, The Netherlands. (josb@win.tue.nl)

J.A. Bergstra[1,2]
Programming Research Group, University of Amsterdam,
Kruislaan 403, 1098 SJ Amsterdam, The Netherlands
and
Department of Philosophy, Utrecht University,
Heidelberglaan 8, 3584 CS Utrecht, The Netherlands. (janb@fwi.uva.nl)

R.N.Bol
Department of Computing Science, Eindhoven University of Technology,
P.O.Box 513, 5600 MB Eindhoven, The Netherlands. (bol@win.tue.nl)

Abstract

Systems can be described at various levels of abstraction: automata, processes and behavior. In this paper, we take the ready trace set as a description of the behavior of a process and we present a ready trace model of real time process algebra. We argue that, especially in the real time case, properties of ready trace sets are best formulated in a dedicated logic (as opposed to describing them in an enriched process notation, such as ACP_τ). We present the syntax and semantics of a logic that could serve this purpose and we apply it to study the existence of so-called coordinated attack protocols. A connection is made with the metric temporal logic of Koymans. This paper is an abbreviated version of [BABB93].

1. Introduction.

Computing systems can be described at various levels of abstraction: automata, processes and behavior. Trace theory is the name for a collection of semantic models for the design, description, and analysis of systems, in which the behavior of a system is described by a set of execution traces, or initial fragments of such traces. A trace encodes a possible behavior of the system by a, finite or infinite, sequence of atomic actions, possibly augmented with additional information. In this paper we concentrate on *real time ready traces* (see Section 1.2). By defining operators on trace sets, we obtain algebras of process ready trace sets that are models of certain versions of ACP [BEK84, BAB91].

We argue that, especially in the real time case, properties of ready trace sets are best formulated in a dedicated logic. An alternative would be the description of a property by a characteristic process. This alternative is attractive in the untimed case, using a hiding operator (i.e., an operator that abstracts from internal actions). However, in the real time case, defining a good hiding mechanism is hard, if at all possible: it is usually desirable to hide a part of the timing information, but not all of it.

We propose a behavioral property logic called RdTrL (Ready Trace Logic). Its syntax and semantics are closely related to ready trace theory. We compare the expressivity of RdTrL with the metric temporal logic of KOYMANS [KOY89a,b]. We demonstrate how the logic can help in studying the (non-)existence of coordinated attack protocols (as was done in [HAM90] using epistemic logic). We conclude by listing some shortcomings of the logics, thus pointing out topics for further research.

---

[1] This author received partial support from ESPRIT Basic Research Action 7166, CONCUR2.
[2] This author received partial support from ESPRIT Basic Research Action 6454, CONFER.

1.1. Conceptual Abstraction.

Systems are described at various levels of abstraction. The most basic distinction, on which we elaborate in the rest of this section, is the distinction between automata, processes, and behavior. We shall call this the *conceptual abstraction hierarchy*. Its levels are comparable to the Nets, Terms, and Formulas of [OLD91].

Within each level in this hierarchy, there are different dimensions and levels of *technical abstraction*. As an example, one of these dimensions is the treatment of time, where we can distinguish (from low to high abstraction): real space/time, real time, discrete time, and symbolic (untimed). Where necessary, we can identify intermediate levels as well. Without aiming at completeness, we list the following dimensions of technical abstraction:

- treatment of time,
- treatment of silent steps,
- treatment of divergence and fairness,
- probabilistic vs. non-probabilistic,
- choice abstraction (as described in [vGL90] for processes).

An *automaton* or transition system is a class of states equipped with a transition relation. Transitions are labelled with so-called actions. Labels (attributes) of states are called signals. Automata represent the lowest level of (conceptual) abstraction in system description needed in this document. Automata exist in many levels of (technical) abstraction: with and without roots, termination states, deadlocks, time stamps, probability assignments, fairness conditions on paths etc. A format that allows one to present automata is often called a *program notation*. Operational semantics assigns an automaton to a program. Of course, the program notation reflects the technical level of abstraction at which the automata are to be treated.

A *process* is an abstraction of an automaton. At this level of conceptual abstraction, information about states is hidden, wheras information about actions is kept. Important is that a process has (in general) an operational semantics and allows simulation (but not necessarily implementation). A format that allows one to describe processes can be called a *process notation*. In general, one expects the existence of an algorithm that returns a simulator, when given a term in process notation. Processes are usually shaped as equivalence classes of states of automata.

A *behavior* is a (conceptual) abstraction of a process. Seen as an abstraction of an automaton, it hides even more information about the state space than a process. The usual shape of a behavior is a collection (satisfying certain natural closure conditions) of traces of (system) paths in an automaton, all starting in the same root. The difference between a path and a trace is that a path contains all information of the states that are visited, whereas a trace will abstract (to some extent) from this intermediate state information.

A *behavioral property* is a collection of traces or a logical description of such a collection in some appropriate logical format. The main difference between a behavior and a property is that a property need not satisfy any closure conditions. A format (logic) that allows one to specify properties may be called a *property notation*.

We will concentrate on one behavioral property logic in this paper, called RdTrLρ (RdTrL for Ready Trace Logic, and ρ for real time). This logic provides primitive properties for traces. Using a convention for the implicit presence of a trace variable and a temporal variable, each formula of RdTrLρ can be interpreted as a set of traces (at the corresponding level of technical abstraction).

1.2. Trace Theory.

It may be useful to explain why in our view trace theory is a necessary tool in the systems design and analysis area. Trace theory complements process algebra based on bisimulation semantics by being more abstract and allowing a full exploitation of the expressive power of linear time temporal logic. In addition, trace theory allows a very flexible expression of system properties related to *fairness*. There is an extensive literature about system description in terms of their set of traces, e.g., [BRO92, VIN90, MEY85].

The main degree of freedom in the development of trace theories lies in the additional decoration of the traces with information on how a computation could have proceeded in alternative ways. In Eindhoven, starting with [REM83], a group of researchers has developed a form of trace theory using undecorated traces. Many theoretical results have been found (e.g., [SNE85, KAL86]) and significant applications concerning integrated circuit design and verification, as well as foundational advances in the concepts of selftimed and delay insensitive systems, have been obtained (e.g., [UDD86, EBE89, BER92]).

In [BRHR84, OLH83], two forms of decorated trace theory have been introduced: failure set semantics and ready set semantics. This work provided a basis for theoretical CSP, a language that has been a platform for many subsequent studies (e.g. [HOA85]). Quite related to this work is the refusal sets model of Phillips [PHI87] and work by [HDN84] on testing.

In [PNU85], the barbed wire model was proposed. This is a decorated trace theory in which, for each action, the set of actions that a process might have done alternatively is recorded. In [BABK87] a similar model was proposed under the heading of a *ready trace set* model. Both proposals originate in the observation that for certain system construction techniques (e.g. broadcasting, priority mechanism), the distinguishing power of Rem's trace theory or that of [BRHR84] is not sufficient.

This paper proposes a version of trace theory that uses the ready trace set model of [BABK87], but slightly modified in order to accommodate a precise account of fairness and liveness and to support a mixed term formalism that exploits linear time ready trace logic as a system construction primitive (Definition 4.4). Our contribution is a systematic development of a version of a trace theory for the syntax of ACPρI [BAB91]. In particular, we develop an appropriate property language compatible with ACPρI. No novelty lies in any of the semantic techniques, but our ready trace theory provides a selfcontained explanation of a language for process description, as well as a preferred semantic model.

2. Real Time Ready Traces.

Due to the lack of space, we present here only one version of the theory: *real time ready traces*. Two other versions, *untimed (symbolic)* and *discrete time ready traces* can be derived from it, see [BABB93]. Throughout the paper, let A be a set, whose elements are used as atomic actions, and let $\delta \notin A$.

$RT\rho(A)$, the collection of real time ready traces over A, consists of the functions
f: $\mathbb{R}_{\geq 0} \to \{\surd, \delta, \Omega\} \cup \{(U, a) : U \subseteq A \cup \{wait\}, a \in U\}$, with the conditions:

- $f(t) = \surd$ $\Rightarrow \forall s > t\ f(s) = \surd$ terminated trace
- $f(t) = \delta$ $\Rightarrow \forall s > t\ f(s) = \delta$ deadlocked trace
- $f(t) = \Omega$ $\Rightarrow \forall s > t\ f(s) = \Omega$ trace with undefined tail (see Example 4.3)
- $f(0) = (U, a) \Rightarrow$ a=wait no action at 0.

If $f(t) = (U, a)$, we put $f(t)_1 = U$, $f(t)_2 = a$. Otherwise, these notions are undefined. We call the (U, a) *ready pairs*. We call a ready trace α *terminating* if there is a t with $\alpha(t) = \sqrt{}$, *deadlocking* if there is a t with $\alpha(t) = \delta$, *eventually undefined* if there is a t with $\alpha(t) = \Omega$.

We call a ready trace set V *ready closed* if $\forall \beta \in V \, \forall r \in \mathbb{R}_{\geq 0} \, \forall U \subseteq A \cup \{wait\} \, \forall a,b \in U$:

$$[\beta(r) = (U, a) \Rightarrow \exists \beta' \in V \, \beta'(r) = (U, b) \wedge \forall s<r \, \beta'(s) = \beta(s)]$$

(i.e., if V contains a trace that shows that an action b is ready at time r, then V contains also a trace that deviates from the previous one exactly by taking the action b at time r).

We call a ready trace set V *time deterministic* if

$$\forall \alpha, \beta \in V \, \forall r \in \mathbb{R}_{\geq 0} \, (\forall s < r \, \alpha(s)_2 = \beta(s)_2 = wait) \Rightarrow$$
$$(\alpha(r)_1 = \beta(r)_1 \vee \alpha(r) = \sqrt{} \vee \beta(r) = \sqrt{})$$

(i.e., if V contains two traces that have only waited until r, then they have the same ready set at r (unless one of them terminated); thus they made no choice by just waiting.)

We call a ready trace α *sufficiently defined* if

$$\forall t \in \mathbb{R}_{\geq 0} \, (\alpha(t) = \Omega \Rightarrow \{t' < t \mid \alpha(t') = (U, a) \wedge a \neq wait\} \text{ is infinite})$$

(i.e., a trace can only be eventually undefined after an infinite number of actions). A ready trace set is sufficiently defined if all its traces are sufficiently defined.

We call a ready trace set V *right closed* if $\forall W \subseteq V \, \forall t_1 \, \forall t_2 \geq t_1$ (in particular $t_2 = \infty$)

$(\,(\forall \alpha, \beta \in W \, (\forall t < t_1 \quad \alpha(t) = \beta(t) \wedge$
$\qquad \forall t \in [t_1, t_2) \, ((\forall s \in [t_1, t) \, \alpha(s)_2 = \beta(s)_2 = wait) \Rightarrow \alpha(t)_1 = \beta(t)_1)) \wedge$
$\forall t \in [t_1, t_2) \, \exists \alpha \in W \, \forall s \in [t_1, t) \, \alpha(s)_2 = wait)$
$\Rightarrow (\exists \alpha \in V \, (\forall t \in [t_1, t_2) \, \alpha(t)_2 = wait \wedge$
$\qquad \forall \beta \in W \quad (\forall t < t_1 \, \alpha(t) = \beta(t) \wedge$
$\qquad\qquad \forall t \in [t_1, t_2) \, (\forall s \in [t_1, t) \, \beta(s)_2 = wait \Rightarrow \alpha(t)_1 = \beta(t)_1))) \,)$

(i.e., if a process can wait until a time arbitrarily close to t_2, then it must contain a trace that can wait until (but not including) t_2. This also holds for a 'subprocess' of V: a subset W of traces of V that are the same until time t_1, after which they obey time-determinism). See Section 3 for further explanation. A *process ready trace set* is a ready trace set that is ready closed, time deterministic, sufficiently defined and right closed.

3. An Algebra of Process Ready Trace Sets.

Let $\gamma: A \cup \{\delta\} \times A \cup \{\delta\} \to A \cup \{\delta\}$ be a commutative and associative function with $\gamma(\delta, a) = \delta$ for all $a \in A \cup \{\delta\}$. γ is called a *communication function*, and $\gamma(a,b)$ represents the action that results if a and b occur simultaneously. Let PRTS be the class of process ready trace sets. As we only deal with the behavior of processes (i.e., their ready trace set) in this paper, we will usually call a process ready trace set a *process*.

We define the following operators on PRTS:

- U: PRTS $\to \mathbb{R}_{\geq 0} \cup \{\infty\}$ ultimate delay; ∞ denotes infinite delay: $\sup(\mathbb{R}_{\geq 0})$.
- $a(t), \delta(t) \in$ PRTS atomic action and deadlock; ($a \in A$, $t \in \mathbb{R}_{\geq 0} \cup \{\infty\}$).
- $+, \cdot, \|$: PRTS $\times$ PRTS $\to$ PRTS alternative, sequential and parallel composition.
- ∂_H: PRTS $\to$ PRTS encapsulation. ($H \subseteq A$).
- $\int_V$: (V $\to$ PRTS) $\to$ PRTS ($V \subseteq \mathbb{R}_{\geq 0}$; for $\int_V \lambda v.x(v)$, we write $\int_{v \in V} x(v)$): the infinite alternative composition of $x(v)$, $v \in V$.

These operators have the following interpretation [.] (where a,b range over $A \cup \{wait\}$).

- $U(x) = \sup\{t \in \mathbb{R}_{\geq 0} : \exists \alpha \in [x] \; \forall t' < t \; \alpha(t')_2 = wait\}$
- $[\delta(t)] = \{\alpha\}$, where $\alpha(s) = (\{wait\}, wait)$ for $s < t$, δ for $s \geq t$.
- $[a(t)] = \{\alpha\}$, where $\alpha(s) = (\{wait\}, wait)$ for $s < t$, $(\{a\}, a)$ for $s = t$ and $\sqrt{}$ for $s > t$.
  In particular, $[\delta(\infty)] = [a(\infty)] = \{\alpha\}$, where for all s: $\alpha(s) = (\{wait\}, wait)$.

The interpretations of $x + y$, $x \cdot y$, etc. are obtained from [x] and [y] in two steps. First, there is a *selection* of traces from [x] and [y], in which traces from one process that are incompatible with the other process are discarded. Second, there is a *combination* of traces. This combination usually involves, for each point in time, the choice of an action from [x] or one from [y], and a combination of ready sets.

- $[x + y] = \{\alpha' : \alpha \in [x] \cup [y] \text{ and } U(\{\alpha\}) < U([x] \cup [y]) \Rightarrow \alpha(U(\{\alpha\})) \neq \delta\}$ *(discard a trace that ends in deadlock after only waiting, if the process allows a longer wait)*, where
  $\alpha'(r) = (Z, a)$ if $\alpha(r)_2 = a$, $U(\{\alpha\}) = r$, and $Z = \cup\{\beta(r)_1 : \beta \in [x] \cup [y] \text{ and } U(\{\beta\}) \leq r\}$
  (if α has only waited so far, then add readies from the other process),
  $\alpha(r)$ in all other cases.

- $[x \cdot y] = \{\alpha \cdot_{rt} \beta : \alpha \in [x], \beta \in [y] \text{ and } \forall r \; (\alpha(r) \neq \sqrt{} \land wait \in (\beta(r)_1) \Rightarrow \beta(r)_2 = wait)\}$
  (discard those traces from [y] that start doing actions too early), where
  $(\alpha \cdot_{rt} \beta)(r) = \alpha(r)$ if $\alpha(r) \neq \sqrt{}$,
  $\qquad\qquad\quad\;\; \beta(r)$ if $\alpha(r) = \sqrt{}$ and $\forall t < r \; (\alpha(t) = \sqrt{} \lor \beta(t)_2 = wait)$,
  $\qquad\qquad\quad\;\; \delta\;\;\;$ if $\alpha(r) = \sqrt{}$ and $\exists t < r \; (\alpha(t) \neq \sqrt{} \land \beta(t)_2 \neq wait)$.
  (β had to start before α was finished, which results in deadlock).

- $[x \parallel y] = \{\alpha \parallel_{rt} \beta : \alpha \in [x], \beta \in [y] \text{ and }$
  $\forall r \; ((\alpha(r) = (U, a) \land \beta(r) = (V, b) \land a,b \in A \land \gamma(a,b) = \delta) \Rightarrow$
  $(wait \notin U \cup V \land \forall c \in U \; \forall d \in V \; \gamma(c,d) = \delta))\}$, *(discard pairs of traces that deadlock because they fail to communicate, if one of the processes could wait, or if communication was possible)*, where

$(\alpha \parallel_{rt} \beta)(r) = \;\delta \qquad\quad$ if $\alpha(r) = \delta$ or $\beta(r) = \delta$ or *(communication failed or fails:)*
$\qquad\qquad\qquad\qquad\qquad \exists t \leq r \; \alpha(t) = (U, a), \beta(t) = (V, b), a,b \in A, \gamma(a,b) = \delta$,
$\qquad\qquad\quad\;\; \Omega \qquad\quad$ if $\alpha(r) = \Omega$ or $\beta(r) = \Omega$,
$\qquad\qquad\quad\;\; \beta(r) \qquad\;$ if $\alpha(r) = \sqrt{}$,
$\qquad\qquad\quad\;\; \alpha(r) \qquad\;$ if $\beta(r) = \sqrt{}$,
$\qquad\qquad\quad\;\; (Z, a) \qquad$ if $\alpha(r) = (U, a), \beta(r) = (V, wait)$,
$\qquad\qquad\quad\;\; (Z, b) \qquad$ if $\alpha(r) = (U, wait), \beta(r) = (V, b)$,
$\qquad\qquad\quad\;\; (Z, \gamma(a,b))$ if $\alpha(r) = (U, a), \beta(r) = (V, b), a,b \in A, \gamma(a,b) \neq \delta$.
In the last three cases, $Z = U \cup V \cup \{\gamma(a,b) : a \in U, b \in V, \gamma(a,b) \neq \delta\}$.

- $[\partial_H(x)] = \{\partial_{H,rt}(\alpha) : \alpha \in [x] \text{ and } \forall r \; (\alpha(r) = (U, a) \land a \in H \Rightarrow U \subseteq H)\}$
  (discard traces showing an encapsulated action, where another action or waiting is possible), where
  $\partial_{H,rt}(\alpha)(r) = \alpha(r) \quad\;\;$ if $\alpha(r) \in \{\sqrt{}, \delta, \Omega\}$ and $\forall t < r \; (\alpha(t) = (U, a) \Rightarrow a \notin H)$,
  $\qquad\qquad\quad\;\; (U-H, a) \;$ if $\alpha(r) = (U, a)$ and $\forall t \leq r \; (\alpha(t) = (U, a) \Rightarrow a \notin H)$,
  $\qquad\qquad\quad\;\; \delta \qquad\qquad$ if $\exists t \leq r \; \alpha(t) = (U, a)$ and $a \in H$.

- $[\int_{v \in V} x(v)] = \{\alpha' : \alpha \in \cup_{v \in V} [x(v)] \cup [\delta(\sup\{U(x(v)) : v \in V\}]$ *(see below)* and
  $U(\{\alpha\}) < \sup\{U(x(v)) : v \in V\} \Rightarrow \alpha(U(\{\alpha\})) \neq \delta\}$ *(see $[x + y]$)*, where

  $\alpha'(r) = (Z, a)$ if $\alpha(r)_2 = a$, $U(\{\alpha\}) = r$, and $Z = \cup\{\beta(r)_1 : \beta \in \cup_{v \in V} [x(v)], U(\{\beta\}) \leq r\}$
  $ \alpha(r)$ in all other cases.

The reason for adding the trace $[\delta(\sup\{U(x(v)) : v \in V\}]$ when considering the integral is that otherwise the definitions would give us e.g. $a(1) \cdot \int_{1<t<2} b(t) + a(1) \cdot \int_{1<t<3} b(t) = a(1) \cdot \int_{1<t<3} b(t)$, which seems undesirable. Namely, all traces of $\int_{1<t<2} b(t)$ would have to do a b-step before time 2, thus no trace could signal that b is no longer ready at time 2. The addition of $\delta(2)$ solves this problem, and also makes the trace set of $\int_{1<t<2} b(t)$ right closed. On the other hand, adding $\delta(2)$ to $\int_{1<t\leq 2} b(t)$ makes no difference, because it already has a trace α with $\alpha(t)_2 = $ wait for all $t<2$ and $\alpha(2) = (\{b\}, b)$.

In this way, we also avoid that $[\int_{t>0} \delta(t)] = \emptyset$, which would mean $\partial_{\{r,s\}}((i(1) \cdot s(2) + j(1)) \parallel \int_{t>0} r(t)) = i(1) \cdot c(2)$ in a context with $\gamma(r,s) = c$. (Compare with Section 6: it should not be possible that a component of a system, that decides internally whether or not to send a message, is forced to send one because another component is waiting to receive it.)

Claim 3.1. PTRS *with the above operators is a model of the process algebra* ACPρI.

Of course, more identities hold in this model, which is based on ready traces, than in ACPρI, which axiomatizes bisimulation.

4. Real Time Ready Trace Logic.

We proceed by giving the syntax and semantics of the primitives of a language describing real time ready trace sets, parametrised by the set of atomic actions A.

Definition 4.1. Let α be a real time ready trace.

    α **sat** $a(t)$ if $\alpha(t) = (U, a)$ for some $U \subseteq A \cup \{\text{wait}\}$
        (In particular, if α **sat** wait(t), then the system does not perform an action at t.)
    α **sat** $R(a, t)$ if $\alpha(t) = (U, b)$ for some $U \subseteq A \cup \{\text{wait}\}$, $a, b \in U$
        (In particular, α **sat** $R($wait$, t)$, then the system need not perform an action at t.)
    α **sat** $\sqrt{}(t)$ if $\alpha(t) = \sqrt{}$
    α **sat** $\delta(t)$ if $\alpha(t) = \delta$
    α **sat** $\Omega(t)$ if $\alpha(t) = \Omega$

The real time ready trace language RdTrLρ(A) consists of the primitives mentioned above, augmented with constants for all objects in its time domain $\mathbb{R}_{\geq 0}$, a total ordering $<$ and binary operators $+$, $\cdot$, $\dot{-}$, and $/$ on it. Furthermore, we have the following constructors from standard predicate logic: $\wedge, \vee, \neg, \rightarrow, \leftrightarrow, \forall$, and $\exists$ (quantification can be over the time domain, the set of atomic actions A, or parts thereof). The semantics of all these language constructs is the standard one. In this way we obtain an explicit time temporal logic. For similar logics see [KOY89a]. (Note that we have some freedom in choosing operators on the time domains: different choices result in different languages with different expressive powers, see Example 5.4).

Definition 4.2. Let V be a real time ready trace set and φ a closed RdTrLρ(A) formula.
- V **sat** φ iff for all $\alpha \in$ V: α **sat** φ.
- RTρ(A, φ) = {$\alpha \in$ RTρ(A) | α **sat** φ}, the set of traces that satisfies φ.

Example 4.3. We present some examples of meaningful RdTrLρ(A) formulas. The following formulas (where a,b range over A $\cup$ {wait}) are satisfied by all ready trace sets.
$\forall$ s,t t < s $\rightarrow$ (δ(t) $\rightarrow$ δ(s) $\wedge$ $\sqrt{}$(t) $\rightarrow$ $\sqrt{}$(s) $\wedge$ Ω(t) $\rightarrow$ Ω(s)).
$\forall$ t a(t) $\rightarrow$ R(a, t).
$\forall$ t a(t) $\wedge$ b(t) $\rightarrow$ a = b.
$\forall$ t (δ(t) $\vee$ $\sqrt{}$(t) $\vee$ Ω(t)) $\rightarrow$ $\neg$R(a, t).

In the following, we use RdTrLρ(A) to formulate an interesting classification of traces. A trace α is a *bounded action frequency trace* if

$$\alpha \text{ sat } \neg \exists t \in \mathbb{R}_{\geq 0} [(\forall s < t \ \exists r \in (s,t) \ \exists a \in A \ a(r))].$$

(This formula says that there is no time point t, such that an infinite sequence of actions occurs, having t as the limit of their time points.) A trace α is a *Zeno trace* if

$$\alpha \text{ sat } \exists t \in \mathbb{R}_{\geq 0} [(\forall s < t \ \exists r \in (s,t) \ \exists a \in A \ a(r))] \wedge$$
$$\forall t \in \mathbb{R}_{\geq 0} [(\forall s < t \ \exists r \in (s,t) \ \exists a \in A \ a(r)) \rightarrow \Omega(t)].$$

(After the unique limit point t, the trace is undefined.) A trace α is a *supertask trace* if

$$\alpha \text{ sat } \exists t \in \mathbb{R}_{\geq 0} [(\forall s < t \ \exists r \in (s,t) \ \exists a \in A \ a(r)) \wedge \neg \Omega(t)].$$

(The trace continues after a limit point.)

Notice that this defines three disjoint sets of traces: each trace either has bounded action frequency, or is Zeno or supertask. A set of traces has bounded action frequency if all traces it contains have bounded action frequency (but see Section 6.3), and a set of traces is called non-supertask if all traces it contains either have bounded action frequency or are Zeno. Usually, we consider non-supertask trace sets only.

Koymans [KOY89a] states in Section 5.3 (page 73) that *syntactical abstractness* imposes the restriction to specify message passing systems solely in terms of their input and output actions. It turns out that the decision to restrict the specification langauge to a syntactically abstract one is both clarifying and mathematically rewarding. We adhere to Koymans' criterion of syntactical abstractness by restricting the languages RdTrL as to use a, R(a) (a $\in$ A $\cup$ {wait}), $\sqrt{}$, δ, and Ω only.

Message passing systems are a special kind of processes. Therefore some compromise with Koymans' requirement on syntactical abstractness is to be expected. We think that our extension of the admitted propositions with readies, termination, deadlock, undefinedness and error captures a meaningful version of the concept of syntactical abstractness for temporal description formalisms in our specific context.

In contrast to [KOY89a], we have included natural numbers, booleans and various common operators in RdTrL. This makes the language more complex than Koymans allows, but on the other hand provides it with a uniform logical complexity. We maintain that even these extensions are consistent with Koymans' request for syntactical abstractness, because the additional mechanisms are fully standard in mathematics.

It is not decidable whether RTρ(A, φ) is ready closed, given a formula φ in RdTrLρ(A) (see [BABB93]). The fact that such a basic aspect of RTρ(A,φ) is not guaranteed and 'even worse' not decidable, justifies that the language RdTrLρ(A) is called a *property language* rather than a process description language (or in programming language terminology: a process notation). So we distinguish between *process notations* such as ACPρI(A) and *property notations* such as RdTrLρ(A). Semantically, both determine subsets of RTρ(A), the difference being that a process notation always denotes a ready closed, time deterministic and right closed trace set (i.e., a process).

It should be noted that in the untimed case, a property notation can often be found very close to the process notation. A typical property notation is $\tau_I(X) = Q$. This asserts of process X that after abstraction from steps in I (i.e. turning X-actions in I into silent ones) X becomes equal to process Q. In principle, this technique can be used in the real time case just as well. The drawback however is that known abstraction operators reduce process complexity much less than in the untimed case. This is due to the fact that timing information cannot easily be suppressed by means of an abstraction operator.

We conclude that in the real time case, a distinction between a property notation and a process notation is justified, if not unavoidable.

Definition 4.4. Having defined ready trace language, we can now define the function $\varphi \Box$: PRTS → PRTS, for $\varphi \in$ RdTrLρ(A): for $x \in$ PTRS, $\varphi \Box x = \{\alpha \in x : \alpha$ **sat** $\varphi\}$, if this set is a process ready trace set (undefined otherwise).

We notice that x **sat** φ iff $\varphi \Box x = x$. An operator similar to $\varphi \Box x$ has been defined on the bisimulation model in Parrow's thesis [PAR85]. "$\varphi \Box x$" imports the property language into the process notation. We can, conversely, add a special primitive to RdTrLρ(A), that imports process notation into it.

Definition 4.5. For an expression P in a process notation, the primitive formula cd(P) denotes the *complete description* of P, that is, a property that is satisfied by a trace α iff α is in the trace set of P.

It depends on the expressibility of the logic and of the process notation, whether for every process P, cd(P) can be expressed in the other primitives of the logic. (Even if this is the case, the complete description primitive may be useful as syntactic sugar.) We expect that in most cases, even for simple processes involving recursion, the complete description is not expressible in the other primitives (see Example 5.4). The results of [KOY89b] point in the same direction. However, our conjecture is that each recursion-free process expression in ACPρI(A) has a complete description in RdTrLρ(A).

Example 4.6. We provide complete descriptions of two process expressions; the intention is to exemplify the difference in nature between ACPρI(A) and RdTrLρ(A). First,

cd(a(7)) = $\forall t<7$ (wait(t) $\wedge \forall b \in A \neg R(b, t)) \wedge$
$\quad$ a(7) $\wedge \neg R(wait, 7) \wedge \forall b \in A$ (R(b, 7) → b=a) $\wedge \forall t>7$ $\sqrt{(t)}$.

Second, let B = $\int_{t>0} \sum_{d \in D} r(d)(t) \cdot s(d)(t + 1)$. Then

cd(B) = $\exists t_0 > 0 \, \exists d \in D \; \forall t \leq t_0 \, \forall a \in A$ (R(a, t) ↔ $\exists e \in D$ a=r(e)) $\wedge$ R(wait, t_0) $\wedge$
$\quad$ r(d)(t_0) $\wedge$ s(d)(t_0 + 1) $\wedge \forall t$ (t ≠ $t_0 \wedge$ t < t_0+ 1 → wait(t)) $\wedge$
$\quad \forall t \, \forall a \in A$ (t_0 < t < t_0+1 → $\neg R(a, t)) \wedge$
$\quad \forall a \in A$ (R(a, t_0+1) → a=s(d)) $\wedge \neg R(wait, t_0+1) \wedge$
$\quad \forall t > t_0 + 1$ $\sqrt{(t)}$
$\vee \; \forall t > 0$ (wait(t) $\wedge \forall a \in A$ (R(a, t) ↔ $\exists e \in D$ a=r(e))).

We conclude that for several simple processes a complete description is fairly complex and as a consequence, uninformative. But our hope is, that a complete construction of cd(B) is not necessary for proving a statement like cd(B) → ∀d,t r(d)(t) → s(d)(t + 1). Even without such a proof system, our logic is valuable as a means to *express* such a statement.

5. Recursion.

Now we consider recursive equations in this framework. As an example, we take $X = \int_{t>0} a(t) \cdot X$. A first solution is the ready trace set determined by the formula:

$\varphi_1 \equiv \forall t>0 \ (R(wait, t) \wedge \forall b \in A \ (R(b, t) \leftrightarrow b=a)) \wedge \neg \exists t \in \mathbb{R}_{\geq 0} \ [(\forall s < t \ \exists r \in (s,t) \ a(r))]$.

This formula determines a ready trace set that is a solution and consists only of bounded action frequency traces. But the ready trace set determined by the formula:

$\varphi_2 \equiv \varphi_1 \wedge \exists t \ \forall r, s \ (t < r < s \wedge a(r) \wedge a(s) \wedge \forall v \ (r < v < s \rightarrow \neg a(v)) \rightarrow s = r + 1)$.

also denotes a process ready trace set. The formula says that after a certain time t, if more a-actions come, then they must be one time-unit apart. Putting an additional a-action in front does not change this property. Thus we find that this process also denotes a solution of the recursion equation. Many more solutions exist, some containing Zeno and supertask traces. Thus this equation comes nowhere near to having a unique solution.

These considerations change if only finite ready traces are considered (thus bringing the approach closer to that of timed CSP [RER88]). The choice to work with complete (usually infinite) traces rather than with incomplete (finite) traces is motivated by the marvellous expressive power concerning different forms of liveness and fairness that is obtained if systems are described by means of these complete traces. This expressive power seems to be mainly responsible for the success of the temporal logic approach to concurrency.

If we want that recursive equations nevertheless define a unique process, then there are at least two options. One option is to define a metric on ready traces, and to allow only closed sets of ready traces as solutions. This leads us to the field of so-called *topological process theory*, initiated with the work of De Bakker & Zucker [BAZ82]. Restricting the class of ready trace sets by topological means in an appropriate way leads to a domain in which guarded equations have unique solutions. In the real time case, the topological techniques become much more complex, unfortunately. To our taste, these techniques are not satisfactory, and we propose a different way to have recursion equations define a process.

Our proposal for the introduction of recursively defined processes in extensional ready trace theory is the following, using techniques similar to [BEK88]. Let $\langle X(t_1, ..., t_k) \mid E \rangle$ denote a mapping from $\mathbb{R}^k$ to PRTS for each guarded recursive specification E involving a process variable X with k real parameters. For particular real values $r_1, ..., r_k$ a process ready trace set

$$P(r_1, ..., r_k) = \langle X(t_1, ..., t_k) \mid E \rangle(r_1, ..., r_k) = \langle X(r_1, ..., r_k) \mid E \rangle$$

is obtained as follows:
i. Determine a real time transition system from E for each of its (parametrised) process variables, following [BAB91]. (For simplicity, we do not consider the φ □ operator in this section.)
ii. Determine the ready trace set of the transition system thus obtained for $X(r_1, ..., r_k)$. This ready trace set is $P(r_1, ..., r_k)$.

We briefly recall the definition of a transition system from [BAB91, Section 4.4]; our presentation here is somewhat simplified.

Definition 5.1. A *state* is a pair $\langle p, t\rangle$, where p is a closed process expression or the termination symbol $\sqrt{}$, and $t \in \mathbb{R}_{\geq 0}$. A *transition* is a triple (source, action, target), usually denoted as source $\xrightarrow{\text{action}}$ target, where source and target are states and action $\in A \cup \{\text{wait}\}$. Intuitively, $\langle p, t\rangle \xrightarrow{a} \langle p', t'\rangle$ means that the process p, when the time has become t, can wait until time t', at which time it performs an a-step and turns into p'.

A *transition system* is a set TS of transitions satisfying for all $a \in A \cup \{\text{wait}\}$:

- $\langle s, t\rangle \xrightarrow{a} \langle s', t'\rangle \in TS \Rightarrow (t < t' \land s \neq \sqrt{} \land (a = \text{wait} \Rightarrow s = s'))$,
- $\langle s, t\rangle \xrightarrow{a} \langle s', t'\rangle \in TS \Leftrightarrow \forall t'' \in (t, t') \; \langle s, t\rangle \xrightarrow{\text{wait}} \langle s, t''\rangle \in TS \land \langle s, t''\rangle \xrightarrow{a} \langle s', t'\rangle \in TS$.

The rules for determining a transition system from a recursive specification, as given in [BAB91], ensure that these properties hold. A *path* through a transition system TS is a *countable* sequence $t_1, t_2, \ldots$ of transitions from TS such that for all $i > 0$:

- t_{i+1} exists if and only if the target of t_i is the source of one or more transitions in TS,
- if t_{i+1} exists, then the target of t_i equals the source of t_{i+1}.
- if, for some $t \in \mathbb{R}_{\geq 0}$, all transitions $t_i, t_{i+1}, \ldots$ have wait as their action and a time $t' < t$ in their target, then there exists an $n \geq i$ such that for all sources s of $t_n, t_{n+1}, \ldots$, TS contains no transition $s \xrightarrow{a} \langle s'', t\rangle$ ($a \in A \cup \{\text{wait}\}$).

The last clause of this definition prevents paths that fail to proceed without being 'trapped'. For example, if TS contains the transition $\langle a(1),0\rangle \xrightarrow{a} \langle \sqrt{},1\rangle$ and all the transitions that come with this one, then $\langle a(1), 0\rangle \xrightarrow{\text{wait}} \langle a(1), 1/2\rangle \xrightarrow{\text{wait}} \langle a(1), 3/4\rangle \xrightarrow{\text{wait}} \langle a(1), 7/8\rangle \xrightarrow{\text{wait}} \ldots$ is not a valid path. In contrast, if TS contains $\{\langle s,0\rangle \xrightarrow{a} \langle \sqrt{},t\rangle \mid 0 < t < 1\}$, but *not* $\langle s,0\rangle \xrightarrow{a} \langle \sqrt{},1\rangle$, nor any other transition $\langle s,0\rangle \xrightarrow{b} \langle s',1\rangle$ ($b \in A \cup \{\text{wait}\}$), then this path is valid. (E.g., $s = \int_{0<t<1} a(t)$.)

Definition 5.2. Let σ be a path through a transition system TS. The *ready trace determined by* σ *in* TS, $RT(\sigma, TS)$, is for each time $t \in \mathbb{R}_{\geq 0}$ defined as $RT(\sigma, TS)(t) =$

$\sqrt{}$	if σ is finite and the target of its last transition is $\langle \sqrt{}, t'\rangle$, with $t' < t$;
δ	if σ is finite and the target of its last transition is $\langle s, t'\rangle$, with $t' < t$ and $s \neq \sqrt{}$ (closed time stop);
δ	if σ is infinite, $t' < t$ for all targets $\langle s, t'\rangle$ of transitions in σ, and finitely many transitions in σ have an action other than wait (open time stop);
Ω	if σ is infinite, $t' < t$ for all targets $\langle s, t'\rangle$ of transitions in σ, and infinitely many transitions in σ have an action other than wait;
(U, a)	for $a \in A \cup \{\text{wait}\}$, if there exist s', t', s such that $\langle s', t'\rangle \xrightarrow{a} \langle s, t\rangle \in \sigma$ and $U = \{b \in A \cup \{\text{wait}\} \mid \exists s'' \; \langle s', t'\rangle \xrightarrow{b} \langle s'', t\rangle \in TS\}$;
(U, wait)	if there exist $a \in A \cup \{\text{wait}\}$, s', t', s'', t'' such that $\langle s', t'\rangle \xrightarrow{a} \langle s'', t''\rangle \in \sigma$, $t' < t < t''$ and $U = \{\text{wait}\} \cup \{b \in A \mid \exists s \; \langle s', t'\rangle \xrightarrow{b} \langle s'', t\rangle \in TS\}$.

Now let p be a closed process expression, and let TS be the transition system associated to p as defined in [BAB91]. Then the ready trace set p is $\{RT(\sigma,TS) \mid \sigma$ is a path through TS and the source of the first transition of σ is $\langle p,0 \rangle\}$.

Proposition 5.3. *Let p be a closed process expression without recursion. Then the ready trace set p defined above coincides with* [p], *as defined in Section 3. (It is especially interesting to check this claim for integrals over right-open intervals.)*

Note that we had *no* intention that unique solutions are obtained. Rather, we have defined a uniform way to select some solution of the system E and to use that to evaluate $\langle X(t_1, ..., t_k) \mid E \rangle$. Other selection mechanisms can lead to other interpretations of this syntax. Also note that the definitions imply that the ready trace defined by a path in a transition system is not a supertask. A special construction is needed for obtaining supertasks; this construction is introduced in Definition 5.5.

Example 5.4. We provide complete descriptions of the three clocks of [BAB91].
The process $\langle C_1(1) \mid C_1(t) = \text{tick}(t) \cdot C_1(t+1) \rangle$ (a perfect clock) has the following complete description:

$\forall n \in \mathbb{N} \, (n \neq 0 \rightarrow \text{tick}(n) \wedge \forall a \in A \, (R(a, n) \rightarrow a = \text{tick})) \wedge$
$\forall t \geq 0 \, (\forall n \in \mathbb{N} \, t \neq n \vee t = 0 \rightarrow \text{wait}(t) \wedge \forall a \in A \, \neg R(a, t)).$

The process $\langle C_2(1) \mid C_2(t) = \int_{v \in [t-0.01, t+0.01]} \text{tick}(v) \cdot C_2(t+1) \rangle$ (a clock allowing some fluctuation of the ticks) has the following complete description:

$\forall t < 0.99 \; \text{wait}(t) \wedge \forall a \in A \neg R(a, t) \wedge$
$\forall n \in \mathbb{N} \, (n \neq 0 \rightarrow$
$\quad \exists t \in [n-0.01, n+0.01]$
$\quad\quad (\text{tick}(t) \wedge (t \neq n+0.01 \leftrightarrow R(\text{wait}, t)) \wedge \forall a \in A \, (R(a, t) \rightarrow a = \text{tick})) \wedge$
$\quad \forall r \in [n-0.01, t) \, (\text{wait}(r) \wedge R(\text{tick}, r) \wedge \forall a \in A \, (R(a, r) \rightarrow a = \text{tick})) \wedge$
$\quad \forall r \in (t, n+0.99) \, (\text{wait}(r) \wedge \forall a \in A \neg R(a, r))).$

Whether or not the process $\langle C_3(1) \mid C_3(t) = \int_{v \in [t-0.01, t+0.01]} \text{tick}(v) \cdot C_2(v+1) \rangle$ (a clock cumulating the errors) has a finite complete description, depends on the expressiveness of the logic. An infinite series of consecutive choices, each one depending on the previous choice, has to be made. We can suggest the following higher order description:

$\exists t_0, t_1, t_2, t_3, ... \quad t_0 = 0 \wedge \forall t < 0.99 \; \text{wait}(t) \wedge \forall a \in A \neg R(a, t) \wedge$
$\forall n \in \mathbb{N} \, (t_{n+1} \in [t_n+0.99, t_n+1.01] \wedge \text{tick}(t_{n+1}) \wedge (t_{n+1} \neq t_n+1.01 \leftrightarrow R(\text{wait}, t_{n+1})) \wedge$
$\quad \forall a \in A \, (R(a, t_{n+1}) \rightarrow a = \text{tick})) \wedge$
$\quad \forall t \in [t_n+0.99, t_{n+1}) \, (\text{wait}(t) \wedge R(\text{tick}, t) \wedge \forall a \in A \, (R(a, t) \rightarrow a = \text{tick})) \wedge$
$\quad \forall t \in (t_{n+1}, t_{n+1}+0.99) \, (\text{wait}(t) \wedge \forall a \in A \neg R(a, t))).$

But higher order logic is not needed here: we can encode the sequence $t_0, t_1, t_2, ...$ by one real number r. If we find an encoding such that the corresponding function decode, satisfying $\forall n \in \mathbb{N} \; \text{decode}(r,n) = t_n$, is expressible in the arithmetic of the logic, then we can replace in the above description $\exists t_0, t_1, t_2, ...$ by $\exists r$, and each t_n by decode(r,n).

The normal task axiom (NTA) excludes supertasks:

$\text{NTA} \equiv \forall t \, ((\forall s < t \, \exists r \in (s,t) \, \exists a \in A \; a(r)) \rightarrow \Omega(t))$, or equivalently:
$\neg \exists t \in \mathbb{R}_{\geq 0} \, [(\forall s < t \, \exists r \in (s,t) \, \exists a \in A \; a(r)) \wedge \neg \Omega(t)].$

Suppose one intends to allow traces that do not satisfy the NTA, i.e. supertask traces. This can be useful for the conceptual analysis of certain communication protocols (see Section 6). We provide an operator that introduces supertask processes (i.e., ready trace sets that may contain supertasks).

Definition 5.5. The operator $\sqrt{}_\Omega$ is defined on ready traces by: $\sqrt{}_\Omega(\alpha)(t) = \sqrt{}$ if $\alpha(t) = \Omega$, $\alpha(t)$ otherwise. On ready trace sets, $\sqrt{}_\Omega$ is defined by applying it to each element of the set. A supertask is then obtained, e.g., as follows:

$$P = \sqrt{}_\Omega(\langle X(1) \mid X(t) = a(t)\cdot X(1 + t/2)\rangle)\cdot a(3).$$

It should be noticed that this operator is meaningless on transition systems. Consequently, a semantic model for recursion equations involving this operator requires more sophicated techniques than the ones outlined above.

6. (Non-)Existence of Coordinated Attack Protocols.

In this section, we look at real time ready trace theory. The protocol we consider, is the so-called *Coordinated Attack Protocol:* via communication through unreliable media M_{12} and M_{34}, processes P and Q should synchronise on a certain action they each perform independently (or at least, the execution of the two actions should be close enough in time). For more information, see [HAM90]. See Fig. 1.

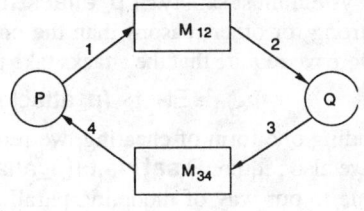

Fig. 1.

The story that goes with this picture is the following: P and Q are two generals that want to synchronise an attack on an army located between them, because only by working together they can beat this army. Their only means of communication is sending messengers that have to pass enemy lines. The messenger may arrive safely at the other army, or may be captured en route.

In the following, we use the convention, common in ACP, that s denotes a send action, r a receive action and i an internal action. Internal actions are subscripted for reference only. Send and receive actions are subscripted by a channel number; they communicate to a communication action c, also subscripted by the channel number. Thus the communication function γ satisfies: $\gamma(r_i(d), s_i(d)) = c_i(d)$ for $i = 1,2,3,4$, $d \in D$, In the interleaving framework of ACPρ, only one action at a time can occur. In order to allow simultaneous actions a and b, we make them 'communicate' to the so-called multi-action a&b. For simplicity, we shall not treat actions like a&b separately: we write X **sat** $a(t) \wedge b(t)$ instead of X **sat** (a&b)(t), violating the statement $\forall t\ a(t) \wedge b(t) \rightarrow a = b$ from Example 4.3.

6.1. The classical case.

We can describe the media as follows:

$$M_{12} = \int_{t>0} \sum_{d \in D} r_1(d)(t) \cdot (i_1(t+1)\cdot s_2(d)(t+2) + i_2(t+2)) \cdot M_{12}$$

$$M_{34} = \int_{t>0} \sum_{d \in D} r_3(d)(t) \cdot (i_3(t+1)\cdot s_4(d)(t+2) + i_4(t+2)) \cdot M_{34}$$

We may assume that the choice in the media is fair, in particular (and more specifically), let
$$\varphi_k = \forall t \, \exists s > t \, R(i_k, s) \rightarrow \forall t \, \exists s > t \, i_k(s),$$
then both media together can be written as
$$M = (\varphi_1 \Box M_{12}) \| (\varphi_3 \Box M_{34}).$$

In order to let the intended communication channels, *and only those*, function properly, we must

1. disallow all actions $r_1(d), s_2(d), r_3(d)$ and $s_4(d)$ by P and Q ($d \in D$),
2. encapsulate the system by the set $H = \{r_i(d), s_i(d) : i = 1, 2, 3, 4, d \in D\}$,
3. define: $\gamma(r_i(d), s_i(d)) = c_i(d)$ for $i = 1, 2, 3, 4, d \in D$,
   $\gamma(a, b) = a\&b$ otherwise.

Thus actions occurring at the same time, other than corresponding reads and sends, are completely independent.

The aim is that P and Q perform actions p_attack and q_attack at the same time: the final system must satisfy $\exists t \; p\_attack \& q\_attack(t)$. This requirement, however, might be too strong for other reasons than the one we aim at (such as relativistic considerations); therefore we require that the attacks take place at most one time unit apart:
$$\Phi_{syn} \equiv \exists t_1, t_2 \; (p\_attack(t_1) \wedge q\_attack(t_2) \wedge |t_1 - t_2| < 1).$$

Excluding one form of cheating, we require P **sat** $\forall t \neg q\_attack(t)$. Excluding another one, we also require P **sat** $\forall s, t \; ((p\_attack(s) \wedge p\_attack(t)) \rightarrow s = t)$. Similarly for Q.

Due to our way of modeling parallelism, all components of a system share a global clock. We need to incorporate some mechanism that excludes using this clock for achieving synchronization. For example, the solution $P = p\_attack(1)$, $Q = q\_attack(1)$ must be avoided. What we want to express is that $U(P) = \infty$, that is
$$\forall t > 0 \; \exists \alpha \in V_P \; \forall t' < t \; \alpha(t')_2 = \text{wait}$$
(where V_P denotes the ready trace set of P), but this statement cannot be rephrased as P **sat** φ for any formula φ, because here the quantification over traces is *existential*, whereas the definition of P **sat** φ employs *universal* quantification. So let us say that the generals must first arrive at their positions, and that they cannot tell in advance how long this will take. Until (and included) their time of arrival, messages sent to them are lost:

$$P\_init(t) = \int_{0 < u < t} (\sum_{d \in D} r_4(d)(u) \cdot P\_init(t)) + \sum_{d \in D} (r_4(d) \& p\_arrive)(t) + p\_arrive(t)$$

$$Q\_init(t) = \int_{0 < u < t} (\sum_{d \in D} r_2(d)(u) \cdot Q\_init(t)) + \sum_{d \in D} (r_2(d) \& q\_arrive)(t) + q\_arrive(t).$$

Now we can write the whole system as:
$$R = \partial_H(\int_{t > 0} (P\_init(t) \cdot P(t)) \| M \| \int_{t > 0} (Q\_init(t) \cdot Q(t))).$$

This formulation works only under the additional assumption that for all t, the ready trace set of P(t) is non-empty, and the same for Q.

Theorem 6.1. *For all processes P and Q satisfying the requirements above,*
$$R \; \textbf{sat} \; \Phi_{syn} \; \textit{is false.}$$

PROOF. See [BABB93].

6.2. Supertasks.

We see that we cannot obtain global synchronisation, even if the processes can have arbitrary form. Now we show that, if we relax the definition of the channel by allowing infinitely many inputs in a finite amount of time, we can obtain global synchronisation. Consider the following channels:

$$M^*_{12} = \int_{t>0} \sum_{d \in D} r_1(d)(t) \cdot (M^*_{12} \parallel (i_1(t+1) \cdot s_2(d)(t+2) + i_2(t+2)))$$

$$M^*_{34} = \int_{t>0} \sum_{d \in D} r_3(d)(t) \cdot (M^*_{34} \parallel (i_3(t+1) \cdot s_4(d)(t+2) + i_4(t+2)))$$

Immediately after receiving an input, a new input can be received. Still, each attempted communication, whether successful or not, takes two time units. We redefine M (and thus R) by:

$$M = (\varphi_1 \square M^*_{12}) \parallel (\varphi_3 \square M^*_{34}).$$

We can get global synchronisation, if we allow supertasks (see Example 4.3 and Definition 5.5). Before we define P and Q, we define two auxiliary processes. The first process, $\text{Send}_i(t)$, is a supertask: it sends within one second, beginning at t, an infinite number of messages through channel i; then it terminates. We do not need any content in the messages: $D = \{d\}$.

$$\text{Send}_i(t) = \sqrt{\Omega}(\langle S_i(t,1) \mid S_i(u,n) = s_i(d)(u) \cdot S_i(u+2^{-n}, n+1)\rangle).$$

Channel fairness ensures that from such a burst of messages, at least one arrives. The second process, Deaf_i, simply accepts and ignores messages coming in through channel i.

$$\text{Deaf}_i = \int_{t>0} r_i(d)(t) \cdot \text{Deaf}_i.$$

The roles of P and Q in this protocol are asymmetric. After P arrives, it sends a burst of messages every second, while listening for an answer. Half a time unit after the first answer comes (meaning that Q has arrived), P attacks. When Q arrives, it waits until it receives a message from P. It answers this message by a one-second burst. If this burst starts at time s, it occupies the interval [s,s+1). Thus at least one message of the burst arrives at P in the interval [s+2,s+3), and P attacks in the interval [s+2.5,s+3.5). So it is safe for Q to attack at s+3. P and Q accept and ignore all incoming messages after the first one.

$$P(t) = \langle X(t+1) \mid X(u) = \text{Send}_1(u) \cdot X(u+1)\rangle \parallel \int_{u>t} (r_4(d)(u) \cdot (\text{Deaf}_4 \parallel p\_attack(u+0.5))).$$

$$Q(t) = \int_{u>t} (r_2(d)(u) \cdot (\text{Deaf}_2 \parallel (\text{Send}_3(u+1) \cdot q\_attack(u+4)))).$$

Theorem 6.2. *If we do* not *allow supertasks, then for all processes* P *and* Q *satisfying the requirements of Section 6.1 (recall that* R *is now defined using* M^*_{12} *and* M^*_{34}*),*

$$R \text{ sat } \Phi_{\text{syn}} \text{ is false.}$$

PROOF. See [BABB93].

6.3 Variable Transmission Speed.

Instead of allowing infinitely many inputs in a finite amount of time, we can also allow the transmission speed to be variable. In this case, we can even obtain global synchronisation with processes without supertasks. Each trace of the system will be a bounded action frequency trace, but for every ε>0, there is a trace in the system's trace set for which the bound is smaller than ε. Therefore it is not implementable. This protocol was suggested by J.F. Groote [GRO92].

Consider the following channels with programmable transmission speed (the only message to be transmitted is the transmission speed, so we remove D):

$$M''_{12} = \int_{t>0} \int_{r \in (0,1)} r_1(r)(t) \cdot (i_1(t+r) \cdot s_2(r)(t+2r) + i_2(t+2r)) \cdot M''_{12}$$

$$M''_{34} = \int_{t>0} \int_{r \in (0,1)} r_3(r)(t) \cdot (i_3(t+r) \cdot s_4(r)(t+2r) + i_4(t+r2)) \cdot M''_{34}$$

M and R are redefined accordingly.

Again, P and Q have asymmetric definitions. This time every eight time units P sends a message in order to determine if both processes are alive. When Q receives such a message, it will time out after seven time units. In order to make sure P knows this, one succesful message exchange suffices. This exchange, initiated by Q, will be done faster and faster, in order to fit within a six units time frame.

In fact, only a message from Q to P is necessary, but P replies in order to stop Q from sending more messages (otherwise, Q would have to be a supertask, sending an infinite number of messages before attacking). Of course, such a reply can get lost, but the fact that Q can send an arbitrary number of messages, means that an arbitrarily large subset of these messages arrives at P, thus P replies arbitrarily many times, and at least one of these replies must reach P. Formally,

$$P(t) = s_3(1/2)(t+8) \cdot P(t+8) + \int_{v \leq t+7} r_2(r)(v) \cdot s_3(r)(v+r) \cdot P^*(v + 10r) \;\; ,$$

$$P^*(u) = p\_attack(u+1/2) + \int_{v \leq u} r_2(r)(v) \cdot s_3(r)(v+r) \cdot P^*(u) \;\; .$$

$$Q(t) = \int_{v > t} r_4(1/2)(v) \cdot s_1(1/2)(v+1/2) \cdot Q^*(v+1/2, 1/2, 7+v) \;\; ,$$

$$Q^*(t, r, u) = r_4(r)(t+5r) \cdot q\_attack(u)) + s_1(r/2)(t + 6r) \cdot Q^*(t+6r, r/2, u).$$

(Here, t is the time of the previous message sent, r is the transmission speed of that message, and u is the time-out time.)

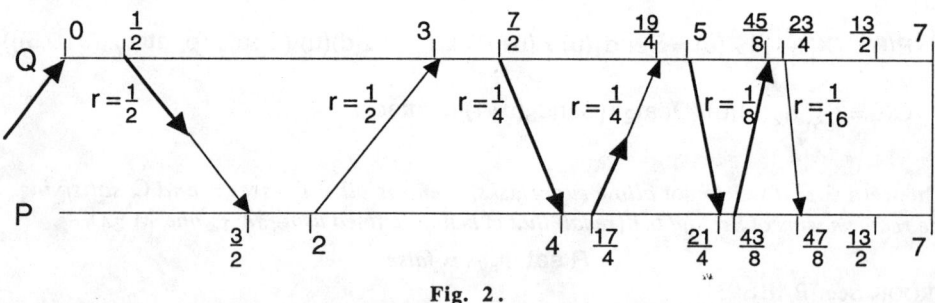

Fig. 2.

Fig. 2 depicts an example trace of the protocol, in which Q receives the first message from P at (relative) time 0 and both processes time out at time 7. The figure shows three attempted message exchanges, not counting the initial messages from P. The first attempted exchange is interrupted in M''_{12} (the thin arrows denote messages that could have occurred, but did not occur, in this example). The second exchange is interrupted in M''_{34} and the third one succeeds. Therefore the fourth and later possible exchanges do not occur. P only replies if the corresponding message from Q arrives. As soon as a pair of corresponding messages succeeds, Q stops sending, and, as a consequence, so does P.

7. Conclusions, Problems, and Future work.

Several problems remain to be solved. We can divide them into two groups: problems concerning the algebra of process ready trace sets and problems concerning the logic.

Ready trace theory has appeared to be more complex than we expected. As a result, Section 3 has grown into a complex definition, of which one would like to prove that it corresponds to the intuitions behind it (for example by verifying Proposition 5.3).

The decision to add a summand $\delta(2)$ to $\int_{1<t<2} b(t)$ seems counterintuitive. But on the other hand it is plausible that $[\partial_{\{b\}}(a(1) \cdot \int_{1<t<2} b(t))] = [a(1) \cdot \delta(2)]$ (rather than $\varnothing$) and also that $a(1) \cdot \int_{1<t<3} b(t)$ differs from $a(1) \cdot \int_{1<t<2} b(t) + a(1) \cdot \int_{1<t<3} b(t)$. Adding a summand $b(2)$ could be an alternative, but that would be at least as counterintuitive, and technically even more complicated. That $\int_{t>0} a(t)$ has an ever waiting trace $\delta(\infty) = a(\infty)$ seems also reasonable.

As it is defined here, Ready Trace Logic can only be used to describe properties of *all* traces of a process. In Section 6.1, we wanted to state that a process has a certain trace (for every t>0). Thus the usefulness of RdTrL could be improved through replacing the implicit universal quantification over the trace variable by arbitrary quantification.

In order to be useful not only for specification, but also for verification, the operators on processes should be translated to 'connectives' in the logic: if P **sat** φ and Q **sat** ψ then P+Q **sat** φ+ψ. Notice that the laws for these connectives will differ from the equations of the process algebra, for example: φ+φ ≠ φ. Defining these connectives in terms of RdTrL is not easy. It is probably worthwhile to study a language of which the primitives are on a higher level.

Apart from facilitating the translation from operators to connectives, this language could also solve another problem of RdTrL, namely that its notation is very explicit. When used on a high level of specification, this is an advantage of RdTrL, but on lower levels it becomes cumbersome to write down not only which actions are ready and take place, but also which actions are not ready, and when the process waits. Perhaps even a non-monotonic logic with a construction for invoking a Closed World Assumption [REI78] could be used (e.g., '¬ R(a,t) holds, unless R(a,t) is explicitly stated.'). But a good understanding of RdTrL is obviously crucial, before this higher level logic can be defined.

Acknowledgements.

We thank J.F. Groote (Utrecht University) for discussions and helpful suggestions on Section 6. We thank J. Katoen and R. Koymans (Philips Research) for suggesting examples from industry.

References.

[BAB91] J.C.M. BAETEN & J.A. BERGSTRA, *Real time process algebra*, Formal Aspects of Computing 3 (2), 1991, pp. 142-188.

[BABB93]   J.C.M. BAETEN, J.A. BERGSTRA & R.N. BOL, *A real time process logic*, report CSN 93/15, Dept. of Computing Science, Eindhoven University of Technology, 1993.

[BABK87]   J.C.M. BAETEN, J.A. BERGSTRA & J.W. KLOP, *Ready trace semantics for concrete process algebra with priority operator*, British Computer Journal 30 (6), 1987, pp. 498-506.

[BAZ82]   J.W. DE BAKKER & J.I. ZUCKER, *Processes and the denotational semantics of concurrency*, I&C 54, 1982, pp. 70-120.

[BEK84]   J.A. BERGSTRA & J.W. KLOP, *Process algebra for synchronous communication*, Inf. & Control 60, 1984, pp. 109-137.

[BEK88]   J.A. BERGSTRA & J.W. KLOP, *A complete inference system for regular processes with silent moves*, in: Proc. Logic Coll. 1986, Hull (F.R. Drake & J.K. Truss, eds.), North-Holland 1988, pp. 21-81.

[BER92]   C.H. VAN BERKEL, *Handshake circuits: an intermediary between communicating processes and VLSI*, Ph.D. Thesis, Eindhoven University of Technology 1992.

[BRHR84]   S.D. BROOKES, C.A.R. HOARE & A.W. ROSCOE, *A theory of communicating sequential processes*, J. ACM 31 (3), 1984, pp. 560-599.

[BRO92]   M. BROY, *Functional specification of time sensitive communication systems*, NATO ASI series, series F: computer and systems sciences, Vol. 88, pp. 325-367.

[EBE89]   J.C. EBERGEN, *Translating programs into delay-insensitive circuits*, Tract 56, CWI Amsterdam 1989.

[VGL90]   R.J. VAN GLABBEEK, *The linear time – branching time spectrum*, in: Proc. CONCUR'90, Amsterdam (J.C.M. Baeten & J.W. Klop, eds.), Springer LNCS 458, 1990, pp. 278-297.

[GRO92]   J.F. GROOTE, Personal communication, 1992.

[HAM90]   J.Y. HALPERN & Y.O. MOSES, *Knowledge and common knowledge in a distributed environment*, J. ACM 37, 1990, pp. 549-587.

[HDN84]   M. HENNESSY & R. DE NICOLA, *Testing equivalences for processes*, TCS 34, 1984, pp. 83-134.

[HOA85]   C.A.R. HOARE, *Communicating sequential processes*, Prentice Hall 1985.

[KAL86]   A. KALDEWAIJ, *A formalism for concurrent processes*, Ph.D. Thesis, Eindhoven University of Technology 1986.

[KOY89a]   R.L.C. KOYMANS, *Specifying message passing and time-critical systems with temporal logic*, Ph.D. Thesis, Eindhoven University of Technology 1989.

[KOY89b]   R.L.C. KOYMANS, *Specifying message passing systems requires extending temporal logic*, in: Proc. Temporal Logic in Specification (B. Banieqbal, H. Barringer & A. Pnueli, eds.), Springer LNCS 398, 1989, pp. 213-223.

[MEY85]   J.-J. CH. MEYER, *Merging regular processes by means of fixed point theory*, TCS 45, 1985, pp. 193-260.

[OLD91]   E.-R. OLDEROG, *Nets, Terms and Formulas*, Cambridge Tracts in Theor. Comp. Sci. 23, Cambridge University Press 1991.

[OLH83]   E.-R. OLDEROG & C.A.R. HOARE, *Specification-oriented semantics for communicating processes,* in: Proc. ICALP 83 (J. Díaz, ed.), Springer LNCS 154, 1983, pp. 561-572.

[PAR85]   J. PARROW, *Fairness properties in process algebra - with applications in communication protocol verification,* Ph.D. Thesis, Uppsala University 1985.

[PHI87]   I.C.C. PHILIPS, *Refusal testing,* TCS 50, 1987, pp. 241-284.

[PNU85]   A. PNUELI, *Linear and branching structures in the semantics and logics of reactive systems,* in: Proc. ICALP 85 (W. Brauer, ed.), Springer LNCS 194, 1985, pp. 15-32.

[REI78]   R. REITER, *On closed world databases,* in: Logic and Databases (H. Gallaire and J. Minker, eds.), Plenum Press, 1978.

[RER88]   G.M. REED & A.W. ROSCOE, *A timed model for communicating sequential processes,* TCS 58, 1988, pp. 249-261.

[REM83]   M. REM, *Partially ordered computations, with applications to VLSI design,* in: Proc. Found. of Comp. Sci. IV.2 (J.W. de Bakker J. van Leeuwen, eds.), MC Tract 159, Math. Centre, Amsterdam 1983, pp. 1-44.

[SNE85]   J.L.A. VAN DE SNEPSCHEUT, *Trace theory and VLSI design,* Springer LNCS 200, 1985.

[UDD86]   J.T. UDDING, *Classification and composition of delay-insensitive circuits,* Ph.D. Thesis, Eindhoven University of Technology 1986.

[VIN90]   E.P. DE VINK, *Designing stream based semantics for uniform concurrency and logic programming,* Ph.D. Thesis, Free University, Amsterdam 1990.

SOMETIMES "TOMORROW" IS "SOMETIME"
Action Refinement in a Temporal Logic of Objects

José Luiz Fiadeiro
ILTEC & Department of Informatics
Faculty of Sciences, University of Lisbon
Campo Grande, 1700 Lisboa, Portugal
llf@di.fc.ul.pt

Tom Maibaum
Department of Computing
Imperial College of Science, Technology and Medicine
180 Queen's Gate, London SW7 2BZ, UK
tsem@doc.ic.ac.uk

Abstract. We address the hierarchical (vertical) decomposition, or abstract implementation, of object specification in temporal logic. Whereas previous approaches to refinement in the context of temporal logic such as those developed by Lamport and by Barringer, Kuiper and Pnueli are based on a single logic that accommodates different levels of action granularity, our approach is based on relating different logics corresponding to different levels of granularity. More precisely, we map abstract actions (propositions) to concrete objects (theories) and, through inference rules that relate the different logics, derive properties of the abstracted actions from the behaviour of the corresponding objects. In this way, we keep a tighter control of action granularity and interference, enabling us to maintain the use of the "next" operator and make the development of reactive systems more tractable.

1 Introduction

In this paper, we address the hierarchical (also called vertical) decomposition, or abstract implementation, of temporal logic specifications of reactive systems. We do so in the context of an object-based approach to system design. In this approach, objects are taken as the basic building blocks of the design process. Each object has a certain number of actions that it can perform, either in reaction to some change in its state or a request from another object. The actions update, atomically, the state of the object. As a means of localising change, each object has a private state structured in terms of a collection of attributes (e.g. program variables) to which no other object can have access except via the actions that were declared for that object.

1.1 Stepwise System Design

In the object-based approach that we have in mind, systems are designed following a stepwise discipline. Each layer of the design process is characterised by a collection of objects that constitute the components out of which more complex objects (systems) may be built. Objects are interconnected to form complex systems by identifying the actions they share, i.e. in which they synchronise. Indeed, we assume, for each layer, a synchronous, multiprocessor architecture, in which, at each execution step, several actions may be performed concurrently. All objects at a given layer of abstraction have the same action granularity: the granularity that is given by the (abstract) processors.

When the actions cannot be considered atomic from the point of view of the (concrete) processors belonging to the intended implementation environment, it is necessary to refine the granularity associated with the abstract processors. Each step in this (vertical) decomposition process can be seen as the "implementation" of each abstract object in terms of a collection of concrete ones that are "assembled" in a configuration that provides the functionality required for implementing the abstract behaviour. Because each such configuration itself defines a (complex) object, the basic construction to be characterised in the formalisation of this refinement discipline is the implementation of an object in terms of another object.

More precisely, what is required is that the actions of the abstract object be decomposed in terms of *transactions* defined over the actions of the concrete object. Each such transaction can be described as an object, e.g. it can have its own private state (auxiliary variables as they are usually called). In fact, a transaction is a special kind of object – one that is guaranteed to terminate. Hence, our approach to refinement will be based on "implementing" abstract actions as concrete (transactional) objects.

1.2 Object Specification in Temporal Logic

Following [Pnueli 77], temporal logic can be used for specifying and reasoning about the behaviour of reactive systems, and object-based systems in particular [Fiadeiro and Maibaum 92]. In the framework that we put forward in [Fiadeiro and Maibaum 92], and which we shall summarise in section 2, an object is specified as a theory presentation of a linear and discrete temporal logic. A theory presentation consists of a signature θ and a set of axioms Φ. The signature θ corresponds to the declaration of the non-logical symbols that are used in the specification, namely the attributes (program variables) and actions of the object. Each axiom in Φ specifies a particular aspect of the behaviour of the object.

The granularity of time in the underlying Kripke structure corresponds to the granularity of the corresponding (abstract) processor. Therefore, it also corresponds to the granularity of the actions that the object can perform. Actions give rise to propositions of the temporal language. An action proposition is true at an instant i iff the action is performed in the transition from i to i+1. Hence, typical axioms which describe the way in which actions change the state of an object are of the form

$$action \wedge condition1 \rightarrow Xcondition2$$

where **X** is the "next" or "tomorrow" temporal operator and the conditions are first-order formulae over the attributes of the object. The use of **X** reflects the fact that actions determine the granularity of "change": actions are atomic in the sense that the object cannot be "caught" (its attributes cannot be accessed) while it is performing an action.

1.3   Action Granularity and Temporal Logic

Because, from the point of view of the object that implements it, an action has a duration (it lasts while the object is executing), the intended notion of action refinement requires us to deal with different action granularities in temporal logic.

The problem of working with different granularities in temporal logic is not new. Lamport's work on hierarchical decomposition and refinement [Lamport 83, 89] addresses this problem by forbidding the use of the "next" operator and introducing stuttering. Lamport addresses action decomposition within a single logic. That is why the use of "next" has to be avoided: it commits the logic to a fixed (multi purpose) granularity. The adoption of the reals in [Barringer et al 86] serves the same purpose: to allow several granularities of action to be accommodated within the same logic.

However, both from the specification and verification points of view, there are advantages in working with a discrete granularity of time. The use of the "next" operator is, indeed, intuitive when associated with the granularity of the actions that the system may perform: the use of "primed variables" as in [Lamport 89] is, in our opinion, a good indication of this fact. Furthermore, as recently pointed out by other researchers [e.g. Jones 92], control of granularity is essential for making the development of concurrent programs more tractable. Together with its control on interference, this is a clear advantage of object-based methods over more traditional ones, which suggests that we seek formal support for action refinement that does not require granularities to be collapsed and maintains the use of "next".

Hence, our approach is radically different from both Lamport and Barringer/Kuiper/ Pnueli. It is not based on a single logic but on the decomposition of the logic itself. That is, rather than accommodating different levels of granularity within the same logic, we propose to establish relationships between different logics, each corresponding to a different level of granularity.

As a result, the use of "next" is allowed at each level of abstraction. From a model-theoretic point of view, such relationships have to provide for the abstraction of atomic transitions (caused by abstract actions) from sequences of concrete ones (i.e. from the behaviours of the corresponding transactional objects). This is achieved by relating the Kripke structures of both levels, i.e. by relating abstract and concrete "time", namely, abstract instants to intervals of concrete instants of time.

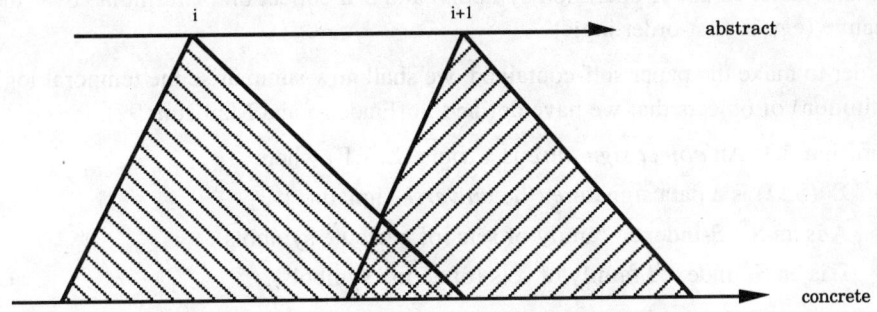

From a proof-theoretic point of view, the required relationships between logics will have to establish a way of deriving what will happen in the "abstract tomorrow" from the "concrete future". That is, we will have to implement the abstract **X** in terms of the concrete **F**. This will be achieved via inference rules that relate the two levels.

There will not be enough room in the paper to discuss structure and compositionality in depth. A summary of the main results on these two important issues will be given in section 4.

2 Temporal Logic of Objects

As already mentioned, the specification of an object at a given level of abstraction is based on a certain action granularity. More precisely, each level of abstraction is characterised by a collection of data types that are assumed to be available for manipulating the attributes. In other words, each level of abstraction defines the primitives that are available for structuring the "memory" of the component (the types of the attributes) and the corresponding operations to manipulate those structures. These operations constrain the nature of the actions that may be declared for the component. For instance, if the data type INT of integers is available with addition as an operation, actions may add integers to integer attributes. If, however, only the successor operation is made available, additions are not possible as (atomic) actions: transactions are required to implement them.

Hence, our formal model of object specification includes, as a means of identifying the level of abstraction to which that object belongs, the specification of the abstract data type (ADT) that is necessary to support its actions and attributes. Because abstract data type specification is already a well-studied subject, we shall not get into detailed descriptions of this aspect of object specification. Instead, we shall assume that ADT specification is given in the traditional algebraic flavour, i.e. we shall take an ADT specification as a pair (Σ,Δ) where $\Sigma=(S,\Omega)$ is a signature (i.e. S is a set of sorts and Ω is an S-indexed set of operation symbols) and Δ a collection of formulae over that signature (e.g. in first-order logic).

In order to make the paper self-contained, we shall now summarise the temporal logic (institution) of objects that we have defined in [Fiadeiro and Maibaum 92].

Definition 2.1: An *object signature* is a triple (Σ,A,Γ) where
- $\Sigma=(S,\Omega)$ is a data signature (the *universe* signature).
- A is an $S^*\times S$-indexed family of sets (of attribute symbols).
- Γ is an S^*-indexed family of sets (of action symbols). □

The families Ω, A, and Γ are assumed to be disjoint and finite (i.e., there are only a finite number of function, attribute and action symbols). Attribute parameters allow us to define complex data structures such as arrays, and action parameters allow us to model exchange of data (input and output).

A semantic interpretation structure for an object signature is given by an algebra that interprets the data parameters, a mapping that gives the values taken by the attributes at each instant, and a mapping that gives the actions that will occur at each instant:

Definition 2.2: A θ–*interpretation structure* for a signature $\theta=(\Sigma,A,\Gamma)$ is a triple $(\mathcal{U},\mathcal{A},\mathcal{G})$ where:
- $\mathcal{U}$ is a Σ-algebra, i.e. to each sort symbol $s\in S$ a set $s_\mathcal{U}$ is assigned, and to each function symbol $f\in\Omega_{<s_1,\ldots,s_n>,s}$ a function $f_\mathcal{U}: s_{1\mathcal{U}}\times\ldots\times s_{n\mathcal{U}}\to s_\mathcal{U}$ is assigned;
- $\mathcal{A}$ maps $f\in A_{<s_1,\ldots,s_n>,s}$ to $\mathcal{A}(f): s_{1\mathcal{U}}\times\ldots\times s_{n\mathcal{U}}\times\omega\to s_\mathcal{U}$;
- $\mathcal{G}$ maps $a\in\Gamma_{<s_1,\ldots,s_n>}$ to $\mathcal{G}(a): s_{1\mathcal{U}}\times\ldots\times s_{n\mathcal{U}}\to\wp(\omega)$. □

The natural numbers are being used as a canonical frame for linear temporal logic. Actions are being interpreted as predicates over instants of time (states): the fact that an instant i belongs to the interpretation of an action means that the action occurs in the transition from i to i+1. Notice that we allow for more than one action to be performed during such a transition. Our overall view of an object is, therefore, that of a collection of actions that can be executed concurrently upon a set of attributes that are private to the object. That is, actions provide the granularity of concurrency in a synchronous, multi-processor architecture.

Notice that it is also possible that no action occurs during a state transition. Such transitions correspond to steps performed by the environment and reflect an *open* semantics of behaviour. Such an open semantics is an essential ingredient for structure and compositionality, as argued in [Barringer and Kuiper 84].

The semantics of behaviour is not completely open. Objects have an intrinsic notion of locality or encapsulation according to which only the actions declared for an object can change the values of its attributes. This notion of locality is captured as follows:

Definition 2.3: Given an object signature $\theta=(\Sigma,A,\Gamma)$ and a θ-interpretation structure $(\mathcal{U},\mathcal{A},\mathcal{G})$, let $\mathcal{E}=\{i\in\omega \mid i\in \mathcal{G}(a)(v_1,...,v_n)$ for some $a\in \Gamma_{<s_1,...,s_n>}$, $(v_1,...,v_n)\in s_{1\mathcal{U}}\times...\times s_{n\mathcal{U}}\}$. That is, $\mathcal{E}$ consists of those instants during which an action of the object occurs, i.e. these instants denote state transitions in which the object will be engaged. The θ-interpretation structure $(\mathcal{U},\mathcal{A},\mathcal{G})$ is said to be a *locus* iff, for every instant $i\notin \mathcal{E}$ and every $f\in A$, $\mathcal{A}(f)(i) = \mathcal{A}(f)(i+1)$. □

Attribute symbols generate terms whose denotation varies in time (non-rigid designators) and action symbols generate predicates of a linear and discrete temporal language [Goldblatt 87] which we shall use to specify objects:

Definition 2.4: The set of *terms* for a signature θ is inductively defined as follows: (a) variables of sort s are terms of sort s; (b) every nullary symbol in Ω_s or A_s is a term of sort s; (c) if $t_1,...,t_n$ are terms of sorts $s_1,...,s_n$ and f belongs to $\Omega_{<s_1,...,s_n>,s}$ or $A_{<s_1,...,s_n>,s}$, then $f(t_1,...,t_n)$ is a term of sort s.

The *formulae* for a signature θ are inductively defined as follows: (a) if $t_1,...,t_n$ are terms of sorts $s_1,...,s_n$ and a belongs to $\Gamma_{<s_1,...,s_n>,s}$, then $a(t_1,...,t_n)$ is a formula; (b) if t_1 and t_2 are terms of sort s, $(t_1=_s t_2)$ is a formula; (c) if ϕ is a formula so are $(\neg\phi)$ and $(\mathbf{X}f)$; (d) if ϕ_1 and ϕ_2 are formulae so are $(\phi_1\rightarrow\phi_2)$ and $(\phi_1\mathbf{U}\phi_2)$; (e) if ϕ is a formula and x is a variable of sort s, $(\forall x:s)\phi$ is a formula. We call *state formulae* those which are constructed from equations and the first-order connectives. □

Terms and formulae are interpreted as follows:

Definition 2.5: Given an interpretation structure $\mathcal{S}=(\mathcal{U},\mathcal{A},\mathcal{G})$ for an object signature (Σ,A,Γ) and an assignment $\mathcal{V}$ of values to variables, we define the value $[\![t]\!](i)$ of t at time i for every term t as follows: (a) $[\![x]\!](i)=\mathcal{V}(x)$ for every variable x; (b) if $f\in\Omega_{<s_1,...,s_n>,s}$, $[\![f(t_1,...,t_n)]\!](i) = f_\mathcal{U}([\![t_1]\!](i),...,[\![t_n]\!](i))$; (c) if $f\in A_{<s_1,...,s_n>,s}$, $[\![f(t_1,...,t_n)]\!](i) = \mathcal{A}(f)([\![t_1]\!](i),...,[\![t_n]\!](i),i)$.

Satisfaction of a formula at time i is defined as follows: (a) $(\mathcal{S},\mathcal{V})^i \vDash a(t_1,...,t_n)$ iff $i\in \mathcal{G}(a)([\![t_1]\!](i),...,[\![t_n]\!](i))$; (b) $(\mathcal{S},\mathcal{V})^i \vDash (t_1=_s t_2)$ iff $[\![t_1]\!](i)=[\![t_2]\!](i)$; (c) $(\mathcal{S},\mathcal{V})^i \vDash (\neg\phi)$ if and only if it is not the case that $(\mathcal{S},\mathcal{V})^i \vDash \phi$; (d) $(\mathcal{S},\mathcal{V})^i \vDash (\mathbf{X}\phi)$ if and only if $(\mathcal{S},\mathcal{V})^{i+1} \vDash \phi$; (e) $(\mathcal{S},\mathcal{V})^i \vDash (\phi_1\rightarrow\phi_2)$ if and only if $(\mathcal{S},\mathcal{V})^i \vDash \phi_1$ implies $(\mathcal{S},\mathcal{V})^i \vDash \phi_2$; (f) $(\mathcal{S},\mathcal{V})^i \vDash (\phi_1\mathbf{U}\phi_2)$ if and only if, for some $j\geq i$, $(\mathcal{S},\mathcal{V})^j \vDash \phi_2$ and, for every $i\leq k<j$, $(\mathcal{S},\mathcal{V})^k \vDash \phi_1$; (g) $(\mathcal{S},\mathcal{V})^i \vDash (\forall x:s)\phi$ iff $(\mathcal{S},\mathcal{V}[x:=v])^i \vDash \phi$ for every $v\in s_\mathcal{U}$, where by $\mathcal{V}[x:=v]$ we are denoting the assignment that is equal to $\mathcal{V}$ except that it assigns the value v *tu the variable* x.

A formula ϕ is said to be *true* in $(\mathcal{S},\mathcal{V})$ if it is satisfied at every instant i, and it is said to be valid in $\mathcal{S}$ iff it is true for every assignment of values to variables. □

As usual, we also work with derived operators, namely **F** ("sometimes in the future") which is such that (**F**φ) corresponds to (true**U**φ). That is, $(S,\mathcal{V})^i \models (\mathbf{X}\phi)$ iff $(S,\mathcal{V})^j \models \phi$ for some j≥i.

An object specification is just a theory presentation in this logic:

Definition 2.6: An *object specification* is a pair (θ,Φ) where θ is an object signature and Φ is a (finite) set of θ-formulae (the axioms of the specification). A model of a specification is a θ-locus that makes all the formulae of Φ valid, i.e. true at every instant i. Hence, the axioms specify properties which are to hold throughout the life of the object. □

The fact that only θ-loci are used as models implies that each specification has the following logical axiom, called the locality axiom for θ=(Σ,A,Γ)

$$\left(\bigvee_{a \in \Gamma} (\exists x_a) a(x_a) \right) \vee \left(\bigwedge_{f \in A} (\forall x_f)(\mathbf{X} f(x_f) = f(x_f)) \right)$$

where, for each symbol u, x_u is a tuple of variables (all distinct) of the appropriate sorts. As required, this formula imposes that either the next transition will be performed by one of the actions of the object, or else every attribute will remain invariant. Notice that this is a logical axiom for the given object signature and, hence, is not required to be explicitly stated in a specification. Further discussion on locality axioms can be found in [Fiadeiro and Maibaum 92].

As an example, consider the specification of the behaviour of a counter:

COUNTER	
data sorts	nat, bool
data functions	usual functions and constants for the booleans plus
	zero: nat; succ, pred: nat→nat; <: nat,nat→bool
attribute symbols	rem: nat
action symbols	set(nat), count
axioms	usual axioms for the booleans and natural numbers plus
	c1 (∀n:nat)(set(n) → **X**(rem=n))
	c2 (∀n:nat)(set(n) → rem=zero)
	c3 (∀m:nat)(count∧(rem=m) → **X**(rem=pred(m)))
	c4 (count → zero<rem)
	c5 (zero<rem → **F**(count))

This specification allows us to illustrate the three types of axioms which are usually used for describing the behaviour of objects.

    1.   action∧condition1 → **X**condition2

where **X** is the "next" or "tomorrow" temporal operator and the conditions are state formulae. Such axioms (e.g. c1 and c3) specify the effects of the actions on the attributes. Given an attribute f and a term t, $\mathbf{X}(f=t)$ means that whatever next action is performed, it will find f equal to t. This entails the "instantaneity" of actions in the sense that an object cannot be "caught" while performing an action.

2. action → condition

where the condition is a state formula. Such axioms (e.g. c2 and c4) specify restrictions on the occurrence of the actions. For instance, c2 specifies that the counter can only be set when the attribute *rem* has value zero. Such conditions are sometimes called pre-conditions or enabling conditions.

3. condition → **F**action

where **F** is the "sometime" temporal operator and the condition is a state formula. Such axioms specify requirements on the occurrence of the actions (e.g. c5 specifies that *count* must occur while the attribute is different from zero). That is, they specify the liveness properties of the object. For instance, it is possible to derive from these axioms the following liveness property:

$(\forall n{:}nat)(set(n) \rightarrow \mathbf{F}(rem=zero))$

Verification in this setting, including inference rules that take into account the encapsulation of change implied by the locality axioms, is discussed in more detail in [Fiadeiro and Maibaum 92, 93].

3 Refining Actions over Objects

As motivated in the introduction, our overall aim is to support a stepwise, object based discipline for reactive system design. This process goes through a succession of levels of abstraction each of which is characterised by the operations that the (abstract) processors execute, i.e. by an abstract data type. The object specifications which "sit" at each such level are based on a fixed granularity of change: each action updates the attributes *atomically*, i.e. each occurrence of an action lasts for a single state transition – the state of an object cannot be read or written while a state transition is taking place. On the other hand, a (synchronous) multi-processor architecture is assumed: several actions may occur during the same state transition as long as they are enabled and do not have conflicting post-conditions. This was not the case for the example of the counter: for instance, c2 and c4 imply that $set(n)$ and $count$ are never simultaneously enabled. All the actions which occur during the same state transition will read the same values of the attributes; the values read at step i+1 are those that were written at step i.

An implementation step, moving the design process from one layer of abstraction to a more concrete one, is necessary when the granularity of change at the abstract level is too coarse. This happens when the operations of the underlying ADT or the abstract actions are not available in the intended implementation environment.

In order to motivate and illustrate our approach to action refinement we have chosen an example which is very simple but still representative of the situations which typically arise. Consider, then, the following specification of an object which can reset, add to and subtract from an integer variable:

ADDER

data sorts	int
data functions	zero: int; succ, pred: int→int; +, –: int,int→int
attribute symbols	var: int
action symbols	reset, add(int), sub(int)
axioms	usual axioms for the integers plus
	a1 (reset → **X**(var=zero))
	a2 ($\forall$i,j:int)(add(i)$\wedge$(var=j) → **X**(var=j+i))
	a3 ($\forall$i,j:int)(sub(i)$\wedge$(var=j) → **X**(var=j–i))

Assume, for the sake of argument, that + and – are not available in the target implementation environment, but only *succ* (for successor) and *pred* (for predecessor). As a result, the actions *add(i)* and *sub(j)* do not have the required granularity and, hence, have to be refined; the object ADDER has to be implemented over simpler objects.

The specification of the integer variable which is available at the concrete level is:

STEP

data sorts	int
data functions	zero: int; succ, pred: int→int
attribute symbols	var: int
action symbols	creset, inc, dec
axioms	usual axioms for the integers plus
	d1 (creset → **X**(var=zero))
	d2 ($\forall$j:int)(inc$\wedge$(var=j) → **X**(var=succ(j)))
	d3 ($\forall$j:int)(dec$\wedge$(var=j) → **X**(var=pred(j)))

The refinement of the abstract actions must be given in terms of "transactions" over this concrete object. For instance, the refinement of *add(i)* will required a "loop" over the action *inc*. In order to define such a loop, we need an "auxiliary variable" which counts the number of cycles necessary for the required increment. Therefore, we need another (concrete) object for the implementation: a counter! Indeed, the implementation of an abstract object may require several concrete objects (the implementation *kernel*) which must be assembled into a configuration which provides the functionality that is required by the abstract object. Hence, ADDER will be implemented over STEP and COUNTER.

The refinement of each abstract action a gives rise to another object ITF_a: an object which controls the execution of the transaction which implements a (the *transactional interface* of a). Each such object consists of two actions – beg_a and end_a, to denote the beginning and the end of the transaction, respectively, and an attribute in_a which is true while the transaction is taking place. (Such start and end points have been used in the literature, e.g., [Aceto and Hennessy 89] in the context of Process Algebra.)

ITF_a

data sorts	bool
data functions	usual boolean functions
attribute symbols	in_a: bool
action symbols	beg_a, end_a
axioms	usual axioms for the booleans plus
i1	$beg_a \rightarrow \neg in_a \wedge (\mathbf{X} in_a \vee end_a)$
i2	$end_a \rightarrow (in_a \vee beg_a) \wedge \mathbf{X} \neg in_a$

In order to complete the implementation of ADDER, it remains to relate these transactional interfaces to the kernel objects. This is done through axioms which state how the actions of the transactional interfaces are related to the actions of the kernel. For instance, for the action *add(i)* we have:

$beg_{add(i)} \rightarrow set(i)$
$in_{add(i)} \rightarrow (count \leftrightarrow inc)$
$in_{add(i)} \wedge rem = succ(zero) \rightarrow (count \leftrightarrow end_{add(i)})$

That is, the beginning of the transaction sets the counter to the parameter of the action, during the transaction *count* (from COUNTER) and *inc* (from STEP) are synchronised, and the last *count* is synchronised with the end of the transaction. (Notice that when an action has parameters, the corresponding transactional interface inherits these parameters.)

These axioms establish the interconnections between the various kernel objects and between these and the transactional interface so that the functionality required by the abstract action is achieved (i.e. the implementation is correct). We shall see later on how to prove correctness of the implementation.

There is, however, an axiom which is missing in the implementation of *add(i)*. Intuitively, the correctness of the implementation rests on the fact that *count* and *inc* are synchronised so that the necessary number of increments is performed. However, in order to make sure that, in the end, *var* has the right value, it is necessary to forbid that decrements be performed:

$$in_{add}(i) \rightarrow (\neg dec)$$

The need for this axiom becomes obvious when the correctness proof for the implementation is done, namely when an appropriate invariant is chosen.

The implementation of *sub(i)* is very similar except that *count* synchronises with *dec* instead of *inc*:

$$beg_{sub}(i) \rightarrow set(i)$$
$$in_{sub}(i) \rightarrow (count \leftrightarrow dec)$$
$$in_{sub}(i) \land rem=succ(zero) \rightarrow (count \leftrightarrow end_{sub}(i))$$
$$in_{sub}(i) \rightarrow (\neg inc)$$

The implementation of *reset* is much easier because it already has the required granularity. Hence, it can be implemented directly over the concrete reset:

$$beg_{reset} \leftrightarrow end_{reset} \leftrightarrow creset$$

Summarising,

Definition 3.1: Given an (abstract) object specification $S=(\theta_S, \Phi_S)$, an *implementation body* for S consists of:
- a collection KER of (concrete) object specifications
- for each action symbol a, a transactional interface ITF_a
- a collection INT of axioms in the language of the kernel and the interfaces

Together, these items define a specification BDY_S as follows:
- the signature of BDY_S consists of the union of the signatures of the kernel objects and of the transactional interfaces (this union being taken componentwise)
- the axioms of BDY_S consist of the union of the axioms of the kernel objects and of the transactional interfaces (including their locality axioms), together with INT.

If the signature of the body is $(\Sigma_B, A_B, \Gamma_B)$, we denote by A^*_B the union of the sets of attributes of the kernel objects, i.e. A_B minus the in_a attributes. □

4 Semantics and Proof-Theory

How, then, can we relate such an implementation body to the abstract specification? In particular, how can we prove that the implementation is *correct*?

First of all, we need a way of establishing a relationship between the implementation body and the abstract specification – an *implementation mapping*, which we shall denote by ι. At the level of the actions, this implementation mapping simply associates every action symbol with its transactional interface. With respect to the abstract data types and the attributes, the implementation mapping relies on notions of refinement in the algebraic theory of abstract data type specification. We take here the approach that has been put forward in [Veloso et al 85]. (Other approaches, such as those that have been put forward for Hoare's data refinement, may be adopted.)

According to this approach, we build a conservative extension $(\Sigma_B, \Delta_B) \to (\Sigma', \Delta')$ where (Σ_B, Δ_B) is the universe specification (the specification of the underlying abstract data type) for BDY_S and an interpretation $\iota:(\Sigma_S, \Delta_S) \to (\Sigma', \Delta')$ mapping the abstract data sorts and operations to the concrete ones.

The extension $(\Sigma_B, \Delta_B) \to (\Sigma', \Delta')$ may be required for defining the abstract data operations in terms of the concrete ones. For instance, in the example we have been using, it is necessary to define + and − in terms of *succ* and *pred*. These definitions would extend the specification (Σ_B, Δ_B) of the data types of the implementation body to produce (Σ', Δ'). We denote by x-BDY_S the extension of BDY_S with the new data type specification.

The interpretation $\iota:(\Sigma_S, \Delta_S) \to (\Sigma', \Delta')$ gives the translation between the abstract and the concrete data types. We extend this translation to the attributes of the abstract specification so that, for every attribute f of S, $\iota(f)$ is a term of x-BDY_S. We also require that $\iota(f)$ be built over A^*_B i.e. the *in* attributes are not allowed in the representation of the abstract attributes; only the attributes of the kernel objects and the data type operations. In this way, every state formula φ gets translated to a state formula $\iota(\varphi)$ in the language of the body. We have chosen our example such that ι is the identity, allowing us to concentrate on action refinement only.

The mapping of the abstract actions to the transactional interfaces also has to respect the translation ι between the data types: if an abstract action has parameters $<s_1,...,s_n>$, its transactional interface should be such that the *beg* and *end* actions, as well as the *in* attribute, all have parameters $<\iota(s_1),...,\iota(s_n)>$.

Given an implementation body and mapping, we can abstract from it interpretation structures for the abstract signature:

Definition 4.1: Given an interpretation structure $(\mathcal{U}_B, \mathcal{A}_B, \mathcal{G}_B)$ for the body x-BDY_S, we can abstract a set of interpretation structures for θ_S. Each abstraction consists of:

1) a function $\rho:\omega \to 2^\omega$ (from abstract to concrete "time") such that for every $i \in \omega$,

   a) $\rho(i)$ is an interval [beg(i),end(i)]

   b) beg(i)<beg(i+1) and end(i)<end(i+1)

   c) for every beg(i+1)≤m<end(i) and attribute $f \in A^*_B$, $\mathcal{A}_B(f)(m) = \mathcal{A}_B(f)(m+1)$.

That is to say, we identify for each abstract step the interval of concrete steps that realise it. For flexibility and generality, we allow for such intervals to overlap but require that attributes remain invariant during intersections.

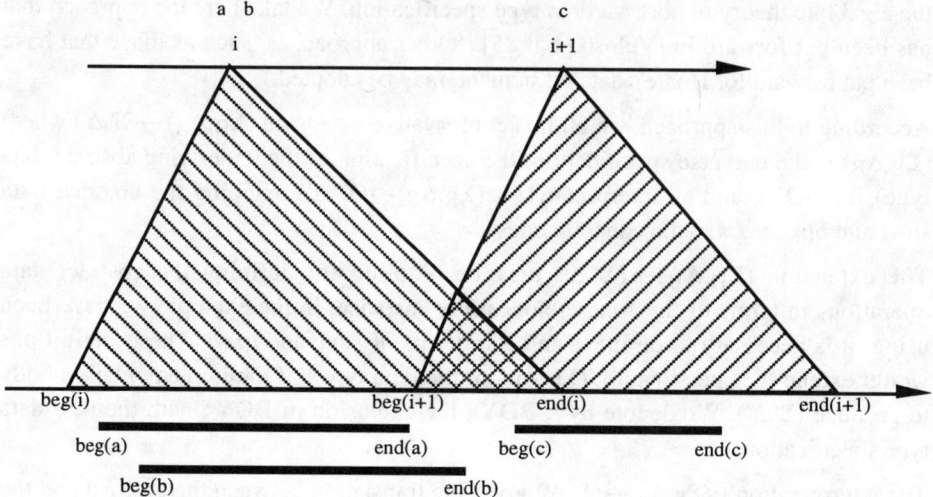

2) a mapping α (from abstract actions to abstract "time") such that, for every $a \in \Gamma_{S<s_1,\ldots,s_n>}$, $\alpha(a): s1_{\mathcal{U}_B} \times \ldots \times sn_{\mathcal{U}_B} \to \wp(\omega)$ is such that, for every $<v_1,\ldots,v_n> \in s1_{\mathcal{U}_B} \times \ldots \times sn_{\mathcal{U}_B}$ and $i \in \alpha(a)(v_1,\ldots,v_n)$, there exist exactly one $j \in \rho(i)$ and one $k \in \rho(i)$, $j \leq k$, such that

a) $j \in \mathcal{G}_B(beg_a)(v_1,\ldots,v_n)$

b) $k \in \mathcal{G}_B(end_a)(v_1,\ldots,v_n)$

c) $\mathcal{A}_B(f)(m) = \mathcal{A}_B(f)(m+1)$ for every attribute $f \in A^*_B$ and $beg(i) \leq m < j$ (i.e. no action may start once the values of the attributes have changed)

d) $\mathcal{A}_B(f)(m) = \mathcal{A}_B(f)(m+1)$ for every attribute $f \in A^*_B$ and $k < m \leq end(i)$ (i.e. after the end of an action the values of the attributes cannot change until the end of the interval).

That is to say, the occurrence of an action at (abstract) time i corresponds to the execution of the interface for that action once and only once during the interval $\rho(i)$.

3) if $j \in \mathcal{G}_B(a)$ for some action a of Γ_B then there is $i \in \omega$ such that $j \in \rho(i)$ and $i \in \alpha(a)$. That is, actions of the interfaces occur only as part of executions of the abstract actions. □

Definition 4.2: Given an interpretation structure $(\mathcal{U}_B, \mathcal{A}_B, \mathcal{G}_B)$ for the body x-BDY$_S$ and an implementation mapping ι, a θ_S-interpretation structure $(\mathcal{U}_S, \mathcal{A}_S, \mathcal{G}_S)$ abstracted through (ρ, α) is defined as follows:

- $\mathcal{U}_S$ is an ι-reduct of $\mathcal{U}_B$.

- $\mathcal{A}_S(f)(i) = \mathcal{S}_B(\iota(f))(\text{beg}(\rho(i)))$
- $\mathcal{G}_S = \alpha$ □

That is to say, attributes take at abstract time i the value that their images take at the beginning of the interval $\rho(i)$, and the abstracted actions occur in the intervals during which the corresponding interfaces execute.

Definition 4.3: An implementation is said to be correct if every θ_S-interpretation structure abstracted from a model of x-BDY$_S$ is also a model of the abstract specification S, i.e., if it is a locus for θ_S and if it validates all the axioms of S. □

The locality requirement is actually enforced by the construction thanks to condition 3 of definition 4.1:

Proposition 4.4: Each interpretation structure abstracted in this way is a locus for θ_S, i.e. it satisfies the locality condition for θ_S. □

Indeed, being a logical notion at the specification level, locality preservation should also be made logical in the definition of an implementation step. That is, it is up to the implementation environment to see to it that the kernel objects are used only for that abstract object.

Hence, in order to prove correctness, it remains to prove that the axioms Φ_S of S remain valid. We can do this proof-theoretically by defining the specification ABS$_S$ which is abstracted from the implementation body, and prove that every axiom of S is translated by ι to a theorem of ABS$_S$, i.e. that ABS$_S \vdash \phi$ for every $\phi \in \Phi_S$.

Definition 4.5: Given an implementation body x-BDY$_S$ and an implementation mapping ι, we define the specification ABS$_S = (\theta_S, \Psi)$ where Ψ consists of all the formulae which are valid in all the θ_S-interpretation structures asbtracted from models of x-BDY$_S$ as defined in 4.2. □

Notice that ABS$_S$ is a theory, i.e. it is already closed under consequence. Hence, ABS$_S \vdash \phi$ iff $\phi \in \Psi$. In the absence of an axiomatic presentation for ABS$_S$, we resort to a set of proof rules that allow us to use the presentation of x-BDY$_S$ instead (which, recall from definition 3.1, is built from the presentations of the kernels, the transactional interfaces and the interaction axioms, and hence finite assuming that the kernel presentations are themselves finite).

Theorem 4.6: The following proof rules are sound with respect to 4.5:

- change: $\dfrac{\text{x-BDY}_S \vdash (\text{beg}_a \wedge \iota(\varphi)) \to \mathbf{F}(\text{end}_a \wedge \mathbf{X}\iota(\psi)))}{\text{ABS}_S \vdash (a \wedge \varphi \to \mathbf{X}\psi)}$

- safety: $\dfrac{\text{x-BDY}_S \vdash (\text{beg}_a \to \iota(\varphi))}{\text{ABS}_S \vdash (a \to \varphi)}$

- liveness: $\dfrac{\text{x-BDY}_S \vdash (\iota(\varphi) \to \mathbf{F}\text{beg}_a)}{\text{ABS}_S \vdash (\varphi \to \mathbf{F}a)}$

where φ and ψ are state formulae. □

Each of these rules corresponds to one of the classes of formulae that are used to describe the behaviour of objects as discussed in Section 2. The first rule states that the effects of the occurrence of an abstracted action are recovered at the end of the corresponding transaction. Notice the role of conditions 2c and 2d of definition 4.1 in establishing the soundness of this rule: we must make sure that the values of the attributes at the start of the interval are still the same at the start of the action, and that their value at the end of the action will be the same at the end of the interval. The second and third rules transfer the safety and liveness requirements on the abstracted actions to the beginning of the corresponding transaction.

Applications of the rules on safety and liveness are usually straightforward. Application of the rule on the effects of the attributes (change) usually requires traditional techniques known from the literature, namely proof of invariants. For instance, in order to prove

$$\text{x-BDY}_S \vdash (\text{beg}_a \wedge \iota(\varphi) \to \mathbf{F}(\text{end}_a \wedge \mathbf{X}\iota(\psi)))$$

we usually look for an invariant *inv* such that:

$$\text{x-BDY}_S \vdash (\text{beg}_a \wedge \iota(\varphi) \to \mathbf{X}inv)$$
$$\text{x-BDY}_S \vdash (\text{in}_a \wedge inv \to \mathbf{X}inv)$$
$$\text{x-BDY}_S \vdash (\text{end}_a \wedge inv \to \mathbf{X}\iota(\psi))$$

For instance, in the case of the refinement of *add(i)* we can prove axiom *a2* by choosing as an invariant the following formula:

(i–rem)=(var–j)

where, recall, φ is (var=j).

The proofs of each of the new proof obligations can be conducted as explained in [Fiadeiro and Maibaum 93], namely by using case-analysis techniques which are allowed by encapsulation. These allow one to break the proof of ($\text{in}_a \wedge inv \to \mathbf{X}inv$) into the proofs of ($b \wedge \text{in}_a \wedge inv \to \mathbf{X}inv$) for every action *b* of the body.

Notice that the "next" operator is, indeed, used at the concrete level to prove a property about the concrete **F** which, in turn, is used to prove the property about the abstract **X**. The two levels are, however, never mixed. Each consequence relation ($\vdash$) operates at a fixed level of granularity. (Meta-) inference rules as above allow us to operate between different levels, i.e. between different logics.

5 Structure and Compositionality

One of the advantages of adopting an object-based discipline for reactive system design is modularity. Having provided separate specifications of the objects which we want to have as components of the system, we should just have to specify how they are interconnected in order to define a (logical) configuration of the envisaged system. Hence, from the point of view of implementing the system, we should be able to start from separate implementations of the components and build from them an implementation of the system. This property of the development method is usually called compositionality.

In the approach which we have been following, objects interact by sharing actions. That is, we specify interconnections by identifying pairs of actions, one in each of the objects. Formally, such interconnection mechanisms are best formalised in terms of Category Theory [Fiadeiro and Maibaum 92]. Because it is not possible to define this approach in the space that is left, we will have to limit our discussion of compositionality to a more informal level.

Consider the situation where we have two (abstract) object specifications $S_1=(\theta_1,\Phi_1)$ and $S_2=(\theta_2,\Phi_2)$ interconnected via a simple channel (one action) $a=<a_1,a_2>$ where $a_i \in \Gamma_i$. The composed (aggregated system) $S_1\|_a S_2$ is defined as follows: its attributes consist of the union of attributes of the components and its actions consist of the union of the sets of actions of the components except that a_1 and a_2 are replaced by a new symbol, say a, which corresponds to the joint execution of a_1 and a_2. Such an aggregation defines two translations κ_1 and κ_2 between the signatures of the components and the new signature which replace a_i by a.

Consider now implementation bodies for S_1 and S_2. If the kernels of the implementation bodies are at the same level of granularity, we want to aggregate them in such a way that an implementation body is obtained for the composite system $S_1\|_a S_2$.

The idea is to aggregate the given implementation bodies making sure that only one interface is provided for the joint action $a=a_1\|a_2$. That is, we have to "merge" the interfaces ITF_{a_1} and ITF_{a_2}. This merging will identify the *beg* and *end* actions of each of the interfaces, i.e. we shall take $beg_a=beg_{a_1}\|beg_{a_2}$ and $end_a=end_{a_1}\|end_{a_2}$, as well as the corresponding *in* attributes.

That is, given implementation bodies (KER_1,ITF_1,INT_1) and (KER_2,ITF_2,INT_2) we form the implementation body $(KER_1 \cup KER_2, ITF_1\|ITF_2, \kappa_1(INT_1)\cup\kappa_2(INT_2))$ where $(ITF_1\|ITF_2)_a$ is ITF_{i_a} if a is an action of S_i which is not shared, and $(ITF_1\|ITF_2)_a$ is as defined above in case a is indeed shared.

Similar constructions will take care of the situations in which the given objects are interconnected via more than one action.

One should check that the new implementation body satisfies the requirements set up in section 3. The theorems required on the interface ITF_a are the exact translations of those of the primitive interfaces. The conditions on the new body are also trivially satisfied.

On the other hand, one should also check that every behaviour abstracted from this implementation body is a behaviour abstracted from the two primitive implementation bodies. Conditions 1 are trivially checked: only 1c depends on the actual objects and invariance of B-attributes implies trivially invariance of B_i attributes. Conditions 2 are also trivially checked because each α defines two restrictions α_1 and α_2 on the actions of the components which satisfy the conditions (c and d) on the attributes. The same applies to the third condition.

Hence, our implementation method is compositional in the sense that we can build an implementation of a composite system from implementations of its components in a universal way, i.e. without having to know exactly how the implementations were made.

6 Concluding Remarks

In this paper, we addressed the refinement of temporal logic specifications of reactive systems. More concretely, we have shown how action granularity could be refined by implementing actions at a given layer of abstraction in terms of objects of a lower level.

This approach differs from other approaches that have been put forward for refinement in the object-oriented paradigm, namely [Ehrich and Sernadas 90], in that it is "proof-theoretic". It seems worthwhile investigating how the monadic approach developed in [Costa et al 92] can be lifted to propositions and theories in order to obtain a closer relationship between the two approaches.

This approach also differs from other techniques for action refinement in temporal logic that have been proposed in the literature [Lamport 83, 89, Barringer et al 86] in that refinement takes place between different logics (each corresponding to a level of granularity) and not within a single logic that accommodates all levels of granularity. In this sense, we agree with [Boudol 91] in that the "correct" interpretation of action refinement is not by means of "syntactic" substitution of the action by a process term in the context of another process term. We believe that it is such substitution-based techniques that lead to stuttering or dense structures for interpreting actions. Our aim has been to preserve the notion of level of abstraction and, hence, it makes no sense to have actions of different levels of granularity coexisting or for any substitution to be operated. We intend to investigate more closely the relationship between our approach and the operational model developed in [Boudol and Castellani 88].

We were thus able to retain the use of the next operator (which is the logical manifestation of the atomicity of actions) to specify the effects of actions. In fact, the refinement procedure we advocated consists in implementing an abstract **X** operator in terms of a concrete **F** operator. Hence, "tomorrow" is, sometimes, "sometime". We are currently refining the process of building the implementation body to guarantee that, in some sense, if the kernel objects have already been implemented, their implementation does not have to be changed. In the case of ADT implementation, this requirement corresponds to the conservativeness of the extension of the concrete data type specification.

Another direction for future work is to assess the impact that these principles may have in approaches to parallel system design such as UNITY [Chandy and Misra 88] and Action Systems [Back and Kurki-Suonio 88], which do not rely on an explicit control of the granularity of the actions involved. A more detailed comparison with other recent approaches, such as [Jones 92], which explore the use of object-based techniques, should also allow us to assess the impact of the object-oriented paradigm in concurrent system development, namely in the dimensions that we explored in the paper: control of granularity and interference.

Acknowledgments

This work was partially supported by the Esprit BRA 6071 (IS-CORE) and 8013 (MODELAGE), by the Eureka Project ESF and by JNICT and the British Council under an exchange protocol. Discussions on action refinement with Félix Costa, Grit Denker, Hans-Dieter Ehrich, Martin Honnen and Gunter Saake are gratefully acknowledged. The example which we used for illustration was developed with Jean-François Liesenborghs and Philippe Mottet.

References

[Aceto and Hennessy 89]
    L.Aceto and M.Hennessy, "Towards Action Refinement in Process Algebras", in *Proc. 4th LICS*, IEEE 1989, 138-145

[Back and Kurki-Suonio 88]
    R.Back and R.Kurki-Suonio, "Distributed Cooperation with Action Systems", *ACM TOPLAS* 10, 1988, 513-554

[Barringer and Kuiper 84]
> H.Barringer and R.Kuiper, "Hierarchical Development of Concurrent Systems in a Temporal Framework", in S.Brookes, A.Roscoe and G.Winskel (eds) *Seminar on Concurrency*, LNCS 197, Springer-Verlag 1984, 35-61

[Barringer et al 86]
> H.Barringer, R.Kuiper and A.Pnueli, "A Really Abstract Concurrent Model and its Temporal Logic", in *Proc. 13th ACM Symposium on Principles of Programming Languages,* ACM 1986, 173-183

[Boudol 91]
> G.Boudol, *Atomic Actions*, Research Notes, INRIA Sophia Antipolis, 1991

[Boudol and Castellani 88]
> G.Boudol and I.Castellani, "Concurrency and Atomicity", *Theoretical Computer Science* 59, 1988, 25-84

[Chandy and Misra 88]
> K.M.Chandy and J.Misra, Parallel Program Design - A Foundation, Addison Wesley 1988

[Costa et al 92]
> J.F.Costa, A.Sernadas and C.Sernadas, *Inductive Objects,* Research Report, DMIST/INESC July 1992

[Ehrich and Sernadas 90]
> H.-D.Ehrich and A.Sernadas, "Algebraic Implementation of Objects over Objects", in *Stepwise Refinement of Distributed Systems: Models, Formalisms, Correctness*, J. de Bakker, W.-P. de Roever and G. Rozenberg (eds), Springer Verlag, 1990, 239-266.

[Fiadeiro and Maibaum 92]
> J.Fiadeiro and T.Maibaum, "Temporal Theories as Modularisation Units for Concurrent System Specification", *Formal Aspects of Computing* 4(3), 1992, 239-272

[Fiadeiro and Maibaum 93]
> J.Fiadeiro and T.Maibaum, *Verifying for Reuse*, Technical Report,FCUL, 1993

[Jones 92]
> C.Jones, *An Object-Based Design Method for Concurrent Programs*, Technical Report, University of Manchester, 1992

[Lamport 83]
> L.Lamport, "Specifying Concurrent Program Modules", *ACM TOPLAS* 6(2), 1983

[Lamport 89]
> L.Lamport, "A Simple Approach to Specifying Concurrent Systems", *Communications ACM* 32(1), 1989, 32-45

[Sernadas et al 92]
> C.Sernadas, P.Gouveia, J.Gouveia, A.Sernadas and P.Resende, "The Reification Dimension in Object-oriented Data Base Design", in *Specification of Data Base Systems*, D. Harper and M. Norrie (eds), Springer Verlag, 1992, 275-299

[Veloso et al 85]
> P.Veloso, T.Maibaum and M.Sadler, "Program Development and Theory Manipulation", *Proc. Third International Workshop on Software Specification and Design,* London, IEEE Computer Society Press 1985, 228-232.

Applications of Transaction Logic to Knowledge Representation

Anthony J. Bonner[*][1] and Michael Kifer[**][2]

[1] University of Toronto
Department of Computer Science
Toronto, Ontario M5S 1A4, Canada
bonner@db.toronto.edu
[2] SUNY at Stony Brook
Department of Computer Science
Stony Brook, NY 11790, U.S.A.
kifer@cs.sunysb.edu

Abstract. We present applications of the recently proposed *Transaction Logic*—an extension of classical logic that accounts in a clean and declarative fashion for the phenomenon of state changes in knowledge bases, including logic programs and databases. Transaction Logic has a natural model theory and a sound-and-complete proof theory, but unlike many other logics, it allows users to *program* transactions. Its semantics also leads naturally to features whose amalgamation in a single logic has proved elusive in the past. These features include both hypothetical *and* committed updates, static and dynamic constraints, nondeterministic actions, nested transactions, bulk updates, view updates, active databases, subjunctive queries, and more. Finally, Transaction Logic holds promise as a logical model of hitherto non-logical phenomena, including *procedural knowledge* in AI, and the *behavioral* aspect of object-oriented databases, especially methods with side effects. This paper outlines the model theory of Transaction Logic, and then focuses on some of its applications to AI, including action definition and execution, planning, and dynamic constraints.

1 Introduction

Transaction Logic (abbreviated $\mathcal{TR}$) accounts in a clean, declarative fashion for the phenomenon of updating arbitrary logical theories, most notably, databases and logic programs. Unlike most logics of action, $\mathcal{TR}$ is a declarative formalism for specifying and executing procedures that update and permanently change a database, a logic program or, more generally, a logical theory. As a special

---

[*] Work supported in part by an Operating Grant from the Natural Sciences and Engineering Research Council of Canada and by a Connaught Grant from the University of Toronto.

[**] Supported in part by NSF grant CCR-9102159. Part of this work is done during sabbatical year at the University of Toronto. Support of Computer Systems Research Institute of University of Toronto is gratefully acknowledged.

case, transactions can be defined as logic programs. This is possible because, like classical logic, $\mathcal{TR}$ has a "Horn" version that has *both* a procedural and a declarative semantics, as well as an efficient SLD-style proof procedure. Since the formal aspects of $\mathcal{TR}$ can be found in [6, 5], this paper focuses on applications of $\mathcal{TR}$ to AI. Numerous other applications can be found in [5, 8]

$\mathcal{TR}$ was designed with several applications in mind, especially in databases, logic programming, and AI. It was therefore developed as a general logic, so that it could solve a wide range of update-related problems. Individual applications can be carved out of different fragments of the logic. These applications, both practical and theoretical, are discussed in great detail in [5]. For instance, in logic programming, $\mathcal{TR}$ leads to a clean, logical treatment of the *assert* and *retract* operators in Prolog, which effectively extends the theory of logic programming to include updates as well as queries. In object-oriented databases, $\mathcal{TR}$ can be combined with object-oriented logics, such as F-logic [15], to provide a logical account of *methods*, *i.e.*, procedures hidden inside objects that manipulate these objects' internal states. Thus, while F-logic covers the structural aspect of object-oriented databases, its combination with $\mathcal{TR}$ would account for the behavioral aspect as well. In AI, $\mathcal{TR}$ suggests a logical account of planning. STRIPS-like actions [11], for instance, and many aspects of hierarchical and non-linear planning are easily expressed in $\mathcal{TR}$. In spite of the previous efforts to give these phenomena declarative semantics, until now there has been no unifying *logical* framework to account for all of them.

On the surface, there would seem to be many other candidates for a logic of transactions, since many logics reason about time or action. However, despite a plethora of logics, researchers continue to complain that there is no clear declarative semantics for updates, whether in databases or in logic programming [4, 3, 20]. In fact—in stark contrast to classical logic—no logic of change has ever become a core of databases or logic-programming, in theory or in practice. There appear to be a few simple reasons for this unsuitability of existing logics of change. These reasons are discussed at length in [5], and we discuss some of them briefly here.

First, most logics of time or action are *hypothetical*. For instance, some systems can infer that if action A precedes B, and B precedes C, then A must precede C. Others can infer, say, that if a student took History400, then he could graduate. Such systems were intended to be observers of action, not participants. They are therefore useful for reasoning about alternatives, or for analyzing programs and plans; but they are not very useful for defining procedures that actually *accomplish* the state changes being reasoned about. In $\mathcal{TR}$, actions can be carried out hypothetically or they can be executed and have a permanent effect on the database, depending on the intent. Furthermore, the proof theory of $\mathcal{TR}$ is not only a verifier of truth, but also an *executor* of transactions. That is, given a transaction and an initial database state, $\mathcal{TR}$ will materialize the final database state.

Second, many logics insist on strict separation between queries and updates. However, this distinction is blurred in object-oriented systems, where both queries and updates are special cases of a single idea: method invocation. In such systems, an update can be thought of as a query with side effects. We would like to model this behavior and thereby provide a logical foundation for object-oriented databases. $\mathcal{TR}$ achieves this by allowing every logical formula to have not only a truth value, but also a "side effect" on the database. In this way, one can account for the *behavior* of object-oriented databases—something that most formalisms do not do. By integrating $\mathcal{TR}$ with F-logic [15], the structural aspect of object-oriented systems can be accounted for as well.

In [5], we provide a detailed comparison of $\mathcal{TR}$ with many other works. The system that comes closest in *spirit* to $\mathcal{TR}$ is Prolog. Unfortunately, updates in Prolog are non-logical operations and, as a result, state-changing procedures are often the most awkward of Prolog programs, and the most difficult to understand, debug, and maintain. $\mathcal{TR}$ provides a general solution to the aforementioned limitations, both of Prolog and of action and temporal logics.

2 Overview of Transaction Logic

$\mathcal{TR}$ is an extension of first-order logic, both syntactically and semantically. It has a natural model theory and a sound-and-complete proof theory. This section gives an overview of the syntax and the model theory. A complete development of $\mathcal{TR}$, including proof theory, can be found in [5].

Like classical logic, $\mathcal{TR}$ has a "Horn" version that is of particular interest for logic programming and deductive databases. In Horn $\mathcal{TR}$, a transaction is defined by Horn-style rules, where the premise specifies a *sequence* of queries and updates. $\mathcal{TR}$ is, thus, a logical language for programming database transactions, just as classical logic is a logical language for programming queries. Furthermore, $\mathcal{TR}$ has an inference system and two natural proof procedures: an efficient SLD-style procedure, and a dual, bottom-up procedure [6, 5]. These proof procedures both answer queries, execute transactions, *and* update the database. Because of its importance, much of this paper focuses on applications of Horn $\mathcal{TR}$. First, though, we describe full $\mathcal{TR}$, without the Horn restriction.

2.1 Syntax

The syntax of $\mathcal{TR}$ distinguishes two kinds of formulas: *transaction formulas* and *elementary transitions*. The former define composite transactions, and the latter define elementary updates.

Transaction formulas are used to define transactions (or complex actions) and to formulate queries. Transaction formulas extend first-order formulas with a new connective, $\otimes$, called *serial conjunction*. Thus, if ϕ and ψ are transaction formulas, then so are $\phi \vee \psi$, $\phi \wedge \psi$, $\phi \otimes \psi$, $\neg \phi$, $(\forall X)\phi$, and $(\exists X)\phi$, where X is a variable. The expression $a(X) \vee \neg[b(X) \otimes c(X,Y)]$ is a transaction formula. Informally, $\psi \otimes \phi$ says, "Do ψ and then do ϕ." A dual connective, *serial disjunction*, is also useful: $\psi \oplus \phi$ is equivalent to $\neg(\neg \phi \otimes \neg \psi)$.

Serial conjunction provides a basic way to *sequence* transactions, where $\phi \otimes \psi$ means "do ϕ then do ψ." In contrast, classical conjunction, "$\wedge$", *constrains* the non-determinism of a transaction. For instance, $\phi \wedge \psi$ means, "do ϕ in a way compatible with doing ψ." This use of "$\wedge$" is further discussed in Section 3.2. Apart from this, "$\wedge$" also has the traditional role of forming logic programs: in $\mathcal{TR}$, as in classical logic, any finite set of rules (called a *transaction base*) is equivalent to a conjunction of all the rules in the set.

A transaction base defines complex formulas in terms of simpler ones. However, we also need a way to specify *elementary changes* to a database. One way to define such transitions is to build them into the semantics as in [17, 18, 9, 1]. The main problem with this approach is that adding new kinds of elementary transitions requires redefining the very notion of a model and, hence, entails an overhaul of the entire proof theory. This is a serious drawback since there appears to be no small, single set of elementary transitions that is best for all purposes [5]. Indeed, in [5, 8], we introduce two new kinds of elementary update. Thus, rather than committing $\mathcal{TR}$ to a fixed set of elementary transitions, we have chosen to treat the elementary transitions as a *parameter* of $\mathcal{TR}$. To achieve this, elementary transitions are defined by logical axioms.

Elementary transitions are formulas of the form $\langle\phi, \psi\rangle u$, where ϕ, ψ are (sets of) closed first-order formulas and u is an atomic formula, called the *name* of the transition. Intuitively, this formula says that u is an update that transforms database ϕ into database ψ. For instance, if the atoms $q.ins(t)$ and $q.del(t)$ stand for the insertion and deletion of atom $q(t)$, then they would be defined by an enumerable set of elementary transitions consisting of the following formulas:

$$\langle \mathbf{D}, \mathbf{D} + \{q(t)\}\rangle\, q.ins(t) \qquad\qquad \langle \mathbf{D}, \mathbf{D} - \{q(t)\}\rangle\, q.del(t)$$

for every relational database $\mathbf{D}$.[3] Enumerable sets of elementary transitions are called *transition bases*. In practice, these formulas would not be materialized all at once, but would be generated on demand by an algorithm. The reader is referred to [5] for a more detailed discussion of transition bases.

As seen from the syntax, there is no strict distinction in $\mathcal{TR}$ between predicates that query the database and predicates that update it. As in classical logic, every predicate has a truth value, but in addition, every predicate may have a side effect by changing the state of the database. This uniformity of representation is important for modeling *methods* in object-oriented databases, where one generally does not distinguish between information-retrieving and state-changing methods. Nevertheless, if desired, $\mathcal{TR}$ can make such a distinction by using different sorts of predicates, one for updates and one for queries. In fact, it may be a good programming practice to reserve a special set of predicates for certain basic updates. This paper uses just such a convention: for each predicate symbol p, we use another predicate symbol, $p.ins$, to represent insertions of tuples into p. Likewise, we represent deletions from p by the predicate $p.del$. Thus the formula $p.ins(a) \otimes p.ins(b) \otimes p.ins(c)$ represents an updating transaction that inserts $p(a)$ into the database, then $p(b)$, and then $p(c)$.

[3] For relational databases, the operators $+$ and $-$ can be thought of as union and set-difference. However, if $\mathbf{D}$ is a general first-order formula, then defining insertion and deletion is more involved [14].

2.2 Example

Before presenting the semantics, we illustrate $\mathcal{TR}$ through an example of a robot arm moving blocks around a table top [19]. Planning of robot actions is discussed in Section 3.1, and in more detail in [5]. The example shows how updates are combined with queries to define complex transactions. In the example, the body of each rule is a sequence of atomic formulas, some of which are queries and some of which are updates. We say that such rules are *serial-Horn*, or simply *Horn*. The example also illustrates the use of transaction subroutines and shows how $\mathcal{TR}$ improves upon Prolog's *assert* and *retract* operators.

Example 1. Non-Deterministic Robot Actions A state in our blocks world is defined via three database predicates: $on(x, y)$, which says that block x is on top of block y; $isclear(x)$, which says that nothing is on top of block x; and $color(x, c)$, which says that c is the color of block x. The rules, below, define six actions that change the state of the world. Together, these rules form a transaction base. For each action, the premises are evaluated in the order given, and the action fails if any of its premises fails (in which case the database is left in its original state). All variables are universally quantified outside the rules.

$stackSameColor(Z) \leftarrow color(Z, C) \otimes stack2colors(C, C, Z)$
$stack2colors(C_1, C_2, Z) \leftarrow color(X, C_1) \otimes color(Y, C_2) \otimes stack2blocks(X, Y, Z)$
$stack2blocks(X, Y, Z) \leftarrow move(Y, Z) \otimes move(X, Y)$
$move(X, Y) \leftarrow pickup(X) \otimes putdown(X, Y)$
$pickup(X) \leftarrow isclear(X) \otimes on(X, Y) \otimes on.del(X, Y) \otimes isclear.ins(Y)$
$putdown(X, Y) \leftarrow wider(Y, X) \otimes isclear(Y) \otimes on.ins(X, Y) \otimes isclear.del(Y)$

The basic actions are $pickup(X)$ and $putdown(X, Y)$, meaning, "pick up block X," and "put down block X on top of block Y," respectively. Both are defined in terms of elementary inserts and deletes to database relations. The remaining rules combine simple actions into more complex ones. For instance, $move(X, Y)$ means, "move block X to the top of block Y," and $stack2blocks(X, Y, Z)$ means, "stack blocks X and Y on top of block Z." These actions are deterministic: Each set of argument bindings specifies only one robot action.

In contrast, the two actions $stack2colors$ and $stackSameColor$ are *non-deterministic*. For instance, $stack2colors(C1, C2, Z)$ means, "stack two blocks, of colors C_1 and C_2, on top of block Z." The action does not say which two blocks to use, only their colors. To perform the action, the inference system searches the database for blocks of the appropriate color that can be stacked. If several such blocks are available, the system chooses any two arbitrarily. The action $stackSameColor(Z)$ means, "stack two blocks on top of Z that are of the same color as Z." Again, the inference system searches the database for appropriate blocks. In this way, by defining non-deterministic actions, a user can specify what to do (declarative knowledge) and how to do it (procedural knowledge).

Note that the rules in Example 1 involve queries as well as updates. In the last rule, for instance, the atom $isclear(Y)$ (which itself may be defined by other deductive rules) is a Boolean test that must return *true* in order for the transaction to succeed. In the first rule, the atom $color(Z, C)$ is a query that retrieves the color C of the block Z. The second rule is, perhaps, the most interesting. Here, the atoms $color(X, C_1)$ and $color(Y, C_2)$ are non-deterministic queries. They retrieve two blocks X and Y, of colors C_1 and C_2, respectively. The particular blocks retrieved by these queries then determine the *future* course of action taken in the rest of the transaction.

Example 1 easily extends to recursively defined actions. For instance, $stack(N, X)$ can be defined as an action that recursively stacks N blocks on top of block X [5].

Finally, observe that the rules in Example 1 can easily be rewritten in Prolog form, by replacing "$\otimes$" with ",", and by replacing the elementary state transitions with *assert* and *retract*. However, the resulting, apparently innocuous, Prolog program does not execute correctly! The problem is that Prolog updates are not undone during backtracking. For instance, suppose that during a *move* action, the robot picked up $blkA$, the widest block on the table. The *move* action would then fail, since the robot cannot put $blkA$ down on the stack, since $blkA$ is too wide. In $\mathcal{TR}$, the inference system simply backtracks and then tries to find another block to pick up.[4] Prolog, too, will backtrack, but it will leave the database in an incorrect state, since it will not undo the *pickup* action. Thus, if $blkA$ was previously on top of $blkB$, then $on(blkA, blkB)$ would remain deleted and $isclear(blkB)$ would stay in the database.

2.3 Model Theory

This section gives an overview of the model theory of $\mathcal{TR}$. The reader is referred to [5, 7, 6] for a full treatment, including the proof theory. Just as the syntax of $\mathcal{TR}$ is based on two basic ideas—serial conjunction and elementary transitions—semantics is also based on a few fundamental principles, described below.

Transaction Execution Paths: When the user executes a transaction, the database may change, going from some initial state to some final state. In doing so, the database may pass through any number of intermediate states. For example, execution of the transaction $a.ins \otimes b.ins \otimes c.ins$ takes the database from an initial state, $\mathbf{D}$, through the intermediate states $\mathbf{D} + \{a\}$ and $\mathbf{D} + \{a, b\}$, to the final state $\mathbf{D} + \{a, b, c\}$. This idea allows us to model a wide range of constraints. For example, we may require that every intermediate state satisfy some condition, or we may forbid certain sequences of states. Note that a path of length 1 corresponds to a single database state. In this way, one model-theoretic device, paths, accounts for databases, updates, queries and more general transactions. This focus on paths is related to the version of Process Logic in [13], but the

---

[4] The important point here is that this backtracking behavior of the proof theory is grounded in the model theory, which is unlike the various ad hoc ideas about "fixing" Prolog's *assert* and *retract*.

two logics are fundamentally different [5]. Because of the emphasis on paths, we refer to semantic structures in $\mathcal{TR}$ as *path structures*.

Truth on Paths: In $\mathcal{TR}$, truth is defined on paths, where a path is a finite sequence of states. For example, we would say that the formula $a.ins \otimes b.ins$ is true on the path $\mathbf{D}, \mathbf{D}+\{a\}, \mathbf{D}+\{a,b\}$, since the formula represents an insertion of the proposition a followed by an insertion of b. Note that this formula is *false* on the path $\mathbf{D}, \mathbf{D}+\{a,b\}$, on the path $\mathbf{D}, \mathbf{D}+\{b\}, \mathbf{D}+\{a,b\}$, and on $\mathbf{D}, \mathbf{D}+\{a,c\}, \mathbf{D}+\{a,b,c\}$. All logical connectives are interpreted on paths, i.e., in terms of action. For instance, $\phi \vee \psi$ is true on a path iff ϕ is true or ψ is true. This gives rise to non-deterministic actions, since intuitively, $\phi \vee \psi$ means, "Do ϕ or do ϕ." This idea is illustrated in [5, 8]. Likewise, $\phi \wedge \psi$ is true on a path iff ϕ and ψ are both true. This provides a way of constraining non-deterministic actions. For instance, $\phi \wedge \neg \psi$ intuitively means, "Do ϕ but without doing ψ in the process." Section 3.2 illustrates this idea. Finally, note that $\otimes$ and $\wedge$ are identical on paths of length 1, and both coincide with classical conjunction. Thus, on states, $\mathcal{TR}$ reduces to classical logic.

Database States: In $\mathcal{TR}$, a database state is a class of equivalent first-order classical formulas. The need to consider equivalence classes rather than just formulas arises because syntactically distinct formulas may represent the same database state. For instance, we do not normally think of $a \wedge b$ and $b \wedge a$ as being two different databases. For convenience, we will let a single member of a class stand for the entire class. Treating states as formulas provides a lot of flexibility when defining elementary updates. Such flexibility is needed since, for general databases, the semantics of elementary updates is not obvious, not even for relatively simple updates like insert and delete. For example, what does it mean to insert an atom b into a database that entails $\neg b$, especially if $\neg b$ itself is not explicitly present in the database? There is no simple answer to this question, and many solutions have been proposed (see [14] for a comprehensive discussion). For these reasons, we take a general approach to elementary updates. For us, an elementary update is a mapping that takes one database, $\mathbf{D}_1$, to some other database, $\mathbf{D}_2$, where a database is any (equivalence class of) first-order formulas. More generally, an elementary update may be non-deterministic, so it is not just a mapping, but a *binary relation* on databases.

The rest of this section presents the Herbrand semantics of $\mathcal{TR}$. A detailed discussion and development of this semantics is given in [7]. A general, non-Herbrand semantics and a sound-and-complete proof theory are given in [5, 6]. For convenience, we define a *split* of a path $\pi = \langle s_1, ..., s_n \rangle$ to be any pair of subpaths, $\pi_1 = \langle s_1, ..., s_i \rangle$ and $\pi_2 = \langle s_i, ..., s_n \rangle$. In this case, we write $\pi = \pi_1 \circ \pi_2$. We also define $\top$ to be a special classical Herbrand structure that satisfies every first-order formula.

Definition 2.1 [**Structures**] *A Herbrand path structure is a mapping*, $\mathbf{M}$, *from paths to classical Herbrand structures. For every state, s, the structure $\mathbf{M}(\langle s \rangle)$ must satisfy all the formulas in s.*

A *variable assignment* is a mapping that takes a variable as input, and returns a Hebrand term as output. We extend the mapping from variables to function terms in the usual way. As usual in logic, we omit the variable assignment for *sentences*, *i.e.*, for formulas with no free variables.

Definition 2.2 [**Satisfaction**] *Let* $\mathbf{M}$ *be a Herbrand path structure,* π *be a a path, and* ν *be a a variable assignment. Then,*

- $\mathbf{M}, \pi \models_\nu p$ *iff* $\mathbf{M}(\pi) \models^c_\nu p$, *for any atomic formula* p, *where* "$\models^c_\nu$" *denotes classical satisfaction.*
- $\mathbf{M}, \pi \models_\nu \neg \phi$ *iff it is not the case that* $\mathbf{M}, \pi \models_\nu \phi$.
- $\mathbf{M}, \pi \models_\nu \phi \wedge \psi$ *iff* $\mathbf{M}, \pi \models_\nu \phi$ *and* $\mathbf{M}, \pi \models_\nu \psi$. $\vee$ *is the dual of* $\wedge$.
- $\mathbf{M}, \pi \models_\nu \phi \otimes \psi$ *iff* $\mathbf{M}, \pi_1 \models_\nu \phi$ *and* $\mathbf{M}, \pi_2 \models_\nu \psi$ *for some split* $\pi_1 \circ \pi_2$ *of path* π.
- $\mathbf{M}, \pi \models_\nu \phi \oplus \psi$ *iff* $\mathbf{M}, \pi_1 \models_\nu \phi$ *or* $\mathbf{M}, \pi_2 \models_\nu \psi$ *for every split* $\pi_1 \circ \pi_2$ *of path* π.
- $\mathbf{M}, \pi \models_\nu (\forall X)\phi$ *iff* $\mathbf{M}, \pi \models_\mu \phi$ *for every variable assignment* μ *that agrees with* ν *everywhere except on* X. $\exists X$ *is the dual of* $\forall X$.

Definition 2.3 [**Models – Transaction Formulas**] *A path structure* $\mathbf{M}$ *satisfies a transaction formula* ϕ *iff* $\mathbf{M}, \pi \models \phi$ *for every path* π. $\mathbf{M}$ *is a model of a set of formulas iff it satisfies every formula in the set.*

Definition 2.4 [**Models – Elementary Transitions**] *A path structure* $\mathbf{M}$ *satisfies an elementary transition* $\langle \mathbf{D}_1, \mathbf{D}_2 \rangle u$ *if and only if* $\mathbf{M}, \langle \mathbf{D}_1, \mathbf{D}_2 \rangle \models u$. $\mathbf{M}$ *is a model of a set of transitions iff it satisfies every transition in the set.*

Definition 2.5 [**Executional Entailment**] *Let* $\mathcal{B}$ *be a set of elementary transitions,* $\mathbf{P}$ *be a set of transaction formulas,* ϕ *be a transaction formula, and* $\mathbf{D}_0, \mathbf{D}_1, \ldots, \mathbf{D}_n$ *be a sequence of first-order formulas. Then,*

$$\mathcal{B}, \mathbf{P}, \mathbf{D}_0, \mathbf{D}_1, \ldots, \mathbf{D}_n \models \phi$$

if and only if $\mathbf{M}, \langle \mathbf{D}_0, \mathbf{D}_1, \ldots, \mathbf{D}_n \rangle \models \phi$ *for every model,* $\mathbf{M}$, *of* $\mathcal{B}$ *and* $\mathbf{P}$.

3 Applications

A wide variety of interesting and useful formulas can be constructed in $\mathcal{TR}$, formulas that capture many of the novel and important features of database and knowledge-base systems. These features include action definition and execution, planning, consistency maintenance, static and dynamic constraints, non-deterministic actions, random sampling, nested transactions, view updates, heterogeneous databases, scientific databases, simulation of systems with state, rule-based inference, and more. This section describes some of these applications. Additional applications and further details of the applications presented herein can be found in [5] and [8]. In addition, $\mathcal{TR}$ can be easily extended with a modal necessity operator, □, which captures a whole new range of applications. These applications include hypothetical reasoning, subjunctive queries, counterfactuals, imperative programming constructs, active databases, software verification, and more [5, 8]. $\mathcal{TR}$ thus provides a wide range of features whose amalgamation in a single declarative formalism has proved elusive in the past. Furthermore, these features all follow naturally from $\mathcal{TR}$'s path-based semantics.

3.1 Planning

Planning of robot actions is carried out by representing various planning regimes as formulas in $\mathcal{TR}$. This is possible because rules in $\mathcal{TR}$ can be used to represent two types of knowledge about actions: action-definitions, which describe how to execute actions; and planning strategies, which describe how and when to execute actions.

Unlike many planning systems, $\mathcal{TR}$ does not need special mechanisms or special syntax to deal with these two types of knowledge. In $\mathcal{TR}$, both types of knowledge can be represented as rules. Planning itself is carried out by executing plan-generating transactions using the general proof theory of [5]—no need to invent specialized search algorithms for each new planning strategy. Moreover, in $\mathcal{TR}$, the infamous *frame* problem is *not an issue* when it comes to planning and execution of actions, just as it is not an issue in STRIPS planning systems or in Pascal programming. $\mathcal{TR}$ captures this attractive property within a formal logic. The frame problem in the context of $\mathcal{TR}$ is extensively discussed in [5].

In [5], we consider three planning regimes. In the first regime, which we call *naive* planning, the inference system searches blindly for any sequence of actions that achieves the goal. Naive planning corresponds roughly to planning using a forward production system as described in [19]. The naive planner is simple to formulate, but is very inefficient, and so its utility is limited to a mere demonstration that planning is possible in $\mathcal{TR}$. However, due to space limitation, this paper gives a detailed description of naive planning only.

The other two regimes improve upon the naive system by incorporating knowledge about how goals are to be achieved. In the second regime, planning is described in terms of a hierarchy of "scripts" or "skeletal plans." Scripts suggest ways of achieving goals, and often they summarize known methods in a particular problem domain. This kind of planning, which is exemplified by systems such as NOAH [21] and MOLGEN [12, 22], is natural for $\mathcal{TR}$, since each script corresponds to a high-level, non-deterministic action. For example, a script for making coffee might be as follows: grind coffee, boil water, put coffee in filter, pour water into filter [10, Article XV.D1]. In $\mathcal{TR}$, this script could be represented by a rule like the following:

$$makeCoffee \leftarrow grindCoffee \otimes boilWater \otimes fillFilter \otimes pourWater \quad (1)$$

As this example illustrates, scripts and skeletal plans often have the form of a procedure (or recipe). For this reason, they are often specified in a procedural language. As the rule above illustrates, however, they can also be specified *in a logic*, $\mathcal{TR}$, that integrates procedural and declarative knowledge in a single framework.

The third planning regime is exemplified by the well-known STRIPS system [11, 16], which plans movements for a robot arm. STRIPS is naturally represented in $\mathcal{TR}$ by a script consisting of just four deductive rules and rule-schemas. It can be shown that STRIPS's planning strategy is sound, albeit incomplete, in $\mathcal{TR}$'s semantics. This incompleteness is responsible for certain failures of STRIPS, *e.g.*, its inability to exchange the contents of two registers [19]. This problem was fixed in RSTRIPS, a successor of STRIPS, by using a better search strategy. The point, however, is that, since $\mathcal{TR}$ has a *complete* proof theory [5], it provides a means of programming powerful planning systems that are free from the problems created by the use of rigid, ad hoc search algorithms. Other problems with STRIPS-like planning (which extend to many other planning systems) are pointed out and solved in [5].

All examples in this paper are based on the insertion and deletion of tuples from a database. They thus bear a conceptual resemblance to STRIPS-actions. Several differences are worth noting:

- Unlike STRIPS-actions, $\mathcal{TR}$-rules are formulas in a rigorous *logical* formalism. They are therefore declarative as well as executable.
- Rules in $\mathcal{TR}$ are *hierarchical* and can be defined at many levels of abstraction. The six rules in Example 1, for instance, represent six different levels of abstraction. In contrast, STRIPS only allows actions to be defined at one level, directly in terms of database inserts and deletes; *i.e.*, it does not support intermediate-level actions (subroutines).
- Unlike STRIPS, actions in $\mathcal{TR}$ can be *non-deterministic*. As Example 1 shows, non-determinism can make actions simpler and easier to formulate, reducing the amount of detail that a user must specify.
- STRIPS-actions are relatively simple; they consist of a pre-condition (which is a series of tests), followed by a set of deletes, followed by a set of inserts, in that order. In contrast, in $\mathcal{TR}$, inserts, deletes, and tests can be sequenced in any order. Thus, apart from pre-conditions and post-conditions, any number of intermediate tests can also be specified. In fact, even more general actions are possible in $\mathcal{TR}$, since in addition to sequential ordering, formulas may be combined via classical conjunction, disjunction and negation.
- Finally, lest this is forgotten, $\mathcal{TR}$ is a general-purpose logic for which STRIPS, planning, and other issues considered here and in [5] are only some of the many applications.

Naive Planning. Naive planning is based on primitive planning actions (or operators). These primitive *actions* differ from elementary *updates* in that they correspond to real-world activities, such as robot actions. In Example 1, for instance, there are two primitive robot actions, $pickup(X)$ and $putdown(X, Y)$, in terms of which the other robot actions are defined. These primitive actions are in turn defined in terms of low-level database inserts and deletes, which are elementary updates that do not correspond to robot actions.

To see how naive planning is done in $\mathcal{TR}$, consider a transaction base with n primitive actions, $a_1, \ldots, a_n$. We define a *plan* to be any sequence of these actions. Such sequences are defined by the following $\mathcal{TR}$ rules:

$$plan \leftarrow act \otimes plan \qquad act \leftarrow a_1$$
$$plan \leftarrow act \qquad\qquad act \leftarrow a_2$$
$$\cdots$$
$$act \leftarrow a_n$$

These rules state that *act* can be any primitive action, and *plan* is any sequence of such actions. Now, suppose we want to achieve a particular goal, such as $isclear(b)$, which states that block b must be clear. The transaction $?-plan \otimes isclear(b)$ would then achieve this goal, and a sequence of primitive actions that results in $isclear(b)$ being true would be recorded in the proof. More generally, we will want to achieve complex goals. For instance, suppose we want to stack block b on top of c on top of d, i.e., the planning goal is $on(b,c) \wedge on(c,d)$. The transaction $?-plan \otimes [on(b,c) \wedge on(c,d)]$ achieves this goal. Since the predicate *on* is a query, and thus does not change the database, we can replace the classical conjunction by a serial one and write $?-plan \otimes on(b,c) \otimes on(c,d)$.

These ideas are easily generalized. If ψ is an arbitrary planning goal, then $?-plan \otimes \psi$ is a transaction that achieves the goal. This transaction, therefore, specifies a sequence of primitive actions such that ψ is true in the final database state. The most common situation is when ψ is a conjunction of queries, i.e., ψ has the form $g_1 \wedge \ldots \wedge g_m$, where each g_i is a query. In this case, the transaction $?-plan \otimes g_1 \otimes \ldots \otimes g_m$ achieves the planning goal. The plan itself can be extracted from the proof generated by the inference system. Indeed, it suffices to retain only those steps in which the primitive actions, $a_1, \ldots, a_n$, were used in the proof of $plan \otimes g_1 \otimes \ldots \otimes g_m$, disregarding lower-level actions used to define the a_l.

Note that this formulation of planning relies on several specific features of $\mathcal{TR}$ actions:

– *Serial conjunction*, which allows primitive actions to be combined sequentially.
– *Post-conditions* on actions, which allow planning goals to be stated.
– *Recursion*, which allows the inference system to generate sequences of actions of arbitrary length, by applying the same rule(s) over and over again.
– *Non-determinism*, which allows actions to be carried out in several different ways. Together with recursion, non-determinism allows us to define a plan to be *any* sequence of primitive actions.

The approach to planning which we have just illustrated is conceptually simple, but is too unfocussed in its blind search for plans. The inference systems of $\mathcal{TR}$ [5, 6], for instance, will simply string actions together at random until a sequence is found that achieves the goal. This behavior is often called forward (or bottom-up) planning. If the plan does not achieve the goal, then the inference system simply backtracks and generates another plan. This process continues until a plan that achieves the goal is finally found.

It is due to this unfocussed behavior that we call the above planning strategy "naive." In contrast, script-based planning strategies described in [5] are highly focussed, as they augment action definitions with "scripts" or "skeletal plans" that are known to be useful in achieving certain kinds of goals. In $\mathcal{TR}$, one can easily represent the planning strategies of STRIPS and NOAH via skeletal plans consisting of a small number of rules. The actual plan is then obtained via the general proof procedure of $\mathcal{TR}$, which *executes* these script. (See [5] for details.)

3.2 Constraints on Execution of Complex Actions

Because transactions are defined on paths, it is possible to express a large variety of constraints on the way they execute. For instance, we can place conditions on the state of the database during transaction execution, or we may forbid certain sequences of states. We refer to such conditions as *path constraints*, or *dynamic constraints*. Such constraints are particularly well suited to areas such as planning and design, where it is common to place constraints on the way things are done. This section illustrates a variety of dynamic constraints expressible in $\mathcal{TR}$. These include temporal constraints in the style of James Allen [2], such as, "immediately after," "some time after," "before," etc.

There are several important problems related to constraints. One such problem is *constraint satisfaction*. That is, given a transaction and a constraint, we want to execute the transaction in such a way that it satisfies the constraint. For example, we might ask a robot to carry out a task while not entering restricted areas and not executing certain undesirable or dangerous sequences of action. In general, starting from the current database, we want to find *some* way of executing a transaction while satisfying constraints.

Constraint satisfaction problems are particularly easy to express in $\mathcal{TR}$ because they correspond to classical conjunction. That is, if ψ and ϕ are transaction formulas, then $\psi \wedge \phi$ can be interpreted as, "Do transaction ψ so that ϕ will be satisfied on the execution path."[5] Intuitively, if ψ is a non-deterministic transaction, then ϕ acts as a filter, removing unwanted execution paths, and reducing the non-determinism of the transaction. If ψ is deterministic, then ϕ acts as a guard, forbidding execution unless the constraints are satisfied. Note that in either case, it is execution *paths* that are constrained.

The examples center around a planning system for robot navigation. The system is composed of rules defining an action, $goto(Y)$, that instructs the robot to go to location Y. This action is highly non-deterministic since there may be many routes that the robot can take. Dynamic constraints can force the planner to reject certain routes or to focus its attention on others.

A simple constraint might require the robot to do something while en route to some location, such as passing certain check points. There are two natural cases to this problem. In the first case, the constraint imposes an order on the way the robot does things. The constrained transaction can then be expressed as a sequence of goals. For instance, suppose we request the robot to go to room A in a building and to pass through room B along the way. This request can be

---

[5] Note that "$\wedge$" is symmetric, so other interpretations of this conjunction are also possible.

expressed as the goal $goto(roomB) \otimes goto(roomA)$, i.e., "go to room B, then go to room A."

In the second, and more interesting case, the constraint does not imply an order on the way things are done. For instance, suppose we request the robot to go to room A, passing through rooms A_1 and A_2 on the way. This does not commit the robot to visiting the rooms in a particular order, and so it cannot be expressed as a single, serial goal. Instead, it is properly expressed as a *conjunction* of serial goals, where each conjunct constrains the robot to pass through a particular room.

To express such constraints, we define a new proposition, path, which is true on every path, i.e., $\text{path} \equiv \phi \vee \neg \phi$, for any formula ϕ. The formula $\text{path} \otimes at(X)$, thus, specifies a path in which the robot ends up at location X. Likewise, the formula $at(X) \otimes \text{path}$ specifies a path in which the robot starts off at location X, while $\text{path} \otimes at(X) \otimes \text{path}$ specifies a path in which the robot passes through location X. For convenience, we abbreviate this latter formula as $go\_thru(X)$, by adding the following rule to the transaction base:

$$go\_thru(X) \leftarrow \text{path} \otimes at(X) \otimes \text{path}$$

It is now easy to specify paths in which the robot must pass through any number of locations, without specifying an order. For instance, the following formula specifies that the robot must go to room A, passing through rooms A_1 and A_2 along the way:

$$goto(roomA) \wedge go\_thru(roomA_1) \wedge go\_thru(roomA_2)$$

The use of classical conjunction ensures that this formula is true only on paths where all three conjuncts are true. In this way, the formulas $go\_thru(X)$ constrain the way in which the transaction $goto(roomA)$ may execute.

We can build up more complex constraints by combining classical and serial conjunction. For instance, the following formula requests that the robot go to room A and that along the way it pass first through rooms B_1 and B_2, in any order, and then through rooms C_1 and C_2, in any order:

$$goto(roomA) \wedge \begin{bmatrix} [go\_thru(roomB_1) \wedge go\_thru(roomB_2)] \\ \otimes \\ [go\_thru(roomC_1) \wedge go\_thru(roomC_2)] \end{bmatrix}$$

Many more examples of constraints can be found in [5].

Acknowledgments: We wish to thank Mariano Consens, Alberto Mendelzon, and Ray Reiter for many insightful discussions. Comments by Greg Meredith, Gösta Grahne, Peter Revesz, Fangzhen Lin, Javier Pinto, Dimitris Lagouvardos, Jan Van den Bussche, and Roel Wieringa are also gratefully acknowledged.

References

1. S. Abiteboul and V. Vianu. Procedural and declarative database update languages. In *ACM SIGACT-SIGMOD-SIGART Symposium on Principles of Database Systems (PODS)*, pages 240–250, 1988.

2. J.F. Allen. Towards a general theory of action and time. *Artificial Intelligence*, 23:123–154, July 1984.
3. F. Bancilhon. A logic-programming/Object-oriented cocktail. *SIGMOD Record*, 15(3):11–21, September 1986.
4. C. Beeri. New data models and languages—The challenge. In *ACM SIGACT-SIGMOD-SIGART Symposium on Principles of Database Systems (PODS)*, pages 1–15, San Diego, CA, June 1992.
5. A.J. Bonner and M. Kifer. Transaction logic programming (or a logic of declarative and procedural knowledge). Technical Report CSRI-270, University of Toronto, April 1992. Revised: February 1994. Available in *csri-technical-reports/270/report.ps* by anonymous ftp to *csri.toronto.edu*.
6. A.J. Bonner and M. Kifer. Transaction logic programming. In *Intl. Conference on Logic Programming (ICLP)*, Budapest, Hungary, June 1993.
7. A.J. Bonner and M. Kifer. An overview of transaction logic. *Theoretical Computer Science*, 133, October 1994.
8. A.J. Bonner, M. Kifer, and M. Consens. Database programming in transaction logic. In A. Ohori C. Beeri and D.E. Shasha, editors, *Proceedings of the International Workshop on Database Programming Languages (DBPL)*, Workshops in Computing, pages 309–337. Springer Verlag, February 1994. Workshop held on Aug 30–Sept 1, 1993, New York City, NY.
9. W. Chen. Declarative specification and evaluation of database updates. In *Intl. Conference on Deductive and Object-Oriented Databases (DOOD)*, volume 566 of *Lecture Notes in Computer Science*, pages 147–166. Springer Verlag, December 1991.
10. P.R. Cohen and E.A. Feigenbaum, editors. *The Handbook of Artificial Intelligence*, volume III. Addison-Wesley Publishing Co., 1986.
11. R.E. Fikes and N.J. Nilsson. STRIPS: A new approach to the application of theorem proving to problem solving. *Artificial Intelligence*, 2:189–208, 1971.
12. P.E. Friedland. *Knowledge-Based Experiment Design in Molecular Genetics*. PhD thesis, Computer Science Department, Stanford University, 1979. Report Number 79-771.
13. D. Harel, D. Kozen, and R. Parikh. Process Logic: Expressiveness, decidability, completeness. *Journal of Computer and System Sciences*, 25(2):144–170, October 1982.
14. H. Katsuno and A.O. Mendelzon. On the difference between updating a knowledge database and revising it. In *Proceedings of the International Conference on Knowledge Representation and Reasoning (KR)*, pages 387–394, Boston, Mass., April 1991.
15. M. Kifer, G. Lausen, and J. Wu. Logical foundations of object-oriented and frame-based languages. Technical Report 93/06 (a revision of 90/14), Department of Computer Science, SUNY at Stony Brook, April 1993. To appear in Journal of ACM. Available in *pub/TechReports/kifer/flogic.ps.Z* by anonymous ftp to *cs.sunysb.edu*.
16. V. Lifschitz. On the semantics of STRIPS. In *Reasoning about Actions and Plans: Proceedings of the 1986 Workshop*, Timberline, OR, 1987. Reprinted in *Readings in Planning*, J. Allen, J. Hendler, A. Tate (eds.), Morgan-Kaufmann, 1990, 523–530.
17. S. Manchanda and D.S. Warren. A logic-based language for database updates. In J. Minker, editor, *Foundations of Deductive Databases and Logic Programming*, pages 363–394. Morgan-Kaufmann, Los Altos, CA, 1988.

18. S. Naqvi and R. Krishnamurthy. Database updates in logic programming. In *ACM SIGACT-SIGMOD-SIGART Symposium on Principles of Database Systems (PODS)*, pages 251–262, March 1988.
19. N.J. Nilsson. *Principles of Artificial Intelligence*. Tioga Publ. Co., Paolo Alto, CA, 1980.
20. G. Phipps, M.A. Derr, and K.A. Ross. Glue-Nail: A deductive database system. In *ACM SIGMOD Conference on Management of Data*, pages 308–317, 1991.
21. E.D. Sacerdoti. The non-linear nature of plans. In *Intl. Joint Conference on Artificial Intelligence (IJCAI)*, pages 206–214, 1975. Also appears in *Readings in Planning*, pp. 162–170. Morgan Kaufmann, San Mateo, CA, 1990.
22. M.J. Stefik. *Planning with Constraints*. PhD thesis, Computer Science Department, Stanford University, 1980. Report Number 80-784.

Circumscribing Features and Fluents

Patrick Doherty[1] and Witold Łukaszewicz[2]

[1] Department of Computer and Information Science
Linköping University
S-58183 Linköping, Sweden
patdo@ida.liu.se
[2] Institute of Informatics
Warsaw University
00-913 Warsaw 59, Poland

Abstract. Sandewall has recently proposed a systematic approach to the representation of knowledge about dynamical systems that includes a general framework in which to assess the range of applicability of existing and new logics for action and change and to provide a means of studying whether and in what sense the logics of action and change are relevant for intelligent agents. As part of the framework, a number of logics of preferential entailment are introduced and assessed for particular classes of action scenario descriptions. This paper provides syntactic characterizations of several of these relations of preferential entailment in terms of standard FOPC and circumscription axioms. The intent is to simplify the process of comparison with existing formalisms which use more traditional techniques and to provide a basis for studying the feasibility of compiling particular classes of problems into logic programs.

1 Introduction

Sandewall has recently proposed a systematic approach to the representation of knowledge about dynamical systems that provides a general framework which can be used to assess the range of applicability of existing and new logics for action and change and to provide a means of studying whether and in what sense the logics of action and change are relevant for intelligent agents. This is made possible in part by introducing the formal notion of an inhabited dynamic system (IDS). An IDS can be related to a number of standard models used in control theory for modeling traditional dynamic systems, while at the same time serving as an underlying semantics in which to assess the applicability of logics of action and change proposed to solve various reasoning tasks.

Because both the logics and the IDS are formal structures, and the ontological and epistemological assumptions about the world and agents acting in it are made explicit, one can prove that a proposed logic is guaranteed to obtain the correct set of conclusions for a particular class of dynamical systems. Recent work in the area reinforces the importance of such an approach where the assessment of logics is less dependent on representative examples and is instead studied in terms of particular classes of action scenarios ([7],[8]). Details of Sandewall's work may be found in several papers ([10],[12]), and more recently, in a book manuscript [11].

1.1 Action Scenario Descriptions

A majority of reasoning problems involving action and change can be represented in terms of *(action) scenario descriptions*. A scenario description is a partial specification in a logical language of the initial state of a system, combined with descriptions of some of the actions that have occurred together with their timing. The "Yale shooting problem" or "Stanford murder mystery" may be considered scenario descriptions.

The formal syntax for specifying scenario descriptions is defined in terms of a surface language $\mathcal{L}(SD)$, consisting of action occurrence statements (ac1,ac2), action (law) schemas (acs1,acs2), and observation statements (obs1). In what follows, all expressions occurring in scenario descriptions will be prefixed. We will use the symbols "obs", "ac" and "acs" to denote observation statements, action occurrence statements and action schemas, respectively.

Example 1. The following is the Yale shooting scenario (below a and l are fluent constants standing for *alive* and *loaded*, respectively, while *Load* and *Fire* are action symbols).

obs1 $[0]$ $a \wedge \neg l$
ac1 $[2,4]$ $Load$
ac2 $[5,6]$ $Fire$
acs1 $[t_1, t_2]$ $Load \leadsto [t_1, t_2]\, l := T$
acs2 $[t_1, t_2]$ $Fire \leadsto ([t_1]\, l \longmapsto [t_1, t_2](a \wedge l) := F)$.

In the initial state, the gun is not loaded and the turkey is alive. The gun is loaded during the period 2 to 4, and fired during the period 5 to 6. In the interim between the two actions is a waiting period. The action schema for "Load" states that if a gun is loaded during the period t_1 to t_2, then at the end of the duration the gun will in fact be loaded at time t_2. A similar type of reading may be given to the action schema for "Fire". Note that during the duration of an action, the values of influenced fluents are not known. Reassignment expressions of type $[t, t']\gamma := T$ and $[t, t']\gamma := F$ are used to assert that a fluent is reassigned a new value of *true* or *false*. The idea is that during the duration t to t', we do not know when γ becomes *true* or *false*, but only that it is *true* or *false* at time t'.

The surface syntax of a scenario description is translated into wffs in $\mathcal{L}(FL)$, the language for the base logic FL, using a two step process. We will say more about this translation in the next section.

1.2 The Reasoning Task

After translating a scenario description into a set of wffs in $\mathcal{L}(FL)$, the reasoning task is to derive additional statements about the course of events. Temporal prediction and postdiction, among others, are two types of reasoning tasks one might be interested in. In a sense, this distinction is unimportant. Essentially, what we would like to do is to conclude as much as possible about the values of fluents at different points in time. Such sets of conclusions will be referred to as *chronicle completions*. Of course, our interest is in intended completions, i.e. those completions containing desirable conclusions constrained by an assumption of inertia. For instance, the Yale shooting scenario has exactly one intended completion given by

[0,2] $a \wedge \neg l$
[3] a
[4,5] $a \wedge l$
[6,∞) $\neg a \wedge \neg l$

Within Sandewall's framework, the intent is to find the proper relation of preferential entailment for the class that this scenario description belongs to, and show that for any scenario in this class, only the conclusions in the intended completions for each scenario description are in fact preferentially entailed. We will consider a number of different ontological classes and their corresponding syntactic characterizations in this paper. A good description of the assessment technique may be found in [12].

1.3 Structure of Paper

The structure of the paper is as follows: In Section 2, the base logic FL is introduced and in Section 3, it is shown how to reduce scenario descriptions in $\mathcal{L}(SD)$ to sets of formulas in $\mathcal{L}(FL)$. In Section 4, a brief account of the logics of preferential entailment considered, and their corresponding ontological classes, is provided. In Section 5, a brief summary of circumscription is provided. In Section 6, the circumscription axioms for four of the preferential entailment relations are given. In Section 7, the concept of occlusion and nochange axioms are introduced and in Section 8, circumscription axioms for three more preferential entailment relations, which use the occlusion concept, are given. In Section 9, we conclude with a discussion.

2 The Base Logic FL

There are a number of different possibilities for choosing a base logic in which to compile scenario descriptions. Sandewall uses a specialized logic called discrete fluent logic with a number of interesting nonstandard features which can potentially aid in developing efficient implementations for the various action classes. In this section, we introduce a logic of fluents FL, which will be used as a basis for formalizing reasoning about action and change. We will use a two-sorted FOPC with a sort $\mathcal{T}$ for temporal entities and a sort $\mathcal{F}$ for truth-valued fluents. In what follows, $\mathcal{T}$ should be interpreted as a linear discrete time line where $\mathcal{T}$ is considered isomorphic to the natural numbers. We work with an interpreted theory, assuming the standard interpretation for the natural numbers. The logical language is similar to the logic of persistence used by Kautz [4] in one of the first solutions to the Yale Shooting Problem. We will show an example after introducing the language.

2.1 Syntax

The language $\mathcal{L}(FL)$, is a sorted first-order language with equality. We use the standard connectives ¬ (negation), ∧ (conjunction), → (implication), ≡ (equivalence), ∨ (disjunction), and the standard quantifiers ∀ and ∃. In addition, scoping will be indicated by parentheses or the dot notation.

There are two domain independent sorts, $\mathcal{T}$ for time-points and $\mathcal{F}$ for propositional fluents. A propositional fluent is a function of time with the boolean truth values as range.

The letters t and s, possibly subscripted and/or primed, will be used to denote variables of type $\mathcal{T}$, while the numerals (0, 1, 2, ...) will be used to denote constants of type $\mathcal{T}$.

The letters f and g, possibly subscripted and/or primed, will be used to denote variables of type $\mathcal{F}$, while italic lowercase letters of the alphabet or combinations of these, will be used to denote constants of type $\mathcal{F}$.

We include the predicate symbols $Holds$, $Occlude$, and $Clip$, of type $\mathcal{T} \times \mathcal{F}$, and the predicate symbols $<$ and $\leq$ (interpreted as the usual "less than" and "less than or equal to" relations on natural numbers) of type $\mathcal{T} \times \mathcal{T}$, and the equality predicate $=$.

We include the function symbols $+$, and $-$ (interpreted as the usual "plus" and "minus" functions on natural numbers) of type $\mathcal{T} \times \mathcal{T} \mapsto \mathcal{T}$.

In the following, the symbols **t** and **s**, possibly subscripted and/or primed, will serve as meta variables denoting temporal terms and the symbols **f** and **g**, possibly subscripted and/or primed, will serve as meta variables denoting fluent terms. The letters α, β, possibly subscripted and/or primed, will serve as meta variables ranging over formulas.

We define the *set of temporal terms* as the smallest set satisfying: 1) Each temporal variable and each temporal constant (i.e. a numeral denoting a natural number) is a temporal term; 2) if **t** and **t'** are temporal terms, then so is **t** + **t'** and $t - t'$. We define the *set of fluent terms* as the smallest set satisfying: 1) Each fluent variable and each fluent constant is a fluent term.[3]

The set of *atomic formulas* is defined to be the smallest set satisfying:

1. If $\mathbf{f}$ is a fluent term and $\mathbf{t}$ is a temporal term, then $Holds(\mathbf{t,f})$, $Clip(\mathbf{t,f})$, and $Occlude(\mathbf{t,f})$ are atomic formulas;
2. if $\mathbf{t}$ and $\mathbf{t'}$ are temporal terms, then $\mathbf{t} = \mathbf{t'}$, $\mathbf{t} < \mathbf{t'}$ and $\mathbf{t} \leq \mathbf{t'}$ are atomic formulas.

We define the *set of formulas* as the smallest set satisfying:

1. Each atomic formula is a formula;
2. if α and β are formulas, then so are $\neg\alpha$, $\alpha \rightarrow \beta$, $\alpha \equiv \beta$, $\alpha \vee \beta$, and $\alpha \wedge \beta$;
3. if α is a formula, t is a temporal variable, f is a fluent variable then $\forall t.\alpha$ and $\exists t.\alpha$, $\forall f.\alpha$ and $\exists f.\alpha$ are formulas.

In what follows, $\mathbf{t} < \mathbf{t'} < \mathbf{t''}$, $\mathbf{t} \leq \mathbf{t'} < \mathbf{t''}$, $\mathbf{t} < \mathbf{t'} \leq \mathbf{t''}$ and $\mathbf{t} \leq \mathbf{t'} \leq \mathbf{t''}$, stand for $\mathbf{t} < \mathbf{t'} \wedge \mathbf{t'} < \mathbf{t''}$, $\mathbf{t} \leq \mathbf{t'} \wedge \mathbf{t'} < \mathbf{t''}$, $\mathbf{t} < \mathbf{t'} \wedge \mathbf{t'} \leq \mathbf{t''}$ and $\mathbf{t} \leq \mathbf{t'} \wedge \mathbf{t'} \leq \mathbf{t''}$, respectively.

2.2 Kautz's Logic of Persistence

Kautz [4] presented one of the first solutions to the Yale shooting problem based on the notion of chronological minimization, preferring violations of persistence at later rather than earlier points in time. Due to the similarities between Kautz's language and $\mathcal{L}(FL)$, we will describe his original axiomatization of the Yale shooting problem in order to provide some intuitions as regards the meaning of the $Hold$ and $Clip$ predicates and of a more conventional representation of an action scenario. Axiom (1) serves as the persistence or frame rule and $Clip$, representing violations of persistence, is minimized chronologically:[4]

$$Holds(t,f) \rightarrow (Holds(t+1,f) \oplus Clip(t+1,f)) \qquad (1)$$

$$Holds(t, FIRE) \wedge Holds(t, LOADED) \rightarrow$$
$$\neg Holds(t+1, LOADED) \wedge \neg Holds(t+1, ALIVE) \qquad (2)$$

$$Holds(1, LOADED) \qquad (3)$$

---

[3] The notion of a fluent term can be extended in the standard way to include more complex terms, if necessary.
[4] The symbol "$\oplus$" denotes the exclusive-or operator and will also be used later in the paper.

$$Holds(1, ALIVE) \qquad (4)$$

$$Holds(1, FIRE) \qquad (5)$$

In addition, the axiomatization contains the unique names axioms $LOADED \neq ALIVE$, etc.

Kautz provides a definition of preferential entailment and a circumscriptive axiom sound and complete relative to the minimal models. We show the definition of preferential entailment. Let $M1$ and $M2$ be two models, where $M[P]$ is the extension of predicate P in the model M. $M1 \leq M2$ iff

1. $M1$ and $M2$ have the same domain
2. Every constant, function, and predicate symbol other than $Clip$ and $Hold$ receives the same interpretation in $M1$ and $M2$.
3. The following (meta-theoretic) statement is true:

$$\langle f,t \rangle \in M1[Clip] \rightarrow (\langle f,t \rangle \in M2[Clip] \vee$$
$$\exists t', f'. \langle f',t' \rangle \in M2[Clip] \wedge \langle f',t' \rangle \notin M1[Clip] \wedge \langle t',t \rangle \in M1[<])$$

where the final clause states that t' is before t.

Kautz's solution will be considered in more detail later in the paper. We will use many different notions of entailment and provide additional extensions to our language, such as the *Occlude* predicate.

3 Reducing Scenario Descriptions to $\mathcal{L}(FL)$

Given a scenario description Υ, consisting of statements in the surface language $\mathcal{L}(SD)$, these statements can be translated into formulas in the language $\mathcal{L}(FL)$ via a two-step process. In the first step, action schemas in Υ are instantiated with action occurrence statements, resulting in what are called *schedule statements*. The resulting schedule statements replace the action schemas and action occurrence statements. The result is an *expanded (action) scenario description* Υ', consisting of both schedule and observation statements. In the second step, abbreviation definitions are used to translate statements in Υ' into formulas in $\mathcal{L}(FL)$.

3.1 Action Instantiation in $\mathcal{L}(SD)$

To represent actions in a scenario description, we introduce a set of *action symbols*. These will usually be denoted by everyday English words such as *Load*, *Fire*, etc. An *action occurrence statement* is any expression of the form [s,t]**A**, where **A** is an action symbol and **s** and **t** are temporal terms. An *action (law) schema* is any expression of the form $[s,t]\mathbf{A} \leadsto \alpha$, where **A** is an action symbol, s and t are temporal variables, and α is a restricted formula of FL.

We say that an action occurrence *corresponds* to an action schema if the same action symbol occurs in both expressions. Let $[\mathbf{s},\mathbf{t}]\mathbf{A}$ be an action occurrence statement corresponding to an action schema $[s,t]\mathbf{A}\leadsto \alpha$. The *result* of the action occurrence wrt the action schema is the formula $\alpha(s/\mathbf{s},t/\mathbf{t})$, where $\alpha(s/\mathbf{s},t/\mathbf{t})$ is obtained from α by substituting the terms **s** and **t** for the variables s and t. For instance, the result of the action $[4,6]$ *Load* wrt the action schema $[t_1, t_2]$ *Load* $\leadsto [t_1, t_2]\, l := T$ is the formula $[4,6]\, l := T$.

Let Υ be a scenario description. An *expanded (action) scenario description* Υ', for Υ is the set of statements obtained from Υ by replacing all action occurrence statements in Υ by their results wrt the corresponding action schemas and then deleting all action schemas in Υ. In the sequel, these results will be referred to as *schedule statements*, and will be prefixed by the symbol "scd".

Example 2. The corresponding expanded scenario description for the scenario in Example (1), with abbreviations, is,

obs1 $[0]\, a \wedge \neg l$
scd1 $[2,4]\, l := T$
scd2 $[5]\, l \rightarrow [5,6](a \wedge l) := F$.

3.2 From $\mathcal{L}(SD)$ to $\mathcal{L}(FL)$

Rather than provide an exhaustive list of the abbreviation definitions used in translating statements in expanded scenario descriptions to formulas in $\mathcal{L}(FL)$, we list a few of the more important and refer the reader to Doherty [2] for the details. Since it can be shown that any boolean combination of fluent constants prefixed by $[\mathbf{t}]$ can be reduced to a boolean combination of statements of type $[\mathbf{t}]\delta$, where δ is a fluent constant (possibly negated), the following abbreviations should suffice to make the point.

$[\mathbf{t}]\delta \stackrel{\text{def}}{=} Holds(\mathbf{t}, \delta)$
$[\mathbf{s},\mathbf{t}]\delta \stackrel{\text{def}}{=} \forall t.\ \mathbf{s} \leq t \leq \mathbf{t} \rightarrow Holds(t, \delta)$
$[\mathbf{s},\mathbf{t})\delta \stackrel{\text{def}}{=} \forall t.\ \mathbf{s} \leq t < \mathbf{t} \rightarrow Holds(t, \delta)$
$[\mathbf{s},\mathbf{t}]\delta := T \stackrel{\text{def}}{=} \exists t.\mathbf{s} \leq t < \mathbf{t} \wedge \forall t'(t < t' \leq \mathbf{t} \rightarrow Holds(t', \delta))$
$[\mathbf{s},\mathbf{t}]\delta := F \stackrel{\text{def}}{=} \exists t.\mathbf{s} \leq t < \mathbf{t} \wedge \forall t'(t < t' \leq \mathbf{t} \rightarrow \neg Holds(t', \delta))$
$Holds(\mathbf{t}, \neg\delta) \stackrel{\text{def}}{=} \neg Holds(\mathbf{t}, \delta)$.

Example 3. The corresponding set of labeled formulas in $\mathcal{L}(FL)$ after expanding the abbreviations in Example (2) is,

obs1 $Holds(0,a) \land \neg Holds(0,l)$
scd1 $\exists t.2 \leq t < 4 \land \forall t'(t < t' \leq 4 \rightarrow Holds(t',l))$
scd2 $Holds(5,l) \rightarrow$
  $([\exists t.5 \leq t < 6 \land \forall t'(t < t' \leq 6 \rightarrow \neg Holds(t',a))] \land$
  $[\exists t.5 \leq t < 6 \land \forall t'(t < t' \leq 6 \rightarrow \neg Holds(t',l))]).$

4 Entailment Classes in Features and Fluents

Sandewall's assessment tool identifies the class of scenarios for which a given logic obtains exactly the intended models. Classes of scenarios are characterized by a coding scheme that makes explicit the *epistemological* and *ontological* assumptions of that class. The simplest category in what Sandewall calls the *overview taxonomy* is $\mathcal{K} - IA$. The $\mathcal{K}$ signifies the epistemological completeness of the class, where actions are known by the agent, along with the (possible) effects of the actions. The IA signifies simple inertia and alternative actions: Features do not change value unless explicitly changed by an action, context-dependent actions and actions with random effects are allowed, along with actions with duration. The timing and order of actions can be incompletely specified in a scenario. Partial or complete specifications are permitted at any state in a scenario including the first. Given the outer framework, more specific classes can be defined by the use of *subordinate subspecialties*, constraints on the $\mathcal{K}, I$, or A parts of the coding.

The preferential entailment relations considered in this paper will use combinations of the following subspecialties within the $\mathcal{K} - IA$ framework:

$\mathcal{K}s$: The state of the world at time zero is completely specified by the scenario.
$\mathcal{K}p$: No observation statements for times later than zero.
$\mathcal{K}c$: The scenario is consistent. The set of intended models is not empty.
$\mathcal{K}l$: The scenario is linear. Actions and observations occur in the same order in all intended models.
Is: Actions take a single time step.
Ad: Actions are deterministic.
Ae: Actions are equidurational. The set of possible durations of an action is independent of its start state.
Au: Actions have uniform change. The set of features affected by an action is independent of its start state.
An: Each action has necessary change. Either it is deterministic, or nondeterministic, but changes in one outcome are never a subset of changes in the other.
Ax: For actions with random effects. No preference of choice for effects of equal duration.

In the paper we will consider the following logics of preferential entailment. The following is a list of the classes of scenarios where the associated logics are guaranteed to provide only the intended conclusions relative to the intended models of the underlying semantics.

OCM Original Chronological Minimization of Change. $\mathcal{K}sp - IsAd$.
PCM Prototypical Chronological Minimization of Change. $\mathcal{K}p - IAex$.
PGM Prototypical Global Minimization of Change. $\mathcal{K}cl - IsAun$.
PCMF Prototypical Chronological Minimization of Change with Filtering. $\mathcal{K} - IAex$.
PMON Global Minimization of Occlusion with Nochange Premises. $\mathcal{K} - IA$.
CMON Chronological Minimization of Occlusion with Nochange Premises. $\mathcal{K} - IA$.
CMOC Chronological Minimization of Occlusion and Change. $\mathcal{K} - IAe$.

Two other logics, CAMC and CAMOC, described by Sandewall [11], can also be given circumscriptive representations, but will not be considered in this article. McCarthy's original proposal for solving the frame problem is similar to PGM. Kautz's original solution to the frame problem is similar to OCM. Various benchmarks used in the literature to focus on particular types of reasoning problems can also be classified within the various classes. For example, the Yale Shooting Problem is of type $\mathcal{K}sp - IAde$, while the Stanford Murder Mystery is of type $\mathcal{K} - IAde$. For a more detailed study of the various classes, benchmark examples, and assessment techniques, we refer the reader to Sandewall [11].

5 Brief Summary on Circumscription

We will assume familiarity with both second-order and generalized pointwise circumscription [6] and use the notation in Lukaszewicz [9]. Briefly,

Definition 1 Second-Order Circumscription. Let $\bar{P}$ be a tuple of distinct predicate constants, $\bar{S}$ be a tuple of distinct function and/or predicate constants disjoint from $\bar{P}$, and let $\Gamma(\bar{P}, \bar{S})$ be a theory. The *second-order circumscription* of $\bar{P}$ in $\Gamma(\bar{P}, \bar{S})$ with variable $\bar{S}$, written $Circ_{SO}(\Gamma; \bar{P}; \bar{S})$, is the sentence

$$\Gamma(\bar{P}, \bar{S}) \land \forall \bar{\Phi}, \bar{\Psi}. \neg [\Gamma(\bar{\Phi}, \bar{\Psi}) \land \bar{\Phi} < \bar{P}]$$

where $\bar{\Phi}$ and $\bar{\Psi}$ are tuples of variables similar to $\bar{P}$ and $\bar{S}$, respectively.

Observe that this can be rewritten as,

$$\Gamma(\bar{P}, \bar{S}) \land \forall \bar{\Phi}, \bar{\Psi}. [\Gamma(\bar{\Phi}, \bar{\Psi}) \land [\bar{\Phi} \leq \bar{P}] \rightarrow [\bar{P} \leq \bar{\Phi}]]$$

where $\bar{U} \leq \bar{W}$ is equivalent to $\bigwedge_{i=1}^{n} \forall \bar{x}. [U_i(\bar{x}) \rightarrow W_i(\bar{x})]$.

Definition 2 Generalized Pointwise Circumscription. Let P be a predicate constant of arity n, $\bar{S} = \{S_1, \ldots, S_m\}$ be a tuple of distinct function and/or predicate constants different from $\bar{P}$, and let $\Gamma(P, \bar{S})$ be a theory. Let V be a $2 \times n$-ary predicate expression and suppose that $\bar{U} = \{U_1, \ldots, U_m\}$, where each $U_i (1 \leq i \leq m)$ is a predicate expression whose arity is the arity of S_i plus n. It is assumed that $V, U_1, \ldots, U_m$ have no parameters and contain none of $P, S_1, \ldots, S_m$. The *pointwise* *circumscription of P in $\Gamma(P,\bar{S})$ with P allowed to vary on $V(\bar{x})$ and S_i allowed to vary on $U_i(\bar{x})(1 \leq i \leq m)$, written $Circ_{PW}(\Gamma; P/V; \bar{S}/\bar{U})$, is the sentence*

$$\Gamma(P,\bar{S}) \wedge \forall \bar{x}, \Phi, \bar{\Psi}. \neg \left(P(\bar{x}) \wedge \neg \Phi(\bar{x}) \wedge EQ_{V(\bar{x})}(P,\Phi) \wedge \bigwedge_{i=1}^{m} [EQ_{U_i(\bar{x})}(S_i, \Psi_i)] \wedge \Gamma(\Phi, \bar{\Psi}) \right)$$

where $EQ_V(U, W) = \forall \bar{x}. \neg V(\bar{x}) \rightarrow [U(\bar{x}) \equiv W(\bar{x})]$ and where Φ and $\bar{\Psi}$ correspond to P and $\bar{S}$, respectively.

The intuition behind the use of $EQ_V(U,W)$ is that only points that have property V are allowed to vary. Consequently, when minimizing a point in a predicate P, it is possible to vary other points in P, or points in additional predicates $S_i \in \bar{S}$. Just as one can control the variation of predicate extensions in second-order circumscription, here one can control variation of points *within* a predicate extension. This increase in granularity of the minimization policy is in fact the property that enables direct translations of a number of Sandewall's notions of preferential entailment.

5.1 Some additional Notation

In what follows, we will need some additional notation to distinguish between different types of formulas within a scenario description. Let $\Gamma_C = \Gamma_{OBS} \cup \Gamma_{SCD} \cup \Gamma_{UNA}$ denote the set of formulas in $\mathcal{L}(FL)$ corresponding to a scenario description Υ, where Γ_{OBS} are the observation axioms, Γ_{SCD} are the schedule axioms and Γ_{UNA} are the appropriate unique name axioms for the fluent constants in Υ. In addition, Γ_{PER} will denote the set containing the two persistence axioms,

$$\forall f, t. Holds(t, f) \rightarrow [Holds(t+1, f) \oplus Clip(t+1, f)] \qquad (6)$$

$$\forall f, t. \neg Holds(t, f) \rightarrow [\neg Holds(t+1, f) \oplus Clip(t+1, f)]. \qquad (7)$$

We will use Γ^+_{SCD} to denote $\Gamma_{SCD} \cup \Gamma_{PER}$ and Γ^+_C to denote $\Gamma_{OBS} \cup \Gamma^+_{SCD} \cup \Gamma_{UNA}$. This notation will be useful in later sections.

6 Circumscription Axioms for OCM, PCM, PGM and PCMF

6.1 Original Chronological Minimization – OCM

OCM is similar to Kautz's [4] original solution to the Yale Shooting Problem, but not quite the same thing. We will consider the difference in a later section.[5]

---
[5] In the following $\langle t, f \rangle = \langle t', f' \rangle$ is an abbreviation for $t = t' \wedge f = f'$.

Let $\Gamma(Clip, Holds) = \Gamma_C^+$.
Let $V(t,f) = \lambda t', f'.[\langle t, f \rangle = \langle t', f' \rangle \vee t < t']$.

$Circ_{PW}(\Gamma(Clip, Holds); Clip/V(t,f); Holds) =$

$\quad \Gamma(Clip, Holds) \wedge$
$\quad \forall t, f.\forall \Phi, \Psi.\neg[Clip(t,f) \wedge \neg\Phi(t,f) \wedge EQ_{V(t,f)}(Clip, \Phi) \wedge \Gamma(\Phi, \Psi)]$ (8)

When minimizing $Clip$ at $\langle t, f \rangle$, the only part of $Clip$'s extension that varies is the point $\langle t, f \rangle$ itself, or the points $\langle t', f' \rangle$ where $t' > t$ and f' is arbitrary. This logic is suitable for the ontological class $\mathcal{K}sp - IsAd$.

6.2 Prototypical Chronological Minimization – PCM

Let $\Gamma(Clip, Holds) = \Gamma_C^+$.
Let $V(t,f) = \lambda t', f'.[\langle t, f \rangle = \langle t', f' \rangle \vee t < t']$ and $U(t,f) = \lambda t', f'.[t \leq t']$.

$Circ_{PW}(\Gamma(Clip, Holds); Clip/V(t,f); Holds/U(t,f)) =$

$\quad \Gamma(Clip, Holds) \wedge$
$\quad \forall t, f.\forall \Phi, \Psi.\neg[Clip(t,f) \wedge \neg\Phi(t,f) \wedge EQ_{V(t,f)}(Clip, \Phi) \wedge$
$\quad\quad EQ_{U(t,f)}(Holds, \Psi) \wedge \Gamma(\Phi, \Psi)]$. (9)

When minimizing $Clip$ at $\langle t, f \rangle$, the only part of $Clip$'s extension that varies is the point $\langle t, f \rangle$ itself, or the points $\langle t', f' \rangle$ where $t' > t$ and f' is arbitrary. In addition, $Holds$ should remain fixed at all points $\langle t', f' \rangle$ where $t' < t$ and f' is arbitrary. This logic is suitable for the ontological class $\mathcal{K}p - IAex$.

6.3 Prototypical Global Minimization – PGM

PGM is essentially the minimization policy that started all the problems. Rather than minimize change chronologically, the extension of $Clip$ is minimized globally. This is essentially McCarthy's original suggestion for solving the frame problem that results in a weak disjunction for the Yale shooting problem. Here, we use standard second-order circumscription.

Let $\Gamma(Clip, Holds) = \Gamma_C^+$.

$Circ_{SO}(\Gamma(Clip, Holds); Clip; Holds) =$

$\quad \Gamma(Clip, Holds) \wedge \forall \Phi, \Psi.\neg[\Gamma(\Phi, \Psi) \wedge \Phi < Clip]$. (10)

$Clip$ is minimized at all points simultaneously while $Holds$ is allowed to vary at all points simultaneously. This logic is suitable for the ontological class $\mathcal{K}cl - IsAun$.

6.4 Prototypical Chronological Minimization with Filtering – PCMF

Filtered preferential entailment is a technique originally introduced by Sandewall [10] for dealing with postdiction. The filtering technique is based on distinguishing between different types of formulas in a scenario, in this particular case, between schedule and observation axioms. Given a scenario description Γ_C, the basic idea is to minimize only the schedule axioms Γ_{SCD} and then use the intersection of these models with the models for the observation axioms Γ_{OBS} as the class of preferred models. We use the same pointwise circumscription axiom as for PCM with some minor modifications for our restricted set of axioms.

Let $\Gamma(Clip, Holds) = \Gamma^+_{SCD}$.
Let $V(t, f) = \lambda t', f'.[\langle t, f \rangle = \langle t', f' \rangle \vee t < t']$ and $U(t, f) = \lambda t', f'.[t \leq t']$.

$Circ_{PW}(\Gamma(Clip, Holds); Clip/V(t, f); Holds/U(t, f)) =$

$\Gamma(Clip, Holds) \wedge$
$\forall t, f. \forall \Phi, \Psi. \neg[Clip(t, f) \wedge \neg \Phi(t, f) \wedge EQ_{V(t,f)}(Clip, \Phi) \wedge$
$\quad EQ_{U(t,f)}(Holds, \Psi) \wedge \Gamma(\Phi, \Psi)].$ (11)

The PCM filtered circumscription is then defined to be

$$\Gamma^+_C \wedge Circ_{PW}(\Gamma(Clip, Holds); Clip/V(t, f); Holds/U(t, f)).$$

Note that we filter with Γ^+_C rather than Γ_{OBS}. This is explained by the fact that $\Gamma^+_C = \Gamma_{OBS} \cup \Gamma^+_{SCD}$ and Γ^+_{SCD} is true in each of the $Circ_{PW}$ preferred minimal models, so it makes no difference. This logic is suitable for the ontological class $\mathcal{K} - IAex$.

7 Occlusion and Nochange Axioms

In this section, we provide syntactic characterizations of a class of preferential entailment definitions that are defined using what Sandewall refers to as *occlusion* and *nochange axioms*.

7.1 Occlusion

Associated with each action type is a subset of fluents that are influenced by the action. If the action has duration, then during its performance, it is not known in general what value the influenced fluents have. Since the action performance can potentially change the value of these fluents at any time, all that can generally be asserted is that at the end of the duration the fluent is assigned a specific value.

To specify such behavior, an *Occlude* predicate is introduced and the definition of reassignment expressions is modified accordingly. The occlusion predicate is used as part of the definition of a reassignment expression which in turn is used as part of the definition of an action schema.

The predicate *Occlude* takes a fluent and timepoint as argument. For example, if the $[t,t']Load$ action is performed then the formula $\forall t''.t < t'' \leq t' \rightarrow Occlude(t'', loaded)$ represents the fact that if *Load* is performed from t to t', then *load* will be occluded. The modified definition for the reassignment expression $[\mathbf{s},\mathbf{t}]\delta := T$ is

$$(\exists t.\mathbf{s} \leq t < \mathbf{t} \wedge \forall t'(t < t' \leq \mathbf{t} \rightarrow Holds(t', \delta)))$$
$$\wedge (\forall t''(\mathbf{s} < t'' \leq \mathbf{t} \rightarrow Occlude(t'', \delta)))$$

The definition for $[\mathbf{s},\mathbf{t}]\delta := F$ is similar, but the $Hold$'s atom is negated. Technically, occlusion is a device which is used to mask fluent changes from influencing various minimization of change policies. We will use the following abbreviation in the examples. Let f be a fluent constant.

$$(\mathbf{s},\mathbf{t}]f^* \stackrel{\text{def}}{=} \forall t''(\mathbf{s} < t'' \leq \mathbf{t} \rightarrow Occlude(t'', f)).$$

Example 4. Below we show the scenario corresponding to the Yale shooting scenario, where the schedule axioms are constructed according to the above definition.

obs1 $Holds(0, a) \wedge \neg Holds(0, l)$
scd1 $([\exists t.2 \leq t < 4 \wedge \forall t'(t < t' \leq 4 \rightarrow Holds(t', l))] \wedge (2, 4]l^*)$
scd2 $Holds(5, l) \rightarrow ([[(\exists t.5 \leq t < 6 \wedge \forall t'(t < t' \leq 6 \rightarrow \neg Holds(t', a))) \wedge (5, 6]a^*]$
$\wedge [(\exists t.5 \leq t < 6 \wedge \forall t'(t < t' \leq 6 \rightarrow \neg Holds(t', l))) \wedge (5, 6]l^*]).$

7.2 Nochange Axioms

We next introduce an additional axiom set, Γ_{NCG},

$$\{\forall f, t.Holds(t, f) \oplus Holds(t+1, f) \rightarrow Occlude(t+1, f)\}, \tag{12}$$

which we will call the *nochange axiom set*. This axiom states that for any fluent f and timepoint t, if the value of the fluent changes value from t to $t+1$ then f must be *occluded* at $t+1$. Consequently, it is axiomatized that a fluent may only change value when it is occluded. The nochange axiom implicitly includes a persistence assumption which is made clear by taking the contraposition of (12),

$$\{\forall f, t.\neg Occlude(t+1, f) \rightarrow Holds(t, f) \equiv Holds(t+1, f)\}. \tag{13}$$

Both the occlusion predicate and the use of nochange axioms provide a means of importing parts of our model-theoretic preference policy into the object language itself.

7.3 The Relation between *Clip* and *Occlude*

There is a great deal of interaction between the *Occludes* and *Clip* concepts. In the purely model-theoretic characterization of entailment using these predicates, the idea will be to first minimize the extent of *Occludes* and then minimize the extent of *Clip* in non-occluded regions, the intent being that the only change we are concerned with is that change which occurs in non-occluded regions. The only occluded durations should be those that are explicitly encoded in the axiomatization for a scenario description. One method for avoiding the prioritized minimization of *Occludes* and *Clips* is to add an explicit set of nochange axioms to each scenario description stating that the only changes allowed must occur in occluded regions of the time line. Using nochange axioms, we need only minimize *Occludes*. This policy is similar to, but somewhat different from the prioritized minimization of *Occlude* and *Clips*. We discuss this issue in a later section.

8 Circumscription Axioms for PMON, CMON, and CMOC

8.1 Pointwise Minimization of Occlusion with Nochange Premises – PMON

The PMON minimization policy uses the filtering technique in the following manner. Given a scenario description Γ_C, we would like to minimize *Occludes* globally relative to Γ_{SCD} and then filter with Γ_{NCG} and Γ_{OBS}. Note that in this policy, we exclude the persistence axioms in Γ_{PER}. We will have more to say about this after introducing CMON. The following circumscription axiom will be used for PMON, where *Occlude* is globally minimized and *Holds* is fixed.

Let $\Gamma(Occlude) = \Gamma_{SCD}$.

$Circ_{SO}(\Gamma(Occlude); Occlude) =$

$$\Gamma(Occlude) \wedge \forall \Phi. \neg [\Gamma(\Phi) \wedge \Phi < Occlude]. \tag{14}$$

PMON circumscription is then defined as

$$\Gamma_{NCG} \wedge \Gamma_C \wedge Circ_{SO}(\Gamma(Occlude); Occlude).$$

Note that the circumscription policy is extremely simple. What is even more surprising is that this particular minimization policy covers the very broad ontological class $\mathcal{K} - IA$, yet it is one of the simpler policies considered. This particular policy appears to be a good candidate for implementing efficiently.

8.2 Chronological Minimization of Occlusion with Nochange Premises – CMON

The CMON policy is a chronological version of PMON and also uses the filtering technique. Given a scenario description Γ_C, we would like to minimize *Occludes* chronologically relative to Γ_{SCD}, while keeping *Holds* fixed up to the point where *Occludes* is minimized.

Let $\Gamma(Occlude, Holds) = \Gamma_{SCD}$.
Let $V(t,f) = \lambda t', f'.[\langle t,f \rangle = \langle t', f' \rangle \vee t < t']$ and $U(t,f) = \lambda t', f'.[t \leq t']$.

$Circ_{PW}(\Gamma(Occlude, Holds); Occlude/V(t,f); Holds/U(t,f)) =$

$\Gamma(Occlude, Holds) \wedge$
$\forall t, f. \forall \Phi, \Psi. \neg[Occlude(t,f) \wedge \neg \Phi(t,f) \wedge EQ_{V(t,f)}(Occlude, \Phi) \wedge$
$EQ_{U(t,f)}(Holds, \Phi) \wedge \Gamma(\Phi, \Psi)]$ (15)

CMON filtered circumscription is then defined as

$\Gamma_{NCG} \wedge \Gamma_C \wedge Circ_{PW}(\Gamma(Occlude); Occlude/V(t,f); Holds/U(t,f)).$

This logic is suitable for the ontological class $\mathcal{K} - IA$.

8.3 The relation between *Clip* and *Occlude*

Before considering the CMOC minimization policy, it will be useful to consider the relation between the *Occlude* predicate and the *Clip* predicate. Recall that Kautz's original logic of persistence and the OCM, PCM, PGM and PCMF logics used just the *Clip* predicate. We then introduced the *Occlude* predicate and used it for the PMON and CMON logics in the context of the nochange axiom. In the next section we will reintroduce the *Clip* predicate in combination with the *Occlude* predicate while excluding the nochange axioms. Since PMON, CMON and CMOC are in general applicable to the same class $\mathcal{K} - IA(e)$, one can assume that there are strong similarities between the various combinations of using *Occlude*, *Clip*, and nochange axioms. Let's look at both Γ_{NCG} and the two persistence axioms Γ_{PER}:

$\forall f, t. Holds(t,f) \oplus Holds(t+1, f) \rightarrow Occlude(t+1, f)$ (16)

$\forall f, t. Holds(t,f) \rightarrow [Holds(t+1,f) \oplus Clip(t+1,f)]$ (17)

$\forall f, t. \neg Holds(t,f) \rightarrow [\neg Holds(t+1,f) \oplus Clip(t+1,f)].$ (18)

There are four interesting cases involving the value of a fluent at timepoints t and $t+1$ and what can be derived relative to the axioms listed above:

1. Suppose $Holds(t, f)$ and $\neg Holds(t+1, f)$. Then $Occlude(t+1, f)$ and $Clip(t+1, f)$.
2. Suppose $\neg Holds(t, f)$ and $Holds(t+1, f)$. Then $Occlude(t+1, f)$ and $Clip(t+1, f)$.
3. Suppose $\neg Holds(t, f)$ and $\neg Holds(t+1, f)$. Then $\neg Clip(t+1, f)$.
4. Suppose $Holds(t, f)$ and $Holds(t+1, f)$. Then $\neg Clip(t+1, f)$.

Based on cases 1 and 2, it is easily observed that positive instances of $Clip$ and $Occlude$ coincide, or would if the persistence axioms (17) and (18) were included in Γ_C while using PMON and CMON. $Clip$ would be fixed because both Γ_{PER} and Γ_{NCG} would be excluded from the minimization. As regards the negative cases of $Clip$, in some cases, there will be a negative $Occlude$ and in some not. This is dependent on the way the axioms in Γ_{SCD} are written and the effect of minimizing $Occludes$. For cases 3 and 4, the minimization effect should be that $\neg Occlude(t+1, f)$ holds. Consequently, $Occlude$ and axiom 16 plus filtering (PMON) would basically be doing a job similar to minimizing $Clip$ in PCM if minimizing $Clip$ was suspended in action occurrence durations. It would be straightforward to provide a circumscriptive definition of such a logic and interesting to assess it in the context of this discussion. Note that Kautz's original formulation lacked axiom (18), so that case 2 would not result in a clip. This points out the fact that OCM is in fact not quite the same policy as in Kautz's original formulation.

It is important to point out that although occlusion can sometimes be interpreted technically as behaving in a manner similar to clipping, it plays a much larger conceptual role in theories of action and change. For example, it is essential for modeling actions with duration and certain types of ambiguity specific to the action and change domain. In conclusion, since $Clip$ is subsumed by $Occlude$ and Γ_{NCG} in PMON and CMON, we can simply do without it.

8.4 Chronological Minimization of Occlusion and Change – CMOC

Is it possible to achieve the same effect as PMON or CMON, without using the nochange axioms? Sandewall presents an additional policy CMOC that comes close to doing that. For the CMOC policy, we reintroduce $Clip$. What we will do is to minimize both $Clip$ and $Occlude$ chronologically, but $Occlude$ will have a higher priority than $Clip$. In order to encode such a policy, we will need prioritized pointwise circumscription.

There is one additional modification that must be made. The notion of $Clip$ is somewhat more constrained under CMOC than it was for the previous policies. Here, the extension of $Clip$ will include only those $\langle f, t \rangle$ tuples that are unoccluded. Instead of using Γ_{PER}, we will use the modified persistence axioms,

$$\forall f, t. (\neg Occlude(t, f) \wedge Holds(t, f)) \rightarrow [Holds(t+1, f) \oplus Clip(t+1, f)] \quad (19)$$

$$\forall f, t.(\neg Occlude(t,f) \land \neg Holds(t,f)) \rightarrow [\neg Holds(t+1,f) \oplus Clip(t+1,f)] \quad (20)$$

We denote these persistence axioms as the set Γ^+_{PER}. Simply by viewing these new persistence axioms, it is easily observable how minimizing $Occlude$ while varying $Clip$ and then minimizing $Clip$ with $Occlude$ fixed will serve to maximize persistence.

A result of Lifschitz [5], informs us that prioritized pointwise circumscription can be reduced to a conjunction of pointwise circumscriptions. We will do just that. The first pointwise circumscription axiom minimizes $Occlude$ and varies $Clip$ and $Hold$, while the second minimizes $Clip$ and varies $Hold$ while leaving $Occlude$ fixed.

Let $\Gamma(Occlude, Holds, Clip) = \Gamma_{SCD} \cup \Gamma^+_{PER}$ and Let

$$V(t,f) = \lambda t', f'.[\langle t,f \rangle = \langle t',f' \rangle \lor t < t']$$

$$U_1(t,f) = \lambda t', f'.[t \leq t'].$$

$$U_2(t,f) = \lambda t', f'.[true].$$

Let $O = Occludes$, $H = Holds$, and $C = Clip$. The following circumscription axioms will be used for CMOC.

$Circ^1_{PW}(\Gamma(O,H,C); O/V(t,f); H/U_1(t,f), C/U_2(t,f)) =$

$\Gamma(Occlude, Holds, Clip) \land$
$\forall t, f.\forall \Phi, \Psi, \Psi'.\neg[Occlude(t,f) \land \neg \Phi(t,f) \land EQ_{V(t,f)}(Occlude, \Phi) \land$
$\land EQ_{U_1(t,f)}(Holds, \Psi) \land EQ_{U_2(t,f)}(Clip, \Psi') \land \Gamma(\Phi, \Psi, \Psi')]. \quad (21)$

Let $\Gamma(Clip, Holds) = \Gamma_{SCD} \cup \Gamma^+_{PER}$.

$Circ^2_{PW}(\Gamma(C,H); C/V(t,f); H/U_1(t,f)) =$

$\Gamma(Clip, Holds) \land$
$\forall t, f.\forall \Phi, \Psi.\neg[Clip(t,f) \land \neg \Phi(t,f) \land EQ_{V(t,f)}(Clip, \Phi) \land$
$EQ_{U_1(t,f)}(Holds, \Psi) \land \Gamma(\Phi, \Psi)]. \quad (22)$

CMOC filtered circumscription is then defined as

$\Gamma_C \land Circ^1_{PW}(\Gamma(O,H,C); O/V(t,f); H/U_1(t,f), C/U_2(t,f))$
$\land Circ^2_{PW}(\Gamma(C,H); C/V(t,f); H/U_1(t,f)). \quad (23)$

This logic is suitable for the ontological class $\mathcal{K} - IAe$.

9 Discussion

We have provided syntactic characterizations of a majority of Sandewall's definitions of preferential entailment in terms of both generalized pointwise- and second-order circumscription. The logic used is standard sorted FOPC with the method of reification for propositional types. This work contributes to ongoing research in the area in two ways. Firstly, it is our goal to implement a number of these ontological classes in a larger body of software being developed in our research group for the construction of autonomous systems. One approach that may be taken is to compile our circumscriptive characterizations into logic programs along lines similar to that of Lifschitz and Gelfond [3]. One class of particular interest is PMON, which according to Sandewall covers the $\mathcal{K} - IA$ class of reasoning problems. As noted previously, not only is this one of the broadest classes studied in Features and Fluents, but it also has one of the simpler circumscription axioms. Preliminary results by Doherty [1] show that we can in fact reduce the circumscription axiom to a first-order formula in the context of certain restrictions on action schemas.

Secondly, Sandewall provides an extremely robust framework for not only generating new logics of action and change, but also using it as a tool for analyzing previously proposed logics. By reformulating many of the ideas in a more conventional and well understood format, we hope this will contribute to accelerating the process of comparison and perhaps introduce Features and Fluents to a broader audience. As one instance of this, we have already considered the relation between the *Clip* and *Occludes* concepts, the former well known in traditional approaches, the latter new on the scene and quite flexible in its various uses. In addition, just as the Features and Fluents framework may be used to analyze other approaches, the circumscription axioms and the body of results associated with them can be used to provide verification of relations between the various ontological classes. For example, it should be possible to show the equivalence between PMON and CMON based on manipulation of the circumscription axioms in the context of restrictions on action schemas, or equivalence between PMON and the proposed modification to PCM.

Acknowledgements

This work is an outgrowth of research by Erik Sandewall on Features and Fluents. We are grateful for the large number of stimulating discussions we have had with him concerning his work. We also thank Lars Karlsson for his comments on earlier drafts.

References

1. P. Doherty. Reasoning about action and change using occlusion. In *Proceedings of the 11th European Conference on Artificial Intelligence, Aug. 8-12, Amsterdam*, pages 401–405, 1994.
2. P. Doherty. Reasoning about action and change using occlusion: Extended report. Technical report, Linköping University, 1994.
3. M. Gelfond and V. Lifschitz. Representing actions in extended logic programming. In K. Apt, editor, *Proc. Joint Int'l Conf. and Symp. on Logic Programming*, pages 559–573, 1992.
4. H. Kautz. The logic of persistence. In *Proc. National Conf. on Artificial Intelligence, (AAAI-86)*, pages 401–405, 1986.
5. V. Lifschitz. Computing circumscription. In *Proc. IJCAI-85*, volume 1, pages 121–127, 1985.
6. V. Lifschitz. Pointwise circumscription. In M. Ginsberg, editor, *Readings in Non-monotonic Reasoning*, pages 179–193. Morgan Kaufmann, 1988.
7. V. Lifschitz. Toward a metatheory of action. In *Proc. International Conf. on Knowledge Representation and Reasoning, (KR-91)*, pages 376–386, 1991.
8. F. Lin and Y. Shoham. Provably correct theories of action (preliminary report). In *National Conference on Artificial Intelligence (AAAI-91)*, pages 349–354, 1991.
9. W. Lukaszewicz. *Non-Monotonic Reasoning – Formalization of Commonsense Reasoning*. Ellis Horwood Series in Artificial Intelligence. Ellis Horwood, 1990.
10. E. Sandewall. Filter preferential entailment for the logic of action and change. In *Proc. Int'l Joint Conf. on Artificial Intelligence, (IJCAI-89)*, 1989.
11. E. Sandewall. Features and fluents: A systematic approach to the representation of knowledge about dynamical systems. Technical Report LITH-IDA-R-92-30, Department of Computer and Information Science, Linköping University, September 1992. Second Review Version.
12. E. Sandewall. The range of applicability of nonmonotonic logics for the inertia problem. In *Proc. Int'l Joint Conf. on Artificial Intelligence, (IJCAI-93)*, 1993.

Dealing with Time Granularity in a Temporal Planning System

Silvana Badaloni[*], Marina Berati[+]

(*)Department of Electronics and Computer Science
University of Padova - Italy
e-mail: badaloni@ladseb.pd.cnr.it

(+) LADSEB-CNR - Italy
e-mail:. berati@sun2.ladseb.pd.cnr.it

Abstract. We have introduced the notion of temporal granularity in a planning system based on a temporal model. The use of different time scales together with the semantic effect of unit time change allows the planning of actions defined at different levels of abstractions. The advantage of a hierarchical planner consists in the fact that it can work, at each moment, at a granularity level as coarse as required, thus reducing the complexity of the problem we are addressing.

Introduction

In complex domains, there may be the need for planning systems to have the capability of reasoning abstractly about tasks that they are called upon to solve, at different levels of abstraction. The method of abstraction may allow one to reason hierarchically [Tenenberg 86] in order to build differently refined plans. In this way, a planner can reason about actions and their effects at different levels of detail. Our idea is to introduce in automated planning systems the notion of granularity and, particularly, that of temporal granularity.

A general definition of granularity can be the level of abstraction at which knowledge can be represented, the resolution power of a certain knowledge representation [Hobbs 85, Montanari 92]. A road can be viewed as a line, if we are planning a trip, as a surface, if we are driving on it and as a volume, if we hit a pothole [Hobbs 93]. A proper characteristics connected to this notion is that of distinguishability; if we view a road as a line, we are not distinguishing between two points that are at the same place along its length, even though they are, for example, on different sides of the road. The distinguishability comes out when we change grain.

The process of abstraction from the domain may take into account and represent in a stratified way only those aspects which are relevant for the problem solver goal. Let us speak of layers of knowledge and of different patterns of reasoning, depending on the level of granularity; when passing from a coarser to a finer grain, an extra knowledge can be acquired. This semantic effect is particularly interesting in order to build up a hierarchical planner. From this new perspective, the term "hierarchical planner" is then used in tight connection with the way in which knowledge has been structured and with the notions of decomposition and refinement of actions.

We propose to extend our planning system, based on a temporal model [Badaloni

93a, Badaloni 93b], by including the temporal granularity intended as the resolution power of the temporal domain. The use of different time scales together with the semantic effect of unit time change allows to plan actions defined at different levels of abstraction. When passing from a coarser temporal grain to a finer one, it may happen that an action can be decomposed into other sub-actions and, when an action is refined, the temporal relations between sub-actions may be specified too. In other words, an action seen at a certain level of abstraction may be an atomic event while, at a finer grain, it may be decomposed into some steps. Indeed, in our temporal planner the process of knowledge stratification and the planning mechanism depend strongly on the temporal structure itself.

Besides, an action can be decomposed into a set of sub-actions at the same level of time granularity. A stack action, for instance, can be decomposed into the following ones: pick-up the block, move the arm, put down the block. That is, one way of accomplishing this action can be represented by the above listed sub-actions (taking into account their relative temporal constraints), while remaining within the same temporal domain. Thus, in the proposed extension of our planning system, we have taken into account that an action can be dealt with as a complex structured object: on the ground of the notion of refinement, an action can be split at different levels of time granularity (vertical decomposition). On the other hand, on the ground of the notion of decomposition, an action can be decomposed into a set of sub-actions at the same level of granularity (horizontal decomposition).

The two main aims of this work are: on the one hand, to achieve the possibility to plan actions at different levels of abstraction (granularity and time granularity); on the other hand, to reduce the computational complexity of the planning system, since, in many cases, only coarse planning is needed. The paper is organised as follows. After a brief description of the temporal reasoner and of the planning system without taking into account granularity (sections 1 and 2), in section 3 granularity will be defined and section 5 will show how this definition alters the semantics of Allen's logic. In sections 4 and 6 what will be described is how the notion of granularity has been dealt with within the temporal reasoner and within the planning system, respectively.

1 The Temporal Reasoner TEMPNET

In our temporal reasoning system, both qualitative and quantitative temporal information is represented in a constraint network. Nodes represent temporal intervals, and the directed arcs are labelled by temporal constraints. Unary and binary constraints can be defined: unary constraints represent possible durations for one interval; binary constraints represent the set of temporal relations that can hold between two intervals.

A binary constraint is the disjunction of relations belonging to a sub-algebra that we have called IDSA (Interval Distance Sub-Algebra). Each relation $R \in$ IDSA between two intervals A and B is a tuple of four distances between the endpoints of the two intervals. This information defines the mutual position of the intervals A and B. When the relation is given as an input to the temporal reasoner, it is also possible to specify a qualitative relation $RQ \in$ SAc between the two intervals (SAc is a sub algebra of Interval Algebra, [vanBeek 90, vanBeek92]). Thus, one can specify either only the qualitative relation RQ, or only some distances between the four endpoints (when a distance is left undefined, it is intended to vary from $-\infty$ to $+\infty$), or, finally, RQ plus some distances (it is worth noting, however, that, when the four distances are specified, the qualitative relation RQ is completely defined, since it can be inferred from the tuple of distances).

Qualitative information can be expressed as the set of possible relations between two intervals, that is:
- atomic relations of Allen's interval algebra IA [Allen 83];
- pointizable relations belonging to SAc (obviously, SAc contains the thirteen atomic relations of IA), that is, those relations of the interval algebra that can be translated into a conjunction of relations of PAc (continuous point algebra) between the beginning and the ending points of the intervals; e.g. *(before, overlaps)* is allowed, but *(before, after)* is not [vanBeek 90, vanBeek 92]. To this set (SAc) belong the relations proposed by [Freksa 92], where Allen's thirteen mutually exclusive relations are grouped to process directly coarse temporal knowledge (e.g. *precedes* represent *(before, meets)*).

In this way, qualitative constraints are expressed as disjunction of relations belonging to SAc. To use SAc sub-algebra, instead of IA, to express qualitative relations between intervals, means to compact the knowledge associated with each edge, thus reducing considerably the computational effort (as we will explain later).

Quantitative information can be expressed as metric constraints representing the duration of intervals (unary constraints), namely, the temporal distances between the beginning and the ending points of an interval (e.g. I_1 (1,5) means that the duration of I_1 is between 1 to 5 (referring to a given time scale)). Moreover, as previously illustrated, binary relations can be specified by defining some (or all) of the four distances between the four extremes belonging to the pair of constrained intervals (e.g. I_1 is *before* I_2 of 10, that is, the distance between the endpoint of I_1 and the startpoint of I_2 is 10, referring to a given time scale).

When new intervals or new constraints are added to the network, the temporal reasoning system has to compute how this new information propagates in the network, that is, all the logical consequences of the input data. This is done by computing the *minimal network*: the minimal network is defined as the complete graph deriving from the propagation of each constraint over the whole network; it is characterised by the set of the tightest temporal constraints (that is, all the inconsistent relations are eliminated from the network). To compute the minimal network is a NP-hard problem [Vilain 90]: polynomial algorithms can be applied to the sub-networks obtained by considering only one relation for each constraint and only one duration for each interval in the whole network (singleton labellings). Each sub-network thus obtained is translated into a new network, where nodes represent time points (which are the extremes of the intervals belonging to the initial network). The computation of the minimal network for each point-network has a complexity of $O(n^3)$ (where n is the number of nodes), because it is sufficient to impose path-consistency over the network itself [Decther 91]. Each minimal network over time points is then translated again into a network over intervals (this is possible because the SAc sub-algebra has been used: the set of four relations between four points, which represent the endpoints of two intervals, specifies a relation between intervals belonging to SAc) and the union of all minimal networks is then executed. The number of these networks, derived by singleton labellings, increases exponentially according to the number of edges and to the number of relations associated to each edge in the IDSA network (k^e, where e is the number of edges, and k is the maximum number of relation for each edge). This shows the advantage, previously mentioned, of having a small number of disjunctive qualitative relations between intervals, obtained by using SAc qualitative relations instead of IA qualitative relations.

The temporal reasoner has been implemented by using the CLP(R) language [Jaffar 87], a logic programming language that is able to handle arithmetical constraints in a declarative way, together with normal prolog-like clauses.

2 The Planning System

Planning tasks are performed on the basis of the knowledge derived by the temporal network. Based on a temporal world model, the planner is able to reason about simultaneous actions (non-sequential planning) and to plan robot actions in complex domains. Most of the classical systems are not adequate, to this aim; they are based on a simplified world model, where actions are supposed to be instantaneous and completely independent. Furthermore, since our temporal model is fully constrained, both qualitatively and quantitatively, the planner can make plans involving deadlines and metric constraints. In the previous section, we have shown how the metric information relative to the durations of intervals has been included into the temporal world model, making it possible to construct a plan composed by actions fully constrained.

Let us now briefly describe our planning system, which is an extension of the one proposed by Allen [Allen 91, Allen 93], including metric information; a more detailed description can be found in [Badaloni 93a]. Knowledge about the world is represented by a set of temporally qualified facts referred to intervals (properties, processes, events [Allen 84, Allen 91]). According to Allen's definition, the predicate *holds(P, I)* is used to mean that the property P holds over the interval I. Properties are used to represent "static" knowledge; if a property P is true over an interval I, then it holds over all sub-intervals of I. To represent "dynamic knowledge", events and processes are used: the predicate *occur(A, E, I)* means that an action A has been executed, causing the event E to occur over the interval I. The predicate *occurring(P, I)* means that a process P is occurring along the interval I.

Knowledge about cause-effect relations among facts is expressed by means of planning rules, called *backward rules*, in the form:

$G \leftarrow\!\!-\!\!- [A_1, ...A_n], [constr_1,...constr_k]$

G is true if $A_1, ...A_n$ are true such that $constr_1,...constr_k$ (temporal constraints) are satisfied. G and A_i represent temporally qualified facts, while $constr_i$ represent temporal constraints that may exist between pair of intervals associated to the facts specified in the rule. The goal is a temporally qualified fact, and it is reached by applying the planning rules with a backward chaining strategy. The temporal constraints may be both qualitative and quantitative, so that durations and deadlines can be expressed. This information is maintained by the temporal reasoner in a temporal network, associated with the plan.

Applying a rule in backward chaining means unifying the consequent G with the goal and introducing the set of constraints into the temporal network; if the constraints can be introduced without producing inconsistency, then the antecedents become the new subgoals. A rule may also be applied by the system in forward chaining, in order to compute the consequences of the assumptions made about persistence of properties and action attempts. This is done by the "prediction reasoner" subsystem, which derives new knowledge from "persistence assumptions" and "ability assumptions", by applying planning rules in forward chaining. Two hypotheses have been assumed according to [Allen 91]; the first one is a default persistency hypothesis (a property persists until an event changes it) thus avoiding to deal with the frame problem. The second one represents the fact that, if an agent attempts to perform an action (represented by the predicate $try(A, E, I)$), then it would occur an event associated to that action. Thus, if the predicate $try(A, E, I)$ is true, the associated predicate $occur(A, E, I)$ will be true, but not vice versa. Furthermore, knowledge about the world is codified in a set of domain constraints concerning properties and actions which are mutually exclusive.

Given a goal and an initial description of the world, the system works in backward chaining by instantiating actions and making assumptions, the aim being to find a set of actions that verify the initial goal. A plan is a set of temporally qualified actions, that is, a set of actions which are not strictly ordered, but that are linked together by complex temporal constraints. This means that the ordering of the actions in the plan satisfies the temporal constraints between intervals in which actions take place.

We have applied the planning system to plan and schedule the actions of a manipulator mounted on a mobile platform, in the case of the maintenance intervention over a hydraulic circuit (experiment under investigation in the framework of the Special Research Project on Robotics (PFR) of the Italian National Council of Research (CNR)) . Starting from an incomplete temporal lay-out of actions and events, the ordering of actions in the final plan derives from maintaining the temporal constraints consistency and from computing the minimal network [Badaloni 93a,b].

The obtained results are certainly satisfactory and show the importance of using a fully expressive temporal model as the basis of a planning system. However, an analysis of computational complexity, even though reduced with respect to other temporal constraint propagating systems, leads to the need of conceiving hierarchical planning, as it has been explained in the introduction.

But, before discussing about granularity, let us now consider a quite simple problem, just to show, now, some examples of planning rules and to show, later, how the planning system endowed with the treatment of granularity actually works. Suppose that two objects have to be transported from one place to another within a maximum time of five days. Each object can be either carried by a truck (with the constraint that the truck can transport only one object at a time) or sent by mail. It is known that the truck travel takes three days, as well as the mail dispatch, but the transport by truck is preferable because it is cheaper. Relative to this scenario, the two possible actions are: mail_send(X) and truck_transport(X). Knowledge about actions is coded into a set of planning rules; we will illustrate some of them:

R1)   holds(all_transported,T)<---
           [holds(transported(1),T1),holds(transported(2),T2)],
           [constr(T1,T,pr),constr(T2,T,pr)].

where the constraint "pr" is the relation "precedes"; the two intervals T1 and T2 have positive durations (pinf stands for +∞):

           int(T1, d(1,pinf)).
           int(T2, d(1,pinf)).

The fact that an object can be transported by truck or sent by mail is expressed by:

R2)   holds(transported(X), T) <---
           [occur (truck_transport(X), E, etime(E)],
           [constr(eff1(E), T, e))].

R3)   holds(transported(X), T) <---
           [occur (mail_send(X), E, etime(E))],
           [constr(eff1(E), T, e)].

where etime(E) denotes the intervals in which the events associated with the actions take place, eff1(E) is the name of the interval during which the effect of the event occurs and "e" is the relation "equal".

The condition that the two actions, mail_send and truck_transport, cannot occur simultaneously for an object can be expressed by the following domain constraint:

D1)   dom_constr([try(truck_transport(X), E, etime(E)),
                        try(mail_send(X), E1, etime(E1))],
                       [constr(etime(E),etime(E1),[pr, sd])]).

where "sd" is the constraint "succeeds".

It is self-evident that a plan to transport the two objects by truck while satisfying the supposed deadline, does not exist whereas there exist a plan composed by the transport by truck of an object and by the dispatch by mail of the other object. Obviously, this result follows from the inferential mechanism.

3 About Granularity

The introduction of granularity allows one to use different abstraction levels, in such a way that the world model can be based only on what is actually needed to reach a certain goal, disregarding all the irrelevant aspects that would unnecessarily complicate the model itself.

It appears quite natural to associate a temporal domain with the various granularity levels: in this way, each temporally qualified fact can be described with respect to the most suitable temporal domain. At each granularity level, a determined temporal scale is defined (seconds, minutes, hours, etc.): the knowledge "visible" at a fixed level is the knowledge relative to the current temporal domain, and to coarser ones. Passing to a finer grain, new information may be visible and it is linked to that defined in coarser domains.

To change abstraction level does not always imply to change temporal domain: generally, it is possible to pass to a finer grain, intended as the degree of detail of the model, without passing to a finer temporal domain; on the other hand, to change time granularity often implies to change abstraction level. We can conclude, therefore, that "granularity" and "time granularity" are tightly connected, even if there is not a one-to-one correspondence between them.

The advantage deriving from the use of granularity consists in the ability of changing the complexity of the model we are dealing with, so that we can work, in every situation, at a granularity level as coarse as required, thus reducing the complexity of the reasoning problem we are addressing. Of course, the coarser the granularity level is, the more "approximate" the solution will be, because the available knowledge is less detailed.

4 Time Granularity in the Temporal Reasoner

To provide the planner with time granularity requires that the temporal reasoning system is able to manage different temporal domains. Let us define a set of disjoint temporal domains (as in [Montanari 92]), $T_1, \ldots , T_n$, each characterised by a metric m_i, the union of which constitutes the temporal universe. For each i, with $1 \leq i \leq n$, the domain T_{i+1} is finer than T_i. All domains are discrete, possibly except the finest one, which can be dense. All domains must be discrete because a dense domain would always be at the finest granularity, for, in such a domain, each number could be expressed with the desired degree of precision, and hence it would be meaningless to create a hierarchy of different grained domains. In fact [Corsetti 91], "mapping, say, a set of reals into another set of reals would only mean changing the unit of measure with no semantic effect. Just in the same way, one could decide to describe geometric facts by using, say, Kmeters and centimetres. However, if Kmetres are measured by real numbers, the same level of precision as with centimetres can be achieved. Instead, the key point in time granularity is that saying that something holds for all days in a given interval does not imply that it holds every second within the 'same' interval".

Let us consider a scenario consisting of seconds, minutes, hours and days; we assume that the metric value of the finer domain is 1. Thus, m=1 corresponds to the domain of seconds and m=60, m=3600, m=3600*24 correspond to the domains of minutes, hours and days, respectively. Given a couple of domains T_i and T_{i+1} (with T_{i+1} finer than T_i), we define a function which maps each point of T_i into an interval of T_{i+1}, as shown in figure 1 (days to hours conversion). For example, the point P=2, within the domain of days, is mapped, within the domain of hours, into the interval [25, 48], that is the time interval which begins as soon as the day 1 has finished and terminates with the end of day 2.

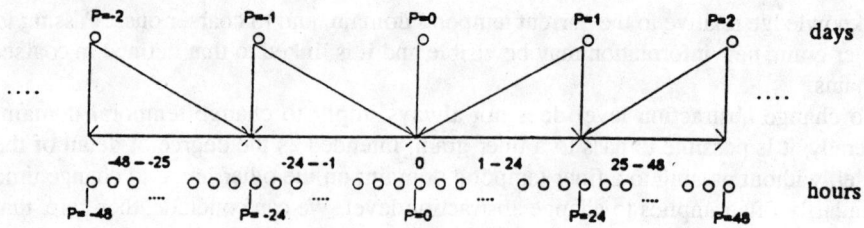

- Figure 1: relation between the domain of days and the domain of hours -

Let us consider now intervals, instead of points, and let us see how they are mapped when passing to a finer domain: the interval [A, B] (which can represent the duration of an interval, or the distance between a couple of points) can be viewed as the set of time points enclosed between A and B, extremes included. So, if we want to translate d(A, B), expressed with respect to a fixed domain, into a finer domain, we will obtain the interval d(A',B') expressed with respect to the finer domain such that:
- A' is the lower bound of the interval A_f corresponding to the projection of the point A in the finer grain and B' is the upper bound of the interval B_f corresponding to the projection of the point B into the finer grain. A general rule can be given concerning a projection from a coarser domain D_c to a finer domain D_f. Let d(A,B) (hp. A>=0 and B>=0) a duration in D_c and m_c its metric. In the finer domain D_f of metric m_f, d(A',B') can be computed according to:

if A= 0 then A'= 0
else A'=[(A - 1) * m_c / m_f] + 1;
B'= B * m_c / m_f;

For example, the interval d(1,2) in the domain of days becomes, in the domain of hours, d(1,48).

It may happen that A or B or both have a negative value; then the rule has to be modified as follows:

if A= 0 then A'= 0
else if A > 0
    then A' = [(A - 1) * mc / mf] +1
    else A' = A * m_c / m_f ;
if B = 0 then B' = 0
else if B<0
    then B' = [(B + 1) * m_c / m_f] - 1
    else B' = B* m_c / m_f ;

When passing from a finer grain to a coarser one, it is sufficient to translate both A' and B' (expressed with respect to the finer grain with metric m_f) into their coarse grain equivalent (metric m_c). The duration d(A,B) in the domain D_c will be given by:

$$A = ceil\,(abs\,(A') * m_f/m_c) * sign(\,A'\,);$$
$$B = ceil\,(abs\,(B') * m_f/m_c) * sign(\,B'\,);$$

where : sign(x)=1 if x>0, 0 if x=0, -1 if x<0
ceil(x) gives the smallest integer y such that y>= x.

By using these rules, all metric information represented in the temporal network can be translated, when passing from a domain to another one. It is worth noting that, in this way, we obtain something more than a simple time scale change: when moving to a finer grain, a new interval is obtained with a new meaning, which it could not have had before, at the coarser grain. For instance, if in the domain of days a distance D is expressed as [1, 2], it indicates that the interval D lasts from 1 to 2 days; but, within the domain of hours, a more general statement is asserted: D is translated as [1, 48], thus indicating that the interval D can have a duration of, say, 1 hour, 2 hours, etc., and this knowledge could not be specified in the domain of days. Vice versa, when passing to a coarser grain, some information is lost: for instance, if in the domain of hours, we know that a duration D varies from 3 to 26 hours, [3, 26], in the domain of days we can only assert that D varies from 1 to 2 days, [1, 2].

For each node, in the temporal network, a maximum domain is defined in this way:
Def.: we define maximum domain of an interval I, the temporal domain Dmax such that the interval itself is visible within that domain and in finer ones, while in coarser domains the interval is not defined.

This definition is connected to the notion of granularity in its general meaning: at a certain level of abstraction, which is associated with a given temporal domain, a fact, temporally qualified by an interval I, is defined. At finer levels, the fact is still visible, and so is the interval I, but a temporal domain will exist (the maximum domain) such that, at coarser levels the fact itself will not be visible, and, consequently, the interval I will not be visible, too: as soon as a domain coarser than the maximum domain of I is reached, the interval is deleted from the temporal network which represents the current level of abstraction.

5 Semantic Effects

But, how does the meaning of properties, events and processes change with time granularity?

• Properties

Let us consider the predicate *holds(P, I)*. If we take into account time granularity, the interval I, referred to a given domain D_c, will be mapped into an interval I', referred to a finer domain D_f, by using the above reported rules. Is the property still true in every sub-interval of I', that is, is *holds(P, I')* still true? The definition itself yields the answer: if P is true over a given interval, this means that it is true during the whole interval (we could say at *each* point of that interval), at any level of granularity. Otherwise, it would be a process (see below). Vice versa, when passing from a domain D_f to a coarser one D_c, the interval I (relative to D_f) is transformed into I' (relative to D_c), which can be specified with a minor degree of detail, but which represents the same interval. Thus, if *holds(P, I)* was true, *holds(P, I')* is surely true.

If D_c is coarser than the maximum domain of a property P, the property is no longer visible, and a simpler world model is obtained.

• Processes

According to Allen's definition [Allen 84, Allen 93], processes describe *dynamic* aspects of the world: they describe activities which have not a precise ending point (or accomplishment). For example "I am running" is a process. A process is represented by a predicate P, and is associated with the interval I, over which it is true, by using the predicate *occurring*. The predicate *occurring (P, I)* is true if the process P is true over at least a sub-interval of I. This means that, if *occurring(P, I)* is true, *occurring(P, I')* is true for each I' that contains I. It is clear that this is not a formal definition of process, because processes are recursively defined in terms of themselves, as noticed by [Trudel 93].

In general, we can say that properties are a particular type of processes: in fact, if P is true over an interval I and over each sub-interval of I, so that P is a property and *holds(P, I)* is true, we can surely say that P is true over at least a sub-interval of I, so that *occurring(P, I)* is true. On the other hand, in [Trudel 93, Galton 90] it has been shown that a process can be seen as an event in progress, and events in progress are a particular type of properties.

If granularity is changed downward and upward, the interval I related to a process is translated from a temporal domain to another, thus obtaining an interval I'. When passing to a finer domain, the predicate *occurring(P, I')* is still true, because I' represents I in a more detailed way. Passing to a coarser grain, *occurring(P, I')* is still true because I' contains I (generally, when passing to a coarser domain, an interval I cannot be exactly represented, but it is approximated to the smallest interval I' that contains I). Thus, in any case, the predicate *occurring (P, I')* is still true when changing time granularity.

It may happen that, passing to a finer grain, the acquired extra-knowledge makes it possible to specify the process P: the process P can be decomposed into a set of processes and properties temporally linked together. For instance, let P be the predicate *work(X)*. In the domain of days, the assertion that X has worked a day I_0 can be written as: *occurring(work(X), I_0)*. Furthermore, in the domain of hours, one would assert that the property *work(X)* is true over I_1 and I_2, being I_1 and I_2 in a certain temporal relation with I_0. Thus, the process P can be better specified by considering: *holds(work(X),I_1)* and *holds(work(X),I_2)*. Passing to a coarser domain, these two properties are no more visible, while the process is still visible. Besides, if this coarser domain is coarser than the maximum domain of the process P, the process itself is no more visible.

- Events

An event describes an activity that involves an accomplishment. The predicate *occur(A, E, I)* indicates that an event E associated with the action A, occurs over the interval I. By definition, an event E is true over an interval I only if the event happened over the time interval I, and there is no sub-interval of I over which the event happened. Passing to a finer grain, this property is maintained: I is translated into I', in the new domain, and we can still assert *occur(A, E, I')*. The same holds for the passage to a coarser grain, if the maximum domain is not got over, so that the event is still visible.

If we change temporal domain, from a coarser to a finer one, an action A can be refined into a set of sub-actions: each sub-action generates a new event, and a new set of temporally qualified facts is obtained, describing the world model in a more detailed way. The sub-actions are related to the action A by means of a set of temporal constraints and the new events refer to new temporal intervals with new temporal qualitative and metric constraints that have to be included in the temporal model at the finer grain. On the contrary, passing to a coarser domain, the sub-actions are not still visible, thus obtaining a simpler model.

On the other hand, an action can be decomposed into a set of sub-actions at the same level of time granularity, because we can change granularity (intended as abstraction level) without changing time granularity (intended as temporal domain). A stack action, e.g., can be decomposed into the the following ones: pick-up the block, move the arm, put down the block.

Thus, an action can be dealt with as a complex structured object in two different ways: if the passage to a finer granularity level is due to the refinement of the temporal domain (i.e. time granularity), we speak of vertical decomposition or action refinement. While, if the abstraction level is changed within the same temporal domain, we speak of horizontal decomposition.

To sum up, it is possible to pass to a finer granularity level (that is, to change abstraction level), both by changing temporal domain and by remaining within the same temporal domain. In the former case, the metric information of temporal constraints relative to each fact can be specified in a more detailed way. In both cases, when changing abstraction level, properties are not affected by transformation (except their maximum domain is got over), while, for processes and actions, it makes sense to define a decomposition, therefore reaching a more detailed model.

6 Time Granularity within the Planning System

As different temporal domains are available, the knowledge to be provided to the planning system can be structured in a "stratified" way, so that different layers of knowledge are formed, one for each temporal domain. At each level, only facts that have as "natural" domain the domain associated with that level can be described, that is, at each level, only a part of the knowledge related to the problem is represented and "visible".When moving to a finer domain, the previous database is kept, and new knowledge (that is, new facts and new planning rules) is added to it.

In the same way as it has been done for temporal intervals, a maximum domain must be defined for each planning rule and each temporally qualified fact. In order to specify an action decomposition, a new type of planning rule has been defined: the "conversion rules". They assume a form similar to the previously defined backward rules:

$$G <\!\!-\!\!-\!\!- [A_1, ...A_n], [constr_1,...constr_k], D_f, D_c$$

Two temporal domains D_c and D_f have to be specified: they represent, respectively, the domain in which the action G is an atomic one, and the domain in which G is split into a set of more elementary actions. D_c is coarser than D_f if an action refinement (vertical decomposition) is described in the conversion rule. If an action decomposition is specified within the same domain D (horizontal decomposition), both D_f and D_c are equal to D. G is an action, while A_i can be any kind of temporally qualified fact. Generally, the facts A_i will be more elementary events than G, and they represent the elementary steps that build up the action G. Temporal constraints $constr_i$ represent the way in which elementary actions are related to each other, and to G. The metric information included in them is expressed with respect to the unit of measure of the domain D_f.

Thus, we can distinguish two types of planning rules: one that defines a cause-effect relationship among facts, with a given maximum domain D (backward rules) and one that defines an action refinement from a domain D_c to a domain D_f (conversion rules).

Granularity, within the planning system, allows to define a hierarchy of actions on different temporal domains by using the conversion rules: they are rules that map an atomic action, with respect to a certain temporal domain, into some elementary actions, with respect to a finer domain. This characteristic is important, for it is natural that, within coarser domains, more elementary actions cannot be visible, because they have "time constants" with a different order of magnitude (with respect to coarser ones). Furthermore, an action can be decomposed within a domain (i.e. without changing temporal domain): this kind of knowledge can be still translated into a conversion rule referring to a unique domain [Badaloni 93c].

Let us see now how the planning mechanism works. To reach a certain goal, a plan is found by the planning system, within a certain temporal domain, chosen a priori. When the plan is obtained, some (or all) actions can be refined by changing time granularity. Both a passage to a finer time granularity and the definition of the initial temporal grain, in order to deal with the adequate temporal information linked to the actions, may be triggered by data coming from the sensor world.

When passing to a finer domain, all knowledge kept in the planner database is translated into the new domain, along with the associated temporal network. Then, new planning rules, visible in the new domain, are added to the database, and coarser actions can be refined. In practice, while within a coarser domain these actions were added to the plan by an "ability assumption" (i.e. the agent is supposed to be able to perform these actions) and were considered as atomic, now they are planned using the conversion rules. When the new plan is done, one can return to the coarser domain,

D_c: new facts with maximum domain finer than D_c will no more be visible, and the same will happen for the associated temporal intervals (they will be deleted from the temporal network). But the new temporal network (at the domain D_c) will be modified, because new information was propagated at the finer level: constraints between coarser interval can be tightened because of new knowledge deduced at finer level.

It should be noted that explicit relations among facts (including actions) at different temporal granularity levels are defined only within the conversion rules. When passing to a finer domain, only those actions whose refinement is specified in a conversion rule can be replanned. If a certain action A is decomposed into two facts G_1 and G_2, from the domain D_c to the domain D_f, the temporal relation between, e.g., G_1 and another fact G' defined in the coarser domain D_c is determined indirectly via the relation between G_1 and A and the relation between A and G' (by propagating the constraints over the network).

Let us now consider the problem presented in section 3; indeed, in that problem, at that level of knowledge, the temporal information referred exclusively to days. Suppose now that a more detailed information is available, i.e., that it is known that the truck travel takes 53 hours (which viewed with the filter of days corresponds to 3 days). Let us show, now, more in detail, how the treatment of granularity has been introduced into the planning system and how it actually works when dealing with knowledge represented at different temporal domains. The planning rules relative to this problem (some of them are reported in section 3) remain unchanged except for the fact that, in each rule and in each domain constraint, the reference temporal domain (the maximum domain) has to be specified. As an example, the rule R2 is modified as follows:

R2')   holds(transported(X), T) <---
           [occur (truck_transport(X), E, etime(E)],
           [constr(eff1(E), T, e)], day.

In the domain of hours, the actions can be decomposed into a set of elementary sub-actions, and new conversion rules from the domain of days to that of hours have to be added to the data base; as an example, consider the action truck_transport(X) that can be decomposed into the two actions refuel and drive:

C1)   try(truck_transport(X), E, etime(E)) <---
           [occur(refuel, E1, etime(E1)), occur(drive, E2, etime(E2))],
           [constr(etime(E1),etime(E), s), constr(etime(E2),etime(E), f),
           constr(etime(E1),etime(E2), m)], hour, day.

where s, f, m represent the temporal relations starts, finishes and meets between the corresponding intervals. New planning rules relative to the domain of hours are defined, for example:

R4)   occur(refuel, E, etime(E)) <---
           [try(refuel, E, etime(E))], [], hour.

R5)   occur(drive, E, etime(E)) <---
         [try(drive, E, etime(E))], [], hour.

and similar rules for other actions.

There may be two cases:
1) temporal planning is performed at the domain of days and then, triggered by some events, the plan is refined at the level of hours,
2) temporal planning is performed starting directly from the domain of hours.

In the first case, the plan is exactly that proposed in section 3, that is, a plan composed by the transport by truck of an object and by the dispatch by mail of the other object (in a time of less then 5 days). When passing to a finer grain, the plan can be refined, as the actions can be decomposed, and more detailed temporal constraints can be represented. Thus, while the plan obtained in the domain of days was made of two actions, truck_transport(1) and mail_send(2), linked together by a set of temporal relations (of course the two actions cannot be executed sequentially), in the domain of hours the plan is made of four actions: refuel, drive, make_package(2), post(2). It can be noticed that, in the finer domain, the set of facts describing the world model is richer, so that the temporal network contains more nodes. The processing of temporal information, then, takes a greater amount of time, because the more complex the network (and the world model), the more complex the computation of the minimal network and the check of temporal consistency.

Coming back to the coarser domain, the original plan remains unchanged and it is associated with a temporal network that is, in general, characterised by more restrictive unary and binary constraints with respect to the original one.

In the second case, the deadline becomes 120 hours. Then, reasoning within this domain, it is possible to conclude that a plan exists to reach the goal by using only the truck transport for the two objects, as it was preferably expected (two travels by truck take 106 hours, on the whole). The constraint that the two corresponding temporal intervals T1 and T2 have to be disjoint is also satisfied, within the domain of hours. Passing to the coarser domain of days, this plan is no more possible since, in the updated temporal network, the intervals T1 and T2 can no longer be disjoint. This inconsistency is inherent to the notion of granularity itself: it is natural that, at a coarser temporal grain, knowledge becomes coarser too.

A hierarchy of plans is formed: initially, there may be a coarse plan, which can become more and more detailed, as one decides to refine some parts of it, or it is also possible to start planning directly at a certain granularity level.

Conclusion

The hierarchical planner presented in this paper allows one to use different temporal domains and to describe the different components of the scenario in which a problem solution has to be planned within the most suitable domain. Even if referring only to a simple explanatory examples, it has been shown which is the main advantage of using a hierarchical planner based on decomposition and refinement of actions: it can always work at a (temporal) granularity level as coarse as required, thus reducing the complexity of the problem one is addressing (especially from the point of view of the temporal reasoner). In many cases, only coarse planning is needed. In this way, a tradeoff between expressiveness and computational efficiency can be achieved.

As for future developments, we intend to test the multi-grain temporal planner on robotics applications.

Acknowledgements

This work has been developed at LADSEB - CNR of Padova, in the framework of the Italian Special Program on Robotics of the CNR (Progetto Finalizzato Robotica). This research has been partially granted by the Italian Special Program on Automated Planning of the CNR and by the Italian Ministry of University, Science and Technology. We would like to thank A.Montanari for many helpful discussions.

References

[Allen 83] J.F. Allen: "Mantaining Knowledge about Temporal Intervals", Communications of the ACM, number 11, volume 26, November 1983.
[Allen 84] J.F. Allen: "Towards a general theory of action and time", Artificial Intelligence 23, pgg. 123-154, 1984.
[Allen 91] J.F. Allen, H.A. Kautz, R.N. Pelavin, J.D. Tenenberg : "Reasoning about plans", Morgan Kaufmann, San Mateo, Calif., 1991.
[Allen 93] J.F. Allen, G.Ferguson: "Actions and events in interval temporal logic", submitted, 1993.
[Badaloni 93a] S.Badaloni, E.Pagello, L.Stocchiero, A.Zanardi: "Making an autonomous robot plan temporally constrained maintenance operations", in P.Torasso (ed.) "Advances in artificial Intelligence", Lectures Notes in Artificial Intelligence n.728, pg. 290-301, Springer Verlag, 1993.
[Badaloni 93b] S.Badaloni, E.Pagello, L.Stocchiero, A.Zanardi: "Planning temporally qualified robot actions", Proc. of the Int. Conf. on Advanced Robotics ICAR'93, Tokyo, Japan, November 1993.
[Badaloni 93c] S.Badaloni, M.Berati: "Decomposition and refinement of actions", In A.Cesta, S.Gaglio (Eds) Proc. of the Italian Planning Workshop 1993 (IPW'93), Roma, Italy, September 1993.
[Corsetti 91] E.Corsetti, E.Crivelli, D.Mandrioli, A.Montanari, A.Morzenti, P.San Pietro, E.Ratto: "Dealing with different time scales in formal specification", Proc. 6th Int. Workshop on Software Specification and design, Italy, 1991.
[Freksa 92] C.Freksa: "Temporal Reasoning based on Semi-Intervals", Artificial Intelligence vol.54, pgg. 199-227, 1992.
[Galton 90] A.Galton "A critical examination of Allen's theory of action an time", Artificial Intelligence, 42 (2-3), pgg. 159-188, 1990.
[Hobbs 85] J. Hobbs: "Granularity", Proc. of the 9th IJCAI, Los Angeles, USA, 1985.
[Hobbs 93] J. Hobbs: "Sketch of a proposed ontology that underlies the way we talk about the world", Proc. Workshop on Formal Ontology, Padua, 1993.
[Jaffar 87] J.Jaffar, S.Michaylov: "Methodology and implementation of a CLP

system", Proc. of the 4th International Conference on Logic Programming, Melbourne, pgg. 196-218, 1987.

[Montanari 92] A. Montanari, E. Maim, E. Ciapessoni, E. Ratto: "Dealing with Time granularity in the Event Calculus", Proc. of the International Conference on Fifth Generation Computer System, Tokyo, Japan, June 1992.

[Tenenberg 86] J.Tenenberg: "Planning with abstraction", Proc. AAAI-86, Philadelphia, 1986.

[Trudel 93] A.Trudel: "A formal specification of Allen's processes", Proc. of the Second Symposium on Logical Formalization of Commonsense Reasoning, Austin, Texas, 1993.

[vanBeek 90] P. van Beek, Robin Cohen: "Exact and approximate reasoning about temporal relations", Comput. Intelligence 6, pgg. 132-144, 1990.

[vanBeek 92] P. van Beek: "Reasoning about qualitative temporal information", Artificial Intelligence 58, pg 297-326, 1992.

[Vilain 90] M. Vilain, H. Kautz, P. van Beek, "Constraint propagation algorithms for temporal reasoning: a revised report", Readings in qualitative Reasoning about physical objects, pgg. 373-381, 1990.

Axiomatizing U and S over Integer Time

M Reynolds[1]

Imperial College
LONDON SW7 2AZ.

Abstract. We give a Hilbert style axiomatization for the set of formulas in the temporal language with *Until* and *Since* which are valid over the integer number flow of time. We prove weak completeness for this orthodox axiom system.

1 Introduction

We continue a long tradition of axiomatizing temporal logics. Variations on the theme have been achieved by varying the language used to talk about events in time and by varying the assumptions made about the nature of time. In this paper we consider the particularly interesting case of the language with "until" and "since" over integer number time.

Early axiomatization results were for the language with two temporal connectives: F for "will" and P for "was". Completeness of axiomatizations for this language over various flows of time can be proved by tinkering with Henkin constructions (see for example [11]).

The stronger, more expressive, language with "until" (U) and "since" (S) presents more difficulties. Burgess [1] gives a more complicated Henkin construction which is sufficient to prove completeness when we consider certain whole classes of flows of time such as linear flows.

Axiomatization over the rational numbers fall out of Burgess' work easily. However, it is generally more difficult again to get completeness results for specific flows of time. Although axioms for U over the more "useful" natural numbers flow appear with an involved proof in [4] an axiomatization of U and S over the natural numbers flow (along with other classes of well-orderings) has only just been given in [12].

In this paper, we use the techniques developed in [9] to provide an axiomatization for U and S over the integers. After defining the logic in the next section we will show that only a weak completeness result can be expected. Thus we can only find a syntactic analogue of the consistency of a single formula rather than being able to find an analogue of the general consequence relation. It will be seen that the weak form of completeness is still very useful: in fact the distinction between weak and strong here is subtle enough to be often ignored.

---

[1] The author would like to thank the temporal logic group at Imperial College for suggesting many improvements. The work was supported by the U.K. Science and Engineering Research Council under the Metatem project (GR/F/28526).

Then, we present the axiomatization, and comment on its obvious soundness.

It is important to note that like our result in [9] for the reals and unlike that in [3], our axiomatization does not use any unorthodox rules of inference such as the IRR rule of [7]. For reasons discussed in [9] and [12], this rule is slightly controversial and it is important to show that it is not necessary.

In most of the rest of the paper we prove the completeness result devoting sections to finding a rational-flowed model, proving expressive completeness, proving Dedekind completeness for a certain type of definable equivalence, finding a integer-flowed model and concluding the proof.

The integer numbers flow of time is obviously a useful model for situations in which we have discrete moments of time but have neither a beginning or an end of time. We see an example of a such a situation in the temporal database work of [2] where the axiom system presented here is modified into an axiom system for resoning with updates of databases.

2 The Logic

In this paper temporal structures, which we will often just call structures, will be linear. Thus they consist of a domain T, an irreflexive linear order $<$ on T and a valuation h assigning each p of a countable set of atoms to a subset $h(p) \subseteq T$. The underlying linear order $(T, <)$ of a structure $(T, <, h)$ is called the flow of time of $(T, <, h)$.

The temporal language will usually be that generated by the connectives U and S. The set of formulas is defined recursively to contain the atoms, $\top$ and $\bot$ and for formulas A and B, we include $\neg A$ and $A \wedge B$ along with $U(A, B)$ and $S(A, B)$. $U(A, B)$ is read "until A, B" or "B until A". Similarly S is to be read as "since".

Formulas are evaluated at points of the flow of time. The readings above suggest the semantics but more formally truth is defined recursively as follows. Suppose that, inductively, we have defined the truth of formulas A and B in a structure $(T, <, h)$ at all points t: now define

$(T, <, h) \models p(t)$ iff $t \in h(p)$, for p atomic
$(T, <, h) \models \top(t)$
$(T, <, h) \not\models \bot(t)$
$(T, <, h) \models (\neg A)(t)$ iff $(T, <, h) \not\models A(t)$
$(T, <, h) \models (A \wedge B)(t)$ iff $(T, <, h) \models A(t)$ and $(T, <, h) \models B(t)$
$(T, <, h) \models U(A, B)(t)$ iff there is $s > t$ such that
                   $(T, <, h) \models A(s)$ and
                   for all $u \in T$, if $t < u < s$ then $(T, <, h) \models B(u)$
$(T, <, h) \models S(A, B)(t)$ iff there is $s < t$ such that
                   $(T, <, h) \models A(s)$ and
                   for all $u \in T$, if $s < u < t$ then $(T, <, h) \models B(u)$

Often, because of the symmetry of their definitions, results involving U and/or S have *dual* or *mirror* versions which can be stated and proven by simply swapping U and S for each other and swapping $<$ and $>$ in the original. We mention these frequently but, of course, never bother to prove them.

As usual we have all sorts of abbreviations: the classical ones $\vee$, $\rightarrow$ and $\leftrightarrow$ along with

FA for $U(A, \top)$ - A will be true,
PA for $S(A, \top)$ - A was true,
GA for $\neg F \neg A$ - A will always be true,
HA for $\neg P \neg A$ - A was always true,
$K^+ A$ for $\neg U(\top, \neg A)$ - A will be true arbitrarily soon and
$K^- A$ for $\neg S(\top, \neg A)$ - A was true arbitrarily recently.

For formula A and set Γ of formulas we write $\Gamma \models A$ iff for all valuations h into $\mathbb{Z}$, for all $t \in \mathbb{Z}$, if for all $B \in \Gamma$, $(\mathbb{Z}, <, h) \models B(t)$ then $(\mathbb{Z}, <, h) \models A(t)$. We are trying to find a syntactic equivalent to this consequence relation.

In the proof we will have to consider flows of time other than integers but throughout this paper all temporal structures will be assumed to have linear flows.

3 Weak versus Strong Completeness

We are only going to be able to prove a weak completeness result for this logic. Let us review the concepts involved before seeing why. Suppose that we are dealing with a class $\mathcal{K}$ of flows of time and an axiom system Z for the logic of some temporal language $\mathcal{L}$ over $\mathcal{K}$. Assume the usual definition of a *proof* (in Z) of a formula A from a set Γ of formulas. Write $\Gamma \vdash_Z A$ if there is such a proof. Since each inference rule only has one formula as a conclusion in such axiom systems we will call them *finitary* axiom systems.

Assume all the usual definitions for

- syntactic concepts like theorems, consistent sets, consistent formulas and maximal consistent sets for Z,
- semantic concepts like $\models_\mathcal{K}$, models of sets of formulas, satisfiability and validity for $\mathcal{K}$ and
- soundness of Z for $\mathcal{K}$.

Recall that Z is *strongly complete* (often just written *complete*) for $\mathcal{K}$ iff one of the following two equivalent conditions hold:

- if Γ is Z-consistent then Γ has a model in $\mathcal{K}$,
- for all Γ, A, if $\Gamma \models_\mathcal{K} A$ then $\Gamma \vdash_Z A$.

To show that we can not have a strongly complete axiomatization in the case of U and S over the integers we will use the concept of compactness. We say that a logic has the *compactness property* iff every unsatisfiable set of formulas has an unsatisfiable finite subset.

Theorem 1. *If a logic has a sound and strongly complete finitary axiomatization then it also has the compactness property.*

This follows easily from the definitions as a proof of inconsistency of a set only uses a finite number of formulas from that set.

It follows that there can be no sound and strongly complete finitary axiomatization of the logic of U and S over integer time. This is because we can see that the logic is not compact by consisdering the set $\{A_n \mid n \in \mathbb{N}\}$ where $A_0 = F(q \wedge H(\neg q))$ and for each $n \in \mathbb{N}$, $A_{n+1} = F(p \wedge A_n)$. The set is clearly unsatisfiable while all its finite subsets are.

Fortunately, there is a weaker notion of complteness which is still useful. We say Z is *weakly complete* for $\mathcal{K}$ iff one of the following two equivalent conditions hold:

- if A is Z-consistent then A has a model in $\mathcal{K}$,
- for all finite Γ, for all A, if $\Gamma \models_\mathcal{K} A$ then $\Gamma \vdash_Z A$.

In the rest of this paper we will show that this is indeed a weaker notion by showing that the logic of U and S over the integers does admit a weakly complete axiomatization.

4 The Axiomatization

Our system **US/Z** has the usual rules for a temporal logic: i.e. modus ponens, generalizations and substitution:

$$\frac{A, A \to B}{B} \qquad \frac{A}{GA} \qquad \frac{A}{HA} \qquad \frac{A(q)}{A(q/B)}$$

The axioms of **US/Z** are:

- all classical tautologies,
- the six Burgess-Xu axioms
  $G(p \to q) \to (U(p,r) \to U(q,r))$
  $G(p \to q) \to (U(r,p) \to U(r,q))$
  $p \wedge U(q,r) \to U(q \wedge S(p,r), r)$
  $U(p,q) \to U(p, q \wedge U(p,q))$
  $U(q \wedge U(p,q), q) \to U(p,q)$
  $U(p,q) \wedge U(r,s) \to$
    $U(p \wedge r, q \wedge s) \vee U(p \wedge s, q \wedge s) \vee U(q \wedge r, q \wedge s)$
  along with each of their duals,
- plus axioms for discreteness and no end points:
  $U(\top, \bot)$ and $S(\top, \bot)$,
- and suitable versions of the Prior axioms:
  Prior-UZ: $Fp \to U(p, \neg p)$
  Prior-SZ: $Pp \to S(p, \neg p)$

Soundness is clear: we are going to spend the rest of the time proving *weak* completeness. The next few sections establish a variety of useful preliminary results.

Notions such as maximal consistent and $\vdash$ are assumed to refer to the logic of U and S over the integers and our system **US/Z** unless otherwise specified.

5   The Burgess-Xu Result

Our task in this section is to find a model of a consistent formula which has a vaguely integer-like flow of time and is one in which $U(\top, \bot)$, $S(\top, \bot)$ and all substitution instances of the Prior axioms are valid. Fortunately, most of the work has been done already. Burgess in [1] proves soundness and completeness of a set of axioms for linear time. Xu, in [13], simplifies the set of axioms and the proof.

Theorem 2. *The Burgess-Xu system (the six axioms and duals, propositional tautologies and the four rules) is sound and strongly complete for the US logic on the class of all linear frames.*

Although neither Burgess nor Xu mention *strong* completeness their proofs do establish that. This is just as well for we need strong completeness.

Let us see how we can use this theorem.

Now suppose that we have a set Γ of formulas consistent with the system **US/Z**. Knowing Lindenbaum's lemma, we can, without loss of generality, assume that Γ is maximal consistent.

By the theorem, since Γ is also consistent with the Burgess-Xu system, there will be a linear model for Γ: i.e. a linear structure in which there is a point, t say, at which all the formulas in Γ hold.

By looking at Burgess's construction (or using Löwenheim-Skolem) we can suppose that the structure is countable.

Since $GU(\top, \bot)$ and its mirror must be true at t, the order does not have end points and the order is discrete.

Because it says so in Γ, all the substitution instances of the other axioms hold everywhere so we have...

Corollary 3. *For every **US/Z**-consistent set Γ of formulas, there is a temporal structure M and $t \in M$ such that*

1. *the flow of time of M is countable, discrete and without end points,*
2. *for all $A \in \Gamma$, $M \models A(t)$ and*
3. *all substitution instances of the axioms Prior-UZ and Prior-SZ are valid in M.*

6   Expressive and Dedekind Completeness

An important technique in our proof is that of switching between the temporal language and an associated first-order one. Let us introduce the concepts and results needed.

We will associate a temporal language with a first-order one called the *monadic* language because it is built from a signature containing only 1-ary predicate symbols along with the binary $<$ predicate symbol. Each atom p in the temporal language corresponds to a predicate symbol P. We can make a temporal structure $(T, <, h)$ into a first-order structure in the monadic language, by interpreting $<$ as $<$ and each P as being true of exactly those points in $h(p)$.

If, as will later be the case, we restrict to a temporal language with a finite number of atoms then the monadic signature is finite but otherwise it will contain a countable number of 1-ary predicate symbols.

It turns out, unsurprisingly, that the temporal formula $U(p,q)$ is true at exactly those points in a structure where the monadic formula

$$\psi_{U(p,q)}(t) = \exists s > t(P(s) \land \forall u(t < u \land u < s \to Q(t)))$$

holds. A simple induction, (see for example [5]), establishes that all temporal formulas A have a corresponding monadic formula ψ_A in one free variable such that, for all structures $(T, <)$, for all valuations h, for all $t \in T$,

$$(T, <, h) \models A(t) \text{ iff } (T, <, h) \models \psi_A(t).$$

We call ψ_A the *table* of A. The induction generalises to show that provided the connectives of the language have first-order tables, as U and S do, then all the temporal formulas of any temporal language have first-order tables.

One may ask whether all first-order formulas with one free variable can be got as tables of temporal formulas. This, of course, depends on the temporal connectives used in the language, but it also depends on what class of structures we restrict attention to. Let us be more precise. Suppose that $\mathcal{S}$ is a class of temporal structures. We say that a temporal language is *expressively complete* over $\mathcal{S}$ if and only if for each monadic formula $\phi(t)$ with one free variable, there is a temporal formula A of the temporal language such that for all $(T, <, h) \in \mathcal{S}$, for all $t \in T$,

$$(T, <, h) \models \phi(t) \text{ iff } (T, <, h) \models A(t).$$

Note the uniformity of the translation over the whole of $\mathcal{S}$.

One of the first expressive completeness results was that of Kamp's in [8]. Kamp showed expressive completeness for (the language with) U and S over the class of all structures whose underlying flow of time is Dedekind complete.

As shown in [5], U and S are still expressively complete even if we allow isolated gaps in the structure but as has been known for while, and as is shown in [5], lemma 3, over the whole class of structures whose underlying flow of time is linear, U and S are not expressively complete. To achieve expressive completeness for the class of all structures with linear flows we need to use the so called Stavi connectives U' and S' which were defined in [4]. $U'(A, B)$ holds if B is true from now until a gap in time after which B is arbitrarily soon false but after which A is true for a while: $U'(A, B)$ is as pictured

$$\underset{\text{now}}{\rule{3cm}{0.4pt}}\overset{B}{}\underset{\text{a gap}}{()}\overset{\leftarrow \cdots \quad \neg B}{\rule{3cm}{0.4pt}}A$$

S' is defined dually. Despite involving a gap, U' is in fact a first-order connective and its table is given by:

$$U'(p,q) \equiv$$
$$\exists s \quad t < s$$
$$\wedge \forall u \ (\qquad\qquad\qquad t < u < s \to$$
$$([\quad \exists v(u < v \wedge \forall w(t < w < v \to q(w)) \ ]$$
$$\vee [\qquad\qquad \forall v(u < v < s \to p(v))$$
$$\wedge \qquad \exists v(t < v < u \wedge \neg q(v)) \qquad\qquad]))$$
$$\wedge \exists u[t < u < s \wedge \neg q(u)]$$
$$\wedge \exists u[t < u < s \wedge \forall v(t < v < u \to q(v))]$$

We have

Theorem 4. *The language with $\{U, S, U', S'\}$ is expressively complete for the class of structures with linear flow of time.*

This result is mentioned in [4] without proof. The first published proof - a direct proof - is in [5]. In [6] is a proof using the separation technique of Gabbay.

Obviously there is some connection between definable gaps and our Prior axioms. Call a linear temporal structure a *Prior structure* if it satisfies all substitution instances of

$$\text{Prior-U:} U(\top, p) \wedge F\neg p \to U(\neg p \vee K^+(\neg p), p)$$

and its dual Prior-S. It is easy to see that then there are no definable gaps. Note that this result also holds for our stronger Prior axioms Prior-UZ and Prior-SZ (the weaker axioms are useful in non-discrete structures). It is now not hard to prove the following (see [3], proposition 4.2).

Theorem 5. *The language with U and S is expressively complete for the class of Prior structures.*

Proof. By the expressive completeness of $\{U, S, U', S'\}$ over all linear structures, it suffices to prove that for any $\{U, S, U', S'\}$-formula B', there is a $\{U, S\}$-formula B such that $B' \leftrightarrow B$ is valid in all Prior structures.

This can be achieved by a simple induction on the construction of such B'. The cases of atoms and $\wedge$, $\neg$, U and S are immediate. Let us look at $U'(A, B)$ when, by induction, we can suppose that A and B are US-formulas. We claim that $U'(A, B) \leftrightarrow \bot$ is valid in all Prior structures.

Suppose for contradiction that $M \models U'(A, B)(t)$ in some Prior structure M. Thus B holds for a while up until a gap after which $\neg B$ is true arbitrarily soon. By Prior-U applied to B we have $M \models U(\neg B \vee K^+(\neg B), B)(t)$ which is the contradiction.

The case of S' is similar.

7 No gaps between equivalence classes

We know that the Prior axioms ensure that there will not be any definable gaps in a model. To show that our model can be made into a model over the integers we actually need a stronger result. We need to know that a certain type of definable equivalence relation also does not have its equivalence classes ending at gaps. First some definitions.

The *intervals* of a structure M are just the convex subsets of M and we will use the usual $(a, b]$, etc., notation for them.

If S is a subset of the domain of a temporal structure M (usually S will be an interval here) then we write $M \mid S$ for the temporal structure with domain S, ordering as in M and interpretation of atoms just the restrictions of the interpretations in M to S. $M \mid S$ is called the *substructure* of M with domain S.

Suppose that $\varepsilon(x, y)$ is a monadic formula with two free variables x and y. We say that ε defines a *contemporaneous* equivalence relation if and only if on any temporal structure M, if we define the binary relation $\sim_M$ by

$$a \sim_M b \text{ iff } \models \varepsilon(a, b),$$

then

- $\sim_M$ is an equivalence relation on the domain of M,
- $\sim_M$ partitions M into intervals and
- ε depends only on contemporary properties: i.e. for all $a, b \in M$,

$$\models \varepsilon(a, b) \text{ iff } \mid [a, b] \models \varepsilon(a, b).$$

A binary relation $\sim$ on a structure M is called a *contemporaneous* equivalence relation if and only if it is defined as $\sim_M$ by such an ε.

We prove that no so defined contemporaneous equivalence relation has equivalence classes ending at gaps in any Prior structures.

Given such an ε, define $\rho(x)$ as

$$\exists y > x \; \neg\varepsilon(x, y)$$
$$\wedge \neg\exists z(\varepsilon(x, z) \wedge \forall y > z \; \neg\varepsilon(x, y))$$
$$\wedge \neg\exists z(x < z \wedge \neg\varepsilon(x, z) \wedge \forall y(x < y < z \to \varepsilon(x, y))).$$

This says that x's $\sim$-class ends in a gap on the right. Dually we can define $\lambda(x)$ about left ends. Note that the end of the whole structure is not a gap and that ρ will not hold of points in the last $\sim_M$ class (if there is such a class).

Now by the expressive completeness of U and S there is temporal R true in any Prior structure exactly where $\rho(x)$ is.

Lemma 6. *Suppose that ε defines the contemporaneous equivalence relation $\sim_N$ on any structure N.*

Then there is an US-formula R which holds in any Prior structure N exactly at those points whose $\sim_N$-class ends in a gap on the right.

Dually L.

Now suppose that M is a Prior structure and that $\sim = \sim_M$ is a contemporaneous equivalence relation defined by ε.

Lemma 7. *The maximal intervals in which R holds are open intervals which, if bounded, have elements of M as their (excluded) end points.*

Proof. Suppose that R holds at $t \in M$. Clearly ρ holding at t implies that R will hold for a while after t: up until a gap in fact. Thus t is in a non-singleton interval of R. It is possible that R holds for ever after t.

If R does not hold for ever after t then Prior-U applied to R implies that M contains a last point of this stretch of R (plainly impossible given ρ) or a first point of $\neg R$. This is as claimed.

Now look to the left of t. Looking back from just after t we can use Prior-S and see that either R is true always before t, there is a last point of $\neg R$ just before this stretch of R or there is no last point of $\neg R$, but instead a first point of R. We must rule out the third case. Note that in the case of M not being dense there may be both a last point of $\neg R$ and a first point of R: this possibility, subsumed in the second case above, is acceptable in that it implies an excluded end point.

Suppose, for contradiction, that s is this first point of R so that $M \models (R \wedge K^-(\neg R))(s)$. The $\sim$-class containing s can not stretch for ever into the future for then it does not end in a gap. Neither can it stretch to the end of the maximal interval of R as it would again not end at a gap.

Thus there are other classes in this interval continuing on the other side of the gap which ends s's. And for a while after the gap R continues to be true: we have not reached the end of the interval yet. Thus $R \wedge K^-(\neg R)$ does not hold at the left hand end of any of these classes.

Let B be the temporal formula saying that the $\sim$-class we are now in begins with a point satisfying $R \wedge K^-(\neg R)$. B exists by expressive completeness. B holds in s's class up to the gap and is false arbitrarily soon after the gap. This contradicts Prior-U applied to B.

Lemma 8. *There is no last class and no first class in any maximal interval of R.*

Proof. The last class in a maximal interval of R wouldn't end in a gap.

By expressive completeness, the formula

$$\rho(x) \wedge \forall y < x(\neg \varepsilon(x, y) \to \exists z(y < z < x \wedge \neg \rho(z)))$$

has a temporal equivalent which is true only in the first classes of maximal intervals of R. If there is a first class then no immediately subsequent classes satisfy this and so we have this formula holding up to a gap and false arbitrarily soon afterwards. This contradicts Prior-U.

Lemma 9. *If a temporal formula holds somewhere in one $\sim$-class in a maximal interval of R, then it holds somewhere in each $\sim$-class in the interval.*

Furthermore, each pair of the $\sim$-classes in a maximal interval of R are elementarily equivalent (taken as substructures of M).

Proof. For a contradiction to the first statement, suppose that A holds in one class but not anywhere in some other class in the same maximal interval of R.

Using expressive completeness and ε, find B which is true at points only if A occurs somewhere in their $\sim$-class. By using $\neg B$ instead if necessary we may suppose that we have B holding throughout one $\sim$-class in our maximal interval of R and false throughout a later class. Choose a point t in this former class in which B holds. B holds in the whole of a class if it is true anywhere at all in the class so it continues for a while after t. By Prior-U there is either a last point where B holds after t (not possible as B must continue for a while) or a first point $s > t$ where $\neg B \wedge K^-(B)$ holds.

So s must be the left hand end point of its $\sim$-class. Look at the gap at right hand end of this class. We can not have B arbitrarily soon after the gap because of Prior-U. Thus for a while after this class B stays false.

Let C be the temporal formula saying that we are now in a class whose left hand end point is also in the class and at that point $K^-(B)$ holds. Now C is true in s's class but false afterwards contradicting Prior-U.

Now consider the second statement in the lemma. Given a monadic sentence ϕ, we relativise it by restricting quantifiers to where $\varepsilon(x,-)$ holds. We get a formula $\phi(x)$ of one free variable. By expressive completeness this is equivalent to a temporal formula. This is true exactly throughout $\sim$-classes which model ϕ. Then, by the first part of the lemma, it can't be true somewhere and false elsewhere in the interval.

We define a *bad* point to be where $R \vee L$ holds. We define a *bad* interval as a non-empty and maximal one in which $R \vee L$ holds throughout.

Lemma 10. *Bad points only occur in non-singleton bad intervals.*

In any bad interval both R and L hold throughout. Any bad interval, if bounded, has excluded end points in M (neither R nor L holds at these end points).

Proof. We first show that L holds wherever R does. Suppose for contradiction that we have a maximal interval of R in which L fails to hold somewhere. So $\neg L$ holds throughout at least one $\sim$-class. By the definition of L, there are two cases. Either this particular $\sim$-class is one which includes its left hand end point or it is one which begins just after some point of M. The class can not be unbounded below for then it would be first in this bad interval.

In fact we can not have a class beginning just after a point r of M. Since the class can not be first in the bad interval r itself must be in a $\sim$-class in the bad interval. But r's class can not end in a gap on the right when r must be its right hand end point.

Thus we have a class in the bad interval which includes its left hand end point. Its not hard to use the previous result to show that throughout the bad interval all classes include their left hand end points.

Let B be a temporal formula true at times which are not left hand end points of their $\sim$-classes. B is then true continuously in any class from just after the left hand end point up until the gap at the right hand end point. B must be false arbitrarily soon after the gap contradicting Prior-U.

Using mirror images of the above and previous results we get our proof.

Lemma 11. *If a formula B is true for a while at the start of a $\sim$-class in a bad interval then it holds throughout the bad interval. Similarly at the end.*

If a formula is true anywhere in a bad interval it is true arbitrarily close to each end of each class in the interval.

Proof. Suppose that $\gamma < \delta$ are gaps and that (γ, δ) is a $\sim$-class within a bad interval.

Suppose that B holds for a while after γ but that $\neg B$ holds somewhere in the bad interval. By lemma 4, $\neg B$ also holds somewhere in (γ, δ).

Using ε and expressive completeness we can find a temporal formula C which is true only at points within a $\sim$-class after some $\neg B$ in that class. C will be false for a while at the beginning of each class and then true for a while at the end.

In fact C is true for a while up to the gap at the end and false arbitrarily soon after the gap. This contradicts Prior-U.

Applying the above to the negation of a formula gives us the second part.

Let us see what happens if we interfere with M by replacing a whole bad interval by one of its $\sim$-classes.

Let Q^- be the subset of the domain of M being all that precedes the bad interval. Let Q^+ be all that follows. Either or both of these may be empty. Let Q_0 be the bad interval itself and I be any one of its $\sim$-classes.

We look at N, the substructure of M whose domain is just $Q^- \cup I \cup Q^+$.

Lemma 12. *For all temporal formulas A, for all $t \in N$,*

$$\models A(t) \text{ iff } N \models A(t)$$

Proof. We proceed by induction on the construction of A. The cases of atomic and boolean A are immediate. Now consider $U(A, B)$: $S(A, B)$ is similar.

($\Rightarrow$): Consider then when $M \models U(A, B)(t)$ with $t \in N$. Say that $s \in M$, that $t < s$, that $M \models A(s)$ and for all $u \in M$, if $t < u < s$ then $M \models B(u)$.

There are several cases.

1. $t < s \in Q^-$: Apply the induction hypothesis to A and B at s and at all points in between.
2. $t \in Q^-$ and $s \in Q_0$: A holds somewhere in Q_0 so somewhere in I (by lemma 6). So holds there in I in N. B holds for a while into Q_0 so, by lemma 6, holds everywhere in Q_0. By the induction hypothesis, B holds everywhere in I in N. Hence result.
3. $t \in Q^-$ and $s \in Q^+$: We can deduce that B holds throughout I in both M and N and get the result.
4. $t < s \in I$: Straight forward use of inductive hypothesis.

5. $t \in I$ and s later in Q_0: Again by lemma 6 we have B true throughout I in M and so in N. Since A is true somewhere in Q_0 in M, lemma 6 tells us that A is true arbitrarily close to the end of I in M and so in N. This gives us our result.
6. $t \in I$ and $s \in Q^+$: B is true throughout I and we have our result.
7. $t < s \in Q^+$: Apply induction hypothesis to A and B at s and at all points in between.

($\Leftarrow$) : Consider then when $N \models U(A,B)(t)$. Say that $t < s$, that $N \models A(s)$ and for all $u \in N$, if $t < u < s$ then $N \models B(u)$.

Again there are several cases:

1. $t < s \in Q^-$: Apply induction hypothesis to A and B at s and at all points in between.
2. $t \in Q^-$ and $s \in I$: B holds from t up until the end of Q^- in both M and N. B holds at the beginning of I in N and so in M. By lemma 6 B holds throughout Q_0. A holds in I in N and so in M and we have our result.
3. $t \in Q^-$ and $s \in Q^+$: B holds throughout I in N and so in M. Lemma 6 tells us B holds throughout Q_0 in M.
4. $t < s \in I$: Straight forward use of inductive hypothesis.
5. $t \in I$ and $s \in Q^+$: B is true throughout I and we have our result.
6. $t < s \in Q^+$: Apply induction hypothesis to A and B at s and at all points in between.

Lemma 13. *In fact there can't have been any bad points anyway.*

Proof. By lemma 7, R holds in I in N.

But by lemma 1, R holds at a point in any Prior structure (not just M) if and only if the $\sim$-class of the point ends in a gap (where $\sim$ is the appropraite equivalence relation for the structure). And N is a Prior structure: we still have all the instances of Prior-U/S continuing to hold as any counterexample point in N is also one in M.

By the contemporaneity of ε, I as a subset of N, like I as a subset of M, is all in one $\sim_N$-class. Could the class be bigger now?

R is true of this class so that it is bounded above amongst other things. Thus Q^+ is non-empty and by lemma 5 begins with a point q say. Also by lemma 5 $\neg R$ holds at q in M and so in N. Clearly q is not in the class of I in N. Thus the class ends just before q.

R can not have been true in this class after all.

Thus we have proven...

Theorem 14. *Suppose that $\sim$ is a contemporaneous equivalence relation on a Prior structure M.*

Then the $\sim$-classes do not end at gaps.

8 Using Contemporaneity on the Integers

Let us see how we can use the theorem above in our proof. Note that here we are working in a language with only a finite number of atoms though.

Theorem 15. *Suppose that M is a temporal structure in a finite language such that*

- *the flow of time of M is countable, discrete and without end points,*
- *all substitution instances of the axioms Prior-UZ and Prior-SZ are valid in M.*

Then for all $k < \omega$, there is a temporal structure with flow of time the integers satisfying the same monadic first-order sentences of quantifier depth at most k as M does.

Proof. First some preliminaries. Fix $k \geq 3$.

Here a *structure* will mean a linear temporal structure in our finite language.

If M and N are structures we write $M \equiv_k N$ if and only if M and N agree on the truth of monadic sentences of quantifier depth at most k. Note that since $k \geq 3$, if $M \equiv_k N$ then M and N either both have a right(respectively left) hand end point or both do not have a right(resp. left) hand end point. Discreteness is also preserved.

We assume familiarity with lexicographic sums of linear orders and with the fact that $\equiv_k$ is preserved under such sums. See [10], [3] or [9] for details.

If a is an element of a discrete structure M then we write $a - 1$ for its immediate predecessor if it has one and $a + 1$ for its immediate successor, if it has one.

Say that M is *good* if and only if there is some $N \equiv_k M$ such that the flow of time of N is an interval of the integers.

Say that M is *very good* if and only if, for all $t \leq u$ in M, the substructure $M \mid [t, u]$ is good.

Lemma 16. *If N is countable and very good then it is good.*

Proof. All finite structures are good so suppose that N has countably infinite domain. If N has two end points then it is clearly good. First consider the case when N has a beginning a_0 but no (right hand) end.

Choose $a_i \in N$ for each positive integer i such that $i < j$ implies $a_i < a_j$ and for all $t \in N$, there is j such that $t < a_j$. Since N is very good, $N \mid [a_i, a_{i+1} - 1]$ is good. For $i = 0, 1, ...$, take $Z_i \equiv_k N \mid [a_i, a_{i+1} - 1]$ with a finite interval of $\mathbb{Z}$ as a flow.

Because $\equiv_k$ is preserved under lexicographic sums,

$$N \equiv_k \Sigma_{i \in \mathbb{N}}(Z_i)$$

the latter having flow isomorphic to a (half) subinterval of $\mathbb{Z}$.

If N has an end but no begining then the proof is similar. If N has no end points then choose $a_0 \in N$, use the above arguments on $(-\infty, a_0]$ and $[a_0 + 1, +\infty)$, and then use the lexicographic sum result to add appropriate structures together.

Define $\sim_M$ on a temporal structure M by for any $a, b \in M$, $a \sim_M b$ if and only if

- $a = b$,
- $a < b$ and $M \mid [a, b]$ is very good or
- $b < a$ and $M \mid [b, a]$ is very good.

Lemma 17. $\sim_M$ *is a contemporaneous equivalence relation on the domain of any M.*

Proof. Clearly there are only finitely many logically inequivalent maximal consistent conjunctions γ of sentences of quantifier depth $\leq k$. Any structure is a model of just one such γ, so if $N_1 \models \gamma$ then $N_2 \equiv_k N_1$ iff $N_2 \models \gamma$. Only some will be true of good structures- $\{\gamma_1, ..., \gamma_s\}$ say. N is good iff $N \models \bigvee_{i \leq s} \gamma_i$.

Let $\gamma(z, t)$ be the result of relativizing the quantifiers of $\bigvee_{i \leq s} \gamma_i$ to $[z, t]$, where z and t are new variables.

Then
$$\varepsilon(x, y) = \quad x < y \to \forall zt(x \leq z \leq t \leq y \to \gamma(z, t))$$
$$\land \, y < x \to \forall zt(y \leq z \leq t \leq x \to \gamma(z, t))$$

is a formula defining $\sim_M$.

To show that $\sim$ is contemporaneous, we first show that it is an equivalence relation. The difficult part is transitivity. Suppose that $a < b < c$ are in M and $a \sim_M b$ and $b \sim_M c$. We show that $M \mid [a, c]$ is very good by showing that if $a \leq t < u \leq c$ then $M \mid [t, u]$ is good.

If t and u are on the same side of b then this is clear. If $b = t$ or $b = u$ then use a lexicographic sum.

So assume that $a \leq t < b < u \leq c$. First note that since $M \mid [b, c]$ is very good then it is also good (even if it is not countable). Thus its flow is discrete and there is a successor $b + 1$ of b. Now $M \mid [t, b]$ and $M \mid [b + 1, u]$ are both good. Choose $Z_1 \equiv_k M \mid [t, b]$ and $Z_2 \equiv_k M \mid [b + 1, u]$ each with flow a subset of $\mathbb{Z}$. Then we know that $M \mid [t, u] \equiv_k Z_1 + Z_2$ whose flow is isomorphic to an interval of $\mathbb{Z}$ itself.

That the $\sim_M$ classes are intervals follows from the fact that very goodness is inherited by substructures on subintervals.

Contemporaneity then follows from the fact that the definition of $\sim_M$ is in terms of exactly the right substructure.

Now let us turn to the proof of the main theorem.

If M is good then we are done. So suppose not. Thus M is not very good. So there is $a < b \in M$ such that $M \mid [a, b]$ is not good. Thus $M \mid [a, b]$ is not very good and we have two disjoint $\sim$ classes.

Now a's class can not end at a gap on the right (by theorem 5 and the fact that Prior-UZ and dual imply Prior-U and dual) so it must include a point c but not the successor $c+1$ of c. This can not be because $M \mid [c, c+1]$, like all finite structures is very good and $\sim$ is transitive.

9 Completeness

Finally

Theorem 18. *The system* **US/Z** *is sound and weakly complete for the semantics over structures with integers flow.*

Proof. Soundness is straightforward.

To show weak completeness, we suppose that we are given a formula A_0 consistent with **US/Z**. We will find a model of it with flow of time the integers.

First use Burgess-Xu Corollary 1 to furnish us with a structure M_0 and $t_0 \in M_0$ such that

1. the flow of time of M_0 is countable, discrete and without end points,
2. $M_0 \models A_0(t_0)$ and
3. all substitution instances of the axioms Prior-UZ and Prior-SZ are valid in M_0.

By ignoring all the atoms which don't appear in A_0 we have a temporal structure M from a finite language. M is still a model of A_0.

Thus we can apply theorem 6.

Let k be one greater than the quantifier depth of the table $\alpha(t)$ of A_0. We have a temporal structure $\mathcal{Z}$, with flow of time the integers, satisfying the same monadic sentences of quantifier depth at most k as M does.

Thus $\mathcal{Z}$ like M is a model of $\exists t \alpha(t)$. Say $b \in \mathcal{Z}$ and $\mathcal{Z} \models \alpha(b)$.

We have $\mathcal{Z} \models A_0(b)$ as promised.

Axiomatizing U and S over the natural numbers can be done in a similar manner.

References

1. J P Burgess. Axioms for tense logic I: "since" and "until". *Notre Dame J Formal Logic*, 23(2):367–374, 1982.
2. Marcelo Finger. Handling database updates in two-dimensional temporal logic. *J. of Applied Non-Classical Logic*, pages 201–224, 1992.
3. D M Gabbay and I M Hodkinson. An axiomatisation of the temporal logic with until and since over the real numbers. *Journal of Logic and Computation*, 1(2):229 – 260, 1990.
4. D.M. Gabbay, A. Pnueli, S. Shelah, and J. Stavi. On the temporal analysis of fairness. In *7th ACM Symposium on Principles of Programming Languages, Las Vegas*, pages 163–173, 1980.

5. D M Gabbay, I M Hodkinson, and M A Reynolds. Temporal expressive completeness in the presence of gaps. In *Proceedings ASL European Meeting 1990*, Lecture Notes in Logic. Springer-Verlag, 1993.
6. D Gabbay, I Hodkinson, and M Reynolds. *Temporal Logic: Mathematical Foundations and Computational Aspects, Vol. 1.* OUP, to appear 1994.
7. D M Gabbay. An irreflexivity lemma with applications to aximatizations of conditions on tense frames. In U Monnich, editor, *Aspects of Philosophical Logic*, pages 67–89. Reidel, Dordrecht, 1981.
8. J Kamp. *Tense Logic and the theory of linear order.* PhD thesis, Michigan State University, 1968.
9. M Reynolds. An axiomatization for Until and Since over the reals without the IRR rule. *Studia Logica*, 51:165–194, 1992.
10. J G Rosenstein. *Linear orderings.* Academic Press, New York, 1982.
11. J van Bentham. *The Logic of Time.* Reidel, Dordrecht, 1983.
12. Y Venema. Completeness via completeness. In M de Rijke, editor, *Colloquium on Modal Logic, 1991.* ITLI-Network Publication, Instit. for Lang., Logic and Information, University of Amsterdam, 1991.
13. Ming Xu. On some U, S-Tense Logics. *Journal of Philosophical Logic*, 17:181–202, 1988.

Temporal Logic with Reference Pointers

Valentin Goranko *

Department of Mathematics, University of the North, QwaQwa Campus,
Private Bag X13, Phuthaditjhaba 9866, South Africa

Abstract. An extension of the propositional temporal language is introduced with a simple syntactic device, called "reference pointer" which provides for making references within a formula to "instants of reference" specified in the formula. The language with reference pointers $\mathcal{L}_{trp}$ has a great expressive power (e.g. Kamp's and Stavi's operators as well as Prior's clock variables are definable in it), especially compared to its frugal syntax, perspicuous semantics and simple deductive system. The minimal temporal logic $\mathbf{K}_{trp}$ of this language is axiomatized and strong completeness theorem is proved for it and extended to an important class of extensions of $\mathbf{K}_{trp}$. The validity in $\mathcal{L}_{trp}$ is proved undecidable.

1 Introduction

The rapidly expanding scope of applications (actual or potential) of temporal logic to theoretical computer science and artificial intelligence demands, inter alia, strengthening of the expressive power of the temporal language to make it a really appropriate tool for adequate treatment of the various temporal phenomena, while keeping a relatively simple and efficient mechanism for derivations, convenient for applications and for implementation of automated deduction systems. This demand is particularly relevant for the propositional temporal languages. Their most valuable assets are the perspicuity of the syntax and deductive apparatus on the one hand, and on the other hand their *intensional* semantic nature which allows for representation of sophisticated first- or higher-order schemata on a propositional syntactic level. These two assets are in mutual controversy, reflecting the fundamental controversy in logic: "expressiveness vs. tractability". A number of temporal languages and systems have been devised in seeking the best compromise in this controversy.

This article is intended as a further contribution in this trend. It proposes a rather simple but particularly strong extension of the propositional temporal language $\mathcal{L}_t$. As it usually happens, this enterprise was motivated from dissatisfaction with the expressiveness of $\mathcal{L}_t$. One of the major drawbacks, at least from point of view of natural language, is its lack of means to make references to

---

* This article was supported by a research grant GUN 2019536 of the Foundation for Research Development of South Africa. The author is also grateful to Patrick Blackburn for the stimulating discussion at the beginning of this research, to Maarten de Rijke for some suggestions, and to Tinko Tinchev for supplying him with LaTeX/EmTeX.

particular points in time, somehow specified in formal context. Simply speaking, one just cannot say "now", "then", "when" in the classical temporal language. (For thorough discussion on "now" and "then" the reader is referred to e.g. [13] and [19]). This is sometimes particularly annoying because, as a consequence, very basic and natural features of the temporal frame cannot be expressed syntactically. To give just one example, the simple fact that "*now* will not occur again", or formally, that the flow of time is irreflexive, is beyond the expressive abilities of $\mathcal{L}_t$. Various enrichments of this language have been proposed to improve its expressiveness. Let us mention three basic types of enrichments. The first one extends the language with new particular *operators* intended to express and formalize specific temporal phenomena, e.g. Kamp's binary temporal modalities *Since* and *Until* and the more sophisticated Stavi's connectives $\mathbf{U}'(p,q)$ and $\mathbf{S}'(p,q)$ (see e.g. [4, 5] and [7]; see the latter as well for a general approach to this type of enrichments of the temporal language). The second one provides the language with new *sorts* of syntactic objects, (constants, variables etc.) having specific interpretation in the temporal framework and thus increases generally the expressive power of the language. A characteristic example are Prior's "clock-variables" ([14]; see also [3], the *nominals* in [2], and the *names* in [9]) which are bound to be true at *exactly one instant* of the flow of time. And the third one increases the expressiveness of the language by adding more rules of inference intended to depict semantic features not expressible by means of formulae. A notable example here is Gabbay's "irreflexivity rule" (see [6]).

What is proposed in this article can be atributed to each of these types of enrichments of the language. We extend the language with a specific syntactic device intended to enable making references to points in time, but the result turns out to be a significant general improvement of the expressiveness of the language. The idea in nutshell is this: we want to specify an instant in our formal temporal expression, to which we want to be able to make references further. In order to do that, when reach our *instant of reference*, we say "*now*" (i.e. fix a *point of reference*), and further, when we want to make reference to that "*now*", we put a *reference pointer* i.e. say "*then*". Schematically it looks like this: ...now(... then ... then ... then ...)..., which we formally express like this: ... ↓ (... ↑ ... ↑ ... ↑ ...) ... . This construction can be iterated, e.g ... ↓ (... ↑ ... ↓ (... ↑ ...) ... ↑ ...) ... etc. The language can be further extended with more than one sets for reference "now-then" but even as such it is rich enough both to express Kamp's and Stavi's operators, and to simulate Prior's clock-variables. This makes it an appropriate medium for formalization of various particular systems for linear and branching time; a few basic examples are given at the end.

In the article we introduce the extended language $\mathcal{L}_{trp}$, its semantics, briefly comment on the expressiveness and prove that the satisfiability problem is undecidable. Then we give an axiomatization of the basic temporal logic $\mathbf{K}_{trp}$ for which we prove a strong completeness theorem and generalize this result to an important class of extensions of $\mathbf{K}_{trp}$. The reader is assumed to have a background in propositional temporal logics (syntax, semantics, deductive systems and completeness theorem) within the bounds of either of e.g. [1, 5, 11]. Following a referee's suggestion, a reference is also made to [16] (unfortunately not accessible to the author) where reportedly similar ideas have been discussed.

2 Syntax

The language $\mathcal{L}_t$ of the propositional temporal logic contains a countable set $P = \{p_1, p_2, \ldots\}$ of propositional variables, logical constants $\bot, \top$, connectives $\neg$ and $\wedge$, and temporal modalities **G** ("always in the future") and **H** ("always in the past"). The symbols $\vee, \rightarrow, \leftrightarrow, \mathbf{F}$ ("sometime in the future") and **P** ("sometime in the past") are definable in a standard way.

We extend the language $\mathcal{L}_t$ with three additional symbols:

- *universal modality* **A** ("always"), whose dual $\neg\mathbf{A}\neg$ is denoted by **E** ("sometime");
- *reference pointer* $\uparrow$ ("then"), and
- *point of reference* $\downarrow$.

The resulting language is denoted by $\mathcal{L}_{trp}$. The symbol $\uparrow$ behaves syntactically like a propositional variable, while $\downarrow$ is an unary connective which resembles a quantifier binding $\uparrow$. The recursive definition of formulae in $\mathcal{L}_t$ is extended for $\mathcal{L}_{trp}$ with the following clauses:

- ($\uparrow$) $\uparrow$ is a formula;
- (**A**) If φ is a formula then $\mathbf{A}\varphi$ is a formula;
- ($\downarrow$) If φ is a formula then $\downarrow\varphi$ is a formula.

We need a few syntactic notions borrowed from first-order logic:

The first occurrence of $\downarrow$ in the formula $\downarrow\varphi$ has a *scope* φ.

An occurrence of $\uparrow$ in a formula φ is *bound* if it is in the scope of an occurrence of $\downarrow$; otherwise it is *free*.

If φ and ψ are formulae, $\varphi(\psi/\uparrow)$ denotes the result of simultaneous substitution of all free occurrences on $\uparrow$ in φ by ψ.

A formula φ is *closed* if there are no free occurrences of $\uparrow$ in φ.

Complexity of a formula φ of $\mathcal{L}_{trp}$ is the number of logical connectives (including **A**, $\uparrow$ and $\downarrow$) in φ.

Reference depth of a formula φ is the largest number $r(\varphi)$ of nested occurrences of $\downarrow$ in φ.

3 Semantics

We shall deal only with relational semantics for propositional temporal logic. The notions of (temporal) *frame*, *valuation* and *model* are the standard ones. Given a model $\mathcal{M} = <T, R, V>$ and a point (*instant*) $t \in T$ we have a recursive definition of truth $\mathcal{M} \models \varphi[t]$ for all formulae of $\mathcal{L}_t$, which we want to extend over the new symbols. However, we have no suitable way to define truth at an instant for non-closed formulae, since this truth would depend on the "instant of reference" which is not determined if the formula is not closed. To avoid this obstacle we choose another way to define truth at an instant, viz. by translation of the formulae of $\mathcal{L}_{trp}$ into the first-order language L_1 containing a binary predicate R and a countable set of unary predicates $\{P_1, P_2, \ldots\}$. For convenience we list the set of individual variables of L_1 as $\{x, w, y_0, y_1, y_2, \ldots\}$ since x and w will play special roles, viz. x will represent the actual point in time (the current "now"), and w will represent the instant of reference ("then"). Now we extend the translation denoted by ST in [1] as follows:

1. $ST(p_i) = P_i x$,
2. $ST(\uparrow) = x = w$,
3. $ST(\neg \varphi) = \neg ST(\varphi)$,
4. $ST(\varphi \wedge \psi) = ST(\varphi) \wedge ST(\psi)$,
5. $ST(\mathbf{G}\varphi) = \forall y (Rxy \rightarrow ST(\varphi)(y/x))$,
6. $ST(\mathbf{H}\varphi) = \forall y (Ryx \rightarrow ST(\varphi)(y/x))$,
7. $ST(\mathbf{A}\varphi) = \forall y (ST(\varphi)(y/x))$,
8. $ST(\downarrow \varphi) = ST(\varphi)(x/w)$.

In 5, 6, and 7 above y is the first variable different from x and w, which does not occur in $ST(\varphi)$; u/v means uniform substitution of u for all free occurrences of v.

Note that x and w can only have free occurrences in $ST(\varphi)$, where they are the only possibly free variables. Furthermore, φ is closed if and only if w does not occur in $ST(\varphi)$.

The model $\mathcal{M} = <T, R, V>$ can be regarded as an L_1-model where R is interpreted by R and P_i by $V(p_i), i = 0, 1, 2, \ldots$. In order to distinguish validity in $\mathcal{M}$ as an L_1-model from validity in $\mathcal{M}$ as a temporal model we shall use the symbol $\Vdash$ for the former case and $\models$ for the latter. Now we define for any *closed* formula φ:

$$\mathcal{M} \models \varphi[t] \text{ if } \mathcal{M} \Vdash ST(\varphi)(t/x),$$

and

$$\mathcal{M} \models \varphi \text{ if } \mathcal{M} \models \varphi[t] \text{ for every } t \in T, \text{ i.e. if } \mathcal{M} \Vdash \forall x ST(\varphi).$$

Finally, φ is *valid in a frame* if it is valid in every model on the frame, and φ is (*universally*) *valid* if it is valid in every temporal frame.

We only define validity for closed formulae since only they have a determined meaning. Hereafter we shall not be interested in non-closed formulae.

4 Notes on Expressiveness and Definability in $\mathcal{L}_{trp}$

The language $\mathcal{L}_{trp}$ has a great expressive power. Here we shall not give complete characterization of its expressiveness (this will be done elsewhere) but shall only present a few eloquent testimonials for it.

1. Various postulates for temporal frames, which are beyond the scope of $\mathcal{L}_t$ are readily expressed in $\mathcal{L}_{trp}$. Just two simple examples:
   - irreflexivity: $F \Vdash \forall x \neg Rxx \Leftrightarrow F \models \downarrow\mathbf{G}\neg\uparrow$,
   - antisymmetry: $F \Vdash \forall x \forall y (x < y \to \neg y < x) \Leftrightarrow F \models \downarrow\mathbf{GG}\neg\uparrow$.
2. A number of modalities not definable in $\mathcal{L}_t$ can be easily defined in $\mathcal{L}_{trp}$. A simple but important example is the *difference modality* $[\neq]$ (see e.g. [17]):

$$[\neq]\varphi = \downarrow\mathbf{A}(\neg\uparrow \to \varphi).$$

Kamp's $\mathbf{S}(p,q)$ (*Since*) and $\mathbf{U}(p,q)$ (*Until*) and Stavi's $\mathbf{U}'(p,q)$ and $\mathbf{S}'(p,q)$ are explicitly definable here, too:

$$\mathbf{U}(p,q) = \downarrow\mathbf{F}(p \wedge \mathbf{H}(\mathbf{P}\uparrow \to q)) \text{ and } \mathbf{S}(p,q) = \downarrow\mathbf{P}(p \wedge \mathbf{G}(\mathbf{F}\uparrow \to q));$$

$\mathbf{U}'(p,q) =$

$\downarrow\mathbf{FH}(\mathbf{P}\uparrow \to q) \wedge \neg\mathbf{U}(\neg q \vee \neg\mathbf{U}(\top, q), q) \wedge \downarrow\mathbf{F}(\neg q \wedge p \wedge \mathbf{H}(\mathbf{P}\uparrow \wedge \mathbf{P}(\mathbf{P}\uparrow \wedge \neg q) \to p))$

and likewise for $\mathbf{S}'(p,q)$.
3. The idea of *clock-variables* or *names for instants* can be adequately formalized in $\mathcal{L}_{trp}$ without introducing a separate sort for variables: The formula $\downarrow\mathbf{A}(p \leftrightarrow \uparrow)$ says "p is valid only now", and accordingly, $\mathbf{E}\downarrow\mathbf{A}(p \leftrightarrow \uparrow)$ means "p is valid at exactly one instant". Thus, in the consequent of the formula $\mathbf{E}\downarrow A(p \leftrightarrow \uparrow) \to \varphi(p, \ldots)$ the variable p plays a role of a clock-variable. This fact will be essentially exploited in our axiomatic system.

As we showed above, $\mathcal{L}_{trp}$ is at least as strong as a temporal language with difference modality or clock-variables (these two are equivalent with respect to definability, shown in [9]). Therefore some results about these two languages (see [9, 17]) hold for $\mathcal{L}_{trp}$, too:

- every finite frame is described up to isomorphism in $\mathcal{L}_{trp}$.
- all universal sentences in the monadic second-order language for R and $=$ are definable in $\mathcal{L}_{trp}$.

In fact, $\mathcal{L}_{trp}$ is stronger than any of the above mentioned languages. Indeed, the formula

$$\mathbf{A}(\mathbf{F}\top \wedge \downarrow\mathbf{G}\neg\uparrow) \to \mathbf{E}\downarrow\mathbf{E}(\mathbf{FF}\uparrow \wedge \mathbf{G}\neg\uparrow)$$

says that if a frame is irreflexive and every point has a successor then it is not transitive. It is valid in every *finite* model but not in the frame $<\mathcal{N},<>$. Therefore this condition is not definable in any of those languages, since their minimal logics enjoy the finite model property. Moreover, as one could expect about such a powerful language, the set of valid formulae in $\mathcal{L}_{trp}$ is not recursive.

Theorem 1. *The satisfiability problem in $\mathcal{L}_{trp}$ is Π_1^0-complete.*

Proof. We show that the *unbounded tiling problem* for $\mathcal{N} \times \mathcal{N}$, known to be Π_1^0-complete (see [12]), is reducible to the satisfiability problem in $\mathcal{L}_{trp}$. The idea for doing this we borrow from [18].

First we define a formula GRID which is suppose to set the grid for tiling:

$$GRID = (p \wedge q) \wedge \varphi_1 \wedge \varphi_2 \wedge \varphi_3$$

where:

$$\begin{aligned}
\varphi_1 = \mathbf{A}(&(p \wedge q \to \mathbf{F}(p \wedge \neg q) \wedge \mathbf{F}(\neg p \wedge q) \wedge \mathbf{G}((p \wedge \neg q) \vee (\neg p \wedge q))) \wedge \\
&(p \wedge \neg q \to \mathbf{F}(p \wedge q) \wedge \mathbf{F}(\neg p \wedge \neg q) \wedge \mathbf{G}((p \wedge q) \vee (\neg p \wedge \neg q))) \wedge \\
&(\neg p \wedge q \to \mathbf{F}(\neg p \wedge \neg q) \wedge \mathbf{F}(p \wedge q) \wedge \mathbf{G}((\neg p \wedge \neg q) \vee (p \wedge q))) \wedge \\
&(\neg p \wedge \neg q \to \mathbf{F}(\neg p \wedge q) \wedge \mathbf{F}(p \wedge \neg q) \wedge \mathbf{G}((\neg p \wedge q) \vee (p \wedge \neg q)))),
\end{aligned}$$

$$\begin{aligned}
\varphi_2 = \mathbf{A} \downarrow (&(p \wedge q \to \mathbf{A}(\mathbf{F}\uparrow \to \mathbf{G}(p \wedge q \to \uparrow))) \wedge \\
&(p \wedge \neg q \to \mathbf{A}(\mathbf{F}\uparrow \to \mathbf{G}(p \wedge \neg q \to \uparrow))) \wedge \\
&(\neg p \wedge q \to \mathbf{A}(\mathbf{F}\uparrow \to \mathbf{G}(\neg p \wedge q \to \uparrow))) \wedge \\
&(\neg p \wedge \neg q \to \mathbf{A}(\mathbf{F}\uparrow \to \mathbf{G}(\neg p \wedge \neg q \to \uparrow)))),
\end{aligned}$$

$$\begin{aligned}
\varphi_3 = \mathbf{A} \downarrow (&(p \wedge q \to \mathbf{A}((\neg p \wedge \neg q \wedge \mathbf{FF}\uparrow) \to \mathbf{GG}(p \wedge q \to \uparrow))) \wedge \\
&(p \wedge \neg q \to \mathbf{A}((\neg p \wedge q \wedge \mathbf{FF}\uparrow) \to \mathbf{GG}(p \wedge \neg q \to \uparrow))) \wedge \\
&(\neg p \wedge q \to \mathbf{A}((p \wedge \neg q \wedge \mathbf{FF}\uparrow) \to \mathbf{GG}(\neg p \wedge q \to \uparrow))) \wedge \\
&(\neg p \wedge \neg q \to \mathbf{A}((p \wedge q \wedge \mathbf{FF}\uparrow) \to \mathbf{GG}(\neg p \wedge \neg q \to \uparrow)))).
\end{aligned}$$

The formula $\varphi_1 \wedge \varphi_2 \wedge \varphi_3$ says that every point of the model has exactly two successors; at one of them the valuation of p changes and the valuation of q remains the same (that would be the move "to the right"), while at the other (the move "upwards") the opposite happens. Moreover, by φ_3, the routes "right;up" and "up;right" converge. That will be enough to embed a copy of $\mathcal{N} \times \mathcal{N}$ into any model of GRID.

Now, consider a tiling problem with a set of tiles $T = \{t_1, ..., t_m\}$ and colours $C = \{c_1, ..., c_k\}$. Every tile has four sides: "up", "down", "left" and "right", each coloured in one of the colours from C. To every colour c_i we assign four propositional variables u_i ("up"), d_i ("down"), l_i ("left"), and r_i ("right"). Each tile t with sides "up", "down", "left" and "right" coloured respectrively in $c_{i_1}, c_{i_2}, c_{i_3}$, and c_{i_4}, we represent by the formula

$$\theta_t = (u_{i_1} \wedge \bigwedge_{j \neq i_1} \neg u_j) \wedge (d_{i_2} \wedge \bigwedge_{j \neq i_2} \neg d_j) \wedge (l_{i_3} \wedge \bigwedge_{j \neq i_3} \neg l_j) \wedge (r_{i_4} \wedge \bigwedge_{j \neq i_4} \neg r_j).$$

Now we define the formulae

$$COVER_T = \mathbf{A}(\bigvee_{i=1}^{m} \theta_i),$$

which says that the model is properly tiled, i.e. every point in the model is covered by exactly one tile (note that θ_i and θ_j are incompatible when $i \neq j$);

$$MATCHUP = \mathbf{A}(\bigwedge_{i=1}^{k} (u_i \to (p \wedge q \to \mathbf{G}(p \wedge \neg q \to d_i)) \wedge$$
$$(p \wedge \neg q \to \mathbf{G}(p \wedge q \to d_i)) \wedge$$
$$(\neg p \wedge q \to \mathbf{G}(\neg p \wedge \neg q \to d_i)) \wedge$$
$$(\neg p \wedge \neg q \to \mathbf{G}(\neg p \wedge q \to d_i)))),$$

which says that the colour "up" of each tile of the cover matches the colour "down" of the one above it;

$$MATCHRIGHT = \mathbf{A}(\bigwedge_{i=1}^{k} (r_i \to (p \wedge q \to \mathbf{G}(\neg p \wedge q \to l_i)) \wedge$$
$$(p \wedge \neg q \to \mathbf{G}(\neg p \wedge \neg q \to l_i)) \wedge$$
$$(\neg p \wedge q \to \mathbf{G}(p \wedge q \to l_i)) \wedge$$
$$(\neg p \wedge \neg q \to \mathbf{G}(p \wedge \neg q \to l_i)))),$$

which says that the colour "right" of each tile of the cover matches the colour "left" of the one to the right of it.

Finally, we put

$$\Phi_T = GRID \wedge COVER_T \wedge MATCHUP \wedge MATCHRIGHT.$$

We claim that Φ_T is satisfiable if and only if $\mathcal{N} \times \mathcal{N}$ can be properly tiled by T.

Indeed, if $\mathcal{N} \times \mathcal{N}$ can be tiled by T we can define a model $\mathcal{M} = <\mathcal{N} \times \mathcal{N}, R, V>$ where

$<m_1, n_1> R <m_2, n_2>$ iff $m_2 = m_1 + 1, n_2 = n_1$ or $m_2 = m_1, n_2 = n_1 + 1$;

$$V(p) = \{<2m, n>: m, n \in \mathcal{N}\}, \quad V(q) = \{<m, 2n>: m, n \in \mathcal{N}\},$$

$$V(u_i) = \{<m, n>: \text{ the "up" colour of the tile at } <m, n> \text{ is } c_i\},$$

and likewise for $d_i, l_i,$ and r_i.
Then,
$$\mathcal{M} \models \Phi_T[<0,0>].$$

Conversely, if for some model $\mathcal{M} = <W, R, V>$, $\mathcal{M} \models \Phi_T[x]$, we define a mapping $f: \mathcal{N} \times \mathcal{N} \to W$ as follows: $f(0,0) = x$; suppose that $f(m,n)$ is defined, then $f(m+1,n)$ is the unique "right"-successor of $f(m,n)$, and $f(m,n+1)$ is the unique "up"-successor of $f(m,n)$. The tiling of $\mathcal{N} \times \mathcal{N}$ is now determined by COVER: for any $<m,n>$ there is a unique θ_i which is true at $f(m,n)$; then we put the tile t_i at $<m,n>$. Due to MATCHUP and MATCHRIGHT the tiling is a proper one, and this completes the proof. □

5 Deductive Systems in $\mathcal{L}_{trp}$

Here we axiomatize the basic logic $\mathbf{K}_{trp}$ of $\mathcal{L}_{trp}$ as follows:
Axioms:

$A1$) The axioms of the basic temporal logic $\mathbf{K}_t$, written over propositional variables. .
$A2$) $\mathbf{S5(A)}$
$A3$) $\mathbf{A}p \to (\mathbf{G}p \wedge \mathbf{H}p)$
$A4$) $\downarrow\uparrow$
$A5$) $\downarrow \mathbf{A}(\uparrow \leftrightarrow p) \to (q \to \mathbf{A}(p \to q))$
$A6$) $\downarrow \mathbf{A}(\uparrow \leftrightarrow p) \to (\downarrow\psi \leftrightarrow \psi(p/\uparrow))$,

where p, q are propositional variables.
Rules:

1. **MP:**
$$\frac{\varphi, \varphi \to \psi}{\psi};$$

2. **NEC$_\mathbf{A}$:**
$$\frac{\varphi}{\mathbf{A}\varphi};$$

3. **CLSUB:**
$$\frac{\varphi}{clsub(\varphi)},$$

where $clsub(\varphi)$ is a result of uniform substitution of *closed* formulae for propositional variables in φ.

4. **WITNESS:**
$$\frac{\downarrow \mathbf{A}(\uparrow \leftrightarrow p) \to \varphi \text{ for every propositional variable } p}{\varphi}.$$

Note the following:

- Only closed formulae are deducible in $\mathbf{K}_{trp}$.
- In the presence of **CLSUB** the infinitary rule **WITNESS** can be replaced by a finitary version:

$$\textbf{WITNESS}_f : \frac{\downarrow \mathbf{A}(\uparrow \leftrightarrow p) \to \varphi \text{ for some prop. variable } p \text{ not occurring in } \varphi}{\varphi}$$

- The rules
$$\text{NEC}_\mathbf{G} : \frac{\varphi}{\mathbf{G}\varphi} \quad \text{and} \quad \text{NEC}_\mathbf{H} : \frac{\varphi}{\mathbf{H}\varphi}$$
are derivable from $\text{NEC}_\mathbf{A}$, MP and A3.

Here are some important theorems of $\mathbf{K}_{trp}$:

- $(t1)\ \downarrow\varphi \leftrightarrow \varphi$ for every closed formula φ;
- $(t2)\ \downarrow\neg\varphi \leftrightarrow \neg\downarrow\varphi$;
- $(t3)\ \downarrow(\varphi \wedge \psi) \leftrightarrow (\downarrow\varphi \wedge \downarrow\psi)$;
- $(t4)\ \downarrow\mathbf{A}(\uparrow\leftrightarrow p) \to p$.

We exemplify derivations in $\mathbf{K}_{trp}$ sketching a proof of $(t2)$:

1. $\downarrow\mathbf{A}(\uparrow\leftrightarrow p) \to (\downarrow\neg\varphi \leftrightarrow \neg\varphi(p/\uparrow))$ by A6,
2. $\downarrow\mathbf{A}(\uparrow\leftrightarrow p) \to (\downarrow\varphi \leftrightarrow \varphi(p/\uparrow))$ by A6,
3. $\downarrow\mathbf{A}(\uparrow\leftrightarrow p) \to (\neg\downarrow\varphi \leftrightarrow \neg\varphi(p/\uparrow))$ by 2 and contrapositions,
4. $\downarrow\mathbf{A}(\uparrow\leftrightarrow p) \to (\neg\downarrow\varphi \leftrightarrow \downarrow\neg\varphi)$ by 1 and 3,
5. $\neg\downarrow\varphi \leftrightarrow \downarrow\neg\varphi$ by 4 and **WITNESS**.

The other derivations are similar.

Remark. 1. The reason we design our axiomatic system for closed formulae is that we have defined validity only for these formulae. Besides that, however, the formulae which are not closed behave rather irregularly. For instance, the rule for equivalent replacement

$$\frac{\varphi \leftrightarrow \psi}{\theta(\varphi/p) \leftrightarrow \theta(\psi/p)},$$

applied to such formulae, does not always preserve validity in a frame. The same happens with the substitution rule $\frac{\varphi}{sub(\varphi)}$, so that special provisions must be made, like "a formula free for substitution in a formula" etc. which would lead to the typical complications of the first-order machinery.

2. The intuition behind the rule **WITNESS** (which has a number of ancestors, e.g. some versions of rules for quantifiers in first-order logic, the "irreflexivity rule" in [6], COV in [9] etc.) is the following. If a formula is not valid, then it is false at some instant t of some model $\mathcal{M}$. Then, a propositional variable p can be made a "t o'clock-variable" i.e. evaluated to be true exactly at that instant t, and then p will be a "witness" of the falsity of φ. Therefore if all clock-variables testify that φ is true at the instants in which they live, then φ must be valid.

3. Although **WITNESS** and **WITNESS**$_f$ infer the same theorems, they yield deductive systems different with respect to logical consequence which is compact in the system with **WITNESS**$_f$, but not in the system with **WITNESS**.

Amongst the basic syntactic properties of $\mathbf{K}_t$ which are transferred, mutatis mutandis, to $\mathbf{K}_{trp}$ we only mention the following

Lemma 2. *For any closed formulae α, β, φ and variable p, if*

$$\mathbf{K}_{trp} \vdash \alpha \leftrightarrow \beta$$

then

$$\mathbf{K}_{trp} \vdash \varphi(\alpha/p) \leftrightarrow \varphi(\beta/p).$$

Proof. (Sketch) Induction on the reference depth $r(\varphi)$.

If $r(\varphi) = 0$, the proof repeats the standard one for $\mathbf{K}_t$.

If $r(\varphi) = r > 0$ we assume that for all formulae with reference depth less than r the statement holds, and then do induction on the complexity of φ. The interesting case is $\varphi = \downarrow \theta$. Here we apply axiom $A6$ (with a variable q not occurring in φ, α, β) which replaces $\varphi(\alpha/p)$ by $\{\theta(\alpha/p)\}(q/\uparrow) = \{\theta(q/\uparrow)\}(\alpha/p)$ and likewise for $\varphi(\beta/p)$. The resulting formulae have reference depths lesser than r. Applying the inductive hypothesis for them, followed by application of **WITNESS**$_f$, completes the induction. □

Theorem 3 (Soundness theorem). *1. All axioms of $\mathbf{K}_{trp}$ are valid.*
2. All rules of $\mathbf{K}_{trp}$ preserve validity in a frame, and hence preserve universal validity.

Proof. 1) For A1 the result comes from $\mathbf{K}_t$; for A2 from $\mathbf{S5}$; for A3 and A4 it is quite simple. As for A5, it is enough to note that
$\forall x ST(\downarrow \mathbf{A}(\uparrow \leftrightarrow p_i) \to (p_j \to \mathbf{A}(p_i \to p_j))) =$
$\forall x (\forall y (y = x \leftrightarrow P_i y) \to (P_j x \to \forall y (P_i y \to P_j y)))$ which is universally valid.
Finally, take $A6$. $ST(\psi(p/\uparrow))$ is obtained from $ST(\psi)$ by replacing all occurrences of the kind $y = w$ (which are the only possible occurrences of w in $ST(\psi)$) by $P_i y$. Due to the antecedent $\forall y (y = x \leftrightarrow P_i y)$ this is equivalent to replace all occurrences of $y = w$ by $y = x$. The result of this substitution is exactly $ST(\psi(x/w)) = ST(\downarrow \psi))$.

2) The only interesting case is the rule **WITNESS**. Suppose that for some model $<T, V>$ and instant $t \in T$, $<T, V> \not\models \varphi[t]$. Choose a variable p not occurring in φ and change the valuation V to V' as follows: $V'(p) = \{t\}$, and V' coincides with V elsewhere. Then $<T, V'> \models \downarrow \mathbf{A}(\uparrow \leftrightarrow p)[t]$ and $<T, V'> \not\models \varphi[t]$, hence $<T, V'> \not\models \downarrow \mathbf{A}(\uparrow \leftrightarrow p) \to \varphi[t]$. □

Now we set ourselves to prove completeness of $\mathbf{K}_{trp}$. Basically we follow an elaborated version of the traditional in modal and temporal logic "canonical model technique", further developed in [15, 9]. The basic steps of the proof will be scrupulously outlined in a series of lemmata, but the standard details in their proofs will be usually omitted.

First we introduce another syntactic notion (originating from the *admissible forms* in [10]; see also [9]). Let $*$ be a symbol not belonging to $\mathcal{L}_{trp}$. We define recursively *universal forms* of $*$ as follows:

1. $*$ is a universal form of $*$.
2. If $u(*)$ is a universal form of $*$ and φ is a closed formula then $\varphi \to u(*)$, $\mathbf{G}u(*)$, $\mathbf{H}u(*)$ and $\mathbf{A}u(*)$ are universal forms of $*$.

Every universal form of $*$ can be represented (up to tautological equivalence) in a uniform way:

$$u(*) = \varphi_0 \to \mathbf{L}_1(\varphi_1 \to \ldots \mathbf{L}_k(\varphi_k \to *)\ldots)$$

where $\mathbf{L}_1, \ldots, \mathbf{L}_k \in \{\mathbf{A}, \mathbf{G}, \mathbf{H}\}$ and some of $\varphi_1, \ldots, \varphi_k$ may be $\top$ if necessary.

For every universal form $u(*)$ and a formula θ we denote by $u(\theta)$ the result of substitution of θ for $*$ in $u(*)$. Obviously, if θ is a closed formula then $u(\theta)$ is a closed formula, too.

Now we introduce the rule

$$\mathbf{WITNESS}_U : \frac{u(\downarrow\mathbf{A}(\uparrow\leftrightarrow p) \to \varphi)}{u(\varphi)} \quad \text{for every propositional variable } p$$

where u is an arbitrarily fixed universal form.

Although $\mathbf{WITNESS}_U$ seems much stronger than $\mathbf{WITNESS}$, they are in fact equivalent. The key observation for that (see [8]) is the following: given a universal form

$$u(*) = \varphi_0 \to \mathbf{L}_1(\varphi_1 \to \ldots \mathbf{L}_k(\varphi_k \to *)\ldots)$$

we define a form

$$u'(*) = \neg * \to (\varphi_k \to \mathbf{L}'_k(\varphi_{k-1} \to \ldots \mathbf{L}'_1 \neg \varphi_0)\ldots),$$

where $\mathbf{A}' = \mathbf{A}, \mathbf{G}' = \mathbf{H}$ and $\mathbf{H}' = \mathbf{G}$. Now, for every closed formula θ, $u(\theta)$ is deductively equivalent to $u'(\theta)$ in the sense that $\mathbf{K}_{trp} + u'(\theta) \vdash u(\theta)$ and $\mathbf{K}_{trp} + u(\theta) \vdash u'(\theta)$.

Definition 4. 1. A *theory* in $\mathcal{L}_{trp}$ is a set of closed formulae of $\mathcal{L}_{trp}$, which contains all theorems of $\mathbf{K}_{trp}$ and is closed with respect to $\mathbf{MP}$.
2. A *W-theory (witnessed theory)* is a theory in $\mathcal{L}_{trp}$ which is closed with respect to $\mathbf{WITNESS}_U$.

Note that for every set of closed formulae Γ there is a minimal W-theory $\mathrm{WTh}(\Gamma)$ /a minimal theory $\mathrm{Th}(\Gamma)$/ containing Γ. Indeed, the set of all closed formulae is a W-theory. Furthermore, the intersection of every family of W-theories is a W-theory. Then $\mathrm{WTh}(\Gamma)$ is the intersection of all W-theories containing Γ. Likewise for theories.

Definition 5. A theory /W-theory/ is *consistent* if it does not contain $\bot$. A set of closed formulae Δ is *W-consistent* if $\mathrm{WTh}(\Delta)$ is consistent.

The following property, well-known for theories, hold for W-theories, too.

Lemma 6 (Deduction theorem for W-theories). *If Γ is a W-theory and φ, ψ are closed formulae then $\varphi \to \psi \in \Gamma$ iff $\psi \in \mathrm{WTh}(\Gamma \cup \{\varphi\})$.*

Proof. If $\varphi \to \psi \in \Gamma$ then, by **MP**, $\psi \in \text{WTh}(\Gamma \cup \{\varphi\})$. Vice versa, suppose that $\psi \in WTh(\Gamma \cup \{\varphi\})$ and consider the set

$$\Delta = \{\theta : \theta \text{ is a closed formula and } \varphi \to \theta \in \Gamma\}.$$

We shall prove that Δ is a W-theory containing $\Gamma \cup \{\varphi\}$. The proof goes as in the standard deduction theorem, with one additional step: closedness with respect to **WITNESS**$_U$, which follows form the fact that Γ is a W-theory, and $\varphi \to u(*)$ is a universal form whenever $u(*)$ is. □

Lemma 7. *If Γ is a set of closed formulae in which infinitely many propositional variables have no occurrences then $WTh(\Gamma) = Th(\Gamma)$.*

Proof. It is a standard fact that

$$\text{Th}(\Gamma) = \{\theta : \gamma_1 \wedge \ldots \wedge \gamma_k \to \theta \in \mathbf{K}_{trp} \text{ for some } \gamma_1, \ldots, \gamma_k \in \Gamma\}.$$

Let us show that $\text{Th}(\Gamma)$ is closed with respect to **WITNESS**$_U$. Suppose that for some universal form u, $u(\downarrow \mathbf{A}(\uparrow \leftrightarrow p) \to \varphi) \in \text{Th}(\Gamma)$, i.e. $\gamma_1^i \wedge \ldots \wedge \gamma_{k_i}^i \to u(\downarrow \mathbf{A}(\uparrow \leftrightarrow p_i) \to \varphi) \in \mathbf{K}_{trp}$ for every propositional variable p_i. We can choose a propositional variable p_j which does not occur in either of u, φ, or Γ. Then, substituting in $\gamma_1^j \wedge \ldots \wedge \gamma_{k_j}^j \to u(\downarrow \mathbf{A}(\uparrow \leftrightarrow p_j) \to \varphi)$ any variable p for p_j we obtain $\gamma_1^j \wedge \ldots \wedge \gamma_{k_j}^j \to u(\downarrow \mathbf{A}(\uparrow \leftrightarrow p) \to \varphi) \in \mathbf{K}_{trp}$ for every p. Therefore $\gamma_1^j \wedge \ldots \wedge \gamma_{k_j}^j \to u(\varphi) \in \mathbf{K}_{trp}$ by **WITNESS**$_U$, hence $u(\varphi) \in \text{Th}(\Gamma)$. □

As a corollary, every set of formulae which satisfies the condition of Lemma 7 is W-consistent iff it is consistent.

Definition 8. A W-theory Γ is *maximal* if for every closed formula φ, either $\varphi \in \Gamma$ or $\neg\varphi \in \Gamma$ but not both.

Every maximal W-theory is consistent and cannot be extended to another consistent W-theory. Moreover, every maximal W-theory Γ contains a "witness" $\downarrow\mathbf{A}(\uparrow\leftrightarrow p)$ for some propositional variable p ("Γ o'clock variable"). For otherwise all $\neg\downarrow\mathbf{A}(\uparrow\leftrightarrow p)$ would be in Γ, and hence, by **WITNESS**$_U$, $\bot$ would belong to Γ.

Lemma 9 (Lindenbaum lemma). *Every W-consistent set Γ_0 can be extended to a maximal W-theory.*

Proof. First, note that $\Gamma = \text{WTh}(\Gamma_0)$ is a consistent W-theory. Let $\psi_1, \psi_2, \ldots$ be a list of all closed formulae of $\mathcal{L}_{trp}$ and $u_1, u_2, \ldots$ be a list of all universal forms in $\mathcal{L}_{trp}$. Then we can list all combinations $\{u_i(\psi_j)\}_{i,j=1}^{\infty}$ in a sequence $\theta_1, \theta_2, \ldots$.(Obviously, there will be repetitions in this sequence, but that does not matter). Now we shall define a sequence of consistent W-theories $T_0 \subseteq T_1 \subseteq \ldots$ as follows: $T_0 = \Gamma$; suppose that T_n is defined and consider $\text{WTh}(T_n \cup \{\theta_n\})$. If it is consistent, this is T_{n+1}. Otherwise let $\theta_n = u_i(\psi_j)$. Then $\neg u_i(\psi_j) \in T_n$ by the deduction theorem. Therefore $u_i(\downarrow\mathbf{A}(\uparrow\leftrightarrow p) \to \psi_j)$ does not belong to T_n for some propositional variable p. Then put

$$T_{n+1} = \text{WTh}(T_n \cup \{\neg u_i(\downarrow\mathbf{A}(\uparrow\leftrightarrow p) \to \psi_j)\}). \text{ Finally, put } T = \bigcup_{n=0}^{\infty} T_n.$$

By virtue of the construction, T is a maximal W-theory. □

For any set of formulae Δ we define

$$\mathbf{G}\Delta = \{\varphi : \mathbf{G}\varphi \in \Delta\}, \mathbf{H}\Delta = \{\varphi : \mathbf{H}\varphi \in \Delta\} \text{ and } \mathbf{A}\Delta = \{\varphi : \mathbf{A}\varphi \in \Delta\}.$$

Lemma 10. *If Δ is a maximal W-theory then $\mathbf{G}\Delta, \mathbf{H}\Delta$ and $\mathbf{A}\Delta$ are W-theories.*

Proof. We shall do the proof for $\mathbf{G}\Delta$, the others are analogous.
That $\mathbf{G}\Delta$ contains all theorems of of $\mathbf{K}_{trp}$ and is closed with respect to **MP** is nothing new. $\mathbf{G}\Delta$ is also closed with respect to **WITNESS**$_U$ since Δ is closed and $\mathbf{G}u(*)$ is a universal form whenever $u(*)$ is. □

Lemma 11. *If Δ is a maximal W-theory and $\mathbf{F}\theta \in \Delta$ / $\mathbf{P}\theta \in \Delta$ / $\mathbf{E}\theta \in \Delta$ respectively then there is a maximal W-theory Δ' such that $\theta \in \Delta'$ and $G\Delta \subseteq \Delta'$ / $H\Delta \subseteq \Delta'$ / $A\Delta \subseteq \Delta'$ respectively.*

Proof. By Lemma 10 $\mathbf{G}\Delta$ is a W-theory. Moreover, $\mathbf{G}\neg\theta \notin \Delta$ since Δ is consistent. Therefore $\neg\theta \notin \mathbf{G}\Delta$, hence $\text{WTh}(\mathbf{G}\Delta \cup \{\theta\})$ is consistent. Then, by Lemma 9 it can be extended to a maximal W-theory Δ'. The other cases are analogous. □

Definition 12. A model $< T, R, V >$ is called *clock-model* if for every $t \in T$ there is a "t o'clock-variable" p_t such that $V(p_t) = \{t\}$.

Lemma 13 (Strong completeness theorem for W-consistent sets in $\mathbf{K}_{trp}$).
Every W-consistent set Γ_0 is satisfiable in a clock-model.

Proof. First, we extend Γ_0 to a maximal W-theory Γ. Then we define a *canonical* model $\mathcal{M} = < T, R, V >$ as follows:

- $T = \{\Delta : \Delta \text{ is a maximal } W\text{-theory and } \mathbf{A}\Gamma \subseteq \Delta\}$;
- for any $\Delta_1, \Delta_2 \in T$, $R\Delta_1\Delta_2$ if $\mathbf{G}\Delta_1 \subseteq \Delta_2$;
- for any propositional variable $p, V(p) = \{\Delta \in T : p \in \Delta\}$.

It is a standard task to prove that for any $\Delta_1, \Delta_2 \in T$, $R\Delta_1\Delta_2$ iff $\mathbf{H}\Delta_2 \subseteq \Delta_1$ and that $\mathbf{A}\Delta_1 \subseteq \Delta_2$.

Now we are going to prove that Γ_0 is satisfied at the point Γ of the model $\mathcal{M}$. This will follow from the following sub-lemma:

Truth-lemma: For every closed formula θ and $\Delta \in T$,

$$\mathcal{M} \models \theta[\Delta] \text{ iff } \theta \in \Delta.$$

Proof of the Truth-lemma: Induction on $r(\theta)$. when $r(\theta) = 0$, i.e. θ contains no reference pointers, the proof goes by induction on the complexity of θ and repeats, mutatis mutandis, the proof the the truth-lemma for $\mathbf{K}_t$, as the universal modality **a** is dealt with in the same way as the temporal modalities. Now, let $r(\theta) > 0$ and assume that for all closed formulae with reference depth less than $r(\theta)$ the statement holds. Then we do again an induction on the complexity of θ. The only non-standard case is $\theta = \downarrow\psi$. Let $\downarrow\mathbf{A}(\uparrow\leftrightarrow p)$ be a "witness" in Δ. Then, by axiom $A6$, $\downarrow\psi \leftrightarrow \psi(p/\uparrow) \in \Delta$, i.e. $\downarrow\psi \in \Delta$ iff $\psi(p/\uparrow) \in \Delta$. Since $r(\downarrow\psi) > r(\psi(p/\uparrow))$, by the inductive hypothesis $\mathcal{M} \models \psi(p/\uparrow)[\Delta]$ iff $\psi(p/\uparrow) \in \Delta$. To complete the proof of the lemma it remains to show that $\mathcal{M} \models \psi(p/\uparrow)[\Delta]$ iff $\mathcal{M} \models \downarrow\psi[\Delta]$. Again by axiom $A6$ $\mathcal{M} \models \downarrow\mathbf{A}(\uparrow\leftrightarrow p) \to (\downarrow\psi \leftrightarrow \psi(p/\uparrow))$, hence it is enough to show that $\mathcal{M} \models \downarrow\mathbf{A}(\uparrow\leftrightarrow p)[\Delta]$. $ST(\downarrow\mathbf{A}(\uparrow\leftrightarrow p)) = \forall y(y = x \leftrightarrow Py)$ where P is the unary predicate symbol corresponding to p. Thus, $\mathcal{M} \models \downarrow\mathbf{A}(\uparrow\leftrightarrow p)[\Delta]$ iff $\mathbf{M} \Vdash \forall y(y = \Delta \leftrightarrow Py)$ which means that $V(p) = \{\Delta\}$, i.e. p is a "Δ o'clock variable". Indeed, this is so: First, $p \in \Delta$ by the $\mathbf{K}_{trp}$-theorem (t2) and **MP**. Now, suppose that $p \in \Delta'$ for some $\Delta' \in T$. Take any $\chi \in \Delta$. According to axiom $A4$, $\mathbf{A}(p \to \chi) \in \Delta$, hence $p \to \chi \in \Delta'$, so $\chi \in \Delta'$. Thus $\Delta \subseteq \Delta'$, which implies $\Delta = \Delta'$. So, $p \in \Delta'$ iff $\Delta' = \Delta$. The sub-lemma is proved, which completes the proof of Lemma 13.

□

Theorem 14 (Strong completeness theorem for $\mathbf{K}_{trp}$). *Every consistent set Γ of closed formulae in $\mathcal{L}_{trp}$ is satisfiable.*

Proof. With a simple trick we reduce the theorem to Lemma 13. Let ρ be a "renaming" of the propositional variables in $\mathcal{L}_{trp}$ as follows: $\rho(p_i) = p_{2i+1}, i = 1, 2, \ldots$. If φ is a formula, denote by $\rho(\varphi)$ the result of uniform substitution $\rho(p_i)$ for each p_i in φ, and then put $\rho(\Gamma) = \{\rho(\varphi) : \varphi \in \Gamma\}$. Now, $\rho(\Gamma)$ is a consistent set, since consistency is not affected by the renaming ρ. Furthermore, since the variables with even indices do not occur in formulae of $\rho(\Gamma)$, it is W-consistent by Lemma 7, hence satisfiable at some instant t of a clock-model $<T, R, V>$. Now we define a valuation V' in $<T, R>$ as follows: $V'(p) = V(\rho(p))$. The resulting model $<T, R, V'>$ (which is not necessarily a clock-model) satisfies Γ at t.

□

Definition 15. $\mathcal{L}_{trp}$-*logic* is every simple *closed* extension (extension by means of closed axioms only) of $\mathbf{K}_{trp}$.

The strong completeness theorem is provable, mutatis mutandis, for all $\mathcal{L}_{trp}$-logics:

Theorem 16. *For each $\mathcal{L}_{trp}$-logic $\mathbf{L}$, every consistent in $\mathbf{L}$ set of closed formulae is satisfied in some $\mathbf{L}$-model.*

Of course, a valuable completeness theorem would guarantee satisfiability in a model *over an* **L**-*frame*. Very few general results in that direction exist in modal and temporal logic, but for $\mathcal{L}_{trp}$-logics there is an important one, stated in Theorem 18 below.

For any formula θ we denote $w(\theta) = \mathbf{E}\downarrow\mathbf{A}(\uparrow \leftrightarrow \theta)$.
$w(\theta)$ says that θ is true at exactly one instant of the model.

Definition 17. 1. A formula φ is *witnessed* if it is of the kind

$$\varphi = w(q_1) \wedge \ldots \wedge w(q_k) \to \psi(q_1, \ldots, q_k),$$

where ψ is a closed formula which only contains propositional variables amongst $q_1, \ldots, q_k$. In particular, φ can be a formula without propositional variables.
2. An $\mathcal{L}_{trp}$-logic is *witnessed* if it is axiomatized over $\mathbf{K}_{trp}$ by means of witnessed formulae only.

Theorem 18. *Let* **L** *be a witnessed $\mathcal{L}_{trp}$-logic. Every* **L**-*consistent set of closed formulae is satisfied in some model based on an* **L**-*frame.*

Proof. Given a witnessed formula (which we can assume to be written over $p_1, \ldots, p_k$) $\varphi = w(p_1) \wedge \ldots \wedge w(p_k) \to \psi(p_1, \ldots, p_k)$ we define the first-order formula $\alpha(\varphi) = \forall x \forall z_1 \ldots \forall z_k \sigma(ST(\psi))$, where $z_1, \ldots, z_k$ are variables not occurring in $ST(\psi)$ and $\sigma(ST(\psi))$ is the result of uniform substitution in $ST(\psi)$ of all occurrences of atomic formulae of the kind $P_i y$ by $y = z_i$ respectively, for $i = 1, \ldots, k$. It is not difficult to verify the following: for every temporal frame F, $F \models \varphi$ iff $F \Vdash \alpha(\varphi)$. Now, let φ be an axiom of **L** and $\mathcal{M} = <T, R, V>$ be a clock-model of **L**. Then $\mathcal{M}$ satisfies all variants $w(p_{i_1}) \wedge \ldots \wedge w(p_{i_k}) \to \psi(p_{i_1}, \ldots, p_{i_k})$ of φ since they are theorems of **L**. This implies that $<T, R> \Vdash \alpha(\varphi)$ and hence $<T, R> \models \varphi$. Thus, every clock-model of a witnessed $\mathcal{L}_{trp}$-logic **L** is based on a frame for **L**. Therefore, by Lemma 13 and Theorem 14, every consistent in **L** set of closed formulae is satisfied in a model based on an **L**-frame. □

We conclude with a few simple examples of basic kinds of temporal logics readily axiomatized as witnessed $\mathcal{L}_{trp}$-logics.

Corollary 19. *The following extensions of* $\mathbf{K}_{trp}$ *are strongly complete.*

1. *The* Logic of Linear (irreflexive) Time: $\mathbf{LT}_{trp} = \mathbf{K}_{trp}+$
    - *(irreflexivity)* $\downarrow \mathbf{G}\neg \uparrow$,
    - *(transitivity)* $\downarrow \mathbf{A}(\mathbf{FF} \uparrow \to \mathbf{F} \uparrow)$,
    - *(linearity)* $\downarrow \mathbf{A}(\mathbf{F} \uparrow \vee \uparrow \vee \mathbf{P} \uparrow)$.
2. *The* Logic of Forward Branching Time: $\mathbf{FBT}_{trp} = \mathbf{K}_{trp}+$
    - *(common histories)* $\downarrow \mathbf{APF} \uparrow$,
    - *(linear past)* $\downarrow \mathbf{GH}(\mathbf{F} \uparrow \vee \uparrow \vee \mathbf{P} \uparrow)$,
    - *(proper branching in the future)* $\downarrow \mathbf{HF}\neg \uparrow$.
3. *The* Logic of Discrete Linear Time: $\mathbf{DLT}_{trp} = \mathbf{LT}_{trp}+$
    - *(immediate predecessor)* $\downarrow \mathbf{PGG}\neg \uparrow$,
    - *(immediate successor)* $\downarrow \mathbf{FHH}\neg \uparrow$.

References

1. van Benthem J.F.A.K. *The Logic of Time*, Reidel, Dordrecht, 1983.
2. Blackburn P. (1989), *Nominal Tense Logic*, Dissertation, Centre for cognitive science, University of Edinburgh, 1989.
3. Bull R. (1970), An Approach to Tense Logic, *Theoria*, **36**(3),1970, 282-300.
4. Burgess J., Axioms for Tense Logic I: "Since" and "Until", *Notre Dame Journal of Formal Logic*, **23**(2), 1982, 367-374.
5. Burgess J. Basic Tense Logic, in: *Handbook of Philosophical Logic*, D. Gabbay and F. Guenthner (eds.), Reidel, Dordrecht, vol.II, 1984, 89-133.
6. Gabbay, D., An Irreflexivity Lemma with Applications to Axiomatizations of Conditions on Tense Frames. in: *Aspects of Philosophical Logic*, U. Monnich (ed.). Reidel, Dordrecht, 1981, 67-89.
7. Gabbay, D., Expressive Functional Completeness in Tense Logic, Ibid., 91-117.
8. Gabbay D. & I. Hodkinson, An Axiomatization of the Temporal Logic with Since and Until Over the Real Numbers. *Journal of Logic and Computation*, **1**, 1990, 229-259.
9. Gargov G. & Goranko, V., Modal Logic with Names, *Journal of Philosophical Logic*, **22**(6), 1993, 607-636.
10. Goldblatt R.I., *Axiomatizing the Logic of Computer Programming*, Springer LNCS **130**, 1982.
11. Goldblatt R.I., *Logic of Time and Computation*, CSLI Lecture Notes, No. 7., 1987.
12. Harel D., Recurring dominoes, in: *Foundations of Computing Theory*, LNCS 158, Springer, Berlin,177-194.
13. Kamp J.A.W., Formal Properties of "Now", *Theoria*, **37**, 1971, 227-273.
14. Prior A., *Past, Present, and Future,* Clarendon Press, Oxford, 1967.
15. Passy S. & Tinchev T., An Essay in Combinatory Dynamic Logic, *Information and Computation*, **93**(2), 1991, 263-332.
16. Richards B. et al (?), *Temporal Representation and Inference*, Academic Press, 1989.
17. de Rijke M., The Modal Theory of Inequality, *Journal of Symbolic Logic*, **57**(2), 1992, 566-584.
18. de Rijke M., *Extending Modal Logic*, ILLC Dissertation Series 1993-4, Amsterdam, 1993.
19. Vlach F., *Now and then: a formal study in the logic of tense anaphora*, Dissertation, UCLA, 1973.

Completeness through Flatness in Two-Dimensional Temporal Logic

Yde Venema[1,2]

[1] Faculteit Wiskunde en Informatica, Vrije Universiteit, Amsterdam
[2] Centrum voor Wiskunde en Informatica, P.O. Box 94079, 1090 GB Amsterdam;
e-mail: yde@cwi.nl

Abstract. We introduce a temporal logic TAL and prove that it has several nice features. The formalism is a two-dimensional modal system in the sense that formulas of the language are evaluated at pairs of time points. Many known formalisms with a two-dimensional flavor can be expressed in TAL, which can be seen as the temporal version of square arrow logic.

We first pin down the expressive power of TAL to the three-variable fragment of first-order logic; we prove that this induces an expressive completeness result of 'flat' TAL with respect to monadic first order logic (over the class of linear flows of time).

Then we treat axiomatic aspects: our main result is a completeness proof for the set of formulas that are 'flatly' valid in well-ordered flows of time and in the flow of time of the natural numbers.

1 Introduction

Two-dimensional temporal logic In the last twenty years, various disciplines related to logic have seen the idea arising to develop a framework for modal or temporal logic in which the possible worlds are *pairs* of elements of the model instead of the elements themselves. Often the motivation for developing such *two-dimensional* formalisms stems from a dissatisfaction with the expressive power of ordinary one-dimensional modal or temporal logic. Let us mention three examples of two-dimensional modal logics (we refer to VENEMA [23] for a more substantial overview):

First, in tense logic, there is a research line inspired by linguistic motivations. In the seventies, the development of formal semantics and the strive to give a logical foundation for it, lead people like Gabbay, Kamp and Åqvist (cf. [5] for an overview) to develop two-dimensional modal logics taking care of the linguistic phenomenon that the truth of a proposition may not only change with the time of reference, but also with the time of utterance by the speaker. Second, in artificial intelligence it has been argued that from a philosophical or psychological point of view, it is more natural to consider temporal ontologies where periods of time are the basic entities instead of time points. In some approaches, for instance in HALPERN & SHOHAM [7], modal logics of time intervals have been studied where the possible worlds are intervals; and an interval is identified with the pair consisting of its beginning point and its endpoint. Our last example concerns arrow logic (VENEMA [26]) which is based on the idea that in its semantics, transitions (arrows) do not link the possible worlds, they *are* the possible worlds. Two-dimensional arrow logic arises if we see transitions as a pair consisting of an input and an output state.

Diverse as all these two-dimensional modal formalisms may be in background and nature, they have many aspects in common, one of which is the close connection with algebraic logic (for an introductory overview of algebraic logic we refer to NÉMETI [17]). For instance arrow logic can serve as a tool to study the theory of Tarski's *relation algebras*. Second, and partly related to the first point, the axiomatics of two-dimensional modal logics is not a trivial matter. In particular, if one is interested in a *full square semantics*, i.e. models where *all* pairs of points are admissible as possible worlds, there are few interesting logics that are decidable or finitely axiomatizable by a standard Hilbert-style derivation system.

There are various ways to get around these negative results: for instance in arrow logic, an interesting approach is to drop the constraint that the universe of a model should be a *full* square. The theory of such *relativized* squares may be both decidable and nicely axiomatizable (cf. MARX ET ALII [15] for some examples). For the axiomatizability problem, a different solution was found by Gabbay (cf. [3]); by introducing so-called irreflexivity rules, various logics which are not finitely axiomatizable by standard means, do allow a finite derivation system. Obviously, such rules do not affect the undecidability of the logic. In order to deal with the latter issue, one may take up a third idea, viz. to restrict the *interpretation* of the two-dimensional semantics. A very common constraint is to make the truth of an *atomic* formula at a pair dependent on *one* coordinate of the pair only. Such valuations are called *weak* or *flat*. It is not very difficult to see that with this restriction, two-dimensional modal logic corresponds to *monadic* first-order logic instead of to *dyadic* predicate calculus, and hence decidability follows immediately for various classes of models. This approach originates with the literature on two-dimensional tense logic and is also followed in formalisms like process logic, cf. HAREL, KOZEN & PARIKH [8].

The system TAL Let us now say a few words about the formalism TAL that we investigate in this paper. It is a temporal logic, i.e. its intended models will be flows of time $\mathfrak{T} = (T, <)$, and it is a two-dimensional system: the basic declarative statement of TAL will be of the form

$$\mathfrak{T}, V, s, t \Vdash \phi,$$

where ϕ is a formula, V is a (possibly flat) interpretation function for the propositional variables, and (s,t) is a *pair* of time points.

The position of TAL within the landscape of two-dimensional modal and temporal logics is best explained by noting that it is the *temporal version of arrow logic*. To be precise, the language of TAL is that of arrow logic, extended with a *constant* λ which refers to the ordering relation, i.e.

$$\mathfrak{M}, s, t \Vdash \lambda \iff s < t.$$

An advantage of the presence of this constant is that it makes it easy to express properties of the ordering relation in the language. Compared to other two-dimensional temporal logics, *TAL* is a quite expressive formalism, being able to express the operators of most of the systems that are known from the literature.

Overview The aim of the paper is to look at the expressive power and the axiomatics of *TAL*.

After giving the necessary formal definitions in the next section, we will see in section 3 that, as a consequence of results known from the theory of relation algebras, *TAL* is expressively equivalent to the three-variable fragment of first-order logic with dyadic predicates; as an interesting corollary we can prove that over the class of linear orders *TAL* is *expressively complete* with respect to first order logic with *monadic* predicates. Concerning axiomatics, in section 4 we first prove some completeness results for derivation systems having an irreflexivity rule. These results are easy consequences of similar results obtained for arrow logic, cf. VENEMA [23].

In section 5 we pay special attention to the well-ordered flows of time and in particular, to the flow of time ω of the natural numbers. There are two reasons to do so: first of all, for these structures we can prove a completeness result for *flat* validity of a system *without any non-orthodox derivation rules*. An interesting aspect of the proof is that it essentially uses the expressive completeness of *TAL* over the class of linear orderings; this completeness-by-completeness argument was first used in GABBAY & HODKINSON [6] for Kamp's one-dimensional functional complete logic with S ('Since') and U ('Until'). Second, well-ordered models for *TAL* have close connections with relation algebras defined in MADDUX [13], but for limitations of space we cannot go into detail here.

To motivate of the special attention for ω, let us note that the set of *intervals* $\{(s,t) \in N \times N \mid s \leq t\}$ over ω can be seen as to represent *finite computation paths*.[3] Hence our results may have applications in the theory of program verification.

2 Definitions

Definition 1. *TAL* is the similarity type having, besides the boolean connectives, the following set of modal operators: a dyadic operator $\circ$, a monadic $\otimes$ and two constants δ and λ. The set of *TAL*-formulas are defined as usual, i.e. given a set *VAR* of propositional variables, a *TAL*-formula is either atomic (i.e. in the set $VAR \cup \{\delta, \lambda\}$), or it has the form $\neg\phi$, $\phi \vee \psi$, $\otimes\phi$ or $\phi \circ \psi$, where ϕ and ψ are formulas.

As abbreviations we will use the usual boolean connectives and constants, and the following *compass diamonds*

---

[3] Infinite computation paths could come into the picture if we would consider the ordering of the successor ordinal of ω; for this flow of time $(\omega + 1, <)$ results can be obtained that are similar to the ones reported on in this paper.

$$\Diamond_N \phi = \phi \circ \otimes \lambda \qquad\qquad \Diamond_W \phi = \otimes \lambda \circ \phi$$
$$\Diamond_S \phi = \phi \circ \lambda \qquad\qquad \Diamond_E \phi = \lambda \circ \phi$$
$$\Diamond_V \phi = \Diamond_N \phi \vee \phi \vee \Diamond_S \phi \qquad \Diamond_H \phi = \Diamond_W \phi \vee \phi \vee \Diamond_E \phi$$
$$\Diamond \phi = \Diamond_H \Diamond_V \phi$$

together with their obvious duals $\Box_N$, $\Box_S$, $\Box_V$, etc. Besides these, we define the following diamond D:

$$D\phi \equiv \neg \delta \circ (\phi \circ \top) \vee (\top \circ \phi) \circ \neg \delta.$$

Definition 2. A *frame* for TAL is a pair $\mathfrak{T} = (T, <)$ with $<$ a binary relation on T. A *flow of time* is a frame where $<$ is a transitive, irreflexive relation on T. A *model* is a triple $\mathfrak{M} = (T, <, V)$ such that $(T, <)$ is a frame and V is a valuation, i.e. a function mapping propositional variables to subsets of $T \times T$.

Truth of a formula ϕ at a pair (s,t) in a model $\mathfrak{M}$, notation: $\mathfrak{M}, s, t \Vdash \phi$, is defined as follows:

$\mathfrak{M}, s, t \Vdash p$ if $(s,t) \in V(p)$,
$\mathfrak{M}, s, t \Vdash \delta$ if $s = t$,
$\mathfrak{M}, s, t \Vdash \lambda$ if $s < t$,
$\mathfrak{M}, s, t \Vdash \neg \phi$ if $\mathfrak{M}, s, t \not\Vdash \phi$,
$\mathfrak{M}, s, t \Vdash \phi \vee \psi$ if $\mathfrak{M}, s, t \Vdash \phi$ or $\mathfrak{M}, s, t \Vdash \psi$,
$\mathfrak{M}, s, t \Vdash \otimes \phi$ if $\mathfrak{M}, t, s \Vdash \phi$,
$\mathfrak{M}, s, t \Vdash \phi \circ \psi$ if there is a u such that $\mathfrak{M}, s, u \Vdash \phi$ and $\mathfrak{M}, u, t \Vdash \psi$.

Validity is defined and denoted as usual. For instance, a formula is valid in a class K of frames, notation: $\mathsf{K} \models \phi$, if for every frame $\mathfrak{T}$ in K, every valuation V on $\mathfrak{T}$, and every pair s, t of points in $\mathfrak{T}$ we have $\mathfrak{T}, V, s, t \Vdash \phi$. A formula ϕ is *satisfiable* in a model if $\neg \phi$ is not valid in the model.

A valuation is called *flat* if for for every propositional variable p and every s, t and u in T, we have $(s, t) \in V(p)$ iff $(s, u) \in V(p)$. *Flat validity*, notation: $\models_\flat$, is defined like ordinary validity, but with the restriction to flat valuations.

Note that informally, a valuation is flat if the truth of a propositional variable at a pair (s, t) only depends on the first coordinate s.

In the sequel, we will consider *linear* flows of time mainly, i.e. flows of time $(T, <)$ where $<$ is a total relation. Such frames allow a nice, two-dimensional representation, cf. the pictures below. The set of pairs where δ holds consists of the *diagonal* elements of the universe; λ is true precisely in the 'north-western' halfplane. The operator $\otimes$ corresponds to *mirroring* in the diagonal. A formula $\phi \circ \psi$ holds at a pair a, if we can draw a *rectangle abcd* such that: b lies on the vertical line through a and ϕ holds at b, d lies on the horizontal line through a and ψ holds at d; and c lies on the diagonal.

The subscripts N, S, W, E, H and V are mnemonics for respectively north, south, west, east, horizontal and vertical. Note that according to the truth definition given above, these compass operators receive their natural interpretation,

Fig. 1. *TAL*'s operators in linear frames.

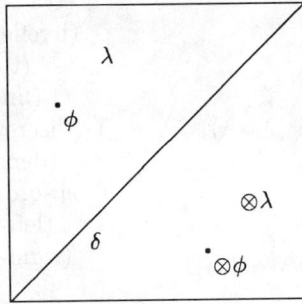 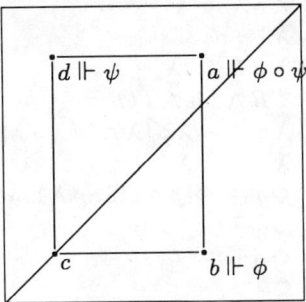

e.g.

$\mathfrak{M}, s, t \Vdash \Diamond_S \phi \iff$ there is a point (s, u) *south* of (s, t) with $s, u \Vdash \phi$,
$\mathfrak{M}, s, t \Vdash \Diamond_V \phi \iff$ there is a point (s, u) with $s, u \Vdash \phi$,

The diamond D is a so-called *difference operator*, i.e. its 'accessibility relation' is the inequality relation:

$$\mathfrak{M}, (s, t) \Vdash D\phi \text{ iff there are } s', t' \text{ with } (s, t) \neq (s', t') \text{ and } \mathfrak{M}, (s', t') \Vdash \phi. \quad (1)$$

Finally, $\Diamond$ is a *universal* operator, i.e.

$$\mathfrak{M}, (s, t) \Vdash \Diamond \phi \text{ iff there are } s', t' \text{ with } \mathfrak{M}, (s', t') \Vdash \phi. \quad (2)$$

We leave it to the reader to verify (1) and (2).

A central role in our paper is played by the class of well-orderings and the flow of time of the natural numbers.

Definition 3. Let ω denote the flow of time of the natural numbers, i.e. ω is the frame $(N, <)$ where N is the set of natural numbers and $<$ is the usual ordering on N. WO denotes the class of well-ordered flows of time, i.e. linear frames such that every non-empty set of time-points has a *smallest* element.

3 Expressiveness

In this section we investigate the expressive power of *TAL*. First we will see how some properties of temporal frames can be expressed in the language; then we turn to the level of models, where we compare the expressive power of *TAL* to that of first-order logic. It turns out that on the model level, *TAL* is as expressive as the three variable fragment of first-order logic; as a consequence, we obtain an expressive completeness result for flat *TAL* over the class of linear orders.

So let us start with some correspondence theory on the frame level. Note that as a nice consequence of having an explicit referent to the (ordering) relation in the object language, it becomes very easy to *characterize* properties of $<$:

Definition 4. Consider the following *TAL*-formulas:

(TR)	$\lambda \circ \lambda \to \lambda$	(transitivity)
(IR)	$\lambda \to \neg \delta$	(irreflexivity)
(TO)	$\lambda \vee \delta \vee \otimes \lambda$	(totality)
(LN)	$TR \wedge IR \wedge TO$	(linearity)
(DI)	$\lambda \circ \lambda \to \lambda \circ (\lambda \wedge \neg(\lambda \circ \lambda)) \wedge (\lambda \wedge \neg(\lambda \circ \lambda)) \circ \lambda$	(discreteness)
(DE)	$\lambda \to \lambda \circ \lambda$	(denseness)
(W)	$\Diamond p \to \Diamond(p \wedge \Box_S \neg p \wedge \Box_W \Box_V \neg p)$	(well-orderings)
(UL)	$\Diamond_W \top$	(left-serial)
(UR)	$\Diamond_N \top$	(right-serial)

Proposition 5. *Let* $\mathfrak{T} = (T, <)$ *be a frame. Then*

(i) $\mathfrak{T} \models TR$ $\iff$ $<$ *is transitive,*
(ii) $\mathfrak{T} \models IR$ $\iff$ $<$ *is irreflexive,*
(iii) $\mathfrak{T} \models TO$ $\iff$ $<$ *is total,*
(iv) $\mathfrak{T} \models LN$ $\iff$ $<$ *is linear,*

Now suppose $<$ *is linear. Then*

(v) $\mathfrak{T} \models DI$ $\iff$ $<$ *is discrete,*
(vi) $\mathfrak{T} \models DE$ $\iff$ $<$ *is dense,*
(vii) $\mathfrak{T} \models W$ $\iff$ $<$ *is well-ordered,*
(viii) $\mathfrak{T} \models UL$ $\iff$ T *is left-serial,*
(ix) $\mathfrak{T} \models UR$ $\iff$ T *is right-serial,*
(x) $\mathfrak{T} \models DI \wedge W \wedge UR$ $\iff$ $T \cong \omega$.

Proof. As an example, we prove **(v)**, one direction of **(vii)**, and **(x)**. Let $\mathfrak{T}$ be linear.

For **(v)** first assume that $\mathfrak{T}$ is discrete, and that $\mathfrak{M}$ is a model on $\mathfrak{T}$ with $\mathfrak{M}, s, t \Vdash \lambda \circ \lambda$. Clearly then t is a successor of s, but not the immediate one. So let u be the immediate successor of s. By linearity of $<$ we have $s < u < t$, and as u is the immediate successor of s: $s, u \Vdash \lambda \wedge \neg(\lambda \circ \lambda)$. So $s, t \Vdash (\lambda \wedge \neg(\lambda \circ \lambda)) \circ \lambda$. The other conjunct in the consequent of DI is treated likewise.

For the other direction, assume that $\mathfrak{T} \models DI$ and let $s < t$. We have to find an immediate successor for s. If t is the immediate successor of s, we are finished. Otherwise, $s, t \Vdash \lambda \circ \lambda$ (in every model on $\mathfrak{T}$), so $s, t \Vdash (\lambda \wedge \neg(\lambda \circ \lambda)) \circ \lambda$ by assumption. By the truth definition, there is a u with $s, u \Vdash \lambda \wedge \neg(\lambda \circ \lambda)$ and $u, t \Vdash \lambda$. It is then straightforward to verify that this u is the immediate successor of s.

Now we will show the direction $\Leftarrow$ of **(vii)**. Assume that $<$ is well-ordering of T, and suppose that V is a valuation on $\mathfrak{T}$ such that for some $s, t \in T$, we have $\mathfrak{T}, s, t \Vdash \Diamond p$. This implies that $V(p) \neq \varnothing$, so we obtain that the set X, defined by $X = \{s \in T \mid \exists t \, (s, t) \in V(p)\}$, is not empty. As $<$ is a well-ordering, X has a *smallest element* x. Now let Y be the set $\{t \in T \mid (x, y) \in T\}$, then $Y \neq \varnothing$ by definition of X. So also Y has a *smallest* element y. It is then straightforward to verify that $\mathfrak{T}, V, x, y \Vdash (p \wedge \Box_S \neg p \wedge \Box_W \Box_V \neg p)$, whence $\mathfrak{T}, V, s, t \Vdash \Diamond(p \wedge \Box_S \neg p \wedge \Box_W \Box_V \neg p)$.

For **(x)**, note that ω is the only well-ordering which is discrete and right-serial. Hence, (x) follows from (v), (vii) and (ix). □

Compared to the existing two-dimensional tense logics, we feel that TAL has the advantage of being both quite expressive and perspicuous. In fact, concerning the first point, all of the systems known to us can be seen as subsystems of TAL. For example, the system studied by Åqvist in [27] uses a set of operators all of which can be defined in TAL:

$$\{bf = \lambda, id = \delta, af = \otimes\lambda,$$
$$\langle P\rangle\phi = \Diamond_W\phi, \langle F\rangle\phi = \Diamond_E\phi, \langle O\rangle\phi = \Diamond_H(\delta \wedge \phi), \langle X\rangle\phi = \otimes\phi\}.$$

As a second example, one of the systems discussed by Gabbay in [5] has two modal operators, F and P, with F having the following semantics:

$$\mathfrak{M}, s, t \Vdash F\phi \iff \text{either } s = t \text{ and for some } t' > t, \mathfrak{M}, s, t' \Vdash \phi$$
$$\text{or} \quad s < t \text{ and } \mathfrak{M}, t, t \Vdash \phi$$
$$\text{or} \quad s > t \text{ and for some } s < u < t, \mathfrak{M}, u, u \Vdash \phi.$$

It is a straightforward exercise to show that $F\phi$ can be defined in TAL as

$$(\delta \to \Diamond_N\phi) \wedge (\lambda \to \Box_H(\delta \to \phi)) \wedge (\otimes\lambda \to \Diamond_W\Diamond_N(\delta \wedge \phi)).$$

Of course, for practical purposes such operators may be necessary: Gabbay's motivation for the introduction of F is to capture the future perfect tense in English. However, we feel that it is better to use a formalism where the *primitive* operators have a more perspicuous semantics, provided that this clarity does not stand in the way of the system's expressive power.

In the second part of this section we compare the expressive power of TAL to that of first-order logic on the level of models. To start with, let us define the first-order language used to describe our models.

Definition 6. Let $L^{2<}$ be a signature of first-order logic with a designated dyadic predicate symbol R and the following set of *dyadic* predicates $\{P_i \mid p_i \in VAR\}$. $L^{1<}$ is defined as $L^{2<}$, but now all predicates P_i are *monadic*.

Let N be a set of L-formulas, k a natural number and X a set of variables in L. We define
$N(X) = \{\phi \in N \mid \text{all } \textit{free} \text{ variables of } \phi \text{ are in } X\}$,
$N_k = \{\phi \in N \mid \text{all variables of } \phi \text{ are among } x_0, \ldots, x_{k-1}\}$.

Now we can view the models of our modal formalism as structures for $L^{2<}$ as follows. Let $\mathfrak{M} = (T, <, V)$ be a model for TAL, then we can define an interpretation I for $L^{2<}$ on T by putting $I(R) = <$ and $I(P_i) = V(p_i)$. If $\mathfrak{M}$ is *flat*, its induced $L^{1<}$-structure will be the pair (T, I) where $I(R) = <$ and

$$I(P_i) = \{t \in T \mid (t, u) \in V(p_i) \text{ for some } u \in T\}.$$

In the sequel, we will identify TAL-models with their induced L-structures.

The basic result concerning the expressive power of our system is that TAL has the same expressive power as $L_3^2(x_0, x_1)$, a fragment of L which we will call 'the three variable fragment of first-order logic', by a slight abus de langue. We hasten to remark that this claim is an immediate consequence of a well-known result in algebraic logic (cf. for instance TARSKI & GIVANT [21]).

Proposition 7. (i) *There is an effective translation τ from* TAL- *formulas to* $L^2(x_0, x_1)$-*formulas such that for every model $\mathfrak{M}$ for* TAL, *every* TAL-*formula ϕ and every pair $(t_0, t_1) \in T \times T$*

$$\mathfrak{M}, t_0, t_1 \Vdash \phi \iff \mathfrak{M} \models \tau(\phi)[x_0 \mapsto t_0, x_1 \mapsto t_1].$$

(ii) *There is an effective translation μ from $L^2(x_0, x_1)$-formulas to* TAL-*formulas such that for every L^2-structure $\mathfrak{M}$ every $L^2(x_0, x_1)$-formula ϕ, and every pair $(t_0, t_1) \in T \times T$*

$$\mathfrak{M} \models \phi[x_0 \mapsto t_0, x_1 \mapsto t_1] \iff \mathfrak{M}, t_0, t_1 \Vdash \mu(\phi).$$

Omitting the proof of this theorem, we just mention the definition of τ:

$$\begin{aligned}
\tau p_i &= P_i x_0 x_1 \\
\tau \delta &= x_0 = x_1 \\
\tau \lambda &= R x_0 x_1 \\
\tau \neg \phi &= \neg \tau \phi \\
\tau(\phi \vee \psi) &= \tau \phi \vee \tau \psi \\
\tau \otimes \phi &= \tau \phi[x_0/x_1, x_1/x_0] \\
\tau(\phi \circ \psi) &= \exists x_2 \, [\tau \phi[x_2/x_1] \wedge \tau \psi[x_2/x_0]],
\end{aligned} \qquad (3)$$

where we assume that we have a suitable devise to perform the variable substitutions ($[x_0/x_1, x_1/x_0]$, $[x_2/x_1]$ and $[x_0/x_1]$) *within* the three variable fragment of $L^{2<}$.

As a straightforward consequence of this proposition, we see that over the class of *flat* models, every TAL-formula ϕ has an equivalent $\tau_\flat \phi$ in $L_3^{1<}(x_0, x_1)$ and vice versa. The crucial (and only) difference in the translation $\tau_\flat$ lies in the atomic clause:

$$\tau_\flat p_i = P_i x_0 \wedge x_1 = x_1. \qquad (4)$$

A fortiori, every TAL-formula has an equivalent in $L^{1<}(x_0, x_1)$. We will now show the converse of this fact to hold as well, and thus establish an *expressive completeness* theorem, in the style of Kamp's famous result concerning the operators S ('Since') and U ('Until') (cf. KAMP [10])[4].

Theorem 8 (Flat expressive completeness over linear orders). *Over the class of flat models based on a linear frame, every $L^{1<}(x_0, x_1)$-formula has an equivalent in* TAL, *and vice versa.*

---

[4] Kamp's theorem states that that over the class of Dedekind-complete linear orderings, every formula in $L^{1<}(x_0)$ has an equivalent in the one-dimensional formalism with operators S and U; we refer to GABBAY E.A. [5] for a more accessible proof.

Proof. Let ϕ be a formula in $L^{1<}(x_0, x_1)$. By results in GABBAY [4], resp. IMMERMAN & KOZEN [9], $L^{1<}$ has *Henkin-dimension* three, resp. the *three-variable property* over the class of linear orderings, both implying that ϕ has an equivalent in $L_3^{1<}(x_0, x_1)$. Then by proposition 7, ϕ has a *TAL*-equivalent over the class of flat linear models. □

Note that the restriction to *flat TAL/monadic* predicates is essential here, as it is shown in VENEMA [22] that no *finite* system of two-dimensional temporal operators can be as expressive as $L_3^{2<}(x_0, x_1)$. Finally, we should mention that *TAL* is not the only two-dimensional expressively complete system, and that our notion of two-dimensional expressive completeness is not the only one possible. We refer to GABBAY E.A. [5] for more details.

4 Axiomatics: the general case

In this section we develop the axiomatics of *TAL*. First we will define a derivation system which is sound and complete with respect to the class of all frames; then we will define complete axiom systems for various classes of flows of time, like the linear, dense and discrete ones. At the end of the section we treat axiom systems for flat validity.

The basic systems in our axiomatics are the following:

Definition 9. The axiom system AR is given by the following sets of axioms (where $\phi \circ \psi$ and $\otimes \phi$ abbreviate $\neg(\neg \phi \circ \neg \psi)$ and $\neg \otimes \neg \phi$, respectively):

(CT) all classical tautologies
(DB) $(p \to p') \circ q \to (p \circ q \to p' \circ q)$
        $p \circ (q \to q') \to (p \circ q \to p \circ q')$
        $\otimes(p \to q) \to (\otimes p \to \otimes q)$
$(A1)$ $\neg \otimes p \leftrightarrow \otimes \neg p$
$(A2)$ $\otimes \otimes p \to p$
$(A3)$ $\otimes(p \circ q) \leftrightarrow \otimes q \circ \otimes p$
$(A4)$ $\delta \circ p \leftrightarrow p$
$(A5)$ $p \circ \neg(\otimes p \circ q) \to \neg q$
$(A6)$ $p \circ (q \circ r) \leftrightarrow (p \circ q) \circ r$

and the following set of derivation rules: Modus Ponens, Universal Generalization and Substitution:

(MP) $\phi, \phi \to \psi \ / \ \psi$
(UG) $\phi \ / \ \phi \circ \psi, \psi \circ \phi$
         $\phi \ / \ \otimes \phi$
(SUB) $\phi \ / \ \sigma \phi$

where σ is a map uniformly substituting formulas for propositional variables in formulas.

The derivation system AR^+ is the extension of AR with the Irreflexivity Rule for D:

(IR_D) $(p \wedge \neg Dp) \to \phi \ / \ \phi$, provided p does not occur in ϕ.

Definition 10. Let $\Lambda = (A, R)$ be a derivation system with A being the set of axioms and R the set of derivation rules. A *derivation* is a sequence $\phi_0, \ldots, \phi_n$ such that every ϕ_i is either an axiom or the result of applying a rule to formulas of $\{\phi_0, \ldots \phi_{i-1}\}$. A formula ϕ is a *theorem* of Λ, notation: $\Lambda \vdash \phi$, if ϕ appears as the last item of a derivation in Λ. A formula ϕ is a Λ-*consistent* if $\neg \phi$ is not a theorem. A derivation system Λ is *sound* with respect to a class K of frames if every Λ-theorem is valid in K and *complete* if every K-valid formula is a theorem of Λ.

We refer the reader to VENEMA [26] for more information on the axioms. The meaning of the Irreflexivity Rule for D is perhaps best understood by reading it as follows: 'if ψ is consistent and p does not occur in ϕ, then $(p \wedge \neg Dp) \wedge \psi$ is consistent'. Note that the formula $p \wedge \neg Dp$ ('p is true here, and nowhere else') can be seen as a *name* for a world (s, t). So (IR_D) states that every consistent formula remains consistent if we take its conjunction with a name. For more information concerning rules like IR_D, which originate with GABBAY [3], we refer to VENEMA [25].

Theorem 11. AR^+ *is sound and complete with respect to the class of all frames.*

Note that Theorem 11 is the TAL-version of a square completeness theorem for arrow logic. For a *proof* we refer to VENEMA [23].

Definition 12. Let $AL^{(+)}$ be the derivation system $AR^{(+)}$ extended with the set $\{TR, IR, TO\}$ as axioms (cf. Definition 4). For $X \subseteq \{DI, DE, UL, UR, W\}$, we define $AL^{(+)}X$ as the derivation system $AL^{(+)}$ extended with the set X as axioms.

Theorem 13. *If $X \subseteq \{DI, DE, UL, UR\}$, then AL^+X is sound and complete with respect to the class K_X of frames where X is valid.*

Proof. An inspection of the proof of Theorem 11 reveals that it can be easily modified to prove Theorem 13. The basic observation is that the theorem only mentions sets X of *closed* axioms, i.e. formulas without any propositional variables. The proof method of Theorem 11 yields, for an arbitrary consistent formula ϕ, a model $\mathfrak{M}$ such that (i) ϕ is satisfiable in $\mathfrak{M}$ and (ii), X is true in every world of $\mathfrak{M}$. It is then a straightforward consequence of Proposition 5 that the underlying frame of $\mathfrak{M}$ belongs to the intended class. □

As instances of Theorem 13, we obtain that AL^+ is sound and complete with respect to the class of all linear flows of time, and $AL^+\{DE, UL, UR\}$ with respect to the class of unbounded, dense flows of time, and thus with respect to the flow of time of the rational numbers.

Finally, we turn to the matter of flat validity. Note that the set of formulas which are flatly valid in a class of frames cannot be a logic in the ordinary sense of the word, since it will not be closed under substitution. For, the formula $p \to \Box_V p$ will be flatly valid, but for instance the formula $\Diamond_H \otimes q \to \Box_V \Diamond_H \otimes q$ can easily be falsified in a flat model. Therefore, we have to use a trick.

Definition 14. Let Ω be one of the axiom systems defined above, and ϕ a *TAL*-formula. Let VAR_ϕ be the set of propositional variables that occur in ϕ. We say that ϕ is *flatly* derivable in Ω, notation: $\Omega_b \vdash \phi$, if

$$\Omega \vdash (\bigwedge_{p \in VAR_\phi} \Box(p \leftrightarrow \Box_V p)) \to \phi.$$

Corollary 15. *If $X \subseteq \{DI, DE, UL, UR\}$, then $\mathrm{AL}^+ X_b$ is 'flatly sound and complete' with respect to the class K_X, i.e. for any formula ϕ:*

$$\Omega_b \vdash \phi \iff \mathsf{K}_X \models_b \phi.$$

Proof. For a change, we only prove soundness. Suppose that $\Omega_b \vdash \phi$, and let $\mathfrak{M}$ be a flat model $(T, <, V)$; we have to show that $\mathfrak{M} \models \phi$.

To start with, by definition of flat derivability we have that

$$\Omega \vdash (\bigwedge_{p \in VAR_\phi} \Box(p \leftrightarrow \Box_V p)) \to \phi.$$

So by our soundness assumption, for *any* model $\mathfrak{N}$:

$$\mathfrak{N} \models (\bigwedge_{p \in VAR_\phi} \Box(p \leftrightarrow \Box_V p)) \to \phi.$$

As it is straightforward to verify that the formula $\Box(p \leftrightarrow \Box_V p)$ is valid in any flat model, validity of ϕ in $\mathfrak{M}$ follows immediately. $\square$

5 Completeness for well-orderings.

In this section we prove our main result, viz. soundness and completeness of ALW_b with respect to flat validity in the class of well-ordered frames. In some sense, this result is the best we can get for WO, for ordinary validity in this class does not allow a recursive axiomatization[5]. Neither do we have *strong* completeness for flat validity in WO, as an easy compactness argument shows.

On the other hand, what makes the results in this section interesting is that the complete axiom systems are orthodox in the sense that they do not use an irreflexivity rule. This is interesting from a theoretical point of view and may also have applicational virtues: note that derivations involving the irreflexivity rule use material (the proposition letter p which does not occur in ϕ) in a very 'resource-unconscious' way.

Theorem 16 (Flat soundness and completeness for well-orderings).
For any TAL-*formula ϕ*

$$ALW_b \vdash \phi \iff \mathsf{WO} \models_b \phi.$$

---

[5] In HALPERN & SHOHAM [7] the authors show how to code the behavior of Turing machines in a subsystem HS of TAL. It easily follows from their results that the TAL-theory of WO is not recursively enumerable.

Proof. We leave it to the reader to establish soundness. For completeness, let ϕ be an $ALW_\flat$-consistent formula. We will find a flat well-ordered model $\mathfrak{M}$ in which ϕ is satisfiable. Define

$$\phi' \equiv (\bigwedge_{p \in VAR_\phi} \Box(p \leftrightarrow \Box_V p)) \to \phi,$$

then by definition of $ALW_\flat$, ϕ' is ALW-consistent. Now we need the following lemma:

Lemma 17. *For every formula ψ:*

$$ALW \vdash \psi \iff AL^+W \vdash \psi.$$

Proof. It suffices to show that the irreflexivity rule for D is conservative over ALW. So, let us assume that

$$ALW \vdash (p \land \neg Dp) \to \phi. \tag{5}$$

Then we have to prove that $ALW \vdash \phi$. Abbreviate $ALW \vdash \chi$ by $\vdash \chi$ and let $\texttt{first}(\chi)$ denote the formula

$$\texttt{first}(\chi) \equiv (\chi \land \Box_S \neg \chi \land \Box_W \Box_V \neg \chi),$$

then the well-ordering axiom W reads: $\Diamond p \to \Diamond \texttt{first}(p)$. From (5) it follows by the rule of substitution that

$$\vdash (\texttt{first}(\neg\phi) \land \neg D\texttt{first}(\neg\phi)) \to \phi. \tag{6}$$

We leave it to the reader to verify that for every q

$$\vdash \texttt{first}(q) \to \neg D\texttt{first}(q). \tag{7}$$

From (6) and (7) it follows that

$$\vdash \texttt{first}(\neg\phi) \to \phi.$$

On the other hand, by definition of $\texttt{first}$, we have

$$\vdash \texttt{first}(\neg\phi) \to \neg\phi,$$

so we find that

$$\vdash \neg\texttt{first}(\neg\phi),$$

whence an application of *NEC* gives

$$\vdash \neg\Diamond\texttt{first}(\neg\phi).$$

But then by the instantiation $W(\neg\phi)$ of the well-ordering axiom W we find the desired

$$\vdash \neg\phi \to \bot.$$

□

It follows from the lemma that $AL^+ \not\vdash \neg\phi'$, so by Theorem 13[6] ϕ has a *linear* model $\mathfrak{M} = (T, <, V)$ such that V is *flat* and $\mathfrak{M} \models AL^+W$, i.e. every theorem of AL^+W is valid in $\mathfrak{M}$. Unfortunately $\mathfrak{M}$ need not be well-ordered, as not *every* subset of T needs to have a smallest element. Fortunately however, $\mathfrak{M}$ is definably well-ordered: call a first-order structure $\mathfrak{N}$ for $L^{1<}$ *definably well-ordered* if every first-order definable subset of the domain has a smallest element, or to be more precise, if $\mathfrak{N}$ satisfies the condition, that for every first-order formula $\psi(x_0) \in L^{1<}$, the set

$$S_\psi = \{t \in T \mid \mathfrak{M} \models \psi(x_0)[x_0 \mapsto t]\}$$

has a smallest element (provided that $S \neq \varnothing$).

Lemma 18. $\mathfrak{M}$ *is definably well-ordered.*

Proof. Assume that ψ is such that $S_\psi \neq \varnothing$. First note that by our expressive completeness result Theorem 8, the formula $x_1 = x_1 \wedge \psi(x_0)$ has an equivalent ψ' in TAL, whence we have

$$\mathfrak{M}, s, t \Vdash \psi' \iff s \in S_\psi.$$

Second, as $\mathfrak{M}$ is an AL^+W-model, the ψ'-instantiation of the well-orderedness axiom W is valid in $\mathfrak{M}$, so

$$\mathfrak{M} \models \Diamond\psi' \to \Diamond(\psi' \wedge \Box_W \Box_V \neg\psi').$$

and as $S_\psi \neq \varnothing$, this immediately gives a smallest element for $S = V(\psi)$. □

We now consider the first-order equivalent $\tau_b(\phi) \in L_3^{1<}(x_0, x_1)$ of ϕ', cf. (3) and (4). Note that, translated into first-order logic, our problem is that $\tau_b(\phi)$ is satisfiable in a definably well-ordered model, while we need to satisfy it in a truely well-ordered model. The solution of this problem is given by the lemma below, for which we need some terminology.

Let $\equiv_n$ denote the following relation between two structures of first-order logic: $\mathfrak{M} \equiv_n \mathfrak{M}'$ iff $\mathfrak{M}$ and $\mathfrak{M}'$ satisfy the same first-order sentences of quantifier depth $\leq n$.

Lemma 19 (Doets). *Let $\mathfrak{N}$ be a definably well-ordered model for $L^{1<}$ and n a natural number. Then $\mathfrak{N}$ has a well-ordered n-equivalent, i.e. there is a well-ordered structure $\mathfrak{N}'$ such that $\mathfrak{N}$ is well-ordered and $\mathfrak{N} \equiv_n \mathfrak{N}'$.*

For a *proof* of this lemma we refer the reader to DOETS [2], Corollary 4.4.

Now let n be the quantifier depth of the formula $\tau_b(\phi)$; lemma 19 supplies us with a well-ordered structure $\mathfrak{M}'$ such that $\mathfrak{M}' \equiv_{n+2} \mathfrak{M}$. It is then immediate to verify that $\mathfrak{M}' \models \exists x_0 x_1 \tau_b(\phi)(x_0, x_1)$.

This implies that $\mathfrak{M}'$ (now seen as a TAL_b-model) is a well-ordered model for ϕ. □

---

[6] Actually, we need a strong completeness theorem (which for lack of space we could not state or prove here).

Corollary 20 (Flat soundness and completeness for ω). *Let $AL\omega$ be the axiom system AL extended with the axioms W, DI and UR. Then for any TAL-formula ϕ*

$$AL\omega_\flat \vdash \phi \iff \omega \models_\flat \phi$$

Proof. By a suitable adaptation of the proof of Theorem 16, one can satisfy any $AL\omega_\flat$-consistent formula in a well-ordering which is discrete and unbounded to the right. The underlying frame of the model must then be isomorphic to the ordering of the natural numbers (cf. Proposition 5(x)). □

6 Questions

We finish the paper with mentioning two open problems:

1. In the introduction we already mentioned the fact that the proof method applied in section 5 stems from GABBAY & HODKINSON [6] for the uni-dimensional case. In the cited paper the authors prove a completeness result for the flow of time of the real numbers. However, their derivation system crucially uses an *irreflexivity rule*. In VENEMA [24] we applied the method to an *orthodox* system, to obtain a complete axiomatization for the S,U-logic of the class of well-orders and the flow of time of the natural numbers. Reynolds [19] solved the more difficult problem to find an orthodox complete axiomatization for the S,U-logic of the reals.

    Concerning two-dimensional temporal logics, it is an intrigueing question whether there is a derivation system Λ such that (i) Λ gives a complete enumeration of the flat[7] TAL-logic of the real number flow of time. Similar questions may be asked for other flows of times and other classes of (linear) frames. Note that the method applied in this paper cannot be extended to other linear flows of time in a straightforward manner, as our proof of Lemma 17 crucially depends on the presence of the well-orderedness axiom W in the logic.

2. One may read the completeness results of Theorem 16 (and Theorem 20) as follows: $L_3^{1<}(x_0, x_1)$ is sufficiently expressive to contain a Hilbert-style derivation system which is complete for the class WO (resp. for ω). The question is whether it is also expressive enough to define complete Gentzen-style calculi in it (not necessarily only for well-orderings)[8]. In particular, it would be interesting to have cut-free Gentzen-style calculi with nice properties like the subformula property or decidability.

---

[7] The related question for ordinary validity is solved in the negative, as the TAL-theory of the reals is not recursively enumerable, cf. footnote 5 and HALPERN & SHOHAM [7].

[8] Again, the restriction to *monadic* first-order logic is essential here: it follows from results in algebraic logic that there is no upper bound on the number of variables needed to prove theorems of first-order logic with *dyadic* predicates. There are calculi known that are complete for $L_3^{2<}(x_0, x_1)$, cf. MADDUX [12] or ORŁOWSKA [18], but these calculi essentially use formulae of $L_n^{2<}(x_0, x_1)$, n arbitrary large.

References

1. H. Andréka, J.D. Monk & I. Németi (eds.), *Algebraic Logic; Proceedings of the 1988 Budapest Conference on Algebraic Logic*, North-Holland, Amsterdam, 1991.
2. K. Doets, "Monadic Π_1^1-theories of Π_1^1- properties", *Notre Dame Journal of Formal Logic*, 30 (1989) 224-240.
3. D.M. Gabbay, "An irreflexivity lemma with applications to axiomatizations of conditions on linear frames", in: [16], pp. 67–89.
4. D.M. Gabbay, "Expressive functional completeness in tense logic", in: [16], pp. 91–117.
5. D.M. Gabbay, I.M. Hodkinson & M. Reynolds, *Temporal Logic: Mathematical Foundations and Computational Aspects*, Oxford University Press, Oxford, 1992, to be published.
6. D.M. Gabbay & I.M. Hodkinson, "An axiomatization of the temporal logic with Since and Until over the real numbers", *Journal of Logic and Computation*, 1 (1990) 229–259.
7. J.Y. Halpern & Y. Shoham "A propositional modal logic of time intervals", *Proceedings of the First IEEE Symposium on Logic in Computer Science, Cambridge, Massachusettes*, 279-292.
8. D. Harel, D. Kozen & R. Parikh, "Process Logic: expressiveness, decidability, completeness, *Journal of Computer and System Sciences*, 26 (1983) 222–243.
9. N. Immerman and D. Kozen, "Definability with bounded number of bound variables", in: [11], pp. 236–244.
10. J.A.W. Kamp, *Tense Logic and the Theory of Linear Order*, PhD Dissertation, University of California, Los Angeles, 1968.
11. *Proceedings of the Symposium on Logic in Computer Science 1987, Ithaka, New York*, Computer Society Press, Washington, 1987.
12. R. Maddux, "A Sequent Calculus for Relation Algebras", *Annals of Pure and Applied Logic*, 25 (1983) 73–101.
13. R.D. Maddux, "Finitary Algebraic Logic", *Zeitschrift für Mathematisch Logik und Grundlagen der Mathematik*, 35 (1989) 321–332, 39 (1993) 566–569.
14. M. Masuch, L. Pólós & M. Marx (eds.), *Arrow Logic*, Oxford University Press, to appear.
15. M. Marx, Sz. Mikulás, I. Németi & I. Sain, "Investigations in arrow logic", to appear in [14].
16. U. Mönnich, ed., *Aspects of Philosophical Logic*, Reidel, Dordrecht, 1981.
17. I. Németi, "Algebraizations of quantifier logics: an introductory overview", *Studia Logica* 50 (1991) 485-570.
18. E. Orłowska, "Relational interpretation of modal logics", in: [1], 443–471.
19. M. Reynolds, "An axiomatization for Since and Until over the reals without the IRR rule", *Studia Logica*, to appear.
20. M. de Rijke (ed.), *Diamonds and Defaults*, Kluwer Academic Publishers, Dordrecht, 1993.
21. A. Tarski and S. Givant, *A formalization of set theory without variables*, AMS Colloquium Publications 41, 1987.
22. Y. Venema, "Relational Games", in [1], 695–718.
23. Y. Venema, *Many-Dimensional Modal Logic*, Doctoral Dissertation, University of Amsterdam, 1992, submitted to Oxford University Press.
24. Y. Venema, "Completeness by completeness: Since and Until", in: [20], 349–359.

25. Y. Venema, "Derivation rules as anti-axioms in modal logic", *Journal of Symbolic Logic*, **58** (1993) 1003–1054.
26. Y. Venema, "A crash course in arrow logic", Logic Group Preprint Series **107**, February 1994, Department of Philosophy, Utrecht University; to appear in [14].
27. L. Åqvist, "A conjectured axiomatization of two-dimensional Reichenbachian tense logic", *Journal of Philosophical Logic*, **8** (1982) 315–333.

Efficient Computation of Nested Fix-Points, with Applications to Model Checking

B. Vergauwen, J. Lewi, I. Avau, A. Poté

K.U.Leuven, Department of Computer Science,
Celestijnenlaan 200A, 3001 Leuven, Belgium

Abstract. The paper presents a general algorithm for computing nested fix-points over complete lattices of finite height. The method presented relies on techniques familiar from the realm of functional programming languages, such as e.g. lazy evaluation. The algorithm is constructed in a stepwise fashion: We start with a schema based on some simple facts of fix-point theory. As such this schema is easily seen to be correct. It is, however, rather inefficient. We then trace the sources of inefficiency and refine the basic schema resulting in a correct and more efficient algorithm. After presenting the general algorithm, we apply it, by means of illustration, to the field of model checking.

1 Preliminaries: Blocks

Assume a set $\mathcal{V}$ of values and a partial order relation $\sqsubseteq$ on $\mathcal{V}$ such that $Lat = \langle \mathcal{V}, \sqsubseteq \rangle$ forms a complete lattice of finite height[1]. The bottom element of Lat is denoted by $\bot$, the top element by $\top$. Also assume a (countable) set $\mathtt{X}$ of variables.

Environments An *environment* $\theta : \mathtt{X} \to \mathcal{V}$ maps variables into values. The set of environments is denoted by Θ. Updating of environments is as follows: Define $\theta[x \mapsto v]$ as the environment θ' that agrees with θ except that x is updated to v, i.e., $\theta'(x) = v$ and $\theta'(y) = \theta(y)$ for $y \neq x$. Furthermore for $\rho : \{x_1, \ldots, x_1\} \to \mathcal{V}$ define $\theta[\rho]$ as the environment $\theta[x_1 \to \rho(x_1)] \ldots [x_n \to \rho(x_n)]$. The partial order $\sqsubseteq$ can be extended pointwise towards environments in the usual manner: $\theta_1 \sqsubseteq \theta_2$ iff $\theta_1(x) \sqsubseteq \theta_2(x)$ for every $x \in \mathtt{X}$.

Equations An *equation* is of the form $\langle x \stackrel{\sigma}{=} f \rangle$ where $x \in \mathtt{X}$, $\sigma \in \{\ominus, \oplus\}$ and $f : \Theta \to \mathcal{V}$ is a monotone (isotone) function, i.e, $f(\theta_1) \sqsubseteq f(\theta_2)$ for every θ_1, θ_2 such that $\theta_1 \sqsubseteq \theta_2$. Let $E = \langle x \stackrel{\sigma}{=} f \rangle$ be an equation. Given an environment θ for interpreting variables other than x, one can solve E for x. A value $v \in \mathcal{V}$ is a *solution* of E iff $v = f(\theta[x \mapsto v])$. As f is monotone, E has at least one solution. There even is a unique least solution as well as a unique greatest solution $[\top]$. The role of σ is precisely to specify which of both extremal solutions is meant: A $\ominus$ label denotes the least solution, a $\oplus$ label denotes the greatest solution. In this

---
[1] I.e., there do not exist infinite sequences of the form $v_1 \sqsubset v_2 \sqsubset v_3 \sqsubset \ldots$ or $v_1 \sqsupset v_2 \sqsupset v_3 \sqsupset \ldots$

paper we are not interested in solving single equations but in *nested* equations. The term *Block* is used to refer to a list of nested equations.

Blocks The set **B** of Blocks is defined inductively as the least set such that:

- nil $\in$ **B**, and $lv(\text{nil}) = \emptyset$ (*lv* for left hand side variables)
- Let $B \in $ **B** and let $E = \langle x \stackrel{\sigma}{=} f \rangle$ be an equation such that $x \notin lv(B)$. Then $E \bullet B$ is also in **B**, and $lv(E \bullet B) = \{x\} \cup lv(B)$.

Note that all left hand side variables of a Block are required to be distinct. This is for technical reasons only. A Block can be seen as a recursive definition for its left hand side variables. The order of the equations determines the relative importance of variables. The role of the $\ominus/\oplus$ labels is as explained above. Formally the semantics of Blocks is defined by the following semantic function $[\![\_ \| \_]\!] : (\mathbf{B} \times \Theta) \to \Theta$:

- $[\![\text{nil} \| \theta]\!] = \theta$
- $[\![\langle x \stackrel{\ominus}{=} f \rangle \bullet B \| \theta]\!] = [\![B \| \theta[x \mapsto \mu\Omega]]\!]$
  where $\mu\Omega$ is the least fix-point of $\Omega = \lambda v \in \mathcal{V}.f([\![B \| \theta[x \mapsto v]]\!])$
- $[\![\langle x \stackrel{\oplus}{=} f \rangle \bullet B \| \theta]\!] = [\![B \| \theta[x \mapsto \nu\Omega]]\!]$
  where $\nu\Omega$ is the greatest fix-point of $\Omega = \lambda v \in \mathcal{V}.f([\![B \| \theta[x \mapsto v]]\!])$

Note that the above definition is sound: As the right hand side functions of Blocks are monotone, functional Ω is also monotone and hence unique extremal fix-points of Ω do exist [T].

The following definitions will prove useful later on. For $B \in $ **B** and $x \in lv(B)$ define:

- $eq(B,x)$ as the equation of B having x as its left hand side variable, i.e., $eq(((\langle x \stackrel{\sigma}{=} f \rangle \bullet B'), x) = \langle x \stackrel{\sigma}{=} f \rangle$, and $eq(((\langle y \stackrel{\sigma}{=} f \rangle \bullet B'), x) = eq(B', x)$ if $y \neq x$.
- $B\triangledown$ as the Block obtained from B by stripping off top-equations until an equation carrying a different label is encountered, i.e., $\text{nil}\triangledown = (E \bullet \text{nil})\triangledown = \text{nil}$, $((\langle x_1 \stackrel{\sigma}{=} f_1 \rangle \bullet \langle x_2 \stackrel{\sigma}{=} f_2 \rangle \bullet B')\triangledown = (\langle x_2 \stackrel{\sigma}{=} f_2 \rangle \bullet B')\triangledown$, and $((\langle x_1 \stackrel{\sigma_1}{=} f_1 \rangle \bullet \langle x_2 \stackrel{\sigma_2}{=} f_2 \rangle \bullet B')\triangledown = \langle x_2 \stackrel{\sigma_2}{=} f_2 \rangle \bullet B'$ if $\sigma_1 \neq \sigma_2$.
- $tv(B)$ as the set of toplevel variables of B, i.e., $tv(B) = lv(B) \setminus lv(B\triangledown)$.
- $nd(B, x)$ as the nesting depth of x in B (top-level variables have a nesting depth of 1), i.e., $nd(B, x) = 1$ if $x \in tv(B)$, and $nd(B, x) = 1 + nd(B\triangledown, x)$ if $x \notin tv(B)$.
- $B \setminus x$ as the Block obtained from B by deleting the equation associated with x, i.e., $(\langle x \stackrel{\sigma}{=} f \rangle \bullet B') \setminus x = B'$, and $(\langle y \stackrel{\sigma}{=} f \rangle \bullet B') \setminus x = \langle y \stackrel{\sigma}{=} f \rangle \bullet (B' \setminus x)$ if $y \neq x$.

2 Goals and Overview

The primary goal of this paper is to construct an efficient, general algorithm SOLVE(B, θ, x) for computing component $[\![B \parallel \theta]\!](x)$. We shall concentrate on *sparse* Blocks, i.e. Blocks whose right hand side functions are sparse. Sparseness is a rather qualitative notion: A function $f : \Theta \to \mathcal{V}$ is said to be *sparse* if for 'most' environments θ only a 'small' fragment of θ is actually needed in order to compute $f(\theta)$.

The construction of SOLVE proceeds in a stepwise fashion: In section 3 we first give the basic schema for SOLVE. This basic schema is obviously correct (by construction) but it is rather inefficient. In sections 4 and 5 we then subsequently refine this schema resulting in a correct and efficient algorithm. In section 6 the algorithm is applied, by means of illustration, to the field of model checking.

3 Basic Schema for SOLVE

The following simple facts form the basis for computing $[\![B \parallel \theta]\!](x)$.

(1) $\forall B \in \mathbf{B}, \theta \in \Theta, x \in X : x \notin lv(B) \Rightarrow [\![B \parallel \theta]\!](x) = \theta(x)$

(2) Let $B \in \mathbf{B}$ and let $x_1, x_2 \in lv(B)$ such that $nd(B, x_1) = nd(B, x_2)$. Let B' be the Block obtained from B by switching equations $eq(B, x_1)$ and $eq(B, x_2)$. Then $[\![B \parallel \theta]\!] = [\![B' \parallel \theta]\!]$ for any $\theta \in \Theta$.

(3) $\forall B \in \mathbf{B}, \theta \in \Theta, x \in lv(B) : [\![B \parallel \theta]\!] = [\![B \setminus x \parallel \theta[x \mapsto [\![B \parallel \theta]\!](x)]]\!]$

Based upon the above facts, $[\![B \parallel \theta]\!](x)$ can be computed along the following lines: If $x \notin lv(B)$ then obviously, by fact (1), $[\![B \parallel \theta]\!](x) = \theta(x)$. If $x \in lv(B)$ and $nd(B,x) = 1$ then x is a top-level variable of B. Let $eq(B,x) = \langle x \stackrel{\sigma}{=} f \rangle$. Assume that $\sigma = \ominus$. From fact (2) together with the definition rule for $[\![\_ \parallel \_]\!]$ it then immediately follows that $[\![B \parallel \theta]\!](x) = \mu\Omega$, where $\Omega = \lambda v \in \mathcal{V}.f([\![B \setminus x \parallel \theta[x \mapsto v]]\!])$. As the lattice *Lat* is of finite height, $\mu\Omega$ is effectively computed as the limit point of the chain $\Omega^0(\bot) \sqsubseteq \Omega^1(\bot) \sqsubseteq \Omega^2(\bot) \sqsubseteq \Omega^3(\bot) \sqsubseteq \ldots$ where $\Omega^0(y) = y$ and $\Omega^{i+1}(y) = \Omega(\Omega^i(y))$. In case $\sigma = \oplus$ we compute $\nu\Omega$ as the limit point of the chain $\Omega^0(\top) \sqsupseteq \Omega^1(\top) \sqsupseteq \Omega^2(\top) \sqsupseteq \Omega^3(\top) \sqsupseteq \ldots$ Finally if $x \in lv(B)$ but $nd(B,x) > 1$ then we first compute the values of the top-level variables of B (as explained above), we then eliminate these variables from B using elimination fact (3) and then compute the x-component of the simplified Block.

The basic schema for SOLVE is listed below. We assume a function EVAL for evaluating right hand side functions of Blocks. (Right hand side functions are assumed to be computable in finite time).

```
function SOLVE₁(B, θ, x) = solve₁(B, θ, x)
-- return 〚 B ‖ θ 〛(x)

function solve₁(B, θ, x) =
-- return 〚 B ‖ θ 〛(x)
begin
  case
  ▯ x ∉ lv(B) : return θ(x)
  ▯ x ∈ lv(B) and nd(B, x) = 1 :
      Let eq(B, x) = ⟨x =^σ f⟩
      v_cur := if σ = ⊖ then ⊥ else ⊤ fi
      loop
        -- compute f(〚 B \ x ‖ θ[x ↦ v_cur] 〛)
        θ' := λy ∈ X.solve₁(B \ x, θ[x ↦ v_cur], y)
        v_new := EVAL(f, θ')
        -- v_new = f(〚 B \ x ‖ θ[x ↦ v_cur] 〛)
        if v_cur = v_new then exit loop fi
        v_cur := v_new
      endloop
      return v_cur
  ▯ x ∈ lv(B) and nd(B, x) > 1 :
      ρ := λy ∈ tv(B).solve₁(B, θ, y)
      solve₁(B▽, θ[ρ], x)
  endcase
end
```

Theorem SOLVE₁(B, θ, x) correctly computes 〚 B ‖ θ 〛(x)

The nice thing about SOLVE₁ is that it is extremely simple because it was directly derived from a few simple facts. On the other hand it is impractical because it is also extremely inefficient, the worst-case running time being exponential in $|lv(B)|$. There are two main sources of inefficiency:

1. *Needless computation* : As Blocks are sparse, lots of information may be computed that is actually never used.
2. *Needless re-computation* : The same/related information is computed more than once.

In the next section we refine SOLVE by eliminating needless computations. Needless re-computation is addressed in section 5.

4 Avoiding Needless Computation

As Blocks are assumed to be sparse, lots of information that is computed by solve₁ may turn out not to be needed at all. The key idea to avoid such needless computation is to *postpone* the computation of information until this information is actually needed (i.e. call by need). In this way no time is wasted computing 'irrelevant' information. There are two points in solve₁ where postponing comes into play. We discuss each in turn.

Lazy evaluation of $f(\theta')$. Consider the following code fragment of solve_1:

```
-- compute f([[ B \ x ‖ θ[x ↦ v_cur] ]])
θ' := λy ∈ X.solve₁(B \ x, θ[x ↦ v_cur], y)
v_new := EVAL(f, θ')
-- v_new = f([[ B \ x ‖ θ[x ↦ v_cur] ]])
```

As f is sparse, in general only a small fragment of θ' will actually be needed in order to compute $f(\theta')$. Hence, instead of first computing all components of θ' and thereafter evaluating $f(\theta')$, it seems worth-while to integrate both steps, i.e., to evaluate $f(\theta')$ in a *lazy* (demand driven) way. In order to do this, function EVAL is slightly extended in the following way: We allow the second argument of EVAL to be a *partial* environment $\rho : Dom(\rho) \to \mathcal{V}$ with $Dom(\rho) \subseteq X$. If the values of variables from $Dom(\rho)$ suffice to evaluate f, then $\text{EVAL}(f, \rho)$ returns the desired value; Otherwise a variable $y \in (X \setminus Dom(\rho))$ is returned, indicating that the value of y is needed in order to evaluate f. Lazy evaluation of $f(\theta')$ is now implemented as follows (ϵ denotes the empty environment, i.e. $Dom(\epsilon) = \emptyset$):

```
-- lazy computation of f([[ B \ x ‖ θ[x ↦ v_cur] ]])
ρ := ε    r := EVAL(f, ρ)
while r ∉ V do
   v := solve₁(B \ x, θ[x ↦ v_cur], r)   ρ := ρ[r ↦ v]   r := EVAL(f, ρ)
od
v_new := r
-- v_new = f([[ B \ x ‖ θ[x ↦ v_cur] ]])
```

Delayed computation of ρ. Consider the following code fragment of solve_1:

$(\star)$ $\quad \rho := \lambda y \in tv(B).\text{solve}_1(B, \theta, y) \quad \text{solve}_1(B\nabla, \theta[\rho], x)$

It makes little sense to compute all components of ρ a priori. Instead we would like to delay the computation of ρ-components until they are actually needed by $\text{solve}_1(B\nabla, \theta[\rho], x)$. It is however unclear at what points ρ-components may be needed during execution of $\text{solve}_1(B\nabla, \theta[\rho], x)$, if they will ever be needed at all. Therefore in order to delay the computation of ρ, we have to keep track of the *context* in which components of ρ should be computed when needed later on. To keep track of this context information, we will use extended environments (ξ) instead of ordinary environments. Extended environments differ from ordinary environments in that a variable may also be bound to a *context (closure)* which is just a tuple consisting of a Block together with, again, an extended environment.

Definition The set Ξ of extended environments is the smallest set such that

- $\Theta \subseteq \Xi$,
- Let $\xi \in \Xi, x \in X, v \in \mathcal{V}$. Then $\xi[x \mapsto v]$ is also in Ξ,

– Let $\xi_1, \xi_2 \in \Xi, x \in \mathbf{X}, B \in \mathbf{B}$. Then $\xi_1[x \mapsto (B, \xi_2)]$ is also in Ξ.

Note Because of the possible nesting of equations, delayed computations can be nested too. This is why in $\xi_1[x \mapsto (B, \xi_2)]$ we must allow ξ_2 to be an extended environment and not just an ordinary environment.

Using extended (ξ) instead of ordinary environments (θ), code fragment ($\star$) becomes:

$$\texttt{solve}_1(B\nabla, \xi[tv(B) \mapsto (B, \xi)], x)$$

Eliminating needless computation as explained above, function $\texttt{SOLVE}_1$ is refined into the function $\texttt{SOLVE}_2$ listed below.

Notation We shall simply write $[\![\, B \parallel \xi \,]\!]$ instead of $[\![\, B \parallel \overline{\xi} \,]\!]$, where $\overline{\xi}$ is the ordinary environment 'equivalent' with ξ, i.e., $\overline{\theta} = \theta$, $\overline{\xi[x \mapsto v]} = \overline{\xi}[x \mapsto v]$, and $\overline{\xi_1[x \mapsto (B, \xi_2)]} = \overline{\xi_1}[x \mapsto [\![\, B \parallel \overline{\xi_2} \,]\!](x)]$.

---

function $\texttt{SOLVE}_2(B, \theta, x) = \texttt{solve}_2(B, \theta, x)$
-- **return** $[\![\, B \parallel \theta \,]\!](x)$

function $\texttt{solve}_2(B, \xi, x) =$
-- **return** $[\![\, B \parallel \xi \,]\!](x)$
begin
  case
    [] $x \notin lv(B)$ and $\xi(x) \in \mathcal{V}$: **return** $\xi(x)$
    [] $x \notin lv(B)$ and $\xi(x) \equiv (B_d, \xi_d)$: $\texttt{solve}_2(B_d, \xi_d, x)$ -- compute delayed value
    [] $x \in lv(B)$ and $nd(B, x) = 1$:
      Let $eq(B, x) = \langle x \stackrel{\sigma}{=} f \rangle$
      $v_{cur} := $ **if** $\sigma = \ominus$ **then** $\bot$ **else** $\top$ **fi**
      loop
        -- lazy computation of $f([\![\, B \setminus x \parallel \xi[x \mapsto v_{cur}] \,]\!])$
        $\rho := \epsilon \quad r := \texttt{EVAL}(f, \rho)$
        while $r \notin \mathcal{V}$ **do**
          $v := \texttt{solve}_2(B \setminus x, \xi[x \mapsto v_{cur}], r) \quad \rho := \rho[r \mapsto v] \quad r := \texttt{EVAL}(f, \rho)$
        od
        $v_{new} := r$
        -- $v_{new} = f([\![\, B \setminus x \parallel \xi[x \mapsto v_{cur}] \,]\!])$
        if $v_{cur} = v_{new}$ **then exit loop fi**
        $v_{cur} := v_{new}$
      endloop
      return v_{cur}
    [] $x \in lv(B)$ and $nd(B, x) > 1$: $\texttt{solve}_2(B\nabla, \xi[tv(B) \mapsto (B, \xi)], x)$
  endcase
end

Note We obviously don't want to compute the delayed value $[\![\, B_d \parallel \xi_d \,]\!](x)$ anew each time it is needed. An actual implementation should first check whether or not this delayed value has already been computed previously. If so, then this previously computed value is returned instead of calling $\texttt{solve}_2(B_d, \xi_d, x)$.

The following definition and fact are useful for proving termination of SOLVE_2.

Definition Let $B \in \mathbf{B}$ and let $\theta \in \Theta$. Then define the closure $cl(B, \theta) \subseteq \mathbf{B} \times \Xi$ as the least set such that:

- $(B, \theta) \in cl(B, \theta)$,
- Let $(B', \xi') \in cl(B, \theta)$, let $x \in tv(B')$ and let $v \in \mathcal{V}$. Then $(B' \setminus x, \xi'[x \mapsto v])$ is also in $cl(B, \theta)$.
- Let $(B', \xi') \in cl(B, \theta)$. Then $(B'\nabla, \xi'[tv(B') \mapsto (B', \xi')])$ is also in $cl(B, \theta)$.

It is clear that $(B', \xi') \in cl(B, \theta)$ for any call $\texttt{solve}_2(B', \xi', \_)$ that is entered during execution of $\text{SOLVE}_2(B, \theta, \_)$.

Fact (4) Let $(B_1, \xi_1) \in cl(B, \theta)$ and let $\xi_1(x) = (B_2, \xi_2)$. Then:

$$\{y \notin lv(B_2) \,|\, \xi_2(y) \in \mathcal{V}\} = \{y \notin lv(B_1) \,|\, \xi_1(y) \in \mathcal{V} \text{ and } nd(B, y) \leq nd(B, x)\}$$

Theorem $\text{SOLVE}_2(B, \theta, x)$ correctly computes $[\![B \parallel \theta]\!](x)$

The advantage of postponed evaluation is that only information that is really needed is computed. As such SOLVE_2 has the same worst-case time complexity as SOLVE_1, i.e., exponential in $|lv(B)|$. For practical applications, however, SOLVE_2 is expected to perform significantly better than SOLVE_1, especially if Blocks are really sparse.

5 Avoiding Needless Re-Computation

The reason for the bad worst-case time complexity of SOLVE_2 is that each $\texttt{solve}_2$-call is treated isolated, i.e., there is no re-use of information between $\texttt{solve}_2$-calls.[2] Take for example call $\texttt{solve}_2(B \setminus x, \xi[x \mapsto v_{cur}], y)$ in the body of the inner while-loop. This call may be executed for different values of y and v_{cur}. There is, however, not the slightest re-use of information between any two such calls. The goal of this section is precisely to refine (extend) function $\texttt{solve}_2$ such that, apart from returning $[\![B \parallel \xi]\!](x)$, it also returns some additional information. This additional information should be *useful*, i.e., it should allow a considerable speed-up for subsequent $\texttt{solve}_2$-calls, and it should be *easy to compute*. The

---

[2] Function $\texttt{solve}_2$ may even be called more than once with exactly the same argument values. This, of course, could easily be alleviated by dynamically keeping track of a table in which all results computed so far are stored, and then performing a look-up before actually computing a component. This would guarantee that the same information is computed at most once. However, it would not improve upon the worst case behaviour as the size of this table may grow exponentially in the number of variables.

additional information that we will consider takes the form of lower and upper bounds. We first indicate how knowledge of such lower/upper bounds can be exploited to speed up the computation process. Then we discuss some (simple) rules that can be used to compute lower/upper bounds.

Using upper/lower bounds. The cornerstone of SOLVE is iterative extremal fix-point computation. For example $\mu\Omega$ is computed as the limit point of the increasing sequence $\Omega^0(\bot) \sqsubseteq \Omega^1(\bot) \sqsubseteq \Omega^2(\bot) \sqsubseteq \Omega^3(\bot) \sqsubseteq \ldots$. Now if it is known in advance that v_l (resp. v_u) is a lower (resp. upper) bound for $\mu\Omega$, then the following more refined schema can be used: Iteration starts at v_l and terminates when a fix-point of Ω is reached or when v_u is reached, whatever occurs first. Hence knowledge of upper/lower bounds allows to reduce the number of iteration steps needed to compute extremal fix-points.

```
-- v_l ⊑ μΩ ⊑ v_u
v_cur := v_l
loop
    if v_cur = v_u then exit loop fi
    v_new := v_cur ⊔ Ω(v_cur)
    if v_cur = v_new then exit loop fi
    v_cur := v_new
endloop
-- v_cur = μΩ
```

The next definition is just to make lower/upper bound information of Blocks more explicit.

Definition Let $\mathcal{L}, \mathcal{U} \in \Theta$ such that $\mathcal{L} \sqsubseteq \mathcal{U}$. Then define the Block $B\lfloor\mathcal{L},\mathcal{U}\rceil$:

$$\texttt{nil}\lfloor\mathcal{L},\mathcal{U}\rceil = \texttt{nil} \quad \text{and} \quad ((\langle x \stackrel{\sigma}{=} f \rangle \bullet B')\lfloor\mathcal{L},\mathcal{U}\rceil = \langle x \stackrel{\sigma}{=} f' \rangle \bullet (B'\lfloor\mathcal{L},\mathcal{U}\rceil)$$

where $f' = \lambda\theta.(f(\theta) \sqcup \mathcal{L}(x)) \sqcap \mathcal{U}(x)$.

Observation Let $x \in lv(B)$. Then $\mathcal{L}(x) \sqsubseteq [\![B\lfloor\mathcal{L},\mathcal{U}\rceil \parallel \theta]\!](x) \sqsubseteq \mathcal{U}(x)$

Computing lower/upper bounds. The computation of lower/upper bound information is based upon elimination fact (3) together with the following facts:

(5) $\forall B \in \mathbf{B}, \theta_1, \theta_2 \in \Theta : \theta_1 \sqsubseteq \theta_2 \Rightarrow [\![B \parallel \theta_1]\!] \sqsubseteq [\![B \parallel \theta_2]\!]$
(6) $\forall B \in \mathbf{B}, \theta, \mathcal{L}, \mathcal{U} \in \Theta : \mathcal{L} \sqsubseteq [\![B \parallel \theta]\!] \sqsubseteq \mathcal{U} \Rightarrow [\![B \parallel \theta]\!] = [\![B\lfloor\mathcal{L},\mathcal{U}\rceil \parallel \theta]\!]$

Elimination fact (3) comes into play upon exit from the loop-construct in $\texttt{solve}_2$ because at that point we have that $v_{cur} = [\![B \parallel \xi]\!](x)$. Hence, by fact (3), lower and upper bounds obtained for $[\![B \setminus x \parallel \xi[x \mapsto v_{cur}]]\!]$ can be fully re-used for the subsequent computation of $[\![B \parallel \xi]\!]$-components.

The monotonicity fact (5) allows a partial re-use of information in the following way: In $\sigma = \ominus$ (resp. $\oplus$) then the successive values obtained by variable v_{cur} form an increasing (resp. decreasing) chain $v_0, v_1, v_2, \ldots$. Hence if $\sigma = \ominus$ (resp.

⊕) then lower (resp. upper) bound information obtained for $[\![\, B \setminus x \parallel \xi[x \mapsto v_i]\,]\!]$ can be re-used later on when computing components of $[\![\, B \setminus x \parallel \xi[x \mapsto v_j]\,]\!]$ with $j > i$.

Fact (6) comes into play when computing several different components of the same Block. This is for example the case inside the while-loop of solve$_2$ where different components of $[\![\, B \setminus x \parallel \xi[x \mapsto v_{cur}]\,]\!]$ are computed for the same value of v_{cur}. To see how fact (6) can speed up the computation process, assume that we have just computed $[\![\, B \parallel \xi \,]\!](x)$. As a result of this computation we also obtained lower bounds $\mathcal{L}$ and upper bounds $\mathcal{U}$ such that $\mathcal{L} \sqsubseteq [\![\, B \parallel \xi \,]\!] \sqsubseteq \mathcal{U}$. Hence, by fact (6), $[\![\, B \parallel \xi \,]\!] = [\![\, B \lfloor \mathcal{L}, \mathcal{U} \rceil \parallel \xi \,]\!]$. I.e., the computation of subsequent components of $[\![\, B \parallel \xi \,]\!]$ is reduced to computing components of $[\![\, B \lfloor \mathcal{L}, \mathcal{U} \rceil \parallel \xi \,]\!]$, which can be done more efficient as the 'range' of the right hand sides of $B \lfloor \mathcal{L}, \mathcal{U} \rceil$ is smaller compared to right hand sides of B.

Note 1 Fact (6) is a generalization of fact (3). This generalization in necessary because fact (3) only allows to eliminate a variable x from B on condition that $[\![\, B \parallel \theta \,]\!](x)$ is known exactly. However, as a result of applying monotonicity fact (5), we do not always know the exact value of components. Unlike fact (3), fact (6) allows to simplify Blocks if only *partial* information on components (under the form of lower/upper bounds) is available.

Note 2 The only point where lower/upper bound information is actually *generated* is upon exit from the loop-construct in solve$_2$. There we have that $v_{cur} = [\![\, B \parallel \xi \,]\!](x)$ and hence v_{cur} acts as a lower and an upper bound for $[\![\, B \parallel \xi \,]\!](x)$. Facts (3) and (5) only allow to *pass (propagate)* lower/upper bound information from $[\![\, B_1 \parallel \theta_1 \,]\!]$ to $[\![\, B_2 \parallel \theta_2 \,]\!]$ where (B_1, θ_1) and (B_2, θ_2) are different but related.

Incorporating lower/upper bound information as explained above, function SOLVE$_2$ is refined into a function SOLVE$_3$ (see appendix). Thanks to the re-use of information, the average run time behaviour of SOLVE$_3$ is significantly better than for SOLVE$_2$. Under the assumption that delayed values are never needed (i.e. the second case-arm of solve$_3$ is never applicable), we can even prove that SOLVE$_3$ runs in time polynomial in $|lv(B)|$ and only exponential in the alternation depth $ad(B)$, where $ad(\texttt{nil}) = 0$ and $ad(B) = 1 + ad(B\triangledown)$ if $B \neq \texttt{nil}$. The above assumption is trivially satisfied in case B is alternation free, i.e. $ad(B) = 1$. Hence SOLVE$_3$ runs in time polynomial in $|lv(B)|$ for alternation free Blocks.

6 Application : Model Checking

In this section we apply algorithm SOLVE to the field of automated system verification, in casu *model checking*. We briefly sketch how SOLVE can be turned into a local model checker, i.e. an algorithm for automatically deciding whether a concurrent system satisfies a desired property.

Modeling Behaviour . We use (labeled) finite state machines for modeling the operational behavior of (concurrent) systems. A finite state machine (fsm) is a triple $\mathcal{M} = \langle \mathcal{S}, \mathcal{A}, \rightarrow \rangle$ where $\mathcal{S}$ is a finite set of system states, $\mathcal{A}$ is a finite set of actions that the system is capable of performing, and $\rightarrow \subseteq \mathcal{S} \times \mathcal{A} \times \mathcal{S}$ is the transition relation, representing the state transitions resulting from the execution of actions. We shall write $s \xrightarrow{a} s'$ instead of $(s, a, s') \in \rightarrow$. Hence $s \xrightarrow{a} s'$ expresses that the system may perform action a when in state s and in doing so it moves to state s'.

Specifying System Properties . Various temporal and modal logics have been proposed for describing system properties. One particular expressive logic is the propositional modal mu-calculus [K]. This calculus allows a wide range of system properties to be expressed, including liveness, safety and fairness properties. Examples of properties expressible as modal mu-formulas are: 'eventually action a will happen', 'it is always possible to perform action a, 'action a can happen infinitely often', etc. The expressive power of the mu-calculus mainly stems from the possibility of nested (alternating) fix-points. Lots of other logics have uniform encodings in the mu-calculus.

Fix a fsm $\mathcal{M} = \langle \mathcal{S}, \mathcal{A}, \rightarrow \rangle$. Define $\mathbf{B}_{mu}$ as the set of Blocks over the lattice $\langle 2^{\mathcal{S}}, \subseteq \rangle$ with right hand side functions of the form:[3]

$$\lambda \theta. \bigcup_{x \in Y} \theta(x) \quad | \quad \lambda \theta. \bigcap_{x \in Y} \theta(x) \quad | \quad \lambda \theta. \langle a \rangle \theta(x) \quad | \quad \lambda \theta. [a] \theta(x)$$

where Y ranges over 2^{X}, x over X and a over $\mathcal{A}$. Furthermore, for any $Q \subseteq \mathcal{S}$, $\langle a \rangle Q = \{s \in \mathcal{S} | \exists s' \in Q : s \xrightarrow{a} s'\}$ and $[a] Q = \{s \in \mathcal{S} | \forall s' \in \mathcal{S} : s \xrightarrow{a} s'$ implies $s' \in Q\}$.

Note that $\mathbf{B}_{mu}$ grosso modo corresponds to the modal mu-calculus. It is fairly straightforward to transform modal mu-formulas into Blocks of $\mathbf{B}_{mu}$. Hence $\mathbf{B}_{mu}$ can be used for specifying properties of $\mathcal{M}$, in the same way as modal mu-formulae. As an illustration consider the modal mu-formula $\nu x_1.(\mu x_2.([a] x_1 \wedge [b] x_2))$ with $\mathcal{A} = \{a, b\}$. This formula, taken from [C], states that it is always the case that an action a is infinitely often possible, assuming that $\mathcal{M}$ has no deadlock states. The x_1-component of the following Block corresponds to the above modal mu-formula:

$$\langle x_1 \stackrel{\oplus}{=} \lambda \theta. \theta(x_2) \rangle \bullet \langle x_2 \stackrel{\ominus}{=} \lambda \theta. \theta(x_3) \cap \theta(x_4) \rangle \bullet \langle x_3 \stackrel{\ominus}{=} \lambda \theta. [a] \theta(x_1) \rangle \bullet B$$

where $B \equiv \langle x_4 \stackrel{\ominus}{=} \lambda \theta. [b] \theta(x_2) \rangle \bullet \texttt{nil}$

More formally properties of $\mathcal{M}$ can be specified by a triple (B, θ, x) where $B \in \mathbf{B}_{mu}$, $x \in lv(B)$ indicates the component of B we are actually interested in, and $\theta : X \rightarrow 2^{\mathcal{S}}$ gives an interpretation for the free variables of B, i.e., variables not in $lv(B)$ that occur in the right hand sides of B. (Free variables of B play the same role as atomic propositions of modal mu-formulas).

---

[3] We use normal forms. Furthermore $\bigcup_{\emptyset} = \emptyset$ and $\bigcap_{\emptyset} = \mathcal{S}$.

SOLVE as a (local) model checker. Given a system $\mathcal{M}$ and a property specification (B, θ, x) for $\mathcal{M}$, we can use algorithm SOLVE to compute $[\![\,B \parallel \theta\,]\!](x)$. This type of model checking is usually referred to as *global* model checking [EL,CKS,CES,AC,CS1, CS2,LP,A1,VL1] because *all* states of $\mathcal{M}$ satisfying a specified property are computed. This application of SOLVE is however not very interesting. The reason is that the Blocks involved in most property specifications are not really sparse.

In practice, however, one is not so much interested in all states that satisfy a given property, but only in a few states, typically the initial system states. It then seems overwhelming to compute the set of states satisfying a property just to check whether the state of interest is in this set. This idea is central to the development of *local* model checkers [A1,C,SW,W,VL2]. Given a system $\mathcal{M}$, a state s of $\mathcal{M}$ and a property specification (B, θ, x), a local model checker will try to check whether $s \in [\![\,B \parallel \theta\,]\!](x)$ without having to compute the complete set $[\![\,B \parallel \theta\,]\!](x)$. The rest of this section sketches how such a local model checker for $\mathbf{B}_{mu}$ can be obtained using SOLVE.

The key idea is to 'merge' $\mathcal{M}$ and B in order to obtain an 'equivalent' boolean Block. A boolean Block is a Block over the lattice $\langle \{\text{false, true}\}, \leq \rangle$ with the usual ordering $\text{false} < \text{true}$. More precisely define a translation $\mathbf{Tr}$ from $\mathbf{B}_{mu}$ into boolean Blocks as follows:[4]

$$\mathbf{Tr}(\text{nil}) = \text{nil} \quad \mathbf{Tr}(\langle x \stackrel{\sigma}{=} f \rangle \bullet B') = \langle x.s_1 \stackrel{\sigma}{=} f \star s_1 \rangle \bullet \ldots \bullet \langle x.s_n \stackrel{\sigma}{=} f \star s_n \rangle \bullet \mathbf{Tr}(B')$$

where $\mathcal{S} = \{s_1, \ldots, s_n\}$ and $f \star s$ is defined as follows:[5]

$(\lambda \theta. \bigcup_{x \in Y} \theta(x)) \star s = \lambda \theta'. \bigvee_{x \in Y} \theta'(x.s)$
$(\lambda \theta. \bigcap_{x \in Y} \theta(x)) \star s = \lambda \theta'. \bigwedge_{x \in Y} \theta'(x.s)$
$(\lambda \theta. \langle a \rangle \theta(x)) \star s = \lambda \theta'. \bigvee_{s' \in \mathcal{S} \mid s \stackrel{a}{\to} s'} \theta'(x.s')$
$(\lambda \theta. [a] \theta(x)) \star s = \lambda \theta'. \bigwedge_{s' \in \mathcal{S} \mid s \stackrel{a}{\to} s'} \theta'(x.s')$

It can now be proved that Block B is equivalent with $\mathbf{Tr}(B)$ in the following sense:

$$\forall x \in \mathbf{X}, \forall s \in \mathcal{S} : \quad s \in [\![\,B \parallel \theta\,]\!](x) \iff [\![\,\mathbf{Tr}(B) \parallel \theta'\,]\!](x.s) = \text{true}$$

where $\theta' = \lambda(x.s).$ if $s \in \theta(x)$ then true else false.

Hence in order to check wether a state s is in $[\![\,B \parallel \theta\,]\!](x)$, we simply compute $[\![\,\mathbf{Tr}(B) \parallel \theta'\,]\!](x.s)$ using SOLVE. In this way SOLVE is turned into a local model checker for $\mathbf{B}_{mu}$.

This use of SOLVE is much more interesting because, as opposed to B, Block $\mathbf{Tr}(B)$ usually exhibits a high degree of sparseness. There is

- *Semantic* sparseness: To evaluate a boolean expression usually only the values for a subset of variables occurring in that expression are needed,

---

[4] Note that variables of $\mathbf{Tr}(B)$ are of the form $x.s$, where $x \in \mathbf{X}$ and $s \in \mathcal{S}$.
[5] $\bigvee_\emptyset = \text{false}$ and $\bigwedge_\emptyset = \text{true}$.

- *Syntactic* sparseness: A right hand side of $\mathbf{Tr}(B)$ usually only contains a small number of boolean variables. This is a result of the fact that for most application systems the number of transitions leaving a state is small compared to number of states. I.e. $|\rightarrow|$ is of the order $\mathcal{O}(|\mathcal{S}|)$, rather than $\mathcal{O}(|\mathcal{S}|^2)$.

Note There is no need to explicitly store and construct $\mathbf{Tr}(B)$ from B in order to compute $[\![\,\mathbf{Tr}(B)\parallel\theta'\,]\!](x.s)$. Instead we use the rules for $f\star s$. Hence we only have to store B together with the fragment of $\mathcal{M}$ that has been explored so far.

7 Discussion

We have presented a new local algorithm for computing nested fix-points over complete lattices of finite height. Several improvements of the basic algorithm have been proposed. As an illustration, we sketched how the proposed algorithm could be used as a local model checker for (essentially) modal mu-formulas. Lots of other application areas can be thought of, e.g. data flow analysis.

Local model checkers for the modal mu-calculus have been presented, under the form of tableau systems, by [SW,C] and, under the form of a rewrite relation, by [W]. A fundamental difference between SOLVE and [SW,C,W] is that SOLVE is based upon lazy fix-point approximation, whereas [SW,C,W] are based upon an implicit use of fix-point induction. Nevertheless, there seems to be some interesting similarities. For example in [SW] definition lists are used and fresh constants are introduced for renaming recursive propositions each time they are unwound. The role of this definition list and constants seems to be similar to our closures: Both serve to save the necessary context information needed to compute delayed values when they are needed later on. In [C] hypotheses sets are used. This seems to be yet another equivalent way for representing closures. The set $H(B_1,\xi_1)$ of hypotheses associated with a closure $(B_1,\xi_1)\in cl(B,\theta)$ is roughly given by $\{(x,\xi_1(x))\,|\,x\in lv(B)\text{ and }x\notin lv(B_1)\text{ and }\xi_1(x)\in\mathcal{V}\}$. Note that knowledge of $H(B_1,\xi_1)$ is sufficient to construct (B_1,ξ_1) from (B,θ). Hence SOLVE can be rewritten solely in terms of hypotheses sets. Furthermore let $\xi_1(x)=(B_2,\xi_2)$. Then rephrasing fact (4) in terms of hypotheses sets yields: $H(B_2,\xi_2)=H(B_1,\xi_1)\setminus\{(y,v)\in H(B_1,\xi_1)\,|\,nd(B,x)<nd(B,y)\}$, explaining in a way the removal of hypotheses in rules R7,R8 and DR3 of [C].

Thanks to the re-use of information, the average run time behaviour of SOLVE$_3$ is significantly better than for SOLVE$_2$. For the special case where B is alternation free, i.e. $ad(B)=1$, SOLVE$_3$ even runs in time polynomial in $|lv(B)|$. In [L1] Larsen also gives a polynomial time local model checker for alternation free boolean Blocks. The improvement of [L1] over [L2] is essentially based on facts (3) and (5). These two facts indeed suffice in the boolean case. However for lattices other than the boolean lattice, fact (6) is also needed, because fact (6) allows to simplify Blocks when only partial information on component values is known.

Despite the re-use of information, the worst-case behaviour of SOLVE$_3$ is still exponential in $|lv(B)|$ if $ad(B) > 1$. The reason is that the re-use of information doesn't work for delayed values. Intuitively, this can be seen as follows: Consider the recursive call $\mathtt{solve}_3((B\triangledown)\lfloor\mathcal{L},\mathcal{U}\rfloor,\xi[tv(B) \mapsto (B\lfloor\mathcal{L},\mathcal{U}\rceil,\xi)], x)$. in function solve$_3$. If the value of a variable $y \in tv(B)$ turns out to be needed during execution of this call, then the computation of x is temporarily suspended and the delayed value $[\![B\lfloor\mathcal{L},\mathcal{U}\rfloor \parallel \xi]\!](y)$ is computed. Hence at this point we are computing different components (x and y) of the same Block $[\![B\lfloor\mathcal{L},\mathcal{U}\rfloor \parallel \xi]\!]y$ without there being any re-use of information. This causes the exponential blow-up. Local algorithms that are polynomial in $|lv(B)|$ and only exponential in $ad(B)$ are contained in [X] for the full mu-calculus, and in [A2] for Blocks of alternation depth 2. We believe that it should be possible to refine SOLVE$_3$ into a polynomial time algorithm. This could probably be done along the following lines: If during execution of $\mathtt{solve}_3((B\triangledown)\lfloor\mathcal{L},\mathcal{U}\rfloor,\xi[tv(B) \mapsto (B\lfloor\mathcal{L},\mathcal{U}\rfloor,\xi)], x)$ the value of a variable $y \in tv(B)$ is needed, then we do *not* compute this value, because this caused the exponential blow-up. Instead we *assume* a value $\mathcal{L}(y)$ for y if y is a $\ominus$ variable, and a value $\mathcal{U}(y)$ if y is a $\oplus$ variable. When $\mathtt{solve}_3((B\triangledown)\lfloor\mathcal{L},\mathcal{U}\rfloor,\xi[tv(B) \mapsto (B\lfloor\mathcal{L},\mathcal{U}\rfloor,\xi)], x)$ then returns, we check whether the assumption for y was justified. If not then x has to be recomputed using the correct value for y. The point is that in such a modified schema there is an optimal re-use of information: The lower/upper bounds information obtained by $\mathtt{solve}_3((B\triangledown)\lfloor\mathcal{L},\mathcal{U}\rfloor,\xi[tv(B) \mapsto (B\lfloor\mathcal{L},\mathcal{U}\rfloor,\xi)], x)$ can be fully re-used when checking the y-assumption, and the information obtained from checking this y-assumption can be fully re-used when re-computing x with the correct y-value. The above ideas are of course very informal and need be further explored. This is an interesting topic for future research.

References

[A1] Andersen, H. R.: *Model Checking and Boolean Graphs*, ESOP'92, LNCS 582, 1992

[A2] Andersen, H. R. : *Verification of Temporal Properties of Concurrent Systems*, PhD thesis, Aarhus University, DAIMI PB - 445, 1993

[AC] Arnold, A., Crubille, P.: *A linear algorithm to solve fixed-points equations on transition systems*, Information Processing Letters, vol.29, 57-66, 1988

[CES] Clarke, E.M., Emerson, E.A., Sistla, A.P.: *Automatic verification of finite-state concurrent systems using temporal logic specifications*, ACM Transactions on Progr. Languages and Systems, Vol.8, No. 2, pp. 244-263, April 1986

[C] Cleaveland, R.: *Tableau-based model checking in the propositional mucalculus*, Acta Informatica, 1990

[CKS] Cleaveland, R., Klein, M., Steffen, B.: *Faster Model Checking for the Modal MuCalculus*, CAV'92, LNCS 663

[CS1] Cleaveland, R., Steffen, B.: *Computing Behavioural Relations, Logically*, ICALP 91, pp. 127-138, LNCS 510

[CS2] Cleaveland, R. and Steffen, B.: *A Linear-Time Model-Checking Algorithm for the Alternation-Free Modal Mu-Calculus*, CAV'91, LNCS 575, 1992

[EL] Emerson, E.A., Lei, C.-L.: *Efficient model checking in fragments of the propositional µ-calculus*, LICS, 267-278, 1986

[K] Kozen, D.: *Results on the propositional mu-calculus*, TCS 17, 1983

[L1] Larsen, K.G.: *Efficient Local Correctness Checking*, CAV'92, LNCS 663

[L2] Larsen, K.G.: *Proof systems for Hennessy-Milner logic with recursion*, CAAP, 1988, see also TCS, 72, 1990

[LP] Lichtenstein, O., Pnueli, A.: *Checking that finite state concurrent programs satisfy their linear specification*, (Proc.) 12th ACM annual Symposium on Principles of Programming Languages, pp. 97-107, 1985

[SW] Stirling, C., Walker, D.: *Local model checking in the modal mu-calculus*, TCS, October 1991, see also LNCS 351, 369-383, CAAP 1989

[S] Stirling, C.: *Modal and Temporal Logics*, in Handbook of Logic in Computer Sciences, Volume 2. Edited by S. Abramsky, M. Gabbay and T.S.E. Maibaum; Oxford Science Publications, 1992

[T] Tarski, A.: *A Lattice-Theoretical Fixpoint Theorem and its Applications*, Pacific Journal of Mathematics, 5: 285-309, 1955

[VL1] Vergauwen, B., Lewi, J.: *A linear algorithm for solving fixed points equations on transition systems*, CAAP'92, LNCS 581, 322-341

[VL2] Vergauwen, B., Lewi, J.: *A Linear Local Model Checking Algorithm for CTL*, CONCUR'93, LNCS 715

[W] Winskel, G.: *A note on model checking the modal nu-calculus*, ICALP, LNCS 372, 1989, see also TCS 83, 1991

[X] Xinxin, L.: *Specification and Decomposition in Concurrency*, PhD thesis, Aalborg University, 1992, R 92-2005

Appendix

function SOLVE$_3$(B, θ, x) =
begin
  $(v, \_, \_) := \text{solve}_3(B\lfloor \lambda y \in \mathbf{X}.\bot, \lambda y \in \mathbf{X}.\top \rceil, \theta, x)$
  return v
end

function solve$_3$($B\lfloor \mathcal{L}, \mathcal{U} \rceil, \xi, x$) =
-- return $(v_{cur}, \mathcal{L}', \mathcal{U}')$ such that
-- $v_{cur} = [\![B\lfloor \mathcal{L}, \mathcal{U} \rceil \parallel \xi]\!](x)$
-- $\mathcal{L} \sqsubseteq \mathcal{L}' \sqsubseteq \mathcal{U}' \sqsubseteq \mathcal{U}$ and $[\![B\lfloor \mathcal{L}, \mathcal{U} \rceil \parallel \xi]\!] = [\![B\lfloor \mathcal{L}', \mathcal{U}' \rceil \parallel \xi]\!]$
begin
  case
    [] $x \notin lv(B)$ and $\xi(x) \in \mathcal{V}$: **return** $(\xi(x), \mathcal{L}, \mathcal{U})$
    [] $x \notin lv(B)$ and $\xi(x) \equiv (B_d \lfloor \mathcal{L}_d, \mathcal{U}_d \rceil, \xi_d)$:
      $(v, \_, \_) := \text{solve}_3(B_d \lfloor \mathcal{L}_d, \mathcal{U}_d \rceil, \xi_d, x)$
      return $(v, \mathcal{L}, \mathcal{U})$
    [] $x \in lv(B)$ and $nd(B, x) = 1$:
      Let $eq(B, x) = \langle x \stackrel{\sigma}{=} f \rangle$
      $v_{cur} := $ **if** $\sigma = \ominus$ **then** $\boxed{\mathcal{L}(x)}$ **else** $\boxed{\mathcal{U}(x)}$ **fi**
      $\boxed{v_{bnd} := \text{if } \sigma = \ominus \text{ then } \mathcal{U}(x) \text{ else } \mathcal{L}(x) \text{ fi}}$
      $\boxed{\mathcal{L}' := \mathcal{L} \ \ \mathcal{U}' := \mathcal{U}}$
      loop
        -- $[\![(B \setminus x) \lfloor \mathcal{L}, \mathcal{U} \rceil \parallel \xi[x \mapsto v_{cur}]]\!] = [\![(B \setminus x) \lfloor \mathcal{L}', \mathcal{U}' \rceil \parallel \xi[x \mapsto v_{cur}]]\!]$
        $\boxed{\text{if } v_{cur} = v_{bnd} \text{ then exit loop fi}}$
        -- lazy computation of $f([\![(B \setminus x) \lfloor \mathcal{L}', \mathcal{U}' \rceil \parallel \xi[x \mapsto v_{cur}]]\!])$
        $\rho := \epsilon \quad r := \text{EVAL}(f, \rho)$
        while $r \notin \mathcal{V}$ **do**
          $(v, \boxed{\mathcal{L}', \mathcal{U}'}) := \text{solve}_3(\boxed{(B \setminus x) \lfloor \mathcal{L}', \mathcal{U}' \rceil}, \xi[x \mapsto v_{cur}], r)$
          $\rho := \rho[r \mapsto v]$
          $r := \text{EVAL}(f, \rho)$
        od
        $v_{new} := r$
        -- $v_{new} = f([\![(B \setminus x) \lfloor \mathcal{L}', \mathcal{U}' \rceil \parallel \xi[x \mapsto v_{cur}]]\!])$
        $\boxed{v_{new} := \text{if } \sigma = \ominus \text{ then } v_{new} \sqcup v_{cur} \text{ else } v_{new} \sqcap v_{cur} \text{ fi}}$
        if $v_{cur} = v_{new}$ **then exit loop fi**
        $v_{cur} := v_{new}$
        $\boxed{\text{if } \sigma = \ominus \text{ then } \mathcal{U}' := \mathcal{U} \text{ else } \mathcal{L}' := \mathcal{L} \text{ fi}}$
      endloop
      return $(v_{cur}, \boxed{\mathcal{L}'[x \mapsto v_{cur}], \mathcal{U}'[x \mapsto v_{cur}]})$
    [] $x \in lv(B)$ and $nd(B, x) > 1$:
      $\text{solve}_3((B \triangledown) \lfloor \mathcal{L}, \mathcal{U} \rceil, \xi[tv(B) \mapsto (B \lfloor \mathcal{L}, \mathcal{U} \rceil, \xi)], x)$
  endcase
end

How Linear Can Branching-time Be?

Orna Grumberg[1] and Robert P. Kurshan[2]

[1] The Technion, Department of Computer Science, Haifa 32000, Israel.
Email: orna@cs.technion.ac.il
[2] AT&T Bell Laboratories, Murray Hill, NJ 07974, USA Email: k@research.att.com

Abstract. We suggest a new characterization that draws finer lines between branching-time and linear-time formulas of the logic CTL*. We define three types of linearity, *strong linearity*, *sub-linearity* and *equi-linearity*, each of which contains all LTL formulas. We prove that these notions are distinct. Moreover, strong linearity implies sub-linearity which implies equi-linearity. We investigate these notions over Kripke structures with and without fairness and show that they do not coincide. We give a syntactic characterization for linear ∀CTL* formulas. Finally, we discuss the practical implication of the new characterization.

1 Introduction

Finite-state concurrent systems arise in several applications of computer science, like hardware designs and communicating protocols. Such systems might easily become very complicated and require formal methods to check their correctness. Fortunately, for finite-state system, efficient verification algorithms were developed and implemented.

There are two main approaches to verifying finite-state systems. One approach describes the specification by means of a formula in some propositional temporal logic. The system is verified by applying a model checking procedure that checks whether the system model satisfies the specification formula. This approach supports both linear-time and branching-time specifications. LTL [16] is a linear-time temporal logic that assumes a unique future at any specific time. CTL [3] is a branching-time temporal logic that assumes several possible futures at any moment. In addition to the usual temporal operators CTL contains path quantifiers, ∃ ("there is a path") and ∀ ("for every path"). CTL* [3] is a powerful logic that is strictly more expressive then LTL and CTL [7]. Model checking procedures have been suggested for each one of these logics [13], [4], [18], [8].

The other approach describes both the system and the required specification as finite-state models. Verification is done by checking that every behavior of the system model is a behavior of the specification model. When linear-time behaviors are considered, both models are treated as finite automata on infinite tapes and their ω-regular languages are compared. Such an approach is taken, for instance, in [12], [20]. When branching-time behaviors are considered, the models are compared by means of bisimulation relation [15] or simulation relation [14].

There is a close relationship between the two approaches. A CTL* formula will have the same value (true or false) on all models that are bisimilar to each other [1]. Similarly, a LTL formula will have the same value on all models that have identical languages. The latter, however, is not an accurate characterization of LTL formulas. Consider the CTL* formula $f = \forall \mathbf{G}\, a \vee \forall \mathbf{G}\, \neg a$. It is true of a model if the model either satisfies a everywhere (along every path, in any state along the path) or satisfies $\neg a$ everywhere. This formula is not in LTL and no LTL formula is equivalent to it. However, it has the same value on all models which have identical languages.

Thus, we conclude that the distinction between branching and linear formulas within CTL* is not fine enough. We define three types of linear properties that relate CTL* formulas to the language of models rather than to their branching structure. The new characterization draws finer lines between subclasses of CTL* formulas. In the conclusion we refer also to some practical implications of the characterization.

Each type of linearity is identified by a property of the models that agree on any formula of that type. The "most linear" type, called strong linearity, requires that a formula f will be true of any structure whose language is contained in some fixed language $\mathcal{L}_f$. The next one, called sub-linearity, requires that if f is true of some structure, then it is true of any structure whose language is contained in the first one. The third type, called equi-linearity, requires that f will have the same value for any two structures that have identical languages.

Clearly, LTL formulas have all three properties. However, we show that only the set of strong linear formulas coincides with LTL. We also show strict inclusion between the three notions and between CTL*.

We examine the three properties with respect to models that include fairness constraints. Fairness constraints are often used to model a system at some level of abstraction. For such models, both the language induced by a model and bisimulation relations are defined with respect to fair paths. These models are a special form of Kripke structures, over which temporal logics are usually interpreted. We also examine these notions with respect to regular Kripke structures, in which every path is considered fair. We show that, every formula that has some property when fairness is considered, has the same property when no fairness is assumed. Surprisingly, we found a formula that has different linearity property when considered over fair structures and over regular Kripke structures.

We pay special attention to a special subset of CTL*, called ∀CTL*, in which only universal path quantifiers are allowed. This subset is strictly more expressive than LTL. ∀CTL* formulas are preserved by simulation relation, i.e., if they are true of a structure, then they are also true of structures that are smaller by the simulation relation. When simulation is used to relate a system model to some abstract model, correctness of the system model can be deduced by checking ∀CTL* formulas on the abstract model. This approach has been taken for instance in [9], [19], [5], [6]. Consequently, ∀CTL* is useful in techniques that combine model checking and abstraction. We show that for ∀CTL*, sub-linearity and equi-linearity coincide. We give a syntactic characterization of a special subset of the (sub- and equi-) linear ∀CTL* formulas.

The paper is organized as follows: Section 2 describes the model and the logics. Section 3 defines simulation and bisimulation relations and states some of their properties. Two special structures - the union structure and the intersection structure are defined in this section. The next section describes the three types of linearity and proves their relationship. Section 5 gives a syntactic characterization of ∀CTL* linear formulas. In Section 6 we conclude by discussing some practical implications of this characterization.

2 Basic Definitions

We assume a fixed finite set of atomic propositions AP over which we define structures, languages and logics.

2.1 Fair structures

Definition 2.1: A fair structure is a tuple $P = (S, s_0, R, L, F)$ where, S is a finite set of states; $s_0 \in S$ is the initial state; $R \subseteq S \times S$ is a transition relation; $L : S \to \mathcal{P}(AP)$ is the labeling function that maps each state to the set of atomic propositions true in that state; and $F \subseteq S$ is the Büchi fairness constraints.

When $F = S$ we say that P has *trivial fairness constraints*. We denote by (P, s) the structure which is identical to P except that s is its initial state. A structure P is *deterministic* iff for every $s \in S$, if $(s, t) \in R$ and $(s, u) \in R$ then $L(t) \neq L(u)$.

An infinite sequence of states $\pi = s_1, s_2, \ldots$ is a path from s in P iff $s_1 = s$, and for every $i \geq 1$, $(s_i, s_{i+1}) \in R$. For a sequence of states $\pi = s_1, s_2, \ldots$ we define $\inf(\pi) = \{s \mid s = s_i \text{ for infinitely many } i\}$ and $word(\pi) = L(s_1), L(s_2), \ldots$. A path $\pi = s_1, s_2, \ldots$ is *fair* iff $\inf(\pi) \cap F \neq \emptyset$. The notation π^i will denote the suffix of π that begins at s_i. Similarly, $word(\pi)^i = word(\pi^i)$.

Definition 2.2: The language of P from s, $\mathcal{L}(P, s)$, is the set of all sequences $\alpha \in AP^\omega$ for which there is a fair path π in P from s, such that $word(\pi) = \alpha$.

The language of P, is defined by $\mathcal{L}(P) = \mathcal{L}(P, s_0)$.

2.2 Temporal logics

In this section we present several subsets of the temporal logic CTL* over the given set AP of atomic propositions. We will assume that formulas are expressed in positive normal form, in which negations are applied only to atomic propositions. This facilitates the definition of universal and existential subsets of CTL*. The logics are interpreted over a form of Kripke structure with fairness constraints, as defined in the previous section. Path quantifiers range over the fair paths of the structures. Since negations are not allowed anywhere, we need to have both conjunction and disjunction. The next-time operator **X** is the dual of itself and the operator **V** is the dual of **U**. Intuitively, f **V** g will hold of a path if g holds as long as f is false. Only if f becomes true then g may become false at a later point.

Definition 2.3: The logic CTL* is the set of state formulas given by the following inductive definition.

1. If $p \in AP$, then p and $\neg p$ are state formulas.
2. If f and g are state formulas, then so are $f \wedge g$ and $f \vee g$.
3. If f is a path formula, then $\forall f$ and $\exists f$ are state formulas.
4. If f is a state formula, then f is a path formula.
5. If f and g are path formulas, then so are $f \wedge g$, $f \vee g$, $\mathbf{X} f$, $f \mathbf{U} g$, and $f \mathbf{V} g$.

The abbreviations *true*, *false* and $\rightarrow$ are defined as usual. We also use the following abbreviations: $\mathbf{F} f$ and $\mathbf{G} f$, where f is a path formula, denote $(true\ \mathbf{U}\ f)$ and $(false\ \mathbf{V}\ f)$, respectively.

CTL is a branching-time subset of CTL* in which path quantifiers may only precede a restricted set of path formulas. More precisely, CTL is the logic obtained by eliminating rules 3 through 5 above and adding the following rule.

3'. If f and g are state formulas, then $\forall \mathbf{X} f$, $\forall (f \mathbf{U} g)$, and $\forall (f \mathbf{V} g)$ are state formulas. Also, $\exists \mathbf{X} f$, $\exists (f \mathbf{U} g)$, and $\exists (f \mathbf{V} g)$ are state formulas.

LTL is a linear-time subset of CTL*. It consists of formulas that have the form $\forall f$ where f is a path formula in which the only state subformulas permitted are atomic propositions. This definition is due to [7]. Standard definition of LTL usually does not include path quantifiers. More precisely, a path formula of LTL (in positive normal form) is either:

1. If $p \in AP$ then p and $\neg p$ are path formulas.
2. If f and g are path formulas, then so are $f \wedge g$, $f \vee g$, $\mathbf{X} f$, $f \mathbf{U} g$, and $f \mathbf{V} g$.

$\forall$CTL* and $\exists$CTL* (universal and existential CTL*) are subsets of CTL* in which the only allowed path quantifiers are $\forall$ and $\exists$, respectively. $\forall$CTL and $\exists$CTL are the restriction of $\forall$CTL* and $\exists$CTL* to CTL. We now consider the semantics of the logic CTL* with respect to a fair structure.

Definition 2.4: Satisfaction of a state formula f by a state s ($s \models f$) and of a path formula g by a fair path π ($\pi \models g$) is defined inductively as follows.

1. $s \models p$ iff $p \in L(s)$; $s \models \neg p$ iff $p \notin L(s)$.
2. $s \models f \wedge g$ iff $s \models f$ and $s \models g$. $s \models f \vee g$ iff $s \models f$ or $s \models g$.
3. $s \models \forall(f)$ iff for every fair path π starting at s, $\pi \models f$.
   $s \models \exists(f)$ iff there is a fair path π starting at s, $\pi \models f$.
4. $\pi \models f$, where f is a state formula, iff the first state of π satisfies f.
5. $\pi \models f \wedge g$ iff $\pi \models f$ and $\pi \models g$. $\pi \models f \vee g$ iff $\pi \models f$ or $\pi \models g$.
6. (a) $\pi \models \mathbf{X} f$ iff $\pi^1 \models f$.
   (b) $\pi \models f \mathbf{U} g$ iff there is $n \in \mathcal{N}$ such that $\pi^n \models g$ and for all $i < n$, $\pi^i \models f$.
   (c) $\pi \models f \mathbf{V} g$ iff for all $n \in \mathcal{N}$, if $\pi^i \not\models f$ for all $i < n$, then $\pi^n \models g$.

$P \models f$ indicates that $s_0 \models f$.

3 Simulations and Language Containment

Following [9], we define a *fair simulation relation* H from a structure P to a structure P'. Simulation and bisimulation for trivial fairness constraints were defined by Milner in [15], [14].

Definition 3.1: $H \subseteq S \times S'$ is a fair simulation relation from P to P', iff

1. $H(s_0, s'_0)$ and,
2. For every $s \in S, s' \in S'$, $H(s,s')$ implies
   (a) $L(s) = L'(s')$, and
   (b) For every fair path $\pi = s_1, s_2, \ldots$ where $s_1 = s$ there is a fair path $\pi' = s'_1, s'_2, \ldots$ where $s'_1 = s'$ such that $\forall i : H(s_i, s'_i)$.

Definition 3.2: $H \subseteq S \times S'$ is a bisimulation relation between P and P' iff H is a simulation relation that satisfies in addition:

2.(c) For every fair path $\pi' = s'_1, s'_2, \ldots$ where $s'_1 = s'$ there is a fair path $\pi = s_1, s_2, \ldots$ where $s_1 = s$ such that $\forall i : H(s_i, s'_i)$.

We use $P \preceq P'$ to denote that there is a simulation relation from P to P' (P simulates P'). $P \equiv P'$ denotes that there is a bisimulation relation between P and P'. Note that having two simulation relations H_1 from P to P' and H_2 from P' to P does not imply that there is a bisimulation between P and P'. A relation is a bisimulation iff it is symmetric and it is a simulation from P to P' and from P' to P.

It is easy to see that if two structures are bisimilar then they have exactly the same language. Similarly, if $P \preceq P'$ then $\mathcal{L}(P) \subseteq \mathcal{L}(P')$. The other direction, however, is not true in general. Only if P' is deterministic, language inclusion implies simulation and equality of languages implies bisimulation [9].

Theorem 3.3:[[9]]

1. If $P \preceq P'$ then for every $\forall CTL^*$ formula f, $P' \models f \implies P \models f$.
2. If $P \preceq P'$ then for every $\exists CTL^*$ formula f, $P \models f \implies P' \models f$.
3. If $P \equiv P'$ then for every CTL^* formula f, $P' \models f \iff P \models f$.

We now define two new structures: the *intersection structure* and the *union structure*. The new structures have the property that their languages are the intersection and the union, respectively, of the languages of their components. Moreover, the intersection structure simulates each of its components and each component simulates the union structure. We will see in next sections that these structures are useful in proving linearity properties.

Definition 3.4: Let P_1 and P_2 be two structures with identically labeled initial states, i.e., $L_1(s_1^0) = L_2(s_2^0)$. The intersection of P_1 and P_2 is the structure $P_1 \cap P_2 = (S, s_0, R, L, F)$ where,

1. $S = \{(s_1, s_2, c) \mid s_1 \in S_1, s_2 \in S_2, c \in \{0, 1\} \text{ and } L_1(s_1) = L_2(s_2)\}$.
2. $s_0 = (s_1^0, s_2^0, 1)$ where s_0^1, s_0^2 are the initial states of P_1, P_2, respectively.
3. $((s_1, s_2, c), (s'_1, s'_2, c')) \in R$ iff $(s_1, s'_1) \in R_1$, $(s_2, s'_2) \in R_2$ and

$$c' = 0 \Leftrightarrow (c = 1 \wedge s_1 \in F_1) \vee (c = 0 \wedge s_2 \notin F_2)$$

$$c' = 1 \Leftrightarrow (c = 0 \wedge s_2 \in F_2) \vee (c = 1 \wedge s_1 \notin F_1).$$

4. $L((s_1, s_2, c)) = L_1(s_1) = L_2(s_2)$.
5. $F = F_1 \times S_2 \times \{1\}$.

Theorem 3.5: $\mathcal{L}(P_1 \cap P_2) = \mathcal{L}(P_1) \cap \mathcal{L}(P_2)$; $P_1 \cap P_2 \preceq P_1$ and $P_1 \cap P_2 \preceq P_2$.

Definition 3.6: Let P_1 and P_2 be two structures such that S_1 and S_2 are disjoint, and $L_1(s_1^0) = L_2(s_2^0)$. The union structure of P_1 and P_2, $P_1 \cup P_2 = (S, s_0, R, L, F)$ is defined as follows.

1. $S = \{s_0\} \cup S_1 \cup S_2$ where s_0 does not appear in S_1 or in S_2.
2. s_0 is the initial state.
3. $R = R_1 \cup R_2 \cup \{(s_0, s') \mid (s_1^0, s') \in R_1 \text{ or } (s_2^0, s') \in R_2\}$.
4. $L(s) = L_1(s)$ if $s \in S_1$; $L(s) = L_2(s)$ if $s \in S_2$; $L(s_0) = L(s_1^0) = L(s_2^0)$.
5. $F = F_1 \cup F_2$.

For any transition from either s_1^0 or s_2^0 to some s, we add a transition in $P_1 \cup P_2$ from s_0 to s. However, there is no transition in $P_1 \cup P_2$ from some s to s_0. This is done to ensure that each path in the language of $P_1 \cup P_2$ is in the language of either P_1 or P_2 and no path is an interleaving of paths in P_1 and P_2. The transitions to and from s_1^0 or s_2^0 are not eliminated so that paths that go through an initial state more than once will not be truncated.

It should be noted that, for any state $s \neq s_0$ in $P_1 \cup P_2$, $(P_1 \cup P_2, s) \equiv (P_1, s)$, if $s \in S_1$ and $(P_1 \cup P_2, s) \equiv (P_2, s)$, if $s \in S_2$. This implies that the corresponding languages are also identical.

Theorem 3.7: $\mathcal{L}(P_1 \cup P_2) = \mathcal{L}(P_1) \cup \mathcal{L}(P_2)$; $P_1 \preceq P_1 \cup P_2$ and $P_2 \preceq P_1 \cup P_2$.

Corollary 3.8 1. For every $\forall CTL^*$ formula f, if $P_1 \cup P_2 \models f$ then both P_1 and P_2 satisfy f. Moreover, if $P_1 \cap P_2$ is not empty (i.e. it has reachable states) then if either $P_1 \models f$ or $P_2 \models f$ then $P_1 \cap P_2 \models f$.
2. For every $\exists CTL^*$ formula f, if $P_1 \cap P_2 \models f$ then both P_1 and P_2 satisfy f. Moreover, if either P_1 or P_2 satisfies f then $P_1 \cup P_2 \models f$.

4 Linear Properties

We investigate three types of linearity for CTL* formulas.

Definition 4.1: A formula $f \in CTL^*$ is equi-linear iff for every structures P_1, P_2,

$$\mathcal{L}(P_1) = \mathcal{L}(P_2) \Longrightarrow [P_1 \models f \Longleftrightarrow P_2 \models f].$$

Definition 4.2: A formula $f \in CTL^*$ is sub-linear iff for every structures P_1, P_2,

$$\mathcal{L}(P_1) \supseteq \mathcal{L}(P_2) \Longrightarrow [P_1 \models f \Longrightarrow P_2 \models f].$$

Definition 4.3: $f \in CTL^*$ is strong linear if there is a ω-regular language $\mathcal{L}_f$ such that for every structure P,

$$P \models f \Longleftrightarrow \mathcal{L}(P) \subseteq \mathcal{L}_f.$$

Lemma 4.4: If f is strong linear then f is sub-linear. Furthermore, If f is sub-linear then f is equi-linear.

To emphasize that we consider arbitrary fairness constraints we denote by EQL_F, SBL_F and STL_F the sets of equi-linear, sub-linear and strong linear CTL* formulas, respectively.

In the sequel we identify formulas that have different type of linearity when considered over structures with *trivial fairness constraints* $(F = S)$. A formula will be equi-linear for trivial constraints if for every two structures with trivial fairness constraints, whenever they have the same languages then they either both satisfy the formula or both falsify it. Thus, if a formula is equi-linear for the general case, it is also equi-linear for the trivial case. The definitions of sub-linearity and strong linearity are changed accordingly. EQL, SBL and STL will denote the subsets of CTL* formulas that are equi-linear, sub-linear and strong linear for trivial fairness constraints. Clearly, $EQL_F \subseteq EQL$, $SBL_F \subseteq SBL$, and $STL_F \subseteq STL$.

4.1 CTL*, equi-linearity and sub-linearity

We now examine the relationships between the sets of CTL*, equi-linear and sub-linear formulas. To show that CTL* is strictly larger than EQL_F, we first present a CTL* formula that is not equi-linear. We also show a ∀CTL formula which is not equi-linear. We then define a subset of ∃CTL* formulas which are all equi-linear. We show that most of these formulas are not sub-linear. Thus, the set of equi-linear CTL* formulas is strictly larger than the set of sub-linear CTL* formulas. On the other hand, we show that for ∀CTL* formulas the sets of equi-linear formulas and sub-linear formulas are identical.

Example 4.5 *The CTL formula* $\forall \mathbf{G}\, \exists \mathbf{F}\, \neg a$ *is not equi-linear. To see that, consider the structures* P_1 *and* P_2 *of Figure 1. The initial states are pointed to by double arrows and the fair states are marked by double circles. In this example both structures are trivially fair. The label* a *in a state denotes that the proposition* a *is true of this state.*

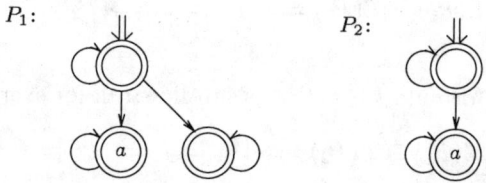

Fig. 1. $\mathcal{L}(P_1) = \mathcal{L}(P_2) = \overline{a}^\omega + \overline{a}^+ a^\omega$, however, $P_1 \not\models \forall \mathbf{G}\, \exists \mathbf{F}\, a$ and $P_2 \models \forall \mathbf{G}\, \exists \mathbf{F}\, a$.

Example 4.6 *The $\forall CTL$ formula $\forall \mathbf{X}(\forall \mathbf{X}\, a \vee \forall \mathbf{X}\, \neg a)$ is not equi-linear. To see that, consider the structures P_1 and P_2 of Figure 2 (with trivial fairness constraints). This example is due to Milner [15]. It has been used to demonstrate the difference between bisimulation and language equivalence.*

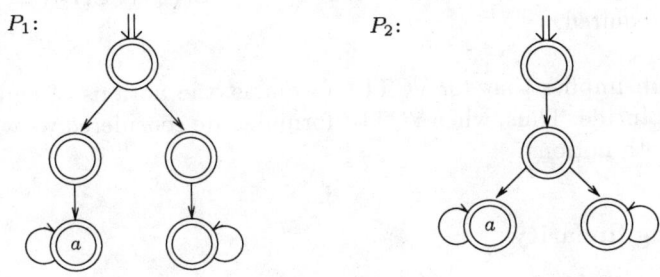

Fig. 2. $P_1 \models \forall \mathbf{X}(\forall \mathbf{X}\, a \vee \forall \mathbf{X}\, \neg a)$ and $P_2 \not\models \forall \mathbf{X}(\forall \mathbf{X}\, a \vee \forall \mathbf{X}\, \neg a)$

Lemma 4.7: If f is a LTL path formula then $\exists f$ is equi-linear.

This is true since if a LTL formula is satisfied by some path π then it is satisfied by any path π' with $word(\pi) = word(\pi')$. Thus, every two structures with identical languages agree on $\exists f$.

Lemma 4.8: If $\exists f$ is sub-linear, then f is equivalent to either *true* or *false* or to a boolean combination of atomic propositions.

Proof: Assume by way of contradiction that $\exists f$ is sub-linear but f is not equivalent to either *true* or *false* or a boolean combination of atomic propositions. Then, there are two structures such that $P_1 \models \exists f$ and $P_2 \not\models \exists f$.

We first show that there are such P_1, P_2 that have identically labeled initial states. To see that, consider the case in which for any two structures, if their initial states are identically labeled then the structures either both satisfy f or both falsify f. Then f is equivalent to a boolean combination of atomic propositions. Since f is assumed not to have this form, we conclude that there are structures P_1, P_2 as required, and for such structures, $P_1 {\cup} P_2$ is defined.

$\exists f \in \exists CTL^*$ and $P_1 \models \exists f$, hence $P_1 {\cup} P_2 \models \exists f$. Moreover, $\exists f$ is sub-linear and $\mathcal{L}(P_2) \subseteq \mathcal{L}(P_1 {\cup} P_2)$ therefore $P_2 \models \exists f$. A contradiction. Thus, if $\exists f$ is sub-linear then f is either *true* or *false* or a boolean combination of atomic propositions.
□

Corollary 4.9 *The set of equi-linear formulas is strictly larger than the set of sub-linear formulas.*

Theorem 4.10: Let f be a $\forall CTL^*$ formula. Then f is equi-linear if and only if f is sub-linear.

Proof: We showed already that every sub-linear formula is equi-linear.

For the other direction, assume f is equi-linear, $\mathcal{L}(P_1) \supseteq \mathcal{L}(P_2)$ and $P_1 \models f$. Since $\mathcal{L}(P_1) \supseteq \mathcal{L}(P_2)$, the initial states of P_1 and P_2 are identically labeled and therefore their intersection $P = P_1 \cap P_2$ is defined. $P \preceq P_1$ and $P_1 \models f$, therefore $P \models f$. Moreover, f is equi-linear and $\mathcal{L}(P) = \mathcal{L}(P_1) \cap \mathcal{L}(P_2) = \mathcal{L}(P_2)$, thus $P_2 \models f$, as required. □

This theorem implies that for $\forall CTL^*$ formulas, the notions of equi-linear and sub-linear coincide. Thus, when $\forall CTL^*$ formulas are considered we will use linear to denote both notions.

4.2 Strong linearity

In this section, we examine the set of strong linear $\forall CTL^*$ formulas. We show that this set is exactly LTL. It is then easy to show that sub-linearity is strictly weaker than strong linearity.

Theorem 4.11: f is strong linear iff $f \in$ LTL.

We base our proof on the characterization of LTL formulas, given in [2]. For every CTL^* formula f, let f^d denote the LTL formula obtained from f by eliminating all its path quantifiers. Given a path $\pi = xy^\omega$, $P(xy^\omega)$ will denote the single-path structure defined by π.[3]

Lemma 4.12:[[2]] For every formula f in CTL^*, f is expressible in LTL iff f is equivalent to $\forall f^d$.

Lemma 4.13:[[2]] Given a structure P and a fair path π in P, then for every $n \geq 0$ there exists a prefix xy of π such that xy^ω is an infinite fair path in P and such that for every LTL formula f with $length(f) \leq n$, $\pi \models f \iff xy^\omega \models f$.

Proof: [of the Theorem] If f is linear then by [20] there is a Büchi automaton that accepts exactly all sequence satisfying f. The language accepted by the Büchi automaton is the required $\mathcal{L}_f$.

For the other direction, let f be strong linear, then there is a ω-regular language $\mathcal{L}_f$ such that for every structure P, $P \models f \iff \mathcal{L}(P) \subseteq \mathcal{L}_f$.

Assume by way of contradiction that f is not expressible in LTL. Then, by the previous lemma, there is a structure P for which either $P \models f$ and $P \not\models \forall f^d$ or $P \not\models f$ and $P \models \forall f^d$.

Consider the first case. $P \models f$ implies that $\mathcal{L}(P) \subseteq \mathcal{L}_f$. On the other hand, $P \not\models \forall f^d$ means that there is a path π of P such that $\pi \not\models f^d$. By the previous lemma, there is a path xy^ω in P that agrees with π on every formula of length smaller than $length(f)$. In particular, xy^ω agrees with π on f^d and therefore, $xy^\omega \not\models f^d$.

---
[3] details can be found in [2].

Let $P(xy^\omega)$ be the single-path structure defined by xy^ω. $P(xy^\omega) \not\models f^d$, and therefore, since it has a single path, $P(xy^\omega) \not\models f$. Since f is strong linear, and since $\mathcal{L}(P(xy^\omega)) = \{xy^\omega\}$, xy^ω should not be included in $\mathcal{L}_f$. However, $xy^\omega \in \mathcal{L}(P)$ and therefore $xy^\omega \in \mathcal{L}_f$. A contradiction.

The other case is similar. □

Example 4.14 *The $\forall CTL$ formula $\forall \mathbf{G}\, a \vee \forall \mathbf{G}\, \neg a$ is not strong linear.*

4.3 Trivial and non trivial fairness constraints

We show the effect of the fairness constraints on linearity by examining the $\forall CTL$ formula $\forall \mathbf{F}\, \forall \mathbf{G}\, a$. This formula has been shown in [2] to be inexpressible in LTL. Thus, it is certainly not strong linear. Consider Figure 4, it summarizes the relationship between eight subsets of CTL* formulas. On the right hand side we have full CTL*. On the left hand side we have LTL that coincides with the sets of strong linear formulas over non trivial and trivial fairness constraints (STL_F, STL). The sets of equi-linear formulas and sub-linear formulas lie in between, where the sub-linear are "closer" to LTL than the equi-linear formulas. The arrows show inclusion between sets (the smaller set is the one that is pointed to). Each arrow is labeled by a formula that differentiates between the sets: it belongs to the larger set and does not belong to the smaller one.

We see that for arbitrary fairness constraints, $\forall \mathbf{F}\, \forall \mathbf{G}\, a$ differentiates between CTL* and EQL_F. Thus, in this case it is not even equi-linear. On the other hand, for trivial fairness constraints it differentiates only between sub-linear and strong linear. Hence, it does belong to EQL and SBL.

In the rest of the section we prove the properties of $\forall \mathbf{F}\, \forall \mathbf{G}\, a$.

Example 4.15 *The $\forall CTL$ formula $\forall \mathbf{F}\, \forall \mathbf{G}\, a$ is not equi-linear over structures with non trivial fairness constraints. To see that, consider the structures P_1 and P_2 of Figure 3.*

Lemma 4.16: $\forall \mathbf{F}\, \forall \mathbf{G}\, a$ is linear over structures with trivial fairness constraints.
Proof: Let $\mathcal{L}(P_1) = \mathcal{L}(P_2)$ and assume $P_1 \models \forall \mathbf{F}\, \forall \mathbf{G}\, a$. Then, for every $\pi = s_1, s_2, \ldots$ there is the smallest j such that $s_l \models a$ for every $l \geq j$.

We first show that $s_1, \ldots, s_j$ does not contain a cycle. Assume the contrary, i.e., $s_1, \ldots, s_j$ contains a cycle $s_k, s_{k+1}, \ldots, s_k$. Since j is the smallest with the above property, $s_{j-1} \models \neg a$. $s_1, \ldots, s_k, (s_{k+1}, \ldots s_k)^\omega$ is a path in P_1 that contains no state that satisfies $\forall \mathbf{G}\, a$, in contradiction to the fact that $P_1 \models \forall \mathbf{F}\, \forall \mathbf{G}\, a$.

From the above argument we conclude that $j \leq m$, where m is the number of states in P_1 and consequently, $\mathcal{L}(P_1) \subseteq \{a, \neg a\}^m a^\omega$. $\mathcal{L}(P_1) = \mathcal{L}(P_2)$, thus $\mathcal{L}(P_2) \subseteq \{a, \neg a\}^m a^\omega$.

Assume $P_2 \not\models \forall \mathbf{F}\, \forall \mathbf{G}\, a$. Then, $P_2 \models \exists \mathbf{G}\, \exists \mathbf{F}\, \neg a$, i.e., there is a path $\pi = s_1, s_2, \ldots$ in P_2 such that for every i, $s_i \models \exists \mathbf{F}\, \neg a$. That means that for every i, there is a path $t_i, t_{i+1}, \ldots$ from s_i and there is $l \geq i$, such that $t_l \models \neg a$. Consider the path $\pi' = s_1, \ldots, s_{m+1}, t_{m+2}, \ldots t_l$. $word(\pi')$ is not a prefix of any

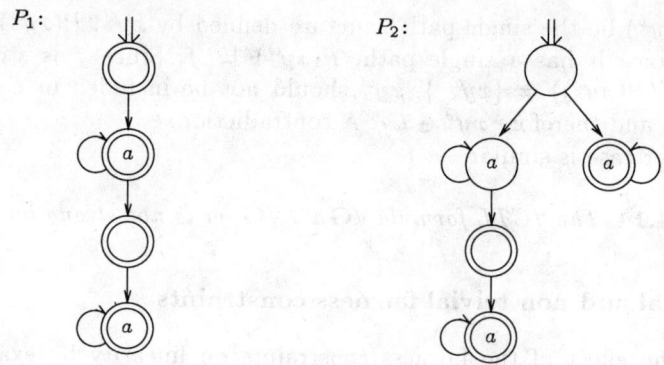

Fig. 3. $\mathcal{L}(P_1) = \mathcal{L}(P_2) = \overline{a}a^\omega + \overline{a}a^+ \overline{a}a^\omega$. However, $P_1 \not\models \forall \mathbf{F}\, \forall \mathbf{G}\, a$ while $P_2 \models \forall \mathbf{F}\, \forall \mathbf{G}\, a$.

word in $\{a, \neg a\}^m a^\omega$. Consequently, this path is not a prefix of any path of P_2, contradicting the assumption that $P_2 \models \exists \mathbf{G}\, \exists \mathbf{F}\, \neg a$.

We therefore conclude that $P_2 \models \forall \mathbf{F}\, \forall \mathbf{G}\, a$ and that $\forall \mathbf{F}\, \forall \mathbf{G}\, a$ is equi-linear over structures with trivial fairness constraints. □

Lemma 4.17: $\forall \mathbf{F}\, \forall \mathbf{G}\, a$ is not strong linear over Kripke structures with trivial fairness constraints.

Proof: By [2], $\forall \mathbf{F}\, \forall \mathbf{G}\, a$ is not in LTL, and therefore it is not strong linear.

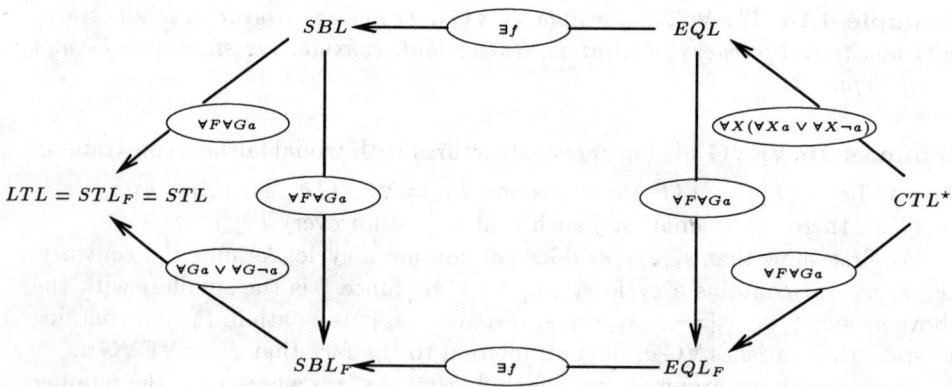

Fig. 4. Sets of linear formulas between CTL* and LTL

5 Syntactic Characterization of ∀CTL* Linear Formulas

In this section we give a syntactic characterization of an interesting subset of the linear ∀CTL* formulas[4]. We note that even for relatively simple formulas like $\forall \mathbf{X} f$, the formula might not be linear even if f is (consider again the formula $\forall \mathbf{X}(\forall \mathbf{X} a \vee \forall \mathbf{X} \neg a)$ which is not linear even though $(\forall \mathbf{X} a \vee \forall \mathbf{X} \neg a)$ is linear). Hence, we define a stronger notion called closed formulas and base on it the characterization of linear formulas. Intuitively, we say that a linear formula is *closed* if whenever it is true of two processes then it is also true of their union.

It is easy to show that all boolean combinations of atomic propositions and all LTL formulas are closed. For the operators $\wedge$, $\forall \mathbf{X}$ and $\forall \mathbf{G}$ we show that if they are applied to closed formulas then the resulting formula is closed.

We show that not in all cases, applying disjunction to closed formulas results in a closed formula. However, we find a necessary and sufficient condition under which disjunction does preserve closedness.

Closedness of formulas that involve the operators $\forall \mathbf{V}$, $\forall \mathbf{U}$ (and $\forall \mathbf{F}$ as a special case of $\forall \mathbf{U}$) are particularly hard to capture. We show examples to demonstrate the difficulty. Intuitively, each of these operators contains an implicit disjunction (this can be viewed in their fixpoint characterization) which often results in a non-closed formula, even when the operator is applied to closed formulas.

We do suggest a sufficient condition for formulas of the form $\forall \mathbf{F} f$ to be closed. However, this condition guarantees closedness only *over structures with trivial fairness constraints*. Based on this condition we can show that if f is closed then $\forall \mathbf{F} \forall \mathbf{G} f$ is closed over structures with trivial fairness constraints. This extends the result of the previous section. Recall that, $\forall \mathbf{F} \forall \mathbf{G} f$, when interpreted over structures with non-trivial fairness constraints, is not linear.

We now turn to a formal presentation of our results.

Definition 5.1: f is closed iff f is linear and for every two structures P_1 and P_2 with $L_1(s_1^0) = L_2(s_2^0)$, if $P_1 \models f$ and $P_2 \models f$ then $P_1 \cup P_2 \models f$.
Note that since $P_1 \preceq P_1 \cup P_2$ and $P_2 \preceq P_1 \cup P_2$ then if f is closed then $P_1 \cup P_2$ satisfies f iff both P_1 and P_2 satisfy f.

Lemma 5.2: If f is closed then $\forall \mathbf{X} f$ is closed.

Proof: We first show that $\forall \mathbf{X} f$ is linear. Let f be closed and let P_1, P_2 be two structures with $\mathcal{L}(P_1) \subseteq \mathcal{L}(P_2)$. Suppose that $P_2 \models \forall \mathbf{X} f$. Then, for every state s such that $(s_2^0, s) \in R_2$, $P_2, s \models f$. Consider a state s', such that $(s_1^0, s') \in R_1$. $\mathcal{L}(P, s') \subseteq \bigcup_{t \in T} \mathcal{L}(P_2, t)$ where $T = \{t \mid (s_2^0, t) \in R_2 \text{ and } L_1(s') = L_2(t)\}$. Since all $t \in T$ satisfy f and since f is closed, the union structure P of all structures $\langle P_2, t \rangle$ satisfies f as well. Since f is linear, $P_1, s' \models f$. This holds for every successor s' of s_1^0. Thus, $P_1, s_1^0 \models \forall \mathbf{X} f$ and $\forall \mathbf{X} f$ is linear.

To see that $\forall \mathbf{X} f$ is also closed consider P_1 and P_2 with $L_1(s_1^0) = L_2(s_2^0)$ and a union structure P, such that $P_1, s_1^0 \models \forall \mathbf{X} f$, $P_2, s_2^0 \models \forall \mathbf{X} f$. Let s be some state in P such that $(s_0, s) \in R$ and suppose $s \in S_1$. By the definition of P,

---

[4] Recall that, for ∀CTL* equi-linearity and sub-linearity coincide.

$(s_1^0, s) \in R_1$, thus $P_1, s \models f$. Moreover, $\mathcal{L}(P, s) = \mathcal{L}(P_1, s)$ and f is linear, thus, $P, s \models f$ and $\forall \mathbf{X} f$ is closed. □

Lemma 5.3: If f_1 and f_2 are closed then the following conditions are equivalent:

1. $f_1 \vee f_2$ is closed.
2. For every $\sigma \in 2^{AP}$, either $f_1 \wedge \sigma \to f_2$ or $f_2 \wedge \sigma \to f_1$.

Lemma 5.4: If f_1 and f_2 are linear then $f_1 \wedge f_2$ is linear; if f_1 and f_2 are closed then $f_1 \wedge f_2$ is closed.

To see that the other direction does not hold, consider $f_1 = \forall \mathbf{X} a \vee \forall \mathbf{X} b$ and $f_2 = \forall \mathbf{X} a$. f_1 is not closed, f_2 is closed and $f_1 \wedge f_2 = f_2$ is closed. (However, $f_2 \to f_1$ in this case. What happens when $f_2 \not\to f_1$?)

Lemma 5.5: If f is closed then $\forall \mathbf{G} f$ is closed.

The other direction does not hold. Consider $f = b \wedge \forall \mathbf{X}(b \wedge (\forall \mathbf{X} b \vee \forall \mathbf{X} \neg b))$. f is not linear and therefore not closed. However, $\forall \mathbf{G} f$ is equivalent to $\forall \mathbf{G} b$. Thus, $\forall \mathbf{G} f$ is closed.

Lemma 5.6: Every formula, constructed from boolean combinations of atomic propositions and from LTL formulas by applications of finitely many conjunctions, $\forall \mathbf{X}$ and $\forall \mathbf{G}$ is closed and therefore linear.

Lemma 5.7: $\forall \mathbf{F} \forall \mathbf{X} a$ is not linear.

Proof: Let P_1 and P_2 be the structures in Figure 5.

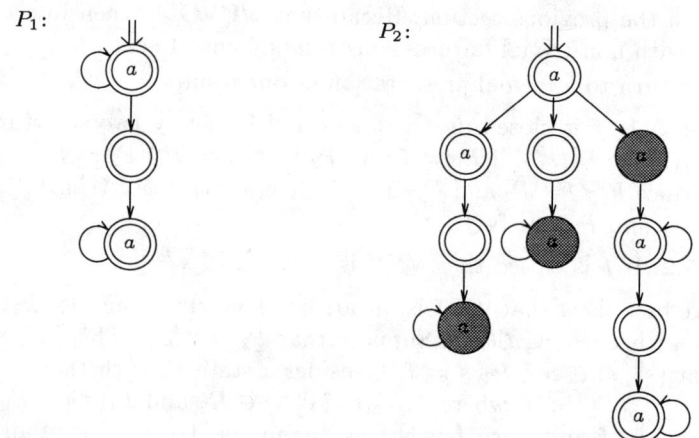

Fig. 5. $\forall \mathbf{F} \forall \mathbf{X} a$ is not linear

$\mathcal{L}(P_1, s_0) = \mathcal{L}(P_2, t_0)$. However, $P_2, s_0 \models \forall \mathbf{F} \forall \mathbf{X} a$ but $P_1, t_0 \not\models \forall \mathbf{F} \forall \mathbf{X} a$, since each state on the path $s_0 s_0 s_0 \ldots$ in P_1 has a successor labeled by $\neg a$. □

The following demonstrates the difficulty in identifying closed formulas of the form $\forall \mathbf{F}\, f$. $\forall \mathbf{F}\, a$, $\forall \mathbf{F}\, \forall \mathbf{G}\, a$ and $\forall \mathbf{F}\, \forall \mathbf{X}\, \forall \mathbf{G}\, a$ are linear while $\forall \mathbf{F}\, \forall \mathbf{X}\, a$ and $\forall \mathbf{F}\, \forall \mathbf{X} (a \wedge \forall \mathbf{X} (a \wedge \forall \mathbf{X}\, \forall \mathbf{G}\, \neg a))$ are not linear even though in all cases the formulas have the form $\forall \mathbf{F}\, f$ where f is closed.

Lemma 5.8: If f is closed and $f \rightarrow \forall \mathbf{X}\, f$ then $\forall \mathbf{F}\, f$ is closed.

Corollary 5.9 *If $\forall \mathbf{G}\, g$ is closed then both $\forall \mathbf{F}\, \forall \mathbf{G}\, g$ and $\forall \mathbf{F}\, \forall \mathbf{X}\, \forall \mathbf{G}\, g$ are closed.*

6 Direction for future research

In this work we suggested three types of linearity for CTL* formulas: equi-linearity, sub-linearity and strong linearity. We showed that the three notions are distinct and can be ordered by inclusion ordering. We also showed that the smallest, strong linearity coincides with LTL and that the largest, equi-linearity is strictly contained in CTL*.

The new characterization has several practical implications. For linear properties (of any type) checking simulation can be replaced by checking language inclusion and bisimulation can be replaced by language equivalence. This will enable verification tools, based on language inclusion (e.g. COSPAN [11], [10]), to handle a wider class of CTL* formulas.

Another implication concerns minimization of structures. Model checking procedures have in general efficient time complexity, however, they often fail to handle large systems due to space limitation. Thus, minimization techniques are valuable.

Minimization should preserve the properties of interest. If these properties are linear, then automata-based minimization can replace bisimulation-based minimization ([1]). Even though the latter is computed more efficiently, the former results in a larger reduction. This is true since two bisimilar states are also language equivalent, but not vice versa. Thus, for a given structure, the minimal structure that is bisimilar to it has, in general, more states than the minimal one that has identical language.

Acknowledgement: We would like to thank Orna Bernholtz, Dennis Dams and Ken McMillan for many useful discussions and helpful suggestions.

References

1. M. C. Browne, E. M. Clarke, and O. Grumberg. Characterizing finite Kripke structures in propositional temporal logic. *Theoretical Computer Science*, 59(1-2), July 1988.
2. E. M. Clarke and I. A. Draghicescu. Expressibility results for linear time and branching time logics. In *Linear Time, Branching Time, and Partial Order in Logics and Models for Concurrency*, volume 354, pages 428-437. Springer-Verlag: Lecture Notes in Computer Science, 1988.

3. E. M. Clarke and E. A. Emerson. Synthesis of synchronization skeletons for branching time temporal logic. In *Logic of Programs: Workshop, Yorktown Heights, NY, May 1981*, volume 131 of *Lecture Notes in Computer Science*. Springer-Verlag, 1981.
4. E. M. Clarke, E. A. Emerson, and A. P. Sistla. Automatic verification of finite-state concurrent systems using temporal logic specifications. In *Proceedings of the Tenth Annual ACM Symposium on Principles of Programming Languages*, January 1983.
5. E. M. Clarke, O. Grumberg, and D. E. Long. Model checking and abstraction. In *Proceedings of the Nineteenth Annual ACM Symposium on Principles of Programming Languages*, January 1992.
6. D. Dams, O. Grumberg, and R. Gerth. Generation of reduced models for checking fragments of CTL. In C. Courcoubetis, editor, *Proceedings of the Fifth Workshop on Computer-Aided Verification*, volume 697 of *Lecture Notes in Computer Science*. Springer-Verlag, July 1993.
7. E. A. Emerson and J. Y. Halpern. "Sometimes" and "Not Never" revisited: On branching time versus linear time. *Journal of the ACM*, 33:151–178, 1986.
8. E. A. Emerson and C.-L. Lei. Modalities for model checking: Branching time strikes back. In POPL85 [17].
9. O. Grumberg and D. E. Long. Model checking and modular verification. To appear in ACM Transactions on Programming Languages and Systems.
10. Z. Har'El and R. P. Kurshan. The COSPAN user's guide. Technical Report 11211-871009-21TM, AT&T Bell Laboratories, 1987.
11. Z. Har'El and R. P. Kurshan. Software for analytical development of communications protocols. *AT&T Technical Journal*, 69(1):45–59, Jan.–Feb. 1990.
12. R. P. Kurshan. Analysis of discrete event coordination. In J. W. de Bakker, W.-P. de Roever, and G. Rozenberg, editors, *Proceedings of the REX Workshop on Stepwise Refinement of Distributed Systems, Models, Formalisms, Correctness*, volume 430 of *Lecture Notes in Computer Science*. Springer-Verlag, May 1989.
13. O. Lichtenstein and A. Pnueli. Checking that finite state concurrent programs satisfy their linear specification. In POPL85 [17].
14. R. Milner. An algebraic definition of simulation between programs. In *Proceedings of the Second International Joint Conference on Artificial Intelligence*, September 1971.
15. R. Milner. *A Calculus of Communicating Systems*, volume 92 of *Lecture Notes in Computer Science*. Springer-Verlag, 1980.
16. A. Pnueli. A temporal logic of concurrent programs. *Theoretical Computer Science*, 13:45–60, 1981.
17. *Proceedings of the Twelfth Annual ACM Symposium on Principles of Programming Languages*, January 1985.
18. J.P. Quielle and J. Sifakis. Specification and verification of concurrent systems in CESAR. In *Proceedings of the Fifth International Symposium in Programming*, 1981.
19. G. Shurek and O. Grumberg. The modular framework of computer-aided verification: Motivation, solutions and evaluation criteria. In R. P. Kurshan and E. M. Clarke, editors, *Proceedings of the 1990 Workshop on Computer-Aided Verification*, June 1990.
20. M. Y. Vardi and P. Wolper. An automata-theoretic approach to automatic program verification. In *Proceedings of the First Annual Symposium on Logic in Computer Science*. IEEE Computer Society Press, June 1986.

First-Order Future Interval Logic

G. Kutty, L. E. Moser, P. M. Melliar-Smith, L. K. Dillon, Y. S. Ramakrishna

Department of Electrical and Computer Engineering
Department of Computer Science
University of California, Santa Barbara, CA 93106

Abstract. Future Interval Logic (FIL) is a linear-time temporal logic that is intended for specification and verification of reactive and concurrent systems. To make FIL useful for specifying and reasoning about practical systems, we present a first-order extension of FIL, including equality and n-ary function and predicate symbols, and a set of sound proof rules for reasoning in the logic. We illustrate the use of the logic by specifying a sliding window protocol and proving that the specifications satisfy a set of correctness requirements.

1 Introduction

Future Interval Logic (FIL) is a linear-time temporal logic, the textual counterpart of Graphical Interval Logic (GIL), that is intended for specification and verification of reactive and concurrent systems [10]. The visual representation [4] of GIL is intended to help programmers and system designers to create, understand and use formal specifications, but logicians may find the textual representation of FIL to be more compact and more appropriate for definition of the logic. Future Interval Logic is derived from the interval logic of [14].

The key construct of FIL is the interval, which provides a context in which properties are asserted to hold. Intervals are defined by sequences of searches. In FIL, a typical formula is $[\rightarrow f_1 \rightarrow f_2 \,|\, \rightarrow f_1 \rightarrow f_2 \rightarrow f_3 \rightarrow f_4)\, g$, which requires g to hold at the first state of the interval $[\rightarrow f_1 \rightarrow f_2 \,|\, \rightarrow f_1 \rightarrow f_2 \rightarrow f_3 \rightarrow f_4)$. The left endpoint of this interval is found by searching for the first state in which f_1 holds and then from that state for the first state in which f_2 holds. The right endpoint of the interval is found by starting at that state and searching for the state in which f_3 holds and then for the state in which f_4 holds. The graphical representation for this formula in GIL is

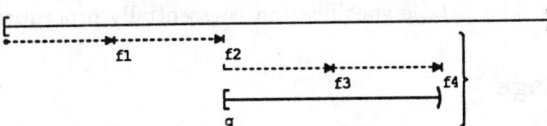

Future Interval Logic is equivalent in expressiveness to Propositional Temporal Logic without the *next* operator, as shown in [7], where we have also presented a deductive system for FIL. In [12] we have given a decision procedure for checking the validity of formulas in FIL. Based on this decision procedure, we have implemented a theorem prover as part of a toolset that also includes a syntax-directed editor and a specification database [6].

This research was supported in part by NSF/ARPA grant CCR-9014382.

Specification and verification of practical systems requires first-order quantification and induction. In this paper we present a first-order extension of FIL, including equality and n-ary function and predicate symbols, and provide a set of sound proof rules for reasoning in the logic. We then use the logic to specify a sliding window protocol for reliable communication over an unreliable communication medium, and prove that the specifications satisfy a set of correctness requirements.

2 Related Work

When a classical first-order theory is extended with modal operators, a number of complications arise. For instance, the rule for substitution of identities in predicate calculus is no longer sound in first-order temporal logic. This is illustrated by $\vdash i = j \Rightarrow (\Box p(i)) \Rightarrow \Box p(j))$, which is not sound because the equality $i = j$ is not necessarily true in all states.

Extending a first-order theory with modal operators requires a large number of decisions to be made regarding the nature of quantification. Many of these decisions can be made independently, resulting in a large number of quantified modal logics. Such logical systems are surveyed in [5], where it is also shown that some of these logical systems are necessarily incomplete.

Manna and Pnueli [9] have developed a proof system for first-order PTL. This system permits quantification over global variables. Abadi and Manna [2] have given a resolution-based proof system for first-order PTL. They have shown that the classical rules for Skolemization are not sound in first-order PTL and that quantifiers cannot necessarily be eliminated by Skolemization as in predicate calculus, and have given modified Skolemization rules for first-order PTL. Abadi [1] has shown that any proof system for first-order PTL is necessarily incomplete by proving that the validity of first-order PTL formulas is Π_1^1-complete. He has proved that arithmetic can be embedded in first-order PTL by representing a number i as the value of a specified variable at time i.

As an example application of first-order FIL, we consider the sliding window protocol. Other formal techniques have been used to specify the sliding window protocol, for instance, HOL [3], CSP [11], and Estelle [13]. The HOL description is axiomatic with explicit quantification over time. The CSP specification is process-algebraic in nature and represents the system using a set of processes that communicate with each other over channels. The Estelle specification is essentially procedural.

3 The Language

The non-temporal portion of FIL corresponds to classical propositional logic. To specify real systems we need to be able to use variables, constants, predicates, functions and quantifiers, as in first-order predicate calculus. We present a first-order extension of FIL augmented with variables, n-ary functions, n-ary predicates and constants. Variables take values from designated domains and are either local or global. A *local variable* may have different values in different states of a computation, while a *global variable* has the same value in all states. Formulas can be quantified over global variables. The syntax and semantics presented below extends the syntax and semantics for FIL presented in [7].

3.1 Syntax

Let $\mathcal{L}$ be a classical first-order theory with a set V of variables, a set F of function symbols, a set P of predicate symbols, and the logical symbols $\neg$, $\vee$ and $\exists$. Each element of F or P has a non-negative integer associated with it called its *arity*; constants are 0-ary function symbols. The set P includes the binary predicate $=$ that denotes equality. The set V is partitioned into two disjoint subsets GV and LV, the global and local variables, respectively. Similarly, P is partitioned into GP and LP, the global and local predicates, respectively. The set GP includes the equality predicate $=$.

We define an extension $FIL(\mathcal{L})$ of $\mathcal{L}$ below, the language of which is defined by the 5-tuple (GV, LV, GP, LP, F). In what follows, we use f to denote either a formula or a function symbol, as is clear from context. We denote an interval by $[\,\alpha\,|\,\beta\,)\,f$, where α is a search pattern that defines the left endpoint of the interval and β is a search pattern that defines its right endpoint. As the notation suggests, an interval is closed on the left and open on the right; this prevents the next operator from being expressed in the logic.

Definition 1 *A term of $FIL(\mathcal{L})$ is defined as follows:*
- *A variable is a term.*
- *If $a_1, a_2, \ldots, a_n$ are terms and $f \in F$ is an n-ary function symbol, then $f(a_1, a_2, \ldots, a_n)$ is a term.*

Definition 2 *A search pattern of $FIL(\mathcal{L})$ is defined as follows:*
- *If f is a formula, then $\rightarrow f$ is a search pattern.*
- *If f is a formula and α is a search pattern, then $\rightarrow f, \alpha$ is a search pattern.*

Definition 3 *A formula of $FIL(\mathcal{L})$ is defined as follows:*
- *true and false are formulas.*
- *If $a_1, a_2, \ldots, a_n$ are terms and $p \in P$ is an n-ary function symbol, then $p(a_1, a_2, \ldots, a_n)$ is a formula.*
- *If f and g are formulas, then $\neg f$ and $f \vee g$ are formulas.*
- *If α and β are search patterns and f is a formula, then $[\,\alpha\,|\,\beta\,)\,f$, $[\,-\,|\,\beta\,)\,f$ and $[\,\alpha\,|\rightarrow)\,f$ are formulas.*
- *If f is a formula and $x \in GV$, then $\exists x\, f$ is a formula.*

The derived operators $\wedge$, $\Rightarrow$, $\equiv$, $\Diamond$, $\Box$ and $\forall$ are defined in the usual manner.

3.2 Semantics

The semantics of an $FIL(\mathcal{L})$ formula in the language (GV, LV, GP, LP, F) are defined below.

Definition 4 *A model M for a formula of $FIL(\mathcal{L})$ is the triple (D, v_G, S), where*
- *D is a non-empty set called the domain of M. For each n-ary function symbol $f \in F$, an n-ary function $f_G : D^n \mapsto D$ is defined. For each n-ary global predicate symbol $p \in GP$, other than $=$, an n-ary predicate p_G on D^n is defined.*
- *$v_G : GV \mapsto D$ is a global valuation function that assigns a value to each global variable.*

- S is a finite or infinite sequence of states with indexing set I, where I is the set of non-negative integers or a finite prefix thereof. The length of such a sequence is denoted by $|S|$, where $|S| = \omega$ if S is infinite. For each n-ary local predicate symbol $p \in LP$ and each $i \in I$, an n-ary predicate $p_L(i)$ on D^n is defined. In addition, a local valuation function $v_L : LV \times I \mapsto D$ is defined that assigns a value to each local variable in each state.

Definition 5 *The value $v(t, i)$ of a term t in state i is defined as follows:*
- *For a global variable x, $v(x, i) = v_G(x)$.*
- *For a local variable y, $v(y, i) = v_L(y, i)$.*
- *For an n-ary function symbol $f \in F$,*
  $v(f(t_1, t_2, \ldots, t_n), i) = f_G(v(t_1, i), v(t_2, i), \ldots, v(t_n, i))$.

The semantics of $FIL(\mathcal{L})$ formulas are defined in terms of the models relation $\models$ using a special value $\perp$ in the case that a search to a formula fails.

Definition 6 *The models relation $\models$ for a model M and a pair of indices i, j, where $i, j \in I$ and $i < |S|$ or $i = \perp$ or $j = \perp$, is defined as follows:*

If $i = \perp$ or $j = \perp$ or $i \geq j$, then $(M, i, j) \models f$.

If $i \neq \perp$ and $j \neq \perp$ and $i < j$, then

$(M, i, j) \models true$ and $(M, i, j) \not\models false$

$(M, i, j) \models t_1 = t_2$ iff $v(t_1, i) \stackrel{D}{=} v(t_2, i)$, where $\stackrel{D}{=}$ denotes identity over the elements of D

$(M, i, j) \models p(t_1, t_2, \ldots, t_n)$ iff $p_G(v(t_1, i), v(t_2, i), \ldots, v(t_n, i))$, where $p \in GP$ is an n-ary predicate symbol other than $=$

$(M, i, j) \models p(t_1, t_2, \ldots, t_n)$ iff $p_L(i)(v(t_1, i), v(t_2, i), \ldots, v(t_n, i))$, where $p \in LP$ is an n-ary predicate symbol

$(M, i, j) \models \neg f$ iff $(M, i, j) \not\models f$

$(M, i, j) \models f \vee g$ iff $(M, i, j) \models f$ or $(M, i, j) \models g$

$(M, i, j) \models [\alpha | \beta) f$ iff $(M, Locate(\alpha, M, i, j), Locate(\beta, M, i, j)) \models f$

$(M, i, j) \models [-|\beta) f$ iff $(M, i, Locate(\beta, M, i, j)) \models f$

$(M, i, j) \models [\alpha | \rightarrow) f$ iff $(M, Locate(\alpha, M, i, j), j) \models f$

$(M, i, j) \models \exists x\, f$ iff $(M', i, j) \models f$ for some model $M' = (D, v'_G, S)$
with $v'_G(y) \stackrel{D}{=} v_G(y)$ for every $y \in GV$ other than x, where

$locate(\rightarrow f, M, i, j) = \begin{cases} k & \text{if } i \neq \perp \text{ and } j \neq \perp \text{ and } i \leq k < j \text{ and } (M, k, j) \models f \\ & \text{and for all } l,\ i \leq l < k,\ (M, l, j) \not\models f \\ \perp & \text{if } i = \perp \text{ or } j = \perp \text{ or } j < i \text{ or } (i \neq \perp \text{ and } j \neq \perp \\ & \text{and for all } l,\ i \leq l < j,\ (M, l, j) \not\models f) \end{cases}$

$locate((\rightarrow f, \alpha), M, i, j) = locate(\alpha, M, locate(\rightarrow f, M, i, j), j)$.

Definition 7 *A well-formed formula f is satisfiable in a model M if and only if $(M, 0, |M|) \models f$. More generally, a formula f is satisfiable if and only if there exists a model M such that f is satisfiable in M.*

Definition 8 *A formula f is valid, denoted $\models f$, if and only if for all models M f is satisfiable in M.*

4 Proof Rules

We now give a set of sound proof rules for reasoning in first-order FIL. In general, proof systems for temporal logics are not complete, and this set of proof rules is not claimed to be complete. We expect, however, that most practically useful theorems can be derived under this framework. Our toolset is based on a decision procedure for the unquantified logic FIL. Thus, when using our toolset, the user must provide instantiations of existentially quantified variables.

The rules given below can be considered to be an extension of the deductive system $\mathcal{D}_{FIL}$ for FIL presented in [7], obtained by augmenting $\mathcal{D}_{FIL}$ with additional axioms and inference rules and by extending the inference rules of $\mathcal{D}_{FIL}$ to first-order FIL formulas. The additional axioms and inference rules can be shown to be sound using the model-theoretic semantics of the logic. Variables in first-order FIL formulas are designated as *free* or *bound* in the usual manner. The notion of *proper substitution* of variables in classical theories must be modified when dealing with modal logics, in particular temporal logics.

Definition 9 *A term t is said to be substitutable for a variable x in a formula f if the substitution of t for each free occurrence of x in f (i) does not create new bound occurrences of global variables, and (ii) does not create new occurrences of local variables in the scope of an interval operator.*

The result of substituting t for x in f is denoted by $f[x \leftarrow t]$.

The following axiom is the standard instantiation axiom of predicate calculus. The second and third axioms are required for reasoning about equality, and express the reflexivity of equality and substitutivity, respectively.

Axiom 1 $\vdash f[x \leftarrow t] \Rightarrow \exists x\, f$, where t is substitutable for x in f.

Axiom 2 $\vdash x = x$ for $x \in V$.

Axiom 3 $\vdash x = y \;\Rightarrow\; (f \Rightarrow f[x \leftarrow y])$ for $x, y \in V$ if f does not contain interval operators.

The three axioms given below define the interaction between interval formulas and quantifiers. In the following, the notation $[\ldots \to f \ldots)$ is used to denote an interval with f as the target formula of one of the search operators used to construct the interval.

Axiom 4 $\vdash \exists x\,[\,\alpha\,|\,\beta\,)\,f \equiv [\,\alpha\,|\,\beta\,)\,\exists x\, f$, where x is not free in the target formulas of any of the search operators in the interval $[\,\alpha\,|\,\beta\,)$.

Axiom 5 $\vdash \forall x\,[\ldots \to f \ldots)\,g \;\Rightarrow\; [\ldots \to \exists x\, f \ldots)\,\exists x\, g$, where x is not free in the target formulas of the other search operators in the interval.

Axiom 6 $\vdash [\ldots \to \exists x\, f \ldots)\,g \;\Rightarrow\; \exists x\,[\ldots \to f \ldots)\,g$, where x is not free in g and x is not free in the target formulas of the other search operators in the interval.

Global variables take values from a designated domain and have the same values in all states, as the following axiom states.

Axiom 7 *For any formula f that is global, i.e. does not involve local variables or predicates,* $\vdash \Box f \vee \Box \neg f$.

The only additional inference rule is the following rule from predicate calculus.

∃-Introduction If $\vdash f \Rightarrow g$ and x is not free in g then $\vdash \exists x\, f \Rightarrow g$.

Note that Axiom 1 and ∃-Introduction, together with Axioms 2 and 3 for equality, provide all theorems of predicate calculus. Theorems of predicate calculus can therefore be used when working with first-order FIL formulas. However, we must be careful when substituting for variables in formulas that contain interval operators.

We now give some useful theorems of first-order FIL.

Theorem 1 $\vdash \forall x\, [\,\alpha\,|\,\beta\,)\, f \equiv [\,\alpha\,|\,\beta\,)\,\forall x\, f$, *where x is not free in* $[\,\alpha\,|\,\beta\,)$.

Theorem 2 $\vdash \forall x\, [\rightarrow(f \vee g)\,|\rightarrow)\, g \equiv [\rightarrow(\exists x\, f \vee g)\,|\rightarrow)\, g$, *where x is not free in g.*

Theorem 3 $\vdash \exists x\, [\rightarrow(f \vee g)\,|\rightarrow)\, g \equiv [\rightarrow(f \vee \exists x\, g)\,|\rightarrow)\, \exists x\, g$, *where x is not free in f.*

Theorem 4 $\vdash \forall x\, \Box f \equiv \Box \forall x\, f$.

Theorem 5 $\vdash \exists x\, \Diamond f \equiv \Diamond \exists x\, f$.

5 An Example Application

To illustrate the use of first-order FIL, we now give a specification for a sliding window protocol that ensures reliable transmission over an unreliable medium. The specification is developed by starting with a simple protocol and by successively introducing modifications and refinements to it. At every stage, the specification is shown to satisfy a set of required properties. Although most of the reasoning has been done using the graphical notation and tools, the specifications are presented here using the textual notation of FIL for conciseness.

The sliding window protocol operates on top of a physical medium which provides only unreliable non-sequenced communication. The operation of the sliding window is divided into that of the *transmitter* and *receiver* processes, which act independently to provide reliable sequenced one-way communication between two higher-level processes *User1* and *User2*.

In the description the *data packets* sent by User1 are assumed to be distinct. A *frame i* consists of a data packet and the sequence number i assigned to it by the transmitter. An *acknowledgment i* sent by the receiver indicates that all frames with sequence numbers up to and including i have been received. A data packet, frame, or acknowledgment is denoted by the unique sequence number assigned to the corresponding frame by the transmitter. Since the data packets are unique and the system does not create any messages, this sequence number uniquely identifies the data packet, frame, or acknowledgment being transmitted. The term *message* is used to denote any of the above.

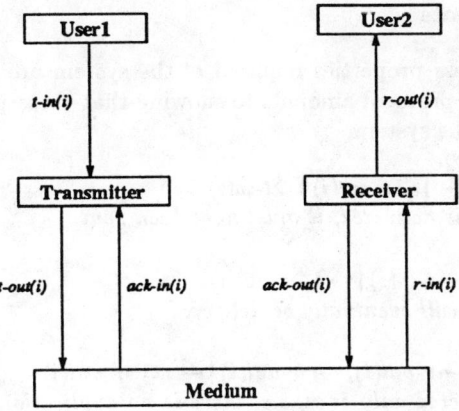

Fig. 1. Sliding Window Protocol Model.

As shown in Figure 1, the system contains six unidirectional channels *t-in*, *t-out*, *ack-in*, *r-in*, *r-out* and *ack-out*. For each message, each channel has an associated predicate that indicates communication of that message over the channel.
- *t-in(i)* : Data packet i is being sent over channel *t-in*.
- *t-out(i)* : Frame i is being sent over channel *t-out*.
- *ack-in(i)* : Acknowledgment i is being sent over channel *ack-in*.
- *r-in(i)* : Frame i is being sent over channel *r-in*.
- *ack-out(i)* : Acknowledgment i is being sent over channel *ack-out*.
- *r-out(i)* : Data packet i is being sent over channel *r-out*.

The values of all variables used in the specification belong to the set of non-negative integers with addition and inequalities. All formulas are assumed to be universally quantified over the free global variables that appear in them.

Channel Specifications

In the following, the predicate $c(i)$ denotes that a message, uniquely identified by i, is being sent over channel c. Koymans [8] has shown that messages must be uniquely identifiable to specify communication mechanisms in temporal logics by axioms like those below. Communication is assumed to be synchronous so that the same predicate can be used for the transmission and reception of a message. In the specification of the protocol, the following properties are instantiated for each channel in the system.

Specification 1 *[Initial]* $\neg c(i)$
Initially, all channels are idle.

Specification 2 $\Box(c(i) \land c(j) \Rightarrow i = j)$
A channel sends one message at a time.

Specification 3 $\Box[\rightarrow c(i) | \rightarrow) \Diamond \neg c(i)$
Every communication event has a finite duration.

Service Requirements

Some of the correctness properties required of the system are given below. Proving the correctness of the protocol amounts to showing that these properties follow from the specifications of the system.

Requirement 1 $[\ -\ |\rightarrow r\text{-}out(i)\,)\,\Diamond t\text{-}in(i)$
Before a data packet is delivered, it must have been sent.

Requirement 2 $[\rightarrow t\text{-}in(i)\,|\rightarrow)\,\Diamond r\text{-}out(i)$
All data packets sent will eventually be delivered.

Requirement 3 $[\rightarrow r\text{-}out(i), \rightarrow \neg r\text{-}out(i)\,|\rightarrow)\,\Box\neg r\text{-}out(i)$
A data packet is delivered only once, i.e. there is no duplication of data packets.

Requirement 4 $[\rightarrow t\text{-}in(i)\,|\rightarrow t\text{-}in(i), \rightarrow t\text{-}in(i+1), \rightarrow r\text{-}out(i+1)\,)\,\Diamond r\text{-}out(i)$
The data packets are delivered in the order in which they were sent.

In the following sections, specifications for the protocol are developed and are shown to satisfy the service requirements given above. The specifications and the proofs are developed in a stepwise fashion. First, a simple protocol is specified and is shown to be correct. Then, the specifications are refined in several steps into a version of the sliding window protocol. All proof steps have been verified using the toolset, except for the application of inductive rules, reasoning involving quantifiers, and reasoning about inequalities, which were done manually.

5.1 Refinement Step 1

First, we consider a simple protocol that does not use acknowledgments. Each frame is transmitted infinitely often and the fairness of the medium ensures that a frame will eventually be delivered.

In the specifications below, the predicate $recd(i)$ indicates that the receiver has completed the reception of frame i. The operator $\rightarrow *$ in Specification 12 is a derived operator, which requires that a state be found in which the target of its search is true.

Medium

The medium does not create any new messages. It may lose or duplicate messages and deliver them out of order. It will eventually deliver a message that is repeatedly sent.

Specification 1 $\Box(\Box\Diamond t\text{-}out(i)\ \Rightarrow\ \Diamond r\text{-}in(i))$
If a frame is sent over the medium infinitely often, then it will eventually be delivered.

Specification 2 $[\ -\ |\rightarrow r\text{-}in(i)\,)\,\Diamond t\text{-}out(i)$
Before a frame is delivered, it must have been sent.

Transmitter

Specification 3 $[\ -\ |\rightarrow t\text{-}in(i+1)\,)\,\Diamond t\text{-}in(i)$
The transmitter assigns sequence numbers to packets in the order in which it receives them.

Specification 4 $[\rightarrow t\text{-}in(i)\,|\rightarrow)\,\Box\Diamond t\text{-}out(i)$
For each data packet received by the transmitter, a frame is sent infinitely often.

Specification 5 $[\ -\ |\rightarrow t\text{-}out(i)\,)\,\Diamond t\text{-}in(i)$
Before a frame is sent, the corresponding data packet must have been received.

Receiver

Specification 6   *[Initial]* $\neg recd(i)$
Initially, no frames have been received.

Specification 7 $[\rightarrow recd(i)\,|\rightarrow)\,\Box recd(i)$
Once a frame is marked as received, it continues to be so marked.

Specification 8 $[\rightarrow r\text{-}in(0)\,|\rightarrow)\,\Diamond recd(0)$
If the first frame arrives at the receiver, then it will be marked as received.

Specification 9 $[\ -\ |\rightarrow recd(0)\,)\,\Diamond r\text{-}in(0)$
The first frame can be marked as received only after it has arrived at the receiver.

Specification 10 $[\rightarrow r\text{-}in(i+1), \rightarrow recd(i)\,|\rightarrow)\,\Diamond recd(i+1)$
If a frame has arrived at the receiver and the previous frame has been marked as received, then that frame will be marked as received.

Specification 11 $[\ -\ |\rightarrow recd(i+1)\,)\,(\Diamond r\text{-}in(i+1)\wedge\Diamond recd(i))$
A frame can be marked as received only after it has arrived at the receiver and the previous frame has been marked as received.

Specification 12 $[\rightarrow recd(i),\rightarrow *r\text{-}out(i),\rightarrow\neg r\text{-}out(i)\,|\rightarrow)\,\Box\neg r\text{-}out(i)$
If a frame is marked as received, the corresponding data packet is delivered exactly once.

Specification 13 $[\ -\ |\rightarrow r\text{-}out(i)\,)\,\Diamond recd(i)$
Before a data packet is delivered, the corresponding frame must have been received.

Specification 14 $[\ -\ |\rightarrow r\text{-}out(i+1)\,)\,\Diamond r\text{-}out(i)$
Data packets are delivered in the order of their sequence numbers.

Proofs of Service Requirements

The service requirements can be proved from the above specifications. The key lemmas used in the proofs follow.

Lemma 1 $[\rightarrow t\text{-}in(i) \,|\, \rightarrow) \Diamond r\text{-}in(i)$
For each data packet received by the transmitter, the corresponding frame eventually reaches the receiver.

Proof: The result follows from Specifications 1, 2, 4, 5 and the initial conditions. ■

Lemma 2 $[\rightarrow t\text{-}in(i) \,|\, \rightarrow) \Diamond recd(i)$
Each data packet received by the transmitter is eventually marked as received by the receiver.

Proof: The proof of this result requires an induction on the global variable i. The basis of the induction, $i = 0$, follows from Specification 8, Lemma $1[i \leftarrow 0]$, Specification $5[i \leftarrow 0]$, Specification $2[i \leftarrow 0]$ and the initial conditions.

Assume that the result holds for $i = n$. We must prove that the result holds for $i = n + 1$, i.e. that $[\rightarrow t\text{-}in(n+1) \,|\, \rightarrow) \Diamond recd(n+1)$. This follows from the inductive assumption, Specification $3[i \leftarrow n]$, Lemma $1[i \leftarrow n+1]$, Specification $7[i \leftarrow n]$, Specification $7[i \leftarrow n+1]$, Specification $10[i \leftarrow n]$ and the initial conditions. ■

The proofs of the service requirements are given below.

Proof of Requirement 1: From Specification 9 and Specification 11 it follows that $[- \,|\, \rightarrow recd(i)) \Diamond r\text{-}in(i)$. The required result follows from this result together with Lemma 1, Specifications 2, 5, 13 and the initial conditions. ■

Proof of Requirement 2: The lemma follows from Requirement 1, Lemma 2, Specification 12 and the initial conditions. ■

Proof of Requirement 3: This follows from Specifications 12, 13 and the initial conditions. ■

Proof of Requirement 4: This follows from Requirements 1 and 2, Specification 14 and the initial conditions. ■

5.2 Refinement Step 2

The previous set of specifications is impractical to implement because each frame must be sent infinitely often. We now modify these specifications to avoid this by introducing acknowledgments. An acknowledgment i effectively acknowledges the receipt of all frames with sequence numbers up to and including i. Thus, every frame need not be individually acknowledged. The predicate $sent(i)$ indicates that the transmitter has successfully completed the transmission of frame i. Once the transmitter has marked frame i as having been correctly sent, that frame need not be sent again.

The additional specifications are given below. Specification 4 is replaced with the new Specification 21, which states that it is not necessary to continue to send a frame repeatedly if $sent(i)$ is true.

Medium

Specification 15 $\Box(\Box\Diamond \text{ack-out}(i) \Rightarrow \Diamond \text{ack-in}(i))$
If an acknowledgment is sent infinitely often over the medium, it will eventually be delivered.

Specification 16 $[-|\rightarrow \text{ack-in}(i))\Diamond \text{ack-out}(i)$
Before an acknowledgment is delivered, it must have been sent.

Transmitter

Specification 17 *[Initial]* $\neg \text{sent}(i)$
Initially, no frames have been marked as sent.

Specification 18 $[\rightarrow \text{sent}(i)|\rightarrow)\Box \text{sent}(i)$
Once a frame is marked as sent, it continues to be so marked.

Specification 19 $\Box[\rightarrow(\text{ack-in}(i) \wedge j \leq i)|\rightarrow)\Diamond \text{sent}(j)$
If an acknowledgment i is received, all frames with sequence numbers less than or equal to i will be marked as sent.

Specification 20 $[-|\rightarrow \text{sent}(i))\Diamond \exists j\,(\text{ack-in}(j) \wedge j \geq i)$
Before a frame can be marked as sent, an appropriate acknowledgment must have been received.

Specification 21 $[\rightarrow t\text{-in}(i)|\rightarrow)(\Box\Diamond t\text{-out}(i) \vee \Diamond \text{sent}(i))$
For every data packet i received by the transmitter, the corresponding frame is sent infinitely often or is marked as sent sometime in the future.

Receiver

Specification 22 $[\rightarrow r\text{-in}(i)|\rightarrow)\Diamond \exists j\,(\text{ack-out}(j) \wedge j \geq i)$
Each frame received must be acknowledged.

Specification 23 $\Box(\text{ack-out}(i) \Rightarrow \neg \text{recd}(i+1) \wedge \text{recd}(i))$
An acknowledgment indicates the frame with the largest sequence number marked as received.

Proofs of Service Requirements

With these refinements, all of the previous proofs still hold, except that the proof of Lemma 1 must be modified to take into account the fact that Specification 4 has been replaced. To this end, we introduce the following lemma.

Lemma 3 $[-|\rightarrow \text{sent}(i))\Diamond \text{recd}(i)$
Before the transmitter marks frame i as sent, the receiver marks the frame as received.

Proof: First we prove that $\Box(recd(n) \Rightarrow (i \leq n \Rightarrow recd(i)))$ using induction on n. For $n = 0$, this holds trivially. To prove the induction step, it suffices to prove that $\Box(recd(n + 1) \Rightarrow recd(n))$. This follows from Specification 7$[i \leftarrow n]$, Specification 11$[i \leftarrow n]$ and the initial conditions.

Next we prove that $\Box(ack\text{-}out(j) \wedge j \geq i \Rightarrow recd(i))$. By Specification 23, $\Box(ack\text{-}out(j) \Rightarrow recd(j))$ and, by the preceding result, $\Box(recd(j) \wedge j \geq i \Rightarrow recd(i))$, which yield the result.

From the preceding result it follows that $\Box(\exists j\, (ack\text{-}out(j) \wedge j \geq i) \Rightarrow recd(i))$ using results from FIL and $\exists$-Introduction.

From Specification 16$[i \leftarrow j]$, it follows that $[\ -\ |\rightarrow ack\text{-}in(j)\,)\,\Diamond ack\text{-}out(j)$. By Axiom 7, $\Box\neg(j \geq i) \vee \Box(j \geq i)$. From the two preceding formulas and the initial conditions, $[\ -\ |\rightarrow ack\text{-}in(j) \wedge j \geq i\,)\,\Diamond(ack\text{-}out(j) \wedge j \geq i)$. By Theorem 5 and Axiom 5, this yields $[\ -\ |\rightarrow \exists j\,(ack\text{-}in(j) \wedge j \geq i)\,)\,\Diamond \exists j\,(ack\text{-}out(j) \wedge j \geq i)$.

The lemma now follows from Specification 20 and the two preceding results. ∎

New Proof of Lemma 1: From Specification 9 and Specification 11, it follows that $[\ -\ |\rightarrow recd(i)\,)\,\Diamond r\text{-}in(i)$. The lemma now follows from this result and Specifications 1, 2, 5, 21, Lemma 3 and the initial conditions. ∎

5.3 Refinement Step 3

The transmitter and receiver are now provided with buffers. This is a simple refinement step in that all of the specifications in the previous version still hold. We assume that the transmitter and receiver both have an infinite number of buffers. We also assume that no restrictions are placed on the window sizes, *i.e.* they could be arbitrarily large. The predicate $t\text{-}buf(i)$ indicates the presence of frame i in the transmitter buffer. Similarly, $r\text{-}buf(i)$ indicates the presence of frame i in the receiver buffer.

Transmitter

The following specifications replace Specifications 5 and 21. Since Specifications 5 and 21 follow from these specifications and the initial conditions, all of the earlier proofs are not affected.

Specification 24   *[Initial]* $\neg t\text{-}buf(i)$
Initially, the transmitter buffers are empty.

Specification 25 $[\rightarrow t\text{-}buf(i), \rightarrow \neg t\text{-}buf(i)\,|\rightarrow)\,sent(i)$
Once frame i is in a buffer, it remains there until it is marked as sent.

Specification 26 $[\rightarrow t\text{-}in(i)\,|\rightarrow)\,\Diamond t\text{-}buf(i)$
For each data packet received by the transmitter, the corresponding frame is eventually constructed and stored in a buffer.

Specification 27 $[\ -\ |\rightarrow t\text{-}buf(i)\,)\,\Diamond t\text{-}in(i)$
Before a frame is stored in a buffer, the corresponding data packet must have been received by the transmitter.

Specification 28 $[\to t\text{-}buf(i) | \to)(\Box\Diamond t\text{-}out(i) \lor \Diamond sent(i))$
Each frame stored in a buffer is transmitted infinitely often or is marked as sent sometime in the future.

Specification 29 $[- | \to t\text{-}out(i))\Diamond t\text{-}buf(i)$
Before a frame is transmitted, it must already be in a buffer.

Receiver

The specifications below replace Specifications 8, 9, 10 and 11. None of the proofs is affected since the replaced specifications can be derived from these.

Specification 30 $[Initial]$ $\neg r\text{-}buf(i)$
Initially, the receiver buffers are empty.

Specification 31 $[\to r\text{-}buf(i), \to \neg r\text{-}buf(i) | \to) recd(i)$
Once frame i is in a buffer, it remains there until it is marked as received.

Specification 32 $[\to r\text{-}in(i) | \to) \Diamond r\text{-}buf(i)$
After a frame is received, it is stored in a buffer.

Specification 33 $[- | \to r\text{-}buf(i))\Diamond r\text{-}in(i)$
Before a frame is stored in a buffer, it must have been received.

Specification 34 $[\to r\text{-}buf(0) | \to) \Diamond recd(0)$
If the first frame is stored in a buffer, then it will be marked as received.

Specification 35 $[- | \to recd(0))\Diamond r\text{-}buf(0)$
The first frame can be marked as received only after it is stored in a buffer.

Specification 36 $[\to r\text{-}buf(i+1), \to recd(i) | \to) \Diamond recd(i+1)$
If a frame is stored in a buffer and the previous frame has been marked as received, then that frame will be marked as received.

Specification 37 $[- | \to recd(i+1))(\Diamond recd(i) \land \Diamond r\text{-}buf(i+1))$
A frame can be marked as received only after it is stored in a buffer and the previous frame has been marked as received.

The refinement proofs for Specifications 5, 8, 9, 10, 11 and 21 are straightforward and are omitted due to lack of space.

5.4   Refinement Step 4

The previous specification is still impractical to implement because it assumes that there are infinitely many buffers available at the transmitter and receiver. The window sizes are now restricted to finite values so that the number of outstanding frames is always bounded. The transmitter and receiver window sizes are TWS and RWS, where $TWS > 0$ and $RWS > 0$.

Specification 38 below further constrains the acceptable behaviors of the transmitter. This does not affect earlier proofs.

Specification 38 $\Box(t\text{-}in(TWS+i) \Rightarrow sent(i))$
When data packet $TWS+i$ is accepted, frame i must be marked as sent.

Specifications 39 and 40 replace Specification 32.

Specification 39 $i < RWS \Rightarrow [\rightarrow r\text{-}in(i) \,|\, \rightarrow) \Diamond r\text{-}buf(i)$
After a frame with a sequence number less than RWS is received, that frame will be stored in a buffer.

From the above, Specification 32 still holds for $i < RWS$.

Specification 40 $[\rightarrow r\text{-}in(RWS+i) \land recd(i) \,|\, \rightarrow) \Diamond r\text{-}buf(RWS+i)$
A received frame that occurs within the current window will be stored in a buffer.

To establish the service requirements, we note that the proof of Lemma 2 no longer holds, since Specification 32 and consequently Specifications 8 and 10 do not hold.

New Proof of Lemma 2: The two cases $i < RWS$ and $i \geq RWS$ are considered separately. For the case $i < RWS$, note that the old proof still holds since Specification 39 yields the old Specification 32 when $i < RWS$.

For the case $i \geq RWS$, assume that the result holds for all $k < i$. It suffices to show that it also holds for i. The basis of the induction is given by the first case above. Let $i = RWS + j$. Assuming $RWS > 0$, it follows that $i > j$ and so the result holds for j.

From Specifications 36 and 37, it follows that $[- \,|\, \rightarrow recd(RWS+j)) \Diamond r\text{-}buf(RWS+j)$. From Specification 3, it follows by induction that $[- \,|\, \rightarrow t\text{-}in(RWS+j)) \Diamond t\text{-}in(j)$. The two preceding results, together with the inductive assumption for $k = j$ and Specifications $1[i \leftarrow RWS+j]$, $2[i \leftarrow RWS+j]$, $5[i \leftarrow RWS+j]$, $7[i \leftarrow j]$, $21[i \leftarrow RWS+j]$, $33[i \leftarrow RWS+j]$, $40[i \leftarrow j]$, Lemma $3[i \leftarrow RWS+j]$ and the initial conditions yield $[\rightarrow t\text{-}in(RWS+j) \,|\, \rightarrow) \Diamond r\text{-}buf(RWS+j)$.

Now let $i = l + 1$ (this is permissible since $i \geq RWS > 0$). The previous result is equivalent to $[\rightarrow t\text{-}in(l+1) \,|\, \rightarrow) \Diamond r\text{-}buf(l+1)$. The induction step now amounts to proving $[\rightarrow t\text{-}in(l+1) \,|\, \rightarrow) \Diamond recd(l+1)$. This follows from the above result together with the inductive assumption for $k = l$, Specifications $2[i \leftarrow l+1]$, $3[i \leftarrow l]$ $5[i \leftarrow l+1]$, $33[i \leftarrow l+1]$, $35[i \leftarrow l+1]$ and the initial conditions. ∎

Refinement Step 4 can be made more concrete by introducing local variables that implement the abstract predicates $sent(i)$ and $recd(i)$. The transmitter uses the local variable lu to denote the frame with the lowest sequence number that has not been acknowledged. The receiver uses the local variable nr to denote the next expected frame. The definitions $sent(i) \equiv i < lu$ and $recd(i) \equiv i < nr$ are compatible with the use of the predicates in the earlier specifications.

6 Conclusion

We have presented a first-order extension of Future Interval Logic, and a set of sound proof rules for reasoning in the logic. The logic is a linear-time temporal logic that is intended for specification and verification of reactive and concurrent systems. We have illustrated the use of the extended logic and our toolset in developing specifications for a sliding window protocol and in verifying its correctness. Some of these proofs have employed the proof rules that we have given for the extended logic.

References

1. Abadi, M., "The power of temporal proofs," *Theoretical Computer Science*, vol. 65, no. 1, pp. 35-83, June 1989.
2. Abadi, M. and Manna, Z., "Non-clausal deduction in first-order temporal logic," *Journal of the ACM*, vol. 37, no. 2, pp. 279-317, April 1990.
3. Cardell-Oliver, R., "The specification and verification of sliding window protocols in higher order logic," Technical Report No. 183, Computer Laboratory, University of Cambridge, October 1989.
4. Dillon, L. K., Kutty, G., Moser, L. E., Melliar-Smith, P. M., Ramakrishna, Y. S., "Graphical specifications for concurrent software systems," *Proceedings of the 14th International Conference on Software Engineering*, Melbourne, Australia, pp. 214-224, May 1992.
5. Garson, J. W., "Quantification in modal logic," in *Handbook of Philosophical Logic*, Gabbay, D. and Guenthner, F. (eds.), vol. II, pp. 249-307, D. Reider Publishing Co., 1984.
6. Kutty, G., Ramakrishna, Y. S., Moser, L. E., Dillon, L. K., Melliar-Smith, P. M., "A Graphical Interval Logic toolset for verifying concurrent systems," *Proceedings of the 5th Conference on Computer Aided Verification*, Elounda, Crete, Greece, Lecture Notes in Computer Science 697, Springer Verlag, pp. 138-153, June 1993.
7. Kutty, G., Moser, L. E., Melliar-Smith, P. M., Ramakrishna, Y. S., Dillon, L. K., "Axiomatizations of interval logics," to appear in *Fundamenta Informaticae*.
8. Koymans, R., "Specifying message passing systems requires extending temporal logic," *Proceedings of the Colloquium on Temporal Logic in Specification*, Lecture Notes in Computer Science 398, pp. 213-223, Springer-Verlag, 1989.
9. Manna, Z. and Pnueli, A., "Verification of concurrent programs: A temporal proof system," *Foundations of Computer Science IV: Distributed Systems*, Mathematical Centre Tracts, 159, Amsterdam, pp. 163-255, 1983.
10. Manna, Z. and Pnueli, A., *The temporal logic of reactive and concurrent systems*, Springer-Verlag, New York, Inc., 1992.
11. Paliwoda, K. and Sanders, J. W., "An incremental specification of the sliding window protocol," *Distributed Computing*, no. 5, pp. 83-94, 1991.
12. Ramakrishna, Y. S., Dillon, L. K., Moser, L. E., Melliar-Smith P. M., Kutty, G., "An automata-theoretic decision procedure for future interval logic," *Proceedings of the Twelfth Conference on Foundations of Software Technology and Theoretical Computer Science*, New Delhi, India, Lecture Notes in Computer Science 652, Springer-Verlag, pp. 51-67, December 1992.
13. Richier, J. L., Rodriguez, C, Sifakis, J. and Voiron, J.,"Verification in Xesar of the sliding window protocol," *Proceedings of the IFIP WG 6.1 7th International Conference on Protocol Specification, Testing and Verification*, Zurich, Switzerland, pp. 235-248, May 1987.
14. Schwartz, R. L., Melliar-Smith, P. M. and Vogt, F., "An interval based temporal logic," *Proceedings of the ACM Workshop on the Logics of Programming*, pp. 443-457, June 1983.

Buy One, Get One Free !!!

Orna Bernholtz and Orna Grumberg [*]
Department of Computer Science
The Technion
Haifa 32000, Israel
orna@cs.technion.ac.il

Abstract

The exponential gap between CTL and LTL model-checking complexity, led to a development of model-checking tools for CTL, while model checkers for LTL have stayed behind. However, users of those tools have to struggle with the limited expressive power of CTL and are often compelled to give up checking many important behaviors. As a matter of course, finding specification languages which are strictly more expressive than CTL and yet maintain its attractive model-checking complexity, is a challenging problem and has been an active area of research. In this paper we introduce such a language.

Our language, CTL^2, is an outcome of a new approach for defining sublanguages of CTL^*. The approach allows a bounded number of linear-time operators within the path formulas of CTL^*. We discuss the expressive power of CTL^2 and, in particular, focus on the relation between CTL^2 and CTL. We show that beyond the increase in the expressive power, a substantial advantage of CTL^2 is the neat and intuitive presentation it provides for specifications whose CTL equivalences are complicated and very hard to understand. We introduce a model-checking procedure for CTL^2. Our model checker is of complexity linear in both the formula and the structure being checked, just as the one for CTL. In addition, we suggest an extension of it that, preserving its complexity, handles fairness.

1 Introduction

Propositional temporal logics, which are propositional modal logics that enable description of occurrence of events in time, serve as a classical tool for verifying concurrent programs [Pnu81]. Two possible views regarding the nature of time induce two types of temporal logics. In linear temporal logics, time is treated as if each moment in time has a unique possible future, while in branching temporal logics, each moment in time may split into various possible futures. Lamport, in [Lam80], was the first to consider this issue. He examined two distinct interpretations of a temporal logic which consists of the temporal operators "always" and "sometimes". The first interpretation, in which the formulas are interpreted over paths, is associated with linear time, and

---

[*]Work carried out under a project on program verification and semantics of programming languages, funded by the Israeli academy of sciences (basic research). The second author was partially supported by the U.S.-Israeli Binational Science Foundation.

the second, in which formulas are interpreted over states, is associated with branching time. The expressive power of those logics, as Lamport showed, is incomparable.

One step forward advanced Clarke and Emerson in [CE81]. They defined the full branching temporal logic CTL* which includes both of Lamport's interpretations and allows specifying both linear-time and branching-time properties. A comprehensive discussion of the expressive power of branching versus linear temporal logics can be found in [EH86]. Emerson and Halpern reexamined Lamport's approach, improved his incomparability result and presented a hierarchy of the relative expressive power of some sub-languages of CTL*. They claimed that, on the one hand, linear temporal logics are suitable for reasoning about concurrent programs where interest is naturally restricted to properties that hold along all computation paths, whereas on the other hand, branching temporal logics have a relatively low model-checking complexity, and enable specifying existential properties. They conclude by noting that "one should use the subset of CTL* most appropriate to the application", where appropriateness is measured by both expressiveness and complexity.

In temporal logic model checking, a given model (representing a program) is checked with respect to a propositional temporal logic formula (specifying a required behavior for the program). Model-checking procedures suggest a computerized mechanism for verifying finite state systems such as communication protocols and hardware designs. Recent methods and heuristics such as BDDs [Bry86], modular model checking [GL91], partial order methods [WG93], and more, cope successfully with the known "state explosion" problem and give rise to model checking not only as a lovely theoretical issue, but also as a practical tool used for formal verification. As far as model-checking complexity is concerned, the following results are known: The model-checking problem for the branching temporal logic CTL is in deterministic linear time [EC82, CES86], and the model-checking problem for the linear temporal logic LTL, so as the one for CTL*, is PSPACE complete [SC85, CES86].

The exponential gap between CTL and LTL model-checking complexity, led to a development of model checking tools for CTL [Bro86, McM93], while model checkers for LTL have stayed behind. However, users of those tools have to struggle with the limited expressive power of CTL. Unfortunately, the restricted branching nature of CTL compels them to give up checking many important behaviors. As a matter of course, finding specification languages which are strictly more expressive than CTL and yet maintain its attractive model-checking complexity, is a challenging problem and has been an active area of research [Eme90]. In this paper we introduce such a language.

Our language, CTL^2, is an outcome of a new approach for defining sub-languages of CTL*. The syntax of CTL* allows branching path quantifiers to be applied to path formulas composed of any combination of linear-time operators. CTL syntax restricts path quantifiers to be applied to path formulas with a single linear-time operator. The idea of our approach is to allow a bounded number of linear-time operators within the path formulas of CTL*. In CTL^2, we investigate the simplest version of this extension: path formulas may contain an assertion composed of two, possibly negated, linear-time operators, either nested or connected by a binary boolean operator. Thus, formulas like $AGFgrant$, $E(X\,req)U\,grant$, $A(F\neg busy \vee Gwait)$, and $E(\neg(busy\ U\ req))U\,grant$, are all CTL^2 formulas. Indeed, CTL^2 is strong enough to express neatly both liveness and safety properties used for reasoning about concurrent programs, such as unconditional fairness, preservation of FIFO order, more accurate specification of accessibility, absence of unsolicited response, and more.

We compare the expressive power of CTL^2 with other branching temporal logics, linear temporal logics, and automata over infinite trees. In particular, we focus on the relation between CTL^2 and CTL. The CTL^2 formula $EGFp$, which is not expressible in CTL [EH86], testifies that CTL^2 is strictly more expressive than CTL. Yet, surprisingly, the increase in the expressive power is not as significant as we might expect. We characterize precisely CTL^2 formulas that have equivalences of CTL and show that beyond the increase in the expressive power, a substantial advantage of CTL^2 is the clear and intuitive presentation it provides for specifications which have clumsy and unreadable equivalences of CTL.

We introduce a model-checking procedure for CTL^2. Given a CTL^2 formula ψ and a structure K, the procedure determines for every state in K whether it satisfies ψ. As known model checkers for CTL, our procedure iteratively labels K with a set of formulas whose satisfaction has to do with the satisfaction of ψ. Unlike in the case of CTL, those formulas are not necessarily sub-formulas of ψ. However, their number is linear in $|\psi|$ and labelling K with each of them is of complexity linear in $|K|$. Thus, model checking of CTL^2 is of complexity linear in both the formula and the structure being checked, just as the one for CTL. We suggest an extension of our CTL^2 model checker that, preserving its complexity, handles fairness for a substantial subset of CTL^2 and, in particular, enables checking strong fairness.

The rest of this paper is organized as follows. In the figure below we give some background and motivation. In section 2 we introduce the languages CTL^2. In section 3 we discuss its expressive power, and in section 4 we present a linear model-checking procedure for it.

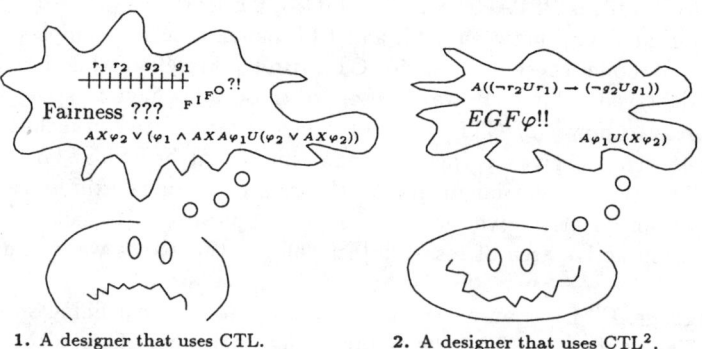

1. A designer that uses CTL.　　2. A designer that uses CTL^2.

2 The Temporal Logic CTL^2

For an introduction to temporal logics see [Eme90]. We introduce a new branching temporal logic, CTL^2. CTL^2 extends CTL by allowing two temporal operators within the path formulas. The temporal operators may be either nested or connected by a binary boolean connective. Negating one or both of the temporal operators is also

allowed. Formally, we distinguish between three types of formulas: state formulas, path formulas of degree 1, and path formulas of degree 2.
Given a set AP of atomic propositions, a CTL^2 state formula is either:

S1. t, f or p, for all $p \in AP$.

S2. $\neg \varphi_1$ or $\varphi_1 \vee \varphi_2$, where φ_1 and φ_2 are CTL^2 state formulas.

S3. $A\varphi_1$ or $E\varphi_1$, where φ_1 is a CTL^2 path formula of degree 1 or 2.

A CTL^2 path formula of degree 1 is either:

P1. $X\varphi_1$ or $\varphi_1 U \varphi_2$, where φ_1 and φ_2 are CTL^2 state formulas.

P2. $\neg \varphi_1$, where φ_1 is a CTL^2 path formula of degree 1.

A CTL^2 path formula of degree 2 is either:

P3. $X\varphi_1$, $\varphi_1 U \varphi_2$ or $\varphi_2 U \varphi_1$, where φ_1 is a CTL^2 path formula of degree 1 and φ_2 is a CTL^2 state formula.

P4. $\varphi_1 \vee \varphi_2$, where φ_1 is either a CTL^2 state formula or a CTL^2 path formula of degree 1, and φ_2 is a CTL^2 path formula of degree 1.

P5. $\neg \varphi_1$, where φ_1 is a CTL^2 path formula of degree 2.

CTL^2 is the set of state formulas generated by the above rules. For example, $AGFgrant$, $E(X req)U grant$, $A(F\neg busy \vee G wait)$, and $E(\neg(busy\ U\ req))U grant$, are all CTL^2 formulas (using the known F and G abbreviations). It is easy to see that as far as syntax is concerned, $CTL \subset CTL^2 \subset CTL^*$, and $CTL^2 \neq BLTL$ (BLTL consists of all CTL^* formulas of the form $A\psi$ for an LTL formula ψ).

3 Expressive Power of CTL^2

In this section we investigate the expressive power of CTL^2 with respect to other branching time logics and automata over infinite trees. We state and prove the following expressiveness relations:

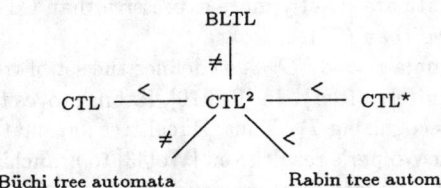

We focus on the relation between CTL^2 and CTL. Definitely, an important advantage of CTL^2 over CTL is the increase in the expressive power. Consider the CTL^2 formula $A(\varphi_1 U \neg(\varphi_2 U \varphi_3))$, which is not expressible in CTL. The formula is a base for many specifications used for reasoning about concurrent programs. Taking for instance $\varphi_1 = req_1$, $\varphi_2 = \neg req_2$, and $\varphi_3 = \neg CS_1$, it specifies the property "in all computation paths, process 1 keeps signalling req_1 until it eventually enters the critical section and stays there as long as process 2 does not signal req_2". However, not less important is the neat presentation CTL^2 suggests for specifications whose CTL equivalences are very hard to understand. The branching nature of CTL, not only makes it difficult to understand the specification, but also makes it difficult for the specifiers to express correctly their intuition. The CTL^2 formula $A((\varphi_1 U \varphi_2) \to (\varphi_3 U \varphi_4))$ can express elegantly a FIFO policy. Taking $\varphi_1 = \neg req_1$, $\varphi_2 = req_2$, $\varphi_3 = \neg grant_1$, and $\varphi_4 = grant_2$, gives "in all computation paths, if req_1 does not precede req_2, then $grant_1$ does not precede $grant_2$". The CTL^2 formula $\neg E(\varphi_1 U(X \varphi_2))$ enables specifying more accurate properties than those enabled without nesting of the next operator. Taking $\varphi_1 = CS_1$ and $\varphi_2 = CS_2$, gives "in all computation paths, if process 2 enters its critical section, then process 1 was not in its critical section two steps earlier". Note that since we usually want to specify behaviors as invariants, we should precede those formulas with AG, resulting in formulas like $AGA\varphi_1 U\neg(\varphi_2 U \varphi_3)$ or $AG\neg E(\varphi_1 U(X \varphi_2))$, all syntactically correct in CTL^2. We start by comparing CTL^2 with CTL^* and BTLT.

Lemma 3.1 *The formula $\psi = AF(p \wedge Xp)$ has no equivalent of CTL^2.*

The proof is a simple extension of Emerson and Halpern's proof from [EH86] for inexpressibility of ψ in CTL.

Claim 3.2 *(1) $CTL^2 \neq BLTL$.*

(2) $CTL^2 < CTL^*$.

Proof: Lemma 3.1 implies that BLTL $\not\leq CTL^2$. The second direction follows from the known CTL $\not\leq$ BLTL result ([Lam80], reproved in [EH86]). $CTL^2 \leq CTL^*$ holds by syntactic containment. Strictness follows from Lemma 3.1. □

Claim 3.3 *(1) $CTL^2 \neq$ Büchi tree automata.*

(2) $CTL^2 <$ Rabin tree automata.

Proof: Büchi tree automata are, as follows from [ES84], more expressive than CTL. We show that Büchi tree automata can not handle an additional temporal operator in the path formulas, and thus, they are not more expressive than CTL^2. On the other hand, as Rabin tree automata are strictly more expressive than CTL^* [EJ88], they are also strictly more expressive than CTL^2.

Consider the CTL^2 formula $\psi = AFG\neg p$. ψ defines the set of trees $T_1 = \{V :$ in all paths of V, p is true only finitely often$\}$. In [Rab70], Rabin proves that there exists no Büchi tree automaton for recognizing T_1. Thus, Büchi tree automata $\not\geq CTL^2$. For the second direction, we extend Wolper's result from [Wol83] to branching time logics, and show that there exists no CTL^2 formula that defines the set of trees $T_2 = \{V :$ in all paths of V, p is true at all even places$\}$. Yet, T_2 is recognizable by Büchi tree automata. It is interesting to note that extending CTL^2 by allowing quantification over atomic propositions [Wol82], results in a language which is strictly more expressive than Büchi tree automata. □

Lemma 3.4 *The CTL^2 formula $\psi = EFGp$ has no equivalent of CTL.*

The Lemma is proved in [EH86] and the following claim is a straightforward corollary.

Claim 3.5 $CTL^2 > CTL$.

In the rest of this section we show that the CTL^2 formula $A\varphi_1 U(\neg(\varphi_2 U \varphi_3))$ and its negation $E(\neg(\varphi_1 U(\neg(\varphi_2 U \varphi_3))))$ embody all the expressiveness superiority of CTL^2 over CTL. Namely, every CTL^2 formula that does not have one of those formulas as a sub-formula, has an equivalent CTL formula. We first suggest a normal form for CTL^2 formulas. We use $\varphi_1 \bar{U} \varphi_2$ as an abbreviation for $\neg(\varphi_1 U \varphi_2)$.

Consider the equivalence $\neg X \varphi \equiv X \neg \varphi$. Since negation of state formulas is allowed, rules P1 and P2 that define a path formula of degree 1, can be replaced by a single rule, according to which, a path formula of degree 1 is either $X\varphi_1$, $\varphi_1 U \varphi_2$, or $\varphi_1 \bar{U} \varphi_2$, where φ_1 and φ_2 are CTL^2 state formulas. Consider now the equivalence $A\psi \equiv \neg E \neg \psi$. Again, as negation of state formulas is allowed, and, moreover, as both A and E are allowed as path quantifiers, rules P3, P4, and P5 that define a path formula of degree 2, can be replaced such that the following logic, called *normal-formed CTL^2*, is equivalent to CTL^2. Given a set AP of atomic propositions, a normal-formed CTL^2 formula is either (where Q stands for either A or E and φ_1, φ_2, φ_3, and φ_4 are normal-formed CTL^2 state formulas):

R1. t, f or p, for all $p \in AP$.

R2. $\neg \varphi_1$ or $\varphi_1 \vee \varphi_2$.

R3. $EX\varphi_1$, $E\varphi_1 U \varphi_2$, or $A\varphi_1 U \varphi_2$.

R4. $EXX\varphi_1$, $EX(\varphi_1 U \varphi_2)$, $EX(\varphi_1 \bar{U} \varphi_2)$,
$Q((X\varphi_1)U\varphi_2)$, $Q((\varphi_1 U \varphi_2)U\varphi_3)$, $Q((\varphi_1 \bar{U} \varphi_2)U\varphi_3)$,
$Q(\varphi_1 U(X\varphi_2))$, $Q(\varphi_1 U(\varphi_2 U \varphi_3))$, or $Q(\varphi_1 U(\varphi_2 \bar{U} \varphi_3))$.

R5. $Q(\varphi_1 \vee (X\varphi_2))$, $Q(\varphi_1 \vee (\varphi_2 U \varphi_3))$, $Q(\varphi_1 \vee (\varphi_2 \bar{U} \varphi_3))$,
$Q((X\varphi_1) \vee (X\varphi_2))$, $Q((X\varphi_1) \vee (\varphi_2 U \varphi_3))$, $Q((X\varphi_1) \vee (\varphi_2 \bar{U} \varphi_3))$,
$Q((\varphi_1 U \varphi_2) \vee (\varphi_3 U \varphi_4))$, $Q((\varphi_1 U \varphi_2) \vee (\varphi_3 \bar{U} \varphi_4))$, or $Q((\varphi_1 \bar{U} \varphi_2) \vee (\varphi_3 \bar{U} \varphi_4))$.

In the sequel, we assume that the CTL^2 formulas are given in that normal form. Note that this at most doubles the formula length.

Let us return to formulas of the form $A\varphi_1 U(\varphi_2 \bar{U} \varphi_3)$, mentioned earlier (note that $E(\varphi_1 \bar{U}(\varphi_2 \bar{U} \varphi_3))$ is not a normal-formed formula). We call formulas of that form *sad formulas*. CTL^2 formulas of other forms are called *happy formulas*. Note that the formulas $AFG\neg p$ and $EGFp$, used for proving Claims 3.3 and 3.5, are sad.

For showing that sad formulas are the only additional power of CTL^2 with respect to CTL, we suggest CTL equivalences for all the happy ones. Rule R5 contains formulas that have a boolean connective in their path formulas. Equivalences to formulas of that form are already known [Eme90] and we give them in Appendix A. Below we introduce and prove CTL equivalences for fourteen forms of formulas that cover all the happy formulas in rule R4. We call them CTL equivalences, meaning that if φ_1, φ_2, and φ_3 are CTL formulas, then the equivalent formula is of CTL [1].

---
[1] We note here that the equivalences are valid also in the context of CTL*; i.e. taking φ_1, φ_2, and φ_3 as CTL* state formulas.

Claim 3.6 *(a)* $EXX\varphi_1 \equiv EXEX\varphi_1$.

(b) $EX\varphi_1 U\varphi_2 \equiv EXE\varphi_1 U\varphi_2$.

(c) $EX\varphi_1 \bar{U}\varphi_2 \equiv EXE\varphi_1 \bar{U}\varphi_2$.

Proof: We prove a stronger claim. Let ψ be a CTL* path formula, we prove $EX\psi \equiv EXE\psi$. $w \models EX\psi$ iff there exists a path $\pi = w_0, w_1, \ldots$, with $w_0 = w$, such that $\pi^1 \models \psi$. This holds iff there exists a path $\pi = w_0, w_1, \ldots$, with $w_0 = w$, such that $w_1 \models E\psi$, which holds iff $w \models EXE\psi$. □

Claim 3.7 *(a)* $A(X\varphi_2)U\varphi_3 \equiv A(AX\varphi_2)U\varphi_3$.

(b) $A(\varphi_1 U\varphi_2)U\varphi_3 \equiv A(A\varphi_1 U\varphi_2)U\varphi_3$.

(c) $A(\varphi_1 \bar{U}\varphi_2)U\varphi_3 \equiv A(A\varphi_1 \bar{U}\varphi_2)U\varphi_3$.

Proof: We prove a stronger claim. Let ψ be a CTL* path formula, we prove $A\psi U\varphi_3 \equiv A(A\psi)U\varphi_3$.

Assume first that $w \models A(A\psi)U\varphi_3$. Then, for every path $\pi = w_0, w_1, \ldots$, with $w_0 = w$, there exists $i \geq 0$ such that $w_i \models \varphi_3$, and for every $0 \leq j < i$, $w_j \models A\psi$. In particular, for every $0 \leq j < i$, $\pi^j \models \psi$. Thus $\pi \models \psi U\varphi_3$ and $w \models A\psi U\varphi_3$.

Assume now that $w \models A\psi U\varphi_3$. Then, for every path $\pi = w_0, w_1, \ldots$, with $w_0 = w$, there exists $i \geq 0$ such that $w_i \models \varphi_3$, and for every $0 \leq j < i$, $\pi^j \models \psi$. Let i_π be the minimal index i for which $w_i \models \varphi_3$. We show that for every $0 \leq j < i_\pi$, $w_j \models A\psi$ and thus $\pi \models (A\psi)U\varphi_3$. Assume, by way of contradiction, that there exists some $0 \leq j < i$ for which $w_j \not\models A\psi$. That is, there exists a path $\pi' = w'_j, w'_{j+1}, \ldots$, with $w'_j = w_j$, such that $\pi' \not\models \psi$. Consider the path $\rho = w_0, w_1, \ldots, w_j, w'_{j+1}, w'_{j+2}, \ldots$. Since π and ρ coincide in their first j states, $i_\pi > j$ implies that $i_\rho > j$ and therefore, as $\rho \models \psi U\varphi_3$, all $0 \leq k \leq j$ have $\rho^k \models \psi$. In particular, $\pi' \models \psi$, and we reach a contradiction. □

Claim 3.8 *(a)* $E(X\varphi_2)U\varphi_3 \equiv \varphi_3 \vee EXE\varphi_2 U(\varphi_3 \wedge \varphi_2)$.

(b) $E(\varphi_1 U\varphi_2)U\varphi_3 \equiv \varphi_3 \vee E(\varphi_1 \vee \varphi_2)U((\varphi_2 \wedge EX\varphi_3) \vee (\varphi_3 \wedge E\varphi_1 U\varphi_2))$.

(c) $E(\varphi_1 \bar{U}\varphi_2)U\varphi_3 \equiv \varphi_3 \vee E(\neg\varphi_2)U(((\neg\varphi_1) \wedge (\neg\varphi_2) \wedge EX\varphi_3) \vee (\varphi_3 \wedge E\varphi_1 \bar{U}\varphi_2))$.

Proof: All three formulas are of the form $E\psi U\varphi_3$, for a CTL* path formula ψ. Their CTL equivalences are based on the same idea: We isolate the case where φ_3, the eventuality of the top level until, holds in the present. If it does not, we modify the until-structure to guarantee both an occurrence of φ_3 eventually, and correctness of ψ along the suffixes that precede the first occurrence of φ_3. We give here the details of (a).

Assume first that $w \models \varphi_3 \vee EXE\varphi_2 U(\varphi_3 \wedge \varphi_2)$. Then, either $w \models \varphi_3$, in which case clearly $w \models E(X\varphi_2)U\varphi_3$, or $w \models EXE\varphi_2 U(\varphi_3 \wedge \varphi_2)$. Then, there exists a path $\pi = w_0, w_1, \ldots$, with $w_0 = w$, for which there exists $i \geq 1$ such that $w_i \models \varphi_3 \wedge \varphi_2$, and for every $1 \leq j < i$, $w_j \models \varphi_2$. This implies that $w_i \models \varphi_3$ and for every $0 \leq j < i$, $w_{j+1} \models \varphi_2$. Thus, $\pi \models (X\varphi_2)U\varphi_3$ and $w \models E(X\varphi_2)U\varphi_3$.

Assume now that $w \models E(X\varphi_2)U\varphi_3$. Then, there exists a path $\pi = w_0, w_1, \ldots$, with $w_0 = w$, for which there exists $i \geq 0$ such that $w_i \models \varphi_3$, and for every $0 \leq j < i$,

$w_{j+1} \models \varphi_2$. If $i = 0$ then $w \models \varphi_3$. Otherwise, $i \geq 1$, w_i satisfies both φ_3 and φ_2, and for every $1 \leq j < i$, $w_j \models \varphi_2$. Thus $w \models \varphi_3 \vee EXE\varphi_1 U(\varphi_3 \wedge \varphi_2)$. □

Claim 3.9 *(a)* $A\varphi_1 U(X\varphi_2) \equiv AX\varphi_2 \vee (\varphi_1 \wedge AXA\varphi_1 U(\varphi_2 \vee AX\varphi_2))$.

(b) $A\varphi_1 U(\varphi_2 U\varphi_3) \equiv A\varphi_1 U(A\varphi_2 U\varphi_3)$.

Proof: Both formulas are of the form $A\varphi_1 U\psi$ for a CTL* path formula ψ. Again, the idea is to distinguish between the case where ψ, the eventuality of the until, holds in all paths in the present, and the case it does not. In the latter, modification of the until-structure is required. In (b), as $(A\psi) \vee A\varphi_1 U(A\psi) \equiv A\varphi_1 U(A\psi)$, both cases coincide. Note that the sad formula $A\varphi_1 U(A\varphi_2 \bar{U}\varphi_3)$, that has no CTL equivalent, has also the form $A\varphi_1 U\psi$, for a CTL* path formula ψ. Below we give the details of (b).

Assume first that $w \models A\varphi_1 U(A\varphi_2 U\varphi_3)$. Then, for every path $\pi = w_0, w_1, \ldots$, with $w_0 = w$, there exists $i \geq 0$ such that $w_i \models A\varphi_2 U\varphi_3$ and for every $0 \leq j < i$, $w_j \models \varphi_1$. In particular, $\pi^i \models \varphi_2 U\varphi_3$ and thus $w \models A\varphi_1 U(\varphi_2 U\varphi_3)$.

Assume now that $w \models A\varphi_1 U(\varphi_2 U\varphi_3)$. Then, for every path $\pi = w_0, w_1, \ldots$, with $w_0 = w$, there exists $i \geq 0$ such that $\pi^i \models \varphi_2 U\varphi_3$ and for every $0 \leq j < i$, $w_j \models \varphi_1$. That is, there exists $k \geq i$ such that $w_k \models \varphi_3$ and for every $i \leq j < k$, $w_j \models \varphi_2$. Thus, π has some state, w_k, that satisfies φ_3, and the prefix $w_0, \ldots, w_k$, of π, has some $0 \leq i \leq k$ such that for every $0 \leq j < i$, $w_j \models \varphi_1$ and for every $i \leq j < k$, $w_j \models \varphi_2$. Let k_π be the minimal index k for which $w_k \models \varphi_3$, and let i_π be the maximal index i that satisfies the above conditions; That is, either $i_\pi = k_\pi$, or $w_{i_\pi} \not\models \varphi_1$. We show that $w_{i_\pi} \models A\varphi_2 U\varphi_3$, and thus $\pi \models \varphi_1 U(A\varphi_2 U\varphi_3)$. Assume, by way of contradiction, that $w_{i_\pi} \not\models A\varphi_2 U\varphi_3$. That is, there exists a path $\pi' = w'_{i_\pi}, w'_{i_\pi+1}, \ldots$, with $w'_{i_\pi} = w_{i_\pi}$ such that $\pi' \not\models \varphi_2 U\varphi_3$. Consider the path $\rho = s_0, s_1, \ldots$, where $s_j = w_j$ for $0 \leq j \leq i_\pi$, and $s_j = w'_j$ for $j > i_\pi$. $\rho \models \varphi_1 U(\varphi_2 U\varphi_3)$ and therefore, ρ has some state, s_{k_ρ}, such that $s_{k_\rho} \models \varphi_3$ and for every $0 \leq j < k_\rho$, $s_j \not\models \varphi_3$ (i.e., k_ρ is minimal). Since ρ and π coincide in their first i_π states and $k_\pi \geq i_\pi$, then $k_\rho \geq i_\pi$. Also, there exists some state s_{i_ρ}, such that $0 \leq i_\rho \leq k_\rho$, for every $0 \leq j < i_\rho$, $s_j \models \varphi_1$, for every $i_\rho \leq j < k_\rho$, $s_j \models \varphi_2$, and either $i_\rho = k_\rho$ or $s_{i_\rho} \not\models \varphi_1$ (i.e., i_ρ is the maximal). If $i_\rho > i_\pi$, then $w_{i_\pi} \models \varphi_1$ and as $w_{i_\pi} \not\models \varphi_3$, this contradicts the maximality of i_π. Therefore, $i_\rho \leq i_\pi$. Thus, π' has $k_\rho \geq i_\pi$ such that $w_{k_\rho} \models \varphi_3$, and for every $i_\pi \leq j < k_\rho$, $w_j \models \varphi_2$. Thus $\pi' \models \varphi_2 U\varphi_3$ and we reach a contradiction. Hence, $w \models A\varphi_1 U(\varphi_2 U\varphi_3)$. □

Claim 3.10 *(a)* $E\varphi_1 U(X\varphi_2) \equiv E\varphi_1 U(EX\varphi_2)$.

(b) $E\varphi_1 U(\varphi_2 U\varphi_3) \equiv E\varphi_1 U(E\varphi_2 U\varphi_3)$.

(c) $E\varphi_1 U(\varphi_2 \bar{U}\varphi_3) \equiv E\varphi_1 U(E\varphi_2 \bar{U}\varphi_3)$.

Proof: We prove a stronger claim. Let ψ be a CTL* path formula, we prove $E\varphi_1 U\psi \equiv E\varphi_1 UE\psi$. $w \models E\varphi_1 U\psi$ iff there exists a path $\pi = w_0, w_1, \ldots$, with $w_0 = w$, for which there exists $i \geq 0$ such that $\pi^i \models \psi$ and for every $0 \leq j < i$, $w_j \models \varphi_1$. This holds iff there exists a path $\pi' = w_0, w_1, \ldots, w_i, w'_{i+1}, \ldots$, such that $w_i \models E\psi$ and for every $0 \leq j < i$, $w_j \models \varphi_1$. That is, iff $w \models \varphi_1 UE\psi$. □

In the next section we show that the additional expressive power of CTL^2 is given for free. Namely, the model-checking problem for the logic CTL^2 is of the same complexity as the one for CTL.

4 A Linear Model-Checking Procedure for CTL^2

In this section we introduce an efficient algorithm for the model-checking problem for CTL^2. Given a Kripke structure $K = \langle W, R, L \rangle$ and a CTL^2 formula ψ, our algorithm labels K such that for every state $w \in W$, w is labelled with ψ iff $w \models \psi$. The algorithm is of complexity linear in both the size of K and the length of ψ. Thus, the increase in the expressive power does not effect model-checking complexity which remains linear, just as the one for CTL. In Section 4.1, we extend our algorithm, preserving its complexity, to handle fairness for a subset of CTL^2.

The algorithm extends the efficient model checker for CTL introduced in [CES86]. There, a formula is handled by successively applying a state labelling algorithm to its sub-formulas. In more details, given a CTL formula ψ and a Kripke structure K, the algorithm takes the sub-formulas of ψ, starting with the innermost ones, and, iteratively, labels with each sub-formula φ of ψ, exactly those states of K that satisfy it. Each iteration handles a single sub-formula which may have one of seven forms that together cover CTL modalities. Since ψ has at most $|\psi|$ sub-formulas, the check terminates after at most $|\psi|$ iterations. Each iteration requires $O(|K|)$ steps and hence, the entire check is accomplished after $O(|\psi| * |K|)$ steps.

In CTL^2, there are much more than seven forms. All the possible combinations of one or two temporal operators should be considered. However, as we showed in the previous section, all those forms, except $A\varphi_1 U(\varphi_2 \bar{U} \varphi_3)$, have CTL equivalences. Thus, labelling with happy sub-formulas can be done by labelling their CTL equivalences. In addition, we introduce a procedure that handles sad sub-formulas.

Below we define the *closure of a formula* for every CTL^2 formula. The closure of φ, $cl(\varphi)$, extends the notion of sub-formulas used in [CES86]. It is the set of CTL^2 formulas that contains φ and all the formulas that the labelling of φ depends on. Given a CTL^2 formula φ, $cl(\varphi)$ is inductively defined as follows (φ_1 and φ_2 are CTL^2 formulas):

- If φ is either t or f, then $cl(\varphi) = \{\varphi\}$.

- If φ is a proposition, then $cl(\varphi) = \{\varphi, t, f\}$.

- If φ is $\neg \varphi_1$ or $EX\varphi_1$, then $cl(\varphi) = \{\varphi\} \cup cl(\varphi_1)$.

- If φ is $\varphi_1 \vee \varphi_2$, $E\varphi_1 U \varphi_2$, or $A\varphi_1 U \varphi_2$, then $cl(\varphi) = \{\varphi\} \cup cl(\varphi_1) \cup cl(\varphi_2)$.

- If φ is a happy formula, then $cl(\varphi) = cl(\varphi')$, where φ' is the CTL equivalent of φ.

- If φ is $A\varphi_1 U(\varphi_2 \bar{U} \varphi_3)$, then $cl(\varphi) = \{\varphi, EtU\varphi_3\} \cup cl(\varphi_1) \cup cl(\varphi_2) \cup cl(\varphi_3)$.

Lemma 4.1 *For every CTL^2 formula ψ,*

(a) All the formulas in $cl(\psi)$ are either t, f, propositions, or have the form $\neg \varphi_1$, $\varphi_1 \vee \varphi_2$, $EX\varphi_1$, $E\varphi_1 U \varphi_2$, $A\varphi_1 U \varphi_2$, or $A\varphi_1 U(\varphi_2 \bar{U} \varphi_3)$.

(b) The size of $cl(\psi)$ is linear in $|\psi|$.

Lemma 4.1 is the key for the model-checking procedure, presented in Figure 1.

```
procedure model_check (ψ, K);
begin
  for i := 1 to |cl(ψ)| do
      for every φ ∈ cl(ψ) with |φ| = i do
          case structure of φ is of the form
              An atomic proposition : noop;
              φ₁ ∨ φ₂ : for every w ∈ W do
                          if φ₁ ∈ L(w) or φ₂ ∈ L(w) then add φ₁ ∨ φ₂ to L(w);
              ¬φ₁     : for every w ∈ W do
                          if φ₁ ∉ L(w) then add ¬φ₁ to L(w);
              EXφ₁    : for every w ∈ W do
                          if φ₁ ∈ L(w') for some successor w' of w then add EXφ₁ to L(w);
              Eφ₁Uφ₂  : EU_check (φ₁, φ₂, φ);
              Aφ₁Uφ₂  : AU_check (φ₁, φ₂, φ);
              Aφ₁U(φ₂Ūφ₃) : AUŪ_check (φ₁, φ₂, φ₃, φ);
          end;
end;
```

Figure 1: The model-checking procedure for CTL^2.

Given a CTL^2 formula ψ, the procedure is called with the CTL equivalent of ψ (if exists). As in [CES86], it iteratively labels formulas from $cl(\varphi)$. Formulas of the form $\neg \varphi_1$, $\varphi_1 \vee \varphi_2$, $EX\varphi_1$, $E\varphi_1 U\varphi_2$, or $A\varphi_1 U\varphi_2$ are labelled using the appropriate procedure described there. Formulas of the form $A\varphi_1 U(\varphi_2 \bar{U} \varphi_3)$, are labelled using the procedure $AU\bar{U}\_Check(\varphi_1, \varphi_2, \varphi_3, \varphi)$, presented in Figure 2 and explained below.

The procedure assumes that upon invocation K is labelled with $cl(\varphi) \setminus \{\varphi\}$. In particular, for every state $w \in W$ and for every $\xi \in \{\varphi_1, \varphi_2, \varphi_3, EtU\varphi_3\}$, $\xi \in L(w)$ iff $w \models \xi$. It uses the following procedures:

- *find_MSCCs*(ψ), which, given a formula ψ, finds the maximal strongly connected components (MSCCs) in the sub-structure of K that consists of all the states in K that are labelled with ψ. The output of *find_MSCCs*(ψ) is a set of (disjoint) subsets of W such that every subset is a MSCC. The complexity of *find_ψ-MSCCs* is linear in $|K|$ [Tar72].

- *foreign_successors*(C), which, given a set C of states, returns the number of transitions $\langle w_1, w_2 \rangle \in R$ for which $w_1 \in C$ and $w_2 \notin C$. The complexity of finding *foreign-successors*(C) for all the MSCCs of K is linear in $|K|$.

- *bd*(w), which, given a state $w \in W$, returns the branching degree of w.

- *comp*(w), which, given a state $w \in W$, returns the MSCC to which w belongs.

The procedure maintains three types of counters:

procedure $AU\bar{U}\_check$ $(\varphi_1, \varphi_2, \varphi_3, \varphi)$;
begin
  for $j \in [0 \ldots |W|]$ **do** $A_3^j := \phi$; $A_1^j := \phi$ **end**;
  $Csets := find\_MSCCs(\neg\varphi_3)$;
  for every $C \in Csets$ **do** $count(C) := foreign\_successors(C)$;
  for every $w \in W$ **do**
    $count_1(w) := bd(w)$; $count_3(w) := bd(w)$;
    if $(\varphi_2 \notin L(w)$ and $\varphi_3 \notin L(w))$ or $(EtU\varphi_3 \notin L(w))$ **then**
      add w to A_3^0; add φ to $L(w)$;
  end;
  $j := 0$;
  while $j < |W|$ **do**
    $S := A_3^j$;
    while $S \neq \phi$ **do**
      remove some w from S;
      for every predecessor w' of w **do**
        $count_3(w') := count_3(w') - 1$; $count_1(w') := count_1(w') - 1$;
        if $\varphi \notin L(w')$ and $count_3(w') = 0$ and $\varphi_3 \notin L(w')$ **then**
          add w' to A_3^{j+1}; add φ to $L(w')$;
        if $\varphi \notin L(w')$ and $\varphi_3 \notin L(w')$ and $w \notin comp(w')$ **then**
          $count(comp(w')) := count(comp(w')) - 1$;
          if $count(comp(w')) = 0$ **then for** every $w'' \in comp(w')$ **do**
            if $\varphi \notin L(w'')$ **then** add w'' to A_3^{j+1}; add φ to $L(w'')$ **end**
        if $\varphi \notin L(w')$ and $count_1(w') = 0$ and $\varphi_1 \in L(w')$ **then**
          add w' to A_1^0; add φ to $L(w')$;
      end;
    end;
    $j := j + 1$;
  end;
  $j := 0$;
  while $j < |W|$ **do**
    $S := A_1^j$;
    while $S \neq \phi$ **do**
      remove some w from S;
      for every predecessor w' of w **do**
        $count_1(w') := count_1(w') - 1$;
        if $\varphi \notin L(w')$ and $count_1(w') = 0$ and $\varphi_1 \in L(w')$ **then**
          add w' to A_1^{j+1}; add φ to $L(w')$
        if $\varphi \notin L(w')$ and $\varphi_3 \notin L(w')$ and $\varphi_1 \in L(w')$ and $w \notin comp(w')$ **then**
          $count(comp(w')) := count(comp(w')) - 1$;
          if $count(comp(w')) = 0$ **then for** every $w'' \in comp(w')$ **do**
            if $\varphi \notin L(w'')$ **then** add w'' to A_1^{j+1}; add φ to $L(w'')$ **end**
      end;
    end;
    $j := j + 1$;
  end;
end;

Figure 2: The procedure $AU\bar{U}\_check$.

- For every MSCC, C in K, $count(C)$ keeps track of the number of transitions $\langle w_1, w_2 \rangle \in R$ for which $w_1 \in C$, $w_2 \notin C$, and there exists a path $\pi = w_1, w_2, \ldots$ that still has not detected as satisfying $\varphi_1 U(\varphi_2 \bar{U} \varphi_3)$. $count(C) = 0$ implies that for every state $w \in C$, all the paths that start in w and go out of C, satisfy $\varphi_1 U(\varphi_2 \bar{U} \varphi_3)$. Then, if all the states in C are labelled with $\neg \varphi_3$, it is guaranteed that they all satisfy φ.

- For every state $w \in W$, $count_3(w)$ keeps track of the number of w's successors that are still not detected as satisfying $A\varphi_2 \bar{U}\varphi_3$. $count_3(w) = 0$ implies that all the successors of w satisfy $A\varphi_2 \bar{U}\varphi_3$ and thus, if $w \not\models \varphi_3$, it satisfies $A\varphi_2 \bar{U}\varphi_3$ too.

- For every state $w \in W$, $count_1(w)$ keeps track of the number of w's successors that are not labelled with φ. $count_1(w) = 0$ implies that all the successors of w satisfy φ and thus, if $w \models \varphi_1$, it satisfies φ too.

We now explain why two counters are required for each state. Note that a state w with $count_3(w) = 0$ is more likely to satisfy φ than a state w with $count_1(w) = 0$. In the first case, w can either satisfy $\neg \varphi_3$ or φ_1. In the second, w must satisfy φ_1; it might be the case where some paths that start in w do not satisfy $\varphi_2 \bar{U}\varphi_3$. Rather, they have some suffix satisfying $\varphi_2 \bar{U}\varphi_3$ and a sequence of states satisfying φ_1, that should not be interrupted, leads to this suffix. Thus, it is essential to keep track of both counters, detecting first states that satisfy φ by satisfying $\varphi_2 \bar{U}\varphi_3$, and only then states that satisfy φ yet do not satisfy $\varphi_2 \bar{U}\varphi_3$. Detection is done by keeping, for every $0 \leq j \leq |W|$, two sets A_3^j and A_1^j, preserving the following invariants:

- Consider the formula $A\varphi_2 \bar{U}\varphi_3$. For every state $w \in W$, if $w \models A\varphi_2 \bar{U}\varphi_3$, then every path $\pi = w_0, w_1, \ldots$, with $w_0 = w$, either has a finite $0 \leq i \leq |W|$ such that $w_i \models (\neg \varphi_2) \wedge (\neg \varphi_3)$ (and for all $0 \leq j \leq i$, $w_j \not\models \varphi_3$), or $\pi \models G\neg \varphi_3$. Let $depth_3(w)$ be the minimal integer for which all paths from w that do not satisfy $G\neg \varphi_3$, have $w_i \models (\neg \varphi_2) \wedge (\neg \varphi_3)$ with $i \leq depth_3(w)$.

  For every $w \in W$ and $0 \leq j \leq |W|$, $w \in A_3^j$ iff $w \models A\varphi_2 \bar{U}\varphi_3$ with $depth_3(w) = j$. In particular, $w \in A_3^0$ iff $w \models A\varphi_2 \bar{U}\varphi_3$ with $w \models AG\neg \varphi_3$ or $w \models (\neg \varphi_2) \wedge (\neg \varphi_3)$.

- Consider the formula $\varphi = A\varphi_1 U(\varphi_2 \bar{U}\varphi_3)$. For every state $w \in W$, if $w \models \varphi$, then all the paths that start in w have a finite $0 \leq i \leq |W|$ such that $\pi^i \models \varphi_2 \bar{U}\varphi_3$. Let $depth_1(w)$ be the minimal integer for which all the paths that start in w have $\pi^i \models \varphi_2 \bar{U}\varphi_3$ with $i \leq depth_1(w)$.

  For every $w \in W$ and $0 \leq j \leq |W|$, $w \in A_1^j$ iff $w \models A\varphi_1 U(\varphi_2 \bar{U}\varphi_3)$ with $depth_1(w) = j$. In particular, $w \in A_1^0$ iff $w \models A\varphi_1 U(\varphi_2 \bar{U}\varphi_3)$ with $w \models A\varphi_2 \bar{U}\varphi_3$.

When $AU\bar{U}\_check(\varphi_1, \varphi_2, \varphi_3, \varphi)$ is called, no state is labelled with φ. The counters are initialized accordingly and states that satisfy $(\neg \varphi_2) \wedge (\neg \varphi_3)$ or $\neg EfU\varphi_3$ are inserted into A_3^0 (note that $\neg EfU\varphi_3 \equiv AG\neg \varphi_3$). Then, executing the first while-loop, satisfaction of $A\varphi_2 \bar{U}\varphi_3$ is detected by reasoning backwards along states that satisfy $\neg \varphi_3$. States w that satisfy $A\varphi_2 \bar{U}\varphi_3$ with $depth_3(w) = i$, are labelled with φ and inserted to A_3^i in the i-th iteration.

Since some paths may satisfy $\varphi_2 \bar{U}\varphi_3$ by satisfying $G\neg \varphi_3$, the backwards diffusion described above is not sufficient. Counters of states that belong to a $\neg \varphi_3$-circle never reach zero and thus, satisfaction of φ in such states can not be detected due to a circular dependency. The counters of the MSCCs take care of this problem. Once a counter of a MSCC, C, whose all states satisfy $\neg\varphi_3$, reaches zero during the first while-loop execution, it is guaranteed that all the paths that go out of C satisfy $\varphi_2\bar{U}\varphi_3$. Since all the paths that remain in C satisfy $G\neg\varphi_3$, it implies that all the states in C satisfy $A\varphi_2\bar{U}\varphi_3$. Hence, they are labelled with φ and inserted to the corresponding A_3^j set.

Only after the backwards diffusion of $A\varphi_2\bar{U}\varphi_3$ runs its course, the second while-loop is executed. Here, the backward diffusion proceeds along states that satisfy φ_1. States w that satisfy φ with $depth_1(w) = i$, are labelled with φ and inserted to A_1^i in the i-th iteration. States w that satisfy φ, do not satisfy $A\varphi_2\bar{U}\varphi_3$, and belong to a MSCC of states satisfying $\neg\varphi_3$, may never reach zero in their counters. Therefore, again, the counters of the MSCC are essential for solving circular dependency. In this case, however, $count(C)$ is decreased only due to edges that their nodes in C satisfy φ_1.

We now consider the complexity of the procedure. As $find\_MSCC(\psi)$ and $foreign$-$successors(C)$ are of complexity linear in $|K|$, initialization is performed in linear time. Since all the sets A_3^j and A_1^j are pairwise disjoint and are all contained in W, the body of both while-loops is executed at most $|W|$ times. The inner for-loop is executed at most once for each transition in R, and finally, since the counter of each MSCC may reach zero only once, the body of the for-loop performed then, is executed at most once for each state in W. Hence, the whole procedure terminates after at most $O(|K|)$ steps.

4.1 A Fair Model Checker for CTL^2

In [CES86], fairness is introduced into CTL. A new logic, CTL^F, interpreted over Kripke structures with fairness constrains, is suggested. Formally, a Kripke structure with fairness constrains (a fair structure) is a four-tuple $\langle W, R, L, \mathcal{F}\rangle$, where W, R, and L are as for usual structures, and $\mathcal{F} \subseteq 2^W$ is a set of subsets of W. Given a path π in K, let $inf(\pi)$ be the set of states that appear in π infinitely often. A path π is fair iff for each set $F \in \mathcal{F}$, $inf(\pi) \cap F \neq \phi$. CTL^F has exactly the same syntax and semantics as CTL, except that all path quantifiers range over fair paths. This enables, while verifying concurrent programs, to refer only to fair execution sequences. In this section we show how the fair model checker, introduced in [CES86], can be used to handle a substantial subset of CTL^2.

As in CTL, let $fair$-CTL^2 (CTL_F^2) be the logic CTL^2 when interpreted over fair structures. Given a fair structure, a state w in it, and a CTL_F^2 formula ψ, $w \models_F \psi$ indicates that ψ holds in w, with path quantifiers ranging over fair paths only. Given two formulas ψ_1 and ψ_2, we say that ψ_1 and ψ_2 are $fair$-$equivalent$ ($\psi_1 \equiv_F \psi_2$) if for every fair structure and for every state w in it, $w \models_F \psi_1$ iff $w \models_F \psi_2$. We first show that translating a happy CTL^2 formula into its CTL equivalent, using the equivalences suggested in Section 3, preserves equivalence also with respect to CTL_F^2 and CTL_F.

Claim 4.2 *Given a happy CTL^2 formula ψ and its CTL equivalent ψ', $\psi \equiv_F \psi'$.*

Proof: We have to show that the fourteen equivalences given in Section 3, remain valid when path quantifiers range over fair paths only. Fortunately, the proofs given there remain valid too. That is, readers are encouraged to copy the proofs given there, add the string "fair " before each occurrence of the string "path", and then to observe that indeed, for every such ψ and ψ', the proof remains valid, fits the case where CTL_F^2 and CTL_F are considered, and thus, $\psi \equiv_F \psi'$. □

Given a happy formula ψ, we say that ψ is *full of happiness*, if all the formulas in $cl(\psi)$ are happy. Note that CTL^2 formulas that do not have any sad sub-formula, are full of happiness, and vice versa. Every full of happiness formula ψ, has an equivalent CTL formula ψ', obtained by recursively translating its sub-formulas to CTL. Moreover, as follows from Claim 4.2, ψ and ψ' are fair-equivalent. Hence, model checking of CTL_F^2 formulas which are full of happiness can be done using the extended and linear model checker described in [CES86].

As in CTL, fairness constrains can be used to increase the expressive power of the logic. Consider the CTL* formula $\psi = A(GFp \rightarrow \varphi)$, where p is atomic. Checking the formula $A\varphi$ with respect to a fair structure that has $\mathcal{F} = \{\{p\}\}$ as its fairness constrains, is equivalent to checking ψ in a structure that has no fairness constrains. Using the extended model checker for CTL^2, we can thus check formulas of the form $A(GFp \rightarrow \varphi)$ for which $A\varphi$ is full of happiness. In particular, we can check the formula $A(GFp \rightarrow GFq)$ that specifies strong fairness.

Note that since our extended model checker cannot handle fairness of sad CTL^2 formulas, we cannot use it for formulas which are not expressible in CTL. Nevertheless, it does contribute to a clear presentation of the formulas being checked.

References

[Bro86] M.C. Browne. An improved algorithm for the automatic verification of finite state systems using temporal logic. In *Proceedings of the First Symposium on Logic in Computer Science*, pages 260–266, Cambridge, June 1986.

[Bry86] R.E. Bryant. Graph-based algorithms for boolean function manipulation. *IEEE Transactions on Computers*, C-35(8), 1986.

[CE81] E.M. Clarke and E.A. Emerson. Design and synthesis of synchronization skeletons using branching time temporal logic. In *Proc. Workshop on Logic of Programs*, volume 131 of *Lecture Notes in Computer Science*, pages 52–71. Springer-Verlag, 1981.

[CES86] E.M. Clarke, E.A. Emerson, and A.P. Sistla. Automatic verification of finite-state concurrent systems using temporal logic specifications. *ACM Transactions on Programming Languages and Systems*, 8(2):244–263, January 1986.

[EC82] E.A. Emerson and E.M. Clarke. Using branching time logic to synthesize synchronization skeletons. *Science of Computer Programming*, 2:241–266, 1982.

[EH86] E.A. Emerson and J.Y. Halpern. Sometimes and not never revisited: On branching versus linear time. *Journal of the ACM*, 33(1):151–178, 1986.

[EJ88] E.A. Emerson and C. Jutla. The complexity of tree automata and logics of programs. In *Proceedings of the 29th IEEE Symposium on Foundations of Computer Science*, White Plains, Oct 1988.

[Eme90] E.A. Emerson. Temporal and modal logic. *Handbook of theoretical computer science*, pages 997–1072, 1990.

[ES84] A.E. Emerson and A.P. Sistla. Deciding full branching time logics. *Information and Control*, 61(3):175–201, 1984.

[GL91] O. Grumberg and D. Long. Model checking and modular verification. In *Proc. 2nd Conference on Concurrency Theory*, volume 527 of *Lecture Notes in Computer Science*, 1991.

[Lam80] L. Lamport. Sometimes is sometimes "not never" - on the temporal logic of programs. In *Proceedings of the 7th ACM Symposium on Principles of Programming Languages*, pages 174–185, January 1980.

[McM93] K.L. McMillan. *Symbolic model checking*. Kluwer Academic Publishers, 1993.

[Pnu81] A. Pnueli. The temporal semantics of concurrent programs. *Theoretical Computer Science*, 13:45–60, 1981.

[Rab70] M.O. Rabin. Weakly definable relations and special automata. In *Proc. Symp. Math. Logic and Foundations of Set Theory*, pages 1–23. North Holland, 1970.

[SC85] A.P. Sistla and E.M. Clarke. The complexity of propositional linear time logic. *ACM*, 32(3):733-749, 1985.

[Tar72] R.E. Tarjan. Depth first search and linear graph algorithms. *SIAM Journal of Computing*, 1(2):146-160, 1972.

[WG93] P. Wolper and P. Godefroid. Partial-order methods for temporal verification. In *Proc. 4th Conferance on Concurrency Theory*, volume 715 of *Lecture Notes in Computer Science*, pages 233–246, Hildesheim, August 1993. Springer-Verlag.

[Wol82] P. Wolper. Specification and synthesis of communicating processes using an extended temporal logic. In *Proc. 9th Symposium on Principles of Programming Language-seedings*, pages 20–33, Albuquerque, January 1982.

[Wol83] P. Wolper. Temporal logic can be more expressive. *Information and Control*, 56(1-2):72–99, 1983.

A Equivalences for Rule R5.

The following equivalences show that the formulas in Rule R5 do not add to the expressive power of CTL. Their redundancy is already known, yet the equivalences we suggest have $cl(\psi')$ of size linear in $|\psi|$, where ψ' is the CTL equivalence of ψ. We use the notation $\varphi_1 \tilde{U} \varphi_2 = (\neg \varphi_1) \bar{U} (\neg \varphi_2)$.

1. $Q(\varphi_1 \vee (X\varphi_2)) \equiv \varphi_1 \vee QX\varphi_2$
   $Q(\varphi_1 \vee (\varphi_2 U \varphi_3)) \equiv \varphi_1 \vee Q\varphi_2 U \varphi_3$

2. $Q((X\varphi_1) \vee (X\varphi_2)) \equiv QX(\varphi_1 \vee \varphi_2)$
   $Q((X\varphi_1) \vee (\varphi_2 U \varphi_3)) \equiv \varphi_3 \vee QX\varphi_1 \vee (\varphi_2 \wedge QX(\varphi_1 \vee Q\varphi_2 U \varphi_3))$
   $Q((X\varphi_1) \vee (\varphi_2 \bar{U} \varphi_3)) \equiv ((\neg \varphi_3) \wedge (\neg \varphi_2)) \vee QX\varphi_1 \vee ((\neg \varphi_3) \wedge QX(\varphi_1 \vee Q\varphi_2 \bar{U} \varphi_3))$

3. $E((\varphi_1 U \varphi_2) \vee (\varphi_3 U \varphi_4)) \equiv (E\varphi_1 U \varphi_2) \vee (E\varphi_3 U \varphi_4)$
   $E((\varphi_1 U \varphi_2) \vee (\varphi_3 \bar{U} \varphi_4)) \equiv (E\varphi_1 U \varphi_2) \vee (E\varphi_3 \bar{U} \varphi_4)$
   $E((\varphi_1 \bar{U} \varphi_2) \vee (\varphi_3 \bar{U} \varphi_4)) \equiv (E\varphi_1 \bar{U} \varphi_2) \vee (E\varphi_3 \bar{U} \varphi_4)$

4. $A((\varphi_1 U \varphi_2) \vee (\varphi_3 U \varphi_4)) \equiv$
   $A((\varphi_2 \vee \varphi_4)\tilde{U}(\varphi_3 \vee A\varphi_1 U \varphi_2)) \wedge A((\varphi_2 \vee \varphi_4)\tilde{U}(\varphi_1 \vee A\varphi_3 U \varphi_4)) \wedge AF(\varphi_2 \vee \varphi_4)$
   $A((\varphi_1 U \varphi_2) \vee (\varphi_3 \tilde{U} \varphi_4)) \equiv$
   $A((\varphi_2 \vee \varphi_3)\tilde{U}(\varphi_4 \vee A\varphi_1 U \varphi_2)) \wedge A((\varphi_2 \vee \varphi_3)\tilde{U}(\varphi_2 \vee A\varphi_3 U \varphi_4))$
   $A((\varphi_1 \bar{U} \varphi_2) \vee (\varphi_3 \tilde{U} \varphi_4)) \equiv$
   $A((\varphi_1 \vee \varphi_3)\tilde{U}(\varphi_2 \vee A\varphi_3 \bar{U} \varphi_4)) \wedge A((\varphi_1 \vee \varphi_3)\tilde{U}(\varphi_4 \vee A\varphi_1 \bar{U} \varphi_2))$

Back and forth through time and events

(Extended Abstract)

Patrick Blackburn,[1] Claire Gardent[2] and Maarten de Rijke[3]

[1] Department of Philosophy, Rijksuniversiteit Utrecht,
Heidelberglaan 8, 3584 CS Utrecht.
Email: `patrick@phil.ruu.nl`

[2] GRIL, Université de Clermont Ferrand, France; and
Department of Computational Linguistics, Universiteit van Amsterdam,
Spuistraat 134, 1012 VB Amsterdam.
Email: `claire@mars.let.uva.nl`

[3] CWI, P.O. Box 4079, 1009 AB Amsterdam.
Email: `mdr@cwi.nl`

Abstract. In this extended abstract *back-and-forth structures* are defined and applied to the semantics of natural language. Back-and-forth structures consist of an event structure and an interval structure communicating via a relational link; transitions in the one structure correspond to transitions in the other. Such structures enable us to view temporal constructions (such as tense, aspect, and temporal connectives) as methods of moving systematically between information sources. We illustrate this with a treatment of the English present perfect, and progressive aspect, that draws on ideas developed in Moens and Steedman (1988).

1 Introduction

Formal accounts of temporal constructions in natural language often disagree about the semantic ontology to be assumed — should it be point based, interval based or event based? We think that more adequate analyses of natural language will be obtained by *combining* ontologies, not choosing between them. We illustrate this by combining interval structures with (various forms of) event structures into what we call *back-and-forth structures* (BAFs). These consist of an interval structure and an event structure linked by a relation so that transitions in the one correspond to transitions in the other.

Such combined ontologies enable us to build our analyses round the following intuition: temporal constructions are means of systematically exploiting links between information sources. Consider the English present perfect. It is common to informally gloss this construction as 'a past tense of present relevance'. For example, '*John has gone to the store*' means that at some past time John went to the store and, moreover, that John's excursion is somehow of relevance to the present context. We see two important transitions here: a move backwards in time through an interval structure, and a move to an associated event in an event structure. The English present perfect coordinates these transitions, and BAFs enable us to model this.

Much of this abstract uses BAFs to explore the ideas of Moens and Steedman (1988); indeed, BAFs developed by thinking about the kind of machinery required to formalise their work. Moens and Steedman provide a wide ranging account of temporal semantics (topics considered include tense, temporal reference, aspect and adverbial modification) couched as a Winograd-style procedural semantics. Their work hinges on (at least) the following ideas: that non-temporal relations between events must be admitted if an adequate account is to be given of the semantics of 'when' and various aspectual phenomena; that there are key event configurations (called 'nuclei') underlying the richness of event ontology; and that adverbial (and other forms of) modification are to be accounted for in terms of 'type coercion'. The Moens and Steedman account is attractive because while it is wide ranging, its explanations reduce to the interaction of a handful of intuitive ideas. Its weakness is that it is largely unformalised. We believe BAFs provide a setting in which substantial parts of their account can be made precise. BAFs can be seen as a way of modeling the insight that a systematic interplay between temporal and non-temporal relations is called for, and by progressively enriching the event structures they are built over one can model ever more of the Moens and Steedman system.

We proceed as follows. We first discuss the semantics of the English present perfect, indicating why the use of combined ontologies seems promising. We then introduce *simple BAFs*. These consist of interval structures combined with an extremely simple type of eventuality structure. Although such structures are too simple to cope with all the subtleties of natural language, their use permits the central idea underlying our proposal to be clearly presented. Following this, we (slightly) enrich the eventuality component to form *sorted BAFs*. This enables us to refine our discussion of the present perfect, and to provide an analysis of progressive aspect that does not run foul of the so-called imperfective paradox. We close the abstract by briefly discussing how we are extending this work, and noting other BAF-like proposals we have found in the literature.

2 The Present Perfect

While descriptive work on the English present perfect abounds, the construction has been notoriously resistant to formal analysis. In this section we discuss the problems the present perfect gives rise to, and argue that these indicate the need for combined ontologies.

It is often argued that the English present perfect is used to describe past events of present relevance. Perhaps the most well-known account of this intuition is that described in Reichenbach (1947), where a present perfect is analysed as describing a past event (the event temporally precedes the speech time) whose reference time coincides with the speech time. Reichenbach's reference point is meant to be the time talked about; or, in other words, the temporal perspective from which the described event is viewed. By insisting that reference and speech time coincide, the present perfect is analysed as relevant to the present. This contrasts with the simple past which is viewed as describing a past event whose reference time coincides with the event time rather than with the speech time.

Although Reichenbach's approach goes one step toward capturing the intuition underlying the use of the present perfect, two problems remain. First, what is the nature of reference times, and how are they determined? Second, the Reichenbachian account fails to account for many observations made in the literature concerning the restrictions governing the use of the present perfect. For instance, it does not explain why the sentence in (1) is infelicitous if uttered at a time occurring after the coffee has been cleaned.

(1)   I have spilled my coffee.

Similarly, it does not account for the restrictions placed by verbal aspectual classes on the use of the present perfect, for example:

(2)   a. ? The house has been empty (stative expression)
       b. ? I have worked in the garden (process expression)
       c. ? The star has twinkled (point expression).

Example (2a) shows that the present perfect is awkward in combination with stative expressions; (2b) and (2c) illustrate its awkwardness in combination with process expressions and point expressions, respectively.

As Moens and Steedman (1986) convincingly argue, these problems can be resolved if the internal structure of events is taken into account. Briefly, the idea is that an event (or *nucleus* in Moens and Steedman's terminology) is a tripartite structure consisting of a *preparatory phase*, a *culmination* and a *consequent state*. Given such a structure, the function of the present perfect is to situate the reference time in the consequent state of the core event being described (cf. Moens and Steedman (1986), p.20). Thus instead of the Reichenbach schema

```
        E              R, S
    ---|--------------|-----
```

Moens and Steedman describe the present perfect by means of the following diagram:

```
              E         R, S
    |----------|------------------
        PP    Cul         CS
```

Their account incorporates the central Reichenbachian intuition, while eliminating its problematic aspects:

— The reference point is given a (more) precise and more motivated location in time, namely within the time stretch of the consequent state.
— Example (1) is explained as follows. An obvious consequence of spilling one's coffee is that coffee is spilled. Under the Moens and Steedman theory, uttering a sentence in the present perfect indicates (i) that the reference time coincides with the speech time and (ii) that both these times are included

in the time stretch of the consequent state. Thus by uttering the present perfect (1), the speaker indicates that coffee is still spilled. Hence the oddity of (1) in a context where it isn't.

- The ill-formedness of the examples in (2) is explained by the fact that stative, process and point expressions are used to describe either states (i.e. unstructured entities) or these parts of the event structure which do not include the consequent state.[4] Since these expressions do not involve the notion of consequent state, they cannot be used in the present perfect whose semantics is defined in terms of this very notion.

The Moens and Steedman approach is intuitively appealing: how can it be made precise? We believe this can be done quite straightforwardly by combining ontologies.

Intuitively, their approach demands a mixture of ontologies: at the very least it seems to call for *temporal* structure, *eventuality* structure, and (crucially) a 'sensible fit' between these two ontologies. The 'past tense' component of the present perfect seems to require some notion of temporal structure; at the very least, this will involve some notion of temporal precedence. But this temporal structure does not suffice: in addition we need to invoke some notion of 'eventuality', and some sort of relation of 'relevance' between eventualities (for example, between the act of spilling the coffee, and the presence of the coffee on the floor). Intuitively this relevance relation isn't temporal; nonetheless, capturing the idea that we want an event of *present* relevance seems to presuppose that some sort of 'synchronisation' between the precedence relation on the temporal structure and the relevance relation on the eventuality structure is in force.

Actually, we will need even more structure than this. As examples (2a)–(2c) showed, the present perfect does not willingly combine with all verb types. We will need to work with a suitably fine-grained view of eventuality structure to capture these restrictions; in particular, by using eventuality structures *sorted* in a manner that reflects verbal aspectual classes we can model more of the Moens and Steedman account.

In the following two sections we will present simple formal models that capture these intuitions. We first present *simple* BAFs. These combine interval structures with a very simple notion of eventuality structure in a way that permits the intuition of 'present relevance' to be directly captured. (Or, to put it in the terminology of Moens and Steedman, they enable us to model the intuition that the present perfect works by locating the reference point in the run-time of consequent state induced by the eventuality being described.) We then refine this simple picture by enriching the eventuality structures used to make BAFs. This allows us to model the aspectual restrictions governing the use of the present perfect, and yields a simple solution to the imperfective paradox.

---

[4] These aspectual notions are discussed in more detail in section 4.

3 Simple BAFs

Simple BAFs consist of four components: an *interval structure*, an *eventuality structure*, and (most importantly) two *links* between them.

An *interval structure* **I** is a triple $\langle I, <, \sqsubseteq \rangle$ as defined in van Benthem (1991). Here I is a set of intervals, $<$ is the precedence relation, and $\sqsubseteq$ is the subinterval relation. We work with *linear*, *atomic* interval structures. That is, we assume that given any two intervals either one precedes the other or they overlap, and that our structures contain minimal, 'point-like' intervals.

An *eventuality* structure of signature $\mathcal{E}$ is (for the purposes of the present section) a triple $\mathbf{O} = \langle O, \texttt{GRiTo}, \{P_e\}_{e \in \mathcal{E}}\rangle$. Here O is a non-empty set, the set of *eventuality occurrences*; GRiTo is a binary relation on O; and all the P_e are unary relations on O. We assume $\mathcal{E} \neq \emptyset$. If e GRiTo e' then we say e *gives rise to* e'. The unary relations P_e can be thought of as 'eventualities' for example *runnings*, *jumpings* and *recitings of poems*.

Now the crucial step. A *back-and-forth structure* (BAF) of signature $\mathcal{E}$ is a quadruple $\langle \mathbf{O}, z, \mathcal{Z}, \mathbf{I} \rangle$, where **O** is an eventuality structure of signature $\mathcal{E}$, **I** is an interval structure, z is a function from O to I that returns the runtime or temporal extent of an eventuality and that preserves the relation GRiTo: if e GRiTo e' then $z(e) < z(e')$. That is, z is an order-preserving morphism from the eventuality structure to the interval structure; it is this morphism that synchronizes the two ontologies. $\mathcal{Z}$ is the relation with domain O and range I defined by $e\mathcal{Z}i$ iff $i \sqsubseteq z(e)$. That is, we assume that all eventualities are *downward persistent* to subintervals.

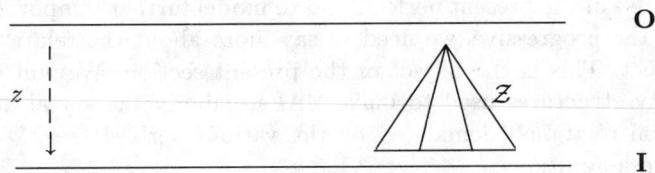

We now formulate a toy language for talking about BAFs: its vocabulary consists of all the items in $\mathcal{E}$, which we shall write as $p, q, r, \ldots$ etc., and call *eventuality symbols*, and an operator PERF. If α is an eventuality symbol then PERF α is well formed (and nothing else is). Obviously it would be possible to add the Boolean operators and allow arbitrary embeddings of PERF; but while this leads to fairly interesting logical territory, it has little relevance to the semantics of natural language.

Now for the semantics. Let **B** ($= \langle \mathbf{O}, z, \mathcal{Z}, \mathbf{I} \rangle$) be a BAF. Then, for all intervals i, and all eventuality symbols q, we define:

$$\mathbf{B}, i \models \text{PERF } q \text{ iff } \exists i' \exists e' \exists e (i' < i \ \&$$
$$i' = z(e') \ \&$$
$$e' \in P_q \ \&$$
$$e' \text{ GRiTo } e \ \&$$
$$e\mathcal{Z}i).$$

Consider what this does. Suppose we have a sentence in the present perfect, say '*Fire has broken out on the oil rig*'. In our toy language this takes the form:

$$\text{PERF}(\textit{Fire breaks out on the oil rig}).$$

If we evaluate this at an interval i in **B**, then we must 'complete a square' in a BAF back to the utterance interval i. That is, we move back in time to an interval i' which is the run-time for an event e'; this e' is an eventuality of the correct type (that is, e' is a breaking out of a fire) and moreover e' gives rise to an event e which is $\mathcal{Z}$ related to our utterance interval i. Intuitively, the eventuality of present relevance e would be the ongoing burning of the fire, that is the consequent state of the breaking out of the fire event. Roughly, this semantics relates to Reichenbach and Moens and Steedman's approaches as follows: i is the time of speech (S), i' is the event time (E) and e is the consequent state induced by the event being described, namely e'. The Reichenbachian constraint according to which speech and reference times coincide is replaced by the Moens and Steedman intuition that the time stretch of the consequent state includes the speech time. In this way, we capture the intuition of present relevance which characterises the English present perfect.

4 Sorted BAFs

Simple BAFs have the virtue of making clear the fundamental idea underlying our approach, but they are very crude. To encode the aspectual restrictions placed on the use of the present perfect, and to model further temporal constructions such as the progressive, we need to say more about the relation between time and aspect. This is the object of the present section. We will insist that the eventuality structures used to make BAFs embody the sortal distinctions (and additional relations) demanded by the various verb classes. We start by motivating these additions.

Eventualities

On the basis of the tenses, aspects and adverbials with which they occur, we classify eventualities into five types; our classification is similar to the one of Carlson (1981) and Moens & Steedman (1988). First we distinguish between indefinitely extending eventualities which we call *states*, and eventualities with defined beginnings and ends called *events*. Sentence (3) describes a state:

(3) Her hair is black.

Events are subdivided into *atomic* and *extended* events, depending on whether or not their runtimes are an atomic interval.

To motivate a further subdivision of the extended events, compare sentences (4) and (5) below.

3. $\forall e\,(\texttt{Culm}(e) \to \exists e'\,(\texttt{Culm\_Proc}(e')\ \&\ e'\,\texttt{Compl}\,e))$.

Now, a *sorted BAF* is a BAF $\mathbf{B} = \langle \mathbf{O}, z, \mathcal{Z}, \mathbf{I}\rangle$, where $\mathbf{O}$ is a sorted eventuality structure in which the following additional conditions are satisfied:

4. $\forall e\,(\texttt{Point}(e) \to z(e)$ is an atomic interval).
   $\forall e\,(\texttt{Culm}(e) \to z(e)$ is an atomic interval).
5. $\forall e\,(\texttt{Proc}(e) \to z(e)$ is an non atomic bounded interval).
   $\forall e\,(\texttt{Culm\_Proc}(e) \to z(e)$ is an non atomic bounded interval).
6. $\forall e\,(\texttt{State}(e) \to z(e)$ is an non atomic, non bounded interval).
7. $\forall e, i\,(i \sqsubseteq z(e) \leftrightarrow e\mathcal{Z}i)$.

Item 4 says that points and culminations are atomic events, item 5 that processes and culminating processes are non atomic bounded eventualities and item 6 that states are non atomic, unbounded eventualities; the seventh item ensures that eventualities are *downward persistent*. Note that BAFs *do* distinguish between points and culminations; only culminations can enter into the Culm relation. Similarly, the Compl relation differentiates between processes and culminating processes.

Present perfect and sentence aspect

As we observed in section 2, not all verbs may be naturally used with the present perfect. More specifically, the examples in (2) show that stative, process and point expressions are awkward in combination with the present perfect. Now consider the semantics we propose for this tense: we require that the event talked about gives-rise-to some other eventuality (the consequent state) whose time stretch includes the speech time. Since we also insist that the 'gives-rise-to' partial function is only defined on culminations, this means that no interpretation can be assigned to a natural language sentence which has tense present perfect and aspectual category anything other than a culmination.[5] In this way, we capture the intuition that only those expressions which evoke a consequent state may be used in the present perfect.

Progressive aspect and the imperfective paradox

We will now examine progressive aspect using sorted BAFs. Following Kamp and Reyle (1993), we assume that the function of the English progressive is to focus attention on the (culminating) process of some eventuality. This idea can be captured as follows. First, we enrich our toy language by adding the operators

---

[5] This is clearly too strong, for given sufficient contextual support such combinations may be naturally interpretable. In the full version of the paper these readings are captured by adding a relation to eventuality structures that explicitly codes this contextual dependence between events. This addition seems to reflect the intentions of Moens and Steedman (cf. their discussion of the *enablement* relation) and is also needed to cope with the semantics of *when*.

Typical examples are:

(a) be green, know
(b) recognize, complete a paper
(c) hiccup, twinkle
(d) build a house, write a thesis
(e) play the piano, sleep, waste time

To sum up: the aspectual category of a sentence determines the sort of eventuality being described. Process, state and point expressions refer to some unstructured entity whereby a stative expression describes some unstructured event stretching over an unbounded period of time, a process expression some unstructured event stretching over a bounded period of time and a point expression some unstructured atomic event. In contrast, culminating process and culmination expressions are used to talk about structured events, that is events consisting of a culmination process, a culmination and a consequent state. Furthermore, it has often been argued (see Kamp and Reyle (1993) and Moens and Steedman (1988)) that the function of the English grammatical aspectual markers (such as the perfective and the progressive aspect) is to indicate which parts of the event structure are being referred to. Roughly, a progressive refers to the culminating process of a structured event and a perfect to its consequent state. In what follows, we show how sorted BAFs allow us to capture these intuitions.

Sorting eventuality structures

We now want to formalize the above ideas by extending our earlier simple BAFs. First, a *sorted eventuality structure* is a tuple

$$\mathbf{O} = \langle \texttt{Point}, \texttt{Culm}, \texttt{Proc}, \texttt{Culm\_Proc}, \texttt{State}; \texttt{GRiTo}, \texttt{Compl}; \{P_e\}_{e \in \mathcal{E}} \rangle,$$

where Point, Culm, Proc, Culm_Proc and State are mutually disjoint domains whose elements are used to interpret the various aspectual categories described above. GRiTo is a specialization of the 'gives-rise-to' relation defined in Section 3; here we insist that it only relates culminations to other eventualities. Compl is a binary relation between culminating processes and culminations. We think of the completion relation Compl as a partial function: if it is defined for an event e, it picks out a preferred or default consequence among all the consequences of e. (As not all events which have a natural culmination actually reach it, we can only have a *partial* function here.) Conversely, we assume that for every culmination there is a culminating process whose completion is this culmination. (We interpret the relation 'has-as-a-culminating process' using the converse of Compl.) More precisely, sorted eventuality structures should satisfy the following conditions:

1. GRiTo is a function whose domain is Culm.
2. Compl is a partial function whose domain is a subset of Culm_Proc and whose range is Culm.

(4) Bert was writing a thesis.

(5) Bert was sleeping.

The difference between sentences such as (4) and sentences such as (5) has been observed by numerous authors, and is often couched in terms of accomplishments and activities, cf. Vendler (1967). We express this distinction between (4) and (5) by saying that the event reported in (4) has a natural *culmination*, viz. the completion of the thesis; (5) has no such culmination. Processes that tend to have culminations in this sense are said to be *culminating*. Both the accomplishments of Vendler (1967) and the culmina*ted* processes of Moens & Steedman (1988) are composite events, consisting of a culminating process and a culmination; we feel it is more natural to split those composites and refer explicitly to the completion relation between culminating processes and their culminations.

Corresponding to the above distinction between processes and culminating processes, we divide atomic events into points and culminations. They differ in that culminations describe the culmination of a structured event (or nucleus) whereas points simply describe isolated atomic events; as a result a culmination may be associated with a culminating process and a consequent state whereas points cannot. To understand this division consider sentences (6) and (7) below.

(6) Bert completed his thesis.

(7) Bert hiccupped.

Sentence (6) reports a culmination; its culminating process is the writing of the thesis, its consequent state a state where the thesis is completed. Without further 'world knowledge' no natural culminating process or consequent state can be associated with the point event of (7).

Here, then, is a scheme of the eventualities we distinguish:

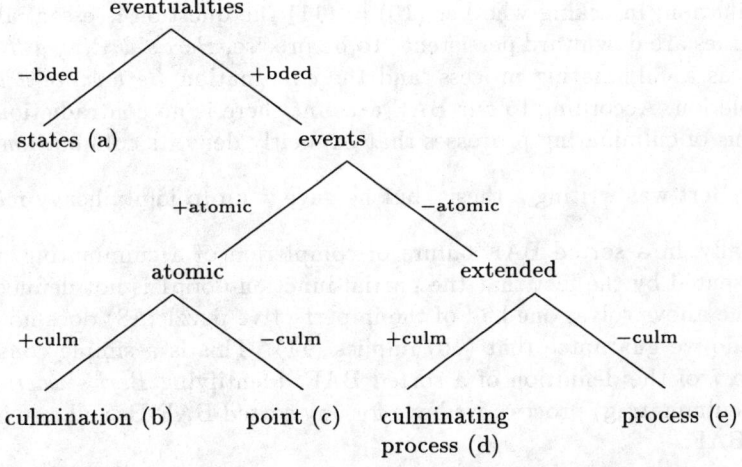

PAST and PROG, and allowing expressions of the form PAST q and PROG q and PAST PROG q to be well formed. As for the semantics, first, define $i \sqsubseteq^+ j$ to hold between two intervals i, j if the following is the case:

$$\frac{\quad j \quad}{\quad\quad i \quad\quad}$$

Let **B** $(= \langle \mathbf{O}, z, \mathcal{Z}, \mathbf{I} \rangle)$ be a sorted BAF. Then, for all intervals i, we define the relation $\mathbf{B}, i \models \phi$ as follows:

$\mathbf{B}, i \models \text{PROG}\, q$ \quad iff $\exists e\, (e \in P_q\ \&\ (\text{Proc}(e) \vee \text{Culm\_Proc}(e))\ \&\ i \sqsubseteq^+ z(e))$
$\mathbf{B}, i \models \text{PAST}\, q$ \quad iff $\exists j, e\, (j < i\ \&\ e \mathcal{Z} j\ \&\ e \in P_q)$
$\mathbf{B}, i \models \text{PASTPROG}\, q$ iff $\exists j\, (j < i\ \&\ \mathbf{B}, j \models \text{PROG}\, q)$.

One of the merits of such a semantics for the progressive is that it yields a simple solution to the so-called 'imperfective-paradox'. Following Dowty (1979), this paradox has been discussed by numerous authors. Briefly, the paradox is this: how can we account for the meaning of a progressive sentence like (8) and (10) in such a way that (8) may be true without (9) ever becoming true, while on the other hand (10) would tautologically imply (11)?

(8) Bert was writing a thesis.

(9) Bert wrote a thesis.

(10) Bert was wasting valuable time and money.

(11) Bert wasted valuable time and money.

The key to a solution to the imperfective puzzle is the observation that there is an important difference between the pair of sentences (8), (9) and (10), (11): in asking whether (8) $\models$ (9) one asks whether a culminating process entails its culmination; in asking whether (10) $\models$ (11) the question is essentially whether processes are downward persistent. To be precise, *Bert's writing a thesis* is classified as a culminating process, and the culmination *Bert wrote a thesis* is its completion. According to our BAF account there is no contradiction in continuations of culminating processes that explicitly deny its culmination:

(12) Bert was writing a thesis, but he gave it up to join a heavy metal band.

Formally, in a sorted BAF failure of completion of a culminating process e is represented by the fact that the partial function Compl is not defined in e.

The above solves one half of the imperfective puzzle: (8) does not imply (9). How do we guarantee that (10) implies (11)? This is a simple consequence of clause 7 of the definition of a sorted BAF. Identifying *Bert's wasting ...* as a (non-culminating) process, we have for any sorted BAF **B**, and any interval i in that BAF:

$\mathbf{B}, i \models \text{PAST PROG}(Bert\ldots)$

iff $\exists j\, (j < i\ \&\ \mathbf{B}, j \models \text{PROG}(Bert\ldots))$

iff $\exists j, e\, (j < i\ \&\ e \in P_{Bert\ldots}\ \&\ \texttt{Proc}(e)\ \&\ j \sqsubseteq^+ z(e))$.

But this means that $j \sqsubseteq z(e)$, and hence $e\mathcal{Z}j$, and thus

$$\mathbf{B}, i \models \text{PAST}\,(Bert\ldots),$$

and (10) implies (11).

5 Conclusion

In this extended abstract we have sketched, in very simple terms, how combined ontologies can be used in the semantics of temporal constructions. To conclude we briefly discuss our ongoing work on richer, more realistic systems, and note other BAF-like proposals we have found in the literature.

Sorted BAFs incorporate some of the Moens and Steedman ideas, but a great deal remains to be done. For example, although the sorts and the GRiTo and Compl relations model something of the Moens and Steedman notion of subevent structure, they don't capture the important idea that this subevent structure is recursively formed out of entities called nuclei. A nuclei is essentially a little 'package' consisting of a culminating process, a culmination, and a consequent state. Sometimes one wants to look at the internal structure of such packages, and sometimes one wants to treat this package simply as a 'lump' which can be linked to other packages. We are currently working with what we term *nucleic* BAFs. These are BAFs in which the eventuality occurrences are recursively generated out of Moens and Steedman style nuclei. Using such structures makes it possible to give analyses of a number of phenomena: in particular, we have given a Moens and Steedman style analysis of adverbial modification, and moreover can account for the interaction of progressive and perfective aspect in a natural way. (This is a topic that Moens and Steedman do not consider.) We are working on the semantics of temporal connectives (such as '*when*' and '*until*') in the setting of nucleic BAFs. An important part of this work is to reconstruct in the (essentially static) BAF framework an analogue of the (essentially dynamic) notion of 'type coercion' used by Moens and Steedman.

But these are topics for the full version of the paper. What can be said at a more general level concerning the idea of using combined ontologies in the study of temporal semantics?[6] We find the approach appealing for a number of reasons. First, it is intuitive. Pre-theoretical talk is often couched in terms of a mixture of different sorts of entities and their interrelations. Rather than ignore these intuitions, it seems better to try and be precise about them. Second, it

---

[6] Actually, the idea of combining ontologies seems of importance in many other areas of applied logic as well; see Blackburn and de Rijke (1994) for further discussion.

seems to work. Formalisations couched in a single ontological setting tend to fare well with a handful of phenomena but can be extended only with difficulty. In contrast, we find the ease with which a wide range of phenomena can be modeled with BAFs striking. (We believe that most of the work of Moens and Steedman can be captured — and extended — in a manner that does no violence to its guiding intuitions.) Thirdly, the approach is, in a very useful sense of the word, conservative. It does not discard the work offered by point based, interval based or event based approaches: rather, it locates them in a richer setting. This retains what is good in earlier analyses, and lets the reasons for their shortcomings become clearly visible. To sum up, while BAFs as we have defined them here are only a crude approximation to the subtlety of temporal discourse, we feel that the underlying idea of combining ontologies will prove useful.

To close the abstract we briefly note some other multiple ontology or BAF-like approaches we are familiar with. First, Oversteegen (1989) analysed the semantics of various English and Dutch expressions in terms of certain moves between an 'objective' and a 'subjective' time flow. Although her structures differ from ours — the 'objective' flow is like an interval structure and the 'subjective' flow is a discrete time line — her approach has many ideas in common with ours. Tense, and perfective and progressive aspect are analysed in terms of a number of basic transition patterns between the structures. Her analysis of Dutch temporal constructions is quite detailed, and we think it would be interesting to formalise her discussion in terms of BAF-like structures.

Second a back-and-forth picture can be found in Seligman and ter Meulen (1992). This aspect of their work may not be immediately obvious, for most of their discussion is devoted to the construction of Dynamic Aspect Trees. Nonetheless, their idea of 'classifying interval frames' involves moving back-and-forth between two structures, and (we would argue) it is this that gives the needed flexibility to drive their dynamic system.

Lastly, our account seems to have affinities with Situation Semantics. This is clear if the Channel Theory initiated by Seligman (1990) is considered. In his terms we are using an interval structure to classify eventuality occurrences. Our treatment of the English present perfect essentially says that the peculiarities of the construction are due to the fact that it exploits this channel in a particularly strong way. More generally, Situation Semantics has long emphasized the importance of ontological diversity, and the way we evaluate formulas in BAFs could be regarded as an instance of their 'relational account' of meaning.

Acknowledgements. Patrick Blackburn and Maarten de Rijke would like to acknowledge the financial support of the Netherlands Organisation for Scientific Research (NWO), project NF 102/62-356 'Structural and Semantic Parallels in Natural Languages and Programming Languages'. Claire Gardent would like to acknowledge the financial support of the LRE 61-062 project 'Towards a Declarative Theory of Discourse'.

References

1. van Benthem, J.: 1991, *The Logic of Time*, sec. edn., Kluwer, Dordrecht.
2. Blackburn, P. and de Rijke, M.: 1994, Zooming in, zooming out. Unpublished manuscript.
3. Carlson, L.: 1981, Aspect and quantification, in Tedeschi, P.J. and Zaenen, A. (eds.): 1981, *Syntax and Semantics 14*, Academic Press, New York.
4. Kamp, H. and Reyle, U.: 1993, *From Discourse to Logic*, Kluwer, Dordrecht.
5. Moens, M, and Steedman, J.: 1986, Temporal information and natural language processing, Research paper EUCCS/RP-2, University of Edinburgh.
6. Moens, M, and Steedman, J.: 1988, Temporal Ontology and Temporal Reference, *Computational Linguistics*, **14**, pp. 15–28.
7. Oversteegen, L.: 1989, *Two track theory of time*, PhD thesis, University of Amsterdam.
8. Reichenbach, H.: 1947, *Elements of symbolic logic*, Free Press, New York.
9. Seligman, J.: 1990, *Perspectives: a Relativistic Approach to the Theory of Information*, PhD thesis, University of Edinburgh.
10. Seligman, J. and ter Meulen, A.: 1992, Dynamic Aspect Trees, in *Logic at Work*, CCSOM, Amsterdam.
11. Vendler, Z.: 1967, Verbs and times. Chapter 4 in *Linguistics and Philosophy*, Cornell U.P., Ithaca.

Interpreting Tense, Aspect and Time Adverbials: A Compositional, Unified Approach*

Chung Hee Hwang & Lenhart K. Schubert
Dept. of Computer Science, University of Rochester
Rochester, New York 14627, U. S. A.

Abstract

We extend our theory of English tense, aspect and time adverbials [Hwang and Schubert, 1992, 1993] to deal with a wider range of time adverbials, including many adverbials of frequency, cardinality, duration, and time span, and adverbials of temporal relation involving subordinating conjunctions such as *after, since,* and *until*. Our theory is fully formal in that it derives indexical (quasi-)logical forms from syntactic-semantic rule pairs of a formal grammar, and nonindexical logical forms via deindexing rules in the form of equivalences and equations. The grammar allows for complex sentences and the semantic rules and deindexing rules are easy to implement computationally, producing formulas in Episodic Logic.

1 Introduction: A Compositional Alternative to Reichenbach

Researchers concerned with higher-level discourse structure, e.g., Webber [1988], Passonneau [1988] and Song and Cohen [1991], have almost invariably relied on some Reichenbach [1947]-like conception of tense. The syntactic part of this conception is that there are nine tenses in English, namely *simple* past, present and future tense, past, present and future *perfect* tense, and *posterior* past, present and future tense (plus progressive variants). The semantic part of the conception is that each tense specifies temporal relations among exactly three times particular to a tensed clause, namely the event time (E), the reference time (R) and the speech time (S). On this conception, information in discourse is a matter of "extracting" one of the nine Reichenbachian tenses from each sentence, asserting the appropriate relations among E, R and S, and appropriately relating these times to previously introduced times, taking account of discourse structure cues implicit in tense shifts. While there is much that is right and insightful about Reichenbach's conception, the lumping together of tense and aspect is out of step with modern syntax and semantics, providing an unsatisfactory basis for a compositional account of intra- and intersentential temporal relations.

---

*Portions of this paper were presented at the DARPA Workshop on Speech and Natural Language, 1990, the 30th Annual Meeting of the ACL, 1992, and the ARPA Workshop on Human Language Technology, 1993. This research was supported in part by NSF Research Grant IRI-9013160 and ONR/ARPA Research Contracts No. N00014-82-K-0193 and No. N00014-92-J-1512.

In particular, we think that the uniform use of E, R, S triples rests on a very dubious basis: first, appeal is made to the intuition that in tensed perfects, there is an implicit reference time involved besides the time of speech and the time of the described event. Then, this extra reference time is also imported into the *simple* tenses, even though for these there is no analogous intuition about the presence of such a reference time. Then some systematic role is sought for these reference times, and different researchers find different uses for them. Often, the "reference time" for a simple past sentence is claimed to be the time of the event introduced by the *previous* sentence, which intuitively tends to be closely aligned with the new event time. But this glosses over the fact that people have quite different intuitions about perfect reference times and these past reference times. More importantly, it glosses over the fact that the *same* "event reference" relations that are felt to exist intersententially for simple pasts like "John picked up the phone. He called Mary" also exist for past perfects like "John had picked up the phone. He had called Mary." In both cases, the "calling" event is felt to be right after the picking up of the phone. But if the time of the "previously reported event" is to be treated as a "reference time" in simple pasts, it ought to be treated as a "reference time" in past perfects as well — in other words, past perfects should have *two* reference times (the perfect reference time and previous event reference time) besides the time of speech and event time! So by the same reasoning, should we then not have two extra reference times for simple tenses, as well?

We think not — rather, we think that the presence of the extra reference time in tensed perfects is due to the presence of the extra *perf* operator (in addition to the *past* operator). More generally, we contend that English past, present, future and perfect are separate morphemes making separate contributions to syntactic structure and meaning. (Note that perfect *have*, like most verbs, can occur untensed; e.g., "She is likely to have left by now.") The corresponding operators *past, pres, futr,* and *perf* contribute separately and uniformly to the meanings of their operands, i.e., formulas at the level of logical form. Thus, for instance, the temporal relations implicit in "John will have left" are obtained not by extracting a "future perfect" and asserting relations among E, R and S, but rather by successively taking account of the meanings of the nested *pres, futr* and *perf* operators in the logical form of the sentence. As it happens, each of those operators implicitly introduces exactly one *episode*, yielding a Reichenbach-like result in this case. (But note: a simple present sentence like "John is tired" would introduce only *one* episode concurrent with the speech time, not two, as in Reichenbach's analysis.) Even more importantly for present purposes, each of *pres, past, futr* and *perf* is treated uniformly in deindexing and context change.[1]

Equally importantly, the clausal structure of sentences (or their logical forms) is not in general "flat," with a single level of constituents and features, but may contain multiple levels of embedding. This substructure can give rise to arbitrarily complex relations among the contributions made by the parts, such

---

[1] Well, almost uniformly; we think there are two variants of *perf* in English. There may also be *generic* variants of *pres* and *past*.

as temporal and discourse relations among subordinate clausal constituents and events or states of affairs they evoke. It is therefore essential that these intrasentential relations be systematically brought to light and integrated with larger-scale discourse structures. Consider, for instance, the following passage.

(1) John will find this note when he gets home.
(2) He *will think*$_{(a)}$ Mary *has left*$_{(b)}$.

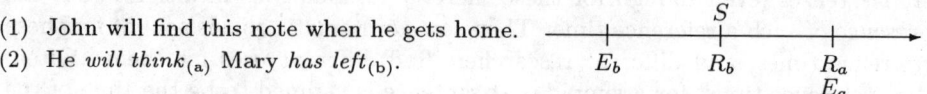

Reichenbach's analysis of (2) gives us $E_b < S, R_b < R_a, E_a$, where $t_1 < t_2$ means t_1 is before t_2, as shown above. That is, John will think that Mary's leaving took place some time before the speaker uttered sentence (2). This is incorrect; it is not even likely that John would know about the utterance of (2). In actuality, (2) only implies that John will think Mary's leaving took place some time before the time of his thinking, i.e., $S < R_a, E_a$ and $E_b < R_b, R_a$. Reichenbach's system fails to take into account the local context created by syntactic embedding. Attempts have been made to refine Reichenbach's theory (e.g., [Hornstein, 1977; Smith, 1978]), but we think these have generally not gone far enough in rebuilding the foundations.

We have developed a uniform, compositional approach to interpretation in which a parse tree leads directly (in rule-by-rule fashion) to a preliminary, *indexical* logical form (LF), and this LF is *deindexed* by processing it in the current *context* (a well-defined structure). The relevant context structures are called *tense trees*. Deindexing simultaneously transforms the LF and the context: context-dependent constituents of the LF, such as operators *past, pres* and *perf* and adverbs like *today* or *earlier*, are replaced by explicit relations among quantified *episodes*; (anaphora are also deindexed, but this is not discussed here); and new structural components and episode tokens (and other information) are added to the context. This dual transformation is accomplished by simple recursive equivalences and equalities. More specifically, they drive the generation and traversal of *tense trees* in deindexing. Our treatment of various kinds of time adverbials is fully compatible and integrated with the treatment of tense and aspect.

We describe tense trees in section 2 and tense-aspect deindexing rules in section 3. We then discuss our compositional approach to the interpretation of temporal adverbials in section 4, and an extension of our system that accommodates aspectual class shifts and the interaction between multiple temporal adverbials in section 5. Concluding remarks are in section 6.

2 Tense Trees

Tense trees provide that part of a discourse context structure which is needed to interpret (and deindex) temporal operators and modifiers within the logical form of English sentences. They differ from simple lists of Reichenbachian indices in that they organize episode tokens (for described episodes and the utterances themselves) in a way that *echoes the hierarchy of temporal and modal operators* of the sentences and clauses from which the tokens arose. Tense trees for successive sentences are "overlaid" in such a way that related episode tokens typically end up as adjacent elements of lists at tree nodes. For instance, tense trees allow the reference times/episodes of the perfect to be automatically identified. The traversal of trees and the addition of new tokens is simply and fully determined by the logical forms of the sentences being interpreted.

The major advantage of tense trees is that they allow simple, systematic interpretation (by deindexing) of tense, aspect, and time adverbials in texts consisting of arbitrarily complex sentences, and involving implicit temporal reference across clause and sentence boundaries. This includes certain relations implicit in the ordering of clauses and sentences. As has been frequently observed, for a sequence of sentences within the same discourse segment, the temporal reference of a sentence is almost invariably connected to that of the previous sentence in some fashion. Typically, the relation is one of temporal precedence or concurrency, depending on the *aspectual class* or *aktionsart* involved (cf., "John closed his suitcase; He walked to the door" *versus* "John opened the door; Mary was sleeping"). However, in "Mary got in her Ferrari. She bought it with her own money," the usual temporal precedence is reversed (based on world knowledge). Also, other discourse relations could be implied, such as *cause-of, explains, elaborates,* etc. Whatever the relation may be, finding the right pair of episodes involved in such relations is of crucial importance for discourse understanding. Echoing Leech [1987], we use the predicate constant `orients`, which subsumes all such relations. Note that the `orients` predications can later be used to make probabilistic or default inferences about the temporal or causal relations between the two episodes, based on their aspectual class and other information. We now describe tense trees more precisely.

The form of a tense tree is illustrated in Figure 1. As an aid to intuition, the nodes in Figure 1 are annotated with simple sentences whose indexical LFs would lead to those nodes in the course of deindexing. A tense tree node may have up to three branches—a leftward *past* branch, a downward *perfect* branch, and a rightward *future* branch. Each node contains a stack-like list of recently introduced episode tokens (which we will often refer to simply as episodes). In addition to the three branches, the tree may have (horizontal) *embedding links* to the roots of embedded tense trees. There are two kinds of these embedding links, both illustrated in Figure 1. One kind, indicated by dashed lines (with the label `mod-sub`), is created by subordinating constructions such as VPs with *that*-complement clauses. The other kind, indicated by dotted lines (and labelled `utt`), is derived from the surface speech act (e.g., telling, asking or requesting) implicit in the mood of a sentence.[2] On our view, the utterances of a speaker (or sentences of a text, etc.) are ultimately to be represented in terms of modal predications expressing these surface speech acts, such as [`Speaker` tell `Hearer` (That Φ)] or [`Speaker` ask `Hearer` (Whether Φ)]. `Speaker` and `Hearer` are indexical constants to be replaced by the speaker(s) and the hearer(s) of the utterance context. The two kinds of embedding links

---

[2]There is also a third kind of link (labelled `sub`), as will be shown in section 4.

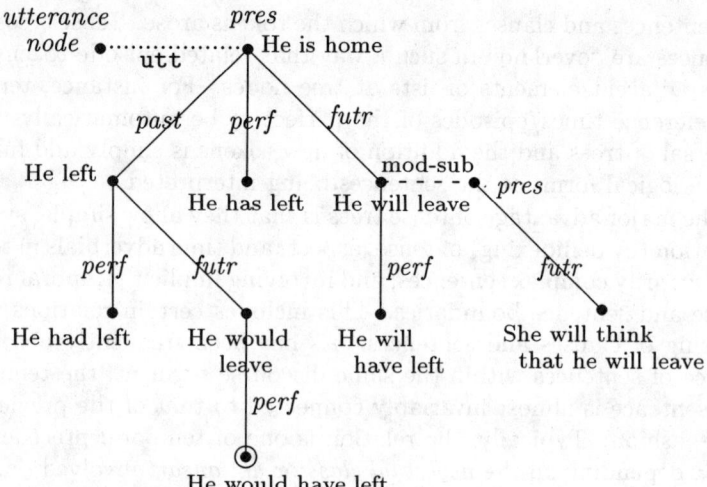

Figure 1. A Tense Tree

require slightly different tree traversal techniques as will be seen later.

A set of trees connected by embedding links is called a *tense tree structure* (though we often refer loosely to tense tree structures as tense trees). At any time, exactly one node of the tense tree structure for a discourse is in focus, and the focal node is indicated by ⊙. Note that the "tense tree" in Figure 1 is in fact a tense tree structure, with the lowest node in focus. By default, an episode added to the right end of a list at a node is "oriented" by the episode which was previously rightmost. For episodes stored at different nodes, we can read off their temporal relations from the tree roughly as follows. At any given moment, for a pair of episodes e and e' that are rightmost at nodes n and n', respectively, where n' is a daughter of n, if the branch connecting the two nodes is a past branch, $[e'$ before $e]$[3]; if it is a perfect branch, $[e'$ impinges-on $e]$ (as we explain later in sections 3, this yields entailments $[e'$ before $e]$ if e' is bounded and $[e'$ extends-to $e]$ if e' is unbounded,[4] respectively illustrated by "John has left" and "John has been busy"); if it is a future branch, $[e'$ after $e]$; and if it is an embedding link, $[e'$ at-about $e]$. These orienting relations and temporal relations are not extracted *post hoc*, but rather are automatically asserted in the course of deindexing using the rules shown later.

As a preliminary example, consider the following passage and a tense tree annotated with episodes derived from it by our deindexing rules:

---

[3]Or, sometimes, *same-time* (cf., "John noticed that Mary *looked* pale" *vs.* "Mary realized that someone *broke* her vase"). This is not decided in an *ad hoc* manner, but as a result of systematically interpreting the context-charged relation bef_T. More on this later.

[4]Technically, *boundedness* is defined for formulas, rather than episodes. However, we can also speak of bounded (or unbounded) episodes, namely those whose characterizing formulas are bounded (or unbounded).

(3) John picked up the phone.
(4) He had told Mary that he would call her.

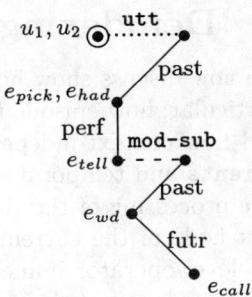

u_1 and u_2 at the root node are utterance episodes for sentences (3) and (4) respectively. Intuitively, the temporal content of sentence (4) is that the event of John's *telling*, e_{tell}, took place *before* some time e_{had}, which is at the same time as the event of John's *picking* up the phone, e_{pick}; and the event of John's *calling*, e_{call}, is located *after* some time e_{wd}, which is the at the same time as the (past perfect) event of John's *telling*, e_{tell}. For the most part, this information can be read off directly from the tree: $[e_{pick}$ orients $e_{had}]$, $[e_{tell}$ before $e_{had}]$ and $[e_{call}$ after $e_{wd}]$. In addition, the deindexing rules yield $[e_{wd}$ same-time $e_{tell}]$. ¿From this, one may infer $[e_{tell}$ before $e_{pick}]$ and $[e_{call}$ after $e_{tell}]$, assuming that the orients relation defaults to *same-time* here.

How does $[e_{pick}$ orients $e_{had}]$ default to $[e_{pick}$ same-time $e_{had}]$? One of the most important features of our account is that the tendency of past perfect "reference time" to align itself with a previously introduced past event is just an instance of a general tendency of atelic episodes to align themselves with their orienting episode. This is the same tendency noted previously for "John opened the door. Mary was sleeping." In the present tense tree, e_{had} is an episode evoked by the past tense operator which is part of the meaning of *had* in (4). It is an *atelic* episode, since this past operator logically operates on a sentence of form (perf Φ), and such a sentence describes a *state* in which Φ has occurred—in this instance, a state in which John has told Mary that he will call her. It is this atelicity of e_{had} which (by default) leads to a *same-time* interpretation of orients.

We remarked that the relation $[e_{wd}$ same-time $e_{tell}]$ is obtained directly from the deindexing rules. We leave it to the reader to verify this in detail (see Past and Futr rules stated in section 3). We note only that e_{wd} is evoked by the past tense component of *would* in (4), and denotes a (possible) *state* in which John *will* call Mary. Its atelicity, and the fact that the subordinate clause in (4) is "past-dominated,"[5] causes $[e_{wd}$ bef$_T$ $e_{tell}]$ to be deindexed to $[e_{wd}$ same-time $e_{tell}]$.

---

[5] A node is *past-dominated* if there is a *past* branch in its ancestry (where embedding links also count as ancestry links).

3 Deindexing with Tense Trees

We now discuss show how tense trees are modified as discourse is processed, in particular, how episode tokens are stored at appropriate nodes of the tense tree, and how context-independent, "deindexed" *episodic logical forms* (ELFs), with **orients** and temporal ordering relations incorporated into them, are obtained. The processing of the (indexical) LF of a new utterance always begins with the root node of the current tense tree (structure) in focus. The processing of the top-level operator immediately pushes a token for the surface speech act onto the episode list of the root node. A typical indexical LF (derivation of indexical LFs is discussed in section 4) looks like:

(*decl* (*past* (¬ [Mary answer])))   "*Mary did not answer.*"

(*decl* stands for *declarative*; its deindexing rule introduces the surface speech act of type "tell"). As mentioned earlier, our deindexing mechanism is a compositional one in which operators *past, futr, perf*, ¬, *That, decl*, etc., contribute separately to the meaning of their operands. As the LF is recursively transformed, the tense and aspect and modal operators encountered, *past, perf* and *futr*, in particular, cause the focus to shift "downward" along existing branches (or new ones if necessary). That is, processing a *past* operator shifts the current focus down to the left, creating a new branch if necessary. The resulting tense tree is symbolized as ↙T. Similarly *perf* shifts straight down, and *futr* shifts down to the right, with respective results ↓T and ↘T. *pres* maintains the current focus. Certain operators embed new trees at the current node, written ↪T (e.g., *That*), or shift focus to an existing embedded tree, written ↪T (e.g., *decl*). Focus shifts to a parent or embedding node are symbolized as ↑T and ←T respectively. As a final tree operation, ⊙T denotes storage of episode token e_T (a new episode symbol not yet used in T) at the current focus, as rightmost element of its episode list. As each node comes into focus, its episode list and the lists at certain nodes on the same tree path provide explicit reference episodes in terms of which *past, pres, futr, perf*, time adverbials, and implicit "orienting" relations are rewritten nonindexically. Eventually the focus returns to the root, and at this point, we have a deindexed ELF, as well as a modified tense tree.

Before we proceed with deindexing rules, we need to mention some basic features of EL, our semantic representation. In EL we take it that utterances characterize situations or episodes, and central to EL are the two episodic operators "*" and "**". Roughly, [Φ * η] means that Φ is true in episode η (or, Φ describes η), and [Φ ** η] means that Φ, and only Φ, is true in episode η (or, Φ characterizes η).[6] As mentioned, each of the deindexing rules for tense-aspect operators introduces an episode into the logical form and predicates that the episode is *characterized* ("**") by the operand (after recursive deindexing). (Use of "*" will be seen later when deindexing of adverbials is discussed.) The

---

[6]Like "situations", "episodes" is a generic term that may stand for events, states, processes, eventualities, etc. Operators similar to our episodic ones are the "support" relation ($\eta \models \Phi$) in situation semantics [Barwise, 1989] and the eventuality "type" condition ($\eta : \boxed{\Phi}$) in DRT [Kamp and Reyle, 1993].

square-bracketed, infixed form is the preferred sentence syntax in EL. In general, $[\tau_n \; \pi \; \tau_1 \ldots \tau_{n-1}]$ is an equivalent way of writing $(\pi \; \tau_1 \ldots \tau_n)$, which is in turn equivalent to $(\ldots((\pi \; \tau_1)\tau_2)\ldots \tau_n)$. Also used in EL are restricted quantifiers of form $(Q\alpha \colon \Phi \; \Psi)$, where Q is a quantifier, α is a variable, and restriction Φ and matrix Ψ are formulas. For details of syntax and semantics of EL, see [Hwang, 1992; Hwang and Schubert, 1993]. We now show some of the basic deindexing rules.[7]

Decl: $(\text{decl } \Phi)_T \leftrightarrow (\exists e_T \colon [[e_T \text{ same-time Now}_T] \wedge [\text{Last}_T \text{ immediately-precedes } e_T]]$
$[[\text{Speaker tell Hearer (That } \Phi_{\hookrightarrow OT})] ** e_T])$
Tree transformation: $(\text{decl } \Phi) \cdot T = \leftarrow (\Phi \cdot (\hookrightarrow OT))$

Pres: $(\text{pres } \Phi)_T \leftrightarrow (\exists e_T \colon [[e_T \text{ at-about Emb}_T] \wedge [\text{Last}_T \text{ orients } e_T]] \; [\Phi_{OT} ** e_T])$
Tree transformation: $(\text{pres } \Phi) \cdot T = (\Phi \cdot (OT))$

Past: $(\text{past } \Phi)_T \leftrightarrow \exists e_T \colon [[e_T \text{ bef}_T \text{ Emb}_T] \wedge [\text{Last}_{\swarrow T} \text{ orients } e_T]] \; [\Phi_{O_{\swarrow}T} ** e_T])$
Tree transformation: $(\text{past } \Phi) \cdot T = \uparrow (\Phi \cdot (O\swarrow T))$

Futr: $(\text{futr } \Phi)_T \leftrightarrow (\exists e_T \colon [[e_T \text{ after Emb}_T] \wedge [\text{Last}_{\searrow T} \text{ orients } e_T]] \; [\Phi_{O_{\searrow}T} ** e_T])$
Tree transformation: $(\text{futr } \Phi) \cdot T = \uparrow (\Phi \cdot (O\searrow T))$

Perf: $(\text{perf } \Phi)_T \leftrightarrow (\exists e_T \colon [[e_T \text{ impinges-on Last}_T] \wedge [\text{Last}_{\downarrow T} \text{ orients } e_T]] \; [\Phi_{O\downarrow T} ** e_T])$
Tree transformation: $(\text{perf } \Phi) \cdot T = \uparrow (\Phi \cdot (O\downarrow T))$

That: $(\text{That } \Phi)_T \leftrightarrow (\text{That } \Phi_{\mapsto T})$
Tree transformation: $(\text{That } \Phi) \cdot T = \leftarrow (\Phi \cdot (\mapsto T))$

As mentioned earlier, **Speaker** and **Hearer** in the Decl-rule are to be replaced by the speaker(s) and the hearer(s) of the utterance. Note that each equivalence pushes the dependence on context one level deeper into the LF, thus deindexing the top-level operator. The symbols Now_T, Last_T and Emb_T refer respectively to the speech time for the most recent utterance in T, the last-stored episode at the current focal node, and the last-stored episode at the current embedding node. When no such stored episodes exist for Last_T, certain other episodes may be substituted for Last_T; and within certain subtrees, Emb_T is interpreted as the embedding node of the "superordinate tree" (see section 4.3). As already mentioned, bef_T in the Past-rule will be replaced by either *before* or *same-time*, depending on the aspectual class of its first argument and on whether the focal node of T is past-dominated. In the Perf-rule, Last_T becomes the analogue of the Reichenbachian reference time for the perfect. The *impinges-on* relation confines its first argument e_T (the situation or event described by the sentential operand of *perf*) to the temporal region preceding the second argument. As in the case of **orients**, its more specific import depends on the aspectual types of its arguments. If e_T is a state or process, *impinges-on* implicates that it persists to the reference time/episode, i.e., $[e_T \text{ extends-to Last}_T]$. If e_T is an event (e.g., an accomplishment), *impinges-on* entails that it occurred sometime

---

[7]See [Hwang, 1992] for the rest of our deindexing rules. Some of the omitted ones are: Fpres ("futural present," as in "John has a meeting tomorrow"), Prog (progressive aspect), Pred (predication), K, K1, Ka and Ke ("kinds"), those for deindexing various operators such as *negation*, etc. Deindexing rules for adverbials are in section 4.

before the reference time/episode, i.e., [e_T before Last_T], and implicates that its main effects persist to the reference time.

To see the deindexing mechanism at work, let us consider sentences (3) and (4) again. The LFs before deindexing are shown in (3a,4a) below (where the labelled arrows mark points we will refer to); the final, context-independent ELFs are in (3b,4b). The transformation from (a)'s to (b)'s and the corresponding tense tree transformations are done with the deindexing rules shown earlier. Anaphoric processing is presupposed here. The snapshots of the tense tree while processing (3a) and (4a) with a null initial context, at points $\uparrow_a - \uparrow_g$, are shown below the formulas.

(3) John picked up the phone.
    a. (decl $\uparrow_a$ (past $\uparrow_b$ [John pick-up Phone])) $\uparrow_c$
    b. ($\exists$ u1:[u1 same-time *Now1*]
        [[*Speaker* tell *Hearer* (That
            ($\exists$ e1:[e1 before u1] [[John pick-up Phone] ** e1]))]
        ** u1])

(4) He had told Mary that he would call her.
    a. (decl (past (perf $\uparrow_d$ [John tell Mary (That $\uparrow_e$ (past (futr $\uparrow_f$ [John call Mary]))))))) $\uparrow_g$
    b. ($\exists$ u2:[[u2 same-time *Now2*] $\wedge$ [u1 immediately-precedes u2]]
        [[*Speaker* tell *Hearer* (That
            ($\exists$ e2:[[e2 before u2] $\wedge$ [e1 *orients* e2]]
            [($\exists$ e3:[e3 impinges-on e2]
              [[John tell Mary (That
                  ($\exists$ e4:[e4 same-time e3]
                    [($\exists$ e5:[e5 after e4] [[John call Mary] ** e5])
                  ** e4]))]
            ** e3])
        ** e2]))]
    ** u2])

What is important here is, first, that Reichenbach-like relations are introduced compositionally. In addition, the recursive rules take correct account of embedding. For instance, the embedded "past-future" in (4) is correctly interpreted as future relativized to John's (past) telling time. But beyond that, episodes evoked by successive sentences, or by embedded clauses within the same sentence, are correctly connected to each other. In particular, note that the orienting relation between John's picking up the phone, e1, and the reference time e2 for the telling event is automatically incorporated into the deindexed formula (4b). Thus we have established inter-clausal connections automatically, which in other approaches require heuristic discourse processing. This was a primary motivation for tense trees.

The `orients` relation is essentially an indicator that there could be a more specific discourse relation between the argument episodes. As mentioned, it can usually be particularized to one or more temporal, causal, or other "standard" discourse relations. Existing proposals for getting these discourse relations right appear to be of two kinds. The first uses the aspectual classes of the predicates involved to decide on discourse relations, especially temporal ones, e.g., [Partee, 1984; Dowty, 1986; Hinrichs, 1986]. The second approach emphasizes inference based on world knowledge, e.g., [Lascarides and Asher, 1993]. Our approach fully combines the use of aspectual class information and world knowledge. For example, in "Mary got in her Ferrari. She bought it with her own money," the successively reported "achievements" are by default in chronological order. Here, however, this default interpretation of `orients` is reversed by world knowledge: one owns things *after* buying them, rather than before. But sometimes world knowledge is mute on the connection. For instance, in "John raised his arm. A great gust of wind shook the trees," there seems to be no world knowledge supporting temporal adjacency or a causal connection. Yet we tend to infer both, perhaps attributing magical powers to John (precisely because of the lack of support for a causal connection by world knowledge). So in this case default conclusions based on `orients` seem decisive. In particular, we would assume that if e and e' are achievements or accomplishments, where e is the performance of a volitional action and e' is not, then $[e$ orients $e']$ suggests $[e$ right-before $e']$ and (less firmly) $[e$ cause-of $e']$.[8]

The tense tree mechanism, and particularly the way in which it automatically supplies orienting relations, is well suited for longer narratives, including ones with tense shifts. For example, in (5) below (from [Allen, 1987] with slight simplification), even though {b–d} would normally be considered a subsegment of the main discourse {a, e}, both the temporal relations *within* each segment and the relations *between* segments (i.e., that the substory temporally precedes the main one) are automatically captured by our rules. For instance, e_1, e_{10} and e_{11} are recognized as successive episodes, both preceded at some time in the past by e_3, e_5, e_7, and e_9, in that order.

(5) a. Jack and Sue went$_{\{e_1\}}$ to a hardware store
  b. as someone had$_{\{e_2\}}$ stolen$_{\{e_3\}}$ their lawnmower.
  c. Sue had$_{\{e_4\}}$ seen$_{\{e_5\}}$ a man take it and had$_{\{e_6\}}$ chased$_{\{e_7\}}$ him down the street,
  d. but he had$_{\{e_8\}}$ driven$_{\{e_9\}}$ away in a truck.
  e. After looking$_{\{e_{10}\}}$ in the store, they realized$_{\{e_{11}\}}$ that they couldn't afford a new one.

---

[8]Our approach to plausible inference in EL in general, and to such default inferences in particular, is probabilistic.

That is not to say that our tense tree mechanism obviates the need for larger-scale discourse structures. For example, many subnarratives introduced by a past perfect sentence may continue in simple past. That is, if *past* is followed by *past*, the latter could be either a continuation of the current perspective and segment (see 6a,b below), or a perspective shift with opening of a new segment (see 6b,c), or closing of the current segment, with resumption of the previous perspective (see 6c,d).

(6) a. Mary found that her favorite vase was broken.
   b. She was upset.
      c. She bought it at a special antique auction,
      d. and she was afraid she wouldn't be able to find anything that beautiful again.

Only plausible inference can resolve these ambiguities. This inference process will interact with resolution of anaphora and introduction of new individuals, identification of spatial and temporal frames, the presence of modal/cognition/perception verbs, and most of all will depend on world knowledge. See [Hwang and Schubert, 1992] for our approaches to this general difficulty.

4 Syntax and Semantics of Time Adverbials

We have shown that tense and aspect can be analyzed compositionally in a way that accounts not only for their more obvious effects on sentence meaning but also, via tense trees, for their cumulative effect on context and the temporal relations implicit in such contexts. We now move on to temporal adverbials.

Previous theoretical work on temporal adverbials has mostly concentrated on adverbials specifying temporal locations (e.g., "yesterday"), durations (e.g., "for a month") and time spans (e.g., "in three hours"). It appears that interest in the first kind of adverbial originated from the desire to correct the erroneous analyses provided by Priorean tense logics (see [Prior, 1967; van Benthem, 1988]), in particular, their treatment of the interaction between time adverbials and tense (see, for example, [Dowty, 1982; Hinrichs, 1988]). The second and third kinds of adverbials were often considered in connection with the aspectual classes of the VPs or sentences those adverbials modify (e.g., durative adverbials may modify only atelic sentences, whereas adverbials of time span may modify only accomplishment sentences). However, other kinds of temporal adverbials have received little attention, including ones specifying *repetition*:

> The engineer shut down the motor *twice* yesterday.
> The engine *frequently* broke down.
> The operator checked the level of oil *every half hour*.
> The inspector visits the lab *every Monday*.

On our analysis, these sentences describe complex events, consisting of a sequence of subevents of specified types, and the given adverbials modify the structure of these complex events: the cardinality of component events ("twice"), the frequency or distribution pattern of component events ("frequently," "regularly," "every hour," etc.), and the temporal location of cyclic events that occur synchronously with other recurrent time frames or events ("every Monday" or "every time the alarm went off").

Other issues that are rarely addressed are the interactions between multiple temporal adverbials, and various kinds of aspectual class shift due to aspectual class constraints on the use of adverbials (occurring singly or jointly with others). The following sentences illustrate these issues.

> John ran *for half an hour every morning for a month.*
> John stepped out of his office *for fifteen minutes.*
> Mary is going to Boston *for three days.*
> Mary won the competition *for four years.*
> John saw Mary *twice in two years.*

Our aim in this section is to provide a uniform analysis for a wide range of temporal adverbials. Our approach is compositional in that the lexicon supplies meanings at the word level (or possibly at the morpheme level, e.g., for '-ly' adverbs), and the meanings of adverbials are computed from the lexical entries by our GPSG-like grammar rules. The grammar rules take care of aspectual compatibility of adverbials with the VPs they modify. The resulting indexical LF is then deindexed by a set of recursive rules. The resultant ELF is formally interpretable and lends itself to effective inference. We now show logical form representations of temporal adverbials, in both indexical and deindexed form, and how to obtain them from the surface structure, together with a brief discussion of semantics.

4.1 The Basic Mechanism

We first discuss the basic interpretive mechanism, using *yesterday* as an example, and then generalize to other types of temporal adverbials. As indicated in the following fragment of a GPSG-like sentence grammar, we treat all adverbial adjuncts as VP-adjuncts at the level of syntax. (Aspectual feature agreement is assumed, but not discussed till section 5.)

> NP ← *Mary*; Mary
> V[1bar, past] ← *left*; <past leave>
> VP ← V[1bar]; V'
> VP ← VP ADVL[post-VP]; (ADVL' VP')
> S ← NP VP; [NP' VP']

However, despite this surface syntax, the semantic rule (ADVL' VP'), specifying functional application of the ADVL-translation to the VP-translation, may lead to either predicate modification or sentence modification at the level of immediate logical form. In particular, manner adverbials (e.g., *with a brush, hastily,* etc.) are uniformly interpreted as predicate modifiers at the level of immediate LF, while temporal (and locative) adverbials are all interpreted as sentence modifiers. How such sentence-modifier interpretations are formed from VP adjuncts is easily seen from rules such as the following:

NP[def-time] ← *yesterday* ; *Yesterday*
PP[post-VP] ← NP[def-time] ; (during NP′)
ADVL ← PP[e-mod, post-VP] ; λPλx((adv-e PP′) [x P]).

(`adv-e` stands for 'episode-modifying adverbial'.[9]) ¿From these rules it is clear that the logical translation of *yesterday*, as an adverbial adjunct, is

λPλx((adv-e (during *Yesterday*)) [x P]).

In the interpretation of a sentence such as "Mary left yesterday," this λ-abstract would be applied to predicate `leave` (initially paired with unscoped tense operator `past`), yielding

λx((adv-e (during *Yesterday*)) [x <past leave>]),

and this in turn would be applied to term `Mary` (translating the NP *Mary*), yielding

((adv-e (during *Yesterday*)) [Mary <past leave>]).

Here, (during *Yesterday*) is a 1-place predicate (the result of applying the 2-place predicate `during` to the indexical constant *Yesterday*, allowable in the "curried function" semantics of EL). `adv-e` maps this 1-place predicate into a *sentence* modifier; i.e., (adv-e (during *Yesterday*)) denotes a function from sentence meanings to sentence meanings. In the present case, the operand is the sentence [Mary <past leave>].

The above indexical LF is obtained quite directly as a byproduct of parsing, and is subsequently further processed—first, by scoping of ambiguously scoped quantifiers, logical connectives, and tense operators, and then by applying deindexing rules, which introduce explicit episodic variables into the LF, and temporally relate these based on tense operators, temporal adverbials, and tense trees. Tense operators are generally assumed to take wide scope over adverbials in the same clause. Thus, after scoping, we get

(past ((adv-e (during *Yesterday*)) [Mary leave])).

Since the deindexing rules "work their way inward" on a given indexical LF, starting with the outermost operator, the past tense operator in the sentence under consideration will already have been deindexed when the `adv-e` construct is encountered. In fact we will have

($\exists e_1$: [e_1 before u_1]
[((adv-e (during *Yesterday*)) [Mary leave])$_T$ ** e_1]),

where, u_1 denotes the utterance event for the sentence concerned, and T denotes the current tense tree. At this point the following deindexing rule for `adv-e` is brought to bear:

For π a monadic predicate, and Φ a formula,
`adv-e`: ((adv-e π) Φ)$_T$ ↔ [$^\vee \pi_T \wedge \Phi_{\pi \cdot T}$]
Tree transformation: ((adv-e π) Φ) · T = Φ · (π · T)

---

[9]Certain feature principles are assumed in the grammar—namely, certain versions of the head feature principle, the control agreement principle, and the subcategorization principle. Notice that in our system, features are treated as trees; e.g., the subtree rooted by feature `mod-vp` has daughters `pre-vp` and `post-vp`, and the subtree rooted by feature `e-mod` has daughters `temp-loc`, `dur`, `time-span`, `freq`, `card`, `cyc-time`, etc., where `temp-loc` in turn has daughters `def-time`, `indef-time`, etc.

This rule essentially splits the formula into a conjunction of two subformulas: one for the adverbial itself, the other for the sentence modified by the adverbial, much as in Dowty's system [1982]. Now the expression $^\vee\pi_T$ on the RHS of the deindexing rule for adv-e is a sentential formula (formed from predicate π_T) which can be read as "π_T *is true of the current episode (i.e., the one at which $^\vee\pi_T$ is evaluated)*." In view of this, the combination $[[^\vee\pi_T \wedge \Phi_{\pi \cdot T}] ** \eta]$ is equivalent to $[[[\eta\ \pi_T] \wedge \Phi_{\pi \cdot T}] ** \eta]$. Note that π_T is now predicated directly of episode η. In the example above, we obtain

$(\exists e_1 : [e_1 \text{ before } u_1]$
  $[[[e_1 \text{ during Yesterday}_T] \wedge [\text{Mary leave}]] ** e_1])$,

and this leaves only Yesterday$_T$ to be deindexed to a specific day (that is, (*yesterday-rel-to* u_1)).

To make the semantics of '$^\vee$', '$*$' and '$**$' a little more precise, we mention two clauses from the truth-conditional semantics of EL:

1. For Φ a formula, and η a term,
$[\![\Phi * \eta]\!]^s = 1$ only if *Actual* $([\![\eta]\!], s)$ and $[\![\Phi]\!]^{[\![\eta]\!]} = 1$;
   $= 0$ only if *Nonactual* $([\![\eta]\!], s)$ or $[\![\Phi]\!]^{[\![\eta]\!]} \neq 1$,
where these conditionals become biconditionals (iffs) and "$\neq 1$" becomes "$= 0$" for s an *exhaustive* (informationally maximal) situation.

2. For $s \in \mathcal{S}$, and π a predicate over situations,
$[\![^\vee\pi]\!]^s = [\![\pi]\!]^{s,s}$, i.e., $[\![\pi]\!](s)(s)$,
where $\mathcal{S}$ is the set of possible situations.

Also, a few relevant axioms are (for π, π' 1-place predicates, η a term, and Φ a formula):

- $[\Phi ** \eta] \leftrightarrow [[\Phi * \eta] \wedge \neg (\exists e : [e \text{ proper-subep-of } \eta] [\Phi * e])]$
- $[^\vee\pi \wedge {}^\vee\pi'] \leftrightarrow {}^\vee\lambda e[[e\ \pi] \wedge [e\ \pi']]$
- $[[^\vee\pi \wedge \Phi] ** \eta] \leftrightarrow [[[\eta\ \pi] \wedge \Phi] ** \eta]$

4.2 Adverbials of Duration, Time-span, and Repetition

Like adverbials of temporal location, durative adverbials are also translated as (adv-e π). For instance, "John slept for two hours" becomes (with tense neglected)

((adv-e (lasts-for (K ((num 2) (plur hour)))))) [John sleep]).

Like during, lasts-for is a 2-place predicate. Here it has been applied to a term (K...), leaving a 1-place predicate.[10] Just as in the case of (during Yesterday), the deindexed LF will contain a predication stating that the episode characterized by *John sleeping* lasts for two hours. Time-span adverbials (as in "John ran the race *in two hours*") are treated in much the same way, using predicate in-span-of.

---

[10]The details of (K...), denoting the abstract kind of quantity, two hours, need not concern us here.

The translation of cardinal and frequency adverbials involves the sentence-modifying construct (`adv-f` π). π is a predicate which applies to a *collection* of temporally separated episodes. It may describe the cardinality of the episodes or their frequency (i.e., their relative density), periodicity or distribution pattern. So, for instance, we have

((adv-f ((num 2) (plur episode))) [John see Movie])

for "John saw the movie twice," and

((adv-f ((attr frequent) (plur episode))) [John call Mary])

for "John called Mary frequently." (`num` is an operator that maps numbers into predicate modifiers, and `plur` (*'plural'*) is a function that maps predicates applicable to individuals into predicates applicable to collections. `attr` (*'attributive'*) is an operator that maps predicates into predicate modifiers.) Table 1 shows lexical rules and PP and ADVL rules handling large classes of frequency adverbials, including periodic ones such as *every two hours* and synchronized cyclic ones such as *every spring*.

The deindexing rule for `adv-f` is as follows:

For π a monadic predicate, and Φ a formula,
`adv-f`: $((\text{adv-f } \pi) \; \Phi)_T \leftrightarrow [{}^\vee \pi_T \wedge (\text{mult } \Phi_{\pi \cdot T})]$
  Tree transformation: $((\text{adv-e } \pi) \; \Phi) \cdot T = \Phi \cdot (\pi \cdot T)$

As illustrated in Table 1, π could take various forms. `mult` on the RHS side of the rule is a function that transforms sentence intensions, and is defined as follows.

For η an episode, and Φ a formula,
$\square \; [(\text{mult } \Phi) \;{**}\; \eta] \leftrightarrow [[\eta \; (\text{plur episode})] \wedge$
  $(\forall e: [e \text{ member-of } \eta]$
    $[[\Phi \;{**}\; e] \wedge \neg (\exists e' \; [[e' \neq e] \wedge [e' \text{ member-of } \eta] \wedge$
    $[e' \text{ overlaps } e]])])]$

Sentences (7)–(9) below illustrate the rules stated in Table 1. The (a)-parts are the English sentences, the (b)-parts their immediate indexical LFs, and the (c)-parts the deindexed ELFs. (7c) says that "some time before the utterance event, there was a 2 month-long (multi-component) episode, that consists three episodes of type 'John date Mary'." (8c) reads as "there was a 10 day-long episode that consists of periodically occurring subepisodes of type 'John take medicine', where the period was 4 hours." (9c) is understood as "at the generic present there is a collection of episodes of type 'Mary swim', such that during each Saturday within the time spanned by the collection,[11] there is such an episode."

(7) a. Mary visited Paris *three times in two months*.
   b. (past ((adv-e (in-span-of (K ((num 2) (plur month)))))
      ((adv-f ((num 3) (plur episode))) [Mary visit Paris])))

---

[11] This constraint on the Saturdays under consideration is assumed to be added by the deindexing process for time- or event-denoting nominals, but has been omitted from (9c).

Table 1: GPSG Fragment

% VP Adjunct Rules
ADVL ← PP[e-mod, post-VP]; λPλx((adv-e PP') [x P])
ADVL ← ADV[e-mod, mod-VP]; λPλx(ADV' [x P])
VP ← VP ADVL[mod-vp]; (ADVL', VP')

% Temporal ADV, PP Rules
NP[def-time] ← yesterday ; Yesterday
PP[post-VP] ← NP[def-time]; (during NP')
  e.g., yesterday' = λPλx((adv-e (during Yesterday)) [x P])

N[time-unit, plur] ← hours; (plur hour)
ADJ[number, plur] ← two; (num 2)
N[1bar, time-length] ← ADJ[number] N[time-unit]; (ADJ' N')
NP ← N[1bar, time-length]; (K N')
P[dur] ← for ; lasts-for
P[span] ← in ; in-span-of
PP[e-mod, post-VP] ← P NP[time-length]; (P' NP')
  e.g., for two hours' = λPλx((adv-e (lasts-for (K ((num 2) (plur hour))))) [x P])
  e.g., in two hours' = λPλx((adv-e (in-span-of (K ((num 2) (plur hour))))) [x P])

ADV[card, post-VP] ← twice; (adv-f ((num 2) (plur episode)))
ADV[freq, mod-VP] ← frequently; (adv-f ((attr frequent) (plur episode)))
ADV[freq, mod-VP] ← periodically; (adv-f ((attr periodic) (plur episode)))
ADV[freq, post-VP] ← Det[every] N[1bar, time-length];
        (adv-f λs[[s ((attr periodic) (plur episode))] ∧ [(period-of s) = (K N')]])
  e.g., twice' = λPλx((adv-f ((num 2) (plur episode))) [x P])
  e.g., frequently' = λPλx((adv-f ((attr frequent) (plur episode))) [x P])
  e.g., every two hours' = λPλx((adv-f λs[[s ((attr periodic) (plur episode))] ∧
                              [(period-of s) = (K ((num 2) (plur hour)))]]) [x P])

N[indef-time] ← spring; spring
NP[cyc-time] ← Det[every] N[1bar, indef-time]; <Det' N'>
PP[post-VP] ← NP[cyc-time]; (during NP')
ADV ← PP[cyc-time, post-VP]; (adv-f λs(∃e [[e member-of s] ∧ [e PP']]))
  e.g., every spring' = λPλx((adv-f λs(∃e [[e member-of s] ∧ [e during <∀ spring>]]))
        [x P])

c. $(\exists e_2 : [e_2 \text{ before } u_2]$
　　　　$[[[e_2 \text{ in-span-of } (\text{K}((\text{num } 2)(\text{plur month})))] \wedge [e_2 ((\text{num } 3)(\text{plur episode}))] \wedge$
　　　　$(\text{mult } [\text{Mary visit Paris}])] ** e_2])$

(8) a. John took medicine *every four hours for ten days*.
　　b. $(\text{past } ((\text{adv-e } (\text{lasts-for } (\text{K }((\text{num } 10)(\text{plur day})))))$
　　　　$((\text{adv-f } \lambda s [[s ((\text{attr periodic})(\text{plur episode}))] \wedge$
　　　　　　$[(\text{period-of } s) = (\text{K }((\text{num } 4)(\text{plur hour})))]])$
　　　　　　[John take (K medicine)]))))
　　c. $(\exists e_4 : [e_4 \text{ before } u_4]$
　　　　$[[[e_4 \text{ lasts-for } (\text{K }((\text{num } 10)(\text{plur day})))] \wedge [e_4 ((\text{attr periodic})(\text{plur episode}))] \wedge$
　　　　$[(\text{period-of } e_4) = (\text{K }((\text{num } 4)(\text{plur hour})))] \wedge (\text{mult } [\text{John take (K medicine)}])]$
　　　　$** e_4])$

(9) a. Mary swims *every Saturday*.
　　b. $(\text{gpres } ((\text{adv-f } \lambda s (\forall d : [d \text{ Saturday}] (\exists e [[e \text{ member-of } s] \wedge [e \text{ during } d]])))$
　　　　[Mary swim]))
　　c. $(\exists e_5 : [e_5 \text{ gen-at } u_5]$
　　　　$[[(\forall d : [d \text{ Saturday}](\exists e [[e \text{ member-of } e_5] \wedge [e \text{ during } d]])) \wedge$
　　　　$(\text{mult } [\text{Mary swim}])] ** e_5])$

We have a tentative account of adverbials such as *consecutively* and *alternately*, but cannot elaborate within the present space limitations. We also set aside certain well-known problems involving temporal adverbials in perfect sentences, such as the inadmissibility of *"John has left yesterday" (for a possible approach, see [Schubert and Hwang, 1990]), and now move on to clausal adverbials.

4.3　SINCE, UNTIL and AFTER: Clausal Adverbials

Since and *until* provide a time "frame" which the episode described by the main clause spans (at least if that main clause is atelic). More specifically, *since* and *until* connect an episode characterized by the main clause to a time indicated by the subordinate clause (or PP) such that the latter specifies the beginning and the end point the episode respectively. For instance, in

　　"Mary has been taking care of John's dog *since* he went off to college,"[12]

the since-clause specifies that the episode of Mary's taking care of John's dog started at the time John went off to college, and in

　　"Mary kept silent *until* John had had his say,"

the until-clause specifies the (earliest) time of John's *having* had his say as the end point of Mary's episode of being silent.[13] On the other hand, *before* and *after* do not provide a time frame but simply specify the temporal ordering between the episodes described by the main and subordinate clauses. In this subsection, we discuss the treatment of adverbial clauses headed by connectives

---

[12] Despite this example and a later one, we do not consider progressives in detail here. Progressives involve syntactic complications as well as appeal to a reference time at the semantics/discourse level, where that reference time is constrained to be within the progressive episode. (It can be "picked" from the tense tree much like other reference times.)

[13] Thus, it is strongly implied that Mary was no longer silent as soon as John finished speaking. However, such an implicature of until-clauses is cancellable as the following example illustrates: "Mary was fine at least until I left."

like *since*, *until* and *after*, which can also serve as temporal prepositions. We do not discuss *while* and *when*, which are not derived from prepositions (but a similar treatment is possible). We also omit discussion of *before*, which requires a slightly more complicated treatment as it can be used hypothetically (even in the past tense), as in "The police arrived before the burglars could run away."

Here are lexical and phrase structure rules we use.

P[since] ← *since*; since
P[until] ← *until*; until
P[after] ← *after*; after
PCONJ[_S[fin]] ← P[pc-temp];
   $\lambda S \lambda e[e\ P'\ <\text{The-earliest}\ \lambda t(\text{sub}\ ((\text{adv-e}\ (\text{at-time}\ t))\ S))>]$
ADVL[post-VP] ← PCONJ[pc-temp] S[fin];
   $\lambda P \lambda x((\text{adv-e}\ (\text{PCONJ}'\ S'))\ [x\ P])$

In the PCONJ rule, feature `pc-temp` ("p-to-conj") is to distinguish CONJ-convertible temporal prepositions from the rest of the temporal prepositions such as *at, for*, etc. The semantic part of the PCONJ rule essentially analyses a phrase of form "since/until/after **S**" as equivalent to "since/until/after the (earliest) time at which **S**." Note that this is a relative-clause analysis of the implicit temporal reference. Thus if the episode corresponding to the main clause is **e**, the adverbial will assert that "e is since/until/after the earliest time at which **S** is true." `sub` ("subordinate") is a sentence operator which is semantically trivial (i.e., an identity operator) but forms a scope island. It is intended to be used for explicit relative clauses as well as in the present case. It also functions as a cue for the deindexer to embed a new tree for the subordinate adverbial clause, as will be seen shortly. **The-earliest** is a definite quantifier much like **The**, except that it strongly (but defeasibly) implicates "earliest."

We regard the entailments of the relational predicate *since* as dependent on the aspectual category of the first argument. For an atelic first argument (as in "The company has thrived since Mary took over"), *since* means "starting at the (earliest) time at which ..." For a telic first argument (as in "John has graduated since you last saw him"), *since* means "after." One might ask how this could account for the implicatures of a sentence like

"Mary has visited John three times since he moved to California."

Here it seems that Mary has visited John three times *only*. We think that the answer lies in the implicatures of cardinal modifiers and of the *impinges* relation we use in the analysis of the perfect. Note that "Mary visited John three times *after* he moved to California" also implicates that he visited her three times *only*. The additional implicature in the present perfect version (due to "impingement" of the three-and-only-three visits on the present reference time) is that the result state of the three-and-only-three visits — viz., that three-and-only-three visits took place — still obtains at present.

A well-known property of *since* is that it requires the main clause to be a perfect sentence. Semantically, though, it appears that since-clauses operate not on the perfect VP or sentence as a whole but rather on "the underlying nonperfect VP" (in Kamp and Reyle's [1993] phrase). In our compositional rule-by-rule approach, we can ensure the right semantics by having since-adverbials combine with phrases of category VP[-en] (a VP headed by a past participle). The underlying VP[-en] is usually atelic (e.g., it would be odd in most contexts to say "Mary bought a book since last December"), but does not have to be as was seen above. In contrast, until-adverbials are normally used with atelic VPs (more exactly, "unbounded" ones — see section 5).

I plan to stay here *until* tomorrow/*until* you return.
*I plan to finish the paper *until* tomorrow/*until* you return.
I plan to finish the paper *by* tomorrow/*by the time* you return.

We now refine the VP rule shown in Table 1 to reflect these properties of *since* and *until*. (*Before* and *after* do not require special rules.)

VP ← VP[-en] ADVL[since]; (ADVL' VP')
VP ← VP[unbounded] ADVL[until]; (ADVL' VP')

For deindexing of adverbial clauses, we assume that the **sub** operator triggers creation of an embedded subtree, much as in the case of that-clauses. Note that there is a slight difference between adverbial clauses and that-clauses. That-clauses signal modal subordination, and the "anchoring" time for them is that of the embedding VP (usually, an attitude VP). As *since*, *until* and *after* signal non-modal subordination, we take the anchoring time for the tense of the subordinate clause to be the same as the anchoring time for the superordinate clause. For instance, in "John arrived after Mary had waited for an hour," the anchoring time for the subordinate clause is the speech time, rather than the arrival-time, and this allows the perfect to be oriented by the arrival-time (as intuitively required). Thus, we use a slightly different kind of embedding link, '= = =', with the label **sub**, which interprets the embedding node (Emb_T) as the "next-highest" embedding node (typically, the "utterance node" at the main tense tree), rather than the immediate embedding node. We now show the relevant deindexing rule.

For Φ a formula,
sub: $(\text{sub } \Phi)_T \leftrightarrow \Phi_{\Rightarrow T}$
Tree transformation: $(\text{sub } \Phi) \cdot T = \leftarrow (\Phi \cdot \Rightarrow T)$

'⇒' indicates: build a new tree, embed it at the current focal node with a double link (i.e., a **sub** link), and move the focus to the root node of the newly embedded tree. Note that the deindexing rule peels off the (semantic) identity operator **sub** from the logical form. We now illustrate the deindexing mechanism using sentence (10) below which we have seen already.

(10) *Mary has been taking care of John's dog since he went off to college.*

With the rules introduced earlier, the adverbial clause "since he [John] went off to college" is translated into

$\lambda P \lambda x((\text{adv-e } \lambda e[e \text{ since } <\text{The-earliest}$
$\lambda t(\text{sub } ((\text{adv-e } (\text{at-time } t)) [\text{John } <\text{past go-college}>]))>]) [x p])$.

This applies to the nonperfect VP "been taking care of John's dog," yielding

λx((adv-e λe[e since <The-earliest
    λt(sub ((adv-e (at-time t)) [John <past go-college>]))>])
        (prog [x take-care-of Dog])).

Applying the perfect auxiliary "has" (λPλx<pres (perf [x P])>) to this and then incorporating the rest of the sentence (i.e., the subject and the punctuation) and scoping the unscoped operators, we get the following indexical logical form for the entire sentence.

(decl (pres (perf
    ((adv-e λe(The-earliest t:(sub (past ((adv-e (at-time t)) [John go-college]))))
        [e since t]))
    (prog [Mary take-care-of Dog])))))

Deindexing is straightforward, and the resultant deindexed EL formula shown below can be easily verified.

(∃u1:[u1 same-time Now1]
    [[**Speaker** tell **Hearer** (That
        (∃e1:[[e1 at-about u1] ∧ [e9 orients e1]]
            [(∃e2:[[e2 impinges-on e1] ∧ [e10 orients e2]]
                [[$^\vee$λe(The-earliest t1: (∃e3:[[e3 before u1] ∧ [e11 orients e3]]
                                    [[$^\vee$(at-time t1) ∧ [John go-college]] ** e3])
                        [e since t1]) ∧
                (prog [Mary take-care-of Dog])]
                ** e2])
            ** e1]))]
    ** u1])

Since we did not provide a context for (10), the deindexer introduces dummy episodes ($e9, e10, e11$) as orienting episodes. These episodes could be resolved against appropriate episodes later if more information becomes available. Finally, by meaning postulates, we get the following logical form.

(∃u1:[u1 same-time Now1]
    [[**Speaker** tell **Hearer** (That
        (∃e1:[[e1 at-about u1] ∧ [e9 orients e1]]
            [(∃e2:[[e2 impinges-on e1] ∧ [e10 orients e2]]
                [[(The-earliest t1: (∃e3:[[e3 before u1] ∧ [e11 orients e3]]
                                    [[[e3 at-time t1] ∧ [John go-college]] ** e3])
                        [e2 since t1]) ∧
                (prog [Mary take-care-of Dog])]
                ** e2])
            ** e1]))]
    ** u1])

This formula says that there is a perfect episode $e1$ which lies at the end of episode $e2$ (Mary's taking care of John's dog); and $e2$ is "since" the earliest time

t at which an event $e3$ of John's going off to college occurred. Note that [e2 impinges-on e1] implies [e2 extends-to e1] (so that $e1$ is at the end of $e2$), as the underlying formula (*prog* [*Mary take-care-of Dog*]) is unbounded.

Below are more sample sentences and their logical forms, followed by some remarks. (For brevity, we omit some orienting relations in the deindexed formulas, and use simplified translations for some irrelevant complex expressions.)

(11) a. John has moved to California since Mary met him last.
    b. (decl (pres (perf ((adv-e λe(The-earliest t1:(sub (past ((adv-e (at-time t1))
                                                      [Mary meet-last John])))
                                      [e since t1])) [John move-to-CA]))))
    c. (∃u1:[u1 same-time Now1]
      [[Speaker tell Hearer (That
        (∃e1:[e1 at-about u1]
        [(∃e2:[e2 impinges-on e1]
          [[(The-earliest t1:(∃e3:[e3 before u1] [[[e3 at-time t1] ∧
                            [Mary meet-last John]] ** e3])
                [e2 since t1]) ∧
            [John move-to-CA]] ** e2])
        ** e1]))]
    ** u1])

(12) a. Mary has visited John three times since he moved to California.
    b. (decl (pres (perf ((adv-e λe(The-earliest t1:(sub (past ((adv-e (at-time t1))
                                                        [John move-to-CA])))
                                      [e since t1]))
        ((adv-f ((num 3) (plur episode))) (mult [Mary visit John]))))))
    c. (∃u1:[u1 same-time Now1]
      [[Speaker tell Hearer (That
        (∃e1:[e1 at-about u1]
        [(∃e2:[e2 impinges-on e1]
          [[(The-earliest t1:(∃e3:[e3 before u1] [[[e3 at-time t1] ∧
                            [John move-to-CA]] ** e3])
                [e2 since t1]) ∧
            [e2 ((num 3) (plur episode))] ∧ (mult [Mary visit John])] ** e2])
        ** e1]))]
    ** u1])

(13) a. Mary had been fine until she had eaten the cake.
    b. (decl (past (perf ((adv-e λe(The-earliest t1:(sub (past ((adv-e (at-time t1))
                                                    (perf [Mary eat Cake]))))
                                      [e until t1]))) [Mary fine])))
    c. (∃u1:[u1 same-time Now1]
      [[Speaker tell Hearer (That
        (∃e1:[[e1 before u1] ∧ [e9 orients e1]]
        [(∃e2:[[e2 impinges-on e1] ∧ [e10 orients e1]]
          [[(The-earliest t1:(∃e3:[[e3 before u1] ∧ [e11 orients e3]]
                    [[[e3 at-time t1] ∧
                          (∃e4:[[e4 impinges-on e3] ∧ [e12 orients e4]]
                          [[Mary eat Cake] ** e4])] ** e3])
                [e2 until t1]) ∧

                    [Mary fine]] ** e2])
                ** e1]))]
        ** u1])

(14) a. John arrived after Mary had left.
   b. (decl (past ((adv-e λe(The-earliest t1: (sub (past ((adv-e (at-time t1))
                                                             (perf [Mary leave]))))
                                        [e after t1])) [John arrive])))))
   c. (∃u1:[u1 same-time Now1]
        [[**Speaker** tell **Hearer** (That (∃e1:[e1 before u1]
            [[(The-earliest t1:(∃e2:[e2 before u1] [[[e2 at-time t1] ∧
                                (∃e3:[[e3 impinges-on e2] ∧ [e11 orients e3]]
                                    [[Mary leave] ** e3])] ** e2])
                        [e1 after t1]) ∧
            [John arrive]] ** e1]))]
        ** u1])

Formula (11b) says that e2 (the event of John's moving) is "since" t, i.e., the time at which Mary met him last, and "impinges-on" e1, i.e., the present time. Since the characterization of e2, [John move-to-CA], is *transition*, the context-charged relation [e2 impinges-on e1] implies that the *result state* of e2 holds until the time of e1. Thus, we can infer that John moved to California some time after Mary met him last and that he is still living in California. Note that the main clause in (12) describes a complex episode whose temporal extent is a multi-interval (that is, the actual times at which the component episodes take place may not be consecutive). And it is this multi-interval that lies in the time frame provided by the since-clause and the present. Formula (13b) says that the episode of Mary's being fine, e1, lasted until the earliest such time at which the cake-eating event can be described as perfect (i.e., the moment she finished eating the cake). Thus, it is strongly implicated that the end points of the cake-eating event and the being-fine state were concurrent and that quite likely Mary was no longer fine right after the eating event. Note that the definite quantifier **The-earliest** makes possible this interpretation. Notice next the orienting relation [e10 orients e2] in (13b). Since an adverbial clause is deindexed with respect to a *new* embedded tree, there is no orienting episode for e2 which would serve as a reference time for the perfect. Thus, the deindexer supplies an unidentified episode (e10 in this case) for the orienting relation. During the subsequent "ampliative" inference stage, this e10 is then identified with e1. Note that we do *not* force the reference time of a subordinate perfect clause uniformly to be identical with the event time of the main clause, in view of sentences like "Mary arrived two hours after John had left." We omit discussion of example (14).

5 Temporal Adverbials & Aspectual Class Shifts

So far, we have assumed aspectual category agreement between temporal adverbials and VPs they modify. We now discuss our aspectual class system and our approach to apparent aspectual class mismatch between VPs and adverbials, based on certain aspectual class transformations. We make use of three aspectual class feature hierarchies, `telicity`, `boundedness` and `temporal-extent` as below:

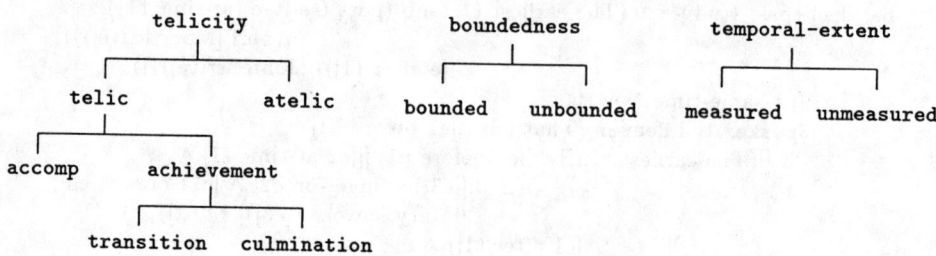

Untensed sentences may be telic or atelic, depending on the type of the predicate (e.g., achievement/accomplishment versus state/process predicates) and on the object and subject (e.g., count versus mass).[14] Sentences or predicates describing achievements or accomplishments are assigned the feature `telic`, while those describing states or processes are assigned the feature `atelic`. Examples of accomplishment VPs are *write a book* and *blink*; transition predicates are *step out, turn off, go to, become, wake up*, etc.; and culmination VPs are *reach the top, win the race*, etc. Intuitively, a formula is `bounded` if the episode it characterizes terminates in a distinctive result state (result states are formally defined in [Hwang, 1992].) That is, episodes with a bounded characterization have a definite end point (in virtue of their characterization), while ones with an unbounded characterization do not. By a co-occurrence restriction, telic formulas are bounded. Atelic formulas are *by default* unbounded. Some atelic episodes are bounded such as an episode of John's being ill, at the end of which he is *not* ill. For instance, *was ill* in "John was ill when I saw him last week" is unbounded as the sentence does not entail that John was not ill right after the described episode. However, when we say "John was ill twice last year," we are talking about bounded "ill" episodes. The `temporal-extent` feature has to do with whether a VP contains a durative adverbial (e.g., *for-* or *throughout-*) or an adverbial of time-span (e.g., *in-*), i.e., whether the temporal extent of an episode is indicated in its characterization. to the persistence of a formula. (E.g., unmeasured atelic formulas are inward persistent (modulo granularity) in general, while telic ones are outward persistent.) See [Hwang, 1992] for further discussion.

As has been discussed by many authors (e.g., in [Moens and Steedman, 1988; Mourelatos, 1981; Vendler, 1967]), VPs and temporal adverbials may not arbitrarily combine. Normally, durative adverbials combine with atelic VPs; cardinal and frequency adverbials with bounded VPs; and adverbials of time-span with

---

[14]Every tensed English sentence, e.g., "Mary left before John arrived," in combination with a *context*, is considered `factual`, where `factual` features are ascribed to atemporal (or, unlocated) sentences whose truth value does not change over space and time. We neglect the `factual` feature in this paper.

telic VPs. Thus, for instance,

> Mary studied *for an hour*.
> *Mary finished the homework *for a second*.
> Mary called John *twice | repeatedly | every five minutes*.
> Mary wrote the paper *in two weeks*.

Note, however, that we also say

> Mary sneezed *for five minutes*.
> Mary stepped out of her office *for five minutes*.
> Mary was ill *twice | repeatedly | every two months*.

The latter group of sentences show that VPs often acquire an interpretation derived from their original, primitive meaning. More specifically, when cardinal, frequency or cyclic adverbials are applied to telic VPs, usually iteration is implied, as in the first sentence. However, in the case of the second sentence, which involves a transition verb, the preferred reading is one in which the adverbial specifies the duration of the resultant episode, i.e., "the *result state* of Mary's stepping out of her office" (i.e., her being *outside* of her office). When cardinal or frequency adverbials (i.e., "bounded" adverbials) are applied to unbounded-atelic VPs, those VPs are interpreted as bounded-atelic. Thus, the third sentence above means that the kind of episode in which Mary becomes ill and then ceases to be ill occurred twice, repeatedly, etc.

To be able to accommodate such phenomena, the syntactic parts of our grammar use **telicity** and **boundedness** as head features. The semantic parts introduce, as needed, operators for aspectual class transformation such as *result-state*, *iter* (iteration), *temporarily* (bounded), etc. (In place of *iter*, we may sometimes use a habitual operator, *H*.)

Adverbials of temporal location like *yesterday* or *last week* may combine with either bounded or unbounded formulas: with bounded ones, these imply a *sometime during* reading; with unbounded ones, by default, a *throughout* reading if they are **unmeasured**, and a *sometime during* reading if they are **measured**. For instance, in "John left last month," the "leaving" episode took place *sometime during* last month, but in case of "Mary was ill for two weeks last month," Mary's "ill" episode is considered to be *sometime during* last month. Synchronized cyclic adverbials like *every spring* or *every time I saw Mary* may combine with either bounded or unbounded and either measured or unmeasured formulas.

Secondly, an application of certain temporal adverbials often induces shifts in the aspectual classes of the resultant VPs. Frequency adverbials and synchronized cyclic ones yield atelic, unmeasured VPs, while while durative adverbials always yield measured VPs. Thus,

> John {{*was ill twice*} *in three years*}.
> ?John {{*was ill twice*} *for three years*}.
> John {{*was frequently ill*} *for three years*}.
> ?John {{*was frequently ill*} *in three years*}.
> John {{*worked for five hours*} *three times*} last week.

We now rewrite the VP adjunct rules introduced earlier to accommodate the interaction between VPs and adverbials and possible shifts in aspectual classes.[15] We also show VP rules that perform aspectual class shifts. Keep in mind that aspectual class features, `telicity`, `bounded`, and `temporal-extent`, are head features, and so are shared between mother and daughter VPs except when explicitly overriden.

VP[measured] ← VP[atelic, unmeasured] ADVL[dur] ; (ADVL' VP')
VP[measured] ← VP[telic, unmeasured] ADVL[span] ; (ADVL' VP')
VP[accomplishment] ← VP[bounded] ADVL[card] ; (ADVL' VP')
VP[atelic, unmeasured] ← VP[bounded] ADVL[freq] ; (ADVL' VP')
VP[atelic, unmeasured] ← VP ADVL[cyc-time] ; (ADVL' VP')

VP[bounded] ← VP[atelic] ; (temporarily VP')
VP[atelic, unmeasured] ← VP[bounded] ; (iter VP')
VP[atelic, unmeasured] ← VP[transition] ; (result-state VP')

These rules allow transitions in aspectual class and VP-adverbial combinations somewhat too liberally. We assume, however, that undesirable transitions and combinations may be ruled out on semantic grounds. We now show some additional sentences and their initial translations (with speech acts neglected) to illustrate some of the above rules.

(15) a. Mary was ill twice in December
    b. (past ((adv-e (during (in-time *December*))))
        ((adv-f ((num 2) (plur episode))) [Mary (*bounded* ill)])))
(16) a. Mary received an award for three years
    b. (past ((adv-e (lasts-for (K ((num 3) (plur year)))))
        [Mary (*iter* $\lambda x (\exists y : [y\ award]\ [x\ receive\ y])$)])))

6 Concluding Remarks

We have shown that tense and aspect can be analyzed compositionally in a way that accounts not only for their more obvious effects on sentence meaning but also, via tense trees, for their cumulative effect on context and the temporal relations implicit in such contexts. Our scheme is easy to implement, and has been successfully used in the TRAINS interactive planning advisor at Rochester [Allen and Schubert, 1991].

Much theoretical work has been done on temporal adverbials (e.g., [Dowty, 1982; Hinrichs, 1988; Kamp and Reyle, 1993; Mittwoch, 1988; Moltmann, 1991; Richards and Heny, 1982]). There is also some computationally oriented work. Moens and Steedman[1988], among others, discussed the interaction of adverbials and aspectual categories. Our work goes further, in terms of (1) the scope of syntactic coverage, (2) interaction of adverbials with each other and with tense and aspect, (3) systematic (and compositional) transduction from syntax

---

[15]Similar kinds of shift in aspectual classes have previously been discussed in the literature; e.g., in [Steedman, 1982; Moens and Steedman, 1988; Smith, 1991].

to logical form (with logical-form deindexing), (4) formal interpretability of the resulting logical forms, and (5) demonstrable use of the resulting logical forms for inference.

Remaining work includes analysis of participial and infinitival adverbials and adverbials involving implicit anaphoric referents. Consider, e.g., "John came back *in ten minutes*" and "*After three years*, John proposed to Mary." These adverbials involve an implicit reference episode. Such implicit referents may often be identified from our tense trees, but at other times require inference. Another important remaining issue is handling of event nominals; e.g., "Mary is angry about the accident. The other driver had been drinking", where the event nominal *accident* serves as reference episode for the subsequent perfect. The interaction between event nominals and frequency adjectives (along the lines of [Stump, 1981]) also calls for further investigation.

References

[Allen and Schubert, 1991] Allen, J. and Schubert, L. K. The TRAINS Project. Technical Report 382, Dept. of Comp. Sci., U. of Rochester, Rochester, NY.

[Allen, 1987] Allen, J. F. *Natural Language Understanding*. Benjamin/Cummings, Reading, MA.

[Barwise, 1989] Barwise, J. *The Situation in Logic*. CSLI, Stanford, CA.

[Dowty, 1982] Dowty, D. Tense, time adverbs and compositional semantic theory. *Linguistics and Philosophy* 5:23–55.

[Dowty, 1986] Dowty, D. The effect of aspectual classes on the temporal structure of discourse: semantics or pragmatics? *Linguistics and Philosophy* 9(1):37–61.

[Hinrichs, 1986] Hinrichs, E. Temporal anaphora in discourses of English. *Linguistics and Philosophy* 9(1):63–82.

[Hinrichs, 1988] Hinrichs, E. W. Tense, quantifiers, and contexts. *Computational Linguistics* 14(2):3–14.

[Hornstein, 1977] Hornstein, N. Towards a theory of tense. *Linguistic Inquiry* 3:521–557.

[Hwang and Schubert, 1992] Hwang, C. H. and Schubert, L. K. Tense trees as the 'fine structure' of discourse. In *Proc. 30th Annual Meeting of the ACL*. Newark, DE. 232–240.

[Hwang and Schubert, 1993] Hwang, C. H. and Schubert, L. K. Episodic Logic: A situational logic for natural language processing. In Aczel, P.; Israel, D.; Katagiri, Y.; and Peters, S., editors, *Situation Theory and its Applications, V. 3*, CSLI, Stanford, CA. 303–338.

[Hwang, 1992] Hwang, C. H. *A Logical Approach to Narrative Understanding*. Ph.D. Dissertation, U. of Alberta, Edmonton, Canada.

[Kamp and Reyle, 1993] Kamp, H. and Reyle, U. *From Discourse to Logic*. Kluwer, Dordrecht, Holland.

[Lascarides and Asher, 1993] Lascarides, A. and Asher, N. Temporal interpretation, discourse relations and common sense entailment. *Linguistics and Philosophy* 16:437–493.

[Leech, 1987] Leech, G. *Meaning and the English Verb (2nd ed)*. Longman, London, UK.

[Mittwoch, 1988] Mittwoch, A. Aspects of English aspect: On the interaction of perfect, progressive and durational phrases. *Linguistics and Philosophy* 11:203–254.

[Moens and Steedman, 1988] Moens, M. and Steedman, M. Temporal ontology and temporal reference. *Computational Linguistics* 14(2):15–28.

[Moltmann, 1991] Moltmann, F. Measure adverbials. *Linguistics and Philosophy* 14:629–660.

[Mourelatos, 1981] Mourelatos, A. P. D. Events, processes and states. In Tedeschi, P. J. and Zaenen, A., editors, *Tense and Aspect (Syntax and Semantics, V.14)*. Academic Press, New York, NY. 191–212.

[Partee, 1984] Partee, B. Nominal and temporal anaphora. *Linguistics and Philosophy* 7:243–286.

[Passonneau, 1988] Passonneau, R. J. A computational model of the semantics of tense and aspect. *Computational Linguistics* 14(2):44–60.

[Prior, 1967] Prior, A. *Past, Present and Future*. Clarendon Press, Oxford, UK.

[Reichenbach, 1947] Reichenbach, H. *Elements of Symbolic Logic*. Macmillan, New York, NY.

[Richards and Heny, 1982] Richards, B. and Heny, F. Tense, aspect, and time adverbials. *Linguistics and Philosophy* 5:59–154.

[Schubert and Hwang, 1990] Schubert, L. K. and Hwang, C. H. Picking reference events from tense trees: A formal, implementable theory of English tense-aspect semantics. In *Proc. Speech and Natural Language, DARPA Workshop*, Hidden Valley, PA. 34–41.

[Smith, 1978] Smith, C. The syntax and interpretations of temporal expressions in English. *Linguistics and Philosophy* 2:43–99.

[Smith, 1991] Smith, C. *The Parameter of Aspect*. Kluwer, Dordrecht, Holland.

[Song and Cohen, 1991] Song, F. and Cohen, R. Tense interpretation in the context of narrative. In *Proc. AAAI-91*. Anaheim, CA. 131–136.

[Steedman, 1982] Steedman, M. J. Reference to past time. In *Speech, Place and Action*. John Wiley & Sons, New York, NY. 125–157.

[Stump, 1981] Stump, G. T. The interpretation of frequency adjectives. *Linguistics and Philosophy* 4:221–257.

[van Benthem, 1988] van Benthem, J. *A Manual of Intensional Logic*. CSLI, Stanford, CA.

[Vendler, 1967] Vendler, Z. *Linguistics in Philosophy*, Chapter 4: Verbs and Times. Cornell U. Press, Ithaca, NY.

[Webber, 1988] Webber, B. L. Tense as discourse anaphor. *Computational Linguistics* 14(2):61–73.

SYNCHRONIZED HISTORIES IN PRIOR-THOMASON REPRESENTATION OF BRANCHING TIME*

MARIA CONCETTA DI MAIO
ALBERTO ZANARDO**

0. Introduction.

English sentences like:

(0.1) If John were to go by bus, he would arrive at the same time as he would were he to go by car.

(0.2) If John were to go by train, he would arrive earlier than he would were he to go by car.

seem to make perfect sense. Moreover, they are also used in inferences which we would intuitively accept. So, for example, the following argument:

a. If John were to go by train, he would arrive earlier than he would were he to go by car.
b. John would only take the transportation which allows him to arrive as soon as possible.
c. It is possible for John to take the train.
d. John would not go by car.

seems intuitively valid.

All the above examples are apparently based on the idea that it makes sense to establish temporal comparisons among different (and incompatible) courses of affairs. What follows is a study about the possibility of expressing temporal comparisons among different courses of affairs in some formal languages whose semantics are based on certain branching-time frames. It will be shown that the usual Prior-Thomason formalism [Prior 67, Thomason 84] does not suffice for this goal (section 2), so that new operators are needed; a relevant part of this work will deal with some such new operators, their semantics and their respective strengths (sections 3 and 4). The results proved in sections 2-4 will be discussed in the last section.

---

* This work was written while the second author was visiting the Center for Philosophy of Science at the University of Pittsburgh. The visit was mainly supported by the Center for Philosophy of Science and it profited also from support of the Italian Consiglio Nazionale delle Ricerche, G.N.S.A.G.A.

** Authors' address: Dipartimento di Matematica Pura ed Applicata - Via Belzoni 7 - I-35131 PADOVA - ITALY.
E-mail: dimaio@pdmat1.math.unipd.it - alberto@pdmat1.math.unipd.it

Besides its role in the formalization of natural languages, the present analysis might help in clarifying the issues in the philosophical debate about simultaneity among possible worlds and it might be useful also as preliminary work for various logics which, even if they do not aim at an explicit investigation of time, are based on a branching time semantics. For instance, the notion 'at this very moment, but on a different evolution of time' is explicitely involved in Belnap's and Perloff's logic of agency [Belnap & Perloff 90], and it also plays a crucial role in several temporal logics used in Computer Science and Artificial Intelligence.

1. Prior-Thomason representation of branching-time.

Let us first consider a language $\mathcal{L}$ containing denumerably many propositional variables $p_0, p_1, \ldots$, the boolean connectives $\wedge$ and $\neg$, the tense operators P (*it was the case that*) and F (*it will be the case that*), and the possibility operator $\Diamond$ (*it is possible that*). The other propositional connectives, as well as the dual temporal and modal operators H ($\neg P \neg$, *it was always the case that*), G ($\neg F \neg$, *it will always be the case that*), $\Box$ ($\neg \Diamond \neg$, *it is necessary that*), are defined in the usual way.

The semantical structures we shall consider for $\mathcal{L}$ are *trees*, that is, pairs $\langle T, \prec \rangle$ where T is a set and $\prec$ is an irreflexive order relation on it such that (tree condition): for $m, m_1, m_2 \in T$, $m_1 \prec m$ and $m_2 \prec m$ imply $m_1 \prec m_2$ or $m_2 \prec m_1$ or $m_1 = m_2$. The intuitive picture behind this kind of structures is that the elements of the structure, which are supposed to represent instantaneous world-slices, are partially ordered by a transitive relation in such a way that each instantaneous world-slice has only one past but it may have more than one future. The elements of T will be referred to as *moments* and, in case $m \prec m'$ we shall say that m is *earlier than* m' or that m' is *later than* m. A *history* is a maximal subset of T linearly ordered by $\prec$. We shall say that the history h *passes through* the moment m whenever h contains m.

Given any tree $\mathbf{T} = \langle T, \prec \rangle$, we shall evaluate the formulas of $\mathcal{L}$ on pairs $\langle m, h \rangle$, where m is a moment and h is a history passing through m (in the sequel, when considering pairs $\langle m, h \rangle$, $m \in h$ will be often understood). Following current use, we shall call this way of evaluating formulas *Ockhamism* [Prior 1967]. The underlying idea of Ockhamist tense logics is that, in general, the truth of a formula depends not only on the moment m in which it is asserted, but also on a given history. So, according to an Ockhamist, an assertion like "John will visit Italy" is true at a pair $\langle m, h \rangle$ if John visits Italy at some $m' \in h$, $m' \succ m$, and false at the pair $\langle m, h \rangle$ if for all $m' \in h$, $m' \succ m$, John does not visit Italy at m'.

A valuation of the propositional variables in the tree $\mathbf{T}$ is a function V assigning to each propositional variable a set of pairs $\langle m, h \rangle$ in $\mathbf{T}$. Valuations of complex formulas at a given pair $\langle m, h \rangle$, are defined inductively by the following rules, where, for every α, $V_{\langle m,h \rangle}(\alpha) = 0$ if and only if $V_{\langle m,h \rangle}(\alpha) \neq 1$

R1. $V_{\langle m,h \rangle}(p_i) = 1$ iff $\langle m, h \rangle \in V(p_i)$

R2. $V_{\langle m,h \rangle}(\neg\alpha) = 1$ iff $V_{\langle m,h \rangle}(\alpha) = 0$

R3. $V_{\langle m,h \rangle}(\alpha \wedge \beta) = 1$ iff $V_{\langle m,h \rangle}(\alpha) = 1$ and $V_{\langle m,h \rangle}(\beta) = 1$

R4. $V_{\langle m,h \rangle}(F\alpha) = 1$ iff $\exists m' \succ m : (m' \in h$ and$) V_{\langle m',h \rangle}(\alpha) = 1$

R5. $V_{\langle m,h \rangle}(P\alpha) = 1$ iff $\exists m' \prec m : V_{\langle m',h \rangle}(\alpha) = 1$

R6. $V_{\langle m,h \rangle}(\Diamond \alpha) = 1$ iff $\exists h'$ passing through $m : V_{\langle m,h' \rangle}(\alpha) = 1$

In case $V_{\langle m,h \rangle}(\alpha) = 1$ [$=0$], we shall say that α is *true* [*false*] at $\langle m, h \rangle$ under V. We shall write $\mathbf{T} \models \alpha$ to mean that, for every valuation V on $\mathbf{T}$, for every history h, and every $m \in h$, $V_{\langle m,h \rangle}(\alpha) = 1$. Valid formulas are those α such that $\mathbf{T} \models \alpha$ for every tree $\mathbf{T}$.

Sometimes, in the course of the present paper, we shall refer to linear-time structures and purely temporal formulas. A linear time structure is an irreflexive linear order $\langle X, < \rangle$ and a valuation on $\langle X, < \rangle$ is a function V assigning a subset of X to each propositional variable. A formula of $\mathcal{L}$ will be said to be purely temporal whenever it does not contain $\Diamond$ and its evaluation is given by the obvious analogue of R1-5.

For every history h in a tree $\mathbf{T}$, the pair $\langle h, \prec \rangle$ is a linear time structure and hence the purely temporal fragment of ockhamist logic is linear-time logic. By R6, the modal logic of $\Diamond$ is S5.

In case $V_{\langle m,h \rangle}(\alpha \equiv \Box \alpha) = 1$ (that is, $V_{\langle m,h \rangle}(\alpha) = V_{\langle m,h' \rangle}(\alpha)$ for all h' passing through m), we shall say that α *is Settled True* or *is Settled false* at m (under the evaluation V) according as to whether $V_{\langle m,h \rangle}(\alpha) = 1$ or $V_{\langle m,h \rangle}(\alpha) = 0$. If either α is settled true at m or α is settled false at m, we shall say that α is *history indepentent* at m (under V) and we shall often write $V_m(\alpha)$ instead of $V_{\langle m,h \rangle}(\alpha)$. Modally closed formulas, that is, formulas α such that $\alpha \equiv \Box \alpha$ is an S5-validity, are history independent formulas at every moment and under any evaluation.

The language $\mathcal{L}$ and the semantics we have given it are the starting point of our analysis, in the sense that we shall now consider semantics which are restrictions of the semantics given above (section 2), and languages which are extensions of $\mathcal{L}$ (sections 3 and 4). However, before going on, two points ought to be mentioned:

i. there are branching time semantics for temporal languages which keep the same frames (trees) as the ones we have spoken about but whose evaluation rules are quite different from the ones we have given.

ii. there are different branching-time frames on which to base a semantics for $\mathcal{L}$.

The Ockhamist semantics we have given evaluates formulas at pairs ⟨moment, history⟩. However, there are semantics which evaluate formulas only at moments. A famous branching-time semantics of this type, is the one adopted in the so-called *Peircean logic* [Prior 67]. The frames of Peircean branching time logic are just the frames of Ockhamist logic, and the two logics differ only with respect to their evaluation rules. In Peircean Logic, the formula $F\alpha$ is true at

m if and only if every history h passing through m contains a moment $m_h \succ m$ at which α is true; $P\alpha$ is true at m if and only if α is true at some $m' \prec m$.

Of course, there is a very close connection between the Peircean reading of the operator F and the Ockhamist reading of the expression $\Box F$.

The Peircean evaluation of the expressions $F\alpha$ and $P\alpha$ is not the only possible way of evaluating such expressions at moments only. For example, we might take $F\alpha$ to be true at m if and only if some history h passing through m contains a moment $m_h \succ m$ at which α is true. This way of evaluating $F\alpha$ is obvioulsy closely connected to the evaluation given to the expression $\Diamond F\alpha$ by what we (following Prior) have called "Ockhamism".

Another way of conceiving frames for $\mathcal{L}$ starts from bundled trees rather than from trees [Burgess 79, Zanardo 85]. Bundled trees can be viewed as pairs $\langle \mathbf{T}, H \rangle$ where H is a set of histories such that every moment in $\mathbf{T}$ belongs to a history in H. If one takes bundled trees as the basic semantic structures, the quantifiers in the evaluation rule of the operator $\Diamond$ quantify over a set of histories in H passing through the moment at hand, *instead of* over the set of *all* such histories.

Bundled trees are very closely related to Kamp frames [Thomason 84, p.147], in fact, every Kamp frame can be shown to be structure-isomorphic to a bundled tree. However, whereas in trees and bundled trees a course of affairs (a history) is seen as a set whose elements are supposed to be instantaneous world-slices, in a Kamp frame a possible course of affairs is taken as a primitive and each course of affairs has its own temporal ordering. This means that the ontological commitments of these two kinds of frames are deeply different.

It is worth noticing that, even though there is an isomorphism between Kamp frames and bundled trees, in a strict sense, Kamp-frame validity and bundled-tree validity do not coincide. The reason is that, in [Thomason 84], the evaluation of propositional variables is assumed to fulfill a condition equivalent to, in our terminology,

(1.1) $\quad \langle m, h \rangle \in V(p) \quad \Leftrightarrow \quad \langle m, h' \rangle \in V(p)$ for all histories $h' \ni m$

so that the propositional variables come out to be history independent. This is equivalent to holding that atomic sentences in natural languages are history independent. A technically unpleasant consequence of (1.1) is that, if it is assumed, *Substitution* is no longer a valid inference rule; for instance, $Pp \to \Box Pp$ is valid under (1.1), but in general the result of substituting an arbitrary formula for p in $Pp \to \Box Pp$ is not valid. This issue is discussed in [Prior 67, pp.123,124], where two kinds of propositional variables are used; propositional variables of the first kind allow for unrestricted substitutions, propositional variables of the second kind allow for substitution only in certain cases. From a semantical point of view, this is equivalent to evaluating the propositional variables of the first kind on arbitrary sets of pairs $\langle m, h \rangle$, and to assume (1.1) for the propositional variables of the second kind.

2. Synchronized histories.

The language $\mathcal{L}$, with the semantics we have given it, is obviously unable to express sentences like the ones we considered in the introduction of this paper:

(0.1) If John were to go by bus, he would arrive *at the same time* as he would were he to go by car.

(0.2) If John were to go by train, he would arrive *earlier than* he would were he to go by car.

In fact, if we denote by m_b, m_c, m_t, the moments of possible arrivals (by bus, by car, by train) spoken about in these examples, and by m the moment in which (0.1) or (0.2) is asserted, we obviously have $m \prec m_b$ and $m \prec m_c$ ($m \prec m_c$ and $m \prec m_t$), but we have also that no history can contain both the moment m_b and the moment m_c (m_c and m_t), because, say, the histories to which m_b belongs and those to which m_c belongs branch out at the moment of departure, that is, before m_b or m_c. This, according to the semantics presented in the preceding section, means that the moments m_b and m_c, and m_c and m_t, are related in no way:

$$(2.1) \quad \begin{array}{c} m_b \neq m_c,\ m_b \not\prec m_c,\ m_b \not\succ m_c \\ m_c \neq m_t,\ m_c \not\prec m_t,\ m_c \not\succ m_t \end{array}$$

In order to make sense of (0.1) and (0.2), what we want is then a semantics endowed with new relations between moments.

One way to go, suggested by example (0.1), is to try and see whether it is possible to define a relation on moments capturing the intuitive idea of "happening at the same time as, on possibly different histories". Given any moments m and m', we will write $m \leftrightarrow m'$ to mean that m happens at the *same time* as m' [1]. Plausible properties of the relation $\leftrightarrow$ are

ST$_1$. $\leftrightarrow$ is an equivalence relation.

ST$_2$. $m \prec m'$ implies $m \not\leftrightarrow m'$ (if m precedes m' in the temporal order, then m cannot happen at the same time as m').

ST$_3$. $\leftrightarrow$ is order-preserving: if $m_1, m_2 \in h$, $m'_1, m'_2 \in h'$, $m_1 \leftrightarrow m'_1$, $m_2 \leftrightarrow m'_2$, and $m_1 \prec m_2$, then $m'_1 \prec m'_2$

---

[1] Our relation $\leftrightarrow$ is defined over moments, which are supposed to represent instantaneous world-slices. This might be considered a deviation from natural languages in that, in English, the relata of "happening at the same time as" are generally taken to be events. If we consider the whole state of the world at a certain time as being just one big event, our relation $\leftrightarrow$ will capture also this aspect of the natural language relation "happening at the same time as". If instead it is thought that the state of the world at a certain time cannot be properly considered an event, our relation $\leftrightarrow$ will differ from the natural language relation "happening at the same time as", still, as we shall see, $\leftrightarrow$ will be very useful when building languages capable of expressing sentences like our examples (0.1) and (0.2).

A consequence of properties ST_1 and ST_2 is that if m bears the relation $\leftrightarrow$ to any moment in a history h', that moment is unique; indeed $m \leftrightarrow m'$ and $m \leftrightarrow m''$ imply $m' \not\prec m''$ by the transitivity of $\leftrightarrow$ and ST_2. Another consequence is that, if $m \in h \cap h'$ then, with respect to h, h', m bears $\leftrightarrow$ to itself. Then, by ST_3, given any two histories h and h', the restriction $\leftrightarrow \cap h \times h'$ of $\leftrightarrow$ to $h \times h'$ is an order-isomorphism $\sigma_{h,h'}$ between a subset of h and a subset of h'. If, in addition to ST_{1-3}, we require that

ST_4. for every moment m and any history h, there is a moment $m' \in h$ such that $m \leftrightarrow m'$,

then, for all histories h, h', there exists an isomorphism $\sigma_{h,h'}$ between h and h'. Properties of these isomorphisms are:

S_1. $\sigma_{h,h'}(m) = m' \Rightarrow \sigma_{h',h}(m') = m$ (symmetry of $\leftrightarrow$).

S_2. $m \in h \cap h' \Rightarrow \sigma_{h,h'}(m) = m$ (reflexivity of $\leftrightarrow$).

S_3. $\sigma_{h,h''}(m) = \sigma_{h',h''}(\sigma_{h,h'}(m))$ (transitivity of $\leftrightarrow$).

We shall say that the tree $\mathbf{T}$ is *synchronizable* whenever a set $\sigma = \{\sigma_{h,h'} : h$ and h' are histories in $\mathbf{T}\}$ with the properties S_{1-3} exists. In this case, σ will be said to be a *synchronism* on $\mathbf{T}$ and the pair $\langle \mathbf{T}, \sigma \rangle$ will be referred to as a *synchronized tree*.

Another way to go is suggested by example (0.2). In this example, a certain event happening on a certain history is said to be earlier than an event happening on a different history.

We then introduce a new relation $\triangleleft$, and we read $m \triangleleft m'$ as saying that m is *weakly earlier than* m, in the sense that m precedes m', but not necessarily on a history to which m' belongs. The *weak precedence* relation $\triangleleft$ has the following properties:

WP_1. $m \prec m' \Rightarrow m \triangleleft m'$ ($\triangleleft$ is an extension of $\prec$)

WP_2. $\triangleleft$ is transitive and irreflexive.

WP_3. for every m and every h', there exists a unique $m' \in h'$ such that $\neg m \triangleleft m'$ and $\neg m' \triangleleft m$.[2]

It can be easily verified that the relations $\leftrightarrow$, $\triangleleft$ and σ, are related to each other in the following ways:

(2.2) $\quad m \triangleleft m' \quad \Leftrightarrow \quad \exists h, h' : \sigma_{h,h'}(m) \prec m' \quad (\Leftrightarrow \quad \exists h, h' : m \prec \sigma_{h',h}(m'))$

(2.3) $\quad\quad\quad m \leftrightarrow m' \quad \Leftrightarrow \quad \neg m \triangleleft m'$ and $\neg m' \triangleleft m$

---

[2] We assume that $\triangleleft$ has also this last property because, in the present paper, we are mainly interested in the notion of synchronism, and hence it is natural to assume that every moment has a correspondent, on each history, which is neither weakly earlier nor weakly later than it. It would be certainly interesting to investigate the logical properties of structures in which only WP_1 and WP_2 hold, or in which WP_3 is relaxed by eliminating the uniqueness of m'

(2.4) $\quad\exists h, h' : m' = \sigma_{h,h'}(m) \quad \Leftrightarrow \quad \neg\, m \triangleleft m'$ and $\neg\, m' \triangleleft m$

Thus, the relations $\triangleleft$, $\leftrightarrow$, σ are interdefinable. This means that the pairs $\langle \mathbf{T}, \sigma \rangle$, $\langle \mathbf{T}, \leftrightarrow \rangle$ $\langle \mathbf{T}, \triangleleft \rangle$ are the same mathematical objects; we shall refer to all of them as synchronized trees and we will shift freely from a notation to another.

Every synchronized tree can be viewed as a $T \times W$-frame [Thomason 84]; given any $T \times W$ frame, however, the corresponding tree-like notion is that of bundled synchronized tree (see section 1).

The class of synchronizable trees is a proper subclass of the class of trees and hence a first problem is whether the former class is definable in $\mathcal{L}$, that is, whether a set Γ of $\mathcal{L}$-formulas exists such that, for every tree $\mathbf{T}$, $\mathbf{T} \models \Gamma$ iff $\mathbf{T}$ is synchronizable. The answer is no; a proof is in [Zanardo 86], where evaluations fulfilling (1.1) are used; another, simpler proof is in [Zanardo 94].

Given the new semantics provided by synchronized trees, let us now see whether it is possible to express in $\mathcal{L}$ a relation among formulas of $\mathcal{L}$ corresponding to our intuitive idea of "happening at the same time as". To that purpose, we first give a metalinguistic characterization of simultaneity [3].

2.5 Definition. *Given a synchronized tree $\langle \mathbf{T}, \sigma \rangle$ and a valuation V on it, we shall say that the (events denoted by the) formulas α and β have a simultaneous occurrence on the histories h and h' (under the evaluation V), whenever there exist moments $m \in h$ and $m' \in h'$ such that $V_{(m,h)}(\alpha) = 1$, $V_{(m',h')}(\beta) = 1$, and $\sigma_{h,h'}(m) = m'$.*

A history-independent definition of simultaneity could be

2.6 Definition. *Given a synchronized tree $\langle \mathbf{T}, \sigma \rangle$ and a valuation V on it, we shall say that the (events denoted by the) formulas α and β have a simultaneous occurrence in $\langle \mathbf{T}, \sigma \rangle$ (under the evaluation V), whenever, for some histories h and h', α and β have a simultaneous occurrence on h and h'*

We can now formulate the problem about the expressibility of the simultaneity relation in $\mathcal{L}$ in several ways. Given any two propositional variables p and q, we can ask whether

(2.7) a formula $\Sigma_1(p, q)$ exists such that, for every synchronized tree $\langle \mathbf{T}, \sigma \rangle$ and every evaluation V, $\langle \mathbf{T}, V \rangle \models \Sigma_1(p, q)$ if and only if p and q have a simultaneous occurrence in $\langle \mathbf{T}, \sigma \rangle$ under V; or

---

[3] The relation "being simultaneous with", as used in physics, has a precise operational definition which, obviously, has nothing to do with what we are talking about in the present paper. However, the notion of simultaneity involved in example (0.1) has nothing to do with physical simultaneity either. At this point one might claim that, since it has no physical meaning, the notion of simultaneity involved in (0.1) is meaningless (and hence, that (0.1) is meaningless too), this, for example, is the position taken in [van Fraassen 89]. We shall instead start from the presupposition that (0.1) makes sense, and we shall see to what extent it can be given a precise logical sense.

(2.8) a formula $\Sigma_2(p,q)$ exists such that, for every synchronized tree $\langle \mathbf{T}, \sigma \rangle$ and every evaluation V, $\Sigma_2(p,q)$ is history independent and $V_m(\Sigma_2(p,q)) = 1$ if and only if p and q have a simultaneous occurrence on two histories passing through m; or

(2.9) a formula $\Sigma_3(p,q)$ exists such that, for every synchronized tree $\langle \mathbf{T}, \sigma \rangle$ and every evaluation V, $V_{\langle m,h \rangle}(\Sigma_3(p,q)) = 1$ if and only if $V_{\langle m,h \rangle}(p) = 1$ and, for some $\langle m', h' \rangle$, $V_{\langle m',h' \rangle}(q) = 1$ and $m' = \sigma_{h,h'}(m)$.

All these questions have a negative answer and it is not difficult to see why.

2.10 Proposition. *The formulas Σ_{1-3} fulfilling (2.7-9) are not $\mathcal{L}$-definable.*

Proof. We have only to observe that 1) the right sides of 'if and only if' in (2.7-9) depend on the synchronism σ whereas the left sides do not and 2) in general, the synchronism on a synchronizable tree is not unique. Consider for instance the synchronized trees $\langle \mathbf{T}, \sigma^1 \rangle$ and $\langle \mathbf{T}, \sigma^2 \rangle$, with $\sigma^1 \neq \sigma^2$, based on the same synchronizable tree $\mathbf{T}$. Let the moments $m \in h$ and $m' \in h'$ be such that $\sigma^1_{h,h'}(m) = m'$ and $\sigma^2_{h,h'}(m) \neq m'$ and let V be an evaluation on $\langle \mathbf{T}, \sigma^1 \rangle$ and V' be an evaluation on $\langle \mathbf{T}, \sigma^2 \rangle$ such that $V(p_0) = V'(p_0) = \{\langle m, h \rangle\}$ and $V(p_1) = V'(p_1) = \{\langle m', h' \rangle\}$.

If $\Sigma_2(p,q)$ existed and m_0 is any moment in $h \cap h'$, then we would have $V_{m_0}(\Sigma_2(p_0,p_1)) = 1$ and $V'_{m_0}(\Sigma_2(p_0,p_1)) = 0$ so that the evaluation of $\Sigma_2(p_0,p_1)$ would depend entirely on the chosen synchronism. But the evaluation rules for $\mathcal{L}$ are such that the evaluation of a formula in $\mathcal{L}$ is *always* independent on the chosen synchronism. Hence, no formula like $\Sigma_2(p_0,p_1)$ exists in $\mathcal{L}$.

Quite similar examples show that $\Sigma_1(p,q)$ and $\Sigma_3(p,q)$ cannot exist either. The same holds for any other fomula of $\mathcal{L}$ which supposedly represents any notion strictly connected with synchronism. ∎

The proof of Proposition 2.10 is strongly based on the possibility of having different synchronisms on a given tree and hence it could suggest that the impossibility of defining Σ_{1-3} is to be ascribed to the lack of a privileged temporal alternative of a given moment in certain synchronizable trees. Theorem 2.12 below shows that this is not true and that $\Sigma_2(p,q)$ cannot be defined in $\mathcal{L}$ even if we consider only synchronizable trees in which the synchronism is unique. The proof can be easily adapted to show that the same holds for $\Sigma_1(p,q)$ and $\Sigma_3(p,q)$.

2.11 Lemma. *Let $\langle X, < \rangle$ and $\langle X', <' \rangle$ be linear-time structures with no first and no last element, and let $\langle X, < \rangle$ be isomorphic to an initial segment of $\langle X', <' \rangle$, that is, there exists a one-to-one function f from X into X' such that, f is order preserving and, for all $x \in X$, $x' <' f(x)$ implies $x' = f(y)$ for a suitable $y < x$. Let V and V' be evaluations on $\langle X, < \rangle$ and $\langle X', <' \rangle$ such that, for every propositional variable p, 1) for all $x \in X$, $x \in V(p)$ iff $f(x) \in V'(p)$ and 2)*

$(*)$ $\quad \exists x \in X : \forall x' >' f(x)(V'_{x'}(p) = 1)$ *or* $\forall x' >' f(x)(V'_{x'}(p) = 0)$

Then, for every $x \in X$ and every purely temporal formula α, $V_x(\alpha) = V'_{f(x)}(\alpha)$.

Proof. By induction of the complexity of α, after having proved that (∗) holds for every purely temporal formula. ∎

2.12 Proposition. *The formula Σ_2 fulfilling (2.8) does not exist even if, in (2.8), **T** is assumed to range over the synchronizable trees on which only one synchronism exists.*

Proof. The definition of a particolar tree we are going to give is rather complex; simpler structures could have been choosen, but in this case complex proofs would have been required. Consider the tree **T** consisting of exactly two histories, h_1 and h_2, such that 1) $h_1 \cap h_2$ is a singleton, say $\{m_0\}$, and both $h_1 - h_2$ and $h_2 - h_1$ are isomorphic to the linear order $\langle X, \prec \rangle$ where 1) X is the set of all pairs $\langle m, n \rangle$ with $m, n \in \omega$ (the set of natural numbers) and 2) $\langle m, n \rangle \prec \langle m', n' \rangle$ if and only if $m > m'$ or $m = m'$ and $n < n'$. Then $\langle X, \prec \rangle$ consists of ω copies of ω; the usual order $<$ between ordinals in each copy agrees with $\prec$, whereas, for $m > m'$, the elements of the m-th copy are in relation $\prec$ with the elements of the m'-th copy. Of course, the identity is the only isomorphism of $\langle X, \prec \rangle$ onto itself and hence there is only one synchronism, σ, on **T**. We shall denote by $\langle m, n \rangle_1$ the representative on h_1 of the pair $\langle m, n \rangle$ and similarly for $\langle m, n \rangle_2$. Since no confusion can arise, $\langle m, n \rangle_i$ will also denote the pair $\langle \langle m, n \rangle_i, h_i \rangle$.

Let V and V' be the evaluations defined by 1) for $i > 1$, $V(p_i) = V'(p_i) = $ the set of all pairs $\langle m, h \rangle$ in **T**, 2) $V(p_1) = V'(p_1) = \langle 0, 0 \rangle_1$, 3) $V(p_0) = \langle 0, 0 \rangle_2$, and $V'(p_0) = \langle 1, 0 \rangle_2$. We shall prove that, for every $\mathcal{L}$-formula α and every history h,

$$(*) \qquad V_{\langle m_0, h \rangle}(\alpha) = V'_{\langle m_0, h \rangle}(\alpha)$$

and this will contradict the existence of Σ_2 because, if such formula existed, we would have $V_{m_0}(\Sigma_2(p_1, p_0)) = 1$ and $V'_{m_0}(\Sigma_2(p_1, p_0)) = 0$.

Consider the right open initial segment $I = [m_0, \langle 0, 0 \rangle_2[$ on h_2. The function f defined by: $f(m_0) = m_0$ and $f(\langle m, n \rangle_2) = \langle m + 1, n \rangle_2$ is an isomorphism between h_2 and I; moreover, for every propositional variable p and every $x \in h_2$, $\langle x, h_2 \rangle \in V(p)$ iff $\langle f(x), h_2 \rangle \in V'(p)$. Then, the hypotheses of Lemma 2.11 are verified for $\langle X, < \rangle = \langle X', <' \rangle = \langle h_2, \prec \rangle$.

This means that every purely temporal formula is true at some $\langle m, n \rangle_2$ under V if and only if it is true at some $\langle m', n' \rangle_2$ under V'. A similar result holds trivially for h_1. Now, if we observe that the modal operator is not vacuous only when we evaluate formulas in m_0, an induction on the complexity of α leads to (∗). ∎

Stronger languages are then needed in order to express a temporal comparison, in a synchronized tree, between moments that do not share a common history, and the semantics for the new operators must involve the syncronism σ or the 'weakly earlier than' relation ⊲.

3. New operators.

Let us call $\mathcal{L}^\sigma$ the language obtained by adding the new operator $\diamond^\sigma$ to $\mathcal{L}$, and $\mathcal{L}^\triangleleft$ the language obtained by adding the operators $F^\triangleleft$ and $P^\triangleleft$ to $\mathcal{L}$. The dual operators $\square^\sigma$, $G^\triangleleft$ and $H^\triangleleft$ are defined in the obvious way. These new operators are meant to be the linguistic correspondents of the semantical relations $\leftrightarrow$ and $\triangleleft$. Thus, we shall interpret them in synchronized trees (viewed as pairs $\langle \mathbf{T}, \sigma \rangle$ or as pairs $\langle \mathbf{T}, \triangleleft \rangle$). Their semantical rules are:

$$V_{\langle m,h \rangle}(\diamond^\sigma \alpha) = 1 \quad \text{iff} \quad \exists m', h' : V_{\langle m',h' \rangle}(\alpha) = 1 \text{ and } m' = \sigma_{h,h'}(m)$$
$$V_{\langle m,h \rangle}(F^\triangleleft \alpha) = 1 \quad \text{iff} \quad \exists m' \triangleright m : V_{\langle m',h' \rangle}(\alpha) = 1 \text{ for some history } h'$$
$$V_{\langle m,h \rangle}(P^\triangleleft \alpha) = 1 \quad \text{iff} \quad \exists m' \triangleleft m : V_{\langle m',h' \rangle}(\alpha) = 1 \text{ for some history } h'$$

A consequence of two of the properties of σ (S2, S3 in section 2) is that the formula $\diamond^\sigma \alpha$ is history independent; history independence holds trivially for $F^\triangleleft \alpha$ and $P^\triangleleft \alpha$, and hence all these formulas can be evaluated on moments only.

A consequence of Theorem 2.10 is that the operator $\diamond^\sigma$ is not definable in $\mathcal{L}$ (because, if it were, the formula $\Sigma_3(p,q)$ would be definable by $p \wedge \diamond^\sigma(q)$). The proof that also $F^\triangleleft$ and $P^\triangleleft$ are not definable in $\mathcal{L}$ is less straightforward, but it is possible to carry it out by means of the same idea as the one on which the proof of Theorem 2.10 is based.

We have already observed that there is a one-to-one correspondence between pairs $\langle \mathbf{T}, \sigma \rangle$ and pairs $\langle \mathbf{T}, \triangleleft \rangle$ and that the relation $\triangleleft$ can be defined by means of the synchronism σ and vice-versa. This fact has a partial linguistic correspondent in the sense that the operators $P^\triangleleft$ and $F^\triangleleft$ are definable in $\mathcal{L}^\sigma$ by

$$(3.1) \qquad P^\triangleleft p \equiv P \diamond^\sigma p \,(\equiv \diamond^\sigma Pp) \qquad F^\triangleleft p \equiv F \diamond^\sigma p \,(\equiv \diamond^\sigma Fp)$$

Proposition 3.3 below shows that $\diamond^\sigma$ is not definable in $\mathcal{L}^\triangleleft$.

3.2 Lemma. *Let $\langle X, < \rangle$ be a linear time structure isomorphic to the set of rational numbers and let $Y \subseteq X$ be such that both Y and $X - Y$ are dense in X. Assume that V is any evaluation such that, for every propositional variable p, and every $m \in X$*

$$(*) \qquad \begin{array}{l} V_m(p) = 1 \text{ and } m \in Y \;\Rightarrow\; \forall m' \in Y, V_{m'}(p) = 1 \\ V_m(p) = 1 \text{ and } m \in X - Y \;\Rightarrow\; \forall m' \in X - Y, V_{m'}(p) = 1 \end{array}$$

Then
(a.) $(*)$ holds for every purely temporal formula α and, in particular,
(b.) if α is $F\beta$ or $P\beta$, then $\{m : V_m(\alpha) = 1\}$ is X or $\emptyset$.

Proof. (a.) is easily proved by induction on the complexity of α, and (b) is an obvious consequence of (a.). ∎

3.3 Proposition. *The operator $\diamond^\sigma$ is not definable in $\mathcal{L}^\triangleleft$.*

Proof. Assume that $\diamond^\sigma$ is definable in $\mathcal{L}^\triangleleft$ and let $S^\triangleleft(p)$ be any $\mathcal{L}^\triangleleft$-formula such that, for every synchronized tree $\langle \mathbf{T}, \triangleleft \rangle$, every evaluation V, and every $\langle m, h \rangle$,

$$V_{\langle m,h \rangle}(S^\triangleleft(p)) = 1 \quad \text{iff} \quad \exists m', h' : V_{\langle m',h' \rangle}(p) = 1 \text{ and } m' = \sigma_{h,h'}(m)$$

Let $\langle\langle T, \prec\rangle, \sigma\rangle$ be a synchronized tree having exactly two histories, h_1 and h_2, and such that 1) $\langle h_1 \cap h_2, \prec\rangle$ is an arbitrary linear order and 2) $\langle h_1 - h_2, \prec\rangle$ and $\langle h_2 - h_1, \prec\rangle$ are both isomorphic to the set of rational numbers. Consider a dense subset X of $h_2 - h_1$ such that $(h_2 - h_1) - X$ is dense too and let V and V' be the evaluations defined by: 1) for $i > 0$, $V(p_i) = V'(p_i) = \{\langle m, h\rangle : m \in T$ and $h \in \{h_1, h_2\}\}$, 2) $V(p_0) = \{\langle m, h_2\rangle : m \in (h_2 - h_1) - X\}$, 3) $V'(p_0) = \{\langle m, h_2\rangle : m \in X\}$. Then, for every $m \in (h_2 - h_1)$, p_0 is true at $\langle m, h_2\rangle$ under V if and only if it is false under V'. The same idea as that of Lemma 3.2 shows that this fact can be extended to the set of all $\mathcal{L}^\triangleleft$-formulas: given any such formula α, for $m_1, m_2 \in X$ and $m'_1, m'_2 \in (h_2 - h_1) - X$,

$$V_{\langle m_1, h_2\rangle}(\alpha) = V_{\langle m_2, h_2\rangle}(\alpha) = V'_{\langle m'_1, h_2\rangle}(\alpha) = V'_{\langle m'_2, h_2\rangle}(\alpha)$$

A consequence of these equalities, and of another induction on the complexity of α, is that

$$\forall m \in h_1 - h_2,\ V_{\langle m, h_1\rangle}(\alpha) = V'_{\langle m, h_1\rangle}(\alpha)$$

This contradicts the existence of $S^\triangleleft(p)$; indeed, if m is any moment of $h_1 - h_2$ such that $\sigma_{h_1, h_2}(m) \in X$, the properties of $S^\triangleleft(p)$ would yield $V_{\langle m, h_1\rangle}(S^\triangleleft(p_0)) = 0$ and $V'_{\langle m, h_1\rangle}(S^\triangleleft(p_0)) = 1$, which, as we have seen, does not happen. ∎

This result shows that, even if the pairs $\langle \mathbf{T}, \sigma\rangle$ are the same mathematical objects as the pairs $\langle \mathbf{T}, \triangleleft\rangle$, the language $\mathcal{L}^\sigma$ has an expressive power greater than that of $\mathcal{L}^\triangleleft$. The following is a discussion about the reasons which lead to this difference in expressive power.

If we didn't as yet know that there is a proof (3.3) showing that the operator $\diamond^\sigma$ is not definable in $\mathcal{L}^\triangleleft$, we might think that a possible candidate for the definition of $\diamond^\sigma$ in $\mathcal{L}^\triangleleft$ would be $P \diamond F\alpha \wedge \neg F^\triangleleft \alpha \wedge \neg P^\triangleleft \alpha$, where the first conjunct merely expresses that α holds somewhere, whereas the second and the third conjuncts constitute the linguistic counterpart of (2.4). However, even if the semantical rules for our new operators imply

(3.4) $$P \diamond F\alpha \wedge \neg F^\triangleleft \alpha \wedge \neg P^\triangleleft \alpha \to \diamond^\sigma \alpha$$

the converse implication in general fails. In fact the left side of (3.4) is true at a given moment m only if all the moments in which α holds are in relation $\leftrightarrow$ with m, but $\diamond^\sigma \alpha$ can be true even if this does not happen.

Similarly, the semantical rules for our new operators imply

(3.5) $$\diamond^\sigma \alpha \to H^\triangleleft F^\triangleleft \alpha \wedge G^\triangleleft P^\triangleleft \alpha$$

but the converse of (3.5) fails because for $H^\triangleleft F^\triangleleft \alpha \wedge G^\triangleleft P^\triangleleft \alpha$ to be true at m, it suffices that α is true at some $m' \triangleleft m$ and at some $m'' \triangleright m$.

The failure of the converse of 3.4 and of 3.5 shows why we cannot define the operator $\diamond^\sigma$ in $\mathcal{L}^\triangleleft$: starting from a history h and a moment $m \in h$, belonging to the synchronized tree $\langle \mathbf{T}, \sigma\rangle$, if the formula $\diamond^\sigma \alpha$ is true at $\langle m, h\rangle$ we can single out a *unique* moment, m', on some history h', at which α holds, and which is *the* moment simultaneous to m on h' relative to $\langle \mathbf{T}, \sigma \rangle$. On the other hand, $F^\triangleleft \alpha$ and $P^\triangleleft \alpha$ allow us to single out a *unique* moment only on very rare occasions, namely, when α is such that $P \diamond F\alpha \wedge \neg F^\triangleleft \alpha \wedge \neg P^\triangleleft \alpha$.

People working in modal logic will recognize the connection between this problem and the general problem of finding a formula capable of expressing the irreflexivity of the accessibility relation in a modal frame ($\prec$ and $\triangleleft$, in our case). It is well known that this cannot be done in propositional modal logic and that, if propositional quantifiers were allowed, the irreflexivity of, for instance, $\prec$, would correspond to the truth, at each $\langle m, h \rangle$ and under any evaluation, of $\exists p \, (p \wedge G \neg p)$. As a matter of fact, (3.6) below (where $p^\#$ denotes $p \wedge G \neg p$) is a definition of $\diamond^\sigma$ in a version of $\mathcal{L}^\triangleleft$ containing propositional quantifiers

(3.6) $\quad \diamond^\sigma \alpha \quad \equiv_D \quad \exists p \, (P \diamond F(\alpha \wedge p^\#) \wedge \neg F^\triangleleft(\alpha \wedge p^\#) \wedge \neg P^\triangleleft(\alpha \wedge p^\#))$

4. Peircean-like simultaneity operators.

Some works based on branching-time semantics use a more relaxed notion of simultaneity than the one we have considered so far. Simultaneity is there explicitly assumed only among events which lie in the future of a given moment; these cases thus involve only the subtree of $\mathbf{T}$ consisting of all histories passing through that moment, rather than the full tree.

For example, the notion of simultaneity "in the future of the moment m_0" plays a crucial role in Belnap's and Perloff's logic of *seeing to it that* [Belnap & Perloff 90]. The operator stit is meant to express this notion and it involves what they call *the instant*, $i_{(m)}$, of the moment m. $i_{(m)}$ is defined as the set $\{m' : \sigma_{h,h'}(m) = m'$, for suitable $h, h'\}$. For $m_0 \prec m$, the *horizon of m_0* at m is the set of all intersections between $i_{(m)}$ and any history passing through m_0. The horizon at m of a suitable m_0 is the set of moments which is taken in consideration for evaluating a stit formula at a moment m [4].

Franz von Kutschera has suggested, in private conversation, the new temporal operators $F_\square$ and $P_\square$ (F_N and P_N in his notation) whose semantics are based on the synchronism between histories holding only in some horizon of the moment of evaluation (the past operator $P_\square$ involves the new notion of 'past' horizon). The formulas $F_\square \alpha$ and $P_\square \alpha$ are history independent and their semantics in a synchronized tree is given by

$V_m(F_\square \alpha) = 1$ iff
$\exists m' \succ m, \exists h' \ni m' : \forall h'' \ni m$, if $m'' = \sigma_{h',h''}(m')$ then $V_{m'',h''}(\alpha) = 1$.

---

[4] In his forthcoming dissertation (University of Pittsburgh), Ming Xu shows that the set of purely stit-validities (not involving temporal operators) does not change if a relaxed notion of instant is adopted. In the correspondent relaxed notion of synchronism we would have that 1) the indexes h, h' in $\sigma_{h,h'}$ do not range over the set of all pairs of histories, but only on a subset of it (including the pairs h, h) and 2) the domain and the image of $\sigma_{h,h'}$ are initial segments of h and h'.

$V_m(P_\square \alpha) = 1$ iff
$\exists m' \prec m, \exists h' \ni m' : \forall h'' \ni m$, if $m'' = \sigma_{h',h''}(m')$ then $V_{m'',h''}(\alpha) = 1$.

$F_\square \alpha$ is then true at m whenever α is true at all $\langle m'', h'' \rangle$ where m'' ranges over a suitable horizon of m and h'' ranges over the set of all histories passing through m; in this sense $F_\square$ can be viewed as a Peircean operator involving a synchronism between histories. The following implications hold in every synchronized tree and clarify the meaning of von Kutschera's operators.

(4.1)
$$F\square^\sigma \alpha \to F_\square \alpha \to F\square \alpha$$
$$P\square^\sigma \alpha \to P\square \alpha \to P_\square \alpha$$

The connections between $F_\square$ and $F^\triangleleft$ can be deduced from the definition of $F^\triangleleft$ in $\mathcal{L}^\sigma$. It is interesting to observe that the dual formula $G_\square \alpha$ is true at m_0 if and only if every horizon of m_0 contains a moment m such that, for some h, α is true at $\langle m, h \rangle$.

If we try to define von Kutschera's operators in $\mathcal{L}^\sigma$ or in $\mathcal{L}^\triangleleft$ we encounter the same kind of problems as we spoke about in connection with the definability of the operators of $\mathcal{L}^\sigma$ in $\mathcal{L}^\triangleleft$. Also in this case, however, propositional quantifiers would get us out of trouble, in fact the definitions of $F_\square$ and $P_\square$ in $\mathcal{L}^\sigma$ would be

$$F_\square \alpha \equiv_D \exists p(F\square^\sigma p^\# \wedge \square F(p^\# \wedge \alpha))$$

$$P_\square \alpha \equiv_D \exists p(P\square^\sigma p^\# \wedge \square P(p^\# \wedge \alpha))$$

and the definition of $\square^\sigma$ in term of $F_\square$ and $P_\square$ would be

$$\square^\sigma \alpha \equiv_D \exists p(p^\# \wedge HF_\square(p^\# \wedge \alpha))$$

The following propositions show that nothing better can be done (we shall only consider $F_\square$).

4.2 Proposition. *The operator $F_\square$ is not definable in $\mathcal{L}^\sigma$.*

Proof. Let $\langle \mathbf{T}, \sigma \rangle$ be the synchronized tree having the following properties. $\mathbf{T}$ consists of three histories, h_1, h_2, h_3, each isomorphic to the set of rational numbers. We will refer to the moments in $\mathbf{T}$ simply as rational numbers and indexes ranging over $\{1, 2, 3\}$ will specify the history to which they belong. Then x_2 denotes the moment in h_2 that corresponds to the rational number x in a fixed isomorphism between h_2 and the set of rational numbers. The synchronism σ is defined by: $\sigma_{h_i,h_j}(x_i) = x_j$. The structure of the tree is described by: 1) for $x \leq -1$ and $i,j \in \{1,2,3\}$, $x_i = x_j$, 2) for $-1 < x \leq 0$, $x_1 \neq x_2 = x_3$, 3) for $x > 0$ and $i \neq j$, $x_i \neq x_j$.

Let X and Y be dense subsets of Q such that $X \cup Y = Q - \{0\}$ and $X \cap Y = \emptyset$; let us denote by X^+ and X^- the sets $\{x : x \in X \text{ and } x > 0\}$, $\{x : x \in X \text{ and } x < 0\}$, respectively, and similarly for Y^+ and Y^-. The evaluations V and V' are defined by: 1) for $k > 0$, $i \in \{1, 2, 3\}$, and every $x \in Q$, $\langle x_i, h_i \rangle \in V(p_k) = V'(p_k)$, and 2)

$$V(p_0) = \{\langle x_1, h_1 \rangle : x \in Y^+\} \cup \{\langle x_2, h_2 \rangle : x \in X^+\} \cup$$
$$\cup \{\langle x_3, h_3 \rangle : x \in X^+\} \cup \{\langle x_i, h_i \rangle : x \in X^-, 1 \leq i \leq 3\}$$

$$V'(p_0) = \{\langle x_1, h_1 \rangle : x \in Y^+\} \cup \{\langle x_2, h_2 \rangle : x \in X^+\} \cup$$
$$\cup \{\langle x_3, h_3 \rangle : x \in Y^+\} \cup \{\langle x_i, h_i \rangle : x \in X^-, 1 \leq i \leq 3\}$$

Then, for $m_0 = 0_2 (= 0_3)$, $V_{m_0}(F_\square p_0) = 1$ and $V'_{m_0}(F_\square p_0) = 0$. We will prove that, for every $\mathcal{L}^\sigma$-formula α and $h \in \{h_2, h_3\}$, $V_{\langle m_0, h \rangle}(\alpha) = V'_{\langle m_0, h \rangle}(\alpha)$, so that $F_\square p_0$ cannot be a formula of $\mathcal{L}^\sigma$.

Both $X^- \cup Y^+$ and $Q - (X^- \cup Y^+)$ are dense in Q and hence Lemma 3.2 can be applied to the histories h_1, h_2, h_3 (linearly ordered by $\prec$) with $X^+ \cup Y^-$ or $X^- \cup Y^+$ playing the role of the set Y in that lemma. For instance, this role is played by $X^- \cup Y^+$ in the history h_3 with the evaluation V', that is, for all $x, y \in X^- \cup Y^+$ and every purely temporal formula α, $V'_{\langle x_3, h_3 \rangle}(\alpha) = V'_{\langle y_3, h_3 \rangle}(\alpha)$, and similarly for $x, y \in X^+ \cup Y^-$. Consider any isomorphism f of Q onto itself, which maps X^+ onto Y^+, Y^+ onto X^+, and is the identity on the rational numbers ≤ 0. Viewed as an isomorphism of h_3 onto itself, f maps the evaluation V to V', that is, $V_{\langle x_3, h_3 \rangle}(p) = V'_{\langle f(x)_3, h_3 \rangle}(p)$ for all $x \in Q$ and every propositional variable p, and hence this equality holds for all purely temporal formulas.

Putting the above facts together, we have that the equalities (a) and (b) below hold for every purely temporal formula α; the distinction between the cases $x, y > 0$ and $x, y < 0$ is due to the particular purposes of this step of the proof.

(a)
$$V_{\langle y_1, h_1 \rangle}(\alpha) = V_{\langle x_2, h_2 \rangle}(\alpha) = V_{\langle x_3, h_3 \rangle}(\alpha) =$$
$$= V'_{\langle y_1, h_1 \rangle}(\alpha) = V'_{\langle x_2, h_2 \rangle}(\alpha) = V'_{\langle y_3, h_3 \rangle}(\alpha)$$

(b)
$$V_{\langle y_i, h_i \rangle}(\alpha) = V_{\langle x_j, h_j \rangle}(\alpha) = V'_{\langle x_j, h_j \rangle}(\alpha) = V'_{\langle y_i, h_i \rangle}(\alpha)$$

where, in (a), x and y are arbitrary rational numbers such that either $x \in X^+$ and $y \in Y^+$ or $y \in X^+$ and $x \in Y^+$ and, in (b), x and y are both in X^- or both in Y^-, and $i, j \in \{1, 2, 3\}$. Moreover, for $i, j \in \{1, 2, 3\}$,

(c)
$$V_{\langle 0_i, h_i \rangle}(\alpha) = V'_{\langle 0_j, h_j \rangle}(\alpha)$$

Now we prove that the restriction to purely temporal formulas in (a), (b) and (c) can be removed; this will conclude the proof because the thesis is that (c) holds for all $\mathcal{L}^\sigma$-formula. We assume that (a), (b) and (c) hold for $\alpha = \beta$ and for $\alpha = \gamma$ and we prove that they hold for $\alpha = \neg\beta$ and for $\alpha = \beta \wedge \gamma$, and so on. The only interesting cases are $\alpha = \square\beta$ and $\alpha = \square^\sigma\beta$. It is worth noticing, however, that (c) cannot be included in (b) because of the case $\alpha = F\beta$: the truth of $F\beta$ at some $\langle x_i, h_i\rangle$ with $x < 0$ does not guarantee that this formula is true also at $\langle 0_i, h_i\rangle$, an example is $F\square^\sigma p_0$.

The case in which $\alpha = \square\beta$ is easy; we have only to observe that, under both V and V', the modal operator is vacuous. This is trivial if we consider pairs $\langle x_i, h_i\rangle$ with $x > 0$, because any such x_i belongs to only one history. For $x \leq 0$, we can use the inductive hypothesis. The inductive hypothesis gives also that $\square^\sigma$ is vacuous for $x \leq 0$.

Assume now that α is $\square^\sigma\beta$ and that $V_{\langle x_i, h_i\rangle}(\alpha) = 1$ with $x > 0$. Observe first that the isomorphism f is the complement of σ_{h_i, h_j} in the sense that $\sigma_{h_i, h_j}(x_i) \in X^+$ if and only if $f(x) \in Y^+$. Then, by (a), β is true at all $\langle x_j, h_j\rangle$ with $x > 0$ and hence this holds also for $\square^\sigma \beta$; in particular (a) is true. ∎

4.3 Proposition. *The operator $\square^\sigma$ is not definable in $\mathcal{L} + \{F_\square, P_\square\}$.*

Proof. Consider the synchronized tree $\langle \mathbf{T}, \sigma\rangle$ consisting of two histories, h_1 and h_2, with the following properties. $h_1 \cap h_2$ is $\{m_0\}$, and both $h_1 - h_2$ and $h_2 - h_1$ are isomorphic to the set Q of rational numbers. As in proposition 4.2, we shall refer to the elements of $h_1 - h_2$ or of $h_2 - h_1$ as rational number and the indexes 1 or 2 will specify the history to which they belong; we also assume $\sigma_{h_1, h_2}(x_1) = x_2$.

Let X be a subset of Q such that both X and $Q - X$ are dense in Q and let the evaluations V and V' be defined by: 1) for $i > 0$, $V(p_i)$ and $V'(p_i)$ coincide and are the set of all pairs $\langle m, h\rangle$ in $\mathbf{T}$, 2) $V(p_0) = \{\langle x_i, h_i\rangle : i \in \{1, 2\}$ and $x \in X\}$, 3) $V'(p_0) = \{\langle x_1, h_1\rangle : x \in X\} \cup \{\langle x_2, h_2\rangle : x > 0$ and $x \in X\} \cup \{\langle x_2, h_2\rangle : x \leq 0$ and $x \in Q - X\}$.

We have $V_{\langle 0_1, h_1\rangle}(\square^\sigma p_0) = 1$ and $V'_{\langle 0_1, h_1\rangle}(\square^\sigma p_0) = 0$ and hence we can derive the thesis by proving that, for all $\langle m, h\rangle$ in $\mathbf{T}$ and every formula α of $\mathcal{L} + \{F_\square, P_\square\}$, $V_{\langle m, h\rangle}(\alpha) = V'_{\langle m, h\rangle}(\alpha)$. The technique of the proof is quite similar to that of previous theorem and it is a further use of Lemma 3.2. We first observe that the operators $F_\square$ and $P_\square$ coincide with F and P on the pairs $\langle m, h\rangle$ with $m \neq m_0$ and that $V_{m_0}(P_\square \alpha) = 0 = V'_{m_0}(P_\square \alpha)$ for all formulas α. Moreover, if $V_{m_0}(F_\square \alpha) = 1$, then we have $V_{\langle x_1, h_1\rangle}(\alpha) = 1 = V_{\langle x_2, h_2\rangle}(\alpha)$ for infinitely many x covering a dense subset of Q and in particular for some $x > 0$; for these x, we have also $V'_{\langle x_1, h_1\rangle}(\alpha) = 1 = V'_{\langle x_2, h_2\rangle}(\alpha)$ and hence $V'_{m_0}(F_\square \alpha) = 1$. If instead $V'_{\langle x_1, h_1\rangle}(\alpha) = 1 = V'_{\langle x_2, h_2\rangle}(\alpha)$, we can consider two cases, namely: $x > 0$ and $x \leq 0$. In the first case, $V_{m_0}(F_\square \alpha) = 1$ is a consequence of an inductive hypothesis. In the second one, we can use the same technique as that of Proposition 4.2 to prove that $V'_{\langle y_1, h_1\rangle}(\alpha) = 1 = V'_{\langle x_2, h_2\rangle}(\alpha)$ for all rational x and y, and hence $V_{m_0}(F_\square \alpha) = 1$ holds again. ∎

5. Concluding remarks.

In section 2, we have seen what are the minimal adjustments to be made on the tree-like structures in order to make sense of those English sentences which, like (0.1) and (0.2), express a temporal comparison between moments in different histories. As we have seen, these adjustments (adding to the semantics either the relation $\leftrightarrow$ or the relation $\triangleleft$) all amount to establishing an isomorphism among the various histories of a tree, in that $\langle \mathbf{T}, \sigma \rangle$, $\langle \mathbf{T}, \leftrightarrow \rangle$, $\langle \mathbf{T}, \triangleleft \rangle$ are the same mathematical entity.

Given a tree with isomorphic histories, we have seen that

1. a language whose only undefined operators are $\diamond, F, P$, is not capable of expressing any temporal relation between moments in different histories;
2. a language with operators $\diamond, F, P, F^\triangleleft, P^\triangleleft$ is capable of expressing temporal precedence between moments in different histories, but it can express across-history simultaneity only on very rare occasions, i.e., when, for some α, $\neg F^\triangleleft \alpha \wedge \neg P^\triangleleft \alpha \wedge \alpha$ is satisfied;
3. a language whose operators are $\diamond, \diamond^\sigma, F, P$ can express both temporal precedence and simultaneity between events in different histories.

Moreover, sentences 0.1 and 0.2 can also be expressed by means of the operators $F_\square$ and $P_\square$ which, as we have seen, are interpreted on trees which might be only partially synchronized.

We saw that $F^\triangleleft$ and $P^\triangleleft$ are definable by means of $\diamond^\sigma$, F and P; all the other operators are not inter-definable. It is important to observe that all the undefinability results would fail if propositional quantifiers were allowed. Of course, adding propositional quantifiers to $\mathcal{L}$ would have deep consequences on the axiomatizability and the decidability of the various notions of validity. This means that the responsibility for the impossibility of defining , for instance, $\diamond^\sigma$ by means of $F^\triangleleft$ and $P^\triangleleft$ is not to be ascribed to the weak expressive power of these operators, but to the inner features of propositional tense logic. As a matter of fact, if we metalinguistically assumed that every event occurs only once on each history, we would have that all the languages considered in this paper have the same expressive power.

Thus, the semantics considered in the previous sections are capable of comparing moments belonging to different histories and some of the languages we have studied are capable of expressing such comparisons. Does this mean that English sentences like (0.1) and (0.2) are fully legitimate? The answer is only partially positive. In fact, as we have seen, we can make sense of sentences like (0.1) and (0.2) only if an isomorphism is chosen among the various histories belonging to a tree. Unfortunately, though, it very often happens that there is more than one isomorphism among the histories of a tree. Thus, with reference to example (0.1), it might very well happen that (0.1) is true relative to a pair $\langle \mathbf{T}, \sigma \rangle$, because, say, relative to isomorphism σ the sentences "I arrive by car" and "I arrive by bus" have simultaneous occurrences in all the histories in which they hold, but (0.1) might be false relative to the pair $\langle \mathbf{T}, \sigma' \rangle$ where $\mathbf{T}$ is as in $\langle \mathbf{T}, \sigma \rangle$, but σ' is an isomorphism which is different from σ and such that, for some histories h and h', "I arrive by car" and "I arrive by bus" are not simultaneous, that is, letting $m \in h$ be the moment in which "I arrive by car" holds, and $m' \in h'$ be the moment in which "I arrive by bus holds", $\sigma'_{h,h'}(m) \neq m'$.

Thus, we can say that sentences like (0.1) are fully legitimate only if the two events spoken about are simultaneous relative to all the isomorphisms among the histories of the given tree (a particular case of this situation is when the isomorphism among the various histories is unique). Probably, though, (0.1) (and the majority of English sentences like (0.1)) does not fall under the above case. If, as it generally happens, we can make sense of them and we can assess their truth-values, it is only because we have a particular, privileged isomorphism in mind relative to which we understand and evaluate them. The extent to which the choice of a privileged isomorphism is arbitrary is an issue which cannot be solved by a logical investigation.

The analysis of the notions of validity considered in the present paper is still studded with open problems. In [Gurevich and Shelah 85] it is proved that, under the hypothesis (1.1), the set of $\mathcal{L}$-formulas valid in the class of all, possibly unsynchronizable trees is decidable, and hence axiomatizable, but no effective axiomatization has been found yet. Apparently, Gurevich's and Shelah's theorem is the only result we have about trees.

The situation looks much better if the semantics for $\mathcal{L}$ is based on bundled trees $\langle \mathbf{T}, H \rangle$ (see section 1) rather than on trees. In [Burgess 79] the set of $\mathcal{L}$-formulas valid in these structures is proved to be decidable and an effective finite axiomatization of this notion of validity is given in [Zanardo 85]. Another axiomatization is quoted in [Thomason 84] and a very simple version of it is given in [Gabbay et al. 94, ch.7]; this latter axiomatization concerns the $\mathcal{L}$-formulas valid under (1.1) and it uses the *Gabbay Irreflexivity Rule* [Gabbay 81].

The sets of $\mathcal{L}$-formulas and of $\mathcal{L}^\sigma$- formulas valid in all synchronized bundled trees are axiomatizable because any synchronized bundled tree is isomorphic to a Thomasons $T \times W$-frame and these latter frames are first order definable [Thomason 84]; no (possibly infinite) axiomatization is known, however. Work in progress makes us confident that such axiomatization can be found. In the case of $\mathcal{L}$-formulas, we think that the axiomatization should consist of infinitely many schemata of the form $\alpha \rightarrow \Box \alpha'$, where α and α' describe, even if only partially, the topological structure of time in the future of the moment at hand. In the case of $\mathcal{L}^\sigma$-formulas, we think that axioms analogous to those of [Zanardo 85] could allow for a finite axiomatization of this notion of validity.

References.

[Belnap & Perloff 90] N. Belnap, M. Perloff, 'Seeing to it that: a canonical form of agentives', in Knowledge representation and defeasible reasonings, H. E. Kyburg, Jr., R. P. Loui and G. N. Carlson (eds.), Kluver Academic Publisher, 1990, pp.175-199.

[Burgess 79] Burgess, J., 'Logic and Time', Journal of Symbolic Logic, 44, 1979, pp. 556-582.

[van Frassen 89] van Frassen, B., Laws and Symmetry, Clarendon Press, Oxford, 1989

[Gabbay et al. 94] D. Gabbay, I. Hodkinson and M. Reynolds, Temporal Logic: Mathematical Foundation and Computational Aspects, Oxford University Press, Oxford, 1994.

[Gabbay 81] Gabbay, D., 'An Irreflexivity Lemma with Applications to Axiomatizations of Conditions on Tense Frames' , in U. Mönnich (ed.), Aspects of Philosophical Logic, D. Reidel, Dordrecht (1981), 67-89.

[Gurevich & Shelah 85] Y. Gurevich, S. Shelah, 'The decision problem for branching time logic', J. of Symbolic Logic, 50 (3) 1985, pp. 668-681.

[Prior 67] Prior, A., Past, Present and Future, Clarendon, Oxford, 1967.

[Thomason 84] Thomason, R., 'Combinations of tense and modality', in D. Gabbay and F. Guenthner (eds.), The Handbook of Philosophical Logic, vol. 2, D. Reidel, Dordrecht (1984), 135-165.

[Zanardo 85] Zanardo, A., 'A Finite Axiomatization of the Set of Strongly Valid Ockhamist Formulas', Journal of Philosophical Logic, 14, 1985, pp. 447-468.

[Zanardo 86] Zanardo, A., 'On the Characterizability of the Frames for the Unpreventability of the Present and the Past', Notre Dame J. of Formal Logic, 1986, 556-564.

[Zanardo 94] Zanardo, A., 'Branching-time logic with quantification over branches. The point of view of modal logic', draft.

On the Completeness of Temporal Database Query Languages

Michael Böhlen and Robert Marti

Institut für Informationssysteme, ETH Zürich
8092 Zürich, Switzerland
Email: (boehlen, marti)@inf.ethz.ch

Abstract. In this paper, we introduce a new definition of completeness for temporal query languages based on the relational model. Our definition relies on the following three notions: relational completeness of non-temporal queries as defined by Codd; the preservation of temporal irreducibility of temporal (valid time) relations, be they stored or returned as results of temporal queries; and the notion of temporal equivalence between temporal and non-temporal queries. Particularly important is the notion of temporal irreducibility which requires that the valid time intervals of two tuples with the same data values must not touch or overlap, since unreduced relations generate incorrect answers to certain types of temporal queries. Finally, we introduce the query language ChronoLog which is a temporally complete extension of Datalog.

1 Introduction

In the past decade, a large number of temporal data models and query languages have been proposed by the database community (see e.g. [Soo91, TCG+93]). The same is true of languages for knowledge representation and temporal reasoning in the realm of artificial intelligence (see e.g. [Rei89]).

A cornerstone of any temporal data model is the representation of time itself. Since ultimately all representations of numeric values on digital computers (integers, rationals, as well as fixed-point and floating-point "reals") are by necessity discrete, we have opted to model time points as integers or fixed point-numbers, depending on the granularity required by the application. Such a value represents the number of seconds which have passed since some reference time point. Moreover, we have decided to use time intervals rather than time points as the basis for modeling the time during which a fact in the real world is perceived to hold. Although point-based models have been shown to be equivalent to interval-based ones in theory, we find that in practice, the interval-based model is more practical for most applications in that it leads to more compact internal representations.

In the area of temporal databases, people typically distinguish between the notions of valid time and transaction time. The *valid time* of some piece of information (e.g. a fact) denotes the time during which this fact is perceived to be true in the real world. Conversely, the *transaction time* of a fact denotes the time during which this fact was recorded in the database. Data models or database systems which support both notions of time are called *bitemporal*. While our research has accommodated both concepts, we will restrict ourselves to valid time in this paper.

Most approaches to manage temporal databases are based on the relational model of data, in particular on first normal form (1NF). As [JSS93] point out, the main reason for this is the simplicity of the relational model. Given the amazing variety of temporal extensions of the relational model, they propose a bitemporal conceptual data model (BCDM) as the least common denominator of the proposed temporal extensions to the relational model. Their BCDM essentially associates complex bitemporal timestamps to each "ordinary" (i.e., non-temporal) tuple. (We remark in passing that this is essentially the same approach as the method of temporal arguments of [Hau87].)

Since we only deal with the aspect of valid time in this paper, the complex timestamps are sets of intervals which represent the valid times of the tuples to which the timestamps are associated. Such timestamps cannot be represented directly in a 1NF relation, though. As in [JSS93], we map the conceptual data model with complex timestamps to a representational data model using single intervals (pairs of integers or fixed-point numbers) as timestamps. As it turns out, there are several ways in which this can be done. However, in order to ensure correct answers to temporal queries, it is important to use a representation in which the time intervals associated with different tuples representing the same fact in the real world at different times do not touch or overlap.

This representation, called the temporally irreducible form of a relation, is introduced in chapter 2.1, together with the concept of the temporal equivalent of a non-temporal relational query. These two definitions are subsequently used to define two interesting classes of temporal query languages, namely, temporally semi-complete and temporally complete languages. In chapter 3, we present the language ChronoLog, which is a temporally complete extension of Datalog. In chapter 4, we review two alternative definitions of completeness for temporal languages due to [CCT93], namely TU-completeness and TG-completeness, and compare our definition to theirs.

2 Completeness of Temporal Query Languages

Before we discuss the completeness of temporal query languages, we have to introduce two auxiliary notions. The first one is *temporal irreducibility* which requires that in a relation, no two value-equivalent tuples [JSS93] exist which overlap or touch in their valid time arguments. The second notion is that of a *snapshot*. Snapshots, a notion that has been used by other authors [Gad86, GY88, Sno93], are needed to define the concept of *temporally equivalent* queries. The final two subsections discuss *temporal semi-completeness* and *temporal completeness*. We will see that temporal semi-completeness is strongly related to the notion of relational completeness that has been coined by [Cod72] and which has been used as a useful metric for the expressive power of database query languages for a long time.

2.1 Temporal Irreducibility of Conceptual Valid Time Relations

Following [JSS93], the basis of our data model and query language is the bitemporal conceptual data model restricted to valid time. Thus, each valid time relation possesses a distinguished attribute *ValidTime* which indicates when the facts in the real world represented by its tuples were considered to be true. On a conceptual level, the values of the *ValidTime* attribute are (finite) unions of time intervals. Hence, a tuple r of an n-ary conceptual valid time relation R has the form

$$r(v_1, \ldots, v_n, \{[t_{b_1}, t_{e_1}), [t_{b_2}, t_{e_2}), \ldots, [t_{b_m}, t_{e_m})\})$$

where the t_{b_j} and $t_{e_j} (1 \leq j \leq m)$ are constants of type time point and the $v_i, 1 \leq i \leq n$ are data values, i.e., constants of type string or number.
Unfortunately, the values of the *ValidTime* attribute cannot be directly represented in the above form using 1NF relations. As a result, we have to decide on a suitable normalized representation. This can be achieved by representing the above tuple as the following collection of tuples:

r

a_1	...	a_n	ValidTime
...	...	...	...
v_1	...	v_n	$[t_{b_1} - t_{e_1})$
v_1	...	v_n	$[t_{b_2} - t_{e_2})$
v_1	...	v_n	...
v_1	...	v_n	$[t_{b_m} - t_{e_m})$
...	...	...	...

As we will see shortly, it is important that the values of t_{b_j} and t_{e_j} in a valid time relation fulfill the following condition: For given data values $v_1, \ldots, v_n$, there must not be any adjacent or overlapping intervals, i.e., for all pairs $j, k (i \neq k)$

$$r(v_1, \ldots, v_n, [t_{b_j}, t_{e_j})) \wedge r(v_1, \ldots, v_n, [t_{b_k}, t_{e_k})) \rightarrow t_{e_j} < t_{b_k} \vee t_{e_k} < t_{b_j}$$

Definition 1. A valid time relation which fulfills the above condition is said to be *temporally irreducible*.

Failing to represent valid time relations in temporally reduced form results in incorrect answers for many queries relating to time. This is demonstrated by the following two simple example queries which for the time being are formulated in "ordinary" first order predicate calculus (similar to Prolog). We assume that the database consists of the following valid time relation[1] which is not in temporally irreducible form:

---
[1] In this and the following examples, we assume a time granularity of a day.

employee

Name	Dept	ValidTime
John	sales	[1988/3/1-1991/1/1]
John	research	[1991/1/1-1992/6/1]
John	research	[1992/6/1-1994/1/1]

Example 1. Who joined the research department after January 1, 1992 and when was this?

$$?-\ \exists T(employee(Person, \text{'research'}, T) \land begin(T) \geq 1992/1/1).$$

This query returns the answer `Person='John'` despite the fact that John had already been working for the research department *before* January 1, 1992.

Therefore, we require that all valid time relations are always temporally irreducible:

employee

Name	Dept	ValidTime
John	sales	[1988/3/1-1991/1/1]
John	research	[1991/1/1-1994/1/1]

Of course, this means that all update operations have to preserve this representation (see [Böh94]). As it turns out, this alone is not good enough: There are queries which compute answer relations which are *not* irreducible. Such answers may be confusing or even misleading for a casual user. More importantly, incorrect answers may be returned if such a query expression is part of a query expression.

Example 2. Who was hired after June 1, 1990 and when was this?

$$?-\ \exists Dept\ employee(Person, Dept, T) \land begin(T) >= 1990/6/1.$$

Again, this query returns John, although he had been continuously working for the company since March 1, 1988. The reason for this is that the "subquery"

$$\exists Dept\ employee(Person, Dept, T)$$

returns a valid time relation – let's call it *is_employed* – which is *not* irreducible:

is_employed

Name	ValidTime
John	[1988/3/1-1991/1/1]
John	[1991/1/1-1994/1/1]

Therefore, if we proceed to select the tuples for which $begin(T) \geq 1990/6/1$ holds, the second tuple will be returned. If the valid time relation *is_employed* had been properly represented, this would not have been the case since the two time intervals would have been merged into a single tuple, namely is_employed

Name	ValidTime
John	[1988/3/1-1994/1/1)

In order to convert a valid-time relation R into its irreducible representation, we have introduced the *temporal reduction operator* ρ. The tuples $r_{red}(Value, Time)$ belonging to the temporal reduction $R_{red} := \rho(R)$ of a relation R can be computed by the following set of rules:

$$r_{red}(V,T) \leftarrow r_{aux}(V,T) \wedge \neg \exists T'(r_{aux}(V,T') \wedge T' \supset T)$$
$$r_{aux}(V,T) \leftarrow r(V,T)$$
$$r_{aux}(V,T) \leftarrow r(V,T_1) \wedge r_{aux}(V,T_2) \wedge T = T_1 \cap T_2 \wedge T \neq \emptyset$$

For an efficient implementation of these rules (using in-place updates) see [BM92].

2.2 Snapshots

Definition 2. The *valid time slice* or *snapshot* $\tau_t(R)$ of a valid time relation R at time point t is a relation R_t which contains all tuples valid at time t without the associated timestamp. More formally, the snapshot $\tau_t(R)$ can be defined in first order predicate calculus the result of evaluating the query

$$\exists T(r(A_1,\ldots,A_n,T) \wedge begin(T) \leq t < end(T)).$$

If the snapshot operation is applied to a whole set of relations $\{R_1,\ldots,R_n\}$ the result consists of the respective snapshots of all relations contained in the set, i.e. $\{\tau_t(R_1),\ldots,\tau_t(R_n)\}$.

The snapshot operation is useful for theoretical considerations because it transforms a set of temporal relations into a set of nontemporal ones. These nontemporal relations can be processed with standard relational techniques. Therefore, we can define temporal operations on a valid time database in terms of nontemporal operations on the set of all possible snapshots. Note however that implementing temporal operations on the basis of snapshots is not practical since it quickly leads to an enormous volume of data. Also, the scheme does not account for temporal built-in predicates (e.g. **overlaps**, **during**, **before**, etc., see [All83]), because they have no nontemporal counterparts.

With these limitations in mind, we use snapshots in order to discuss and explain the completeness of temporal query languages.

Example 3. Let **staff** be the following valid time relation:

staff

Project	Name	ValidTime
project1	Mike	[1990/5-1992/2)
project1	Tom	[1991/1-1992/5)
project1	Bill	[1989/1-1989/5)
project2	Jane	[1992/2-)

The snapshot of staff as of March 18, 1992 (i.e. $\tau_{1992/3/18} staff$) is the (non-temporal) relation

Project	Name
project1	Tom
project2	Jane

2.3 Temporally Semi-Complete Languages

We start with defining the notion of *temporally equivalent* queries. Clearly, non-temporal queries are evaluated over snapshot relations while temporal queries are evaluated over valid time relations. (Note that valid time relations subsume snapshot relations in that snapshot relations can be viewed as valid time relations which appear to hold over the entire time axis. Therefore, if we consider temporal queries we do not have to take into account snapshot relations.)

Definition 3. Let R' be a set of valid time relations, Q' a temporal query, Q a non-temporal relational query, and $Res' := Q'(R')$ the result of applying the query Q' to the set of valid time relations R'. Then, Q' is called a *temporal equivalent* of Q if and only if for all possible time points t, the time-slice $\tau_t(Res')$ is equal to the result of applying the query Q to the time-slices $\tau_t(R')$, i.e., $\tau_t(Q'(R')) = Q(\tau_t(R'))$ (see figure 1).

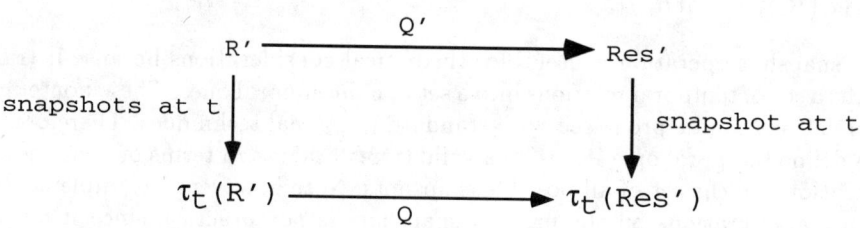

Fig. 1. Graphical illustration of *temporally equivalent* queries

Note that, depending on the actual contents of the valid time relations and the given non-temporal query, there may be more than one temporal query which is temporally equivalent to a given query. However, these queries only differ in their treatment of valid time attributes which disappear when snapshots are taken: All but one of these temporally equivalent queries produce valid time relations in which there are value equivalent tuples with overlapping or adjacent time intervals. Therefore, we require that the temporal equivalent should produce an irreducible valid time relation. This requirement is embodied in the definition of temporal semi-completeness.

Definition 4. A temporal query language is *temporally semi-complete* if, for every relational query Q, the temporal query language allows the formulation of a temporally equivalent temporal query Q' which yields a valid time relation *in temporally irreducible form*.

A temporally semi-complete language is closed since all queries operate on temporally irreducible relations and by definition return temporally irreducible relations.

More importantly, we emphasize that the notion of temporally semi-complete languages serves as a useful guideline for the design of a temporal extension of an existing (relationally complete) query language: A simple syntactic extension to "construct" the temporal equivalent of a non-temporal query makes it possible to not only retrieve information about the current perceived state of the world (as in conventional snapshot databases), but also about the perceived state of the world at any other point in time. Indeed, the language ChronoLog which is introduced in section 3.3 is essentially such a simple temporal extension of Datalog.[2] Such a language is also intuitive for the user: He or she simply decides if a given query formulated in some non-temporal query language should be evaluated in a temporal or non-temporal fashion.

2.4 Temporally Complete Languages

Temporally semi-complete languages constitute an interesting and powerful class of temporal query languages. However, they are limited in that they do not allow the formulation of queries which retrieve information concerning relationships between perceived states of the world at *different* points in time. In his seminal work on representing and reasoning with temporal knowledge, Allen has identified thirteen possible relationships between time intervals [All83], namely *equals, overlaps, overlapped_by, contains, during, precedes, follows, before, after, starts, ends, started_by* and *ended_by*. Temporally semi-complete languages implicitly only account for the first five of them. Moreover, they do not make a distinction between them, treating all cases as T_1 *overlaps* $T_2 \vee T_1$ *overlapped_by* T_2.

If we would like to support all of these temporal relationships explicitly, we have to extend the definition of semi-completeness as follows:

Definition 5. A temporal query language is *temporally complete* if

- a temporally semi-complete formula may be used as an operand in a query expression wherever a valid time relation is permitted
- it supports the thirteen temporal relationships as defined in [All83] between (1) valid time attributes associated with stored valid time relations, (2) implicitly computed valid time attributes associated with semi-complete formulas, and (3) temporal interval constants.

---

[2] There is no reason why the same technique could not be used to design a temporal extension of SQL, for example. Indeed, the design of such a language – to be called ChronoSQL – is currently underway in a joint project with the Union Bank of Switzerland.

It should be obvious that the temporal equivalent of a relational query defines a true temporal relation. From relational theory we know that the result of a query is a relation again. Moreover, according to definition 3, the temporal equivalent of a nontemporal query timestamps every tuple with a valid time argument. Thus, a temporal semi-complete query indeed yields a temporal relation. In contrast to a temporally semi-complete language, a temporally complete language must provide a possibility to use this relation within an arbitrary temporal relational algebra expression. Furthermore, we require the thirteen temporal built-in predicates that have been defined by [All83]. The requirement for these predicates should be obvious as Allen has proved that they exhaustively describe the relationships between two intervals.

3 The Temporal Query Language ChronoLog

3.1 Syntax

The syntax of our temporal query language ChronoLog can be defined as a many-sorted first order logic (see e.g. [HT92, Bur92]). In the following, we will use the term *type* instead of *sort*.

Besides the standard types *integer*, *real* and *string*, we introduce a *time* type to timestamp extensional knowledge. As stated before, we restrict ourselves to *time intervals representing valid time*. However, it would also be possible to choose bitemporal elements [JSS93], bitemporal rectangles [Böh94] or temporal elements [Gad88] instead. A time interval is specified as an ordered pair of time points which denote the lower and upper bound of the interval (start and end point). Time points can be specified to an arbitrary accuracy, but there is an implementation dependent maximum accuracy supported by the system.[3]

As usual, a *term* is a constant or a variable. A term is called temporal if either the constant or the variable is temporal. If $t_1, ..., t_N$ are terms, T is a temporal term and p is an n-ary predicate, then $p(t_1, ..., t_N)@T$ is a *temporal atom*. The meaning of such an atom is that $p(t_1, ..., t_N)$ is valid during the time interval T. T is also called the valid time argument of predicate p. The valid time argument can be existentially quantified like any other argument. In addition, it is possible to omit it for user convenience: Instead of $\exists T(sick(Emp)@T)$ we can simply write $sick(Emp)$.

Besides the (nontemporal) built-in predicates ($<, >, \leq, \geq, =, \neq$) we provide the thirteen *temporal built-in predicates* (*before, after, ...*) that have been proposed and motivated by [All83].

Every temporal atom is a *temporal formula*. Furthermore, if F_1 and F_2 are temporal formulas, x is a variable and T is a temporal term then $F_1 \wedge F_2$, $F_1 \vee F_2$, $\neg F_1$, $F_1 \rightarrow F_2$, $\exists x F_1$, $\forall x F_1$ and $\{F_1\}@T$ are temporal formulas as well. The curly braces in $\{F_1\}@T$ indicate that F_1 has to be evaluated temporally. As a result,

---

[3] It is important to understand that the behaviour of the system is independent of the chosen accuracy. In particular, runtime performance does not suffer even if a very fine time granularity (e.g. nano seconds) is supported.

the temporal term T is unified with the valid time of the formula. Again, T can be simply omitted instead of being existentially quantified. Hence, $\{F\}$ and $\exists T(\{F\}@T)$ are logically equivalent.

A *temporal deductive database* consists of a finite set of temporal facts and temporal rules. A *temporal fact* is a temporal ground atom while a *temporal rule* has the form $H \leftarrow B$ where H is an atom and B is a temporal formula. We call H the *head* and B the *body* of the rule. A *temporal query* has the form ?– Q, where Q is a temporal formula.

3.2 Semantics

The declarative semantics of relational or deductive databases (e.g., deductive databases consisting of Datalog formulas) is defined by assigning truth valuations for all atomic formulas that can possibly be constructed out of the predicate symbols and constants appearing in the database. Typically, the meaning of a (stratified) database (or "program") is defined as the minimal (Herbrand) model of the program which can be computed by determining the least fixpoint of the program operator T_P (see e.g. [Llo87, NT89, CGT90] for a discussion of these terms).

The semantics of formulas which are not subject to temporal normalization (i.e., formulas which are not surrounded by curly braces) are defined in the usual way in the sense that all temporal arguments are treated as if they were "ordinary" arguments: $I(p(X)@T) := I(p(X,T))$.

Temporal formulas of the form $\{\phi\}@[t_b - t_e)$ are interpreted as being true if and only if the following conditions hold:

1. the formula ϕ is true at all time points t with $t_b \leq t < t_e$, and
2. ϕ is false at time point t_e, and
3. ϕ is false at time point preceding t_b.

The second and third condition ensure that the interval $[t_b - t_e)$ is maximal, i.e., the result of a temporal formula is guaranteed to be temporally irreducible. This temporal interpretation is based on a set of interpretations I_t which formalize the notion of "true at time point t":

$I_t(p(\mathbf{x})@[t_b - t_e))$ is true if $p(\mathbf{x})@[t_b - t_e)$ is true and $t_b \leq t < t_e$
$I_t(p(\mathbf{x}))$ is true if $I_t(p(\mathbf{x})@[t_b - t_e))$ is true for some t_b, t_e
$I_t(\phi \wedge \psi)$ is true if $I_t(\phi)$ and $I_t(\psi)$ are true
$I_t(\phi \vee \psi)$ is true if $I_t(\phi)$ and/or $I_t(\psi)$ are true
$I_t(\neg \phi)$ is true if $I_t(\phi)$ is false
$I_t(\exists x \phi)$ is true if $I_t(\phi[x/c])$ is true for some c
$I_t(\forall x \phi)$ is true if $I_t(\phi[x/c])$ is true for all c's

Another possibility to define the semantics of ChronoLog formulas, albeit in a more operational fashion, is a translation to one of the temporal algebras that have been proposed in the literature (see [MS91] for an overview and further pointers to the literature). Unfortunately, most of these temporal algebras are not sufficient for our purposes. In particular, they do not take into account temporal irreducibility. (A notable exception is the temporal algebra in [Sno93] which defines a total of twelve temporal operators. All operators are defined in terms of a tuple calculus and three auxiliary functions: *NotNull*, *Reduce* and *Apply*. However, this algebra is defined over heterogeneous relations with set valued timestamps at the attribute level.)

For this reason, we define our own temporal relational algebra to serve as a target language for the evaluation of ChronoLog queries. The features of our temporal relational algebra are the following: First, it relies on relations with tuples being timestamped with an interval. This allows a straightforward and efficient representation in relational databases. Second, we only introduce one additional operator, namely the temporal reduction operator ρ (see section 2.1). Finally, apart from temporal reduction, our algebra is directly executable by commercial relational database systems.[4] This has several benefits. For one, temporal database systems gain from the work that has been done in the area of relational database systems. Furthermore, due to the common base of temporal and nontemporal database systems, it is possible to smoothly migrate from nontemporal to temporal databases.

In table 1, the first column provides the ChronoLog formula, i.e. a temporal

ChronoLog Formula	Equivalent Temporal Relational Algebra Expression
$\{p(x) \wedge q(x,y)\}@T$	$\pi_{T,x,y}(\sigma_{T=p.vt \cap q.vt \wedge T \neq \emptyset}(P' \bowtie Q'))$
$\{p(x) \vee q(x)\}@T$	$\rho(\pi_{T,x}(\sigma_{T=p.vt}P' \cup \sigma_{T=q.vt}Q'))$
$\{p(x) \wedge \neg q(x)\}@T$	$\pi_{T=p.vt \setminus q.vt, x}(P' \bowtie Q')$
$\{p(x) \wedge F(x)\}@T$	$\pi_{p.vt,x}(\sigma_{F(x)}P')$
$\{\exists y\ p(x,y)\}@T$	$\rho(\pi_{p.vt,x}P')$

Table 1. Translation of temporal logical formulas to temporal relational algebra

logical formula while the second column sketches the translation to temporal relational algebra[5]. Note that the translation is kept rather informal. For example, instead of argument vectors we use predicates with one or two arguments only. In this way, attention is focussed much better on the valid time argument, the key point of the translation. A generalization to argument vectors can be done along the lines of [Mar91] where the translation of logical formulas to relational algebra is discussed.

---

[4] Temporal reduction can be implemented by iterating a SQL statement. The original implementation can be found in [BM93]. A more efficient solution is presented in [BM92].

[5] Temporal negation requires the subtraction of time intervals. It is well known that interval subtraction is not a closed operation (we may get zero, one or two intervals). A thorough discussion of this operation together with a SQL implementation is provided in [BM93].

3.3 Temporal Completeness in ChronoLog

After having discussed the syntax and semantics of ChronoLog we reconsider the issue of temporal completeness which has been investigated theoretically in section 2. In particular, we will have a look at two examples which show that queries which are formulated in a semi-complete subset of ChronoLog are straightforward extensions of Datalog queries. Every example starts with a natural language formulation of a nontemporal problem, the nontemporal ChronoLog solution (i.e. a first order predicate logic formula) and a tabular representation of its "input" and "output". Afterwards, we formulate the equivalent temporal problem. Again, we provide a natural language formulation, the ChronoLog solution and a tabular representation of its input and output.

Our first example deals with projection of relational algebra (existential quantification in first order logic). In the most radical form, all variables of a formula are existentially quantified and we end up with a yes/no query. The relation

staff

Project	Name
project1	Tom
project2	Jane

together with the query "Are there any persons assigned to project 1?" is an example of such a situation. In ChronoLog, the result can be obtained by issuing the query

```
?- exists Person staff('project1', Person).
```

Applied to the above *staff* relation, this query (obviously) returns **yes**. A temporal equivalent query operates on a corresponding relation extended with a valid time argument:

staff

Project	Name	ValidTime
project1	Mike	[1990/5 − 1992/2)
project1	Tom	[1991/1 − 1992/5)
project1	Bill	[1989/1 − 1989/5)
project2	Jane	[1992/2−)

The natural language formulation of the query resembles the original nontemporal one very much. Typically, we would also like to know *when* persons were assigned to the projects, i.e., the valid times associated with the different tuples. Instead of a yes/no query we formulate a query that returns those time intervals: "During which periods were some persons assigned to project 1?". In *ChronoLog*, the resemblance between the nontemporal and the temporal query is even stronger than in the English formulation. We simply enclose the nontemporal query in curly braces and append a variable to it so that it will be unified with the valid time of the formula:

```
?- {exists Person staff('project1', Person)}@ValidTime
```

Evaluation of this query returns the intuitively correct result

ValidTime
[1989/1 − 1989/5]
[1990/5 − 1992/5]

Note that it is necessary to coalesce the valid time of two tuples in order to get this result. Temporal relational query languages not supporting the temporal reduction ρ fail to answer this query correctly because coalescing tuples is out of the scope of relational algebra.

Our second example is a more complex one. It illustrates that *ChronoLog* scales up nicely. The (notational) overhead to handle the temporal dimension is very small and it does not grow with the complexity of the query. Given the relations *department* and *salary*,

department

Name	Dept
Mike	research
Bill	research
Tom	sales
Jane	sales
Tina	research

salary

Name	Amount
Mike	4300
Bill	3800
Tom	6000
Jane	5400
Tina	6700

we issue the query "Which researchers earn more than the best paid salesman?".

```
?- department(Person,'research') &
   salary(Person,Amount) &
   ~exists Person2,Amount2 (department(Person2,'sales') &
                            salary(Person2,Amount2) &
                            Amount2>Amount).
```

The answer is:

Name	Amount
Tina	6700

This query is rather complex in that only systems that support negation and quantification can answer it. Matter becomes even more complicated if we consider the problem in the presence of valid time: "Which researcher earned more than the best-paid salesman did at the same time and when was this?"

department

Name	Dept	ValidTime
Mike	research	[1991/1 − 1992/8]
Bill	research	[1991/1 − 1991/8]
Tom	sales	[1990/8 − 1992/7]
Jane	sales	[1991/6 − 1992/5]
Tina	research	[1991/11 − 1992/9]

salary

Name	Amount	ValidTime
Mike	4300	[1991/1 − 1991/8]
Mike	4300	[1991/8 − 1992/8]
Bill	3800	[1991/1 − 1991/8]
Tom	4600	[1990/8 − 1991/8]
Tom	5400	[1991/8 − 1992/1]
Tom	6000	[1992/1 − 1992/7]
Jane	5400	[1991/6 − 1992/5]
Tina	6700	[1991/11 − 1992/9]

In most of the proposed temporal query languages, this query cannot be expressed, since answering it requires the nontrivial intersection and subtraction of intervals. In *ChronoLog* we simply formulate the temporally equivalent query

```
?- {department(Person,'research') &
    salary(Person,Amount) &
    ~exists Person2,Amount2 (department(Person2,'sales') &
                             salary(Person2,Amount2) &
                             Amount2>Amount)}@ValidTime.
```

Again, we do not have to deal with the temporal dimension at all. Instead, we simply use curly braces in order to indicate that we want to evaluate the query with temporal semantics. The query yields three tuples:

Person	Amount	ValidTime
Bill	5700	[1991/1-1991/6]
Mike	6100	[1992/5-1992/8]
Tina	6100	[1992/5-1992/9]

It can easily be verified that no salesman was earning more than the respective researcher during the computed periods.

Finally, we want to consider an example of a ChronoLog query for which the semi-complete subset is not sufficient. Give are the two valid time relations *employee* and *manager*:

employee

Name	Salary	ValidTime
Mike	4000	[1993/5/1 − 1993/7/1]
Mike	4200	[1993/7/1 − 1994/5/1]
Rick	5250	[1993/8/1 − 1994/2/1]
Bob	4300	[1992/8/1 − 1993/2/1]
Bob	4600	[1993/2/1 − 1993/9/1]
Tom	5000	[1993/11/1 − 1994/5/1]

manager

Mgr	Emp	ValidTime
Bob	Mike	[1993/5/1 − 1993/8/1]
Rick	Mike	[1993/8/1 − 1993/11/1]
Tom	Mike	[1993/11/1 − 1994/5/1]

The question "What was the salary of Bob when he managed Mike and what was the salary of the person who managed Mike immediately after Bob?" is answered in ChronoLog by issuing the query

```
?- { employee('Bob',Sal_Bob) &
     manager('Bob','Mike') }@ValidTime1 &
   { manager(Who,'Mike') &
     employee(Who,Sal_Who) }@ValidTime2 &
   ValidTime2 follows ValidTime1.
```

Note that the query consists of two temporal subqueries. The first one determines Bob's salary at the time he managed Mike while the second determines the salaries and the names of Mike's managers. Finally, we put the two subqueries respectively the valid times of both subqueries in the required temporal relationship. The query yields the expected answer, namely

Sal_Bob	ValidTime1	Who	Sal_Who	ValidTime2
4600	[1993/5-1993/8]	Rick	5250	[1993/8-1993/11]

4 Related Completeness Notions

4.1 TU-Completeness and TG-Completeness

In the following, we briefly review two notions of completeness introduced by [CCT93]. They make a fundamental distinction between *temporally ungrouped* and *temporally grouped* data models. These two terms can be considered as synonyms for *non first normal form (N1NF)* and *first normal form (1NF)* respectively (see e.g. [KS91]). Figure 2 contrasts a temporally grouped relation against

EMPLOYEE			
NAME	DEPT	SALARY	ValidTime
Tom	Sales	20K	[0,3)
Tom	Sales	30K	[3,4)
Tom	Mktg	30K	[4,5)
Thomas	Mktg	27K	[5,6]
Juni	Acctng	28K	[2,6]
Ashley	Engrng	27K	[1,2)
Ashley	Engrng	30K	[2,3)
Ashley	Mktg	30K	[3,4)
Ashley	Engrng	35K	[5,6]

EMPLOYEE			
NAME	DEPT	SALARY	lifespan
[0,5) -> Tom [5,6] -> Thomas	[0,4) -> Sales [4,6] -> Mktg	[0,3) -> 20K [3,5) -> 30K [5,6] -> 27K	{0,1,2,3,4,5,6}
[2,6] -> Juni	[2,6] ->Acctng	[2,6] -> 28K	{2,3,4,5,6}
[1,4) -> Ashley [5,6] -> Ashley	[1,3) -> Engrng [3,4) -> Mktg [5,6] -> Engrng	[1,2) -> 27K [2,4) -> 30K [5,6] -> 35K	{1,2,3,5,6}

Fig. 2. A temporally ungrouped relation (above) and a temporally grouped relation (below) representing the same information

a temporally ungrouped relation to illustrate the difference between the two relation types. In both relations, exactly the same information content is represented. However, in the temporally grouped relation, all information concerning one object is grouped into a single tuple while in the temporally ungrouped relation, the information is in some sense scattered over the whole relation.

TU-Completeness is defined with respect to the calculi TL and TC. The calculus TL is based on a logic without explicit references to time (namely, US logic, a temporal logic with operators *since* and *until*) while the calculus TC is based on a logic with explicit time references (a two sorted first order logic where one sort is time). As [CCT93, p.510] point out, [Gab87] has proven that the two calculi are equivalent. Therefore, it is just a matter of taste whether we rely on a temporal logic with explicit references to time (first order approach) or whether we rely on a temporal logic with operators hiding time (modal approach).

TG-Completeness is defined with respect to a tuple-based historical calculus L_h which is defined in [CCT93, p.514ff]. L_h is a many sorted logic with variables over ordinary values, historical values, and times. It permits quantification over all three sorts of variables. Example 4, which is due to [CCT93, p.516], gives a rough impression of the calculus L_h. However, for a more thorough discussion the interested reader is referred to the original paper cited above.

Example 4. What are the name and the salary histories of those employees in the marketing department at time 6?

$$[e.NAME, e.SALARY : t]\ EMPLOYEE(e) \wedge t \in e.lifespan\ \wedge$$
$$\exists t_1(e.DEPT(T_1) = Mktg\ \wedge$$
$$t_1 \in e.lifespan \wedge t_1 = 6)$$

A language is called *TU-Complete* if it has the same expressive power as the calculi TL respectively TC. Alternatively, a language is called *TG-Complete* if it has the same expressive power as the calculus L_h. Clifford, Croker, and Tuzhilin show that a temporally ungrouped data model can be transformed into a temporally grouped data model by introducing so called *grouping attributes*. (A grouping attribute corresponds to an object identifier in object oriented databases, or to a surrogate in certain extended relational models.) Moreover, TU-Completeness is equivalent to TG-Completeness under this transformation.

4.2 Comparison with our Notion of Completeness

One difference between the notions of completeness in [CCT93] and ours concerns the grouping mechanism. Since our model is purely relational, it does not directly support the grouping of related tuples, i.e., tuples which describe properties of a single object in the real world. As a result, a good database design must include "artificial" identification keys (object identifiers) which may never be changed during the lifetime of an object nor be reused later for another object. This is admittedly a drawback of our model compared to the temporally grouped model of [CCT93]. However, this drawback not only applies to the temporal dimension, but to all situations where one has to deal with multiply valued attributes, and users of relational systems are well aware of this fact.

A more important difference is that in our approach, temporal reduction is at the center of evaluating temporal queries. This is not the case in their approaches. As we have pointed out before, failure to temporally reduce answers (or even intermediate results) can lead to incorrect answers. Such a reduction merges value equivalent tuples with touching and overlapping time intervals. As we will illustrate below, this process is not necessarily restricted to tuples pertaining to a single object and therefore out of the scope of TU- and TG-complete languages. This becomes clear if we remember that the calculi TC, TL, and L_h are all *tuple based calculi* as pointed out in [CCT93, p.513]. Obviously, it is not possible to coalesce tuples with a tuple based approach.

Example 5. A final example shall illustrate this important point in some more detail. We start from the well known employee relation:

employee

Surr	Name	Sal	ValidTime
1874	Mike	4590	[1988/3/1-1991/1/1)
1874	Mike	4700	[1991/1/1-1992/6/1)
2384	Tina	4690	[1992/1/1-1992/8/1)
8472	Jane	5000	[1992/7/1-1993/8/1)

Note that we have introduced the above discussed surrogates in order to approximate temporally grouped data models. A first query is "How have the salaries developed?". For reasons of privacy, we may not want to display the names of the employees. However, we still want to recognize every salary history on its own. The query

?- {exists Name employee(Surr,Name,Sal)}@ValidTime.

yields the relation

employee

Surr	Sal	ValidTime
1874	4590	[1988/3/1-1991/1/1)
1874	4700	[1991/1/1-1992/6/1)
2384	4690	[1992/1/1-1992/8/1)
8472	5000	[1992/7/1-1993/8/1)

However, there are also queries where we want to summarize information pertaining to different objects. For example, the query "During which period(s) was a salary paid?" should return the answer

ValidTime
[1988/3/1-1993/8/1)

In ChronoLog we obtain this answer by issuing the query

?- {exists Surr,Name,Sal employee(Surr,Name,Sal)}@ValidTime

TU-Complete respectively TG-Complete query languages cannot answer this type of queries.

5 Conclusions

We have started with the definition of *temporally semi-complete* and *temporally complete* query languages. Temporally semi-complete languages constitute an interesting class of query languages in that they cover a wide range of temporal queries in which temporal aspects can be handled in an intuitive and consistent way without burdening the user. Explicit manipulations of temporal relationships are only needed if the user wants to relate different statements using special temporal relationships (e.g. *precedes, after*, etc.). This issue is addressed by temporally complete query languages.

In the following, we have introduced ChronoLog, a temporal deductive query language. Its design is based on the theoretical considerations of temporally semi-complete and temporally complete languages. The result is an elegant and small query language which relieves the user from the burden to formulate temporal relationships in most cases. As a result, temporal queries are much more likely to be formulated correctly, leading to more reliable applications. (Note that ChronoLog has been implemented as a front-end to a commercial relational database system (see [BM92, BM93]) and that the ChronoLog system supports typical database services such as persistency, concurrency, and recovery.)

Finally, we have briefly discussed the notions of TU- and TG-completeness introduced by [CCT93]. While their ideas related to the notion of an object are interesting, there are queries that cannot be answered with TU-Complete respectively TG-Complete query languages. We have shown how these queries can be answered in ChronoLog.

References

[All83] J. F. Allen. Maintaining Knowledge about Temporal Intervals. *Communications of the ACM*, 16(11), 1983.

[BM92] M. Böhlen and R. Marti. A Temporal Extension of the Deductive Database System ProQuel. Technical report, Departement Informatik, ETH Zürich, 1992.

[BM93] M. Böhlen and R. Marti. Handling Temporal Knowledge in a Deductive Database System. In A. Oberweis W. Stucky, editor, *Datenbanksysteme in Büro, Technik und Wissenschaft*, 1993.

[Böh94] M. Böhlen. *The Temporal Deductive Database System ChronoLog*. PhD thesis, Departement Informatik, ETH Zürich, 1994.

[Bur92] J. Burse. ProQuel: Using Prolog to Implement a Deductive Database System. Technical report, Departement Informatik, ETH Zürich, 1992.

[CCT93] J. Clifford, A. Croker, and A. Tuzhilin. On the Completeness of Query Languages for Grouped and Ungrouped Historical Data Models. In A. Tansel, J. Clifford, S. Gadia, S. Jajodia, A. Segev, and R. Snodgrass, editors, *Temporal Databases: Theory, Design, and Implementation*, pages 496–533. Benjamin/Cummings Publishing Company, 1993.

[CGT90] S. Ceri, G. Gottlob, and L. Tanca. *Logic Programming and Databases*. Surveys in Computer Science, Springer Verlag, Berlin, 1990.

[Cod72] E. F. Codd. Relational Completeness of Data Base Sublanguages. *Courant Computer Symposia Series*, 6:65–98, 1972.

[Gab87] D. Gabbay. The Declarative Past and Imperative Future: Executable Temporal Logic for Interactive Systems. In B. Banieqbal, H. Barringer, and A. Pnueli, editors, *Temporal Logic in Specification*, pages 409–448. Springer-Verlag, LNCS 398, 1987.

[Gad86] S. K. Gadia. Weak Temporal Relations. In *Proceedings of the International Conference on Principles of Database Systems*, 1986.

[Gad88] S. K. Gadia. A Homogeneous Relational Model and Query Languages for Temporal Databases. *ACM Transactions on Database Systems*, 13(4):418–448, 1988.

[GY88] S. K. Gadia and C. Yeung. A Generalized Model for a Relational Temporal Database. In *Proceedings of the ACM SIGMOD International Conference on Management of Data*, 1988.

[Hau87] B. Haugh. Non-standard Semantics for the Method of Temporal Arguments. In P. Jackson, editor, *Proceedings of the International Joint Conference on Artificial Intelligence*, pages 449–455, 1987.

[HT92] P.M. Hill and R.W. Topor. A Semantics for Typed Logic Programs. In F. Pfenning, editor, *Types in Logic Programming*, chapter 1, pages 1–62. MIT Press, 1992.

[JSS93] C. Jensen, M. Soo, and R. Snodgrass. Unification of Temporal Data Models. In *International Conference on Data Engineering*, 1993.

[KS91] H.F. Korth and A. Silberschatz. *Database system concepts*. McGraw-Hill, 1991.

[Llo87] J. W. Lloyd. *Logic Programming*. Symbolic Computation, Springer Verlag, Berlin, 1987.

[Mar91] R. Marti. Research in Deductive Databases at ETH: The LogiQuel Project. In *SI-DBTA Proceedings Database Research in Switzerland*, pages 130–143, 1991.

[MS91] L. E. McKenzie and R. T. Snodgrass. Evaluation of Relational Algebras Incorporating the Time Dimension in Databases. *ACM Computing Surveys*, 23(4):501–543, 1991.

[NT89] S. Naqvi and S. Tsur. *A Logical Language for Data and Knowledge Bases*. Computer Science Press, New York, 1989.

[Rei89] H. Reichgelt. A Comparison of First Order and Modal Logics of Time. In P. Jackson and F. van Harmelen H. Reichgelt, editors, *Logic-Based Knowledge Representation*. MIT Press, 1989.

[Sno93] R. Snodgrass. An Overview of TQuel. In A. Tansel, J. Clifford, S. Gadia, S. Jajodia, A. Segev, and R. Snodgrass, editors, *Temporal Databases: Theory, Design, and Implementation*, pages 141–182. Benjamin/Cummings Publishing Company, 1993.

[Soo91] M. Soo. Bibliography on Temporal Databases. *SIGMOD RECORD*, 20(1):14–23, 1991.

[TCG$^+$93] A. Tansel, J. Clifford, S. Gadia, S. Jajodia, A. Segev, and R. Snodgrass. *Temporal Databases: Theory, Design, and Implementation*. Benjamin/Cummings Publishing Company, 1993.

The Abductive Event Calculus as a General Framework for Temporal Databases

Kristof Van Belleghem   Marc Denecker   Danny De Schreye

Department of Computer Science, K.U.Leuven,
Celestijnenlaan 200A, B-3001 Heverlee, Belgium.
e-mail : {kristof, marcd, dannyd}@cs.kuleuven.ac.be

Abstract. In earlier work, we have shown that the formalism of abductive logic programs with FOL integrity constraints provides, under a completion semantics, the same declarative expressivity for representing incomplete information as full first order logic. We have shown how the combination of this formalism with a variant of the Event Calculus of Kowalski and Sergot results in a correct and very expressive framework for temporal reasoning and representation. In this paper we demonstrate how this Abductive Event Calculus formalism provides a general framework for the representation and use of temporal databases. On the declarative level, it is particularly convenient for the representation of incomplete knowledge. Complementary, on the procedural level, we are able to provide a number of simple algorithms using abduction and deduction to test the consistency of the base, answer queries, update the database, handle complex formulas and resolve inconsistency. Furthermore, the use of the database for general temporal problem solving is possible using the known Event Calculus and Logic Programming methods. In particular we show how planning is possible in this kind of temporal database.

1 Introduction

The Event Calculus (see [14]) is a well-known formalism for temporal representation and reasoning. The basic concepts of the formalism are events and properties: events initiate and terminate periods of time during which properties hold.

The Event Calculus has been modified in several ways, for example in [20], [9], [17] and [13], mainly to simplify the ontology and to eliminate problems occurring because of bidirectional persistence of properties (forward as well as backward in time). In the context of the Event Calculus, [8], [19] and [17] have introduced abduction to solve planning problems and [7] showed how abduction can be used to solve general temporal postdiction problems in the presence of incomplete information.

Though abduction is clearly a useful computational paradigm, it was not recognized earlier that the formalism of abductive logic programs with first order logic constraints provides the same declarative expressivity for representing incomplete information as full first order logic (FOL). This was shown in [5], where it was exploited to provide an implementation of the A language of [10] in the Situation Calculus formulated as an abductive logic program.

In this paper we demonstrate how a temporal database with incomplete information can be formalized in the Abductive Event Calculus and how a suitable abductive procedure like SLDNFA ([6]), which satisfies sufficiently strong sound- and completeness results, can be used to implement the functionality of the database and to use the data for problem solving.

Kowalski ([13]) has argued earlier that the Event Calculus in Logic Programming can be used to formalize the evolution of a database system. The work presented in our paper addresses different issues than Kowalski's work, in particular the representation of incomplete information in Abductive Logic Programming and the use of an abductive procedure for implementing the functionality of the database.

The paper is organised as follows. In the next section, we briefly introduce the Event Calculus. Section 3 describes the abductive extension to Logic Programming, its semantics and the abductive procedure. In section 4 we specify the considered type of temporal database and in section 5 we show how it can be represented and used with the Abductive Event Calculus.

2 The Event Calculus

In the Event Calculus, information is represented in Horn clauses augmented with *negation as failure*. The following axioms define a simplified version of the Event Calculus, which we use as a basis for our framework:

$$\begin{aligned} holds\_at(P,T) &\leftarrow happens(E), E << T, initiates(E,P), \\ &\quad not\, clipped(E,P,T). \\ clipped(E,P,T) &\leftarrow happens(C), in(C,E,T), \\ &\quad terminates(C,P). \\ in(C,E,T) &\leftarrow E << C, C << T. \end{aligned}$$

$happens(E)$ holds if the event E occurs. Only one event is allowed to occur at any one time point, which makes it possible to represent events by their time of occurrence. In other words, we consider events to be special time points. The relation $<<$ defined on events and other time points is a strict linear order: our theory contains axioms ensuring irreflexivity, antisymmetry, transitivity and linearity, but they are left implicit.

In general, the actions associated with an event determine which properties are initiated or terminated by it. This is formulated in domain dependent axioms, for example

$$\begin{aligned} initiates(E, has(x,b)) &\leftarrow act(E, give(y,b,x)). \\ terminates(E, has(y,b)) &\leftarrow act(E, give(y,b,x)). \end{aligned}$$

3 Abduction

The first order logic theory represented by an Event Calculus program is usually defined as the one corresponding to its Clark completion semantics ([3]). The well-known SLDNF procedure can then be used to make deductive inferences. However, Clark completion semantics requires complete knowledge about the problem domain: incomplete data can not be represented. Furthermore, if we want to provide a wider functionality than the answering of queries (for example, use of the database for planning), deduction alone will not suffice.

These restrictions can be overcome in the following way: on the representational level, our theories are to be interpreted according to Console's completion semantics ([4]) for abductive logic programs, augmented with general first order logic constraints. This semantics allows for the occurrence of *undefined* predicates. Thus, incomplete knowledge can be represented.

On the level of problem solving, we will use *abduction* as well as deduction. This, of course, requires an abductive proof procedure: given a set of logic formulas F (facts and rules about the problem domain) and a number of conclusions G, an abductive procedure attempts to find a set of additional facts Δ such that

- $F + \Delta$ is consistent.
- $F + \Delta \models G$.
- Δ is minimal: no subset of Δ exists that satisfies the first two conditions.

The minimality condition is not always added. The facts allowed in Δ are usually constrained to obtain useful results (for example to avoid the solution $\Delta = \{G\}$).

In general, abduction can be used to deal with incomplete knowledge and to solve diagnosis and planning problems, by constraining the facts in Δ in the appropriate way. In general the predicates allowed in Δ are those we have incomplete knowledge about, in other words the undefined predicates. In the sequel we refer to these as *abducible* or *abductive*. For example, in planning problems we have incomplete knowledge about the actions occurring in the plan and the time relations between them, so actions and $\ll$ will be abducible. We try to find the sequence of actions necessary to prove the goal, which is the desired end state. In a similar way postdiction (diagnosis) and indeterminism can be modeled, as demonstrated in [7]. We return to the topic of planning later, in the context of a temporal database.

3.1 The SLDNFA proof procedure

There have been several attempts to build an abductive proof procedure and to use it in the context of temporal reasoning. Examples can be found in [8], [19], [17] and [11].

In our proposal we use the SLDNFA procedure described in [6] and [7]. This procedure, an extension of the SLDNF resolution of Logic Programming ([12], [3]) that can deal with abductive predicates, can handle deduction as well as abduction and allows for a correct treatment of non-ground abducible atoms, which is necessary in our applications. Its soundness and completeness with respect to Console completion semantics are proven in [6].

The implementation of SLDNFA we are using has a number of features making it more fit for planning in the Event Calculus and for general temporal reasoning. A necessary feature is of course the possibility to indicate which predicates are abducible, depending on the intended kind of problem solving and the available knowledge about the problem domain.

Another feature is the possibility to use constraints "$false \leftarrow A_1, \ldots A_n$". These constraints are handled by adding $A_1, \ldots A_n$ to a list of goals for which finite failure needs to be proven, thus ensuring $\neg false$.

Specifically related to the Event Calculus is the fact that iterative deepening is added, which allows us to find the shortest solution (the one including the smallest number of events) first. It is also possible to indicate a maximum number of events to limit the search. Finally, a special constraint module is implemented that makes sure the $<<$ relation on events constitutes a linear order (actually, a partial order will be returned if all of its possible linearizations are valid solutions).

4 Specification of the considered type of temporal database

4.1 Topology of time

When representing time, we have a choice of several topologies, for example point-based or interval-based time, and numerical or non-numerical time. We allow both time points and intervals in our database, considering intervals as periods of time started and ended by a time point. Currently SLDNFA does not support numerical constraints, so we will not use a numerical time line. The addition of these numerical constraints is one of our further research issues.

4.2 Contents of the database

The data we aim to represent in the database are formulas representing the truth value of properties during intervals and at time points, and the change of truth values at certain time points. Further, formulas representing the order on time points and the relations between intervals will be used. We choose the following set of basic formulas, with P an atom:

- $holds\_at(P, T)$: P is true at time point T.
- $holds(P, int(t_1, t_2))$: P is true throughout the interval $int(t_1, t_2)$. This interval does not need to be "maximal": P may remain true after t_2 or can be true already before t_1.
- $notholds(P, int(t_1, t_2))$: P is false throughout the interval $int(t_1, t_2)$.
- $on(P, T)$: P's value changes from false to true at time point T.
- $off(P, T)$: P changes from true to false at T.

The possible relations between time points and intervals are represented by the following formulas. (In the case of intervals, we distinguish thirteen possible relations, based on those defined in [2], though some names may differ.)

- $T_1 = T_2$: T_1 and T_2 are the same time point.
- $T_1 << T_2$: T_1 is chronologically before T_2.
- $equal(i_1, i_2)$: i_1 and i_2 are the same interval.
- $meets(i_1, i_2)$: the endpoint of i_1 is the starting point of i_2.
- $overlaps(i_1, i_2)$: i_1 starts before i_2, and ends during i_2.
- $starts(i_1, i_2)$: i_1 is an initiating subinterval of i_2.
- $ends(i_1, i_2)$: i_1 is a terminating subinterval of i_2.
- $during(i_1, i_2)$: i_1 is a subinterval of i_2 that is initiating nor terminating.
- $before(i_1, i_2)$: i_1 lies entirely before i_2.
- $after(i_1, i_2)$: inverse of $before$.
- $metby(i_1, i_2)$: inverse of $meets$.
- $overlapped(i_1, i_2)$: inverse of $overlaps$.
- $startby(i_1, i_2)$: inverse of $starts$.
- $endby(i_1, i_2)$: inverse of $ends$.
- $contains(i_1, i_2)$: inverse of $during$.

We can build more complex expressions by combining these basic formulas with logical connectives and quantifiers : if P and Q are valid expressions, then $\neg P$, $P \& Q$, $P \vee Q$, $P \oplus Q$ (exclusive or), $P \Rightarrow Q$, and $P \Leftrightarrow Q$ are valid as well, and if $P(x)$ is an expression, then so are $\forall x : P(x)$ and $\exists x : P(x)$.

4.3 Use of the database

The formulas defined above determine the possible contents of our temporal database. The functionality we require of such database is the following:

- testing whether a database D is consistent.
- answering normal Logic Programming queries as well as more complex ones. Since our data may be incomplete, we distinguish two types of query:
  1. "Is Q necessarily true in the database D ?" (does $D \models Q$ hold?)
  2. "Is Q possible in the database D ?" (is Q + D consistent ?).
- updating the database, with a consistency check of the new data.
- in the case of inconsistency, proposing solutions to restore consistency.
- finally, and maybe most importantly, extending the expressivity to make it possible to use the database for problem solving, in particular planning.

5 A general solution using the Abductive Event Calculus

5.1 Representation of data

We consider our data as a theory that consists of two parts. The first part is a logic program defining all basic formulas in terms of primitive Event Calculus predicates. The second part contains the real data in the base. These are considered integrity constraints on the possible states (models) of the database.

The basic formulas are defined by the following rules:

$$
\begin{aligned}
&holds\_at(P,T) &&\leftarrow happens(E), E \ll T, initiates(E,P),\\
& && \quad not\ clipped(E,P,T).\\
&holds(P, int(T_1,T_2)) &&\leftarrow interval(T_1,T_2), holds\_from(P,T_1),\\
& && \quad not\ clipped(T_1, P, T_2).\\
&notholds(P, int(T_1,T_2)) &&\leftarrow interval(T_1,T_2), notholds\_from(P,T_1),\\
& && \quad not\ started(T_1, P, T_2).\\
&holds\_from(P,E) &&\leftarrow initiates(E,P).\\
&holds\_from(P,E) &&\leftarrow holds\_at(P,E), not\ terminates(E,P).\\
&notholds\_from(P,E) &&\leftarrow terminates(E,P).\\
&notholds\_from(P,E) &&\leftarrow not\ holds\_at(P,E), not\ initiates(E,P).\\
&started(E,P,T) &&\leftarrow happens(C), in(C,E,T), initiates(C,P).\\
&clipped(E,P,T) &&\leftarrow happens(C), in(C,E,T), terminates(C,P).\\
&on(P,E) &&\leftarrow initiates(E,P), not\ holds\_at(P,E).\\
&off(P,E) &&\leftarrow holds\_at(P,E), terminates(E,P).\\
&in(C,E,T) &&\leftarrow E \ll C, C \ll T.
\end{aligned}
$$

It follows from our definitions that the interval $int(t_1,t_2)$ actually denotes the interval $]t_1,t_2]$, containing its endpoint but not its starting point. This choice is made because working with closed intervals can lead to inconsistencies (one time point can belong to two intervals with different values for the same property), while open intervals lead to time points where properties are undefined. A choice between the two types of halfopen intervals is easy: the definition of the Event Calculus naturally leads to the form $]t_1,t_2]$.

The chronological relations between intervals are expressed in terms of relations between their starting points and end points. This allows us to use the linear time constraint module of SLDNFA for reasoning on them.

$$
\begin{aligned}
&equal(int(T_1,T_2), int(T_1,T_2)) &&\leftarrow interval(T_1,T_2).\\
&meets(int(T_1,T_2), int(T_2,T_3)) &&\leftarrow interval(T_1,T_2), interval(T_2,T_3).\\
&overlaps(int(T_1,T_2), int(T_3,T_4)) &&\leftarrow interval(T_1,T_2), interval(T_3,T_4),\\
& && \quad T_1 \ll T_3, T_3 \ll T_2, T_2 \ll T_4.\\
&starts(int(T_1,T_2), int(T_1,T_3)) &&\leftarrow interval(T_1,T_2), interval(T_1,T_3),\\
& && \quad T_2 \ll T_3.\\
&ends(int(T_1,T_2), int(T_3,T_2)) &&\leftarrow interval(T_1,T_2), interval(T_3,T_2),\\
& && \quad T_3 \ll T_1.\\
&during(int(T_1,T_2), int(T_3,T_4)) &&\leftarrow interval(T_1,T_2), interval(T_3,T_4),\\
& && \quad T_3 \ll T_1, T_2 \ll T_4.\\
&before(int(T_1,T_2), int(T_3,T_4)) &&\leftarrow interval(T_1,T_2), interval(T_3,T_4),\\
& && \quad T_2 \ll T_3.\\
&after(int(T_1,T_2), int(T_3,T_4)) &&\leftarrow interval(T_1,T_2), interval(T_3,T_4),\\
& && \quad T_4 \ll T_1.\\
&overlapped(int(T_1,T_2), int(T_3,T_4)) &&\leftarrow interval(T_1,T_2), interval(T_3,T_4),\\
& && \quad T_3 \ll T_1, T_1 \ll T_4, T_4 \ll T_2.
\end{aligned}
$$

$$metby(int(T_1, T_2), int(T_3, T_1)) \leftarrow interval(T_1, T_2), interval(T_3, T_1).$$
$$startby(int(T_1, T_2), int(T_1, T_3)) \leftarrow interval(T_1, T_2), interval(T_1, T_3),$$
$$T_3 << T_2.$$
$$endby(int(T_1, T_2), int(T_3, T_2)) \leftarrow interval(T_1, T_2), interval(T_3, T_2),$$
$$T_1 << T_3.$$
$$contains(int(T_1, T_2), int(T_3, T_4)) \leftarrow interval(T_1, T_2), interval(T_3, T_4),$$
$$T_1 << T_3, T_4 << T_2.$$
$$interval(T_1, T_2) \leftarrow happens(T_1), happens(T_2),$$
$$T_1 << T_2.$$

These rules are interpreted under Console completion semantics ([4]) with happens/1, initiates/2, terminates/2 and $<< /2$ as the undefined predicates. The definitions of other predicates are completed as in Clark completion semantics.

Another important remark is that the "free equality theory", which states that constants and terms with different names are unequal, holds for all terms and constants, except for time points. Time points with different names can be equal and must be treated as skolem constants.

Finally, as indicated earlier, $<<$ is a linear order on time points and events.

The rules defined so far determine the *meaning* of our database. The real data are considered integrity constraints on the possible models of this database. These can be basic formulas as well as complex expressions. Some examples:

$$notholds(has(john, book_1), int(t_1, t_2))).$$
$$\forall(T) : holds\_at(p, T).$$
$$meets(int(T_1, T_2), int(T_3, T_4)) \oplus metby(int(T_1, T_2), int(T_3, T_4)).$$
$$holds\_at(has(X, O), T), holds\_at(has(Y, O), T) \Rightarrow X = Y.$$

The data can be very incomplete, so possibly many different models exist. For example, consider the database containing only two simple constraints:

$$holds(has(john, book_1), int(t_1, t_2)).$$
$$notholds(has(mary, book_2), int(t_3, t_4)).$$

We do not know anything about John having his book outside of the interval $int(t_1, t_2)$. He can own it all the time, or only during the mentioned time period, or during a period that starts at t_1 but continues after t_2, and so on. Likewise for Mary's book we have many possible models. Finally we have no information at all concerning the temporal relation linking $int(t_1, t_2)$ and $int(t_3, t_4)$. These periods can overlap, be disjoint, be equal, etc. Many different models may correspond to different solutions for the undefined predicates.

5.2 Basic functionality of the database

Now that we have expressed the meaning of our data, we can determine how to use them. We first describe how the basic functionality of the database is provided in the special case where only basic formulas are allowed as constraints.

One very important task, useful for consistency testing as well as query answering, is the generation of a logical model for a set of data. To find such models, we use abduction. As indicated earlier, our undefined predicates are happens, $\ll$, initiates and terminates. These predicates are the abductive ones for the procedure. To find a model for a set of data, we collect these data in a goal, and try to build a proof for this goal using the definitions of the basic formulas and a number of abduced new facts. The resulting set of abduced facts (if it exists) forms, in a sense, a model for the data: $Comp(P + \Delta) \models F$, where P is the program consisting of defining rules, Δ is the set of abduced facts, F is the goal to be proven (the data), and $Comp(P)$ is the Clark completion of P. Every abduced solution is an assignment of truth values to the undefined predicates, and corresponds to one possible model for the data. If no solution can be abduced, the data are inconsistent.

Testing the consistency of a database is now straightforward: we check whether $P + F$ is consistent, where P is the program containing the basic definitions and F are the data in the database, by attempting to abduce a model in which F holds. The data are consistent if we find a model, inconsistent if we find failure.

Consider again the database containing the constraints

$$holds(has(john, book_1), int(t_1, t_2)).$$
$$notholds(has(mary, book_2), int(t_3, t_4)).$$

To check its consistency, we form the goal

$\leftarrow holds(has(john, book_1), int(t_1, t_2)), notholds(has(mary, book_2), int(t_3, t_4)).$

and find for example the abduced facts

$happens(t_1).\quad happens(t_2).\quad happens(t_3).\quad happens(t_4).$
$t_1 \ll t_2 \ll t_3 \ll t_4.$
$initiates(t_1, has(john, book_1)).$

which proves consistency of the data.

Answering queries can be done in a similar way. If we want to know whether Q is possible in the database F, we try to abduce a solution that (added to P) entails $F + Q$. For example, using the same data as above, the query "Is it possible that Mary owns book 2 at t_1" will be solved by attempting to prove

$\leftarrow holds(has(john, book_1), int(t_1, t_2)),$
$\quad notholds(has(mary, book_2), int(t_3, t_4)),$
$\quad holds\_at(has(mary, book_2), t_1)).$

which has as a model for example (omitting the *happens*-facts)

$t_{new} \ll t_1 \ll t_2 \ll t_3 \ll t_4.$
$initiates(t_{new}, has(mary, book_2)).$
$initiates(t_1, has(john, book_1)).$
$terminates(t_3, has(mary, book_2)).$

The answer to the query is therefore affirmative.

If the question is whether Q is *necessarily* true given F, we try to abduce a model for $F + \neg Q$. If we find no model ($P + F + \neg Q$ is inconsistent), we know that $P + F \models Q$, which is what we were trying to find out. Using the same query as in the previous example, we would end up trying to prove

$$\leftarrow holds(has(john, book_1), int(t_1, t_2)),$$
$$notholds(has(mary, book_2), int(t_3, t_4)),$$
$$not\ holds\_at(has(mary, book_2), t_1)).$$

which has the solution

$$t_1 \ll t_2 \ll t_3 \ll t_4.$$
$$initiates(t_1, has(john, book_1)).$$

so we can conclude that Mary does not necessarily have book 2 at t_1.

Finally, to update the database, we check whether the resulting data would be consistent, and if so, add the new data item to the base. If inconsistency is detected, a warning results and the update can only be executed through user intervention. For example, if we want to add $holds\_at(has(mary, book_2), t_1))$ to the database, we try to compute a model for

$$\leftarrow holds(has(john, book_1), int(t_1, t_2)),$$
$$notholds(has(mary, book_2), int(t_3, t_4)),$$
$$holds\_at(has(mary, book_2), t_1)).$$

which succeeds as before. The new data item is then added to the base.

This functionality poses no problem for the SLDNFA procedure, except for the treatment of time constants. SLDNFA considers constants in the data to be normal constants, where they are intended to be skolems. We can solve this problem in the following way: we collect all data $(F_1, F_2, \ldots F_N)$ in the conjunction

$$F_1 \& F_2 \& \ldots \& F_N.$$

We write that conjunction in the form $F(t_1, \ldots t_n)$ where the t_i are our time point skolem constants. In short, we call this expression F. We can then "deskolemize" F: we replace all skolem constants by existentially quantified variables, which results in F':

$$\exists T_1, \ldots T_n : F(T_1, \ldots T_n).$$

Skolem's theorem states that for all P and F, with F' the deskolemization of F as defined above: $P + F$ is consistent $\Leftrightarrow P + F'$ is consistent. Therefore, replacing skolem constants by existentially quantified variables does not change the result of a consistency check or a query.

We now do the following: before calling the SLDNFA procedure, we build a table in which we link every time constant to a variable. In the data we pass to the procedure, we replace every constant by its corresponding variable. As indicated, this does not change the consistency results.

To find the solution corresponding to this obtained consistency result, we combine the answer of the SLDNFA procedure with our table of time constants, where some of the variables may be unified by now. In that case the time constants corresponding to these variables are equal in this solution. We will discuss a detailed example later.

The search space can now be limited to solutions with a bounded number of events in a consistency/inconsistency preserving way: it can be shown that, if the data are consistent and contain only N different time points, there exists at least one model for these data containing 2N or less events. In general 2N even is a substantial overestimation of the needed number of events, and in most cases we find solutions with a number of events equal to or a little greater than N.

5.3 Introduction of complex data and queries

In the previous section we showed how SLDNFA can be used to abduce models for data consisting of only basic formulas. However, in the case of complex queries and data a preceding transformation step is required. This transformation step is based on the Lloyd-Topor transformation described in [16].

The transformation provides a method to transform a program containing non-Horn clauses and complex goals, an *extended program*, to a program containing only Horn clauses augmented with negation as failure. The soundness of the transformation under Clark completion semantics is proven in the article, and this soundness result holds for abductive logic programs under Console completion semantics as well. We recall the essence of the transformation here.

An extended logic program is a program consisting of "general clauses" ([15]). These are rules of the form

$$A \leftarrow W.$$

where A is an atom and W an arbitrary first order logic expression. Any variables in A and free variables in W are considered universally quantified at the beginning of the clause.

The transformation to a normal logic program is performed by replacing general clauses by others using a set of transformation rules, until only Horn clauses — possibly with negation in the body — are left. As an example, we include some of the basic transformation rules:

a) Replace $\quad A \leftarrow W_1, W_2, \ldots, (\forall x_1 \ldots x_m : W), \ldots W_n.$
   by $\qquad A \leftarrow W_1, W_2, \ldots, \neg(\exists x_1 \ldots x_m : \neg W), \ldots W_n.$
b) Replace $\quad A \leftarrow W_1, W_2, \ldots, \neg(V \Leftarrow W), \ldots W_n.$
   by $\qquad A \leftarrow W_1, W_2, \ldots, W, \neg V, \ldots W_n.$
c) Replace $\quad A \leftarrow W_1, W_2, \ldots, \neg(\neg W), \ldots W_n.$
   by $\qquad A \leftarrow W_1, W_2, \ldots, W, \ldots W_n.$
d) Replace $\quad A \leftarrow W_1, W_2, \ldots, \neg(\exists x_1 \ldots x_m : W), \ldots W_n.$
   by $\qquad A \leftarrow W_1, W_2, \ldots, \neg p(y_1 \ldots y_k), \ldots W_n.$
   and $\qquad p(y_1 \ldots y_k) \leftarrow \exists x_1 \ldots x_m : W.$

where p is a new predicate symbol not occurring in the program, and $y_1, \ldots y_k$ the free variables in $(\exists x_1 \ldots x_m : W)$.

Similar rules exist for each operator and its negation. A complete list can be found in [16].

The goal of the program can be transformed in the same way:

Replace   ← W.
by   ← $answer(x_1 \ldots x_n)$.
and $answer(x_1 \ldots x_n)$ ← W.
where $x_1 \ldots x_n$ are the free variables in W.

The resulting rule $answer(x_1 \ldots x_n)$ ← W. must be transformed further using the rules described above.

5.4   A detailed example

To illustrate how our system handles complex data and time skolems, we solve a small example query in detail. We have a database DB containing two data items, namely

$$holds(has(john, book_1), int(t_1, t_2)).$$
$$holds(has(mary, book_2), int(t_2, t_3)).$$

We want to know if it is possible that, for arbitrary time points a,b and c,

$$holds(has(john, book_1), int(a, b)) \vee notholds(has(mary, book_2), int(a, c))$$

is true. The following query is used:

← $poss\_query(DB,$
   $or(holds(has(john, book_1), int(a, b)), notholds(has(mary, book_2), int(a, c))))$.

The program collects the data from DB in a list and adds the query to it. All time constants are replaced by variables, and we obtain the following time table:

$$\begin{array}{ll} t_1 - X_1 & a - A \\ t_2 - X_2 & b - B \\ t_3 - X_3 & c - C \end{array}$$

The goal we want to prove becomes

$[holds(has(john, book_1), int(X_1, X_2)), holds(has(mary, book_2), int(X_2, X_3)),$
$or(holds(has(john, book_1), int(A, B)), notholds(has(mary, book_2), int(A, C)))]$

but the data need to be transformed first. In this case only the disjunction is complex. New rules

$$q_0(A, B, C) \leftarrow holds(has(john, book_1), int(A, B)).$$
$$q_0(A, B, C) \leftarrow notholds(has(mary, book_2), int(A, C)).$$

are added to the program, and we try to solve the following query:

← $holds(has(john, book_1), int(X_1, X_2)),$
   $holds(has(mary, book_2), int(X_2, X_3)), q_0(A, B, C)$.

If we ask for a solution with three events, the meta-interpreter replaces time variables by skolem constants, determines the order on these time constants, and abduces the necessary initiations and terminations to prove the goal (using the new rules for q_0 together with the general definitions of the database formulas).

The solution contains three events new_1, new_2 and new_3, where $X_1 = A = new\_1$, $X_2 = B = new\_2$ en $X_3 = C = new\_3$. The abduced order on these events is $new\_1 \ll new\_2 \ll new\_3$. The initiations are

$$initiates(new\_1, has(john, book_1)).$$
$$initiates(new\_2, has(mary, book_2)).$$

and terminations are not necessary. The time table now looks like this:

$$\begin{array}{ll} t_1 - new\_1 & a - new\_1 \\ t_2 - new\_2 & b - new\_2 \\ t_3 - new\_3 & c - new\_3 \end{array}$$

and we can read the following solution

$$t_1 = a,\ t_2 = b,\ t_3 = c,$$
$$t_1 \ll t_2 \ll t_3,$$
$$initiates(t_1, has(john, book_1)),$$
$$initiates(t_2, has(mary, book_2)).$$

which is obviously correct. Of course it is not the only solution, and the program will find many more, for example solutions where g is initiated by t_1 or where f gets terminated by some event. One reason for the many solutions is the occurrence of a disjunction in the query.

5.5 Resolving inconsistency

We have extended this program further to help the user resolve inconsistency in the data. We use abduction to propose solutions for the inconsistency, as illustrated in the following example.

Supppose we have three formulas P, Q and R as data. The program collects these data in a list $[P, Q, R]$, which is given — after transformation — as a goal to the meta-interpreter. If the meta-interpreter returns with a solution, the data are consistent and there is no problem.

If no solution is found, and the user has chosen the "resolve inconsistency" option, control returns to the transformation program. This program will undo all changes it made to the data during transformation, and generates a new transformation, only this time not for $[P, Q, R]$, but for $[reject(P), Q, R]$.

The meta-interpreter then tries to explain $reject(P)$ instead of P. This is always possible, because we make reject/1 an abductive predicate. The result is then that an abduced fact $reject(P)$ is written in the solution, while the program continues trying to find a model for $[Q, R]$. The constraint P is dropped, which possibly resolves the inconsistency.

In further attempts every combination of formulas and rejected formulas is checked until a solution is found. In short then, for any fact P the program can either explain P, or abduce $reject(P)$ and ignore P. Looking at the abduced $reject(P)$ facts, the user sees which constraints have been dropped to restore consistency. Of course more than one solution can exist, and the user can choose the best one, whatever "best" means to him.

As an example, a database containing

$$holds(has(john, book_1), int(t_1, t_2))).$$
$$holds(has(mary, book_1), int(t_1, t_2)).$$
$$holds\_at(has(X, O), T), holds\_at(has(Y, O), T) \Rightarrow X = Y.$$

is inconsistent, and consistency can be restored by deleting any of the three constraints. One of the proposed solutions would be

$$reject(holds(has(john, book_1), int(t_1, t_2)))).$$
$$happens(t_1). \quad happens(t_2). \quad t_1 << t_2.$$
$$initiates(t_1, has(mary, book_1)).$$

This method is of course quite inefficient, since data are selected for rejection in a random way, without looking for the causes of the inconsistency. There exist, however, solutions to this problem. As mentioned earlier, the meta-interpreter maintains a list of constraints (negative goals) to be satisfied. If we keep track of the data items corresponding to a constraint, and determine which fact violates which constraint, we can use this information in a more intelligent method to resolve inconsistency. We will return to this issue briefly in the discussion.

5.6 Planning

If we want to use a database for planning, we of course need to introduce the concept of action. This is a well-known concept in the Event Calculus and we can use it without any modification. To use the database for planning, we define all possible actions and their effects. Then we make the actions abducible (instead of simply the initiation and termination of properties). Initiations and terminations now follow from the occurring actions. In this way, properties can not arbitrarily change their value as they could in the original model. Every change is now caused by an action.

A system for planning using the Abductive Event Calculus has been developed earlier. This system, based on abducible actions, can almost automatically be combined with our database system. We can then use our database for planning without a problem. Furthermore, the use of actions to explain every initiation and termination can be extended to non-planning problems. The actions then define every possible way in which properties can change. As a result, our models not only contain information about which properties changed value, but also about *why* this happened.

As an example, assume we want to build a very simple plan: John owns a certain book, and we want Mary to have it. The only possible action is giving the book to someone. We add the specification of this action's preconditions and effects to our basic definitions:

$$false \quad\quad\quad\quad \leftarrow act(E, give(Y, B, X)), not\ holds\_at(has(Y, B), E).$$
$$initiates(E, has(X, B)) \quad \leftarrow act(E, give(Y, B, X)).$$
$$terminates(E, has(Y, B)) \leftarrow act(E, give(Y, B, X)).$$

We also introduce a special event *start* which occurs before all other events to take care of the first initiations. After this special event, only actions can change the world.

$$happens(start).$$
$$false \leftarrow happens(T), T << start.$$
$$initiates(start, P) \leftarrow initially(P).$$

and we make *happens*, $<<$, *initially* and *act* abducible.

The database *DB* would contain the formulas

$$holds\_at(has(john, book_1), t_1).$$
$$holds\_at(has(X, B), T), holds\_at(has(Y, B), T) \Rightarrow X = Y.$$

and we would try to solve the query

$$\leftarrow poss\_query(DB, [t_1 << t_2, holds\_at(has(mary, book_1), t_2)]).$$

This gives for example the model

$$happens(start).\quad happens(t_1).\quad happens(t_2).\quad happens(t_3).$$
$$start << t_1 << t_3 << t_2.$$
$$initially(has(john, book_1)).$$
$$act(t_3, give(john, book_1, mary)).$$

which explicitly contains the plan.

As a final remark we can indicate that the original idea with abducible initiations and terminations is just a special case of the proposal using actions, in which every action corresponds to one initiation or termination.

6 Discussion

We have demonstrated how the Abductive Event Calculus provides a general framework for the representation and use of temporal databases. Both time points and intervals can be used. The representation of incomplete knowledge is perfectly possible.

Abduction provides a straightforward way to generate models for a set of data. This allows us to check consistency and to answer queries. Complex data can be handled using a preceding transformation step, and deskolemization allows us to represent time points that may be equal to each other.

Using an abductive predicate *reject/1* we introduced a simple method to help us resolve inconsistency by rejecting certain data. Finally, we have shown how the database can be used for planning by introducing actions and making them abducible.

In general, the proposed solutions are not very efficient. We mainly provide a theoretical framework for representing incomplete temporal databases, and give a number of simple algorithms to illustrate how this database can be used. These algorithms may be the basis of research on more efficient implementations.

One of our own further research goals, apart from the introduction of numerical time constraints, is the improvement of the efficiency of our procedures.

On one hand, we hope to obtain this greater efficiency by incorporating CLP techniques and tabulation.

On the other hand, we already find interesting ideas in the literature. For example, in [22] we find an algorithm for resolving inconsistency in a network of interval relations, based on the work in [1]. There, for each pair of intervals a list of possible relations between these intervals is maintained. If ever no possible relations are left between any two intervals, the data are inconsistent. Weigel and Bleisinger have modified and extended this procedure to efficiently derive solutions for the inconsistency.

Their solutions show some similarity to our approach, but work only on interval relations instead of general data. This allows for more efficient algorithms, especially if an incremental consistency checker is used.

Another approach to the representation of temporal databases can be found in [18]. A database is considered a collection of maximal intervals throughout which certain properties hold. For each property a list of such intervals is maintained. Incomplete knowledge can be represented by skolemizing the end points of an interval, and constraints on these end points can be expressed. The framework shows some similarity to ours, though no explicit events are used and only maximal intervals are represented. The system can be mapped to ours, however, and some of its proposed algorithms may be useful to us.

A possible extension for temporal databases is the introduction of a notion of belief. The representation of belief in a theory of time was addressed in [2]. One proposal to incorporate this notion in a temporal database is described in [21]. To incorporate a similar extension in our system, further research will be necessary.

The most important aspect of our framework is probably that it allows for the data in the base to be in the same language as the applications working with them. This is clearly illustrated by the straightforward extension for planning. Thus we hope to show that the Abductive Event Calculus is not only useful in several distinct temporal reasoning domains, but also provides a link between them.

Acknowledgements

Kristof Van Belleghem is partly supported by ESPRIT BR project Compulog II and partly by the Belgian IWONL. Marc Denecker is supported by Dienst Onderzoekscoordinatie, K.U.Leuven. Danny De Schreye is a senior research associate of the Belgian NFWO. We thank anonymous referees for valuable comments.

References

1. J. F. Allen. Maintaining Knowledge About Temporal Intervals. *CACM*, 26(11):832–843, 1983.
2. J. F. Allen. Towards a General Theory of Action and Time. *Artifical Intelligence*, 23(11):123, 1984.

3. K. Clark. Negation as failure. In H. Gallaire and J. Minker, editors, *Logic and databases*, pages 293–322. Plenum Press, 1978.
4. L. Console, D. Theseider Dupre, and P. Torasso. On the relationship between abduction and deduction. *Journal of Logic and Computation*, 1(5):661–690, 1991.
5. M. Denecker. *Knowledge Representation and Reasoning in Incomplete Logic Programming*. PhD thesis, Department of Computer Science, K.U.Leuven, 1993.
6. M. Denecker and D. De Schreye. SLDNFA; an abductive procedure for normal abductive programs. In K. Apt, editor, *Proceedings of the International Joint Conference and Symposium on Logic Programming, Washington*, 1992.
7. M. Denecker, L. Missiaen, and M. Bruynooghe. Temporal reasoning with abductive event calculus. In *Proceedings of ECAI 92, Vienna*, 1992.
8. K. Eshghi. Abductive planning with event calculus. In R. Kowalski and K. Bowen, editors, *Proceedings of the 5th ICLP*, 1988.
9. C. Evans. The Macro-Event Calculus: Representing Temporal Granularity. In *Proceedings of PRICAI, Tokyo*, 1990.
10. M. Gelfond and V. Lifschitz. Describing Action and Change by Logic Programs. In *Proc. of the 9th Int. Joint Conf. and Symp. on Logic Programming*, 1992.
11. A. Kakas and P. Mancarella. Constructive abduction in logic programming. Technical report, Dipartimento di Informatica, University of Pisa, 1993.
12. R. A. Kowalski. *Logic for problem solving*. Elsevier Science Publisher, 1976.
13. R. A. Kowalski. Database updates in the event calculus. *Journal of Logic Programming, 1992*, 1992.
14. R. A. Kowalski and M. Sergot. A logic-based calculus of events. *New Generation Computing*, 4(4):319–340, 1986.
15. J. Lloyd. *Foundations of Logic Programming*. Springer-Verlag, 1987.
16. J. Lloyd and R. Topor. Making prolog more expressive. *Journal of logic programming*, 1(3):225–240, 1984.
17. L. Missiaen. *Localized abductive planning with the event calculus*. PhD thesis, Department of Computer Science, K.U.Leuven, 1991.
18. A. Porto and C. Ribeiro. Temporal inference with a point-based interval algebra. In *Proceedings of ECAI 92, Vienna*, pages 374–378, 1992.
19. M. Shanahan. Prediction is deduction but explanation is abduction. In *Proceedings of IJCAI 89*, page 1055, 1989.
20. M. Shanahan. Representing continuous change in the event calculus. In *Proceedings of the 9th ECAI*, page 598, 1990.
21. S. Sripada. A metalogical programming approach to reasoning about time in knowledge bases. In *Proceedings of IJCAI 93*, 1993.
22. A. Weigel and R. Bleisinger. Support for resolving Contradictions in Time Interval Networks. In *Proceedings of ECAI 92, Vienna*, pages 379–383, 1992.

A Decision Procedure for a Temporal Belief Logic

Michael Wooldridge and **Michael Fisher**

Department of Computing
Manchester Metropolitan University
Chester Street, Manchester M1 5GD
United Kingdom
{M.Wooldridge, M.Fisher}@mmu.ac.uk

Abstract. This paper presents a *temporal belief logic* called $\mathcal{L}_{TB}$. In addition to the usual connectives of linear discrete temporal logic, $\mathcal{L}_{TB}$ contains an indexed set of modal *belief* connectives, via which it is possible to represent the belief systems of resource-bounded reasoning agents. The applications of $\mathcal{L}_{TB}$ in general, and its use for representing the dynamic properties of multi-agent AI systems in particular, are discussed in detail. A tableau-based decision procedure for $\mathcal{L}_{TB}$ is then described, and some examples of its use are presented. The paper concludes with a discussion and future work proposals.

1 Introduction

Temporal logics have been shown to have many applications, in a variety of disciplines. For example: in computer science, temporal logics are used in the specification and verification of reactive systems [16]; in artificial intelligence, they are used as knowledge representation formalisms, and have proved to be a valuable tool in tackling such problems as reasoning about action [18]. For some applications, however, logics containing connectives that operate over just the one modal dimension of time do not provide sufficient expressive power. For such applications, it is necessary to provide connectives that allow us to represent the properties of different modal dimensions *in the same logic*. Logics which contain more than one different type of modality are called *multi-modal* logics [3]. In this paper, we consider a multi-modal logic which contains connectives for representing both time and *belief*.

The obvious approach to defining the semantics of a temporal belief logic involves adapting possible worlds semantics for belief [11]: one might define a world to be a sequence of time points, so that a belief accessibility relation holds between alternative histories (cf. tensed modal logics [20]). Although such an approach is undoubtedly simple, it suffers from at least two drawbacks. The first is that, in common with all normal modal formalisations of belief, it implies that agent's beliefs are closed under logical consequence; this is the logical omniscience problem [17]. While logical omniscience is acceptable in the study of theoretically perfect believers, it is clearly at odds with any reasonable understanding of how belief works in resource-bounded reasoners. The second problem is that belief and time would *interact* in such a way as as to make the development of an automatic proof method awkward [3].

In this paper, we develop a temporal belief logic called $\mathcal{L}_{TB}$, in which the semantics of belief are not based on possible worlds, but on a simple new model of belief which is outlined in §2. The logic $\mathcal{L}_{TB}$ is then developed in §3, which also includes a discussion of its applications. Since time and belief do not interact directly in $\mathcal{L}_{TB}$, it is possible to develop a tableau-based decision procedure for $\mathcal{L}_{TB}$ as a generalisation of the temporal tableau method. Such a decision procedure is presented in §4. Some worked examples, illustrating the decision procedure, are given in §4.1. The paper closes with some comments and future work proposals.

Notational Conventions: If $\mathcal{L}$ is a logical language, then we write $Form(\mathcal{L})$ for the set of (well-formed) formulae of $\mathcal{L}$. We use the lowercase Greek letters φ, ψ, and χ as meta-variables ranging over formulae of the logical languages we consider, and the uppercase Greek letters Δ and Γ as meta-variables ranging over sets of formulae. To give the reader some visual clues, we generally use Δ to denote a set of beliefs, and Γ to stand for an arbitrary set of formulae. We use a VDM-style notation for manipulating sets and functions [13], and use $\emptyset$ for the empty set.

2 Belief Models

In this section we develop the formal framework which will be used in $\mathcal{L}_{TB}$ to give a semantics to belief[1]. This new framework may be used to represent the belief systems of resource bounded reasoning agents, although it is sufficiently rich that it can also be used to represent, for example, the perfect reasoners of possible worlds semantics. In the space available, we can do no more than sketch the properties of the model; for details, see [23].

The structures we use to represent belief systems are called *belief models*. A belief model representing an agent i's belief system is a pair. The first component of this pair is a set of observations that have been made about i's beliefs. These observations are expressed in some *internal language*; throughout this paper we shall call this internal language $\mathcal{L}$. In general, the internal language may be one of rules, frames, semantic nets, or some other kind of KR formalism but, for simplicity, we shall assume that $\mathcal{L}$ is a *logical* language. Thus, the first component of i's belief model is a set of $\mathcal{L}$-formulae representing observations that have been made about i's beliefs. The second component is a relation, which holds between sets of $\mathcal{L}$-formulae and $\mathcal{L}$-formulae. This relation is called a *belief extension* relation, (hereafter abbreviated to 'b.e. relation'), and it is intended to model i's reasoning ability. Let BE_i be the b.e. relation for agent i. Then the way we interpret BE_i is:

if i believes Δ **and** $(\Delta, \varphi) \in BE_i$ **then** i also believes φ.

It is via i's b.e. relation that we are able to make deductions about what other beliefs i has. In [23], we show how a b.e. relation that correctly describes the behaviour of an agent's belief system may be derived in a principled way. We now formally define belief models.

---
[1] Note that *human* belief is *not* the object of study in this paper, and in particular, no claims are made about the validity or usefulness of the model for representing human believers.

Definition 1. A *belief model*, b, is a pair $b = (\Delta, BE)$ where

- $\Delta \subseteq Form(\mathcal{L})$; and
- $BE \subseteq (powerset(Form(\mathcal{L})) \times Form(\mathcal{L}))$ is a countable non-empty binary relation between sets of $\mathcal{L}$-formulae and $\mathcal{L}$-formulae, which must satisfy the following requirements:
  1. *Reflexivity*: if $(\Delta, \varphi) \in BE$, then $\forall \psi \in \Delta, (\Delta, \psi) \in BE$;
  2. *Monotonicity*: if $(\Delta, \varphi) \in BE$, $(\Delta', \psi) \in BE$, and $\Delta \subseteq \Delta'$, then $(\Delta', \varphi) \in BE$;
  3. *Transitivity*: if $(\Delta, \varphi) \in BE$ and $(\{\varphi\}, \psi) \in BE$, then $(\Delta, \psi) \in BE$.

If $b = (\Delta, BE)$ is a belief model, then Δ is said to be its *base set*, and BE its *belief extension relation*. We now define a function bel which takes as its sole argument a belief model, and returns the set of $\mathcal{L}$-formulae representing the *belief set* of that model.

Definition 2.
$$bel((\Delta, BE)) \stackrel{\text{def}}{=} \{\varphi \mid (\Delta, \varphi) \in BE\}$$

Suppose b_i is a belief model which represents agent i's belief system. Then the interpretation of 'belief' in this paper is as follows:

$\varphi \in bel(b_i)$ — i believes φ $\qquad$ $\neg\varphi \in bel(b_i)$ — i believes $\neg\varphi$
$\varphi \notin bel(b_i)$ — i doesn't believe φ $\qquad$ $\neg\varphi \notin bel(b_i)$ — i doesn't believe $\neg\varphi$.

3 The Temporal Belief Logic $\mathcal{L}_{TB}$

In this section, we develop the new temporal belief logic $\mathcal{L}_{TB}$. This logic is essentially a standard linear discrete temporal logic enriched by the addition of a set of unary modal belief connectives, with a semantics given in terms of belief models, as described in the preceding section.

We let time be linear, discrete, bounded in the past, and infinite in the future, giving the temporal model $(I\!N, <)$. We take as primitive just two temporal connectives: $\bigcirc$ ('next'), and $\mathcal{U}$ ('until'). The remaining standard connectives of linear discrete future temporal logic may be derived from these.

3.1 Syntax

In the interests of simplicity, we shall restrict our attention in this paper to propositional languages. We assume an underlying classical propositional language, which we shall call $\mathcal{L}_0$. This language is defined over a set Φ of primitive propositions, and is closed under the unary connective '$\neg$' (not), and the binary connective '$\vee$' (or). The remaining connectives of classical logic ('$\wedge$' (and), '$\Rightarrow$' (implies), and '$\Leftrightarrow$' (iff)) are assumed to be introduced as abbreviations, in the standard way. $\mathcal{L}_0$ is also assumed to contain the logical constants **true** and **false**, and the usual punctuation symbols ')' and '('. Finally, note that $\mathcal{L}_{TB}$ is to be used for representing beliefs expressed in the internal language, $\mathcal{L}$. It follows that $\mathcal{L}$ must appear in $\mathcal{L}_{TB}$ somewhere. For simplicity, we shall assume that $\mathcal{L} = \mathcal{L}_{TB}$, i.e., agents are capable of having beliefs about beliefs, and about how beliefs change over time.

Definition 3. The language $\mathcal{L}_{TB}$ contains the following symbols:

1. All symbols of $\mathcal{L}_0$;
2. The set $Ag = \{1, \ldots, n\}$ of agent names;
3. The symbols ']' and '[';
4. The unary temporal connective $\bigcirc$, and binary temporal connective $\mathcal{U}$.

Definition 4. The set $Form(\mathcal{L}_{TB})$ of (well-formed) formulae of $\mathcal{L}_{TB}$ is defined by the following rules:

1. If $\varphi \in Form(\mathcal{L}_0)$ then $\varphi \in Form(\mathcal{L}_{TB})$;
2. If $\varphi \in Form(\mathcal{L}_{TB})$ and $i \in Ag$ then $[i]\varphi \in Form(\mathcal{L}_{TB})$;
3. If $\varphi \in Form(\mathcal{L}_{TB})$ then $\neg\varphi, \bigcirc\varphi, (\varphi) \in Form(\mathcal{L}_{TB})$;
4. If $\varphi, \psi \in Form(\mathcal{L}_{TB})$, then $\varphi \vee \psi, \varphi \mathcal{U} \psi \in Form(\mathcal{L}_{TB})$.

3.2 Semantics

Before we define the semantics of $\mathcal{L}_{TB}$, we must make a number of assumptions plain. First, we assume that an agent's beliefs can change over time. (If we assumed that beliefs were fixed, then there would be little point in having a temporal component in the language.) Secondly, we assume that an agent's reasoning ability, as represented in its b.e. relation, does not change with time. Although dropping this assumption would be relatively simple in terms of semantics, it would complicate the proof theory of the language considerably, and we do not consider it necessary in practice.

Models for $\mathcal{L}_{TB}$ include a valuation function, giving the truth of each primitive proposition at each time; additionally, they include a function which assigns each agent a base set of beliefs at each moment in time, and an indexed set of b.e. relations.

Definition 5. A model, M, for $\mathcal{L}_{TB}$ is a triple $M = \langle \pi, a, \{BE_i\} \rangle$, where

- $\pi : \mathbb{N} \times \Phi \rightarrow \{T, F\}$ interprets propositions at each time point;
- $a : \mathbb{N} \times Ag \rightarrow powerset(Form(\mathcal{L}_{TB}))$ assigns each agent a base set of beliefs at each time; and
- $\{BE_i\}$ is an indexed set of b.e. relations, one for each agent $i \in Ag$.

As usual, we define the semantics of the language via the satisfaction relation '$\models$'. For $\mathcal{L}_{TB}$, this relation holds between pairs of the form $\langle M, u \rangle$, (where M is a model and $u \in \mathbb{N}$ is a temporal index into M), and $\mathcal{L}_{TB}$-formulae. The rules defining the satisfaction relation are given in Fig. 1. Satisfiability and validity for $\mathcal{L}_{TB}$ are defined as follows: if $\varphi \in Form(\mathcal{L}_{TB})$ and there is some $\langle M, u \rangle$ such that $\langle M, u \rangle \models \varphi$, then φ is said to be *satisfiable*, otherwise φ is said to be *unsatisfiable*. If $\neg\varphi$ is unsatisfiable, then φ is *valid* (notation $\models \varphi$).

The remaining temporal connectives of $\mathcal{L}_{TB}$ are introduced as abbreviations.

$$\diamond \varphi \stackrel{def}{=} \mathbf{true}\, \mathcal{U}\, \varphi$$
$$\square \varphi \stackrel{def}{=} \neg \diamond \neg \varphi$$
$$\varphi \mathcal{W} \psi \stackrel{def}{=} \varphi \mathcal{U} \psi \vee \square \varphi$$

```
⟨M, u⟩ ⊨ true
⟨M, u⟩ ⊨ p           iff    π(u, p) = T (where p ∈ Φ)
⟨M, u⟩ ⊨ ¬φ          iff    ⟨M, u⟩ ⊭ φ
⟨M, u⟩ ⊨ φ ∨ ψ       iff    ⟨M, u⟩ ⊨ φ or ⟨M, u⟩ ⊨ ψ
⟨M, u⟩ ⊨ [i]φ        iff    φ ∈ bel((a(u, i), BE_i))
⟨M, u⟩ ⊨ ○φ          iff    ⟨M, u+1⟩ ⊨ φ
⟨M, u⟩ ⊨ φ U ψ       iff    ∃v ∈ IN s.t. (v ≥ u) and ⟨M, v⟩ ⊨ ψ,
                             and ∀w ∈ IN, if (u ≤ w < v) then ⟨M, w⟩ ⊨ φ
```

Fig. 1. Semantics of $\mathcal{L}_{TB}$

We now informally consider the meaning of the connectives. The formula $[i]\varphi$ is read 'agent i believes φ'; it will be satisfied if φ is present in i's belief set at the current time. The $\bigcirc$ connective means 'at the next time'. Thus $\bigcirc \varphi$ will be satisfied at some time point if φ is satisfied at the *next* time point. The $\mathcal{U}$ connective means 'until'. Thus $\varphi \mathcal{U} \psi$ will be satisfied at some time if ψ is satisfied at that time or some time in the future, and φ is satisfied at all times until ψ is satisfied. Of the derived connectives, $\Diamond$ means 'either now, or at some time in the future'. Thus $\Diamond \varphi$ will be satisfied at some time if either φ is satisfied at that time, or some later time. The $\Box$ connective means 'now, and at all future times'. Thus $\Box \varphi$ will be satisfied at some time if φ is satisfied at that time and at all later times. The binary $\mathcal{W}$ connective means 'unless'. Thus $\varphi \mathcal{W} \psi$ will be satisfied at some time if either φ is satisfied until such time as ψ is satisfied, or else φ is always satisfied. Note that $\mathcal{W}$ is similar to, but weaker than, the $\mathcal{U}$ connective; for this reason it is sometimes called 'weak until'.

3.3 Properties of $\mathcal{L}_{TB}$

Since the propositional connectives of $\mathcal{L}_{TB}$ have standard semantics, all propositional tautologies will be valid; additionally, the inference rule *modus ponens* will preserve validity. In short, we can use all propositional modes of reasoning in $\mathcal{L}_{TB}$. The new logic also inherits the axioms and inference rules associated with its temporal component (see, e.g., [4] for discussion). However, $\mathcal{L}_{TB}$ has some additional properties. To illustrate this, we first establish an analogue of Konolige's attachment lemma [14, pp34–35].

Theorem 6. *The set* $\{[i]\Delta, \neg[i]\Delta'\}$ *is unsatisfiable iff* $\exists \varphi \in \Delta'$ *such that* $(\Delta, \varphi) \in BE_i$.

Proof. This, and all remaining proofs, are omitted due to space restrictions; full details may be found in the associated technical report [23].

This theorem allows us to derive a number of results; for example:

Theorem 7. $\models [i]\varphi_1 \wedge \cdots \wedge [i]\varphi_n \Rightarrow [i]\varphi$, *where* $(\{\varphi_1, \ldots, \varphi_n\}, \varphi) \in BE_i$.

This theorem represents the basic mechanism for reasoning about belief systems: if it is known that i believes $\{\varphi_1, \ldots, \varphi_n\}$, and that $(\{\varphi_1, \ldots, \varphi_n\}, \varphi) \in BE_i$, then this implies that i also believes φ. Note that axiom K and the necessitation rule from classical modal logic do *not* in general hold for belief modalities in $\mathcal{L}_{TB}$, and thus $\mathcal{L}_{TB}$ does not fall prey to logical omniscience. However, $\mathcal{L}_{TB}$ *is* capable of representing logically omniscient believers [23].

3.4 Applications of $\mathcal{L}_{TB}$

Temporal belief logics such as $\mathcal{L}_{TB}$ have a number of applications. For example: formalisms for representing the time-varying properties of multi-agent systems are essential in the emerging discipline of Distributed Artificial Intelligence (DAI) [2]; epistemic temporal logics have been used by researchers in computer science to reason about distributed systems [12]; temporal belief logics have recently found a role in the specification and verification of DAI systems [22, 7]; and ultimately, temporal belief logics may even be *executed*, as in [6]. In the remainder of this section, we consider various properties of agents that may be expressed using $\mathcal{L}_{TB}$ (note that we do not consider the 'standard' axioms of belief — KD45 — in this paper; we are concerned instead with axioms in which time and belief interact).

First, consider the *persistence* of belief. Suppose that at some time an agent believes φ, then how long might the agent persist in this belief? An extreme case is that in which, when an agent comes to believe something, it always believes it:

$$[i]\varphi \Rightarrow \Box[i]\varphi. \tag{1}$$

Agents with property (1) are not very interesting from the point of view of AI. A more reasonable assumption is that beliefs persist until a contradictory belief is held:

$$(([i]\varphi) \land (\neg \bigcirc [i]\neg\varphi)) \Rightarrow \bigcirc [i]\varphi. \tag{2}$$

Property (2) may also be expressed as:

$$[i]\varphi \Rightarrow (([i]\varphi) \, \mathcal{W} \, ([i]\neg\varphi)). \tag{3}$$

We might also like to state that if an agent believes that φ will be true at some point in the future, then at some point in the future the agent will believe φ. This gives the following three axioms.

$$[i]\bigcirc \varphi \Rightarrow \bigcirc [i]\varphi \tag{4}$$
$$[i]\Diamond \varphi \Rightarrow \Diamond [i]\varphi \tag{5}$$
$$[i]\varphi \, \mathcal{U} \, \psi \Rightarrow \Diamond [i]\psi \tag{6}$$

Kraus-Lehmann suggest that (4)–(6) describe a notion of belief closer to the sense of 'religious' belief than the everyday notion of belief as 'readiness to bet' [15, pp166-168]. This weaker notion cannot easily be axiomatized, but (7)–(9) seem reasonable properties.

$$[i]\bigcirc \varphi \Rightarrow [i]\bigcirc[i]\varphi \tag{7}$$
$$[i]\Diamond \varphi \Rightarrow [i]\Diamond[i]\varphi \tag{8}$$
$$[i]\varphi \mathcal{U} \psi \Rightarrow [i]\Diamond[i]\psi \tag{9}$$

Finally, consider the class of agents with the ability to affect the future, perhaps by acting in the world. Divide the set Φ of primitive propositions into two disjoint sets, Φ_e (environment propositions) and Φ_a (agent propositions), and let $\Phi_a = \Phi_{a_1} \cup \cdots \cup \Phi_{a_n}$. The idea is that Φ_e is the set of propositions whose truth or falsity is controlled by the world — propositions in this set are not affected by the actions of agents. The set Φ_a contains propositions under the control of individual agents; Φ_{a_i} is the set of the propositions under the control of agent i. Now consider the following axioms:

$$\left. \begin{array}{l} [i]\bigcirc \varphi \Rightarrow \bigcirc \varphi \\ [i]\Diamond \varphi \Rightarrow \Diamond \varphi \end{array} \right\} \quad \text{where } \varphi \in \Phi_{a_i}. \tag{10}$$

The axioms in (10), which are related to axiom T from classical modal logic, state that an agent's beliefs about the future state of the propositions under its control are reflected in the actual future state of those propositions. Axioms like this have been used to reason about the behaviour of systems in the 'imperative future' paradigm [8, 6]; see [7, 22] for details.

4 A Decision Procedure for $\mathcal{L}_{TB}$

In this section, we present a tableau-based decision procedure for $\mathcal{L}_{TB}$. The procedure consists of two functions: a main function *structure*, and an auxiliary function *tableau*. The function *structure* takes as its input an $\mathcal{L}_{TB}$ formula φ, and systematically searches for a model of φ. If φ is satisfiable, then *structure* returns a graph from which a model for φ can be extracted; if φ is unsatisfiable, then *structure* returns an empty graph. The function *tableau* is used to check the internal consistency of states during graph generation. The algorithm draws on the tableau methods for temporal logic described by Wolper [21], Gough [10], and Ben-Ari [1, pp216–228], and in fact generalises the basic temporal tableau method. Note that the algorithm assumes that we have the belief extension relation of each agent available.

As with all tableau-based decision procedures, our procedure relies upon *alpha* and *beta* equivalences; these equivalences are defined in Fig. 2. (We assume that all input formulae are rewritten into a normal form in which negations are only applied to primitive propositions and belief modalities; see [10].)

Definition 8. If φ is an alpha-formula, (notation is-$alpha(\varphi)$), then its components are given by $\alpha_1(\varphi)$ and $\alpha_2(\varphi)$ respectively. If φ is a beta-formula (notation is-$beta(\varphi)$), then its components are given by $\beta_1(\varphi)$ and $\beta_2(\varphi)$ respectively. The *alpha closure* of a formula is given by the function α^*, which has the signature

$$\alpha^* : Form(\mathcal{L}_{TB}) \to powerset(Form(\mathcal{L}_{TB}))$$

α	α_1	α_2
$\varphi \wedge \psi$	φ	ψ
$\Box \varphi$	φ	$\bigcirc \Box \varphi$

β	β_1	β_2
$\varphi \vee \psi$	φ	ψ
$\Diamond \varphi$	φ	$\bigcirc \Diamond \varphi$
$\varphi \mathcal{U} \psi$	ψ	$\varphi \wedge \bigcirc(\varphi \mathcal{U} \psi)$
$\varphi \mathcal{W} \psi$	$\Box \varphi$	$\varphi \mathcal{U} \psi$

Fig. 2. Alpha and Beta Equivalences

and which is defined by

$$\alpha^*(\varphi) \stackrel{\text{def}}{=} \begin{cases} \alpha^*(\alpha_1(\varphi)) \cup \alpha^*(\alpha_2(\varphi)) & \text{if } \textit{is-alpha}(\varphi) \\ \{\varphi\} & \text{otherwise.} \end{cases}$$

The function α^* is extended to sets of formulae in an obvious way.

The internal consistency of nodes during tableau generation is established by checking whether they are *proper*.

Definition 9. If $\Gamma \subseteq \textit{Form}(\mathcal{L}_{TB})$ then Γ is *proper*, (notation *proper*(Γ)) iff:

1. **false** $\notin \Gamma$;
2. If $\varphi \in \Gamma$, then $\neg \varphi \notin \Gamma$;
3. If $\{[i]\varphi_1, \ldots, [i]\varphi_n, \neg[i]\varphi\} \subseteq \Gamma$, then $(\{\varphi_1, \ldots, \varphi_n\}, \varphi) \notin BE_i$.

Note that proper sets may be unsatisfiable, but improper sets are never satisfiable; the only non-obvious part is (3), which is given by Theorem 6. We now define tableau structures (cf. [19]). Nodes are drawn from some arbitrary set *Node*.

Definition 10. A *tableau*, Υ, is a quad $\Upsilon = (N, T, l_1, l_2)$, where

- $N \subseteq \textit{Node}$ is a set of nodes;
- $T \subseteq N \times N$ is a binary tree over N;
- $l_1 : N \rightarrow \textit{powerset}(\textit{Form}(\mathcal{L}_{TB}))$ labels each node with a set of $\mathcal{L}_{TB}$-formulae; and
- $l_2 : N \rightarrow \{o, c\}$ labels each node in N with either o (open) or c (closed).

Let *Tableaux* be the set of all tableaux. If $\Upsilon = (N, T, l_1, l_2)$ is a tableau, then let *leaves*(Υ) denote the subset of N containing the leaves of T.

The function *tableau*, which is used to check the internal consistency of states during graph generation, is given in Fig. 3. This function has one important property:

Lemma 11. *If tableau($\{\varphi\}$) returns a tableau with no open leaves, then φ is unsatisfiable.*

Definition 12. If $\Upsilon = (N, T, l_1, l_2)$ is a tableau and $n \in \textit{leaves}(\Upsilon)$, then denote by *walk*($\Upsilon, n$) the sequence of nodes obtained by walking from the root n_0 of Υ to n. If *walk*(Υ, n) = $(n_0, \ldots, n_k)$ then let *walk-set*(Υ, n) denote the set $l_1(n_0) \cup \cdots \cup l_1(n_k)$.

```
function tableau(Γ : powerset(Form(ℒ_TB))) : (N, T, l₁, l₂) : Tableaux
vars        n, n', n'' : Node
            flag : 𝔹
begin (* initialise *)
    create new node n
    N := {n}                          (* root *)
    T := ∅                            (* empty tree *)
    l₁ := {n ↦ Γ}                     (* label root with Γ *)
    if proper(Γ) then l₂ := {n ↦ o}
                 else l₂ := {n ↦ c}
    repeat (* main loop *)
        flag := false                 (* no leaves created *)
        for each n ∈ leaves((N, T, l₁, l₂)) s.t. l₂(n) = o do
(* α *)     l₁ := l₁ † {n ↦ α*(l₁(n))}
(* β *)     for each φ ∈ l₁(n) s.t. is-beta(φ) do
                create new nodes n' and n''
                N := N ∪ {n', n''}
                l₁ := l₁ † {n' ↦ (l₁(n) − {φ}) ∪ {β₁(φ)}}
                l₁ := l₁ † {n'' ↦ (l₁(n) − {φ}) ∪ {β₂(φ)}}
                T := T ∪ {(n, n'), (n, n'')}
                if proper(l₁(n')) then l₂ := l₂ † {n' ↦ o}
                                  else l₂ := l₂ † {n' ↦ c}
                if proper(l₁(n'')) then l₂ := l₂ † {n'' ↦ o}
                                   else l₂ := l₂ † {n'' ↦ c}
                if l₂(n') = o or l₂(n'') = o then flag := true
            end-for
        end-for
    until ¬flag
end-function
```

Fig. 3. Function *tableau*

We now move on to the model-like graph structures that will be generated by the procedure; states are drawn from some arbitrary set *State*.

Definition 13. A *structure*, H, is a triple $H = (S, R, L)$, where

- $S \subseteq State$ is a set of states;
- $R \subseteq S \times S$ is a binary relation on S; and
- $L : S \to powerset(Form(\mathcal{L}_{TB}))$ labels each state with a set of $\mathcal{L}_{TB}$-formulae.

Let *Structures* be the set of all structures.

Definition 14. If $\varphi \in Form(\mathcal{L}_{TB})$ is of the form $\chi\, \mathcal{U}\, \psi$ or $\Diamond \psi$ then φ is said to have *eventuality* ψ. If (S, R, L) is a structure, $s \in S$ is a state, R^* is the reflexive transitive closure of R, and $\varphi \in Form(\mathcal{L}_{TB})$, then φ is said to be *resolvable* in (S, R, L) from

s, (notation $resolvable(\varphi, s, (S, R, L)))$, iff if φ has eventuality ψ, then $\exists s' \in S$ s.t. $(s, s') \in R^*$ and $\psi \in L(s')$.

Definition 15. If $\Gamma \subseteq Form(\mathcal{L}_{TB})$, then $next(\Gamma)$ is defined:

$$next(\Gamma) \stackrel{\text{def}}{=} \{\varphi \mid \bigcirc \varphi \in \Gamma\}.$$

The decision procedure is then given by the function *structure*, which is presented in Fig. 4. The following two theorems describe the key properties of this function.

Theorem 16. *If $\varphi \in Form(\mathcal{L}_{TB})$, then φ is satisfiable iff $structure(\varphi)$ returns (S, R, L), and $\exists s \in S$ s.t. $\alpha^*(\varphi) \subseteq L(s)$.*

Theorem 17. *If $\varphi \in Form(\mathcal{L}_{TB})$, then $structure(\varphi)$ terminates.*

4.1 Examples

Example 1: The first example is a purely temporal formula taken from [10].

$$(\Diamond p \wedge \Box(p \Rightarrow \bigcirc p)) \Rightarrow \Diamond \Box p \tag{11}$$

After negating and rewriting into normal form, (11) becomes:

$$\Diamond p \wedge \Diamond \neg p \wedge (\neg p \vee \bigcirc p) \wedge \bigcirc \Box(\neg p \vee \bigcirc p) \wedge \bigcirc \Box \Diamond \neg p. \tag{12}$$

The graph generation stage of *structure* terminates after generating a graph containing seven states, s_1–s_7, labelled as follows.

$s_1 = \{p, \bigcirc \Diamond \neg p, \bigcirc p, \bigcirc \Box(\neg p \vee \bigcirc p), \bigcirc \Box \Diamond \neg p, \neg p \vee \bigcirc p, \Diamond \neg p, \Diamond p\}$
$s_2 = \{\bigcirc \Diamond p, \neg p, \bigcirc \Box(\neg p \vee \bigcirc p), \bigcirc \Box \Diamond \neg p, \neg p \vee \bigcirc p, \Diamond \neg p, \Diamond p\}$
$s_3 = \{\bigcirc \Diamond p, \neg p, \bigcirc p, \bigcirc \Box(\neg p \vee \bigcirc p), \bigcirc \Box \Diamond \neg p, \neg p \vee \bigcirc p, \Diamond \neg p, \Diamond p\}$
$s_4 = \{\bigcirc \Diamond p, \bigcirc \Diamond \neg p, \neg p, \bigcirc \Box(\neg p \vee \bigcirc p), \bigcirc \Box \Diamond \neg p, \neg p \vee \bigcirc p, \Diamond \neg p, \Diamond p\}$
$s_5 = \{\bigcirc \Diamond p, \bigcirc \Diamond \neg p, \bigcirc p, \bigcirc \Box(\neg p \vee \bigcirc p), \bigcirc \Box \Diamond \neg p, \neg p \vee \bigcirc p, \Diamond \neg p, \Diamond p\}$
$s_6 = \{\bigcirc \Diamond \neg p, p, \bigcirc p, \bigcirc \Box(\neg p \vee \bigcirc p), \bigcirc \Box \Diamond \neg p, \neg p \vee \bigcirc p, \Diamond \neg p\}$
$s_7 = \{\bigcirc \Diamond p, p, \bigcirc p, \bigcirc \Box(\neg p \vee \bigcirc p), \bigcirc \Diamond \neg p, \bigcirc \Box \Diamond \neg p, \Diamond \neg p, \neg p \vee \bigcirc p, \Diamond p\}$

The final state of R is summarised in the following adjacency matrix.

From ...

	s_1	s_2	s_3	s_4	s_5	s_6	s_7
s_1		×	×	×	×		×
s_2		×		×			
s_3		×		×			
s_4		×		×			
s_5		×		×			
s_6	×					×	
s_7			×		×		×

To ... (applies to row labels)

Graph contraction then begins; as all states contain unresolved eventualities, they are all deleted. The graph that is returned is therefore empty, so (12) is unsatisfiable, meaning that (11) is valid.

```
function structure(φ : Form(L_TB)) : (S, R, L) : Structures
vars            (N, T, l₁, l₂) : Tableaux
                n : Node
                s, s' : State
                flag : B
begin                      (* stage 1: initialise *)
        S := R := L := ∅
        (N, T, l₁, l₂) := tableau({φ})
        for each n ∈ leaves((N, T, l₁, l₂)) s.t. l₂(n) = o do
                create new state s
                S := S ∪ {s}
                L := L † {s ↦ walk-set((N, T, l₁, l₂), n)}
        end-for
        repeat              (* stage 2: create graph *)
                flag := false
                for each s ∈ S do
                        (N, T, l₁, l₂) := tableau(next(L(s)))
                        for each n ∈ leaves((N, T, l₁, l₂)) s.t. l₂(n) = o do
                                if ∃s' ∈ S s.t. walk-set((N, T, l₁, l₂), n) = L(s') then
                                        R := R ∪ {(s, s')}
                                                    else
                                        create new state s'
                                        S := S ∪ {s'}
                                        L := L † {s' ↦ walk-set((N, T, l₁, l₂), n)}
                                        R := R ∪ {(s, s')}
                                        flag := true
                        end-for
                end-for
        until ¬flag
        repeat              (* stage 3: contract graph *)
                flag := false
                for each s ∈ S do
                        if ∃ψ ∈ L(s) s.t. ¬resolvable(ψ, s, (S, R, L)) or
                                ∃ψ ∈ L(s) s.t. ψ is of the form ○χ
                                and ¬∃s' ∈ S s.t. (s, s') ∈ R and χ ∈ L(s')
                        then
                                S := S − {s}
                                flag := true
                end-for
        until ¬flag
end-function
```

Fig. 4. Function *structure*

Example 2: The formula used in the previous example contained only classical and temporal connectives; it contained no belief modalities. In this section, we present an example in which belief and time interact. Imagine an agent i that is a perfect propositional reasoner; that is, whenever i believes Δ, then i will also believe φ if $\Delta \vdash_{\mathcal{L}_0} \varphi$ (i.e., if there is a proof of φ from Δ in $\mathcal{L}_0$). Thus $(\Delta, \varphi) \in BE_i$ iff $\Delta \vdash_{\mathcal{L}_0} \varphi$. Now consider the following formula.

$$\Box[i](p \wedge (p \Rightarrow q)) \wedge \Diamond \neg [i]q \qquad (13)$$

Intuitively, it is easy to see that (13) is unsatisfiable: it is obvious that $(p \wedge (p \Rightarrow q)) \vdash_{\mathcal{L}_0} q$, and so if i always believes $(p \wedge (p \Rightarrow q))$, then i also always believes q. Hence i can never not believe q. We use the decision procedure to show this formally.

The alpha closure of (13) is:

$$\{[i](p \wedge (p \Rightarrow q)), \bigcirc \Box[i](p \wedge (p \Rightarrow q)), \Diamond \neg [i]q\}.$$

The tableau for this set has two leaves, n_1 and n_2, labelled thus:

$$n_1 = \{[i](p \wedge (p \Rightarrow q)), \bigcirc \Box[i](p \wedge (p \Rightarrow q)), \neg[i]q\}$$
$$n_2 = \{[i](p \wedge (p \Rightarrow q)), \bigcirc \Box[i](p \wedge (p \Rightarrow q)), \bigcirc \Diamond \neg [i]q\}.$$

Node n_1 is closed: it contains both $[i](p \wedge (p \Rightarrow q))$ and $\neg[i]q$, and $(\{p \wedge (p \Rightarrow q)\}, q) \in BE_i$, since $p \wedge (p \Rightarrow q) \vdash_{\mathcal{L}_0} q$. (See Definition 9(3)). Node n_2 is open, so a state s_1 is created in the graph, labelled:

$$\{[i](p \wedge (p \Rightarrow q)), \bigcirc \Box[i](p \wedge (p \Rightarrow q)), \bigcirc \Diamond \neg [i]q, \Diamond \neg[i]q\}.$$

However, the next time formulae of this set correspond to the alpha closure of the input formula, so we need not build another tableau; we simply make a link from s_1 to itself. Graph generation then ends, and graph contraction begins: state s_1 is deleted, as it contains an unresolved eventuality ($\Diamond \neg [i]q$). The structure returned is thus empty, and (13) is therefore unsatisfiable.

Example 3: The *wisest man puzzle* is a classic problem in reasoning about knowledge and belief that is widely used as a benchmark against which formalisms for representing these notions are evaluated. We have used the decision procedure for $\mathcal{L}_{TB}$ to solve a variant of the problem, which involves an element of time. The variant we used, in its most general form, may be stated as follows (see [14, p58] for the original):

> A king wishes to know which of his n advisors is the wisest. He arranges them in a circle, so that they can both see and hear each other, and tells them that he will paint either a white or black dot on each of their foreheads, but that at least one dot will be white. He offers his favour to the one that can correctly identify the colour of his spot. At time 1 he asks advisor 1 if he knows the colour of his spot; the advisor does not know. At time 2 he asks advisor 2 if he knows the colour of his spot; he does not know. The king continues in this way, until at time n he asks advisor n, who correctly identifies that his spot is white.

A solution to the problem (without a temporal component) may be found in [14, pp57–61]; another interesting solution, involving common knowledge and time, and given as a proof in a Hilbert-style axiom system appears in [15, pp168–174][2]. We now give an axiomatisation of the puzzle. We write $w(i)$ for 'agent i's dot is white'. First, we state that every dot is white.

$$\Box \bigwedge_{i=1}^{n} w(i) \qquad (14)$$

Next, we need to state that it is *mutually believed* that at least one dot is white. Since $\mathcal{L}_{TB}$ does not contain a mutual belief operator, we define one: we write $[M]\varphi$ if $[i]\varphi$, and $[i][j]\varphi$ and $[i][j][k]\varphi$, and so on.

$$[M]\varphi \stackrel{\text{def}}{=} ([i_1]\varphi) \wedge ([i_1][i_2]\varphi) \wedge \cdots \wedge ([i_1][i_2]\cdots[i_n]\varphi) \qquad \text{for all } i_1, \ldots, i_n \in \{1, \ldots, n\}$$

The mutual belief that at least one dot is white is represented by the following axiom.

$$\Box [M] \bigvee_{j=1}^{n} w(j) \qquad (15)$$

The following *observation axioms* state that each advisor can see everyone else's dot.

$$\left. \begin{array}{l} \Box w(i) \Rightarrow [j]w(i) \\ \Box \neg w(i) \Rightarrow [j]\neg w(i) \end{array} \right\} \qquad \text{for all } i, j \in \{1, \ldots, n\} \text{ s.t. } i \neq j \qquad (16)$$

The axioms in (16) are mutually believed.

$$\left. \begin{array}{l} \Box [M]w(i) \Rightarrow [j]w(i) \\ \Box [M]\neg w(i) \Rightarrow [j]\neg w(i) \end{array} \right\} \qquad \text{for all } i, j \in \{1, \ldots, n\} \text{ s.t. } i \neq j \qquad (17)$$

Advisors are at least partially consistent: if they believe that some advisor's dot is white, then they do not believe it is not white.

$$\Box [i]w(j) \Rightarrow \neg[i]\neg w(j) \qquad \text{for all } i, j \in \{1, \ldots, n\} \qquad (18)$$

Axiom (18) is mutually believed.

$$\Box [M](([i]w(j)) \Rightarrow \neg[i]\neg w(j)) \qquad \text{for all } i, j \in \{1, \ldots, n\} \qquad (19)$$

At time $u \in \{1, \ldots, n-1\}$, advisor u reveals that he does not know the colour of his dot, making this mutually believed:

$$\bigcirc^u \Box [M] \neg [u] w(u) \qquad \text{for all } u \in \{1, \ldots, n-1\} \qquad (20)$$

where $\bigcirc^u$, for $u \in \mathbb{N}$, means $\bigcirc$ iterated u times. Finally, the aim of the puzzle is to show that at time n, advisor n knows that his spot is white.

$$\bigcirc^n [n] w(n) \qquad (21)$$

---

[2] However, the logic used in [15] is not given a semantics.

We have used the $\mathcal{L}_{TB}$ decision procedure to solve the 2-advisor version of this problem; full details are included in the associated technical report [23]. Extensions to the general n-advisor case are not problematic.

An interesting aspect of the decision procedure, when applied to this and many other problems, is that showing that a set of $\mathcal{L}_{TB}$-formulae containing belief modalities is improper, (and thus unsatisfiable), involves a recursive call on the decision procedure, to show that a simpler set of formulae (essentially the original belief formulae with the outermost belief modality stripped off) is improper (cf. [9]). While this has obvious implications with respect to efficiency, it has the advantage of being conceptually a very simple way of dealing with belief modalities.

5 Concluding Remarks

In this paper, we have developed a temporal belief logic called $\mathcal{L}_{TB}$. By using this logic, it is possible to represent the time-varying properties of systems containing multiple resource-bounded reasoning agents. We have discussed both the applications of the logic, and some properties that might be represented in it, and have presented a tableau-based decision procedure for it.

Future work will focus on the following areas: extensions of the decision procedure to restricted first-order logics; implementing and improving the efficiency of the decision procedure; decision procedures for normal modal temporal belief logics, (and ultimately many-dimensional modal logics in general); and resolution-style calculi for temporal belief logics, (perhaps based on [5]).

References

1. M. Ben-Ari. *Mathematical Logic for Computer Science*. Prentice Hall, 1993.
2. A. H. Bond and L. Gasser, editors. *Readings in Distributed Artificial Intelligence*. Morgan Kaufmann Publishers, Inc., 1988.
3. L. Catach. Normal multimodal logics. In *Proceedings of the National Conference on Artificial Intelligence (AAAI '88)*, St. Paul, MN, 1988.
4. E. A. Emerson. Temporal and modal logic. In J. van Leeuwen, editor, *Handbook of Theoretical Computer Science*, pages 996–1072. Elsevier, 1990.
5. M. Fisher. A resolution method for temporal logic. In *Proceedings of the Twelfth International Joint Conference on Artificial Intelligence (IJCAI '91)*, Sydney, Australia, August 1991.
6. M. Fisher and M. Wooldridge. Executable temporal logic for distributed A.I. In *Proceedings of the 12th International Workshop on Distributed Artificial Intelligence*, Hidden Valley, PA, May 1993.
7. M. Fisher and M. Wooldridge. Specifying and verifying distributed intelligent systems. In M. Filgueiras and L. Damas, editors, *Progress in Artificial Intelligence — Sixth Portuguese Conference on Artificial Intelligence (LNAI Volume 727)*, pages 13–28. Springer-Verlag, October 1993.
8. D. Gabbay. Declarative past and imperative future. In B. Banieqbal, H. Barringer, and A. Pnueli, editors, *Proceedings of the Colloquium on Temporal Logic in Specification (LNCS Volume 398)*, pages 402–450. Springer-Verlag, 1989.

9. C. Geissler and K. Konolige. A resolution method for quantified modal logics of knowledge and belief. In J. Y. Halpern, editor, *Proceedings of the 1986 Conference on Theoretical Aspects of Reasoning About Knowledge*, pages 309–324. Morgan Kaufmann Publishers, Inc., 1986.
10. G. D. Gough. Decision procedures for temporal logic. Master's thesis, Department of Computer Science, Manchester University, Oxford Rd., Manchester M13 9PL, UK, October 1984.
11. J. Y. Halpern and Y. Moses. A guide to completeness and complexity for modal logics of knowledge and belief. *Artificial Intelligence*, 54:319–379, 1992.
12. J. Y. Halpern and M. Y. Vardi. The complexity of reasoning about knowledge and time. I. lower bounds. *Journal of Computer and System Sciences*, 38:195–237, 1989.
13. C. B. Jones. *Systematic Software Development using VDM (second edition)*. Prentice Hall, 1990.
14. K. Konolige. *A Deduction Model of Belief*. Pitman/Morgan Kaufmann, 1986.
15. S. Kraus and D. Lehmann. Knowledge, belief and time. *Theoretical Computer Science*, 58:155–174, 1988.
16. Z. Manna and A. Pnueli. *The Temporal Logic of Reactive and Concurrent Systems*. Springer-Verlag, 1992.
17. H. Reichgelt. Logics for reasoning about knowledge and belief. *Knowledge Engineering Review*, 4(2), 1989.
18. Y. Shoham. *Reasoning About Change: Time and Causation from the Standpoint of Artificial Intelligence*. The MIT Press, 1988.
19. R. M. Smullyan. *First-Order Logic*. Springer-Verlag, 1968.
20. R. H. Thomason. Combinations of tense and modality. In D. Gabbay and F. Guenther, editors, *Handbook of Philosophical Logic Volume II — Extensions of Classical Logic*, pages 135–166. D. Reidel Publishing Company, 1984. (Synthese library Volume 164).
21. P. Wolper. The tableau method for temporal logic: An overview. *Logique et Analyse*, 110–111, 1985.
22. M. Wooldridge. *The Logical Modelling of Computational Multi-Agent Systems*. PhD thesis, Department of Computation, UMIST, Manchester, UK, October 1992.
23. M. Wooldridge. A temporal belief logic. Technical report, Department of Computing, Manchester Metropolitan University, Chester St., Manchester M1 5GD, UK, 1994.

Decidability of Deliberative *Stit* Theories with Multiple Agents

Ming Xu
Department of Philosophy, University of Pittsburgh, Pittsburgh PA 15260, U.S.A.
mxust@vms.cis.pitt.edu

Abstract

The purpose of this paper is to present axiomatizations for some *dstit* (*deliberatively seeing-to-it-that*) logics with multiple agents, and prove the completeness and decidability by way of *finite model property*. The *dstit* logics considered in this paper include the basic *dstit* logic, and for every $n \geq 1$, a logic in which every agent has at most n possible choices at every moment.

For a long time philosophers and logicians have treated agency or action as a modality.[1] In this tradition, agency or action of an individual is represented by, *via* a modal operator, a particular state of affairs that the individual brings about.[2] One aspect of this tradition is to take the phrase "see to it that" (*stit* for short) as a modal operator in various development of logics of agency. Theories of *stit* have been proposed by B. Chellas (see CHELLAS 1969 and 1992), F. von Kutschera (VON KUTSCHERA 1986) and J. Horty (HORTY 1989), and N. Belnap and M. Perloff (starting with BELNAP and PERLOFF 1988).[3] The logics considered in this paper are in accordance with the theory of *stit* proposed independently by von Kutschera and Horty. This theory is now often referred to as *the deliberative stit* (*dstit* for short). This terminology reminds us immediately the notion of deliberative obligation in THOMASON 1981, and goes far back to Aristotle's discussion in *Nichomachean Ethics* about deliberation. Some applications of *dstit* theory in deontic logic have been discussed in BARTHA 1993, BELNAP and BARTHA 1993, and HORTY and BELNAP 1993, etc. In this paper, we will give an axiomatization for the basic *dstit* logic, and prove its completeness and decidability.

---

1. I would like to give thanks to professor N. Belnap for his encouragement and constant help, and to my friend P. Bartha for some useful discussions on the topic.

2. The tradition of treating agency as a modality has been traced at least as far back as St. Anselm. It has been represented by many philosophers and logicians in this century. For a historical review on the subject, see BELNAP 1991 and SEGERBERG 1992.

3. Belnap and Perloff's theory, often referred to as *the achievement stit* (*astit* for short), has been developed in a series of articles, including BELNAP and PERLOFF 1988, 1992, 1993, BELNAP 1991, 1991a, 1991b, PERLOFF 1991, and XU 1993, 1993a, 1994.

A *dstit* sentence is [α *dstit:* A] (read "α deliberately sees to it that A"), where α is any agent term and A any sentence. It is interpreted as, roughly, that A is guaranteed true by a choice made by α. The strict semantics for *dstit* is based on the theory of branching time proposed by A. Prior and R. Thomason (PRIOR 1967 and THOMASON 1970). It seems very natural to combine *dstit* theory with indeterminist tense logics, especially when we consider *deliberately seeing to something* to be connected with what future is like. However, as doing a basic technical work in *dstit* theory, we will use a formal language without tense operators, though we will use the historical necessity operator □ (THOMASON 1970) as a primitive symbol.

Chellas proposed an operator Δ for *stit* in CHELLAS 1992, which is very like the *dstit* operator. In our formal language, we will introduce this operator as an abbreviation. In the presence of □, Chellas' operator and *dstit* operator are interdefinable. In fact, each of the three operators □, [*dstit:*] and Δ can be defined in terms of the other two. The reader familiar with modal logic can easily see that Chellas' theory of *stit* is decidable since he did not propose any condition concerning the relation among different agents. Our proof actually shows that Chellas' theory of *stit* (CHELLAS 1992) is decidable even if he accepts the condition *independence of agents* (defined in §1).[4]

How many possible choices does an agent have at a given moment? For a logic of agency, this is an interesting question. We will show some results about the expressibility of our formal language in this aspect, i.e., for each number $n \geq 1$, we have a scheme that corresponds to the condition that every agent has at most n possible choices at every moment. We will also give, for each $n \geq 1$, an axiomatization for the *dstit* logic in which every agent has at most n possible choices at every moment, and prove its decidability.

1. Syntax and Semantics

Our language contains, as primitive symbols, propositional variables $p_0, p_1, p_2, \ldots$, agent terms $\alpha_0, \alpha_1, \alpha_2, \ldots$, an equation symbol =, truth-functional operators ~ and ∧, the historical necessity operator □, and the *dstit* operator [*dstit:*]. Formulas are constructed in the usual way, except that $\alpha = \beta$ is a formula for any agent terms α and β, and that [α *dstit:* A] is a formula whenever α is an agent term and A is a formula. We will use A, B, C, etc. to range over formulas. As usual, ∨, →, ↔, ⊤, ⊥ and ◇ are introduced as abbreviations. In addition we use $\alpha \neq \beta$ for $\sim \alpha = \beta$ and introduce the following as abbreviations, where A is any formula, and $\alpha, \beta_0, \beta_1, \ldots$ are any agent terms:

1.1. $\Delta \alpha A =_{df} [\alpha \ dstit: A] \vee \Box A$.
1.2. $\textit{diff}(\beta_0) =_{df} \top$, and for any $n \geq 0$,
$\textit{diff}(\beta_0, \ldots, \beta_{n+1}) =_{df} \textit{diff}(\beta_0, \ldots, \beta_n) \wedge \beta_0 \neq \beta_{n+1} \wedge \ldots \wedge \beta_n \neq \beta_{n+1}$.

---

4. Note that Chellas in fact proposed two different operators for *stit*, one was proposed in CHELLAS 1969 and discussed in CHELLAS 1992, and the other was proposed in CHELLAS 1992. I am here talking about the latter, the operator Δ in CHELLAS 1992 (for convenience we use Δ). Note also that the background of Chellas' semantics for *stit* actually falls into the category of T×W account, but his account can be easily translated into a semantics against the background of "tree-like frames".

A *semantic structure for dstit* (*structure* for short) is any quartet $S = \langle T, \leq, Agent, Choice \rangle$ satisfying the following postulates: T is a nonempty set whose elements m, m' etc. are called *moments*, and $\leq$ is a partial order on T subject to *historical connection*, $\forall m \forall m' \exists m'' (m'' \leq m \land m'' \leq m')$, and *no downward branching*, $\forall m \forall m' \forall m'' (m' \leq m \land m'' \leq m \rightarrow m' \leq m'' \lor m'' \leq m')$. For each $m, m' \in T$, we define $m < m'$ iff $m \leq m'$ and $m \neq m'$. For any maximal chains h and h' of moments in T (called *histories* in T), h and h' are *undivided at m*, written $h \equiv_m h'$, iff $\exists m' (m < m' \land m' \in h \cap h')$. For each $m \in T$, we set $H_{(m)} = \{h : m \in h\}$. *Agent* is a nonempty set, whose elements a, a', etc. are called *agents*. *Choice* is a function on $Agent \times T$ such that for each $a \in Agent$ and $m \in T$, $Choice(a, m)$ is a partition of $H_{(m)}$. The elements of $Choice(a, m)$ are called *possible choices for α at m*. h and h' are *choice equivalent for a at m*, written $h \equiv^a_m h'$, if $h, h' \in H$ for some $H \in Choice(a, m)$. The function *Choice* is subject to the postulates *no choice between undivided histories*, $\forall m \forall a \forall h \forall h' (h \equiv_m h' \rightarrow h \equiv^a_m h')$, and *independence of agents*, i.e., for each $m \in T$, and for each function $choice_m$ on *Agent* such that $choice_m(a) \in Choice(a, m)$ for all $a \in Agent$, $\bigcap \{choice_m(a) : a \in Agent\} \neq \emptyset$.[5] When $Choice(a, m) = \{H_{(m)}\}$, we say that a has a *vacuous choice at m*. An agent a has *at most n possible choices at m* iff $|Choice(a, m)| \leq n$.

What the postulate *independence of agent* says is that any combination of choices made by different agents at the same moment must be consistent. The idea behind this postulate is roughly that at each moment, every agent can choose each of his alternatives, no matter what the other agents are doing at this moment. For a discussion about all the postulates mentioned above, see, e.g., BELNAP 1991 or 1991a.

A *model M on S* is a pair $\langle S, V \rangle$ where S is a structure and V is a valuation such that for each propositional variable p, $V(p)$ is a subset of $\{\langle m, h \rangle : m \in h\}$, and for each agent term α, $V(\alpha) \in Agent$. When V is given, we use $h' \equiv^\alpha_m h$ rather than $h' \equiv^{V(\alpha)}_m h$ for every agent term α. Let A be any formula. We define the truth of A at (m/h) in M, written $M \models A$ $[m/h]$, recursively as follows, where $m \in h$, p is any propositional variable, and α and β are any agent terms:

$M \models p$ $[m/h]$	iff	$\langle m, h \rangle \in V(p)$;
$M \models \alpha = \beta$ $[m/h]$	iff	$V(\alpha) = V(\beta)$;
$M \models \sim A$ $[m/h]$	iff	$M \not\models A$ $[m/h]$ (not $M \models A$ $[m/h]$);
$M \models A \land B$ $[m/h]$	iff	$M \models A$ $[m/h]$ and $M \models B$ $[m/h]$;
$M \models \Box A$ $[m/h]$	iff	$M \models A$ $[m/h']$ for all h' with $m \in h'$;
$M \models [\alpha$ *dstit:* $A]$ $[m/h]$	iff	(i) $M \models A$ $[m/h']$ for all h' with $h' \equiv^\alpha_m h$, and (ii) $M \not\models A$ $[m/h'']$ for some h'' with $m \in h''$.

There are two conditions in the definition of the truth of $[\alpha$ *dstit:* $A]$ at m/h. Condition (i) requires that "α should act at m in such a way that the truth of A is guaranteed; α should constrain the histories through m to lie among those on which A is true", while condition (ii) requires that "A should not be settled true, so that α's actions can be seen as having some real effect". (HORTY and BELNAP 1993) It is easy to verify by §1.1 and the definition above that the following hold, where α is any agent term:

---

5. Some disagreement and discussions about the postulate *independence of agents* can be found in BELNAP and PERLOFF 1993 and CHELLAS 1992.

1.3. $M \models \Diamond A \ [m/h]$ iff $M \models A \ [m/h']$ for some h' with $m \in h$;
$M \models \Delta\alpha A \ [m/h]$ iff $M \models A \ [m/h']$ for all h' with $h' \equiv_m^\alpha h$
                              iff $M \models \Delta\alpha A \ [m/h']$ for all h' with $h' \equiv_m^\alpha h$.

Hence $M \models [\alpha \ dstit: A] \ [m/h]$ iff $M \models \Delta\alpha A \land \Diamond \sim A \ [m/h]$. Note that *choice equivalence relation* is an equivalence relation, and hence by the truth definition above, the operators $\Box$, and $\Delta\alpha$, $\Delta\beta$, etc. are just like the operator in modal logic *S5*. Let us define the *validity of A for a model M*, $M \models A$, as $M \models A \ [m/h]$ for every m and h in M with $m \in h$, and the *validity of A for a structure S*, $S \models A$, as $M \models A$ for every model M on S.

Our logic L_0 takes as axioms all substitution instances of truth-functional tautologies as well as the following schemata, where α, β, γ and $\beta_0, ..., \beta_k$ are any agent terms:

A1 $\Box(A \to B) \to (\Box A \to \Box B)$, $\Box A \to A$, $\Diamond A \to \Box\Diamond A$
A2 $\Delta\alpha(A \to B) \to (\Delta\alpha A \to \Delta\alpha B)$, $\Delta\alpha A \to A$, $\sim\Delta\alpha A \to \Delta\alpha \sim \Delta\alpha A$
A3 $[\alpha \ dstit: A] \to \sim \Box A$
A4 $\alpha = \alpha$, $\alpha = \beta \to \beta = \alpha$, $\alpha = \beta \land \beta = \gamma \to \alpha = \gamma$
A5 $\alpha = \beta \to (A \to A(\alpha/\beta))$ where $A(\alpha/\beta)$ is any formula obtained from A by replacing some or all occurrences of α with β,
AIA$_k$ $\textit{diff}(\beta_0, ..., \beta_k) \land \Diamond\Delta\beta_0 B_0 \land ... \land \Diamond\Delta\beta_k B_k \to \Diamond(\Delta\beta_0 B_0 \land ... \land \Delta\beta_k B_k)$ $(k \geq 1)$

and takes as rules of inference *modus ponens* and

RN   from A to infer $\Box A$.[6]

AIA can be called an *axiom scheme for independence of agents*. For each $n \geq 1$, we define L_n to be the axiomatic system obtained by adding the following to L_0 as an extra axiom scheme, where α is any agent term:

APC$_n$ $\Diamond\Delta\alpha A_1 \land \Diamond(\sim A_1 \land \Delta\alpha A_2) \land ... \land \Diamond(\sim A_1 \land ... \land \sim A_{n-1} \land \Delta\alpha A_n) \to A_1 \lor ... \lor A_n$

APC can be called an *axiom scheme for possible choices*. In the next section we show that for every structure S and for every $n \geq 1$, $S \models APC_n$ *iff* every agent (in S) has at most n possible choices at every moment (in S). Other syntactic notions such as theorems in L_n ($\vdash_{L_n}$), L_n-consistency (consistency w.r.t. L_n), maximally consistent sets w.r.t. L_n (L_n-MCS) etc., are defined as usual. It is easy to see that for each i and k with $1 \leq i < k$, $\vdash_{L_i} APC_k$, and hence, identifying L_n with all its theorems, we have $L_0 \subseteq ... \subseteq L_3 \subseteq L_2 \subseteq L_1$. In fact, as the soundness theorem shows, we have $L_0 \subset ... \subset L_3 \subset L_2 \subset L_1$. It is easy to see by §1.1 that the following are derivable in L_0 (and hence in all L_n):

T1 $[\alpha \ dstit: A] \leftrightarrow \Delta\alpha A \land \sim \Box A$ (A3)
T2 $\Box A \leftrightarrow \Delta\alpha A \land \sim [\alpha \ dstit: A]$ (A3)
T3 $\alpha = \beta \to \Box \alpha = \beta$, $\alpha \neq \beta \to \Box \alpha \neq \beta$ (A5, A1, RN)

---

6. When taking $\Box$ and Δ to be primitive, one needs to introduce $[\alpha \ dstit: A] =_{df} \Delta\alpha A \land \sim \Box A$ as an abbreviation, and replace *A3* by $\Box A \to \Delta\alpha A$. When taking [*dstit:*] and Δ to be primitive, one needs to introduce $\Box A =_{df} \Delta\alpha A \land \sim [\alpha \ dstit: A]$ for some particular α, and replace *A3* by $[\alpha \ dstit: A] \to \Delta\alpha A$.

R1	$A / \Delta \alpha A.$	*(RN, T2)*
T4	$\sim [\alpha\ dstit:\ A] \rightarrow \Delta \alpha \sim [\alpha\ dstit:\ A]$	*(T1, A1, A2, T2, R1)*

Note that although the decidability of L_0 will also give us the decidability of the *dstit* theory with a single agent, it would be nice to know an axiomatization of that theory. It is easy to check, along with our proof, that *AA1−3* and *RN* will be sufficient for that axiomatization. Note also that *A2* does not directly show the *a priori* plausibility for *dstit* operator. It is easy to see, however, that the following are derivable in L_0:

T5	$[\alpha\ dstit:\ A] \rightarrow A$	*(T1, A2)*
T6	$[\alpha\ dstit:\ A] \rightarrow [\alpha\ dstit:\ [\alpha\ dstit:\ A]]$	*(A1, A2, T1, T2)*
T7	$[\alpha\ dstit:\ A] \wedge [\alpha\ dstit:\ B] \rightarrow [\alpha\ dstit:\ A \wedge B]$	*(A1, T1)*
T8	$[\alpha\ dstit:\ A] \wedge \Box B \rightarrow [\alpha\ dstit:\ A \wedge B]$	*(T1, A1, A2, RN, R1)*
T9	$[\alpha\ dstit:\ A \wedge B] \rightarrow [\alpha\ dstit:\ B] \vee \Box B$	*(T1, A2, R1)*
T10	$\Diamond [\alpha\ dstit:\ A] \wedge \sim [\alpha\ dstit:\ A] \rightarrow [\alpha\ dstit:\ \sim [\alpha\ dstit:\ A]]$	*(T4, T1)*
T11	$\Box(A \leftrightarrow B) \rightarrow ([\alpha\ dstit:\ A] \rightarrow [\alpha\ dstit:\ B])$	*(A1, A2, T1, T2, RN)*
T12$_k$	$\textit{diff}(\beta_0, ..., \beta_k) \wedge \Diamond [\beta_0\ dstit:\ B_0] \wedge ... \wedge \Diamond [\beta_k\ dstit:\ B_k]$	
	$\rightarrow \Diamond ([\beta_0\ dstit:\ B_0] \wedge ... \wedge [\beta_k\ dstit:\ B_k])$	*(T1, A1, AIA$_k$)*

Let L be the axiomatic system obtained from L_0 by replacing *A2* with *T5−T11*, and replacing AIA_k with $T12_k$. Then L and L_0 are equivalent in the sense that they have the same set of theorems (the proof is omitted).

A philosophically-minded reader may find the following facts interesting: In both *astit* and *dstit* theories, *doing* (w.r.t. the fact that A) is represented by $[\alpha\ stit:\ A]$ ("α sees to it that A"), *refraining from doing* is represented by $[\alpha\ stit:\ \sim [\alpha\ stit:\ A]]$, and *refraining from refraining from doing* is represented by $[\alpha\ stit:\ \sim [\alpha\ stit:\ \sim [\alpha\ stit:\ A]]]$. In *dstit* theory, *being able to do* is represented by $\Diamond [\alpha\ dstit:\ A]$.[7] It is easy to see that the following are all derivable in L_0:

T13	$\Diamond [\alpha\ dstit:\ A] \wedge \sim [\alpha\ dstit:\ A] \leftrightarrow [\alpha\ dstit:\ \sim [\alpha\ dstit:\ A]]$	*(T10, T5, A3)*
T14	$[\alpha\ dstit:\ A] \rightarrow [\alpha\ dstit:\ \sim [\alpha\ dstit:\ \sim [\alpha\ dstit:\ A]]]$	*(A1, T5, T6, A3, T10)*
T15	$[\alpha\ dstit:\ \sim [\alpha\ dstit:\ \sim [\alpha\ dstit:\ A]]] \rightarrow [\alpha\ dstit:\ A]$	*(T13, T10, A1, A3)*

T14 and *T15* give us that *doing* is equivalent to *refraining from refraining*. That is to say, the *refref conjecture* (see BELNAP 1991a) holds for *dstit*.[8] *T13* indicates that the *stit* analysis of *refraining* coincides with von Wright's analysis of *refraining*—not doing conjoined with the ability of doing (see VON WRIGHT 1963).[9]

---

7. See BELNAP and PERLOFF 1988, PERLOFF 1991, or HORTY and BELNAP 1993 for an analysis of refraining and refraining from refraining; and see HORTY and BELNAP 1993 for an analysis of the ability of doing.

8. Note that the equivalence between *doing* and *refraining from refraining* does not in general hold for *astit* (see BELNAP 1991, BELNAP and PERLOFF 1992, and XU 1993).

9. I learned this from HORTY and BELNAP 1993. It is noted there that the left-hand side of *T13* is only an approximation of von Wright's analysis of refraining.

2. Soundness

In this section we prove the validity of AIA_k for all $k \geq 1$ and some results about the expressibility of the language concerning possible choices, and finally we establish the soundness theorem for L_n with all $n \geq 0$.

2.1. Lemma. *For each $k \geq 1$, AIA_k is valid for every structure.*

Proof. Let $S = \langle T, \leq, Agent, Choice \rangle$ be any structure, and let $M = \langle S, V \rangle$ be any model on S, and let $k \geq 1$. Suppose that $M \models \textit{diff}(\beta_0, ..., \beta_k) \wedge \Diamond \Delta \beta_0 B_0 \wedge ... \wedge \Diamond \Delta \beta_k B_k$ [m/h] for some m and h in M with $m \in h$. Then by truth definition and §1.2, all $V(\beta_0), ..., V(\beta_k)$ are different agents in $Agent$, and there are $h_0, ..., h_k$ such that for each i with $0 \leq i \leq k$, $M \models \Delta \beta_i B_i$ [m/h_i]. Hence by §1.3,

(*) $\quad M \models \Delta \beta_i B_i$ [m/h'] for all $h' \equiv_m^{\beta_i} h_i$.

Let $choice_m$ be any function on $Agent$ such that $choice_m(\beta_i) = \{h' : h' \equiv_m^{\beta_i} h_i\}$ for each i with $0 \leq i \leq k$. Then by *independence of agents*, we know that $\bigcap_{0 \leq i \leq k} choice_m(\beta_i) \neq \emptyset$. Hence by (*), there is an h'' with $m \in h''$ such that $M \models \Delta \beta_0 B_0 \wedge ... \wedge \Delta \beta_k B_k$ [m/h''], and then by §1.3, $M \models \Diamond (\Delta \beta_0 B_0 \wedge ... \wedge \Delta \beta_k B_k)$ [m/h]. Hence $M \models AIA_k$. It follows that $S \models AIA_k$. ■

Although all AIA_k with $k \geq 1$ are valid under the condition *independence of agents*, the set $\{AIA_k : k \geq 1\}$ does not fully express this condition. This is so simply because in a structure $S = \langle T, \leq, Agent, Choice \rangle$, it could be true that $|Agent| \geq \aleph_0$.[10] Fortunately, the presence of AIA_k with all $k \geq 1$ is sufficient for our proof of the completeness theorem.

2.2. Lemma. *Let $S = \langle T, \leq, Agent, Choice \rangle$ be any structure, and let $n \geq 1$. Then $S \models APC_n$ iff $Choice(a, m)$ has at most n elements for all $a \in Agent$ and all $m \in T$.*

Proof. Suppose that $|Choice(a, m)| \leq n$ for all $a \in Agent$ and all $m \in T$. Let M be any model on S. If $n = 1$, then it is easy to see that $M \models \Diamond \Delta \alpha A \rightarrow A$. Assume that $n > 1$ and set α to be any agent term, and $A_1, ..., A_n$ any formulas. We show as follows that $M \models APC_n$. Suppose for reductio that $M \not\models APC_n$ [m/h] for some moment m and some history h with $m \in h$. Then $M \models \Diamond \Delta \alpha A_1$ [m/h], $M \models \Diamond (\sim A_1 \wedge \Delta \alpha A_2)$ [m/h], ..., $M \models \Diamond (\sim A_1 \wedge ... \wedge \sim A_{n-1} \wedge \Delta \alpha A_n)$ [m/h], and $M \models \sim A_1 \wedge ... \wedge \sim A_n$ [m/h]. Then there are $h_1, ..., h_n$ such that for each i with $1 \leq i \leq n$, $m \in h_i$ and

---

10. Consider the following example: Let $T = \{m, m_0, m_1, ...\}$, $\leq = \{\langle m', m' \rangle : m' \in T\} \cup \{\langle m, m_k \rangle : k \geq 0\}$, and $Agent = \{a_0, a_1, ...\}$. For each $k \geq 0$, we set $h_k = \{m, m_k\}$. It is easy to see that $H_{(m)} = \{h_0, h_1, ...\}$. Let us define $Choice$ to be a function on $Agent \times T$ such that for each $k \geq 0$, $Choice(a_k, m) = \{H_{k,1}, H_{k,2}\}$ where $H_{k,1} = \{h_0, ..., h_k\}$, $H_{k,2} = \{h_{k+1}, h_{k+2}, ...\}$. Since $\bigcap_{k \geq 0} H_{k,2} = \emptyset$, we know that $Choice$ does not satisfy the postulate *independence of agents*. If, however, we treat $S = \langle T, \leq, Agent, Choice \rangle$ as a structure, and if we keep all other semantic notions the same, then we will still have $S \models AIA_k$ for every $k \geq 0$. Examples of this kind suggest that there are no formulas corresponding to the condition *independence of agents* since we can only talk about finitely many agents in each formula. What $\{AIA_k : k \geq 1\}$ corresponds to is the weaker condition that for each $m \in T$, and for each function $choice_m$ on any finite subset $Agent_f$ of $Agent$ such that $choice_m(a) \in Choice(a, m)$ for all $a \in Agent_f$, $\bigcap \{choice_m(a) : a \in Agent_f\} \neq \emptyset$.

(a) $M \models \Delta \alpha A_1 \, [m/h_1]$, and
    $M \models {\sim} A_1 \wedge \ldots \wedge {\sim} A_{i-1} \wedge \Delta \alpha A_i \, [m/h_i]$ for all i with $1 < i \leq n$.

For each i with $1 \leq i \leq n$, let us set $e_i = \{h' : h' \equiv^\alpha_m h_i\}$. Since $M \models {\sim} A_1 \wedge \ldots \wedge {\sim} A_n \, [m/h]$, we know by (a) and §1.3 that for each i with $1 \leq i \leq n$, $h \not\equiv^\alpha_m h_i$. That is to say, $h \notin e_i$ for all i with $1 \leq i \leq n$. Since $Choice(V(\alpha), m)$ is a partition of $H_{(m)}$, there must be an $e \in Choice(V(\alpha), m)$ such that $h \in e$. But obviously $e \neq e_i$ for all i with $1 \leq i \leq n$. It follows that $|Choice(V(\alpha), m)| \geq n+1$, contrary to our supposition that $|Choice(V(\alpha), m)| \leq n$. We conclude from this reductio that $M \models APC_n$.

Next suppose that $|Choice(a, m)| \geq n+1$ for some $a \in Agent$ and some $m \in T$. We show as follows that there is a model M on S such that for some h with $m \in h$,

(b) $M \not\models \Diamond \Delta \alpha p_1 \wedge \Diamond ({\sim} p_1 \wedge \Delta \alpha p_2) \wedge \ldots \wedge \Diamond ({\sim} p_1 \wedge \ldots \wedge {\sim} p_{n-1} \wedge \Delta \alpha p_n) \to p_1 \vee \ldots \vee p_n \, [m/h]$.

Let $e, e_1, \ldots, e_n$ be $n+1$ different elements of $Choice(a, m)$, let $h \in e$, and let $h_i \in e_i$ for each i with $1 \leq i \leq n$. We define $M = \langle S, V \rangle$, where V is any valuation such that $V(\alpha) = a$, and for each i with $1 \leq i \leq n$, $V(p_i) = \{\langle m, h' \rangle : h' \in e_i\}$. Since $e, e_1, \ldots, e_n$ are all different, it is easy to verify that $M \models {\sim} p_1 \wedge \ldots \wedge {\sim} p_n \, [m/h]$, $M \models \Delta \alpha p_1 \, [m/h_1]$, and $M \models {\sim} p_1 \wedge \ldots \wedge {\sim} p_{i-1} \wedge \Delta \alpha p_i \, [m/h_i]$ for every i with $1 < i \leq n$. It follows that $M \models \Diamond \Delta \alpha p_1 \wedge \Diamond ({\sim} p_1 \wedge \Delta \alpha p_2) \wedge \ldots \wedge \Diamond ({\sim} p_1 \wedge \ldots \wedge {\sim} p_{n-1} \wedge \Delta \alpha p_n) \, [m/h]$ and $M \not\models p_1 \vee \ldots \vee p_n \, [m/h]$. It follows that (b) holds. ∎

Since it is easy to verify that $A1 - A6$ are valid for all structures and RN is validity-preserving, §2.1 and §2.2 give us the following.

2.3. Soundness Theorem. *For every formula A, $\vdash_{L_0} A$ only if $S \models A$ for every structure S; and for each $n \geq 1$, $\vdash_{L_n} A$ only if $S \models A$ for all structure S in which every agent has at most n possible choices at every moment.*

3. Completeness and Compactness

In this section we prove the strong completeness theorem for all L_n with $n \geq 0$, which implies the compactness. Our proof is based on a technology from modal logic, using canonical models. We will assume the reader's familiarity with this technology, and will mainly deal with the general condition *independence of agents* and the conditions about possible choices.

For any $L = L_n$ with $n \geq 0$, let W_L be the set of all L-MCSs, and let R_L be the relation on W_L such that for each $w, w' \in W_L$, wR_Lw' iff $\{A : \Box A \in w\} \subseteq w'$. By $A1$ and RN, we know from modal logic that R_L is an equivalence relation. Let us use X, X' etc. to range over R_L-equivalence classes. Let X be any R_L-equivalence class. We know that the restriction of R_L to X is a universal relation. We define a relation $\cong$ among all agent terms such that for each α and β, $\alpha \cong \beta$ iff $\alpha = \beta \in w$ for some $w \in X$. It is easy to see by $T3$ that $\alpha \cong \beta$ iff $\alpha = \beta \in w$ for all $w \in X$, and $\alpha \not\cong \beta$ iff $\alpha \neq \beta \in w$ for all $w \in X$. We know by $A4$ that $\cong$ is an equivalence relation. For each agent term α, we set $[\alpha]$ to be the $\cong$-equivalence class to which α belongs. Let us assume that $\beta_0, \beta_1, \ldots$ is an enumeration of the representatives of all $\cong$-equivalence classes. *The agent-frame for L on X is the sequence*

$\langle X, [\beta_0], [\beta_1], \ldots \rangle$. Let us define, for each agent term α, a relation $R_{[\alpha]}$ on X such that for every $w, w' \in X$, $wR_{[\alpha]}w'$ iff $\{A: \Delta\alpha A \in w\} \subseteq w'$. By ordinary modal logic, we know that for each α, $R_{[\alpha]}$ is an equivalence relation. We set $E_{[\alpha]}$ to be the set of all $R_{[\alpha]}$-equivalence classes. It is easy to see that if $\alpha \cong \beta$, $R_{[\alpha]}$ is the same relation as $R_{[\beta]}$. We will call the sequence $\langle X, [\beta_0], [\beta_1], \ldots, R_{[\beta_0]}, R_{[\beta_1]}, \ldots \rangle$ *the canonical frame for L w.r.t. X*.

3.1. Lemma. *Let $L=L_n$ for any $n \geq 0$, and let $\langle X, [\beta_0], [\beta_1], \ldots, R_{[\beta_0]}, R_{[\beta_1]}, \ldots \rangle$ be the canonical frame for L w.r.t. X. Then the following hold:*

(i) for each $w \in X$ and each A, $\Box A \in w$ iff $A \in w'$ for all $w' \in X$ iff $\Box A \in w'$ for all $w' \in X$;

(ii) for each $w \in X$ and each α and A, $\Delta\alpha A \in w$ iff $A \in w'$ for all $w' \in X$ with $w' R_{[\alpha]} w$ iff $\Delta\alpha A \in w'$ for all $w' \in X$ with $w' R_{[\alpha]} w$;

(iii) for each $w \in X$ and each α and A, $[\alpha \text{ dstit}: A] \in w$ iff $A \in w'$ for all $w' \in X$ with $w' R_{[\alpha]} w$, and $\sim A \in w''$ for some $w'' \in X$.

Proof. (i) and (ii) are trivial (apply modal logic and T2), and (iii) can be easily obtained by T1, (i) and (ii). ∎

3.2. Lemma. *Let $L=L_n$ for any $n \geq 0$, and let $\langle X, [\beta_0], [\beta_1], \ldots, R_{[\beta_0]}, R_{[\beta_1]}, \ldots \rangle$ be the canonical frame for L w.r.t. X. Suppose that f is any function on $\{[\beta_0], [\beta_1], \ldots\}$ such that $f([\beta_i]) \in E_{[\beta_i]}$ for all $i \geq 0$. Then $\bigcap_{i \geq 0} f([\beta_i]) \neq \varnothing$.*

Proof. Let f be any function on $\{[\beta_0], [\beta_1], \ldots\}$ with $f([\beta_i]) \in E_{[\beta_i]}$ for all $i \geq 0$. We show as follows that $\bigcap_{i \geq 0} f([\beta_i]) \neq \varnothing$. We know by modal logic and §3.1(i,ii) that there are $A_0, A_1, A_2, \ldots$ such that

(a) for every $w \in W$, $w \in X$ iff $\Lambda_\Box = \{\Box A_j: j \geq 0\} \subseteq w$;

and that for each $i \geq 0$, there are $B_{i,0}, B_{i,1}, B_{i,2}, \ldots$ such that

(b) for every $w \in X$, $w \in f([\beta_i])$ iff $\Lambda_i = \{\Delta\beta_i B_{i,j}: j \geq 0\} \subseteq w$.

Since $[\beta_0], [\beta_1], [\beta_2], \ldots$ are different $\cong$-equivalence classes, we have that

(c) for every $k \geq 0$, $\text{diff}(\beta_0, \ldots, \beta_k) \in w$ for all $w \in X$.

We first show that $\Theta = \Lambda_\Box \cup (\bigcup_{i \geq 0} \Lambda_i)$ is L-consistent. To that end, it is sufficient to show that

(d) for each $k \geq 0$, each $\Box A \in \Lambda_\Box$, and each $\Delta\beta_i B_i \in \Lambda_i$ with $0 \leq i \leq k$, $\Box A \wedge \Delta\beta_0 B_0 \wedge \ldots \wedge \Delta\beta_k B_k$ is L-consistent.

It is easy to see by (b) that for each i with $0 \leq i \leq k$, there is a $w_i \in f([\beta_i]) \in E_{[\beta_i]}$ such that $\Delta\beta_i B_i \in w_i$. Then, selecting any $w \in X$, we have by (a), (c) and §3.1(i) that

$$\Box A \wedge \text{diff}(\beta_0, \ldots, \beta_k) \wedge \Diamond\Delta\beta_0 B_0 \wedge \ldots \wedge \Diamond\Delta\beta_k B_k \in w.$$

It follows from AIA_k that $\Box A \wedge \Diamond(\Delta\beta_0 B_0 \wedge \ldots \wedge \Delta\beta_k B_k) \in w$, and hence by modal logic, $\Diamond(\Box A \wedge$

$\Delta\beta_0 B_0 \wedge ... \wedge \Delta\beta_k B_k) \in w$, and consequently, $\Box A \wedge \Delta\beta_0 B_0 \wedge ... \wedge \Delta\beta_k B_k$ is L-consistent. It follows that (d) holds, and hence Θ is L-consistent, and is included in some L-MCS $w^* \in W$. Since $\Lambda_\Box \subseteq w^*$, we know by (a) that $w^* \in X$; and since $\Lambda_i \subseteq w^*$ for each $i \geq 0$, we know by (b) that $w^* \in f([\beta_i])$ for each $i \geq 0$. It follows that $\bigcap_{i \geq 0} f([\beta_i]) \neq \emptyset$. ∎

3.3. Lemma. *Let $L = L_n$ for any $n \geq 1$, and let $\langle X, [\beta_0], [\beta_1], ..., R_{[\beta_0]}, R_{[\beta_1]}, ...\rangle$ be the canonical frame for L w.r.t. X. Then for each $i \geq 0$, there are at most n $R_{[\beta_i]}$-equivalence classes, i.e., $|E_{[\beta_i]}| \leq n$.*

Proof. Suppose for reductio that there is an $i \geq 0$, such that $|E_{[\beta_i]}| \geq n+1$. Setting α to be β_i, we know that there are $e_0, ..., e_n \in E_{[\alpha]}$ such that $e_0, ..., e_n$ are all different. Let us select $w_0 \in e_0, ..., w_n \in e_n$. It follows from §3.1(ii) that there are $A_1, ..., A_n$ such that for each i with $1 \leq i \leq n$, $\Delta\alpha A_i \in w_i$ and $\sim A_i \in w_k$ for all k with $0 \leq k \leq n$ and $i \neq k$. That is to say, $\sim(A_1 \vee ... \vee A_n) \in w_0$, $\Delta\alpha A_1 \in w_1$, $\sim A_1 \wedge \Delta\alpha A_2 \in w_2, ..., \sim A_1 \wedge ... \wedge \sim A_{n-1} \wedge \Delta\alpha A_n \in w_n$. It follows from §3.1(i) that

$$\Diamond\Delta\alpha A_1 \wedge \Diamond(\sim A_1 \wedge \Delta\alpha A_2) \wedge ... \wedge \Diamond(\sim A_1 \wedge ... \wedge \sim A_{n-1} \wedge \Delta\alpha A_n) \wedge \sim(A_1 \vee ... \vee A_n) \in w_0,$$

and hence by APC_n, $\bot \in w_0$, contrary to our assumption of L-consistency on w_0. We conclude from this reductio that $|E_{[\beta_i]}| \leq n$ for all $i \geq 0$. ∎

3.4. Strong Completeness Theorem. *Every L_0-consistent set Θ of formulas is satisfiable in a structure; and every L_n-consistent set Θ of formulas ($n \geq 1$) is satisfiable in a structure in which every agent has at most n possible choices at every moment.*

Proof. Let $L = L_n$ for any $n \geq 0$, and let Θ be any L-consistent set of formulas. Then there is a L-MCS $w \in W_L$ including Θ. Suppose that X is the R_L-equivalence class to which w belongs, and let $\langle X, [\beta_0], [\beta_1], ..., R_{[\beta_0]}, R_{[\beta_1]}, ...\rangle$ be the canonical frame for L w.r.t. X. We want to convert $\langle X, [\beta_0], [\beta_1], ..., R_{[\beta_0]}, R_{[\beta_1]}, ...\rangle$ into a structure $S = \langle T, \leq, \text{Agent}, \text{Choice}\rangle$. Let us first set

$T = \{X\} \cup X$;
$\leq \, = \{\langle X, w\rangle: w \in X\} \cup \{\langle X, X\rangle\} \cup \{\langle w, w\rangle: w \in X\}$;
$\text{Agent} = \{[\beta_0], [\beta_1], ...\}$.

Then we set $h_w = \{X, w\}$ for all $w \in X$. It is clear that h_w is the unique history in $\langle T, \leq\rangle$ to which w belongs, and that there is a one-one correspondence between all $w \in X$ and all h_w in $\langle T, \leq\rangle$. Finally let us set

$\text{Choice}([\beta_i], X) = \{H: \exists e(e \in E_{[\beta_i]} \wedge H = \{h_w: w \in e\})\}$ for each $i \geq 0$; and
$\text{Choice}([\beta_i], w) = \{\{h_w\}\} = \{H_{(w)}\}$ for each $i \geq 0$ and each $w \in X$.

Note that at each w, each agent $[\beta_i]$ has a vacuous choice. It is easy to see that for each $i \geq 0$, $\text{Choice}([\beta_i], X)$ is a partition of $H_{(X)}$, and that the condition *no choice between undivided histories* is vacuously satisfied. To see that *Choice* satisfies *independence of agents*, let choice_X be any function on *Agent* such that $\text{choice}_X([\beta_i]) \in \text{Choice}([\beta_i], X)$ for each $i \geq 0$. Then by definition of *Choice*, for each $i \geq 0$, there is an $e_i \in E_{[\beta_i]}$ such that $\text{choice}_X([\beta_i]) = \{h_w: w \in e_i\}$. Let f be the function on *Agent* such that for each $i \geq 0$, $f([\beta_i]) = e_i \in E_{[\beta_i]}$. Since there is a one-one correspondence between all $w \in X$ and all h_w in $\langle T, \leq\rangle$, it is easy to see that for any $w \in X$ and any $i \geq 0$, $w \in f([\beta_i])$ iff $h_w \in choice_X([\beta_i])$, and hence by §3.2, $\bigcap_{i \geq 0} choice_X([\beta_i]) \neq \emptyset$. It follows that *Choice* satisfies *independence of agents*, and thus S is shown to be a structure.

When $L = L_n$ with $n \geq 1$, we need to show that each agent in S has at most n possible choices at every moment. Since $|Choice([\beta_i], w)| = 1$ for every $i \geq 0$ and every $w \in X$, we only need to show that $|Choice([\beta_i], X)| \leq n$ for every $i \geq 0$. But we know by definition of *Choice* that for each $i \geq 0$ and each $k \geq 1$, $|Choice([\beta_i], X)| = k$ iff $|E_{[\beta_i]}| = k$. It follows from §3.3 that $|Choice([\beta_i], X)| \leq n$.

Let us define a model $M = \langle S, V \rangle$ such that for each agent term α, $V(\alpha) = [\alpha]$; and for each propositional variable p, and each history h_w in S, $\langle X, h_w \rangle \in V(p)$ iff $p \in w$. It is easy to see by our definition that for each h in S, $h \in H_{(X)}$ iff $\exists w \in X(h = h_w)$; and that for agent term α, and every $w, w' \in X$, $h_w \equiv_Y^{[\alpha]} h_{w'}$ iff $\exists e \in E_{[\alpha]}(w, w' \in e)$ iff $w R_{[\alpha]} w'$. Then one can show by induction that $M \models A [X/h_w]$ iff $A \in w$ for every A and every $w \in X$, applying §3.1(i) and §3.1(iii). ■

3.5. Corollary (Compactness) *For every set Θ of formulas, if every finite subset of Θ has a model, then Θ has a model; and if every finite subset of Θ has a model in which every agent has at most n possible choices at every moment, then Θ has a model of the same kind.*

Some general remarks are worth noting. Let us define a *dstit logic* as a set of formulas that contains all truth-functional tautologies, all $A1-A5$ and all AIA_k for $k \geq 1$, and is closed under substitution, modus ponens and *RN*. Let L be any *dstit* logic. A *model for L* is any model M such that $M \models A$ for all $A \in L$. L is *complete for its models* if for every $A \notin L$, there is a model for L such that $M \not\models A$. Going over our §§3.1−2 and the first part of §3.4 (considering L there as an arbitrary *dstit* logic), one can easily see that the following holds.

3.6. Remark. *Every dstit logic is complete for its models.*

A *general structure* is any triple $G = \langle S, \bar{A}, Z \rangle$ such that (i) $S = \langle T, \leq, Agent, Choice \rangle$ is a structure; (ii) $\bar{A} = \{\bar{\alpha}_0, \bar{\alpha}_1, ...\}$, where $\alpha_0, \alpha_1, ...$ are all the agent terms in our language, and $\bar{\alpha}_n \in Agent$ for every $n \geq 0$; and (iii) Z is a subset of the power set of $\{\langle m, h \rangle : m \in h\}$ and is closed under Boolean operations as well as the following, where $z \in Z$ and $\bar{\alpha} \in \bar{A}$:

$$f_\Box(z) = \{\langle m, h \rangle : \forall h'(m \in h' \rightarrow \langle m, h' \rangle \in z\};$$
$$f_{\bar{\alpha}}(z) = \{\langle m, h \rangle : \forall h'(h' \equiv_m^\alpha h \rightarrow \langle m, h' \rangle \in z) \land \exists h''(m \in h'' \land \langle m, h'' \rangle \notin z)\}.$$

Let $G = \langle S, \bar{A}, Z \rangle$ be any general structure. A *model on G* is any model $M = \langle S, V \rangle$ such that $V(\alpha_n) = \bar{\alpha}_n$ for every $n \geq 0$ and $V(p) \in Z$ for every propositional variable p. We set $G \models A$ iff $M \models A$ for all model M on G. Let L be any *dstit* logic. G is a general structure for L if $G \models A$ for all $A \in L$. L is *complete for its general structures* if for every $A \notin L$, there is a general structure G for L such that $G \not\models A$.

3.7. Remark. *Every dstit logic is complete for its general structures.*

Proof. Let L be any *dstit* logic, and let $A \notin L$. We know by §3.6 that there is a model $M=\langle S, V \rangle$ for L such that $M \not\models A$. Let us set $\bar{\alpha}_n = V(\alpha_n)$ for every $n \geq 0$, and $\bar{A} = \{\bar{\alpha}_0, \bar{\alpha}_1, ...\}$; and set $\|B\|^M = \{\langle m, h \rangle : M \models B[m/h]\}$ for every formula B, and $Z = \{\|B\|^M : B \text{ is a formula}\}$. It is easy to show by induction that $G = \langle S, \bar{A}, Z \rangle$ is a general structure. We show as follows that G is a general structure for L. Suppose for reductio that $B \in L$ but $M^* \not\models B$ for some B and some model $M^* = \langle S, V^* \rangle$ on G. Let $B = B(p_0, ..., p_k)$ (all propositional variables occurring in B are among $p_0, ..., p_k$). By definition $V^*(p_0) = \|C_0\|^M, ..., V^*(p_k) = \|C_k\|^M$. Consider the formula $B' = B(C_0/p_0, ..., C_k/p_k)$, i.e., the formula obtained from B by substituting C_i for p_i for each $i \leq k$. We claim that $M \not\models B'$. To justify this claim, it is sufficient to observe that for every formula $D(p_0, ..., p_k)$, $M^* \models D(p_0, ..., p_k) [m/h]$ iff $M \models D(C_0/p_0, ..., C_k/p_k) [m/h]$, which can be easily established by induction. Since $B \in L$, and since L is closed under substitution, we know that $B' \in L$, and hence, since M is a model for L, $M \models B'$, a contradiction. From this reductio we conclude that $B \in L$ implies $M^* \models B$ for every formula B and every model M^* on G, i.e., G is a general structure for L. It is easy to see that M itself is a model on G, and hence, since $M \not\models A$, $G \not\models A$. ∎

4. Finite Model Property

We show in this section that all L_n with $n \geq 0$ have the finite model property, and hence they are all decidable.[11] The basic idea in our proof of the finite model property of L_n is borrowed from the filtration theorem in modal logic. Let A be any formula. We define three sets Σ_A, Λ_A and Π_A of formulas as follows.

$\Sigma_A = \{B: B \text{ is a subformula of } A\}$,
$\quad \Sigma_0 = \Sigma_A \cup \{\Delta\beta B: [\beta \text{ dstit}: B] \in \Sigma_A\} \cup \{\sim \Box B: [\beta \text{ dstit}: B] \in \Sigma_A\}$,
$\quad \Sigma_1 = \Sigma_0 \cup \{\Delta\beta \sim \Delta\beta B: \Delta\beta B \in \Sigma_0\}$,
$\Lambda_A = \{B: B \text{ is a subformula of } D \in \Sigma_1\}$,
$\quad \Sigma_2 = \{\Diamond(\Delta\gamma_0 B_0 \wedge ... \wedge \Delta\gamma_n B_n): n \geq 0, \gamma_0, ..., \gamma_n \text{ are different}, \gamma_i \ (0 \leq i \leq n)$
$\quad \quad \text{occurs in } A, B_i = C_0 \wedge ... \wedge C_j (0 \leq j \leq |\Lambda_A|), \Delta\gamma_i C_0, ..., \Delta\gamma_i C_j \in \Lambda_A\}$,
$\Pi_A = \{B: B \text{ is a subformula of } C \in \Lambda_A \cup \Sigma_2\}$.

It is easy to verify that Π_A is finite. Π_A is particularly designed, in accordance with *AIA*, for the purpose of dealing with the postulate *independence of agents*, with the help of a property of Λ_A stated in the following lemma.

4.1. Lemma. *Let $L = L_n$ for any $n \geq 0$, and let A be any formula. Then for each β occurring in A and each formula $\Delta\beta B \in \Lambda_A$, there is a formula B' such that $\vdash_L \Delta\beta B' \leftrightarrow \Delta\beta \sim \Delta\beta B$, and $\Delta\beta B' \in \Lambda_A$.*

Proof. It is easy to see that if $\Delta\beta B \in \Lambda_A$, then either $\Delta\beta B \in \Sigma_0$ or $\Delta\beta B = \Delta\beta \sim \Delta\beta C$ with $\Delta\beta C \in \Sigma_0$. Suppose that $\Delta\beta B \in \Lambda_A$. If $\Delta\beta B \in \Sigma_0$, then, setting $B' = \sim \Delta\beta B \in \Lambda_A$, we know that

---

[11]. Note that the decidability of *dstit* theories is not a direct consequence of GUREVICH and SHELAH 1985 and 1985a since in their result, the theory on tree-like frames does not contain a quantifier over sets of choice equivalent histories, which is needed because of *Choice* in a structure and the truth conditions for *dstit*.

$\Delta\beta B' \in \Sigma_1$. If $\Delta\beta B \in \Sigma_1 - \Sigma_0$, then $\Delta\beta B = \Delta\beta \sim \Delta\beta C$ for some $\Delta\beta C \in \Sigma_0$, and hence $\Delta\beta \sim \Delta\beta B = \Delta\beta \sim \Delta\beta \sim \Delta\beta C$. Setting $B' = C$ and applying modal logic, we have $\Delta\beta B' \in \Sigma_1$. ∎

Let $L = L_n$ with $n \geq 0$, let X be any R_L-equivalence class, and let $F = \langle X, [\beta_0], [\beta_1], \ldots \rangle$ be the agent-frame for L on X. Consider any formula A. We first define a relation $\approx$ on X as follows: for each $w, w' \in X$, $w \approx w'$ if for every $B \in \Pi_A$, $B \in w$ iff $B \in w'$. $\approx$ is obviously an equivalence relation. Since Π_A is finite, there are only finitely many $\approx$-equivalence classes. Selecting a representative for each of these $\approx$-equivalence classes, we set Y to be the set of all those representatives. Next, since there are only finitely many agent terms occurring in A, there must be finitely many $\equiv$-equivalence classes, say $[\beta_0], \ldots, [\beta_k]$, such that each β occurring in A belongs to $[\beta_i]$ for some i with $0 \leq i \leq k$, and each $[\beta_i]$ with $0 \leq i \leq k$ contains at least some β occurring in A. We may assume, without loss of generality, that all $\beta_0, \ldots, \beta_k$ occur in A. A *filtration of F through Λ_A and Π_A* is a sequence $\langle Y, [\beta_0], \ldots, [\beta_k], \simeq_{[\beta_0]}, \ldots, \simeq_{[\beta_k]} \rangle$, where Y is the set of representatives for $\approx$-equivalence classes, and for each i with $0 \leq i \leq k$, $\simeq_{[\beta_i]}$ is the relation on Y such that

for each $w, w' \in Y$, $w \simeq_{[\beta_i]} w'$ iff for all $\Delta\beta_i B \in \Lambda_A$, $\Delta\beta_i B \in w$ iff $\Delta\beta_i B \in w'$.

Clearly each $\simeq_{[\beta_i]}$ with $0 \leq i \leq k$ is an equivalence relation. Note that the filtration defined here is a little different from those in the area of modal logic. Each filtration there was obtained through a single set of formulas, while the filtration here is defined through *two* different sets: the $\approx$-equivalence classes are determined by Π_A, but the relation $\simeq_{[\beta_i]}$ is defined in terms of Λ_A rather than Π_A.[12] Note also that when we choose different representatives for $\approx$-equivalence classes, we will have different filtrations, but they must be isomorphic to each other. We will thus speak of "the filtration of F through Λ_A and Π_A" as if there is only one such filtration. Consider each i with $0 \leq i \leq k$. Since Y is finite and $\simeq_{[\beta_i]}$ is an equivalence relation on Y, there are finitely many $\simeq_{[\beta_i]}$-equivalence classes. We will use $U_{[\beta_i]}$, for each i with $0 \leq i \leq k$, to denote the set of all $\simeq_{[\beta_i]}$-equivalence classes, and use u, u', etc. to range over its elements.

4.2. Lemma. *Let $L = L_n$ for any $n \geq 0$, let X be any R_L-equivalence class, and let $F = \langle X, [\beta_0], [\beta_1], \ldots \rangle$ be the agent-frame for L on X. Suppose that A is any formula and $\langle Y, [\beta_0], \ldots, [\beta_k], \simeq_{[\beta_0]}, \ldots, \simeq_{[\beta_k]} \rangle$ is the filtration of F through Λ_A and Π_A. Then for any $w \in Y$,*

(i) *if $\Box B \in \Pi_A$, then $\Box B \in w$ iff $B \in w'$ for all $w' \in Y$;*
(ii) *if $\Delta\beta B \in \Lambda_A$, then $\Delta\beta B \in w$ iff $B \in w'$ for all $w' \in Y$ with $w' \simeq_{[\beta]} w$;*
(iii) *if $[\beta\ \text{dstit}: B] \in \Sigma_A$, then $[\beta\ \text{dstit}: B] \in w$ iff $B \in w'$ for all $w' \in Y$ with $w' \simeq_{[\beta]} w$, and $B \notin w''$ for some $w'' \in Y$;*
(iv) *for each $u_0 \in U_{[\beta_0]}, \ldots, u_k \in U_{[\beta_k]}$, $u_0 \cap \ldots \cap u_k \neq \emptyset$.*

Proof. (i) Let $\Box B \in \Pi_A$. Suppose that $\Box B \in w$. Then $B \in w'$ for all $w' \in X$, and hence, $B \in w'$ for all $w' \in Y$. Suppose that $\Box B \notin w$. Then there is a $w'' \in X$ such that $B \notin w''$. Since $B \in \Pi_A$, and

---

12. It is, however, not essential to define our filtrations in terms of two sets of formulas. It is possible to design a single set of formulas such that when we define a filtration in terms of this single set, the filtration still has a property corresponding to the postulate *independence of agents*.

since there is a $w' \in Y$ such that $w' \approx w''$, it follows from our definition of $\approx$ that $B \not\in w'$.

(ii) Let $\Delta\beta B \in \Lambda_A$. Suppose that $\Delta\beta B \in w$. Let γ be the representative of $[\beta]$. Consider any $w' \in Y$ with $w' \simeq_{[\beta]} w$. Then $w' \simeq_{[\gamma]} w$ and $\Delta\gamma B \in w$ by $A5$. Since $\Delta\gamma B \in \Lambda_A$, it follows from the definition of $\simeq_{[\gamma]}$ that $\Delta\gamma B \in w'$, and hence by $A2$, $B \in w'$. It follows that $B \in w'$ for all $w' \in Y$ with $w' \simeq_{[\beta]} w$. Suppose that $\Delta\beta B \not\in w$. Then by $A5$, $\Delta\gamma B \not\in w$. Since $w \in Y \subseteq X$, it follows from §3.1(ii) that there is a $w'' \in X$ such that $B \not\in w''$ and

(*) $\{\Delta\gamma C: \Delta\gamma C \in w\} \subseteq w''$ and $\{\Delta\gamma C: \Delta\gamma C \in w''\} \subseteq w$.

Since there is a $w' \in Y$ such that $w' \approx w''$, it follows from (*) and the definition of $\simeq_{[\gamma]}$ that $w \simeq_{[\gamma]} w'$ (i.e., $w \simeq_{[\beta]} w'$) and $B \not\in w'$.

(iii) Let $[\beta \text{ dstit: } B] \in \Sigma_A$. By $T1$ we know that $[\beta \text{ dstit: } B] \in w$ iff $\Delta\beta B \wedge \sim \Box B \in w$. Since, by definition of Π_A, $\Delta\beta B \in \Lambda_A$ and $\sim \Box B \in \Lambda_A \subseteq \Pi_A$, it follows from (i) and (ii) that $[\beta \text{ dstit: } B] \in w$ iff $B \in w'$ for all $w' \in Y$ with $w' \simeq_{[\beta]} w$, and $B \not\in w''$ for some $w'' \in Y$.

(iv) Consider any $u_0 \in U_{[\beta_0]}, ..., u_k \in U_{[\beta_k]}$. Consider any i with $0 \le i \le k$. It is easy to show by definition of $\simeq_{[\beta_i]}$ that there are $\Delta\beta_i C_0, ..., \Delta\beta_i C_j, \Delta\beta_i D_0, ..., \Delta\beta_i D_l \in \Lambda_A$ such that

(1) $\Delta\beta_i C_0, ..., \Delta\beta_i C_j, \Delta\beta_i D_0, ..., \Delta\beta_i D_l$ are all the formulas in Λ_A that have the form $\Delta\beta_i B$, and
(2) $\Delta\beta_i C_0 \wedge ... \wedge \Delta\beta_i C_j \wedge \sim \Delta\beta_i D_0 \wedge ... \wedge \sim \Delta\beta_i D_l \in w$ for all $w \in u_i$, and
(3) for each $u \in U_{[\beta_i]}$ with $u \ne u_i$, $\sim (\Delta\beta_i C_0 \wedge ... \wedge \Delta\beta_i C_j \wedge \sim \Delta\beta_i D_0 \wedge ... \wedge \sim \Delta\beta_i D_l) \in w'$ for all $w' \in u$.

By (1) we know that $j+l+2 \le |\Lambda_A|$. Since $\Delta\beta_i D_0, ..., \Delta\beta_i D_l \in \Lambda_A$, we know by §4.1 that there are $E_0, ..., E_l$ such that

(4) for each r with $0 \le r \le l$, $\vdash_L \Delta\beta_i E_r \leftrightarrow \Delta\beta_i \sim \Delta\beta_i D_r$ and $\Delta\beta_i E_r \in \Lambda_A$.

Selecting some $w_i \in u_i$, $\Delta\beta_i C_0 \wedge ... \wedge \Delta\beta_i C_j \wedge \Delta\beta_i E_0 \wedge ... \wedge \Delta\beta_i E_l \in w_i$ by (2) and (4). Now let us set $B_i = C_0 \wedge ... \wedge C_j \wedge E_0 \wedge ... \wedge E_l$. Then by modal logic, $\Delta\beta_i B_i \in w_i$. In general, we can define $B_0, ..., B_k$ the same way. Selecting $w_0 \in u_0, ..., w_k \in u_k$, we have

(5) $\Delta\beta_0 B_0 \in w_0, ..., \Delta\beta_k B_k \in w_k$.

It follows from modal logic that $\Diamond \Delta\beta_0 B_0 \wedge ... \wedge \Diamond \Delta\beta_k B_k \in w$ for some (actually, all) $w \in Y$. Since $[\beta_0], ..., [\beta_k]$ are different $\cong$-equivalence classes, it follows from definition of $\cong$ that $\textit{diff}(\beta_0, ..., \beta_k) \wedge \Diamond \Delta\beta_0 B_0 \wedge ... \wedge \Diamond \Delta\beta_k B_k \in w$, and hence by AIA_k,

(6) $\Diamond(\Delta\beta_0 B_0 \wedge ... \wedge \Delta\beta_k B_k) \in w$.

Recall that for each i with $0 \le i \le k$, $B_i = C_0 \wedge ... \wedge C_j \wedge E_0 \wedge ... \wedge E_l$ with $\Delta\beta_i C_0, ..., \Delta\beta_i C_j, \Delta\beta_i E_0, ..., \Delta\beta_i E_l \in \Lambda_A$ and $j+l+2 \le |\Lambda_A|$. Hence $\Diamond(\Delta\beta_0 B_0 \wedge ... \wedge \Delta\beta_k B_k) \in \Pi_A$ by the definition of Π_A. It follows from (6) and (i) that

(7) $\Delta\beta_0 B_0 \wedge ... \wedge \Delta\beta_k B_k \in w'$ for some $w' \in Y$.

In the following, we show that $w' \in u_0 \cap ... \cap u_k$. Consider any i with $0 \leq i \leq k$. Let us assume that the same situation described in (1)−(4) still holds here. We have by (7) that $\Delta\beta_i(C_0 \wedge ... \wedge C_j \wedge E_0 \wedge ... \wedge E_l) \in w'$, and hence by modal logic $\Delta\beta_i C_0 \wedge ... \wedge \Delta\beta_i C_j \wedge \Delta\beta_i E_0 \wedge ... \wedge \Delta\beta_i E_l \in w'$. It follows from (4) that

$$\Delta\beta_i C_0 \wedge ... \wedge \Delta\beta_i C_j \wedge \Delta\beta_i \sim \Delta\beta_i D_0 \wedge ... \wedge \Delta\beta_i \sim \Delta\beta_i D_l \in w',$$

and hence by $A2$, $\Delta\beta_i C_0 \wedge ... \wedge \Delta\beta_i C_j \wedge \sim \Delta\beta_i D_0 \wedge ... \wedge \sim \Delta\beta_i D_l \in w'$. It follows from (1)−(3) that $w' \in u_i$. Thus we can show in general that $w' \in u_0, ..., w' \in u_k$ the same way. It follows that $u_0 \cap ... \cap u_k \neq \emptyset$. ∎

4.3. Lemma. *Let $L = L_n$ for any $n \geq 1$, let X be any R_L-equivalence class, and let $F = \langle X, [\beta_0], [\beta_1], ... \rangle$ be the agent-frame for L on X. Suppose that A is any formula and $\langle Y, [\beta_0], ..., [\beta_k], \simeq_{[\beta_0]}, ..., \simeq_{[\beta_k]} \rangle$ is the filtration of F through Λ_A and Π_A. Then for each i with $0 \leq i \leq k$, there are at most $n \simeq_{[\beta_i]}$-equivalence classes, i.e., $|U_{[\beta_i]}| \leq n$.*

Proof. Suppose for reductio that $|U_{[\beta_i]}| \geq n+1$ for some i with $0 \leq i \leq k$. Let $\alpha = \beta_i$, let $u_0, ..., u_n \in U_{[\alpha]}$ such that $u_0, ..., u_n$ are all different, and let $w_0 \in u_0, ..., w_n \in u_n$. It follows from §4.2(ii) that there are $\Delta\alpha A_1, ..., \Delta\alpha A_n \in \Lambda_A$ such that for each i with $1 \leq i \leq n$, $\Delta\alpha A_i \in w_i$ and $A_i \notin w_j$ for all j with $0 \leq j \leq n$ and $i \neq j$. That is to say, $\sim(A_1 \vee ... \vee A_n) \in w_0$, $\Delta\alpha A_1 \in w_1$, $\sim A_1 \wedge \Delta\alpha A_2 \in w_2, ..., \sim A_1 \wedge ... \wedge \sim A_{n-1} \wedge \Delta\alpha A_n \in w_n$. It follows from §3.1(i) that

$$\Diamond \Delta\alpha A_1 \wedge \Diamond(\sim A_1 \wedge \Delta\alpha A_2) \wedge ... \wedge \Diamond(\sim A_1 \wedge ... \wedge \sim A_{n-1} \wedge \Delta\alpha A_n) \wedge \sim(A_1 \vee ... \vee A_n) \in w_0,$$

and hence by APC_n, $\perp \in w_0$, contrary to our assumption of L-consistency on w_0. We conclude from this reductio that $|E_{[\beta_i]}| \leq n$ for all i with $0 \leq i \leq k$. ∎

4.4. Theorem. *Every L_0-consistent formula A is satisfiable in a finite model; and every L_n-consistent formula ($n \geq 1$) is satisfiable in a finite model in which every agent has at most n possible choices at every moment.*

Proof. Let $L = L_n$ for any $n \geq 0$, let A be any L-consistent formula. Set X to be any R_L-equivalence class such that $A \in w$ for some $w \in X$, and $F = \langle X, [\beta_0], [\beta_1], ... \rangle$ to be the agent-frame for L on X, and $\langle Y, [\beta_0], ..., [\beta_n], \simeq_{[\beta_0]}, ..., \simeq_{[\beta_k]} \rangle$ to be the filtration of F through Λ_A and Π_A. We first construct a structure $S = \langle T, \leq, \text{Agent}, \text{Choice} \rangle$ as follows. Let us set

$T = \{Y\} \cup Y,$
$\leq \, = \{\langle Y, w \rangle : w \in Y\} \cup \{\langle Y, Y \rangle\} \cap \{\langle w, w \rangle : w \in Y\},$
$\text{Agent} = \{[\beta_0], ..., [\beta_k]\}.$

Since Y is finite, T is clearly finite. Similar to our proof of §3.4, for each $w \in T$, we set h_w to be the unique history in $\langle T, \leq \rangle$ to which w belongs, i.e., $h_w = \{Y, w\}$. Note that there is a one-one correspondence between all $w \in Y$ and all h_w in $\langle T, \leq \rangle$, i.e., for every h in $\langle T, \leq \rangle$,

(*) $h \in H_{(Y)}$ iff $\exists w \in Y(h = h_w).$

Finally we set

$Choice([\beta_i], Y) = \{H: \exists u(u \in U_{[\beta_i]} \wedge H = \{h_w: w \in u\})\}$ for each i with $0 \leq i \leq k$; and
$Choice([\beta_i], w) = \{\{h_w\}\} = H_{(w)}$ for each i with $0 \leq i \leq k$ and each $w \in Y$.

The condition *no choice between undivided histories* has been trivially satisfied. We show as follows that *Choice*, so defined, satisfies *independence of agents*. Consider any function $choice_Y$ such that for each i with $0 \leq i \leq k$, $choice_Y([\beta_i]) = H_i \in Choice([\beta_i], Y)$. It is easy to see by definition above that for each i with $0 \leq i \leq k$, there is a $u_i \in U_{[\beta_i]}$ such that $H_i = \{h_w: w \in u_i\}$. Since there is a one-one correspondence between all $w \in Y$ and all h_w in $\langle T, \leq \rangle$, it is then clear that for each $w \in Y$ and each i with $0 \leq i \leq k$, $w \in u_i$ iff $h_w \in H_i$. We know by §4.2(iv) that there is a $w' \in u_0 \cap \ldots \cap u_k$, and hence, $h_{w'} \in H_0 \cap \ldots \cap H_k$, i.e., $\bigcap \{choice_Y([\gamma]): [\gamma] \in Agent\} \neq \varnothing$.

When $L = L_n$ with $n \geq 1$, we need to show that each agent in S has at most n possible choices at every moment. Since $|Choice([\beta_i], w)| = 1$ for every $w \in Y$ and every i with $0 \leq i \leq k$, we only need to show that $|Choice([\beta_i], Y)| \leq n$ for every i with $0 \leq i \leq k$. But we know by definition of *Choice* that for each i with $0 \leq i \leq k$ and each number $n' \geq 1$, $|Choice([\beta_i], Y)| = n'$ iff $|U_{[\beta_i]}| = n'$. It follows from §4.3 that $|Choice([\beta_i], Y)| \leq n$ for every i with $0 \leq i \leq k$.

It is easy to verify that for every $[\gamma] \in Agent$ and every $w, w' \in Y$,

(**) $h_w \equiv_Y^{[\gamma]} h_{w'}$ iff $\exists u \in U_{[\gamma]}(w, w' \in u)$ iff $w \simeq_{[\gamma]} w'$.

Let us now construct a model $M = \langle S, V \rangle$ on S by defining V in such a way that for each agent term α occurring in A, $V(\alpha) = [\alpha]$; and for each propositional variable p occurring in A, and each $\langle Y, h_w \rangle$ with $w \in Y$, $\langle Y, h_w \rangle \in V(p)$ iff $p \in w$. We show by induction that for every formula $B \in \Sigma_A$ and every $w \in Y$,

(***) $M \models B [Y/h_w]$ iff $B \in w$.

It is easy to see by definition of V that (***) holds when B is a propositional variable or is of the form $\alpha = \beta$. The inductive steps for truth functional operators are straightforward. Let $B = \square C$ for some C. We know that $M \models \square C [Y/h_w]$ iff $M \models C [Y/h_{w'}]$ for all $h_{w'} \in H_{(Y)}$, and then by (*), $M \models \square C [Y/h_w]$ iff $M \models C [Y/h_{w'}]$ for all $w' \in Y$, and by induction hypothesis, $M \models \square C [Y/h_w]$ iff $C \in w'$ for all $w' \in Y$, and hence by §4.2(i), $M \models \square C [Y/h_w]$ iff $\square C \in w$. Let $B = [\gamma \, dstit: C]$. Since $M \models [\gamma \, dstit: C] [Y/h_w]$ iff $M \models C [Y/h_{w'}]$ for all $h_{w'} \in H_{(Y)}$ with $h_{w'} \equiv_Y^{[\gamma]} h_w$ and $M \not\models C [Y/h_{w''}]$ for some $h_{w''} \in H_{(Y)}$, we know by (*) and (**) that $M \models [\gamma \, dstit: C] [Y/h_w]$ iff $M \models C [Y/h_{w'}]$ for all $w' \in Y$ with $w' \simeq_{[\gamma]} w$ and $M \not\models C [Y/h_{w''}]$ for some $w'' \in Y$, and then by induction hypothesis, $M \models [\gamma \, dstit: C] [Y/h_w]$ iff $C \in w'$ for all $w' \in Y$ with $w' \simeq_{[\gamma]} w$ and $C \notin w''$ for some $w'' \in Y$, and hence by §4.2(iii), $M \models [\gamma \, dstit: C] [Y/h_w]$ iff $[\gamma \, dstit: C] \in w$. It follows that (***) holds for all $B \in \Sigma_A$.

Since $A \in \Sigma_A$ and $A \in w$ for some $w \in X$, it follows that $A \in w$ for some $w \in Y$. Hence by (*) and (***), $M \models A [Y/h_w]$ for some $h_w \in H_{(Y)}$. This completes the proof. ∎

By theorem §4.4 and a routine argument, the following theorem can be established.

4.5. Theorem.   *For every $n \geq 0$, L_n is decidable.*

References

BARTHA, P.: 1993, "Conditional obligation, deontic paradoxes, and the logic of agency". *Annals of Mathematics and Artificial Intelligence* **9**, 1–23.

BELNAP, N. and BARTHA, P.: 1993, "Marcus and the problem of nested deontic modalities". Forthcoming in *Modality, Morality, and Belief*, Walter Sinnott-Armstrong, Diana Raffman, Nicholas Asher (eds.), Cambridge University Press.

BELNAP, N.: 1991, "Backwards and forwards in the modal logic of agency". *Philosophy and phenomenological research* **51**, 777-807.

BELNAP, N.: 1991a, "Before refraining: concepts for agency". *Erkenntnis* **34**, 137-169.

BELNAP, N.: 1991b, "Agents in branching time". Forthcoming in *Logic and Reality: Essays in Pure and Applied Logic, In Memory of Arthur Prior*, Oxford University Press.

BELNAP, N. and PERLOFF, M.: 1988, "Seeing to it that: a canonical form for agentives". *Theoria* **54**, 175-199. Belnap noted that the informal semantic account of *stit* is garbled in this paper; the account is correct in the version of this paper republished in *Knowledge representation and defeasible reasoning*, H. E. Kyburg, Jr., R. P. Loui, and G. N. Carlson (eds.), Dordrecht/Boston/London, 1990, 167-190; and it is correct in the version in other papers.

BELNAP, N. and PERLOFF, M.: 1992, "The way of the agent". *Studia Logica* **51**, 463–484.

BELNAP, N. and PERLOFF, M.: 1993, "In the realm of agents". *Annals of Mathematics and Artificial Intelligence* **9**, 25–48.

CHELLAS, B. F.: 1969, *The Logical Form of Imperatives*, Perry Lane Press, Stanford.

CHELLAS, B. F.: 1992, "Time and modality in the logic of agency". *Studia Logica* **51**, 485-517.

GUREVICH, Y. and SHELAH, S.: 1985, "The decision problem for branching time logic". *The Journal of Symbolic Logic* **50**, 668–681.

GUREVICH, Y. and SHELAH, S.: 1985a, "To the decision problem for branching time logic". *Foundations of logic and linguistics: Problems and their solutions*, Plenum Press, 1985, 181–198.

HORTY, J. F.: 1989, "An alternative *stit* operator". Unpublished manuscript, Philosophy Department, University of Maryland.

HORTY, J. F. and BELNAP, N.: 1993, "The deliberative *stit*: a study of action, omission, ability, and obligation." Forthcoming in *Journal of Philosophical Logic*.

PERLOFF, M.: 1991, "*Stit* and the language of agency". *Synthese* **86**, 379–408.

PRIOR, A.: 1967, *Past, Present and Future*. Oxford University Press.

SEGERBERG, K.: 1992, "Getting started: Beginnings in the logic of action". *Studia Logica* **51**, 437–378.

THOMASON, R. H.: 1970, "Indeterminist time and truth-value gaps". *Theoria* **36**, 264-281.

THOMASON, R. H.: 1981, "Deontic logic as founded on tense logic". In *New Studies in Deontic Logic*, R. Hilpinen (ed.), D. Reidel Publishing Company, 1981, 165–176.

VON KUTSCHERA, F.: 1986, "Bewirden". *Erdenntnis* **24**, 253–281.

VON WRIGHT, G. H.: 1963, *Norm and Action: A Logical Enquiry*. Routledge and Kegan Paul.

XU, M.: 1993, "Doing and refraining from refraining". Forthcoming in *Journal of Philosophical logic*.

XU, M.: 1993a, "Decidability of *stit* theory with a single agent and *refref* equivalence". Forthcoming in *Studia Logica*.

XU, M.: 1994, "On the basic logic of *stit* with a single agent". Unpublished manuscript, Department of Philosophy, University of Pittsburgh.

Abduction in Temporal Reasoning

Cristina Ribeiro   António Porto

Departamento de Informática
Universidade Nova de Lisboa
2825 Monte da Caparica, Portugal
e-mail:{mcr, ap}@fct.unl.pt
phone: 351-1-2953270

Abstract. Commonsense knowledge often omits the temporal incidence of facts, and even the ordering between occurrences is only available for some of their instances. Reasoning about the temporal extent of facts and their sequencing becomes complex due to this inherent partiality. The generation of hypotheses is adopted here as a natural way to overcome the difficulties in computing answers to temporal queries. The proposed abductive system performs temporal reasoning in a logic programming framework. Queries are taken as goals and the inference system combines deduction with abduction and constraint solving. The convenience of constraints for dealing with temporal information is widely recognized, their interest being twofold: the representation of essential properties of time and the provision for partial information, allowing flexible bounds on times instead of constant bindings. Inference manipulates a language associating propositions with time periods which are maximal intervals for the proposition. The abductive inference procedure is described here, identifying the constraint operations required. It is also shown that the outcome of a derivation is always consistent with the information in the knowledge base.
Keywords: temporal reasoning, abduction, constraint satisfaction, logic programming, deductive databases.

1 Introduction

Temporal reasoning is considered here in the context of the representation and inference of commonsense information. The temporal assertions to be considered associate propositions with time intervals, and at the internal or knowledge base level such statements refer to maximal intervals.

It is common to several well-known approaches to temporal reasoning in Artificial Intelligence the representation of a temporal object as a proposition and an associated temporal extent [All84, McD82, Sho87]. The language considered here also relies on a simple ontology where intervals are associated to propositions. There is however a restriction on the temporal assertions manipulated internally, namely that the intervals are maximal ones.

The present work stresses the usefulness of abduction for answering temporal queries. Two aspects make abductive inference particularly suited for temporal reasoning: the partiality of temporal information and the generalized use of constraints. In what follows, a brief presentation of the knowledge base language is made, followed by the definition of answer to a temporal query. The use of abduction is demonstrated in the description of the inference system and the soundness of computed answers is proved, showing that they comply with the given definition.

2 The knowledge base language

The language of *maximal intervals* will be referred to in the sequel as *MI* and a description of its syntax and model-theoretic semantics can be found in [RP91]. The propositional (plus time) version considers temporal constants, variables, Skolem constants and functions and proposition symbols. Relations $\{=, >, <, \leq, \geq\}$ are used to build constraint formulas, and predicative formulas associate a proposition symbol with a pair of points. In the semantics, a proposition is interpreted as a set of disjoint time periods. Clauses are formed with an atomic formula in the head and a conjunction of formulas in the body. As an illustration of the difference between having maximal and non-maximal intervals, consider a knowledge base (KB) and several facts about proposition p:

$$p[2,4] \qquad p[3,7] \qquad p[10,11]$$

The facts assert intervals where p is known to hold. They are obviously non-maximal. If no further information exists about transitions of p, the same knowledge may be assimilated in terms of maximal intervals as

$$p[\alpha, \beta] \quad \alpha \leq 2 \quad \beta \geq 7 \quad p[\delta, \gamma] \quad \delta \leq 10 \quad \gamma \geq 11$$

where two maximal intervals appear, as well as constraints on their endpoints. Both representations capture the available information but the second is more compact, making clear that the two time periods may even reduce to a single interval, as $\alpha = \delta$ and $\beta = \gamma$ are compatible with the other constraints on the endpoints. Working with maximal intervals can be viewed as an alternative to the incorporation of events in the temporal language, as done for instance in the event calculus [KS86]. Events are used there to characterize the endpoints of intervals. Using maximal intervals results in a simpler ontology, and both time periods and their endpoints are given precise meaning.

To illustrate the language and the inference scheme, the classic Yale Turkey Shoot (YTS) example is used. The scenario description is as follows [HM86]. Initially, the turkey is alive and the gun unloaded. Then loading and shooting take place. If a loaded gun is shot, alive ceases to be true. The formalization in *MI* is stated as

(F1) $alive[0, \alpha]$.
(F2) $load[1, 1]$.
(F3) $shoot[3, 3]$.
(R1) $loaded[L, f(L)] \leftarrow load[L, L]$.
(R2) $Y \leq S \leftarrow shoot[S, S], loaded[A, B], A \leq S, S \leq B, alive[X, Y], X \leq S$.
(C1) $\alpha \geq 0$.
(C2) $f(L) \geq L$.

Note that both events and properties are modelled with temporal propositions. To deal with maximal intervals, Skolem constants and functions are taken for the unspecified endpoints and constraints between them are explicitly stated. Rules with constraints in the head are considered, like R2 above, and contribute to the enforcement of temporal consistency in the propositions involved. Rule R2, for instance, excludes scenarios for the available information such that *shoot* occurs with a loaded gun but the period for *alive* does not terminate. Such rules are treated in the inference as integrity constraints.

According to the adopted logic-based approach queries are expressed as conjunctions of atomic formulas, either predicative or constraints. With such queries it is possible to set up a particular configuration of intervals for a set of propositions, imposing constraints on their endpoints, and then test for its consistency with the KB. In the simplest form, a query is just a tentative retrieval of an interval for a proposition. Possible queries in our example are

Q1- $alive[X, 3]$
Q2- $alive[X, Y], Y > 5$
Q3- $loaded[X, Y], Y < 3$

Q1 asks about an ending of *alive* at point 3, Q2 about the possibility of *alive* extending past 5 and Q3 about the existence of an interval for *loaded* terminating before time point 3.

As we are considering a proposition-based language, only intervals for propositions and constraints on their endpoints appear in the answers. Constraints involve KB constants and functions and query variables; a predicative formula appearing as part of an answer indicates that the query can be satisfied by assuming it and its consequences through the KB rules. The emphasis is on consistency: in the absence of complete knowledge, the answer provides information on scenarios consistent with the query. The constraints and predicative formulas in each answer will be identified with the abductions required to configure the KB in order to derive the corresponding query.

Let us consider a KB K, a query Q and a set of abductions A expected from the derivation of Q by the inference system. The properties of answers are expressed in terms of entailment and consistency in the temporal context. Inconsistency of an answer with the KB is obviously excluded, and it is expected that the answer is a consequence of the knowledge base and the abductions, under interval and point consistency. $\approx$ is used for entailment under the axioms for interval and point consistency and $\Diamond$ for consistency in the same context.

Definition 1 Answer. A conjunction of atoms A is an *answer* to a query Q to a knowledge base K if and only if it is consistent, i.e.

$$\Diamond(K \cup A)$$

and it is sufficient, i.e.

$$K \cup A \models Q$$

The expressions which satisfy the above definition constitute the basic universe of answers, and correspond to the definition of abductive answer as provided in [KKT92]. It is also noted there that this constitutes a weak characterization for answers. In [Rib93] a stronger characterization is provided. This definition is however sufficient for the purpose of the present study.

Answers are nondeterministic, as there is in general more than one way to satisfy a query. For example, if the query reduces to a predicative goal, as Q1 above, we obtain a solution by unifying it with an existing interval in the KB, and another by creating a new (abduced) interval that is made disjoint from all others for the same proposition. As an example, two possible answers to query Q1 are $\{X = 0, \alpha = 3\}$ and $\{X > \alpha, alive[X, 3]\}$.

3 Inference system

In a KB constituted by facts and rules, we want to achieve a query-answering system having the features of goal-directed inference and also incorporate an efficient manipulation of constraints. In a logic-based approach, resolution can be used to work from the query as a goal using the facts and rules of the KB. In a language with constraints, this general approach has the drawback of treating them as regular predicates that have to be axiomatized; as they are central in the system, inference becomes quite inefficient. On the other hand, constraint based approaches to temporal reasoning [DMP91] deal efficiently with constraints but are not query oriented and only consider facts for the temporal relations. In MI, clauses involve predicative and constraint literals, expressing relations between propositions at different times. To have an inference system driven by queries and also deal with constraints in an efficient manner, our strategy is to maintain a constraint machine based on state-of-the-art constraint satisfaction algorithms working under the control of a resolution machine. The abductive resolution is detailed here, and the constraint mechanism is isolated via a set of constraint manipulation primitives which are called where required.

The query is taken as the top goal and resolution proceeds until all subgoals are solved. A goal is solved either by reduction with an existing fact or by abduction. A record of abductions is kept for use in the generation of answers. Abducing a constraint goal is equivalent to choosing a possible partial completion of the temporal order (excluding models of the KB where this constraint is not valid). Abducing a predicative goal corresponds to creating an interval for the proposition that is not supported on existing knowledge—does not unify with any of the existing intervals for this proposition.

3.1 Rules and integrity constraints

The knowledge base comprises rules where the head literal is a constraint, as well as rules having a predicative literal in the head. The rules with a predicative head are considered as rules for the corresponding proposition, and are used deductively for solving goals. Rules with constraints in the head are not useful in this regard and are therefore considered separately. Note that, as the negation of each constraint relation is another constraint relation, the negation of the head literal can be moved to the body. For uniformity, this kind of rules is always represented with empty head and is called an "integrity constraint". In the YSP above, rule R2 is therefore changed to

(R2') $\leftarrow shoot[S,S], loaded[A,B], A \leq S, S \leq B, alive[X,Y], X \leq S, S < Y$

It will be shown that integrity constraints are used to validate the abductions made during a derivation. If all the atoms in the body of an integrity constraint are satisfied then the KB and the abductions are inconsistent and the current derivation does not lead to an answer.

3.2 The constraint network

The inference system does not operate on the KB rules directly, but rather on an alternative partial-evaluated form. The available facts are applied to the body of the rules, producing new facts and reduced versions of the rules. The facts obtained in this process are represented in a constraint network (CN) where nodes exist for points and intervals. A constraint atom appears in the network as an arc between two point nodes and a predicative atom as two arcs between an interval node and a pair of point nodes. Interval nodes are labelled by proposition and therefore the set of intervals for each proposition is directly available in the network.

The motivation for processing the rules into facts comes from the features of the language. Constraints are central in the inference and the insertion and update of constraints is global in nature, so propagations can affect the whole network. The enforcement of maximality, on the other hand, is also a global constraint on the intervals for the same proposition. The forward processing of the KB is computationally very convenient, as it provides at each step a set of facts that can be derived from the KB. Instead of computing them each time consistency has to be enforced, they are maintained in the constraint network, avoiding the repetition of similar tasks for different subgoals relating to the same proposition. When abduction occurs and the network is incremented with some fact, the fact is evaluated in the rules to explore consequences in the form of new facts and new reduced rules. This strategy permits as much of an incremental behaviour as possible in a system requiring the enforcement of constraints. The set of rules and reduced rules is viewed as the complement of the constraint network. The reduced rules facilitate the computation of derived facts when the KB is extended with abductions.

Let us look at the YTS scenario mentioned earlier. Using the single fact for *load* and R1, a new fact for *loaded* is produced

(F4) $loaded[1, f(1)]$.

With the facts for *loaded*, *shoot* and *alive* a reduced rule (a reduced integrity constraint, in fact) is obtained from R2,

(R3) $\leftarrow 1 \leq 3,\ 3 \leq f(1),\ 0 \leq 3,\ 3 < \alpha$

which simplifies to

(R4) $\leftarrow 3 \leq f(1),\ 3 < \alpha$.

The incorporation of the new fact has no further consequences, and the expansion of the KB is completed. The maximality axioms implicit in the semantics are maintained dynamically in the network, by considering the interval nodes partitioned by proposition. The network can be regarded as consisting of a central part of point nodes, plus some mutually disjoint subnetworks of interval nodes, connected to the central part. Constraint propagation takes place in each subnet, and modifications are propagated between them.

3.3 Top level derivation

An inference rule relates a goal with a set of subgoals, and has associated explicit conditions under which it is applicable. The rules also show the evolution of the knowledge base in the course of inference. K designates the KB, and is assumed to be available as a pair $N - R$ where N is the network representation of facts and R the rule expansion. A stands for the set of abductions that represent the computed answer. Auxiliary relations are used for manipulations in the constraint network. $N \vdash g$ expresses that g is derivable from the constraint network, meaning that it is a logical consequence thereof. $\mathcal{P}$ and $\mathcal{C}$ appear in the conditions of some rules and refer to the sets of closed instances of predicative and constraint formulas, respectively.

The top level derivation symbol expresses the relation holding between a KB, a query and a possible answer. Its format is

$$K \vdash g\ \langle A \rangle$$

where g is the query taken as the goal to be derived, K the KB and A a set of abductions which complete K in order to satisfy g. Its definition consists of a single rule.

$$\frac{K \mathrel{\vdash\!\sim} g\ \langle A \mid K' \rangle \qquad \nabla K'}{K \vdash g\ \langle A \rangle}$$

A top level derivation is defined in terms of the abductive derivation plus a global consistency check. The consistency operator ∇ is applied to the potential answer A resulting from the abductive procedure.

3.4 Abductive procedure

Inference is centered on the deductive and abductive procedure. It drives the reduction of the atoms in the query, either unifying them with facts in the knowledge base or abducing conveniently. The following set of rules constitutes the definition of $\mathrel{\mid\!\sim}$.

Empty goal
$$\overline{K \mathrel{\mid\!\sim} \emptyset \, \langle \emptyset \mid K \rangle} \tag{1}$$

Conjunctive goal
$$\frac{K \mathrel{\mid\!\sim} g \, \langle A_1 \mid K' \rangle \qquad K' \mathrel{\mid\!\sim} G \, \langle A_2 \mid K'' \rangle}{K \mathrel{\mid\!\sim} (g, G) \, \langle A_1, A_2 \mid K'' \rangle} \tag{2}$$

Rules (1) and (2) are the structural rules that express the ramification of derivations in the conjunctions and establish termination of a branch when the empty goal is obtained. Abductions are propagated from each branch of a conjunction to the next by updating the KB, so that the initial scenario for the derivation of a goal is the final scenario of the previous one. The abductions for the conjunction are collected from the abductions obtained for each of the conjuncts.

Atomic constraint goal - network retrieval
$$\frac{N \vdash c}{K \mathrel{\mid\!\sim} c \, \langle \emptyset \mid K \rangle} \qquad \{c \in \mathcal{C}\} \tag{3}$$

Atomic predicative goal - network retrieval
$$\frac{N \vdash p[X', Y'] \qquad N-R \mathrel{\mid\!\sim} (X = X', Y = Y') \, \langle A \mid K' \rangle}{N-R \mathrel{\mid\!\sim} p[X, Y] \, \langle A \mid K' \rangle} \tag{4}$$

Rule (3) shows that a constraint is derived when it is a consequence of the constraint network. Rule (4) is the reduction of a predicative goal by unification with an interval in the constraint network. The derivation of the two equality constraints corresponding to the unification is required to solve the goal.

Atomic goal - direct abduction
$$\frac{K \ll g \, \langle A \mid K' \rangle}{K \mathrel{\mid\!\sim} g \, \langle g, A \mid K' \rangle} \qquad \left\{ \begin{array}{c} N \not\vdash g \\ g \in \mathcal{P} \cup \mathcal{C} \end{array} \right\} \tag{5}$$

Rule (5) accounts for the possibility of direct abduction of the atom to be derived. It is required that the goal is not a consequence of the constraint network. If this was the case the abduction would be unnecessary. The consistency of the abduction with the KB is required, and the call to $\ll$ includes this test. $\ll$ adds to the abductions the new consequences A which are produced while enforcing consistency of the abduced goal with K. K' is the expanded KB after incorporation of the abductions.

Atomic predicative goal - reduction through a rule

$$\frac{\begin{array}{l}K \mathrel{\hspace{1pt}\vrule depth 0pt height 1.2ex\hspace{-0.5pt}\sim\hspace{1pt}} (X = X', Y = Y') \langle A_1 \mid K' \rangle \\ N' \gg p[X, Y]\langle A_2 \rangle \\ K' \ll A_2 \langle A_3 \mid K'' \rangle \\ K'' \mathrel{\hspace{1pt}\vrule depth 0pt height 1.2ex\hspace{-0.5pt}\sim\hspace{1pt}} b \langle A_4 \mid K''' \rangle \end{array}}{K \mathrel{\hspace{1pt}\vrule depth 0pt height 1.2ex\hspace{-0.5pt}\sim\hspace{1pt}} p[X, Y] \langle A_1, A_2, A_3, A_4 \mid K''' \rangle} \quad \left\{ \begin{array}{l} K = N - R \\ K' = N' - R' \\ vrt(p[X', Y'] \leftarrow b) \in R \end{array} \right\} \quad (6)$$

Besides the unification with one of the intervals in the network and the direct abduction possible with previous rules, a predicative goal can be derived using the KB rules, as considered in (6). $vrt(h \leftarrow B)$ denotes a variant of a rule, with fresh variables to avoid variable conflicts. The endpoints of the interval in the goal are unified with those for the interval in the head of the rule, with abductions A_1 and resulting scenario K'. As this inference rule should provide an interval not already in the network, the network primitive $\gg$ is used to obtain the set of constraints A_2 forcing $[X, Y]$ to be disjoint from other intervals in the network. These are incorporated in K' producing K''. The body of the KB rule is derived in K'', resulting in the final scenario K'''. This inference rule allows, in the derivation of a goal, the abduction of "upstream" goals. By making abductions in the body of a rule whose head is the goal to be derived, the goal becomes a logical consequence of the extended knowledge base.

The disjointness condition imposed by $\gg$ makes rule (6) exclusive from rule (5). The resulting set of abductions A_4 is not empty, otherwise $p[X, Y]$ would already be a consequence of the KB and belong to its expansion.

3.5 Abductive incorporation

The abductive derivation relies on the abductive incorporation noted $\ll$. The tuples in this relation comprise the initial KB, the set of abductions to be incorporated, a set of new facts A' required for consistency of the former and an updated knowledge base K'. The first rule for $\ll$ corresponds to the base case where the set of abductions is empty and the KB unchanged.

$$\overline{K \ll \emptyset \langle \emptyset \mid K \rangle} \quad (7)$$

The second rule is recursive on the new abductions produced and reduces the incorporation of abductions to the incorporation into the network and the update of rules.

$$\frac{\begin{array}{l} N \ll_n A \langle A_n \mid N' \mid \theta \rangle \\ N' - R\theta \ll_r A^p \langle A_r \mid R' \rangle \\ N' - R' \ll A_r \langle A' \mid K \rangle \end{array}}{N - R \ll A \langle A_n, A_r, A' \mid K \rangle} \quad \{ A \neq \emptyset \} \quad (8)$$

The incorporation into the network $\ll_n$ is one of the constraint primitives, producing an updated network N', abductions A_n and substitution θ. Substitutions allow the replacement of entities which are linked by equality constraints in the network by a single representative token in the rules. The substitution is applied to the set of rules used in the next step.

The rule update $\ll_r$ takes the updated network and the set of rules $R\theta$ and reduces the rules with the predicative part of A, $A^{\mathcal{P}}$. The new set of rules R' is obtained. The constraint network is used in the process to eliminate any satisfiable constraints in the reduced rules. The rule set expansion produces a set of new facts A_r. For a tuple in the $\ll_r$ relation, the reduction of the rules with the facts does not produce any inconsistency. The reduction of the rules is expressed as a function of the initial rules and the network.

$$\frac{\bar{R} : R' + A}{N - R \ll_r F \langle A \mid R' \rangle} \quad \left\{ \begin{array}{l} R_r = \rho(F, R, N) \\ \bar{R} = simplify(R_r, N) \end{array} \right\}$$

The antecedent of the inference rule, $\bar{R} : R' + A$ enforces the fact that no inconsistency is detected in the set of reduced rules $\bar{R}$, while separating it into facts and non-factual rules. The reduction of rules induced by the existing facts appears in the abductive incorporation in the form of a function $\rho(F, R, N)$ described below. $simplify(R, N)$ is the merging of constraints in the body of each rule such that they involving the same entities. An equivalent constraint is computed through the network.

3.6 Rule reduction

The definition of $\rho(F, R, N)$ is supported on two simpler functions, one for the predicative part of the KB and one for the constraint part:

$$\rho(F, R, N) = \rho_c(\rho_p(F, R, N), N)$$

where $\rho_p(F, R, N)$ is the reduction of rules in R with predicative facts in F and $\rho_c(R, N)$ is the reduction of rules in R with constraints in N.

Reduction with predicative facts $\rho_p(P, R, N)$ is defined as the least set satisfying the following recursive equation

$$\rho_p(P, R, N) = R \cup \bigcup_{p \in P} (\sigma(p, R, N) \cup \rho_p(P, \sigma(p, R, N), N))$$

where $\sigma(p, R, N)$ is the reduction of rules triggered by a single predicative fact p. The definition factors it into the reduction of every rule in the set.

$$\sigma(p, R, N) = \bigcup_{r \in R} \sigma_1(p, r, N)$$

Finally the reduction of a single rule by a single predicative fact is defined as follows, with r standing for $h \leftarrow B$:

$$\sigma_1(p, r, N) = \left\{ \begin{array}{ll} \emptyset & \text{if } p = h \\ \{\leftarrow B\} & \text{if } inconsistent(p, h, N) \\ \{red(r, p'\theta, \theta) \mid p' \in B, \theta = mgu(p, p')\} & \text{otherwise} \end{array} \right.$$

The predicative expansion with fact p eliminates the rule if p is the head of the rule and reduces it to an integrity constraint if the interval in the head literal is inconsistent with that in p. If p can be unified with an atom in the body, a reduced version of the rule is produced. The options in the definition of σ_1 are exclusive because there is no recursion in the KB rules. In the substitution θ two situations have to be distinguished, as the terms to be substituted may be universal variables or existential KB entities. For universal variables, it suffices to perform the substitution on the rule atoms, and the universal variables disappear in the reduced rule. But if the rule atom already has any existential entity, this gives rise to an equality constraint in the body, rather than a substitution. This distinction is the basis for defining $red(r, \theta)$. If θ_u is substitution θ restricted to the universal variables and C_θ the rest of the substitution in equational form, the reduced rule is:

$$red(h \leftarrow B, p, \theta) \equiv h\theta_u \leftarrow (B\theta_u \setminus p) \cup C_\theta.$$

Reduction with constraints The reduction of rules with constraints in the network is defined in terms of a reduction of each rule with the network.

$$\rho_c(R, N) = \bigcup_{r \in R} \tau(r, N)$$

where $\tau(r, N)$ is

$$\tau(h \leftarrow B, N) = \begin{cases} \emptyset & \text{if } \exists c \in B, N \vdash \bar{c} \\ \{h \leftarrow (B \setminus \{c \in B \mid N \vdash c\})\} & \text{otherwise.} \end{cases} \quad (9)$$

A rule is eliminated if any of the constraints in the body is contradictory with the network, otherwise it is simplified of those constraints which are satisfied.

3.7 Interleaving of network and reduced rules

It is now possible to see how the use of the constraint network and the set of rules is interleaved. The retrieval of facts from the KB only requires the network, which is considered closed for the factual consequences of the KB. When abduction occurs, both the network and the rules are used, as captured in the definition of the incorporation relation $\ll$. Figure 1 illustrates the process of incorporation, relating the N and R components of the KB during an update.

4 Global Consistency

It is now necessary to evaluate the extent to which the abductive procedure enforces consistency of answers. Supposing that the initial KB is consistent and so are the N and R components obtained from it, the inconsistencies may appear only in the incorporation of new facts when the answer scenario is built. Following the incorporation steps as depicted in figure 1 and the definitions of

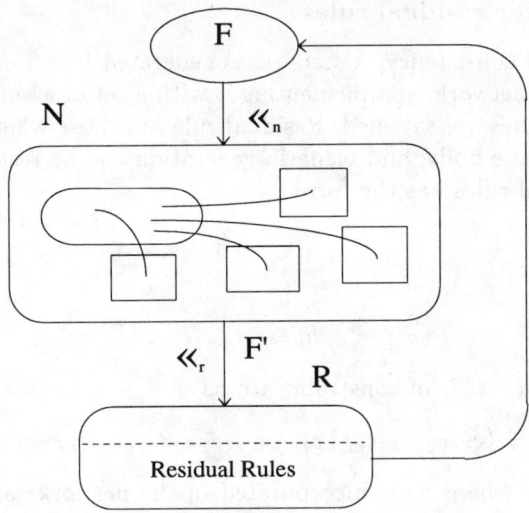

Fig. 1. Interaction between N and R in scenario update.

the derivation relations, it is obvious that the repeated application of the incorporation procedure detects several classes of inconsistency. Consistency of the facts to be incorporated with the network is required in the $\ll_n$ step. With the new configuration of the network, rule reduction is performed. Inconsistencies due to the satisfaction of all constraints in the body of an integrity constraint cause failure of the reduction. The same happens if the body of a residual rule is satisfied, but the head is not consistent.

The consistency enforcement during rule expansion is however not sufficient for consistency of the answers. A set of rules may have implicit consequences which are not obtained via the application of the rule reduction, because this operates on a rule-by-rule basis. It is therefore adequate to explore consequences of conjunctive statements only. We have however disjunctive statements, due to the existence of constraints, and satisfying them involves satisfying one of the constraints in the disjunction. The possibility exists that all scenarios satisfying the KB rules have some consequence which is not already in the network. In case this happens, the fact is also a consequence of the KB, but it is not taken into account by the incorporation procedure. It is also possible that no scenario exists satisfying all the rules, and this is also not taken into account by the abductive procedure.

The border line between consistency enforcement in the inference process and final consistency check has been located according to an incrementality criterium. The incorporation of abductions with associated reduction of rules builds at each abductive step the context in which inference can proceed. This partial consistency enforcement is therefore accompanied by an incremental update of the answer scenario. This is not the case with the generation of a scenario for the integrity constraints, which has to be produced anew for each incorporation of facts.

4.1 Scenario for residual rules

To enforce global consistency, a scenario is generated based on the information in the constraint network, complementing it with a set of assumptions such that all the residual rules are satisfied. Residual rule are those where only constraint atoms remain in the body, and scenario generation can be restricted to them. If the set of residual rules has the form

$$h_1 \leftarrow C_1$$
$$\vdots$$
$$h_m \leftarrow C_m$$

and we consider a set S_c of constraint atoms

$$S_c = \{c_1, c_2, \ldots c_m \mid \bar{c}_1 \in C_1, \bar{c}_2 \in C_2, \ldots \bar{c}_m \in C_m\},$$

then the scenario where S_c is incorporated in the network satisfies all residual rules. The rules with nonempty head can alternatively be satisfied in a scenario by including in it both the head literal and all the constraints in the body.

Global consistency is based on the generation of one consistent scenario. It is now possible to define the meaning of ∇K, the consistency of K, associating it with the existence of a consistent scenario S for K. Let $\underline{R}$ designate the subset of R including all residual rules.

$$\frac{}{\nabla(N-R)} \qquad \{\,\underline{R} = \emptyset\,\} \qquad (10)$$

$$\frac{N-R \ll \bar{c}\,\langle -\mid K'\rangle \qquad \nabla K'}{\nabla(N-R)} \qquad \left\{\begin{array}{l} h \leftarrow B \in \underline{R} \\ c \in B \end{array}\right\} \qquad (11)$$

$$\frac{N-R \ll (h,B)\,\langle -\mid K'\rangle \qquad \nabla K'}{\nabla(N-R)} \qquad \left\{\begin{array}{l} h \leftarrow B \in \underline{R} \\ h \neq \emptyset \end{array}\right\} \qquad (12)$$

Rules (11) and (12) correspond to the increment of the provisional scenario with the atoms that satisfy one of the residual rules. In (11) a constraint complementary to one of the body constraints is considered. In (12) all the constraints in the body and the predicative atom in the head are incorporated. In both, the resulting KB has to be tested for consistency. This is a KB where one of the residual rules has been satisfied and therefore dropped.

5 Constraint handling

The relations among times appearing in temporal assertions can be dealt with by specific constraint handling techniques. The interval algebra proposed by Allen [All83] has triggered a generalized interest in the subject, and different approaches to the problem appeared in recent work. The complexity of constraint reduction in the interval algebra has been analysed in [VK86], where the point algebra is proposed as an alternative. Algorithms for reducing constraints in these algebras [vB89, vB90] benefit from existing results for constraint satisfaction but have to deal with the new ones posed by incomplete knowledge and the features of the temporal relations. The problem can alternatively be considered from a quantitative point of view as a constraint satisfaction problem, as in [DMP91], and its extension to the qualitative relations of interval and point algebras has been treated in [Mei91, KL91].

The constraint primitives appearing in the inference rules can be performed on the network using point algebra on each of the subnetworks. The updates are propagated between them using a systematic translation between interval and point information [Rib93].

6 Consistency of computed answers

To prove that a computed answer is consistent, we show that it admits a model. To this purpose we resort to the notion of scenario given earlier. The scenario generated in the derivation includes the set of abductions obtained during the derivation. If the scenario has a model which is also a model of the KB, then the KB extended with the abductions also has a model, i.e.

$$\Diamond(K \cup S) \Rightarrow \Diamond(K \cup A).$$

Let S be the scenario obtained in the derivation, and N_S its representation as a constraint network. An interpretation I for the scenario is chosen as follows:

- For each existential entity, a value is assigned from the temporal domain, such that the constraints in N_S are satisfied.
- For each proposition, the predicative atoms corresponding to valuated facts in the scenario are true in the interpretation, and all others are false.

It is assumed that the network is consistent, i.e. that there exists an assignment of values in the temporal domain to the time entities in the scenario which satisfies all the constraints. As a consequence of the choice of interpretation, we have that a constraint fact is true if it is subsumed in the constraint network and it is false if its negation is subsumed in the constraint network. We want to prove that the interpretation is a model of the KB; to this end we show that all the facts and rules in the KB are satisfied.

For predicative facts and constraints, this is simply verified. All predicative facts in the KB are in the initial network N and are never discarded during the derivation. Therefore, they belong to the scenario and are true in the interpretation. All constraint facts in the KB are also in N_S and the interpretation satisfies them by hypothesis.

For rules, let us consider all ground instances of KB rules and show they are satisfied. Rule instances have the general form

$$h \leftarrow p_1, \ldots, p_n, c_1, \ldots, c_m$$

where each p_i and c_j are ground instances. All rule instances such that any of the p_i is not in the model are satisfied. This excludes rule instances where predicative atoms in the body are not those in the facts of the scenario.

For the rule instances such that all p_i are true in the interpretation, we can equate their satisfiability to that of corresponding residual rules:

$$h \leftarrow c_1, \ldots, c_m.$$

For each such rule it is now necessary to show that

1. It corresponds to a reduced rule produced in the KB expansion and updated in the incorporation procedure
2. it is true in the interpretation I.

The first part concerns the rule expansion function. It is required that for each tuple of predicative facts in I and each rule having atoms for the corresponding propositions in the body, a residual rule is produced by the rule expansion function. The second part concerns the satisfaction of such reduced rules. The proof for each part is now sketched.

Existence of reduced rule: It is necessary to prove that, for each predicative fact in the network, rule expansion is processed. Consider p_k^g, a ground instance of p_k. One of two cases applies

- If p_k^g is a fact in the initial KB, the initial expansion will use it.
- If p_k^g is abduced during the derivation, there is a step of the incorporation procedure where it is a part of the abductions, and therefore of the set of facts F processed in the call to the rule reduction function $\rho(F, R, N)$.

So, for each predicative fact in the network, there is a rule reduction step where it is used. To show that a reduced rule exists for any tuple of ground facts, we use induction on the number of predicative atoms in the body of the rule. It is easily verified that a rule with no predicative atoms in the body is maintained in the rule set. For the inductive step, we consider that there is a reduction of the rule for $k-1$ atoms. For the kth atom, there is by hypothesis an atom for the same proposition p in the body of the rule. The application of the reduction step $\sigma_1(p, r, N)$ may result in an unsatisfiable constraint in the body, in which case the rule is dropped. Such rules are trivially satisfied and irrelevant in what follows. If this is not the case, a reduced rule is produced and maintained in the set.

Satisfaction of the ground rules: Consider an instance of a rule

$$h \leftarrow p_1^g, \ldots, p_n^g, C^g$$

If an inconsistent constraint is detected during rule reduction, then there is some constraint in C^g which is inconsistent with an intermediate network configuration. As the constraint network information grows monotonically, it is also inconsistent with the network in the scenario, and therefore false in the interpretation. The ground instance is therefore satisfied.

If no inconsistent constraint is detected during rule reduction, the corresponding reduced rule is maintained. In this rule, constraints satisfied in the interpretation by the time the reduction was completed are no longer present. Such constraints are also satisfied in the interpretation, and may therefore be eliminated from the rule. In the reduced rule there are also extra constraints if unification between existential temporal entities has occurred. But we have assumed that all p_i^g are in the scenario; so the unification constraints must be satisfied in the final scenario, and therefore also in the interpretation. As all constraints satisfiable in the final scenario can be discarded from the residual rule, the ground rule is now identifiable with the residual rule. Suppose the ground rule is not satisfied. This happens if all the constraints $C^g = c_1^g, \ldots, c_m^g$ are satisfied but h is not. Suppose that all c_i^g are satisfied in the residual rule. If this is detected during the abductive procedure, then the application of constraint reduction rule (9) would result in a new fact $h \leftarrow$, incorporated in the network according to the $\ll_r$ procedure. But this contradicts the hypothesis. The satisfaction of the c_i^g is therefore not detected until the abductive procedure is completed, and the rule is maintained in the reduced rules. Each residual rule in this set must be satisfied in the scenario generation process. There are two forms of achieving this, according to rules (11) and (12). Either the constraints are all assumed true and incorporated in the network, in which case h is also incorporated. But then h is also true in the interpretation and this contradicts the hypothesis. The alternative is to choose one constraint in the body c and incorporate its converse $\bar{c}$ in the network. Then there is one constraint in the body which is not satisfied in the network and consequently not satisfied in the interpretation. This also contradicts the hypothesis. As scenario generation only terminates when all residual rules are satisfied, the rule instance is satisfied.

The proof above is based on the assumption that there exists an assignment of values from the temporal domain to the existential symbols in the KB scenario such that it is a model of the constraints in N_S. This condition does not hold for arbitrary assignments to the existential entities, due to the semantics of the constraint relations. It can however be proved that it exists for the constraint network provided that constraint propagation and reduction is performed according to some requisites [Rib93].

7 Conclusions

This work intends to show that temporal reasoning can be effectively carried out in a logic programming framework comprising constraint handling, where abduction appears as the natural way of overcoming partial knowledge. Abduction is used in a goal-oriented manner to generate hypotheses on 1) the possible intervals for propositions and 2) the temporal order, such that the query is satisfied. Abduction is therefore always performed in the context of answering a query, and not for completing the KB in some preferred way as is common in nonmonotonic frameworks. Enforcing consistency of the hypotheses is essential in an abductive system. Considerable incrementality is achieved in this task by maintaining an auxiliary representation of facts and reduced rules.

References

[All83] James Allen. Maintaining Knowledge About Temporal Intervals. *Communications of the ACM*, 26(11):832–843, 1983.

[All84] James Allen. Towards a General Theory of Action and Time. *Artificial Intelligence*, (23):123–154, 1984.

[DMP91] R. Dechter, I. Meiri, and J. Pearl. Temporal constraint networks. *Artificial Intelligence*, 49(1), 1991.

[HM86] Steve Hanks and Drew McDermott. Default reasoning, nonmonotonic logics and the frame problem. In *Proceedings of the 5th National Conference on Artificial Intelligence*, pages 328–333. AAAI, 1986.

[KKT92] A. C. Kakas, R. A. Kowalski, and F. Toni. Abductive logic programming. *Journal of Logic and Computation*, 2(6):719–770, 1992.

[KL91] H. Kautz and P. Ladkin. Integrating metric and qualitative temporal reasoning. In *Proceedings of AAAI'91*, 1991.

[KS86] Robert Kowalski and Marek Sergot. A logic-based calculus of events. *New Generation Computing*, 4(1):67–95, 1986.

[McD82] Drew McDermott. A Temporal Logic for Reasoning About Processes and Plans. *Cognitive Science*, (6):101–155, 1982.

[Mei91] I. Meiri. Combining qualitative and quantitative constraints in temporal reasoning. In *Proceedings of AAAI'91*, 1991.

[Rib93] Cristina Ribeiro. *Representation and Inference of Temporal Knowledge*. PhD thesis, FCT-UNL, 1993.

[RP91] Cristina Ribeiro and António Porto. Maximal intervals, an approach to temporal reasoning. In P. Barahona, L. Moniz Pereira, and A. Porto, editors, *EPIA 91—5th Portuguese Conference on Artificial Intelligence*. Springer Verlag, Lecture Notes on Artificial Intelligence 541, 1991.

[Sho87] Yoav Shoham. *Reasonig about Change*. The MIT Press, 1987.

[vB89] Peter van Beek. Approximation algorithms for temporal reasoning. In *Proceedings of the 11th International Joint Conference on Artificial Intelligence*, pages 1291–1296, 1989.

[vB90] Peter van Beek. Reasoning about qualitative temporal information. In *Proceedings of the 8th National Conference on Artificial Intelligence*, pages 728–734, 1990.

[VK86] Marc Vilain and Henry Kautz. Constraint propagation algorithms for temporal reasoning. In *Proceedings of the 5th National Conference on Artificial Intelligence*, pages 377–382, 1986.

A Temporal Logic Approach to Implementation and Refinement in Timed Petri Nets

Miguel Felder and Angelo Morzenti
Politecnico di Milano, Dipartimento di Elettronica e Informazione
email: {felder, morzenti}@elet.polimi.it

Abstract

We define formally the notion of implementation for time critical systems in terms of provability of properties described abstractly at the specification level. We characterize this notion in terms of formulas of the temporal logic TRIO and operational models of timed Petri nets. Refinement steps are often used as a means to derive in a systematic way the system design starting from its abstract specification. We present a method to formally prove the correctness of refinement rules for timed Petri nets and apply it to a few simple cases. We show how the possibility to retain properties of the specification in its implementation can be exploited to greatly simplify the verification of the designed systems by performing incremental analysis at various levels of the specification/implementation hierarchy.

1. Introduction

Real time systems are required to manage their resources in a way that predictably satisfies some given timing constraints. Such systems are often embedded in critical applications such as patient monitoring systems, plant supervision systems, traffic control systems: their correctness is of primary importance, since their failure can have enormous costs and lead to unrecoverable damages. In the past years, the research on formal methods for the specification and verification of real time systems has been particularly active, especially in the field of temporal logic, resulting in the proposal of several specification formalisms and verification methods.

The proposed models are however rarely employed in the industrial development of such systems, where informal and semiformal methods are still largely prevalent. One of the reasons for this unsatisfactory state of the art is that the systematic or algorithmic analysis techniques are very complex so that they cannot be scaled up to realistic systems. For instance the algorithms proposed for system verification and validation are often exponential in the size of the specification [AH90, FM92].

Often, however, the final specification of a (real time) system or its high level design are derived through a sequence of refinement steps. In each of these steps one derives an "implementation", i.e., a more detailed version of the system that includes elements deriving from design choices, starting from a more abstract version that is considered as its specification. If these repeated refinement steps are conducted in a systematic, careful way, the verification activity needs not be repeated from scratch for each implementation step, since the system can be analyzed incrementally. The overall cost of the verification can thus be kept at a reasonable level by reusing in each step the results already obtained in the preceding phases.

In this paper we address the problem of reducing the overall specification and design effort for real time system developed through a sequence of refinement steps. We report here the application of these ideas to the case when the real time system is abstractly modeled with timed Petri nets (a kind of Petri net where each transition is associated with a firing time interval describing its earliest and latest firing time after enabling) and its timing requirements are described by means of formulas in TRIO (a temporal logic providing a metric on time distances, particularly suitable to the specification of real-time systems). We formally define the notion of correct implementation among two timed Petri nets in terms of provability, in the net acting as an implementation, of all the properties that are guaranteed by the net acting as a specification. In previous works [FMM91,FMM94], we defined an axiomatic system for TRIO and an axiomatization of timed Petri nets that adequately copes with the salient features of this operational formalism, such as nondeterministic behavior, multiple simultaneous transition firings, zero-time and infinite-time transitions, and unbounded accumulation of tokens in places. We are therefore able not only to formally characterize the notion of implementation among timed Petri nets, but also to prove (i.e., to derive as a TRIO metatheorem) that such a relation holds for two given nets.

The proof of correct implementation among timed Petri nets, however, may not be performed algorithmically and generally it requires a certain amount of skill and ingenuity, because no general guideline can be provided for it and the TRIO axiomatization is not complete (we recall that the language includes arithmetics over the temporal domain). In a companion paper [FGP93] we introduced a set of refinement rules for timed Petri nets which allow a designer to substitute a place or transition with a net fragment in such a way that the resulting net is a correct implementation of the original one. Here we provide a systematic method to formally prove the correctness of such refinement rules. Once the correctness of a rule has been formally proven, its application to particular nets can be performed systematically or even automatically, since only the topological relations among net elements and the algebraic relations among the time bounds of the involved transitions must be checked.

By adopting this design method based on successive refinements, one can obtain nets that satisfy by construction all properties specified by the initial abstract version of the system. Moreover, the properties "inherited by refinement" can further be used as lemmas in the more detailed analysis of the final version of the system. In this way, when analyzing an implementation obtained through a sequence of refinement steps, the analysis effort can be greatly reduced by performing the proof of intermediate lemmas on the first, more abstract and simplified versions of the system.

The notion of refinement has already been studied in the literature [Vog87, SM83, Mül85], but, to the best of our knowledge, with reference only to untimed Petri nets. Here, and in [FGP93], we propose new techniques specifically devoted to real time systems, thus employing timed Petri nets and a temporal logic with a metric on time such as TRIO. We characterize properties that we wish to be preserved in implementations in a syntactic way, i.e., by means of TRIO formulas, whereas other works [FGP93, Vog90, GG90] adopt more semantic approaches based on execution traces and behaviors. In particular, [FGP93] provides an operational characterization of the correct implementation relation by introducing the notion of timed behaviors. A treatment closer to ours under this respect is adopted in [DDG+90], where the temporal logic MCTL (a modular extension of CTL) is used to describe properties of Petri net modules. Our definitions of implementation and refinement are based on the ability to ensure, at lower levels in the specification/implementation hierarchy, the properties that are specified at the highest specification level. This approach follows similar notions introduced in [AL91] and [Aiz90]. Other approaches to the refinement operation [LA89, GG90] greatly emphasize compositionality and modularity, both in the system structure and in the proof of its properties. The ideas on incremental analysis of refined systems presented in the paper are strongly related to the notions of compositionality and incrementality reported in [YY91].

The paper is structured as follows. Sections 2 and 3 present the TRIO language and its use in the axiomatization of timed Petri nets. Section 4 formally defines the implementation relation among timed Petri nets. Sections 5 and 6 introduce a set of refinement rules and provide a method for proving their correctness. Section 7 presents an example of incremental analysis.

2. The temporal logic TRIO

TRIO is a temporal logic equipped with operators that provide a metric on time, since they express quantitatively the distance in time between events and the length of time intervals. The underlying time is assumed to be linear and the logic easily accommodates both discrete and dense models of time. In the present paper, we assume as time model the set of real numbers, which makes the time domain continuous and unlimited both in the past and in the future. The logic is first order and typed, in that every variable has an associated domain of possible values, and every predicate and function has associated domain and range. The language includes time independent predicates, whose interpretation is independent from the current time instant, and time dependent ones, representing relations that may change with time (elsewhere we defined more complex versions of the language that include time dependent variables and functions: we ignore here such features which are not essential for the presented results). The fundamental modal operator *Dist* is defined in such a way that if A is a formula and t is a term of the temporal type, then $Dist(A,d)$ is a formula meaning that A holds at an instant laying d time units (t.u.) in the future (if d>0) or in the past (if d<0) or at the current time (if d=0).

Several derived temporal operators may be defined starting from Dist, using the propositional operators, conditions on the temporal arguments, and first-order quantification. A sample thereof is given in Table 1, together with short intuitive explanations, whenever needed.

$Futr(\varphi, t)$	$\stackrel{def}{=}$	$t \geq 0 \wedge Dist(\varphi, t)$	Future
$Past(\varphi, t)$	$\stackrel{def}{=}$	$t \geq 0 \wedge Dist(\varphi, -t)$	Past
$Lasts(\varphi, t)$	$\stackrel{def}{=}$	$\forall t'(0<t'<t \rightarrow Dist(\varphi, t'))$	φ holds over an interval
$Lasts_{ie}(\varphi, t)$	$\stackrel{def}{=}$	$\forall t'(0 \leq t'<t \rightarrow Dist(\varphi, t'))$	first extreme included
$Lasts_{ei}(\varphi, t)$	$\stackrel{def}{=}$	$\forall t'(0<t' \leq t \rightarrow Dist(\varphi, t'))$	second extreme included
$SomF(\varphi)$	$\stackrel{def}{=}$	$\exists t\, (t>0 \wedge Dist(\varphi, t))$	sometimes in the future
$AlwF(\varphi)$	$\stackrel{def}{=}$	$\forall t(t>0 \rightarrow Dist(\varphi, t))$	always in the future
$Alw(\varphi)$	$\stackrel{def}{=}$	$\forall t\, Dist(\varphi, t)$	Always φ
$Som(\varphi)$	$\stackrel{def}{=}$	$\exists t\, Dist(\varphi, t)$	Sometimes φ
$WithinF(\varphi, t)$	$\stackrel{def}{=}$	$\exists t'(0 \leq t' \leq t \wedge Dist(\varphi, t'))$	φ will occur within t time units

Table 1 A sample of derived temporal operators.

TRIO has been given a model-theoretical semantics in [GMM90] in a fairly standard way, while [FMM94] reports a sound and (relatively) complete axiomatic system.

3. Timed Petri nets and their aziomatization

Timed Petri nets (TPN) [MF76] differ from traditional Petri nets [Rei85] in that every transition v is associated with a pair of values, usually denoted by $[m_v, M_v]$, belonging to the temporal domain (with $0 \leq m_v \leq M_v \leq \infty$). These are called, respectively, the *lower and upper bound* of v, whereas the pair $[m_v, M_v]$ is called v's *time interval*. Intuitively, the meaning of the pair $[m_v, M_v]$ is that, once v is enabled by the presence of at least one token in each place of its preset, it *can not fire before a time m_v elapses* (we call this property LB, since it originates from the lowerbound of v) and it must fire within M_v unless in the meanwhile it is disabled by the firing of another transition in conflict with it (we refer to this property as UB, since it is related to the upperbound of v). As in traditional Petri nets, tokens are *uniquely* generated and consumed by transition firings. In particular, any firing of a transition consumes one and only one distinct token from each place in its preset (we call this property IU, input unicity), and introduces one and only one token into each place of its postset; that token can contribute to no more than a single transition firing (we call this property OU, output unicity).

For the sake of simplicity and brevity, we consider here a restricted class of timed Petri nets, whose hypotheses, however, are satisfied in many cases of practical interest. A timed Petri net N=<P, T, F, Θ, m_0> is called 1_bounded, simple-place (1-STPN), iff it satisfies the following hypotheses: no transition of N has an empty postset; for each pair of transitions r, s of the net, there is at most one place belonging both to the postset of r and to the preset of s; N is 1-bounded, i.e., no place may contain more than one token at any time. In [FMM94] we proved that in any net satisfying the above hypotheses no multiple simultaneous firings of any given transition can occur.

Formally, a 1-STPN net is defined as a 5-tuple: N=<P_N, T_N, F_N, Θ_N, m_N>, where

- P_N, T_N, F_N denotes, as customary in Petri nets, the set of places, transitions, and arcs of the net (for any transition v∈T we denote its preset and postset as $^\bullet v$ and $v^\bullet$, respectively);
- Θ_N is a function assigning to each transition of the net its *time interval*: Θ_N: T_N → $R_\infty^+ \times R_\infty^+$, where $R_\infty^+ = \{x \mid x \in R \wedge x \geq 0\} \cup \{\infty\}$ is the set of non negative reals enriched with the "infinite" value; for each v∈T_N, $\Theta_N(v)$=<m_v, M_v> is a pair of nonnegative real values such that $0 \leq m_v \leq M_v \leq \infty$. In the following the value of $\Theta_N(v)$ will be indicated as [m_v, M_v] to conform with the literature on the subject.
- $m_N \subseteq P$ is the initial marking of the net; since we describe here only 1-STPN, in the initial state of the net every place may only be marked or unmarked, so that m is just a subset of the places.

In a preceding paper [FMM94] we gave a TRIO axiomatization of timed Petri nets, and in particular, as a significant simplified case, of 1-STPN. As a useful convention, unless otherwise specified we intend that all axioms describing Petri net semantics are preceded by an implicit *Alw* operator, and that identifiers denoting transitions (e.g., r, s, u, v) are constant names, i.e., names of the transitions in the Petri net to which the formula refers, while identifiers for time distances (e.g., d, e) are variables. We also adopt the convention that free variables in axioms are always implicitly universally quantified.

The axiom system for 1-STPNs is based on the following *elementary predicates* (other predicates will be introduced as *derived or auxiliary predicates*).
- fire(r)   states that at the current instant transition r fires.
- tokenF(r, s, d)   the token produced at the current instant by the firing of transition r enters the (unique, and hence implicit) place $p \in r^\bullet \cap ^\bullet s$ and will be consumed by the firing of transition s after d time units.

The following axiom formalizes the intuitive explanation of tokenF and fire presented above.

1. FI (Future Implication):   tokenF(r, s, d) → fire(r) ∧ Futr(fire(s), d)

To simplify some future axioms and deductions, we immediately introduce the derived predicate tokenP(r, s, d) having the same meaning as tokenF but referring to the time of firing of the second transition s:

2. **PF (Past to Future):** $\quad\quad\quad\quad\quad$ tokenP(r, s, d) $\overset{def}{=}$ Past(tokenF(r, s, d), d).

From 1 and 2 the following, symmetric properties can be immediately derived.

3. **PI (Past Implication):** $\quad\quad\quad$ tokenP(r, s, d) $\rightarrow$ fire(s) $\wedge$ Past(fire(r), d)

4. **FP (Future to Past):** $\quad\quad\quad\quad$ tokenF(r, s, d) $\leftrightarrow$ Futr(tokenP(r, s, d), d)

The general form of the axiomatization of a 1-STPN as determined by the net topology and the transitions' time bounds can be found in [FMM94]. For each transition v of the net, the axiomatization includes axioms called LB(v), UB(v), IU(v), and OU(v) describing, respectively, the properties of lowerbound, upperbound, input unicity, and output unicity that hold for v. Here we report, as a concrete example, the axioms related to transition v of figure 1.

LB(v): fire(v) $\rightarrow$ $\exists$d(d$\geq$m$_v$ $\wedge$ (tokenP(r, v, d) $\vee$ tokenP(s, v, d));

UB(v): (fire(r) $\rightarrow$ $\exists$d(d$\leq$M$_v$ $\wedge$ tokenF(r,v,d))) $\wedge$ (fire(s) $\rightarrow$ $\exists$d(d$\leq$M$_v$ $\wedge$ tokenF(s,v,d)));

IU(v): tokenP(x, v, d) $\wedge$ tokenP(y, v, e) $\rightarrow$ x=y $\wedge$ d=e, where x and y are variables;

OU(v): tokenF(v, x, d) $\wedge$ tokenF(v, y, e) $\rightarrow$ x=y $\wedge$ d=e, where x and y are variables.

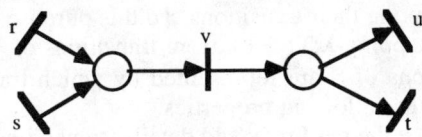

Fig. 1. A simple net.

The initial marking of a 1-STPN is described in terms of firings of a set of special transitions called *initializing transitions* of the net. For each place p$\in$P$_N$, if p$\in$m$_N$ then an initializing transition for p is introduced called 'itp', such that $^\bullet$itp=$\varnothing$ (the preset of itp is empty) and itp$^\bullet$={p} (p is the only place of its postset). Consequently itp does not give rise to any axiom of the kind LB(itp) nor UB(itp) nor IU(itp), but only to axioms of the kind OU(itp), and it appears in the axioms of other transitions having place p in their preset. As a consequence the values of the lower and upper bound m$_{itp}$ and M$_{itp}$ are immaterial: any initializing transition itp fires just once, as specified by the following axiom.

$\quad\quad$ IM(p): $\quad\quad$ fire(itp) $\wedge$ AlwP($\neg$fire(itp)) $\wedge$ AlwF($\neg$fire(itp))

Notice that the above axioms is asserted at the time instant which is assumed as the starting point for the evolution of the net. Starting from the topology and the initial marking of a given net we are thus able to completely axiomatize its temporal behavior by means of the axioms of the kind FI, IU, OU, LB, UB, IM. In the following we call Ax(N) the collection of all such axioms for a net N. Using Ax(N) we are able to prove its temporal properties: for any TRIO formula φ constructed from predicates *fire* and *tokenF*, if Ax(N)$\vdash\varphi$ then every execution of net N satisfies the property described by φ.

As an example of a timing property consider, with reference to the net fragment of Fig. 1, the formula fire(r) $\rightarrow$ WithinF(fire(u)$\oplus$fire(t), M$_v$+max(M$_u$,M$_t$)). It asserts that any firing of r will always be followed by a corresponding firing of either t or u within M$_v$+max(M$_u$,M$_t$) t.u..

4. Implementation relation among 1-STPNs

In this section we characterize the notion of implementation relation among 1-STPNs. An implementation must satisfy, by its very definition, all the requirements expressed in the specification, hence we say that a timed Petri net I correctly implements a timed Petri net S acting as a specification if all the properties satisfied by net S are also ensured by net I. The properties ensured by a timed Petri net are essentially the temporal relations among its observable events, i.e., those temporal relations among transition firings that are satisfied in every execution. Therefore such properties correspond, in our approach, to the TRIO formulas constructed on the time dependent *fire* predicate, which can be derived as valid formulas from the axiomatization of the net. We consider only the *fire* predicate because we assume that the transition firings are the only observable events in the net. The other fundamental predicate in the axiomatization of timed Petri nets, namely predicate 'tokenF', models the topology of the net and the cause-effect relations among transition firings: such information is related to the mechanisms through which the timed Petri net ensures the temporal properties of the systems it models and thus we consider it immaterial for what concerns the implementation relation.

An implementation net I may add (implementation) details, in the form of additional places and transitions, to the specification net S; we require in any case that for every transition in S there is (at least) one transition in I corresponding to it. The implementation relation among nets S and I is therefore defined with reference to a given correspondence among their transitions. To this purpose we introduce the notion of *event function* as a mapping $\lambda: T_I \to T_S$ from transitions of I to transitions of S that specifies which transitions of S are represented by which transitions of I. An event function λ must satisfy the following properties.

- λ may be partial, since the net I may add details (transitions) that are not present in S;
- λ is required to be onto, because *every* property of S must be ensured by I, which implies that every transition of S is represented by some transition in I;
- λ is not required to be one to one: a single transition of S may well be represented by more than one transition of I (e.g. when an action of S is implemented by two exclusive actions of I).

The event function λ allows us to precisely state, for each property ensured by the specification net S, which corresponding property is ensured by the implementation net I. We define a *property function* Λ, uniquely identified by λ, that translates each formula φ constructed on the *fire* predicate applied to transitions of S into a formula that represents the same property in I. Λ is defined by the following clauses.

i. $\Lambda(\text{fire}(v)) = \text{fire}(v_1) \vee \text{fire}(v_2) \vee \ldots \vee \text{fire}(v_n)$, where $\lambda(v_1)=\lambda(v_2)=\ldots=\lambda(v_n)=v$;

ii. $\Lambda(u=v) = u_1=v_1 \vee u_1=v_2 \vee \ldots \vee u_n=v_m$, for $u,v \in T_S$, where $\lambda(v_1)=\ldots=\lambda(v_n)=v$ and $\lambda(u_1)=\ldots=\lambda(u_m)=u$;

iii. $\Lambda(P(t_1\ldots t_n)) = P(t_1\ldots t_n)$ for any other atomic formula where $P \neq \text{tokenF}$;

iv. $\Lambda(\alpha \to \beta)=\Lambda(\alpha) \to \Lambda(\beta)$;   v. $\Lambda(\neg\alpha)=\neg\Lambda(\alpha)$;

vi. $\Lambda(\text{Dist}(\alpha, t))=\text{Dist}(\Lambda(\alpha), t)$;   vii. $\Lambda(\forall x\ \alpha)=\forall x \Lambda(\alpha)$.

The above introduced notations allow us to formally characterize correct implementations among 1-STPNs.

Definition 1 (correct implementation among 1-STPNs). Given two 1-STPNs S and I and an event function $\lambda: T_I \to T_S$, we say that I correctly implements S if, for each TRIO formula φ constructed from the fire predicate applied to transitions of S, $\text{Ax}(S) \vdash \varphi$ implies $\text{Ax}(I) \vdash \Lambda(\varphi)$.

It can be easily shown that the implementation relation as defined above is transitive: given nets N_1, N_2, N_3, if N_2 correctly implements N_1 through event function λ_1 and N_3 correctly implements N_2 through function λ_2, then N_3 correctly implements N_1 through the event function $\lambda_1 \circ \lambda_2$. This allows designers to iterate the implementation process so that, given a sequence of n timed Petri nets $N_1...N_n$ obtained through successive implementations, only $n-1$ proofs are necessary to show that the most detailed net N_n correctly implements the most abstract one N_1.

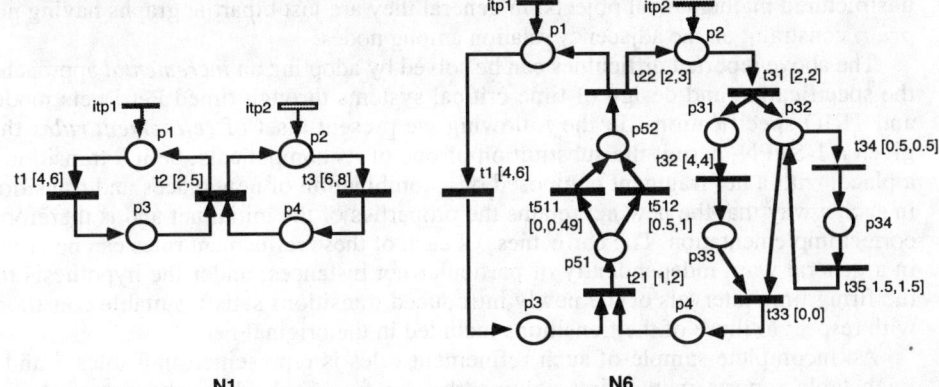

Fig 2. A simple 1-STPN (a) and a more complex net (b) implementing it.

Example 1. As a trivial example of implementation relation among timed Petri nets, let us consider two 1-STPN nets S and I having the same topology and initial marking (i.e., $T_S=T_I$, $P_S=P_I$, $F_S=F_I$, and $m_S=m_I$) but such that the time interval of every transition in I is stricter than that of the same transition in S (i.e., $\forall v \in T_I$, if we let $\Theta_I(v)=[m_{vI}, M_{vI}]$ and $\Theta_S(v)=[m_{vS}, M_{vS}]$ then $m_{vI} \geq m_{vS}$ and $M_{vI} \leq M_{vS}$). Then it can be easily proven that net I correctly implements net S. In fact, every axiom $\alpha_S \in Ax(S)$ for net S is logically implied by the corresponding axiom $\alpha_I \in Ax(I)$ in I, i.e., $\vdash \alpha_I \rightarrow \alpha_S$, because of the stricter time bounds. Then the implementation relation among S and I is satisfied by taking as event function λ the identity function on T_I, so that Λ is the identity function on the temporal properties of the net. The fact that, for any formula φ, $Ax(S) \vdash \varphi$ implies $Ax(I) \vdash \varphi$ (recall that $\Lambda(\varphi)=\varphi$) is an obvious consequence of the following property of the TRIO axiomatic calculus (and in fact of any first order calculus): if $\Gamma; \alpha \vdash \varphi$ and $\vdash \beta \rightarrow \alpha$ then $\Gamma; \beta \vdash \varphi$.

Example 2. As a more complex example of correct implementation, consider the net fragments N_1 and N_6 shown in Fig. 2 (the example is borrowed, with some minor modifications, from [FGP93]). N_1 models a simple rendezvous between a producer and a consumer. The producer gets data (e.g., temperature, pressure) from an external device, in 4 to 6 time units (t.u.), and then, in 2 to 5 t.u., communicates the acquired data to the consumer who is responsible for the elaboration. The elaboration by the consumer takes 6 to 8 t.u. (transition t3). Initially, the producer is ready for the acquisition and the consumer is ready for elaboration (places p1 and p2 are initially marked: notice initializing transitions itp1 and itp2). In the next section we will prove that the net N_6, where several elements modeling design choices have been added to those of N_1, correctly implements it, with event function λ defined as $\lambda(t_1) = t_1$, $\lambda(t_{22}) = t_2$, $\lambda(t_{33}) = t_3$.

5. Implementation through refinements

Proving the existence of a correct implementation relation can be complex and difficult when two arbitrary timed Petri nets (or 1-STPNs) are compared. First of all, the choice of the correct event function may be non trivial, since the space of such function is very large. Furthermore, given an event function, proving the correct implementation relation requires the proof of a metatheorem on the derivability of a rather large set of formulas, and no precise guideline on how to structure such proof can be given if no additional information is available on the relation among the two given nets. This situation is further complicated by the fact that (timed) Petri nets are rather unstructured mathematical objects: in general they are just bipartite graphs having no *a priori* constraint on the adjacency relation among nodes.

The above reported difficulties can be solved by adopting an *incremental* approach to the specification and design of time critical systems through timed Petri nets models and TRIO specifications. In the following we present a set of *refinement rules* that, given a 1-STPN, permit the substitution of one of its components, be it a transition or a place, with a net fragment composed of a combination of new places and transitions in such a way that the new net retains the properties of the initial net and is therefore a correct implementation. The correctness of each of these refinement rules can be proved in a general way, independently of particular net instances, under the hypothesis that the firing time intervals of the newly introduced transitions satisfy suitable constraints with respect to those of the transitions included in the original net.

An incomplete sample of such refinement rules is represented in Tables 2 and 3. Each table reports in the first column the net fragment where the substitution is performed and in the other columns the net resulting from the rule application. For each rule, the table also provides a formula (TC) asserting relations among the time bounds of the involved transitions, and the definition of the event function (EF) for such transitions (since the rest of the net is unchanged, $\lambda(v)=v$ for every other transition).

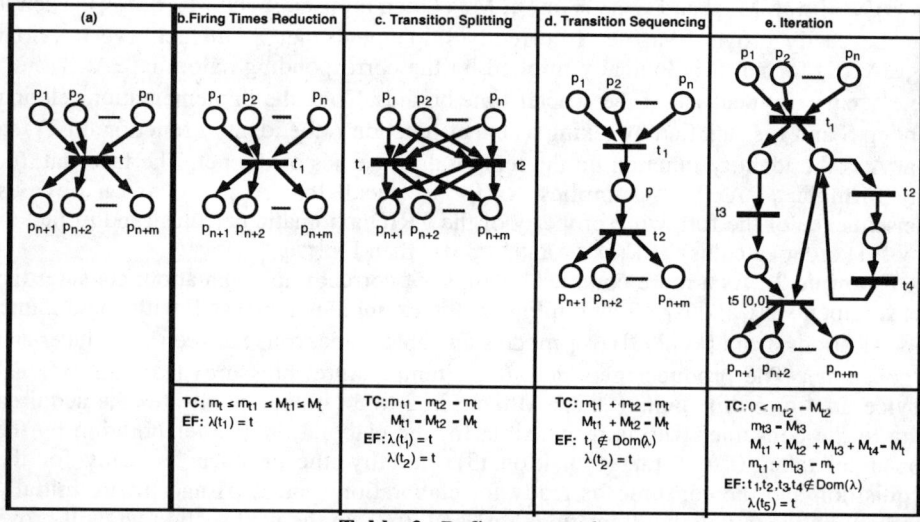

Table 2. Refinement rules

Table 2(d) describes the *transition sequencing* rule, that substitutes a given transition t with a sequence of two transitions t_1 and t_2 and a place p among them representing the start and the end of the action modeled by the original transition t. The rule can be applied only if there are no transitions conflicting with t. We apply this rule to transition t2 of net N_1 in Fig. 2 obtaining the net N_2 of Fig. 3. The action representing the communication between the producer and the consumer is detailed in two actions representing the start and the end of the communication, respectively (transitions t21 and t22). N_2 is a correct implementation of N_1 with the event function λ_1 such that $\lambda_1(t1) = t1$; $\lambda_1(t3) = t3$; $\lambda_1(t22) = t2$.

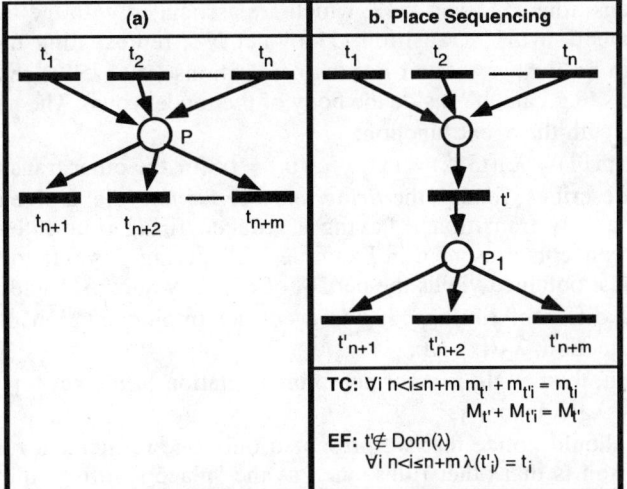

Table 3. Refinement rules (Cont.)

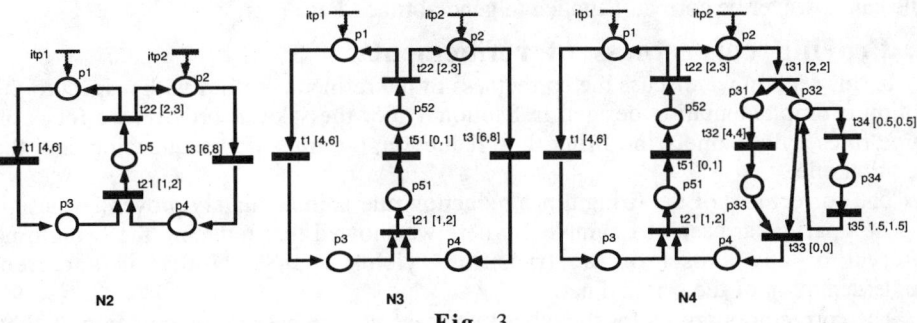

Fig. 3.

Table 3(b) shows the *place sequencing* rule: a place p and the transitions $\{t_{n+1}, \ldots t_{n+m}\}$ in its postset are replaced by places $p1$ and $p2$, transition t' and a set of transitions $\{t'_{n+i}\}$ each corresponding to one of the transitions $t_{n+i} \in p^\bullet$. This rule, applied to place p5 of N_2 in Fig. 3, yields the net N_3. Place p5, representing the communication being held, is refined into two places (p51 and p52) and a transition (t51) in net N_3 to model the action executed during the communication (the body of the rendezvous). N_3 is a correct implementation of N_2 with event function λ_2:

$$\lambda_2(tx) = tx \quad \forall \ tx \in \{t1, t22, t3, t21\}.$$

Table 2(e) presents the *iteration* rule. It consists of substituting a transition t with a set of transition and places modeling repeated firings of a transition with a time upper bound equal to that of the original transition t. Applying this rule to transition t3 of N_3 of Fig. 3 leads to the net N_4. The consumer's behavior is described by two processes: an iteration executing the computation, controlled by a time-out. Now the event function is:

$$\lambda_3(t33) = t3; \qquad \lambda_3(tx) = tx \quad \forall \ tx \in \{t1, t22, t51, t21\}.$$

Table 2(c) presents the *transition splitting* rule, where a transition t is split into two transitions t_1 and t_2 having the same preset and postset and the same firing time interval as t. This rule can be applied to model that the original action, is implemented by two alternatives. The net N_5, obtained by applying this rule to transition t51 in the net of N_4 of Fig. 3, has the same topology as the net N_6 in Fig. 2, with different time intervals for transitions t511 and t512, which are associated with the interval [0,1] as the original transition t51. Transition t51 of net N_4, representing the body of the rendezvous, can be further refined into two transitions (t511,t512) representing two exclusive actions (e.g., an "if" inside the body of the rendezvous). The net N_5 correctly implements N_4 with the event function:

$\lambda_4(t511) = \lambda_4(t512) = t51$; $\lambda_4(tx) = tx$ for the other transitions.

Table 2(b) describes, finally, the *firing times reduction* rule: a given transition t is substituted by a new transition t' having a reduced firing time interval. This rule formalizes the concepts presented in Example 1. Applying this rule to the transitions t511 and t512 just obtained yields the net N_6 of Fig. 2, where t511 and t512 represent actions having different timings. N_6 is a correct implementation of N_5 with the identity event function: $\lambda_5(tx) = tx$.

In conclusion, the net N_6 is a correct implementation of the net N_1 with the event function $\lambda = \lambda_5 \circ \lambda_4 \circ \lambda_3 \circ \lambda_2 \circ \lambda_1$.

The reader should notice that we presented only one refinement rule referring to places. The reason is that other rules, such as the "place splitting rule" presented in [FGP93], yield nets that do not satisfy the 1-STPN requirements; the place splitting rule can however be correctly applied to general timed Petri nets.

6. Proving correctness of refinement

In this section we discuss the correctness of the refinement rules with respect to the notion of implementation defined in Section 4. For the sake of brevity, we focus on three rules only, namely the firing time reduction, the transition sequencing, and the iteration rule.

The correctness of the firing times reduction rule is immediately proven because it is just a particular case of Example 1 where we showed that reducing the firing time interval of any subset of the transitions yields a 1-STPN that is a correct implementation of the original net.

The correctness proofs for the other two cases will be based on a common method that is immediately extendible to all the above described refinement rules and to all rules of the same kind.

To prove the correctness of the transition sequencing and iteration refinement rule we show that for each formula π derivable from the net S there exists in the axiomatization of I a proof of $\Lambda(\pi)$ that can be obtained systematically from the proof of π in Ax(S). The proof of π for net S uses axioms of Ax(S) that describe its topology, initial marking and firing times constraints. The formulas in the proof can therefore contain, in general, not only occurrences of the fire predicate describing the transition firings, but also occurrences of the tokenF predicate that describes the cause-effect relations among transition firings.

We therefore introduce a *proof translation* function Δ that translates TRIO formulas constructed from *both* predicates fire and tokenF applied to transitions of S into other TRIO formulas constructed with the same predicates applied to transitions of I, such that if the proof that $Ax(S) \vdash \pi$ consists of the formulas $\pi_0, \pi_1 ... \pi_n = \pi$, then the proof of $Ax(I) \vdash \Lambda(\pi)$ includes formulas $\Delta(\pi_0) ... \Delta(\pi_1) ... \Delta(\pi_n) = \Delta(\pi)$ and possibly other ones. Notice that the condition $\Delta(\pi) = \Lambda(\pi)$ requires Δ to be an extension of the property function Λ. Δ is therefore essentially characterized by the way it translates the tokenF predicate.

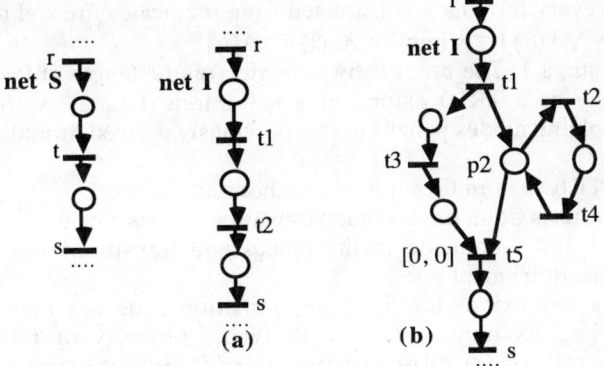

Fig. 4. Transition sequencing (a) and iteration (b) rules applied to a simple net.

6.1 The case of the transition sequencing refinement rule

Let us first apply this proof method to the simple case of the transition sequencing refinement rule; then we will apply it to the iteration rule and will discuss its general applicability to any refinement rule.

For the transition sequencing rule the proof translation function applied to the tokenF predicate is defined by the following clauses

$\Delta(\text{tokenF}(r, t, d)) = \exists d1 \exists d2(d1+d2=d \land \text{tokenF}(r, t1, d1) \land \text{Futr}(\text{tokenF}(t1, t2, d2), d1))$
$\Delta(\text{tokenF}(t, s, d)) = \text{tokenF}(t2, s, d)$
$\Delta(\text{tokenF}(x, y, d)) = \text{tokenF}(x, y, d)$ for every transition x, y: $x \neq t \land y \neq t$

Notice that from this definition it can be easily derived that
$\Delta(\text{tokenP}(s, t, d)) = \exists d1 \exists d2(d1+d2=d \land \text{tokenP}(t1, t, d1) \land \text{Past}(\text{tokenP}(s, t1, d2), d1))$.

For non atomic formulas Δ is defined in the usual compositional way:
$\Delta(\alpha \rightarrow \beta) = \Delta(\alpha) \rightarrow \Delta(\beta)$; $\Delta(\neg \alpha) = \neg \Delta(\alpha)$;
$\Delta(\text{Dist}(\alpha, t)) = \text{Dist}(\Delta(\alpha), t)$; $\Delta(\forall x \alpha) = \forall x \Delta(\alpha)$.

We can now formally prove the correctness of the transition sequencing refinement rule. In such a proof we will consider its application to a transition t having very simple connections with the rest of the net, as shown in Fig. 4: the preset and postset of t consist of a single place, q_1 and q_2, and there is only a single transition, s, having q_1 in its postset, and a single transition, r, having q_2 in its preset. This simple configuration is not necessary for the applicability and correctness of the refinement rule, but it is intuitively clear that the cardinality of the preset and postset of t are immaterial under this respect. In the following we will refer to this simple topology to abbreviate the proof: a similar proof, but with a greater quantity of inessential details can be provided in the case of a general topology. Also notice that in the initial net S, we assume that the places q_1 and q_2 are not marked, so that no initializing transition must be considered: again, the possible inclusion of those places in the initial marking can be taken into account at the price of some complication in the proof, which we do not include for the sake of simplicity (notice instead that the absence of marking in the newly introduced place p is essential). Similar simplifying assumptions will also be taken in the correctness proof of the iteration rule, to be presented subsequently.

Theorem 1 (Correctness of the transition sequencing refinement rule). With reference to nets S and I of Fig. 4(a), for every formula π constructed from the predicate fire applied to transitions of S, $Ax(S) \vdash \pi$ implies $Ax(I) \vdash \Delta(\pi)$.

Proof. The theorem derives directly from the following lemma 1 because Δ is defined as an extension of Λ.

Lemma 1: For every formula π constructed from predicates fire and tokenF applied to transitions of S, $Ax(S) \vdash \pi$ implies $Ax(I) \vdash \Delta(\pi)$.

Proof of Lemma 1. The proof is by induction on the length of the derivation of π. Formula π may be a TRIO axiom, or a net axiom (i.e., $\pi \in Ax(S)$) or it may be obtained by applying modus ponens to two previously derived formulas ψ and $\psi \to \pi$.

Base step.

1. If π is a TRIO axiom then it holds for both nets S and I.

2. If π is a net axiom for S where transition t does not occur, then $\pi \in Ax(I)$, because nets S and I coincide in the places and transitions not involved in the application of the refinement rule.

3. If π is a net axiom for S where transition t occurs then we prove that $Ax(I) \vdash \Delta(\pi)$, i.e., its translation is a derivable property of I (for the reader's convenience we call $IU_N(x)$, $OU_N(x)$, $LB_N(x)$, and $UB_N(x)$ the axioms describing input unicity, output unicity, lower bound and upper bound, respectively for transition x in net N).

- If $\pi = LB_S(t)$ then $\Delta(\pi) = \text{fire}(t2) \to \exists d(d \geq m_t \wedge \exists d1 \exists d2(d1+d2=d \wedge \text{tokenP}(t1,t2,d1) \wedge \text{Past}(\text{tokenP}(r,t1,d2),d1)))$; this formula can be easily derived in $Ax(I)$ using $LB_I(t2)$ and $LB_I(t1)$.

- If $\pi = UB_S(t)$ then $\Delta(\pi) = \text{fire}(r) \to \exists d(d \leq M_t \wedge \exists d1 \exists d2(d1+d2=d \wedge \text{tokenF}(r,t1,d1) \wedge \text{Futr}(\text{tokenF}(t1,t2,d2),d1)))$; this formula can be easily proven in $Ax(I)$ using, in particular, $UB_I(t1)$ and $UB_I(t2)$.

- If $\pi = LB_S(s)$ then $\Delta(\pi)$ is a direct consequence of $LB_I(s)$.

- If $\pi = UB_S(s)$ then $\Delta(\pi)$ is a direct consequence of $UB_I(s)$.

- If $\pi = IU_S(s)$ then $\Delta(\pi) = \exists d1 \exists d2(d1+d2=d \wedge \text{tokenP}(t1,t2,d1) \wedge \text{Past}(\text{tokenP}(s,t1,d1),d1)) \wedge \exists e1 \exists e2(e1+e2=e \wedge \text{tokenP}(t1,t2,e1) \wedge \text{Past}(\text{tokenP}(r,t1,e1),e1)) \to d=e$; we can derive from the premise of $\Delta(\pi)$ and from $IU_I(t2)$ the equality $d1=e1$ and from this, through $IU_I(t1)$, we derive $d2=e2$ and hence $d=e$.

- If $\pi = OU_S(r)$ then $\Delta(\pi)$ is proved in the same way as $\Delta(UB_S(s))$ using $OU_I(r)$ and $OU_I(t1)$.

- If $\pi = IU_S(s)$ then $\Delta(\pi)$ is a direct consequence of $IU_I(s)$.

- If $\pi = OU_S(t)$ then $\Delta(\pi)$ is a direct consequence of $OU_I(t2)$.

- If $\pi = FI_S(r, t)$ then $\Delta(\pi)$ is a direct consequence of $FI_I(r, t1)$ and $FI_I(t1, t2)$.

- If $\pi = FI_S(t, s)$ then $\Delta(\pi)$ is a direct consequence of $FI_I(t2, s)$.

4. Induction step. If π is obtained by applying MP to two previously derived formulas ψ and $\psi \to \pi$, then by the induction hypothesis we may assume that $Ax(I) \vdash \Delta(\psi)$ and $Ax(I) \vdash \Delta(\psi \to \pi)$. But $\Delta(\psi \to \pi) = \Delta(\psi) \to \Delta(\pi)$, so we can now derive $\Delta(\pi)$ by one more application of MP.

QED

The proof of correctness for the transition sequencing refinement rule can be easily modified and adapted to produce similarly structured correctness proofs for the other refinement rules. In fact the statement of the theorem, its structuring into a lemma and point 1, 2, and 4 in the proof of the lemma can be kept without change in the correctness proof of any other refinement rule. Thus, to complete such proofs we only must provide a definition of the proof translation function Δ based on the concrete mechanisms that allows the implementation net I to simulate the specification net S, and a proof that all the net axioms of S hold, under the defined proof translation function, also in net I: formally, we must provide a proof that for each $\varphi \in Ax(S)$, $Ax(I) \vdash \Delta(\varphi)$.

6.2 The case of the transition iteration refinement rule

As a further example of this proof method, let us consider next the transition iteration rule. We consider the following proof translation function:

$\Delta(tokenF(t,s,d)) = tokenF(t5,s,d)$
$\Delta(tokenF(r,t,d)) = \exists dr\ \exists d1\ \exists d3\ (d3+d1+dr = d \land tokenF(r,t1,dr) \land$
$\quad Futr(tokenF(t1,t3,d1) \land Futr(tokenF(t3,t5,d3),d1),dr))$
$\Delta(tokenF(x,y,d)) = tokenF(x,y,d)$ if $x \neq t$ and $y \neq t$

From the above definitions it can be easily proven that

$\Delta(tokenP(t,s,d)) = tokenP(t5,s,d)$
$\Delta(tokenP(r,t,d)) = \exists dr\ \exists d1\ \exists d3\ (d3+d1+dr = d \land tokenP(t3,t5,d3) \land$
$\quad Past(tokenP(t1,t3,d1) \land Past(tokenP(r,t1,dr),d1),d3))$

Theorem 2 (Correctness of the transition iteration rule) With reference to nets S and I of Fig. 4(b), for each axiom $\pi \in Ax(S)$, $Ax(I) \vdash \Delta(\pi)$.

For space reasons we present here only the most relevant steps of the proof.

- if $\pi = LB_S(t)$ then $\Delta(\pi) = fire(t5) \rightarrow \exists d\ (d \geq m_t \land d3+d1+dr = d \land tokenP(t3,t5,d3) \land Past(tokenP(t1,t3,d1) \land Past(tokenP(r,t1,dr),d1),d3))$; this can be derived in $Ax(I)$ using in sequence the $LB_I(t5)$, $LB_I(t3)$, and $LB_I(t1)$.

- if $\pi = UB_S(t)$ then $\Delta(\pi)$: $fire(r) \rightarrow \exists d\ (d \leq M_t \land \exists dr\ \exists d1\ \exists d3\ (d3+d1+dr = d \land tokenF(r,t1,dr) \land Futr(tokenF(t1,t3,d1) \land Futr(tokenF(t3,t5,d3),d1),dr)))$; this can be proved using the lemmas:

    Le.1. $fire(r) \rightarrow \exists d\ \exists d3(d \leq M_{t1} \land d3 \leq M_{t3} \land tokenF(r,t1,d) \land Futr(tokenF(t1,t3,d3),d))$; This describes the firing sequence leading to the firing of t3. Its proof is a simple chain application of the $UB_I(t1)$ and $UB_I(t3)$.

    Le.2. $fire(r) \rightarrow \exists d\ (d \leq M_{t3}+M_{t1} \land Futr(fire(t3),d))$; A direct consequence of Le.1.

    Le.3. $(fire(t4) \land Past(fire(t3),d3)) \rightarrow (tokenP(t4,t5,0) \lor \exists d5(d5 \leq d3 \land (Past(tokenP(t3,t5,d3-d5),d5)))$; This lemma establishes that if transition t4 fires after a firing of t3, then either it is immediately followed by a firing of t5, or else there must have been a precedent firing of that transition. The proof is derived by using $UB_I(t5)$ and by showing (with an argument by contradiction) that the firing of t2 cannot follow the firing of t4. This is easily accomplished considering the axiom $IU_I(t2)$.

    Le.4. $(M_{t2} < X \land Lasts_{ei}(\neg fire(t5),X) \land fire(t4)) \rightarrow Lasts_{ei}(WithinF_{ei}(fire(t4), M_{t4}+M_{t2}), X-M_{t2}))$; If transition t4 fires and transition t5 does not fire in the next X t.u., then t4 will fire periodically within at most $M_{t4}+M_{t2}$ t.u. in the next $X-M_{t2}$ t.u. This formula can be derived using $UB_I(t2)$, $UB_I(t4)$ and the theorem $(a<b \land \varphi \land Lasts_{ie}(\varphi \rightarrow WithinF_{ei}(\varphi,a),b)) \rightarrow Lasts(WithinF_{ei}(\varphi,a),b))$.

Finally, the proof of $\Delta(UB_S(t))$ runs along the following lines:

1. Assume the firing of r.
2. Considering $UB_I(t1)$, derive the firing of t1 and its upper bound M_{t1}.
3. Derive the firing of t3 and its upper bound $M_{t1}+M_{t3}$ from 1. and Le.2.
4. From the firing of t1, it is possible to derive either the firing of t2 or t5. Assume by contradiction that t5 will not fire in the next M_t t.u. (i.e., assume $Lasts_{ei}(\neg fire(t5),M_t)$). From this and Le.4, it follows that t4 fires periodically within $M_{t4}+M_{t2}$ (i.e., $Lasts_{ei}(WithinF_{ei}(fire(t4), M_{t4}+M_{t2}), M_t-M_{t2}))$
5. Consider the instant $M_{t1}+M_{t3}$:
    (a) By Le.2, deduce the occurrence of a firing of t3;

(b) Since $M_{t1}+M_{t3}<M_t$ derive, by step 4, a firing of t4 later than $M_{t1}+M_{t3}$ and earlier than $M_{t1}+M_{t2}+M_{t3}+M_{t4}$.

6. By 5(a), 5(b), and Le.3, derive a firing of t5 before $M_{t1}+M_{t2}+M_{t3}+M_{t4}$.

7. Since $M_{t1}+M_{t2}+M_{t3}+M_{t4}=M_t$ by definition of the iteration rule, 6 contradicts the hypothesis $Lasts_{ei}(\neg fire(t5),M_t)$, whose negation, $WithinF(fire(t5),M_t)$, is therefore true. $\Delta(UB_S(t))$ is now immediately derived using Le.1, the axioms of type OU_I, and DED.

- if π is equal to $LB_S(s)$, $UB_S(s)$, $OU_S(t)$, $IU_S(s)$, or $FI_S(t,s)$, then $\Delta(\pi)$ coincides with the corresponding axiom in I.
- if $\pi = IU_S(t)$ then $\Delta(\pi) = \exists dd1\ \exists dd3\ \exists ddr(dd3+dd1+ddr = d \wedge tokenP(t3,t5,dd3) \wedge Past(tokenP(t1,t3,dd1) \wedge Past(tokenP(r,t1,ddr),dd1),dd3)) \wedge \exists de1\ \exists de3\ \exists der(de3+de1+der = e \wedge tokenP(t3,t5,de3) \wedge Past(tokenP(t1,t3,de1) \wedge Past(tokenP(r,t1,der),de1),de3)) \rightarrow d=e$; we can derive from the premise of $\Delta(\pi)$ and from $IU_I(t5)$ the equality dd3=de3; from this, through $IU_I(t3)$, we derive dd1=de1, and finally, using similarly $IU_I(t1)$, we derive ddr=der and hence d=e.
- if $\pi = OU_S(r)$ then $\Delta(\pi)$ is proved in the same way as $IU_S(t)$, using $OU_I(r),OU_I(t1),OU_I(t3)$.
- if $\pi = FI_S(r,t)$ then $\Delta(\pi)$ is a direct consequence of $FI_I(r,t1)$, $FI_I(t1,t3)$ and $FI_I(t3,t5)$

7. Incremental analysis

Implementation through refinements is not only important as a means of structured development, but it also enhances the analyzability of the specifications by promoting their incremental analysis through inheritance of the properties already proven at more abstract levels. The inherited properties can be used as lemmas for proving new properties at the implementation level that cannot be proven inside the specification, e.g., because they refer to transitions that do not appear at the specification level.

We illustrate these concepts through the system specified and refined in Figures 2 and 3. We outline a simple incremental proof of the following property P for net N_6 of Fig. 2: *AlwF(fire(t1)→WithinF(fire(t511), 6.49))* (i.e., the firing of *t1* will always be followed by a firing of *t511* within 6.49 t.u.).

1. The first lemma in the proof refers to net N_1 and states that after the firing of t1, t3 will eventually fire within 4.t.u.:

P1: $AlwF(fire(t1) \rightarrow WithinF(fire(t3), 4))$

The proof of P1 is based on the LB(t1), UB(t3) and CA rules. The latter rule is applied for considering the firing of either *itp1* or *t2*.

2. Then in net N_2 we are able to prove the property P2: $AlwF(fire(t1) \rightarrow WithinF(fire(t21), 6))$ using $\Lambda_1(P1)$ as a lemma. This property establishes an upper-bound to the firing of transition t2 after that of t1.

3. In the net N_3 we can prove property P3 that establishes an upper bound for the time elapsing from the firing of t21 until the firing of t51.

P3: $Alw(fire(t21) \rightarrow WithinF(fire(t51),1))$

4. Property P3 is inherited in N_6 as

$\Lambda_5(\Lambda_4(\Lambda_3(P3))) = Alw(fire(t21) \rightarrow WithinF(fire(t511) \vee fire(t512), 1))$

5. In N_6 we can now show that the time bounds on transitions *t511* and *t512* resolve the non-determinism introduced by the application of the transition splitting rule.

P4: $Alw(fire(t21) \rightarrow WithinF(fire(t511), 1))$

P4 is easily proved using as lemmas $\Lambda_5(\Lambda_4(\Lambda_3(P3)))$ and property P5: Alw($\neg$fire(t512)), whose proof is reported, for a similar net, in [FMM94].

6. In net N_6 we now derive a stronger constraint on the time bound among the firing of transition t511 after that of t21.

P6: Alw $\big($ fire(t21) $\rightarrow$ WithinF(fire(t511), 0.49) $\big)$

P6 is a straightforward consequence of P5 and of the net axiom UB(t511).

Finally P6, together with $\Lambda_5(\Lambda_4(\Lambda_3(\Lambda_2(P2))))$, allows us to prove the desired property P.

To fully appreciate the simplification introduced by our incremental approach to the proof of net properties the reader should consider, for instance, that the derivation of the first lemma P1, which used the axiom UB(t3) for net N_1, would be quite intricate if conducted at the level of the net N_6, since each application of the net axiom UB(t3) should be substituted by a proof similar to that outlined in section 6.2 for $UB_S(t)$.

8. Conclusions

The specification of a time critical system and its design may contain flaws, which might lead to disastrous consequences if they remain undetected until the production of the system or its delivery. The use of formal specification and verification techniques can improve the reliability of time critical systems by permitting the production of unambiguous specification and by supporting the use of powerful verification methods to detect faults early in the development process. Unfortunately, for most formalisms having high expressive power the verification procedures either cannot be performed mechanically or their computational cost is too high to permit their use in realistic projects. These problems can be addressed by adopting a development methodology based on successive refinements that reduces the overall design and verification effort by reusing, at each phase of the development, the results gathered in the preceding steps.

In the present paper we discussed these issues referring to an operational formalism like timed Petri nets used as abstract system models, and to TRIO temporal logic formulas used as a means to describe desired timing properties. We formally defined the notions of correct implementation and provided a general method to prove that a set of refinement rules are correct with respect to this notion of implementation. We also showed how the development of a system through a series of correct refinement steps greatly facilitates its verification by allowing the designer to prove with a reduced effort intermediate lemmas on the early versions of the system. The described refinement rules are now supported by a tool, called Cabernet, for the incremental specification and design of time critical systems based on timed Petri nets that has been developed at Politecnico di Milano.

For the sake of brevity and simplicity, most of the definitions and proofs were provided with reference to a restricted class of timed Petri nets, namely 1-safe Simple place Timed Petri Nets (1-STPNs), and we occasionally assumed simplified topological configurations. All the reported results hold, however, for unbounded timed Petri nets with unrestricted topology, and we plan to report them in the most general form in a forthcoming paper.

A future research topic will be the investigation of the completeness issue for the set of transformation rules. Given a specification net S and a correct implementation I, a complete set of transformations would permit to obtain, starting from S, the net I or a net equivalent to it.

Finally, we claim that the presented development method, with its related formal definitions and proofs, can be adapted, with suitable modifications, to other formalisms that combine a descriptive language for specifying timing requirements with an operational notation to model system structure, such as, for instance, timed transition systems [HMP91] or ESM/RTTL [Ost89].

Acknowledgement

We would like to thank Prof. Mauro Pezzè for some preliminary insightful discussions.

References

[Apt81] K. Apt, "Ten years of Hoare's Logic: A survey - Part I," *ACM-Transactions on Programming Languages and Systems*, vol. 3, no. 4, pp. 431-483, Oct 1981.

[AH90] R. Alur and T.A. Henzinger, "Real Time Logics: Complexity and Expressivness", Tech. Report no. STANCS901307, Appeared in the 5th IEEE LICS'90 (pp. 390-401), 1990.

[Aiz90] Jacob Itzhack Aizikowitz, "Designing Distributed Services Using Refinement Mappings", Ph.D. Thesis and Tech. Report 89-1040, Cornell University, Ithaca, New York, 1990.

[AL91] M.Abadi and L. Lamport, "The existenece of refinement mappings", Theoretical Computer Science 82 (1991) 253-284, Elsevier Science Publiscers B.V.

[DDG+90] W. Damm, G. Dohmen, V. Gerstner, and B. Josko, "Modular verification of Petri nets, the temporal logic approach," in *Proceedings of Stepwise Refinement of Distributed Systems. Models, Formalisms, Correctnesss,* LNCS 430, Springer Verlag, 1990, pp.181-207.

[End72] H.B. Enderton, *A Mathematical Introduction to Logic.* New York: Academic Press, 1972.

[FGP93] M. Felder, C. Ghezzi, and M. Pezzè, "Analyzing refinements of state based specifications: the case of TB nets," in *Proceedings of ISSTA'93*, Cambridge, 1993, pp. 28-39.

[FMM91] M. Felder, D. Mandrioli, and A. Morzenti, "Proving properties of real-time systems through logical specifications and Petri nets models," Tech. Rep., TR 91-072, Dipariemento di Elettronica e Informazione, Politecnico di Milano, December 1991.

[FMM94] M. Felder, D. Mandrioli, and A. Morzenti, "Proving properties of real-time systems through logical specifications and Petri nets models," *IEEE Transactions on Software Engineering.* vol. 20, no. 2, pp. 127-141, February 1994.

[FM92] M. Felder and A. Morzenti, "Validating real-time systems by executing logic specifications in TRIO," in *Proceedings of 14th International Conference on Software Engineering*, ACM/IEEE, 1992, pp. 199-211.

[GG90] R. Glabbeek and U. Goltz, "Refinement of actions in causality based models", in *Proceedings of Stepwise Refinement of Distributed Systems. Models, Formalisms, Correctnesss,* LNCS 430, Springer Verlag, 1990, pp. 266-300.

[GJM91] C. Ghezzi, M. Jazayeri, and D. Mandrioli, *Fundamentals of Software Engineering.* Englewood Cliffs, N.J.: Prentice-Hall International Editors, 1991.

[GMM90] C. Ghezzi, D. Mandrioli, and A. Morzenti, "TRIO, a logic language for executable specifications of real-time systems," *Journal of Systems and Software*, vol. 12, no. 2, pp. 107-123, May 1990.

[HMP91] T. Henzinger, Z. Manna, and A. Pnueli, "Temporal proof methodologies for real-time systems," in *Proc.of the 18th Annual Symposium on Principles of Programming Languages*, ACM-PRESS, 1991, pp. 353-366.

[LA89] N.A. Lynch and H. Attiya,"Using mapping to prove timing properties" Tech. Report MIT/LCS/TM-412,b Laboratory for Computer Science, MIT, 1989. Appeared in *Proc. PODC'90*

[Men63] E. Mendelson, "Introduction to mathematical logic", Van Nostrand Reinold Company, New York, 1963.

[MF76] P.M. Merlin and D.J. Farber, "Recoverability of communication protocols - Implications of a theoretical study," *IEEE Transactions on Communications*, vol 24, no. 9, pp.1036-1043, September 1976

[Mül85] K. Müller, "Constructable Petri nets", in *Proc. EIK 21*. 1985, pp. 171-199.

[Ost89] J. Ostroff, *Temporal Logic For Real-Time Systems*, Advanced Software Development Series, 1. Taunton, Somerset, England: Research Studies Press LTD., 1989.

[Pnu86] A. Pnueli, "Applications of temporal logic to the specification and verification of reactive systems: A survey of current trends," *LNCS 224*, Springer-Verlag, 1986.

[Rei85] W. Reisig, *Petri Nets: an Introduction.*, EATCS Monographs on Theoretical Computer Science, Springer Verlag, Berlin-New York, 1985.

[SM83] I. Suzuki and T. Murata, "A method of stepwise refinement and abstraction of Petri nets", *Journal of Computer System Sciences,* no. 18, 1979, pp. 35-46.

[Vog87] Walter Vogler, "Behaviour preserving refinements in Petri nets", in *Proc. 12th Int. Worksop on Graph Theoretic Concepts in Computer Sciende*, München ,1986. Springer Verlag, LNCS 246, pp. 82-93.

[Vog90] Walter Vogler, "Failures Semantics based on Interval Semiwords is a Congruence for Refinement", in *Proc. STACS'90,* 1990. Springer Verlag, LNCS 415, pp. 285-297.

[YY91] W.J. Yeh and M. Young, "Compostional Reachability analysis using process algebra", in *4th Int. Workshop on Testing and Verifications,* Victoria, Canada, 1991, ACM Sigsoft, pp. 49-50

A Stuttering Closed Temporal Logic for Modular Reasoning about Concurrent Programs

Abdelillah Mokkedem
Dominique Méry
CRIN-CNRS & INRIA-Lorraine, BP239
54506 Vandœuvre-lès-Nancy, France

Abstract. A simple and elegant formulation of compositional proof systems for concurrent programs results from a refinement of temporal logic semantics. The refined temporal language we propose is closed under w-*stuttering* and, thus, provides a fully abstract semantics with respect to some chosen observation level w. This avoids incorporating irrelevant detail in the temporal semantics of parallel programs. Besides compositional verification, concurrent program *design* and *implementation* of a coarser-grained program by a finer-grained one, turn out to be easily practicable in the setting of the new temporal logic.

1 Introduction

The regular temporal logic [14, 16] provides a powerful tool for *global* specification and *non-compositional* verification of *existing* concurrent programs. However, this logic offers a very poor support for modular specification and verification and, consequently, *systematic* design of concurrent programs is hard (if not impossible) to do in such a setting. The lack of modularity comes from the fact that the semantics of the temporal formalism has been defined in terms of *global states* in such a way that the temporal properties of a given component, viewed within some context, do not make abstraction of invisible state changes performed by other components. In practice, previous work [4, 5, 25] have shown that this complicates the design of compositional proof systems for parallel programs. Now, in our opinion, if one wants to formulate a compositional rules for parallel programs, the first step is to be careful at the stage of the definition of the specification language semantics. *Full abstraction* is one of the main properties that could give an appropriate answer to this problem. This notion means that whenever a model τ satisfies a formula F, any model that is equivalent to τ (modulo some state changes considered irrelevant) also satisfies F. This paper explores of this point and proposes of a new semantics for the temporal logic formalism, which we want to be *fully abstract*. This semantic criterion is used to define an appropriate basis for a compositional theory of modular specification and verification of concurrent programs [3].

This paper is organized as follows. In section 2 problems we want to overcome within the temporal logical TL are stated. In Section 3 the concept of

w-stuttering is introduced. The response to TL's problems is presented; this refines the temporal semantics of the basic operators that cause trouble in abstraction. It is shown that the resulting temporal logic is fully abstract w.r.t w-stuttering. In section 4 we give an axiomatization for the refined temporal logic MTL. In Section 5 a close connection between MTL formulas and a compositional semantics of concurrent modules is presented. Concurrent programs can be modularly specified and composionally verified within the logic MTL. Moreover, *implementation* of a concurrent program by a finer grained one is formalized in an elegant way within this logic. Section 6 concludes the paper, describing future and related work.

2 TL's problem !

The linear discrete temporal logic TL has been perceived to be an appropriate tool for both the semantic description of concurrent (and sequential) programs and the reasoning about them [14, 16]. This relies on the fact that concurrent program behaviour can be easily modeled by all possible totally ordered execution sequences arising from the interleavings of actions in the separate 'sequential' processes of the concurrent program (*interleaving semantics*). One can feel that something is lost by sequentializing a concurrent program in this way, and that a partial ordering among the actions should be used instead. However, as Lamport stated in [7], so long as we consider only safety and liveness properties, there is no loose of generality in considering totally ordered sequences of actions. However, serious problems arise when one wants to apply TL to parallel programs of realistic size. Proofs are not modular and turn out very hard to master. Moreover, one can not develop a program together with its correctness proof. We aim to be able (1) to decompose a proof of a large program into lemmas associated to its components, (lemmas that remain valid for ayn context where these components are used) (2) to ignore details of the reasoning and, if required, to take them into account later without lost proved properties. The logic TL does not provide an appropriate tool to support these notions, it has been strongly criticized from this point of view [6, 7].

- In [6, 7], Lamport objects to the use of the *next* operator as the cause of trouble in abstraction, forcing too much irrelevant detail to be present in the semantic description. It turns out that lowest level of atomicity must be visible, which should not occur in a properly abstract semantics.

- Quantification over flexible variables turned out to be very useful for abstraction, and has been shown to be necessary for reaching compositional completeness. But with its classical definition, flexible quantification does not make abstraction to stuttering [12].

- Manna and Pnueli [12] stated some points of dissatisfaction with the temporal logic presented in [11] due to the *floating* interpretation which does not assign any special significance to the initial state so that satisfiability and validity are evaluated at *all* positions in models. This interpretation needs the generalization rule ($\varphi \vdash \Box\varphi$) in the proof system which violates the *deduction* rule (a powerful tool in the predicate calculus) and, on the other hand, requires the *suffix* closure property for the set of computations when formulas are interpreted over the behaviour of a given program. In fact, they presented an *anchored* temporal logic [12] in which they consider that a formula φ is defined to be *valid* (resp. *satisfiable*) over a set of sequences $\mathcal{C}$, if it holds at position 0 of *every* (resp. *some*) sequence of $\mathcal{C}$.

The following example illustrates these comments.

Example 1.

P_1 : **var** x : **integer**;
$\quad x := x + 1$

P_1' : **var** x : **integer**;
$\quad t$: **integer local**;
$\quad t := 1;\ x := x + t$

P_2 : **var** y : **integer**;
$\quad y := y - 1$

Let us define $p \longrightarrow q =_{df} \Box(p \to \bigcirc q)$, where $\bigcirc$ and $\Box$ denote respectively the operators *next* and *always* of the temporal logic TL [17].

Remark 1. [**abstraction** problem]
$P_1 \models (x = 0) \longrightarrow (x = 1) \qquad P_1' \not\models (x = 0) \longrightarrow (x = 1)$

P_1 and P_1' are (observationally) equivalent but no one satisfies the safety property satisfied by each other. The lesson is that TL, especially its operator $\bigcirc$, is too (operationally) precise— even *w.r.t* invisible changes; TL lacks of abstractness.

Remark 2. [**compositionality** problem]
$P_2 \models (y = 1) \longrightarrow (y = 0) \qquad P_1 \| P_2 \not\models (x = 0) \longrightarrow (x = 1)$
$P_1 \| P_2 \not\models (y = 1) \longrightarrow (y = 0)$

Although the two programs (P_1 and P_3) do not share any variable (i.e. it is a very simple case involving concurrency without communication), in the composition $P_1 \| P_2$ the behaviour of each program disturbs the safety property of each other. TL lacks of compositionality.

Remark 3. [**refinement** problem]
It would be desirable for $P_1' \models \exists t.\ x = 0 \longrightarrow (t = 1 \land x = t)$ to hold since the sole difference between P_1 and P_1' is that P_1' is finer-grained than P_1. The behaviour that concerns the invisible variable t should be completely hidden by the binder $\exists$, if that is so we have success since we can easily check that $\exists t.\ x = 0 \longrightarrow (t = 1 \land x = t)$ implies $x = 0 \longrightarrow x = 1$[1]. But, unfortunately, it is not the case : according to the classical definiton of $\exists$ in the logic TL, $P_1' \not\models \exists t.\ x = 0 \longrightarrow (t = 1 \land x = t)$. The lesson of this is that the logic TL does not provide an adequate mathematical tool for formalizing refinement with implication.

---

[1] This suggests that *refinement* can be formalized by logical *implication*!

3 The logic MTL

The present contribution is concerned with the problems mentioned above. We propose a refined temporal logic MTL[2] in which notions of *abstraction, compositionality* and *refinement* turn out rigourously treated. In this logic we assume an *anchored* version of a *future*-fragment with flexible quantifiers, and the semantics of the *next* operator and the quantifiers are refined in a way to be abstract w.r.t some invisible steps. The temporal semantics of programs shall be formulated in terms of the refined temporal logic MTL. Notice that the design decisions have been especially motivated by the need to reach a sufficient abstraction for the temporal language semantics which should enable the design of composition principles for (compositionally) reasoning about concurrent programs. Moreover, we are interested in an *open* semantic model in which the temporal semantics of a program S describes the execution sequences of S in all (possible) environments. The resulting logic does not require *suffix* closure of program computations, and guarantees *invariance under stuttering* of properties. Besides allowing semantic description of *open* systems, it provides a good abstraction for compositional specification and verification of concurrent systems and also offers a good support for systematic design of concurrent programs.

3.1 Syntax and semantics

We first describe the basic syntax of state formulas and models to define the syntax and semantics of MTL. State formulas (also called *assertions*) are non-quantified formulas expressed in some fragment of the first order language without quantifiers, they describe properties at individual states. We assume an infinite (countable) set of flexible variables $\mathcal{V}_s$ ($x, y, z, \ldots \in \mathcal{V}_s$) and an infinite (countable) set of rigid variables $\mathcal{V}_l$ ($u, v, n, \ldots \in \mathcal{V}_r$). A flexible variable may assume different values in different states of the model, while values of rigid variables does not depend on states. From a computer-science rather than mathematical point of view, rigid variables are intended to represent constants, while flexible variables represent program variables. We assume a set **Val** of *values* including the booleans T and F, natural numbers, strings, ... We will not be precise about what **Val** contains exactly, but will assume that it contains all the values needed for examples considered. In addition to variables we also assume concrete predicates and and concrete functions over their respective domains included in **val**. We agree to view constants as 0-ary functions and *propositions* as boolean variables[3]. We also assume boolean connectives $\neg, \wedge, \vee, \rightarrow, \leftrightarrow$, and equality $=$.

Both variables and rigid variables take values from **Val**, however, since they do not have the same status (i.e. variables may assume different values in different states, while rigid variables may assume only a fixed value), we prefer to interpret them by two different interpretations.

---

[2] MTL stands for Modular Temporal Logic
[3] We call 'flexible variables' simply 'variables' from now on

Definition 1. We define a *state* s over V ($V \subseteq \mathcal{V}_s$) (resp. a valuation ξ over V_r ($V_r \subseteq \mathcal{V}_r$)) to be an assignment of values to variables in V (resp. in V_r) – that is, a mapping from V to **Val** (resp. from V_r to **Val**).

We denote by $\alpha[x]$ the value that the mapping α assigns to the variable x (where x is either a flexible or a rigid variable). Let α, α' be two mappings over V and x be a variable in V, we say that α' is an x-variant of α ($\alpha' =_x \alpha$ in notation) if $\alpha[y] = \alpha'[y]$ for every $y \in V - \{x\}$. State formulas are interpreted by couples (ξ, s) in the usual way, e.g. let $s =< x : 0, y : 3 >$ and $\xi =< u : 1 >$, $(\xi, s)[x = u - 1 \wedge y > u + 1] =_{df} s[x] = \xi[u] - 1 \wedge s[y] > \xi[u] + 1 =_{df} 0 = 0 \wedge 3 > 2 =_{df} \mathrm{T}$.

We denote by $(s, s', s_1, s_2, \ldots \in)\Sigma$ and $(\sigma, \sigma', \sigma_1, \sigma_2, \ldots \in)\Gamma$ the set of all states and the set of all infinite sequences of states respectively. The set of valuations is denoted by $(\xi, \xi', \ldots \in)\Delta$. Let $\sigma : s_0, s_1, \ldots$ and $\sigma' : s'_0, s'_1, \ldots$ be two sequences of states over V, and $x \in V$ be a variable, we say that σ' is an x-variant of σ (we write this $\sigma' =_x \sigma$) if for each $j \geq 0$, $s'_j =_x s_j$. Let σ be a sequence of states over V and w a subset of V, we denote by $\sigma \lceil w$ the sequence $s_1 \lceil w, s_2 \lceil w, \ldots$ where $s_i \lceil w$ denotes the projection of the state s_i onto the set of variables w (i.e. the restriction of the mapping s_i to the subset of variables w). Let $\sigma : s_0, s_1, s_2, \ldots$ be a sequence over V ; σ^i denotes the sequence $s_i, s_{i+1}, \ldots$ and σ_i denotes the ith state s_i in σ. We write σ as an abbreviation of σ^0.

Definition 2. Let $\sigma : s_0, s_1, s_2, \ldots$ be a sequence over V, a step (s_{i-1}, s_i) in σ is called a *stuttering* step iff $s_{i-1} = s_i$. We call *finite stuttering* (resp. *infinite stuttering*) a finite number of stuttering steps $s_i, s_i, \ldots, s_i$ (resp. an infinite number of stuttering steps $s_i, s_i, \ldots$). We define $\natural \sigma$ to be the sequence obtained from σ by removing all finite stutterings.

Definition 3. Let $\sigma^j : s_j, s_{j+1}, s_{j+2}, \ldots$ be a sequence,

$\natural \sigma^j = $ if $\forall i > j,\ s_i = s_j$ then σ^j
    else if $s_j = s_{j+1}$ then $\natural \sigma^{j+1}$
        else $(s_j) \bullet \natural \sigma^{j+1}$

Definition 4 *w-stuttering.* Let σ, τ be two sequences over V and $w \subseteq V$. σ, τ are said to be w-stuttering equivalent (in notation $\sigma \simeq_w \tau$) if $\natural(\sigma \lceil w) = \natural(\tau \lceil w)$. We simply say that σ and τ are stuttering equivalent for the case $w = V$ and we write this $\sigma \simeq \tau$.

Proposition 5. $\forall k \geq 0.\ \exists j \geq k.\ (\natural(\sigma \lceil w), k) = \natural((\sigma, j) \lceil w)$.

Proof. A consequence of definition 3.

Definition 6. A *temporal model* (or a Kripke model) for MTL is a couple (ξ, σ^i) that consists of a valuation ξ of rigid variables and an infinite sequence of states σ^i where the positive index i is used as *now*; we write (ξ, σ) as an abbreviation of (ξ, σ^0).

The new and central concept in the definition of MTL lies in the introduction of a new kind of *next* operator, denoted $\otimes_w$, (and its dual, denoted $\oplus_w$) and the notion of the minimal index. An important feature of $\otimes_w$ is that it is *insensitive* to finite w-stuttering and *sensitive* to infinite w-stuttering (with respect to a given set of variables w), while its dual, $\oplus_w$, is *insensitive* to both finite and infinite w-stuttering. Intuitively, the minimal index of a formula represents the set of (*relevant*) variables which designates the abstraction level of the formula. When applying MTL to programs the minimal index will formalize the notion of observable variables. Another new concept, similar to Lamport's one recently introduced in [9], consists of flexible quantification modulo stuttering steps. We then define the other temporal operators (*always* $\square$, *sometimes* $\diamond$, etc.) according to these new concepts in order to obtain a temporal logic that will enable semantic descriptions which are *invariant* under finite w-stuttering, where w shall represent the set of variables viewed by the component. This is one of the major results to ensure a *desired* level of abstraction necessary for modular specification and compositional verification of concurrent systems. The syntax and semantics of MTL, along the additional notation we use to write MTL formulas, are summarized below. Assuming the meaning of state formulas, which can easily be defined within the predicate logic (see [17]), we provide all one needs to understand MTL formulas.

Syntax Formulas in MTL are indexed by a set of variables. The importance of indexing formulas should be progressively clear throughout the sequel. Intuitively, this will modelize the concepts of *observable* and *non-observable* changes inside the logic itself. We inductively define MTL formulas and their indexes. We shall name MTL formulas by symbols from $\{p, q, f, g, F, G, F_w, G_w, \ldots, \}$, names like F_w precise that w is an index of the formula denoted by F.

- *basic formula*: If f is a state formula and V_f is the set of its (flexible) variables then f is a formula in MTL and every w s.t. $V_f \subseteq w$ is an index of f.
- *strong-next*: If f is a formula and w is an index of f then $\otimes_w f$ is a formula and w is its index
- *always*: If f is a formula and w is an index of f then $\square f$ a formula and w is an index of it
- *rigid quantifier*: If f is a formula, w is an index of f, and u is a rigid variable then $\exists u. f$ is a formula and w is an index of it
- *flexible quantifier*: If f is a formula, w is an index of f, and x is a (flexible) variable then $\exists x. f$ is formula and w is an index of it
- *negation*: If f is a formula and w is an index of f then $\neg f$ is a formula and w is an index of it
- *disjunction*: If f, g are two formulas and w_1, w_2 are respectively two indexes of f, g then $f \vee g$ is a formula and $w_1 \cup w_2$ is an index of it.

When a formula F has more than one index, we shall define w to be the *minimal index* of F iff w is an index of F and for any other index w' of F, $w \subseteq w'$.

Semantics:

$(\xi, \sigma^j) \models p$ iff $(\xi, \sigma_j)[p]$ for a state formula p
$(\xi, \sigma^j) \models \bigotimes_w p$ iff *there is some* $k > j$ s.t $\sigma_k \lceil w \neq \sigma_j \lceil w$ *and* $(\xi, \sigma^k) \models p$
    and for every i, $j \leq i < k$, $\sigma_j \lceil w = \sigma_i \lceil w$
$(\xi, \sigma^j) \models \Box p$ iff $(\xi, \sigma^k) \models p$ *for every* $k \geq j$
$(\xi, \sigma^j) \models \exists u.\, p$ iff *there exists* $\xi' \in \Delta$ *s.t.* $\xi' =_u \xi$ *and* $(\xi', \sigma^j) \models p$
$(\xi, \sigma^j) \models \exists x.\, p$ iff *there exist* $\rho, \tau \in \Gamma$. $\rho \simeq \sigma^j$ *and* $\tau =_x \rho$ *and* $(\xi, \tau) \models p$
$(\xi, \sigma^j) \models \neg p$ iff $(\xi, \sigma^j) \not\models p$
$(\xi, \sigma^j) \models p \vee q$ iff $(\xi, \sigma^j) \models p$ or $(\xi, \sigma^j) \models q$

Additional notation:

1. $p \wedge q =_{df} \neg(\neg p \vee \neg q)$ 2. $p \rightarrow q =_{df} \neg p \vee q$
3. $\Diamond p =_{df} \neg \Box \neg p$ 4. $p \Rightarrow q =_{df} \Box(p \rightarrow q)$
5. $\bigoplus_w p =_{df} \neg \bigotimes_w \neg p$ 6. $\forall u.\, p =_{df} \neg \exists u.\, \neg p$
7. $\forall x.\, p =_{df} \neg \exists x.\, \neg p$ 8. **shuffle**$(w_1, w_2) =_{df} \forall \overline{u}, \overline{v}.\, (w_1 = \overline{u} \wedge w_2 = \overline{v} \wedge$
    $\bigotimes_{w_1 \cup w_2} true) \Rightarrow \bigotimes_{w_1 \cup w_2} (w_1 = \overline{u} \vee w_2 = \overline{v}).$

The predicate **shuffle** will be used in the axiom formalizing conjunction of independent transitions. Intuitively, it asserts that sequence changes leading to the next state may not involve variables from both w_1 and w_2.

3.2 Abstractness

We consider the *abstraction problem*, stated in section 2, that arises when applying temporal logic to describe concurrent program behaviour. Our suggestion aims at solving this problem with the new semantics we have given by making abstraction of state changes of invisible variables (i.e. variables outside the minimal index). Indeed, we will show that the meaning of every formula F is insensitive to w-stuttering, i.e. steps keeping values of all variables in w unchanged, where w is the minimal index of F. This is what we call a *fully abstract* semantics with respect to the minimal index of formulas. The following propositions answer this issue.

Proposition 7. *Let F be a formula, w be the minimal index of F, and σ a sequence of states over V, for every set of variables w' s.t. $w \subseteq w' \subseteq V$, $(\xi, \sigma \lceil w') \models F$ iff $(\xi, \sigma \lceil w) \models F$.*

Proof. By definition, the minimal index of a formula F includes all the variables occurring in F. According to the semantic of formulas, given above, the truth-value of a formula F does not depend on values of variables that appear in the sequence σ but are outside its minimal index.

Proposition 8 Full abstraction. *Given a formula F and a valuation ξ, for any index w of F and every pair of sequences σ, τ such that $\sigma \simeq_w \tau$, $(\xi, \sigma) \models F$ iff $(\xi, \tau) \models F$.*

The proof of this proposition is given in the full paper [24].

Proposition 7 asserts that the *truth-value* of F does not depend of variables outside the minimal index. Proposition 8 asserts that the meaning of any formula F is insensitive to changes preserving the value of all variables in the minimal index of F. This is what should allow the description of temporal semantics of concurrent programs in a modular way.

3.3 Validity and Provability

As in [12, 17] two types of validity are consider. A state formula is defined to be *assertionally valid*, denoted by $\models_A F$, if $s[F]$ for every state $s \in \Sigma$. A temporal formula F is defined to be *temporally valid*, denoted by $\models_T$, if

$$(\xi, \sigma) \models F \text{ for every valuation } \xi \in \Delta \text{ and every sequence } \sigma \in \Gamma.$$

Note that if the formula F contains some temporal operators, then the temporal validity can be applied to it. On the other hand, in the case that F is an assertion, both types of validity apply. Corresponding to these two types of validity, two possible deductive proof systems may be considered. The first proof system supports proving assertional validity of state formulas, while the second system support proving temporal validity of temporal formulas. This leads to two notions of *provability*. We say that a state formula F is *assertionally provable*, denoted by, $\vdash_A F$, if its assertional validity can be proven using the assertional proof system. Similarly, we say that a formula F is *temporally provable*, denoted by $\vdash_T F$, if its temporal validity can be proven by the temporal proof system. Since we are mainly interested in temporal validity and provability, we assume an underlying assertional proof system and we give only axioms and rules dealing with temporal validity. But for the famous results of Gödel [27] the set of valid assertions (allowing quantification and interpretation into concrete structures including natutal numbers) is in general non-recursive and, consequently, any temporal proof system based on it is non-recursive too. To circumvent this situation, we will assume that we have a so called *oracle* to decide whether some assertion of our assertion language is valid or not. The temporal proof system, that we present in section 4, is recursive relative to this oracle, that is, the set of temporally valid formulas may be described by a recursive proof system where we may call upon the oracle to decide the validity of assertions. Focusing on the temporal part, we will omit the subscript T and interpret the simpler $\models, \vdash$ as $\models_T, \vdash_T$ respectively.

3.4 More about quantifiers

Section 2 discussed a problem that concern implementing a program by a finer-grained one. One refers to this problem the *action refinement problem* in [13] which precisely rises when an action in a program is decomposed into two or more actions in another one. Lamport's logic TLA [9] solves this problem by defining the semantics of quantification taking in account possible stuttering steps. A drawback of TLA is that quantifiers do not preserve the laws of classical quantifiers. In particular, the property $\exists x.\, p \Leftrightarrow p$ is no longer valid. We have adopted exactly the same definition for (flexible) quantification in the logic MTL but with a little advantage that the classical laws of quantifiers are conserved. This elegant feature results from the fact that all the temporal operators are insensitive to stuttering (even the *next* operator). That is illustrated first by considering the counter example given by Pnueli in [13] and, then, the theorem that states this property is given. Let the sequence

$$\sigma :<x:-1>,<x:+1>,<x:-1>,<x:+1>,\ldots$$

the expression $\exists y.\, x = x' \Leftrightarrow x = x'$ does not hold on σ in TLA. The formula $x = x'$ is false on σ because $s_0[x] = -1$, while $s_0[x'] = +1$. On the other hand, $\exists y.\, x = x'$ is true on σ because $x = x'$ is true on the following stuttering variant of σ,

$$\sigma' :<x:-1>,<x:-1>,<x:+1>,<x:-1>,<x:+1>,\ldots$$

The equivalent formula in our logic $\exists y.\, x = \bigotimes_x x \Leftrightarrow x = \bigotimes_x x$ (using a simplified notation without introducing rigid variables) is however true on σ because $x = \bigotimes_x x$ holds on σ'. Now let us define $Free(F)$ to be the set of (flexible and rigid) not bounded variables that appear in the formula F, e.g.

- $Free(\bigoplus_{(x,y)}(y = 0)) =_{df} \{x, y\}$
- $Free(\exists x.\, \bigoplus_{(x,y)}(x = u)) =_{df} \{y, u\}$

We have the following theorem.

Theorem 9. *Let x, u be respctively a variable and a rigid variable,*

1. $\models \exists x.\, F \Leftrightarrow F$ *if* $x \notin Free(F)$
   $\models \exists u.\, F \Leftrightarrow F$ *if* $u \notin Free(F)$
2. $\models \exists x.\, (F \vee G) \Leftrightarrow \exists y.\, F \vee G$ *if* $x \notin Free(G)$
   $\models \exists u.\, (F \vee G) \Leftrightarrow \exists y.\, F \vee G$ *if* $u \notin Free(G)$

Proof. Let x, u be a variable and a rigid variable respectively and F be a formula and w its minimal index,

1. $(\xi, \sigma^j) \models \exists x.\, F$ iff $\exists\, \sigma', \tau.(\sigma' \simeq \sigma^j \wedge \tau =_x \sigma')$ and $(\xi, \tau) \models F$
   {because for any ρ, τ s.t $\rho =_x \tau$ and $x \notin Free(F)$, $(\xi, \rho) \models F$ iff $(\xi, \tau) \models F$}
   iff $\exists\, \sigma', \tau.\sigma' \simeq \sigma^j$ and $(\sigma', \tau) \models F$
   {by proposition 8 and $\sigma' \simeq \sigma^j$ implies $\sigma' \simeq_w \sigma^j$}
   iff $(\xi, \sigma^j) \models F$

$(\xi, \sigma^j) \models \exists u. F$ iff there exists $\xi' \in \Delta$ s.t. $\xi' =_u \xi$ and $(\xi', \sigma^j) \models F$
{since $u \notin Free(F)$}
iff $(\xi, \sigma^j) \models F$

2. Similar to 1.

We have presented above the syntax and semantics of the logic MTL and have shown that each formula has a minimal index which, from an abstraction point of view, represent sets of relevant variables, i.e. variables only whose changes to be considered in steps of sequences. Proposition 8 proves the abstractness property of MTL, saying that, for any formula F and any minimal index w of F, the truth value of F does not relate to any interleaving with steps changing values of variables outside w.

4 A proof system for MTL

We give now a system of axioms and rules, namely $\mathcal{G}$, dedicated to mechanizing theorem proving within MTL. An important notion connected to the construction of proofs is *instantiation*.

Definition 10. Let ψ be a formula (scheme) and $p_1, \ldots, p_k$ some of the (propositional) sentence symbols appearing in ψ. A *temporal replacement* $\alpha : [p_1 \mapsto \varphi_1, \ldots, p_k \mapsto \varphi_k]$, specifies for p_i a replacing formula φ_i.

We denote by $\psi[\alpha]$ the formula obtained from ψ by replacing all occurrences of p_i with φ_i, $i = 1, \ldots, k$, respectively. We refer to $\psi[\alpha]$ as an *instantiation* of ψ. For example, the formula $\Box p \vee \Diamond \neg p$ is an instantiation of $p \vee \neg p$ obtained by the replacement $p \mapsto \Box p$.

Working with variables we want to extend temporal replacement to parametrized sentence symbols, but additional restrictions have to be imposed in order to make the instantiation rule, we will give below, sound. The problem that can arise from an uncontrolled temporal replacement of parametrized sentence symbols is clearly stated in [17]. We give here only the restriction undertaken to overcome this problem (for more details see [17]). One restricts the temporal replacement to *rigid* parametrized sentence symbols and we require that all variables appearing in the replacing formulas are not captured by quantifiers in the instantiated formula ψ.

Definition 11. We define a parametrized occurrence of a sentence symbol $p(u_1, \ldots, u_k)$ to be rigid if the variables $u_1, \ldots, u_k$ are rigid. Let ψ be a formula and $p(u_1, \ldots, u_n)$ be a rigid parametrized sentence symbol occurring in ψ, we define a general temporal replacement

$$p(u_1, \ldots, u_m) \mapsto \varphi(u_1, \ldots, u_m), \ m \geq 0,$$

to be *admissible* if $\varphi(u_1, \ldots, u_m)$ does not contain any variable that is quantified in ψ.

Note that by taking $m = 0$, this definition also covers the case of unparametrized replacement $p \mapsto \varphi$. The preceding discussion considered replacements of a sentence symbol p with a formula φ. When dealing with quantifiers and equality we also need replacements of variables by expressions. In the following, we write $p(y)$ to imply that $p(y)$ has one or more *free* occurrence of the variable y, and we use the term '*w-next operator*' to designate either $\bigoplus_w$ or $\bigotimes_w$.

Definition 12. Let x, u be a variable and a rigid variable respectively, e be an expression, and $V(e) = V_s(e) \cup V_r(e)$ where $V_s(e)$ and $V_r(e)$ are respectively the set of variables and rigid variable occurring in e.

1. Let x be a variable and e be an expression. The replacement $x \mapsto e$ is said *compatible* for $p(x)$ if for any *free* occurrence of x within the scope of a *w-next* operator, $V_s(e) \subseteq w$.
2. The replacement $x \mapsto e$ is said *admissible* for $p(x)$ if it is compatible for $p(x)$, and none of the variables appearing in e is quantified in $p(u)$.
3. The replacement $u \mapsto e$ is said *admissible* for $p(u)$ if $V_s(e) = \emptyset$ and none of the variables appearing in e is quantified in $p(u)$.

4.1 The proof system

We write $p(e/v)$ for the instantiated formula $p(v)[v \mapsto e]$ where v is either a variable or a rigid variable.

Axioms for temporal operators :

A1. $\Box p \rightarrow p$
A2. $\bigoplus_{w_1} \neg p \Leftrightarrow \neg \bigotimes_{w_1} p$
A3. $\bigoplus_w (p \rightarrow q) \Leftrightarrow (\bigoplus_w p \rightarrow \bigoplus_w q)$
A4. $\Box(p \rightarrow q) \Rightarrow (\Box p \rightarrow \Box q)$
A5. $\Box p \rightarrow \Box \bigoplus_{w_1} p$
A6. $(p \Rightarrow \bigoplus_{w_1} p) \rightarrow (p \Rightarrow \Box p)$
A7. $\bigoplus_{w_1} p \wedge \bigotimes_{w_1} true \Leftrightarrow \bigotimes_{w_1} p$
A8. $\bigoplus_{w_1} p \vee \bigoplus_{w_2} q \Leftrightarrow \bigoplus_{w_1 \cup w_2} (p \vee q)$
A9. $\bigoplus_{w_1} p \wedge \bigoplus_{w_2} q \Rightarrow \bigoplus_{w_1 \cup w_2} (p \vee q)$
A10. $\bigotimes_\emptyset true \Rightarrow false$
A11. $(\bigotimes_{w_1 \cup w_2} true \rightarrow \bigotimes_{w_1 \cap w_2} true) \Rightarrow (\bigoplus_{w_1} p \wedge \bigoplus_{w_2} q \leftrightarrow \bigoplus_{w_1 \cup w_2} (p \wedge q))$
A12. $\text{shuffle}(w_1, w_2) \rightarrow (p \wedge q \Rightarrow \bigoplus_{w_1 \cup w_2} (p \vee q))$

Axioms for quantifiers :

A13. $\neg \exists y.\, p(y) \Leftrightarrow \forall y.\, \neg p(y)$
A14. $\neg \forall x.\, p(x) \Leftrightarrow \exists x.\, \neg p(x)$
A15. $p(e/x) \Rightarrow \exists x.\, p(x)$ provided $x \mapsto e$ is admissible for $p(x)$
A16. $\neg \forall u.\, p(u) \Leftrightarrow \exists u.\, \neg p(u)$
A17. $p(e/u) \Rightarrow \exists u.\, p(u)$ provided $u \mapsto e$ is admissible for $p(u)$

A18. $\bigoplus_w \exists x.\, p \Rightarrow \exists x.\, \bigoplus_w p$
A19. $\bigoplus_w \exists u.\, p \Leftrightarrow \exists u.\, \bigoplus_w p$

where x, u are respectively a variable and a rigid variable, and e is an expression.

Inference rules :

GEN For any state formula φ, $\dfrac{\models_A \varphi \text{ decided by an oracle}}{\Box \varphi}$

INS $\dfrac{\psi}{\psi[\alpha]}$ where α is an admissible general temporal replacement

MP $\dfrac{p \rightarrow q \,,\; p}{q}$

EXI 1 $\dfrac{p \Rightarrow q}{\exists x.\, p \Rightarrow q}$

EXI 2 $\dfrac{p \Rightarrow q}{\exists u.\, p \Rightarrow q}$

x, y are respectively a variable and a rigid variable which does not occur free in q

Comment. Observe that, from a programming point of view, axioms **A9** and **A12** may be assumed as modeling concurrency without communication and the axiom **A11** as modeling concurrency with communication. In the axioms **A9** and **A12** changes concern variables not shared by the indexes w_1 and w_2, however in **A11** changes involve variables common to w_1 and w_2.

4.2 Theorems and derived rules

Axioms and rules given above are used to derive some additional theorems and rules. A theorem is a statement of the form $\vdash F$, claiming that the formula F is provable in the presented deductive system, and hence is valid (assuming, of course, soundness that is shown below). The proof of a theorem ϕ under assumptions Γ is a finite sequence $\phi_0, \ldots, \phi_n$ so that for all $i \in \{0, \ldots, n\}$:

1. ϕ_i is an assertionally valid assertion, or
2. ϕ_i is an axiom, or ϕ_i is in Γ, or
3. ϕ_i is derived from $\{\phi_0, \ldots, \phi_{i-1}\}$ using a rule of the deductive system $\mathcal{G}$
4. ϕ_n is ϕ

We write $\Gamma \vdash \phi$ a proof of ϕ from Γ, we write simply $\vdash \phi$ when Γ is empty. We say that $\dfrac{p_1, \ldots, p_k}{q}$ is a derived rule of $\mathcal{G}$ if there is a proof $p_1, \ldots, p_k \vdash q$ in $\mathcal{G}$. Once a theorem, or a derived rule, is proven, we may use it in subsequent proofs to justify additional steps. We give some examples of theorems and derived rules in the full paper [24].

4.3 The soundness of the $\mathcal{G}$ system

It can be shown that the axioms and rules dealing only with the propositional fragment of the temporal language are complete. That is, any valid propositional temporal formula can be proven using our proof system. This is because, when we drop variables and quantifiers we obtain exactly the same propositional fragment of the regular temporal logic TL [12, 17] where the *w-next* operator $\bigotimes_w$ is equivalent to $\bigcirc$. For this fragment the proof system provides the same axioms and rules presented in TL which are proven complete for the propositional fragment [12]. However, the axioms and rules we provide to deal with variables and quantifiers, while they allow derivation of a large number of valid formulas, do not lead to a complete proof system. This is not surprising since the underlying assertional language assumes variables that range over concrete structures including integers which no complete deductive system to reason about them exists.

Theorem 13. *The proof system $\mathcal{G}$ is sound, i.e. if $\vdash F$ then $\models F$ for any formula F.*

The proof of this theorem is given in the full paper [24].

5 Design of compositional proof systems for concurrent programs within MTL

In [22, 23], we have shown how a compositional proof system can be easily derived from MTL for modular reasoning about concurrent programs. A modular programming notation (**IPL**) for concurrent modules of a concurrent system is introduced and a computational model based on *Fair Transition Systems* (*FTS*) to represent *open* semantics of IPL is defined. Brievely, in IPL, a module is composed of a *body* wich describes the internal behaviour of the module and an *interface* which stipulates the constraints the environment must satisfy for a correct interaction with the module. The obtained semantics is compositional in the sense that the semantics of a composite system is computed from semantics of its sub-modules. Correctness formulas are defined to state a relation (called *modular validity* and denoted $\sqsupset=$) between programs and MTL formulas [23] :

$$M \sqsupset= F \text{ iff } M\|M' \models F \text{ for all } M' \text{ interface compatible with } M.$$

Then, the logic MTL is augmented by a collection of axioms and rules to derive valid correctness formulas. This extension will permit the derivation, for a given IPL module M, of theorems that are valid over the set of models corresponding to the behaviour of M in an arbitrary context. Clearly, every temporal tautology of the basic logic MTL is a theorem for any program M, but there are formulas that are valid for a given program M but not valid in general. For the search to establish a proof system that should support both compositional verification and incremental (and modular) construction of IPL programs, composition rules are defined in such a way that both program part and specification part of the correctness formulas in premisses reduce in complexity *w.r.t* the conclusion. According to this criterion, given a large specification to be implemented, rules allow the implementor to decompose it into more elementary ones that can be implemented separately. Conversely, given the correctness proofs of some small modules, they allow the verifier to establish the correctness of bigger modules.

Worked examples. Two examples are developed respectively in [22] and [23]. The first one shows how the whole logic (i.e. the pure MTL logic and the program-part logic) provides an elegant theory for incrementally developing concurrent programs from their temporal specifications essentially using rules in backward. The second one shows how large (concurrent) programs can be compositionally verified in the offered theory essentially using rules in forward. Currently, we are developing examples to practice refinement of (concurrent) programs in the setting of this theory. We focus mainly on implementing a program by a finer-grained one.

6 Conclusion and related work

In this paper is presented the preliminary concepts of a refined temporal logic that guarantee a fully abstract semantics *w.r.t* to the chosen level of observation. We then shown how a compositional temporal proof system for concurrent programs can be derived. The resulting full logic should provide a practicable method for both compositional verification and modular construction of concurrent programs. The novelty of the refined temporal logic lies mainly in its ability to express properties with any chosen level of abstraction. Many versions of Pnueli's temporal logic [15] have been proposed to describe a program by a temporal formula [14, 17]. Some differ from the others by expressiveness but all of them represent programs by formulas that are not invariant under stuttering. So a compositional rule for parallel composition was hard to obtain and where it was possible the result was very complex. Moreover a finer-grained program could not implement a coarser-grained one in these logics.

Lamport's TLA [9] is the first logic in which programs are described by formulas that are invariant under stuttering. With the refined semantics for the basic temporal operators proposed here, it is shown that results equivalent to TLA ones may be reformulated in the regular temporal logic with the advantage, in our logic, that temporal quantifiers behave like the first-order quantifiers. Another attempt to tackle the problem of stuttering within the classical temporal logic is done by Pnueli in [13]. The main difference between Pnueli's work and our contribution lies in the fact that, contrary to the *discrete* temporal logic MTL, Pnueli dealt with the temporal logic TLR [13] which is based on a *dense* time domain (isomorphic to reals). Our proposal mainly intends to achieve results equivalent to Lamport's (for TLA) and Pnueli's ones (for TLR) for discrete temporal logic (TL [17]). We thus define a discrete temporal logic that supports refinement and systematic development of concurrent systems.

Finally, this work is undertaken with the idea of refining a previous logic [20, 21] we felt was cumbersome for reasoning about real-size concurrent programs. Invariance to stuttering is aimed at reaching a more modular method (and hence more practical) which can support systematic design of concurrent programs starting from their desired properties.

References

1. M. Abadi and L. Lamport. The existence of refinement mappings. In *Third Annual Symposium on Logic In Computer Science*, pages 165–177, Edinburgh, July 1988.
2. M. Abadi and L. Lamport. Composing specifications. In J. W. de Bakker, W. P. de Roever, and G. Rozenberg, editors, *Stepwise Refinement of Distributed Systems: Models, Formalisms, Correctness.* Springer Verlag, 1990. LNCS 430.
3. H. Barringer. The use of temporal logic in the compositional specification of concurrent systems. In A. Galton, editor, *Temporal logics and their applications*, pages 53–90, London, 1987. Academic Press.
4. H. Barringer, R. Kuiper, and A. Pnueli. Now you may compose temporal logic specifications. In *Sixteenth ACM Symposium on Theory of Computing*, pages 51–63, April 1984. ACM.
5. L. Lamport. The 'Hoare Logic' of concurrent programs. *Acta Informatica*, 14:21–37, 1980.
6. L. Lamport. Specifying concurrent program modules. *ACM Transactions On Programming Languages And Systems*, 2(5):190–222, april 1983.
7. L. Lamport. What good is temporal logic? pages 657–677. IFIP, 1983.
8. L. Lamport. A simple approach to specifying concurrent systems. *Communications of ACM*, 1(32):32–45, January 1989.
9. L. Lamport. The temporal logic of actions. Technical report, DEC Palo Alto, December 1991.
10. L. Abadi and G.D. Plotkin A logical view of composition. Technical report, DEC Palo Alto, May 1, 1992.
11. O. Lichtenstein, A. Pnueli, and L. Zuck. The glory of the past. In *Logics of Programs*, pages 196–218. Spinger Verlag, 1985. LNCS 193.
12. Z. Manna and A. Pnueli. The anchored version of teh temporal framework. In J.W. de Bakker, W.-P. de Roever, and G. Rozenberg, editors, *Linear Time, Branching Time and Partial Order in Logics and Models for Concurrency*, pages 201–284, New York, 1981. Springer Verlag. LNCS 354.
13. A. Pnueli. System Specification and Refinement in Temporal Logic. In LNCS, 1992. Springer Verlag.
14. Z. Manna and A. Pnueli. Verification of concurrent programs: A temporal proof system. In *4th School on Advanced Programming*, pages 163–255, June 1982.
15. Z. Manna and A. Pnueli. The modal logic of programs. Lecture Notes in Computer Science 71, pages 257–289, 1979.
16. Z. Manna and A. Pnueli. Verification of concurrent programs: The temporal framework. In R.S. Boyer and J.S. Moore, editors, *Correctness Problem in Computer science*, pages 215–273, London, 1982. Academic Press.
17. Z. Manna and A. Pnueli. *The Temporal Logic of Reactive and Concurrent Systems.* Springer-Verlag, 1991. ISBN 0-387-97664-7.
18. Z. Manna and A. Pnueli. Verification of concurrent programs: A temporal proof system. In 4th *School on Advanced Programming*, pages 163–255, June 1982.

19. Z. Manna and A. Pnueli. How to cook a temporal proof system for your pet language. In *Proceedings of the Symposium on Principles of Programming Languages*, 1983.
20. D. Méry and A. Mokkedem. A proof environment for a subset of SDL. In O. Faergemand and R. Reed, editors, *Fifth SDL Forum Evolving methods*. North-Holland, 1991.
21. D. Méry and A. Mokkedem. Crocos: An integrated environment for interactive verification of SDL specifications. In G. Bochmann, editor, *Computer-Aided Verification Proceedings*. Springer Verlag, 1992.
22. A. Mokkedem and D. Méry. On using a Composition Principle to Design Parallel Programs. In *Third International Conference on Algebraic Methodology and Software Technology* proceedings, AMAST'93, June 21-25, 1993, University of Twente, The Netherlands.
23. A. Mokkedem and D. Méry. On using temporal logic for refinement and compositional verification of concurrent systems. Technical Report 93-R-324, CRIN, 1993.
24. A. Mokkedem and D. Méry. A Stuttering Closed Temporal Logic for Modular Reasoning about Concurrent Programs. Technical Report, CRIN, 1994.
25. S. Owicki and D. Gries. An axiomatic proof technique for parallel programs I. *Acta Informatica*, 6:319–340, 1976.
26. L. Paulson and T. Nipkow. Isabelle tutorial and users's manual. Technical report, University of Cambridge, Computer Laboratory, 1990.
27. K. Gödel. Über Formal Unentscheidbare Sätze der Principa Mathematica under Verwandeter Systeme, I Monatshefte für Mathematic und Physik, 38, 1931.
28. J. Zwiers and W.P. de Roever. Predicates are predicate transformers : a unified compositional theory for concurrency. *Communications of ACM*, 265–279, 1989.

A Hierarchy of Partial Order Temporal Properties

Marta Kwiatkowska[*][1] and Doron Peled[2] and Wojciech Penczek[**][3]

[1] Dept. of Math. and Comp. Science, University of Leicester
Leicester LE1 7RH, UK
mzk@uk.ac.le.mcs

[2] AT&T Bell Laboratories
600 Mountain Avenue, POB 636, Murray Hill, NJ 07974-0636, USA
doron@research.att.com

[3] Department of Computing Science, Eindhoven University of Technology
P.O. Box 513, 5600 MB Eindhoven, The Netherlands
wojtek@win.tue.nl

Abstract. We propose a classification of partial order temporal properties into a hierarchy, which is a generalization of the safety-progress hierarchy of Chang, Manna and Pnueli. The classes of the hierarchy are characterized through three views: language-theoretic, topological and temporal. Instead of the domain of strings, we take the domain of Mazurkiewicz traces as a basis for our considerations. For the language-theoretic view, we propose operations on trace languages which define the four main classes of properties: safety, guarantee, persistence and response. These four classes are shown to correspond precisely to the two lower levels of the Borel hierarchy of the Scott topology of the domain of traces relativized to the infinite traces. In addition, a syntactic characterization of the classes is provided in terms of a sublogic of the Generalized Interleaving Set Temporal Logic GISTL (an extension of ISTL).

1 Introduction

Recently, Chang, Manna and Pnueli [CMP93] proposed a classification of program properties into classes forming a hierarchy (*safety* and *progress* properties, the latter class including *guarantee, response* and *persistence*). They presented an extensive framework, in which this hierarchy was characterized by means of four views: language-theoretic, topological, temporal and automata-theoretic. The hierarchy is an alternative to the safety-liveness characterization of Alpern and Schneider [AS85].

Central to both approaches is the space Σ^ω of infinite strings (words) over a given alphabet Σ. Each infinite string (of states or events) models a possible

---

[*] Supported in part by CEC grant CIPA3510PL927369. Visiting Academic at the Department of Computing, Imperial College, for the duration of the Nuffield Science Foundation Fellowship SCI/124/528/G.

[**] On leave from Institute of Computer Science, PAS, Warsaw, Poland. Supported in part by CEC grant CIPA3510PL927370 and by the Polish grant No. 2 2047 9203.

computation of a program. A *property* is defined as a set $\Pi \subseteq \Sigma^\omega$ of such computations. A program P is said to *satisfy* the property Π if all computations of P belong to Π. The question then arises as to which classes of extensional properties correspond to the intuitive classification into safety and liveness properties [Lam77], and how they can be characterized.

The framework of [CMP93] presents four different views, which are nevertheless in correspondence with each other. To introduce the language-theoretic view, four operators on string languages (modalities called $\mathcal{A}, \mathcal{E}, \mathcal{R}, \mathcal{P}$) were defined such that the four main classes of properties can be represented in the form $L(\Phi)$, where Φ is a finitary string language (intuitively, some collection of finite observations), and L is one of the above-mentioned operators. The four main classes also turned out to coincide with the two lower levels of the Borel hierarchy in the Cantor topology of infinite strings, with the safety properties corresponding to the *closed* sets, guarantee to their complements (the *open* sets), response to the $\mathbf{G}_\delta$ sets, and, dually, persistence to the $\mathbf{F}_\sigma$ sets. The next characterization is syntactic, as formulas of the linear time temporal logic LTL. Finally, the fourth view complemented the picture by providing a characterization in terms of automata on infinite strings (which serve as models for LTL), in the sense that a property in a given class is accepted by an automaton of the corresponding class.

We are generalizing the results of Chang, Manna and Pnueli concerning the linear temporal logic (these results were also published earlier in [MP89, MP90, MP91]) to the case of *partial order* temporal logic. Now, a natural representation of the behavior of a program is as a certain set of (infinite) Mazurkiewicz traces, that is, a set of equivalence classes of strings with respect to the congruence relation which identifies two strings if they differ in the order of consecutive, independent action symbols. If the independence relation between alphabet letters is empty, then the space of traces is isomorphic to the space of strings. A *property* is a set of such infinite traces.

We give a characterization of the classes of the hierarchy in terms of three views: *language-theoretic, topological* and *temporal*. The hierarchy presented in this paper collapses to the one in [CMP93] if the independence relation is empty.

For the language-theoretic view, we propose suitable definitions of the operators $\mathcal{A}, \mathcal{E}, \mathcal{R}, \mathcal{P}$, such that the intuitively acceptable classes of safety, guarantee, response and persistence properties (as sets of infinite traces) are of the form $\mathcal{A}(\Phi), \mathcal{E}(\Phi), \mathcal{R}(\Phi)$ and $\mathcal{P}(\Phi)$ respectively, for some set of finite traces Φ.

The topological view has been derived on the basis of the statement that *safety properties are the closed sets*. This leads us to rejecting the metric topology on infinite traces [Kwi90, Kwi91a] in favor of the Scott topology relativized to the infinite traces (the two coincide for strings). We find that the four main classes of properties coincide with the two lower levels of the Borel hierarchy. The hierarchy, however, does not have the same structure as the one in [CMP93] – the inclusion between safety and response (and, dually, between guarantee and persistence) does not hold. This is an inherent feature of the topology (it is not Hausdorff – the inclusion between closed and $\mathbf{G}_\delta$ sets holds in all Hausdorff spaces).

This lack of inclusion between certain classes of properties is replicated in the temporal characterization. To this aim, we study the Generalized Interleaving Set Temporal Logic GISTL (an extension of ISTL [KP91]).

2 Mathematical Preliminaries

2.1 Topology

We recall basic definitions and facts, which can also be found in [GHK+80].

A *topological space* is given as a pair $(X, \overline{})$, where X is a set and $\overline{}$ is a *closure* operator assigning to each set $A \subseteq X$ the set $\overline{A} \subseteq X$ satisfying the following axioms:

$$\text{(A1)} \ \overline{A \cup B} = \overline{A} \cup \overline{B} \quad \text{(A2)} \ A \subseteq \overline{A}$$
$$\text{(A3)} \ \overline{\emptyset} = \emptyset \quad \quad \quad \text{(A4)} \ \overline{\overline{A}} = \overline{A}$$

For metric topological spaces a closure operator satisfies also (A5) $\overline{\{x\}} = \{x\}$, for $x \in X$.

A set $A \subseteq X$ is *closed* iff $A = \overline{A}$. A set $A \subseteq X$ is *open* iff its complement $\sim A$ is closed. The family of closed sets (notation **F**) is closed under arbitrary intersection and finite union. Dually, the family of open sets (notation **G**) is closed under arbitrary union and finite intersection.

A family $\mathcal{B}$ of open sets is called a *base* iff each open set can be represented as the union of elements of a subfamily of $\mathcal{B}$.

The *Borel hierarchy* is obtained from the open (i.e., **G**) and closed (i.e., **F**) sets of a topology by taking alternatively countable intersections and unions. Thus, a $\mathbf{G}_\delta$ *set* is a countable intersection of open sets. An $\mathbf{F}_\sigma$ *set* is a countable union of closed sets. A $\mathbf{G}_{\delta\sigma}$ *set* is a countable union of $\mathbf{G}_\delta$ sets, while $\mathbf{F}_{\sigma\delta}$ *set* is a countable intersection of $\mathbf{F}_\sigma$ sets, and so forth.

Many topological notions can be relativized to a subset of the topological space. Let $E \subseteq X$. A set A is *closed relative to* E iff $A = A' \cap E$ for some A' closed. Likewise, a set A is *open relative to* E iff $A = A' \cap E$ for some A' open. More generally, the Borel hierarchy can also be relativized, in the sense that a set A is *Borel relative to* E iff $A = A' \cap E$ for some Borel set $A \subseteq X$.

2.2 Partial Orders and Domains

Let $(P, \leq)$ be a partially ordered set. A subset X of P is *directed* iff it is non-empty and every pair $x, y \in X$ has a bound $z \in X$. For $X \subseteq P$ and $x \in P$ write: $\downarrow X = \{y \in P \mid \exists x \in X : y \leq x\}$, $\uparrow X = \{y \in P \mid \exists x \in X : x \leq y\}$. For $x \in P$ we write $\downarrow x$ in place of $\downarrow \{x\}$, and similarly for $\uparrow x$.

$(P, \leq)$ is called a *complete partial order* (cpo) iff every directed subset X of P has a least upper bound $\bigsqcup X$ and P has the least element $\bot$.

A subset X of P is a *lower set* (also *prefix-* or *downward-closed*) iff $X = \downarrow X$, and an *upper set* (also *upward-closed*) iff $X = \uparrow X$. A subset $X \subseteq P$ is an *ideal* iff X is a directed lower set. Let $\text{Idl}(P)$ denote the set of all ideals of P ordered by inclusion. Then $(\text{Idl}(P), \subseteq)$ is called the *ideal completion* (or the *order completion*) of $(P, \leq)$.

Let $(P, \leq)$ be a cpo. An element $x \in P$ is called *compact* (or *finite*) iff, for every directed set $X \subseteq P$ with $x \leq \bigsqcup X$, there exists $z \in X$ such that $x \leq z$. For $X \subseteq P$ we write $\mathcal{K}(X)$ to denote the set of all compact elements contained in X. P is *algebraic* iff every element $x \in P$ is the least upper bound of the directed set of compact elements below it. P is *ω-algebraic* iff it is algebraic and $\mathcal{K}(P)$ is at most countable.

Let $(P, \leq)$ be an ω-algebraic cpo. The open sets of the *Scott topology* on P consists of all sets U such that U is upward-closed and, for every directed set $M \subseteq P$, if $\bigsqcup M \in U$ then some element of M is in U. One can easily check that the closed sets of the Scott topology consists of all sets U such that U is downward-closed and, for every directed set $M \subseteq U$, $\bigsqcup M \in U$.

3 Traces

This section presents a short summary of basic notions of trace theory; for a complete presentation see [Maz89].

Let Σ be a finite alphabet of action symbols. Define an *independence* relation over Σ to be an irreflexive and symmetric relation $I \subseteq \Sigma \times \Sigma$. The complement $D = (\Sigma \times \Sigma) \setminus I$ of an independence relation is called the *dependency* relation. By a *concurrent alphabet* we mean an ordered pair (Σ, I). We define finitary *trace equivalence* (notation $\equiv^{fin}$) as the least congruence in the monoid of finite strings $(\Sigma^\star, \cdot, \epsilon)$ such that: $a\ I\ b \Rightarrow ab \equiv^{fin} ba$.

An equivalence class $\sigma = [s]$ of a string s is called a *trace*. For example, if $\Sigma = \{a, b, c\}$ where $a\ I\ b$ then: $[abc] = \{abc, bac\}$, $[ac] = \{ac\}$. Denote $\Sigma^\star / \equiv^{fin}$ by $[\Sigma^\star]$. $([\Sigma^\star], \cdot, [\epsilon])$ is a monoid with concatenation satisfying: $[s] \cdot [t] = [s \cdot t]$. The concatenation symbol '$\cdot$' is omitted (i.e., replaced by juxtaposition) when it is clear from the the context (e.g., we write $[s][t]$).

A set $T \subseteq [\Sigma^\star]$ is called a *trace language*. Define a prefix order relation $\sqsubseteq^{fin} \subseteq [\Sigma^\star] \times [\Sigma^\star]$ on traces by: $\sigma \sqsubseteq^{fin} \tau \iff \exists \gamma : \sigma\gamma = \tau$. The pair $([\Sigma^\star], \sqsubseteq^{fin})$ is a partial order, with $[\epsilon]$ its least element. In contrast to strings, a trace may have incomparable prefixes. For example, if $\Sigma = \{a, b\}$ with $a\ I\ b$ we have $[a] \sqsubseteq^{fin} [ab]$ and $[b] \sqsubseteq^{fin} [ab]$, but $[a]$ and $[b]$ are incomparable. Incomparable prefixes arise due to concurrent execution. Finite traces are not satisfactory as a representation of infinite behavior of a concurrent system.

There are a number of equivalent ways of obtaining ω-infinite traces, see *e.g.* [Kwi91, Die91]. For the purpose of this paper we use the ideal completion. We identify the ideals of $[\Sigma^\star]$ with finite and infinite traces. Thus, for $a\ I\ b$ the finite ideal $\{[\epsilon], [a], [b], [ab]\}$ corresponds to $[ab]$, whereas the infinite ideal $\{[\varepsilon], [a], [b], ..., [a^n], [a^n b], ...\}$ to the trace $[ba^\omega]$. The set $\text{Idl}([\Sigma^\star])$ of all ideals is ordered by inclusion. We denote $(\text{Idl}([\Sigma^\star]), \subseteq)$ by $([\Sigma^\infty], \sqsubseteq)$. Let $\Phi \subseteq [\Sigma^\star]$. Define $\uparrow^{fin}\Phi = \{\sigma \in [\Sigma^\star] \mid \exists \sigma' \in \Phi : \sigma' \sqsubseteq^{fin} \sigma\}$, $\uparrow^{inf}\Phi = \{\sigma \in [\Sigma^\omega] \mid \exists \sigma' \in \Phi : \sigma' \sqsubseteq \sigma\}$, and $\downarrow^{fin}\Phi = \{\sigma \in [\Sigma^\star] \mid \exists \sigma' \in \Phi : \sigma \sqsubseteq^{fin} \sigma'\}$.

4 The Language-theoretic View

4.1 Operators on Languages

For the purpose of this paper we shall assume that computations will be represented by *infinite* traces. Denote by $[\Sigma^\omega]$ the set of *infinite* traces, and $\sqsubseteq$ the prefix order on traces. Note that, in contrast to [CMP93], when working with all infinite traces, we shall have to take care to distinguish between a proper prefix of some infinite trace and its finite prefix. This is because infinite traces may have proper infinite prefixes, e.g. $[a^\omega] \sqsubset [ba^\omega]$ for $a\,I\,b$.

A set $\Phi \subseteq [\Sigma^*]$ is called a *finitary property*. An *infinitary property* is a set $\Pi \subseteq [\Sigma^\omega]$. We are ultimately interested in infinitary properties, which will be constructed from the finitary ones by means of operators on trace languages. We define the *complement* of a finitary property $\Phi \subseteq [\Sigma^*]$, and of an infinitary property $\Pi \subseteq [\Sigma^\omega]$, by $\sim\Phi = [\Sigma^*] \setminus \Phi$, $\sim\Pi = [\Sigma^\omega] \setminus \Pi$, respectively.

Following [CMP93], we propose four operators $\mathcal{A}, \mathcal{E}, \mathcal{R}, \mathcal{P}$, which can be applied to a finitary property $\Phi \subseteq [\Sigma^*]$, yielding an infinitary property.

- The property $\mathcal{E}(\Phi)$ consists of all infinite traces σ s.t. *some* finite prefix of σ belongs to Φ, or formally $\mathcal{E}(\Phi) = \{\sigma \in [\Sigma^\omega] \mid \exists \sigma' \sqsubseteq^{\text{fin}} \sigma : \sigma' \in \Phi\} = \uparrow^{\inf}\Phi$.
- The property $\mathcal{A}(\Phi)$ consists of all infinite traces σ s.t. *all* finite prefixes of σ belong to Φ, or formally $\mathcal{A}(\Phi) = \{\sigma \in [\Sigma^\omega] \mid \forall \sigma' \sqsubseteq^{\text{fin}} \sigma : \sigma' \in \Phi\} = \sim(\uparrow^{\inf}\sim\Phi)$.
- The property $\mathcal{R}(\Phi)$ consists of all infinite traces σ s.t. *infinitely many* finite prefixes of σ belong to Φ, or formally $\mathcal{R}(\Phi) = \{\sigma \in [\Sigma^\omega] \mid \text{card}(\downarrow^{\text{fin}}\sigma \cap \Phi) = \omega\}$.
- The property $\mathcal{P}(\Phi)$ consists of all infinite traces σ s.t. *all but finitely many* finite prefixes of σ belong to Φ, or formally $\mathcal{P}(\Phi) = \{\sigma \in [\Sigma^\omega] \mid \text{card}(\downarrow^{\text{fin}}\sigma \cap \sim\Phi) <$

We shall illustrate the definitions by means of a notation based on regular expressions extended with ω-iteration. For example, if $\Sigma = \{a, b\}$ with $a\,I\,b$ then

$$\mathcal{E}([a^*]) = [\Sigma^\omega] \qquad \mathcal{E}([(ab)^+]) = [ab] \cdot [\Sigma^\omega]$$
$$\mathcal{A}([a^*]) = [a^\omega] \qquad \mathcal{A}([(ab)^*]) = \emptyset$$
$$\mathcal{R}([a^*]) = [b^*a^\omega] \cup [(ab)^\omega] \quad \mathcal{R}([(ab)^*]) = [(ab)^\omega] = [(abb)^\omega] = [(aab)^\omega]$$
$$\mathcal{P}([a^*]) = [a^\omega] \qquad \mathcal{P}([(ab)^*]) = \emptyset.$$

(Notice that $[(ab)^*]$ is the set of finite traces that contain strings with an equal number of a's and b's.) The motivation for denoting the latter two operators by $\mathcal{R}$ and $\mathcal{P}$ is that finite prefixes belonging to Φ occur *recurrently* in $\mathcal{R}(\Phi)$, and *persistently* in $\mathcal{P}(\Phi)$. Our formulation is a suitable generalization of the definitions of [CMP93] as it reduces to the linear response class of [CMP93] when $I = \emptyset$. One can give an alternative definition of $\mathcal{R}$, that also reduces to the linear response class when $I = \emptyset$, as follows:

The property $\mathcal{R}(\Phi)$ consists of all infinite traces σ s.t. for each finite prefix ρ of σ there exists a finite prefix ρ' of σ such that $\rho \sqsubseteq^{\text{fin}} \rho'$ and $\rho' \in \Phi$.

However, the latter alternative definition would not give rise to a set $\mathbf{F}_\sigma$ in the Borel hierarchy. The choice between the two alternatives made for the definition of $\mathcal{R}$ will affect the temporal logic view presented in section 6.

4.2 Basic Classes of Properties

We can now define the basic classes of infinitary properties by means of the operators $\mathcal{A}$, $\mathcal{E}$, $\mathcal{R}$ and $\mathcal{P}$.

Definition 1. Let $\Pi \subseteq [\Sigma^\omega]$.

- Π is a *guarantee* property iff $\Pi = \mathcal{E}(\Phi)$ for some finitary property $\Phi \subseteq [\Sigma^*]$.
- Π is a *safety* property iff $\Pi = \mathcal{A}(\Phi)$ for some finitary property $\Phi \subseteq [\Sigma^*]$.
- Π is a *response* property iff $\Pi = \mathcal{R}(\Phi)$ for some finitary property $\Phi \subseteq [\Sigma^*]$.
- Π is a *persistence* property iff $\Pi = \mathcal{P}(\Phi)$ for some finitary property $\Phi \subseteq [\Sigma^*]$.

Note that, by duality, we have that each safety property is a complement of some guarantee property (and vice-versa). Likewise, each response property is a complement of some persistence property (and vice-versa).

4.3 Closure of the Classes

We shall now show that the classes of properties introduced above are closed under intersection and union. Define $\text{Min}(\Phi) = \{\sigma \in \Phi \mid \neg \exists \sigma' \sqsubseteq^{\text{fin}} \sigma : \sigma' \in \Phi \wedge \sigma' \neq \sigma\}$.

Proposition 2. *Let $\Phi_1, \Phi_2 \subseteq [\Sigma^*]$. Then:*

1. $\mathcal{E}(\Phi_1) \cup \mathcal{E}(\Phi_2) = \mathcal{E}(\Phi_1 \cup \Phi_2)$
2. $\mathcal{E}(\Phi_1) \cap \mathcal{E}(\Phi_2) = \mathcal{E}(\uparrow^{\text{fin}} \Phi_1 \cap \uparrow^{\text{fin}} \Phi_2)$
3. $\mathcal{A}(\Phi_1) \cap \mathcal{A}(\Phi_2) = \mathcal{A}(\Phi_1 \cap \Phi_2)$
4. $\mathcal{A}(\Phi_1) \cup \mathcal{A}(\Phi_2) = \mathcal{A}(\sim \uparrow^{\text{fin}} \sim \Phi_1 \cup \sim \uparrow^{\text{fin}} \sim \Phi_2)$
5. $\mathcal{R}(\Phi^1) \cup \mathcal{R}(\Phi^2) = \mathcal{R}(\Phi^1 \cup \Phi^2)$
6. $\mathcal{R}(\Phi^1) \cap \mathcal{R}(\Phi^2) = \mathcal{R}(\Phi)$ where $\Phi = \bigcup_{k \in \omega} \text{Min}(\uparrow^{\text{fin}} \Phi^1_k \cap \uparrow^{\text{fin}} \Phi^2_k)$
   with $\Phi^i_k = \{\sigma \in \Phi^i \mid \exists \sigma_1, ..\sigma_{k-1} \in \Phi^i : \sigma_1 \sqsubset \sigma_2 \sqsubset ... \sqsubset \sigma_{k-1} \sqsubset \sigma\}$
7. $\mathcal{P}(\Phi^1) \cap \mathcal{P}(\Phi^2) = \mathcal{P}(\Phi^1 \cap \Phi^2)$
8. $\mathcal{P}(\Phi^1) \cup \mathcal{P}(\Phi^2) = \mathcal{P}(\Phi)$ where $\Phi = \sim \left(\bigcup_{k \in \omega} \text{Min}(\uparrow^{\text{fin}} \sim \Phi^1_k \cap \uparrow^{\text{fin}} \sim \Phi^2_k)\right)$

Proposition 3.

1. *Safety and persistence are downward-closed.*
2. *Guarantee and response are upward-closed.*

4.4 Characterization of the Lower Classes

We provide an additional characterization of the safety and guarantee properties similar to the characterization of [CMP93].

Proposition 4.

- *An infinitary property $\Pi \subseteq [\Sigma^\omega]$ is a safety property iff $\Pi = \mathcal{A}(\downarrow^{\text{fin}} \Pi)$.*
- *An infinitary property $\Pi \subseteq [\Sigma^\omega]$ is a guarantee property iff $\Pi = \mathcal{E}(\sim (\downarrow^{\text{fin}} \sim \Pi))$.*

4.5 Inclusion among the Classes

In this section, the respective inclusion between the four classes of properties is shown.

Proposition 5. *Safety properties are contained in persistence properties. Guarantee properties are contained in response properties.*

It is easy to see that the inclusion is strict. Observe that safety properties are not included in the response properties. To see this, assume $\Sigma = \{a, b\}$ with $a \, I \, b$. Then $\mathcal{A}([a^\star]) = [a^\omega]$, which is not upward-closed (e.g., it does not contain $[ba^\omega]$). Thus, this property cannot be contained in the response class because all response properties are upward-closed. Likewise, the guarantee properties are not contained in the persistence properties. For example, the guarantee property $\mathcal{E}([b^2])$ over the same concurrent alphabet is not downward-closed, since it contains $[b^2 a^\omega]$ but not $[ba^\omega]$. Hence, according to Proposition 3, it is not a persistence property. This differs from the characterization of [CMP93].

4.6 The Compound Classes

We now define two additional classes, the *obligation* properties, *i.e.* the class obtained by Boolean combinations of guarantee properties, and the *reactivity* properties, *i.e.* the class obtained by Boolean combinations of response properties.

Definition 6.

1. $\Pi \subseteq [\Sigma^\omega]$ is an *obligation* property iff $\Pi = \bigcap_{i=1...m} (\mathcal{A}(\Phi_i) \cup \mathcal{E}(\Psi_i))$
   for some $m > 0$ and finitary properties Φ_i, Ψ_i.
2. $\Pi \subseteq [\Sigma^\omega]$ is a *reactivity* property iff $\Pi = \bigcap_{i=1...m} (\mathcal{P}(\Phi_i) \cup \mathcal{R}(\Psi_i))$
   for some $m > 0$ and finitary properties Φ_i, Ψ_i.

It is easy to see that every safety property is an obligation property, every guarantee property is an obligation property, and that obligation properties are closed under finite union, intersection and complement. However, in contrast to [CMP93], the obligation properties are *not* in general contained in the response and persistence classes. To see this, consider $\Sigma = \{a, b\}$ with $a \, I \, b$. Then $T = \mathcal{A}([a^\star]) \cup \mathcal{E}([b^2])$ is an obligation property, but it is neither a persistence nor a recurrence property; notice that response properties are upward-closed, but T is not upward-closed: it contains the trace $[a^\omega]$, but not $[ba^\omega]$. Persistence properties, on the other hand, are downward-closed, and T violates this since it contains $[b^2 a^\omega]$, but it does not contain $[ba^\omega]$.

5 A Topological View

We shall now characterize the four basic classes of infinitary properties in a topology on $[\Sigma^\omega]$. It is argued that the metric topology [Kwi90] should be rejected in favor of the Scott topology relativized to the infinite traces.

First of all, a suitable closure operator is needed. As we are motivated by the statement that safety properties are the closed sets, we propose the following definition.

Definition 7. Let $X \subseteq [\Sigma^\omega]$ and put: $\overline{X} = \{\sigma \in [\Sigma^\omega] \mid \downarrow^{\text{fin}} \sigma \subseteq \downarrow^{\text{fin}} X\}$.

Proposition 8. *The operator $\bar{}$ is indeed a closure operator (i.e., it satisfies the conditions (A1) – (A4) from Section 2.1).*

Proposition 9 [Kwi91]. $([\Sigma^\infty], \sqsubseteq)$ *is an ω-algebraic cpo.*

Proposition 10. *The topological space $([\Sigma^\omega], \bar{})$ is the Scott topology of $([\Sigma^\infty], \sqsubseteq)$ relativized to $[\Sigma^\omega]$.*

The price to be paid for the choice of the relativized Scott topology is that the statements "response contains safety and guarantee" and "persistence contains safety and guarantee" are no longer true. On the other hand, for the metric topology defined in [Kwi90] or any other metric topology the safety properties are *not* the closed sets. To see this, consider $\Sigma = \{a, b\}$ with $a\ I\ b$, and the infinite trace $[ba^\omega]$. Then, according to the property (A5): $\overline{\{x\}} = \{x\}$, for $x \in X$ of a closure operator in the metric topological spaces, $[ba^\omega]$ is a (relativized) closed set. But, $[ba^\omega]$ is not a safety property since it is not downward-closed (e.g. it does not contain $[a^\omega]$).

5.1 Characterizing Properties

It is easy to see that in the relativized Scott topology safety properties are the closed sets and guarantee properties are the open sets.

Proposition 11.
- *For any set $\Pi \subseteq [\Sigma^\omega]$, Π is a safety property iff Π is a closed set.*
- *For any set $\Pi \subseteq [\Sigma^\omega]$, Π is a guarantee property iff Π is an open set.*

We now characterize the response and persistence properties.

Proposition 12.
- *For any set $\Pi \subseteq [\Sigma^\omega]$, Π is a response property iff Π is a $\mathbf{G}_\delta$ set.*
- *For any set $\Pi \subseteq [\Sigma^\omega]$, Π is a persistence property iff Π is an $\mathbf{F}_\sigma$ set.*

Proof. (Sketch.) Suppose $\Pi = \mathcal{R}(\Phi)$ for some $\Phi \subseteq [\Sigma^\star]$. For $k > 0$ define:

$$\Phi_k = \{\sigma \in \Phi \mid \exists \sigma_1, ..\sigma_{k-1} \in \Phi : \sigma_1 \sqsubset \sigma_2 \sqsubset ... \sqsubset \sigma_{k-1} \sqsubset \sigma\}$$

and define $G_k = \uparrow^{\inf} \Phi_k$. G_k are open, since they are open in the Scott topology relativized to $[\Sigma^\omega]$. One can prove that $\Pi = \bigcap_{k \in \omega} G_k$.

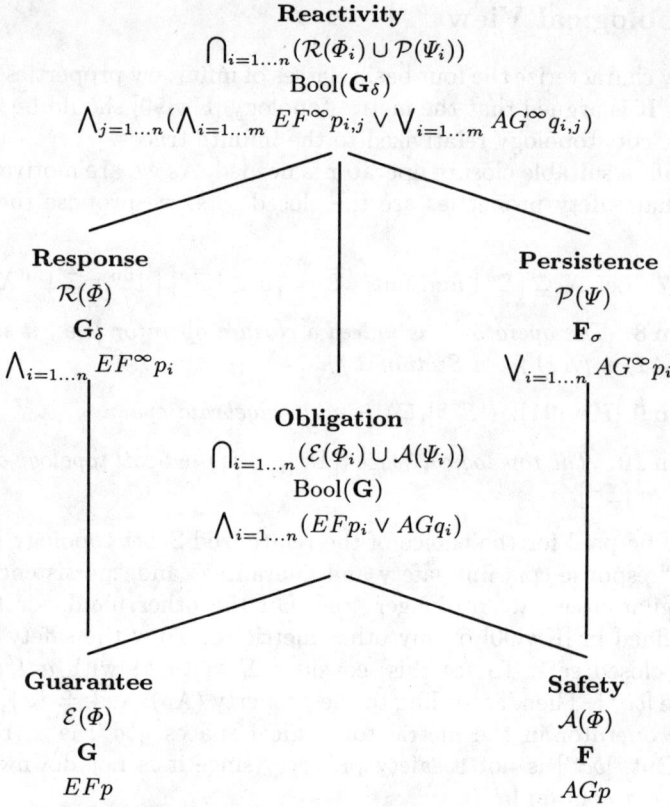

Fig. 1. The hierarchy of partial order temporal properties

6 The Partial Order Temporal Logic View

Each class of the hierarchy introduced in the abstract setting will be shown to correspond to a syntactic class of a temporal logic over partial orders. We call these classes the *partial order temporal logic classes*, as opposed to the classes in [CMP93], which will be referred to as the linear temporal logic classes. We start with introducing the language of a partial order temporal logic, called GISTL (Generalized Interleaving Set Temporal Logic). The language is a variant of the language of ISTL [KP91, PP93] with past modalities and labeled next step operators. It introduces path operators that express that some property can happen *infinitely often* or *eventually always* along a path (these operators can be expressed in ISTL*). These are: $F^\infty p$ (abbreviating GFp and meaning "infinitely often p") and $G^\infty p$ (abbreviating FGp and meaning "almost everywhere p"). These path operators correspond to the linear temporal logic formulas $\Box\Diamond p$ and $\Diamond\Box p$, respectively.

A model for a partial order temporal formula is an ordered pair $M_\sigma = (\downarrow^{\text{fin}}\sigma, \rightarrow)$, where σ is an infinite trace and $\rightarrow$ is the labeled next step relation in $\downarrow^{\text{fin}}\sigma$ such that $\rho \xrightarrow{a} \delta$ iff $\rho[a] = \delta$, for $\rho, \delta \in \downarrow^{\text{fin}}\sigma$.

6.1 Syntax of GISTL

Let Σ be a finite alphabet. The set of GISTL formulas Ξ is the smallest set satisfying the following:

1. $true \in \Xi$.
2. if $\varphi, \psi \in \Xi$ then so are $\neg\varphi$ and $\varphi \wedge \psi$,
3. if $\varphi \in \Xi$ and $a \in \Sigma$ then $EX_a\varphi, EY_a\varphi \in \Xi$.
4. if $\varphi \in \Xi$ then so are $EF\varphi, EP\varphi$.
5. if $\varphi \in \Xi$ then so is $EF^\infty\varphi$.

The other derived temporal operators are defined as follows:
$$AG\varphi \stackrel{def}{=} \neg EF\neg\varphi,\ AH\varphi \stackrel{def}{=} \neg EP\neg\varphi,\ AG^\infty\varphi \stackrel{def}{=} \neg EF^\infty\neg\varphi.$$

6.2 Semantics of GISTL

Let $M = (W, \rightarrow)$ be a model. By a forward (backward) path in M starting at $\rho_0 \in W$ we mean an infinite sequence of traces $x = \rho_0\rho_1\ldots$ such that $\rho_i \xrightarrow{a_i} \rho_{i+1}$ ($\rho_{i+1} \xrightarrow{a_i} \rho_i$, resp.) for some $a_i \in \Sigma$ and $i \geq 0$.

1. $M, \rho \models true$ for each $\rho \in W$,
2. $M, \rho \models \neg\varphi$ iff not $M, \rho \models \varphi$,
   $M, \rho \models \varphi \wedge \psi$ iff $M, \rho \models \varphi$ and $M, \rho \models \psi$,
3. $M, \rho_0 \models EX_a\varphi$ iff $M, \rho_1 \models \varphi$, for some ρ_1 with $\rho_0 \xrightarrow{a} \rho_1$,
   $M, \rho_0 \models EY_a\varphi$ iff $M, \rho_1 \models \varphi$, for some ρ_1 with $\rho_1 \xrightarrow{a} \rho_0$,
4. $M, \rho_0 \models EF\varphi$ iff there is a forward path x starting at ρ_0 with $k \geq 0$ s.t. $M, \rho_k \models \varphi$,
   $M, \rho_0 \models EP\varphi$ iff there is a backward path x starting at ρ_0 with $k \geq 0$ s.t. $M, \rho_k \models \varphi$,
5. $M, \rho_0 \models EF^\infty\varphi$ iff there is a forward path x starting at ρ_0 s.t. for infinitely many i: $M, \rho_i \models \varphi$.

A formula φ holds in a model M (written $M \models \varphi$) if it holds in its initial state $M, [\epsilon] \models \varphi$.

Operators EX_a, EF, and EF^∞, as well as these derived from them, are called the *future* operators; their past counterparts are called the *past* operators. The language does not contain the past counterpart EP^∞ of the operator EF^∞ since interpreting EP^∞ over finite backward does not make sense.

Below, some simple examples of temporal formulas are shown:

- $AG\varphi$ - the infinite trace contains all the prefixes with φ,
- $EF\varphi$ - the infinite trace contains a prefix with φ,

- $AG^\infty \varphi$ - the infinite trace contains only finitely many prefixes with $\neg \varphi$,
- $EF^\infty \varphi$ - the infinite trace contains infinitely many prefixes with φ.

When the independence relation I is empty, the above GISTL formulas correspond to the linear temporal logic formulas $\Box \varphi$, $\Diamond \varphi$, $\Diamond \Box \varphi$ and $\Box \Diamond \varphi$, respectively. This will be used in the sequel to show that some (negative) results between linear temporal logic classes, as proved in [CMP93], can be transferred to the partial order hierarchy. In those cases, the same examples that are used to show that there is no inclusion between the classes in the linear case can be used here, provided we select the empty independence relation.

Let $\Pi \subseteq [\Sigma^\omega]$. Π is said to be *specifiable* if there is a formula φ s.t. Π is exactly the set of all infinite traces which satisfy φ, i.e., $\Pi = \{\sigma \mid M_\sigma \models \varphi\}$. Then, we define $Sat(\varphi) = \Pi$.

The past formulas of ISTL include only the modal operators EY_a, EP, the boolean operators and the constant *true*. The finitary property defined by the past formula p is the set of finite traces whose end states satisfy p (i.e. $\{\rho \mid \rho \in \downarrow^{\text{fin}} \sigma \wedge M_\sigma, \rho \models p$, for some infinite trace $\sigma\}$). Denote by $esat(p)$ the finitary property defined by p. A finitary property Φ is *expressible* in GISTL if it equals to $esat(p)$ for some past formula p.

Definition 13. Define the *linearization* of a trace language T as the set of all strings included in the traces of T, i.e. $\{v \mid [v] \in T\}$.

Lemma 14. *Any past GISTL formula specifies a trace language whose linearization is a regular language.*

Proof. (Sketch.) Construct a finite automaton to recognize the linearization of a trace language defined by a past GISTL formula by structural induction on the past formulas. This automaton recognizes the words backwards, i.e., from the last letter to the first one. It is trivial to construct an automaton for *true*; i.e., just construct a single accepting node with self-loops for all possible letters in Σ. For the boolean operators, use the fact that regular languages are closed under complementation and union.

Now consider formulas of the form $EY_a \varphi$. We assume that there is an automaton A that recognizes the linearization of the language T defined by the past formula φ. Then we can construct a finite automaton that recognizes the linearization of $EY_a \varphi$ in the following way: construct two copies for A. From the first copy, remove all the edges which contain letters b which are dependent on a. Then add an edge marked by a from each state of the first copy to the corresponding state in the second copy. The starting node of the constructed automaton will be the one corresponding to the formerly starting state in the *first* copy, and the accepting states are those corresponding to the formerly accepting states in the *second* copy. It is easy to see that this automaton recognizes the linearization of the concatenation $[a] \cdot T$ (i.e., the linearization of the above $EY_a \varphi$).

The construction of the linearization of the language of $EP\varphi$, obtained from an automaton to recognize the linearization of φ is similar: it should correspond to the linearization of $[\Sigma^\star] \cdot T$, where T is the trace language of φ. Given an automaton A for T, construct a new automaton A' by searching A using depth-first-search. Each node of A will have a corresponding set of nodes in A', each of which is marked by a different subset of the operations Σ. The search is started with $\langle \hat{s}, \Sigma \rangle$ where $\hat{s}$ is the initial node of A. Then, for an already constructed state $\langle s, S \rangle$ in A', for any transition $s \xrightarrow{a} t$ in A, we add the new edge $\langle s, S \rangle \xrightarrow{a} \langle t, S' \rangle$, where $S' = S \cap \{b \mid b \; I \; a\}$. Then add to each node $\langle s, S \rangle$ self-loops marked with all the letters in S.

6.3 The Partial Order Temporal Hierarchy of Properties

Next, we present four temporal classes expressed using the future modalities AG, EF, EF^∞ and AG^∞. These classes correspond to the language operators $\mathcal{A}, \mathcal{E}, \mathcal{R}, \mathcal{P}$. This relation is established through the following equalities:

1) $Sat(EFp) = \mathcal{E}(esat(p))$,
2) $Sat(AGp) = \mathcal{A}(esat(p))$,
3) $Sat(EF^\infty p) = \mathcal{R}(esat(p))$,
4) $Sat(AG^\infty p) = \mathcal{P}(esat(p))$.

It is easy to see from the definitions that the first two equalities hold. The third equivalence follows from the lemma given below. The fourth equivalence stems from duality between the persistence and recurrence class and from the duality between EF^∞ and AG^∞.

Lemma 15. *Every infinite trace σ in a recurrence property $\mathcal{R}(\Phi)$ contains a path of traces that includes an infinite number of traces from Φ.*

The closure of properties expressible by temporal formulas under the operations of union, intersection, and complementation can be translated into disjunction, conjunction, and negation of the formulas expressing the properties.

- $Sat(\varphi) \cup Sat(\psi) = Sat(\varphi \vee \psi)$,
- $Sat(\varphi) \cap Sat(\psi) = Sat(\varphi \wedge \psi)$,
- $\overline{Sat(\varphi)} = Sat(\neg \varphi)$.

6.4 Safety and Guarantee Formulas

A *canonical safety formula* is defined to be a formula of the form AGp, for a past formula p. The class of properties expressible by safety formulas is closed under intersection and union. To prove it, one has to show that the class of safety formulas is closed under conjunction and disjunction. This follows from the following equivalences: $(AGp \wedge AGq) \equiv AG(p \wedge q)$, $(AGp \vee AGq) \equiv AG(AHp \vee AHq)$.

A *canonical guarantee formula* is defined to be a formula of the form EFp, for a past formula p. The class of properties expressible by guarantee formulas is closed under intersection and union. This follows from the fact that EF is dual to AG: $EFp \equiv \neg AG \neg p$, and from the following equivalences: $(EFp \vee EFq) \equiv EF(p \vee q)$, $(EFp \wedge EFq) \equiv EF(EPp \wedge EPq)$.

Notice that the class of guarantee formulas is not closed under negation. The negation of a guarantee formula is a safety formula. Thus, the temporal safety and liveness classes are dual.

6.5 Obligation Formulas

A *canonical simple obligation formula* is defined to be a formula of the form $AGp \vee EFq$, for past formulas p, q. The class of properties specifiable by simple obligation formulas is closed under union. To show this observe the following simple equivalence:

$$[(AGp_1 \vee EFq_1) \vee (AGp_2 \vee EFq_2)] \equiv [(AGp_1 \vee AGp_2) \vee (EFp_1 \vee EFq_2)].$$

Using the closure of both the safety and guarantee formula classes under disjunction, this leads to an equivalent simple obligation formula. However, this class is not closed under intersection. This can be easily shown by reducing our partial obligation class to the linear obligation class in [CMP93] when the independence relation is empty. Therefore, taking conjunctions of simple obligation formulas results in getting a more powerful class.

A *canonical m-obligation formula* is a formula of the following form: $\bigwedge_{i=1}^{m}[AGp_i \vee EFq_i]$.

Now it is easy to show that every finite Boolean combination of safety and guarantee formulas is a general obligation formula. This implies also that the class of general obligation formulas is closed under all Boolean operations.

The class of obligation specifiable formulas strictly contains the class of safety specifiable properties and the class of guarantee-specifiable properties. In fact, it forms an infinite strict hierarchy with classes specifiable by m-conjuncts. The inclusion of the classes is obvious. The strictness of these classes stems from the corresponding strictness of the classes in the linear hierarchy, which is shown in [CMP93], when using the empty independence relation.

6.6 Response Formulas

A *canonical simple response formula* is defined to be a formula of the form $EF^\infty p$, for a past formula p. The class of properties expressible by simple response formulas is closed under union. This is shown by the following equivalence: $(EF^\infty p \vee EF^\infty q) \equiv EF^\infty(p \vee q)$.

Although the class of response properties is closed under intersection, the class of response specifiable formulas needs in its definition the external conjunction. This is because it is not the case that for each conjunction $\varphi = EF^\infty p_1 \wedge EF^\infty p_2$, for some past GISTL formulas p_1 and p_2, there exists some past formula p_3 such that $\varphi \equiv EF^\infty p_3$. To see this, notice that the languages $\mathcal{R}([a^+])$ and $\mathcal{R}([b^+])$ over $\Sigma = \{a, b\}$ with $a\ I\ b$, are easily specifiable in the above form. It will be shown that their intersection is a response property that cannot be built using a finitary trace language with regular linearization. It was shown in Lemma 14 that the past part of GISTL can express only languages with regular linearization. Thus, no such past formula p_3 exists.

Now consider the intersection of recurrence languages $\mathcal{R}([a^+])$ and $\mathcal{R}([b^+])$. It was shown to be equivalent to some recurrence language $\mathcal{R}(\Phi)$. We will show that Φ cannot be regular. The language Φ must contain words with an arbitrary number of a's and an arbitrary number of b's. However, it may not contain a sequence of words where the number of a's (the number of b's, respectively) is bounded while the number of b's (of a's, respectively) grows indefinitely. Otherwise, this class will include words with either a bounded number of a's or a bounded number of b's, which is not in the intersection. Moreover, once some arbitrary word is in the language, then using the independence between a's and b's, one obtains a word which has all the a's before all the b's and is in the language too. Since one can get words with arbitrary numbers of a's (or b's) where all the a's appear first, it follows by the pumping lemma for finite automata that one must have an infinite sequence of words, in which only the number of a's increases, while the number of b's remains the same. However, this contradicts the above requirement for Φ.

We define a *canonical response formula* to be a formula of the following form: $\bigwedge_{i=1}^{m} EF^{\infty} p_i$.

Now, it is easy to show that the class of general response formulas is closed under all positive Boolean operations. The class of response-specifiable formulas strictly contains the class of guarantee-specifiable properties. This is supported by the following equivalence: $EFp \equiv EF^{\infty}(EPp)$.

The class of response-specifiable formulas does not contain the class of safety-specifiable formulas: for the concurrent alphabet $\Sigma = \{a, b\}$ and $a\ I\ b$, the formula $AG(\neg EY_b true)$ specifies the safety language $\mathcal{A}([a^*])$ which was already shown not to be upward-closed, hence it cannot be a response property.

6.7 Persistence Formulas

A *canonical simple persistence formula* is defined to be a formula of the form $AG^{\infty}p$, for a past formula p. The classes of properties specifiable by persistence and response formulas are dual. This follows from the following equivalence: $\neg EF^{\infty}\neg p \equiv AG^{\infty}p$. Therefore, the class of properties expressible by simple persistence formulas is closed under intersection. This is also shown by the following equivalence: $(AG^{\infty}p \wedge AG^{\infty}q) \equiv AG^{\infty}(p \wedge q)$.

We define a *canonical m-persistence formula* to be a formula of the following form: $\bigvee_{i=1}^{m} AG^{\infty} p_i$.

Now it is easy to show that the class of general persistence formulas is closed under all positive Boolean operations. The class of persistence-specifiable formulas strictly contains the class of safety-specifiable properties. This is supported by the following equivalence: $AGp \equiv AG^{\infty}(AHp)$.

The class of persistence-specifiable formulas does not contain the class of guarantee-specifiable formulas: for the concurrent alphabet $\Sigma = \{a, b\}$ and

$a\ I\ b$, the formula $EFEY_b EY_b(\bigwedge_{a\in\Sigma}\neg EY_a true)$ corresponds to the guarantee language $\mathcal{E}([b^2])$ which was already shown not to be downward-closed, and hence it cannot be a persistence property.

6.8 Reactivity Formulas

A *canonical simple reactivity formula* is defined to be a formula of the form $\bigwedge_{i=1}^{m} EF^\infty p_i \vee \bigvee_{i=1}^{m} AG^\infty q_i$, for past formulas p_i, q_i. The class of properties specifiable by simple reactivity formulas is closed under union. To show this observe the following simple equivalence:
$[(\bigwedge_{i=1}^{m} EF^\infty p_{1,i} \vee \bigvee_{i=1}^{m} AG^\infty q_{1,i}) \vee (\bigwedge_{i=1}^{m} EF^\infty p_{2,i} \vee \bigvee_{i=1}^{m} AG^\infty q_{2,i})] \equiv$
$[(\bigwedge_{i=1}^{m} EF^\infty p_1 \vee \bigwedge_{i=1}^{m} EF^\infty p_2) \vee (\bigvee_{i=1}^{m} AG^\infty p_{1,i} \vee \bigvee_{i=1}^{m} AG^\infty q_{2,i})]$.

Using the closure of both the response and persistence formula classes under disjunction, this leads to an equivalent simple reactivity formula. However, this class is not closed under intersection (see [CMP93]). Therefore, taking conjunctions of simple obligations formulas results in getting a more powerful class. In fact, it forms an infinite strict hierarchy with classes specifiable by n-conjuncts. A *canonical n-reactivity formula* is a formula of the following form: $\bigwedge_{j=1}^{n}[\bigwedge_{i=1}^{m} EF^\infty p_{i,j} \vee \bigvee_{i=1}^{m} AG^\infty q_{i,j}]$. Now it is easy to show that every finite Boolean combination of response and persistence formulas is a general reactivity formula. This implies also that the class of general reactivity formulas is closed under all Boolean operations.

The class of reactivity-specifiable formulas strictly contains the class of response-specifiable properties, the class of persistence-specifiable properties, and the class of obligation-specifiable formulas. In fact, it forms an infinite strict hierarchy with classes specifiable by n-conjuncts.

The above three views form a hierarchy of partial order properties: *language-theoretic* as classes of trace languages, *topological* as classes of sets of traces in the Borel hierarchy of the Scott topology relativized to the infinite traces, and *syntactic* as formulas of partial order temporal logic. We obtain the hierarchy of abstract properties as shown in Figure 1.

7 Conclusion

We have provided three characterizations of partial order temporal properties: language-theoretic (in terms of language operators giving rise to properties of a given class), topological (as the lower levels of the Borel hierarchy of the Scott topology relativized to the infinite traces), and temporal (as syntactic forms belonging to a sublogic of GISTL).

Our hierarchy of trace properties is a natural extension of the linear hierarchy proposed in [CMP93], since it collapses to the linear one when the independence relation is empty. The metric topology and the relativized Scott topology over infinite traces can naturally extend the topology used in [CMP93]. However, the Scott topology was chosen since it maintains the classification of the safety properties as the closed sets. The remaining classes were defined according to the two lower levels of the Borel hierarchy over the Scott topology.

The properties definable in the proposed hierarchy correspond to viewing computations of concurrent systems as sets of observations which differ in the ordering of independent actions. Thus, the specified properties assert that there exists at least one possible observation under which the computation evolved in some desired way. Consequently, while the linear guarantee properties can describe an occurrence of some global state in *every execution sequence*, its partial order counterparts describe an occurrence of a global state in at *least one* of the observations of each partial order execution. This corresponds, e.g., to the global states reached in communication closed layers [KP91]. Linear response properties can deal with repeated occurrence of a global state. In the partial order hierarchy, response properties can describe the repeated occurrence of a global state on at least one observation of each partial order execution. This corresponds, e.g., to the global states that describe completed transactions in serializable database systems with repeated occurrence in some observation of each partial order execution.

Automata on infinite traces are only at the early stages of their study (see, e.g., [GP92, DM93]). Hence, no automata theory counterpart was included. The connection between partial order temporal logic and various automata for infinite traces is thus an important issue for further research.

8 Acknowledgements

The first author is grateful to the Theory and Formal Methods Section, Department of Computing, Imperial College for providing an excellent research environment for the duration of the Nuffield Research Fellowship.

References

[AS85] B. Alpern and F. B. Schneider. Defining liveness. *Information Processing Letters*, 21:181–185, 1985.

[CMP93] E. Chang, Z. Manna, and A. Pnueli A Hierarchy of Temporal Properties. Stanford University, Technical Report.

[Die91] V. Diekert. On the concatenation of infinite traces. In Choffrut C. et al., editors, *Proceedings of the 8th Annual Symposium on Theoretical Aspects of Computer Science (STACS'91), Hamburg 1991*, number 480 in Lecture Notes in Computer Science, pages 105–117, Berlin-Heidelberg-New York, 1991. Springer. To appear 1993 in Theoret. Comput. Sci.

[DM93] V. Diekert and A. Muscholl. Deterministic asynchronous automata for infinite traces. In P. Enjalbert, A. Finkel, and K. W. Wagner, editors, *Proceedings of the 10th Annual Symposium on Theoretical Aspects of Computer Science (STACS'93), Würzburg 1993*, number 665 in Lecture Notes in Computer Science, pages 617–628, Berlin-Heidelberg-New York, 1993. Springer.

[GP92] P. Gastin and A. Petit. Asynchronous automata for infinite traces. In W. Kuich, editor, *Proceedings of the 19th International Colloquium on Automata Languages and Programming (ICALP'92), Vienna (Austria) 1992*, number 623 in Lecture Notes in Computer Science, pages 583–594, Berlin-Heidelberg-New York, 1992. Springer.

[GHK+80] G. Gierz, K. Hofmann, K. Keimel, J. Lawson, M. Mislove, and D. Scott. *A Compendium of Continuous Lattices*. Springer-Verlag, 1980.

[Kwi90] M. Kwiatkowska. A metric for traces. *Information Processing Letters*, 35:129–135, 1990.

[Kwi91] M. Kwiatkowska. On the domain of traces and sequential composition. In S. Abramsky and T.S.E.. Maibaum, editors, *Proceedings 16th Coll. on Trees in Algebra and Programming (CAAP'91), Brighton (UK)*, number 493 in Lecture Notes in Computer Science, pages 42–56, Berlin-Heidelberg-New York, 1991. Springer.

[Kwi91a] M. Z. Kwiatkowska. On topological characterization of behavioural properties. In G. Reed, A. Roscoe, and R. Wachter, editors, *Topology and Category Theory in Computer Science*, pages 153–177. Oxford University Press, 1991.

[Lam77] L. Lamport. Proving the correctness of multiprocess programs. *IEEE Transactions on Software Engineering*, SE-3(2):125–143, 1977.

[KP91] S. Katz, D. Peled. Interleaving Set Temporal Logic. *Theoretical Computer Science*, Vol. 75, Number 3, 21–43, also appeared in proceedings of the 6th Annual ACM Symposium on Principles of Distributed Computing, Vancouver, Canada, 178–190, 1987.

[Maz89] A. Mazurkiewicz. Basic notions of trace theory. In *Linear Time, Branching Time and Partial Order in Logics and Models for Concurrency*, volume 354 of *Lecture Notes in Computer Science*, pages 25–34. Springer-Verlag, 1989.

[MP89] Z. Manna and A. Pnueli. The anchored version of the temporal framework. In *Linear Time, Branching Time and Partial Order in Logics and Models for Concurrency*, volume 112 of *Lecture Notes in Computer Science*, pages 201–284. Springer-Verlag, 1989.

[MP90] Z. Manna and A. Pnueli. A hierarchy of temporal properties. In *Proceedings, 9th ACM Symposium on Principles of Distributed Computing*, pages 377–408. ACM Press, 1990.

[MP91] Z. Manna and A. Pnueli. Completing the temporal picture. *Theoretical Computer Science*, 83(1):97–139, 1991.

[PP93] D. Peled and A. Pnueli. Proving partial order liveness properties. *Theoretical Computer Science*, 1993. A preliminary version appeared in the proceedings of the 17th International Colloquium on Automata, Languages and Programming, Warwick University, England, July 1990, Lecture Notes in Computer Science 443, Springer Verlag, 553-571.

[Pen93] W. Penczek. Temporal logics for trace systems: On automated verification. *International Journal of Foundations of Computer Science*, Vol. 4 No. 1, pp. 31:67, 1993.

[SC82] A.P. Sistla and E. Clarke. The complexity of propositional temporal logic. In *14th ACM Symposium on Theory of Computing*, pages 159–167, 1982.

[Sza87] A. Szalas. A complete axiomatic characterization of first-order temporal logic of linear time. *Theoretical Computer Science*, 54:199–214, 1987.

A Graph-Based Approach To Resolution In Temporal Logic*

Clare Dixon[1], Michael Fisher[2] and Howard Barringer[1]

[1] Department of Computer Science, University of Manchester, Oxford Road,
Manchester M13 9PL, UK.
`{dixonc,howard}@cs.man.ac.uk`

[2] Department of Computing, Manchester Metropolitan University, Chester Street,
Manchester M1 5GD, UK.
`michael@sun.com.mmu.ac.uk`

Abstract. In this paper, we present algorithms developed in order to implement a clausal resolution method for discrete, linear temporal logics, presented in [Fis91]. As part of this method, temporal formulae are rewritten into a normal form and both 'non-temporal' and 'temporal' inference rules are applied. Through the use of a graph-based representation for the normal form, "efficient" search algorithms can be applied to detect sets of formulae for which temporal resolution is applicable. Further, rather than constructing the full graph structure, our algorithms only explore and construct as little of the graph as possible. These algorithms have been implemented and have been combined with sub-programs performing translation to normal form and non-temporal resolution to produce an integrated resolution based temporal theorem-prover.

1 Introduction

Although resolution has been widely used as a decision procedure in classical logics, decision procedures in temporal logic have usually been tableau or automata based, for example [Wol83, Gou84, VW86, SV89]. However, some resolution proof procedures for temporal logics have been developed, for example [CdC84, Ven86, AM90].

A naive application of the classical resolution rules to two complementary literals fails as the literals may occur in different moments in time. An obvious extension of the classical resolution rule to temporal logics is to try to resolve complementary formulae appearing within the context of the ' $\Box$ ' (*always in the future*) and ' $\Diamond$ ' (*sometimes in the future*) operators using the rule

$$\frac{A \vee \Box p \quad B \vee \Diamond \neg p}{A \vee B}$$

where A and B are formulae and p is a proposition. This rule is sound since, while ' $\Box p$ ' states that p will always be true, ' $\Diamond \neg p$ ' states that p will be false at some point in the

---

* The work of the first author was supported by SERC under a PhD Studentship.

future. However, due to the interaction (in discrete temporal logics) between '$\Box$' and '$\bigcirc$' (meaning *at the next moment in time*) operators, a formula implying $\Box p$ may be *hidden* within a set of other formulae containing '$\bigcirc$' operators. For example, the formula

$$\Box(a \Rightarrow \bigcirc a) \wedge a \wedge l \wedge \Box(a \Rightarrow \bigcirc l)$$

implies $\Box l$ although this is not immediately obvious, and so this formula should be resolvable with $\Diamond \neg l$.

To address these issues, a normal form for temporal formulae [Fis91, Fis92], together with a resolution method relying upon the structure of the formulae within the normal form [Fis91], have been developed. The resolution method uses both the classical resolution rule, in order to derive contradictions that occur solely within temporal contexts (termed *non-temporal resolution*), together with a new resolution rule, which derives contradictions *across* temporal contexts (termed *temporal resolution*).

In this paper, an implementation of the temporal resolution proof method is described. In particular, a graph-based approach to the detection of sets of formulae to which the temporal resolution rule can be applied is developed.

This paper is structured as follows. In §2, a description of the propositional temporal logic used in this paper is given, while, in §3, an outline of the temporal resolution method and an overview of the graph-based implementation of this method is presented. Details of the algorithms used are given in §4. The soundness and completeness of these algorithms are considered in §5. In §6, both related work and future work of the authors is outlined.

2 A Linear Temporal Logic

2.1 Syntax and Semantics

The logic used in this report is Propositional Temporal Logic (PTL), which is based on a linear, discrete model of time with finite past and infinite future. PTL may be viewed as a classical propositional logic augmented with both future-time and past-time temporal operators. Future-time temporal operators include '$\Diamond$' (*sometime in the future*), '$\Box$' (*always in the future*), '$\bigcirc$' (*in the next moment in time*), '$\mathcal{U}$' (*until*), '$\mathcal{W}$' (*unless* or *weak until*), each with a corresponding past-time operator. Since our temporal models assume a finite past, for convenience, two last-time operators are used namely '$\bullet$' (*weak last*) and '$\boldsymbol{\bullet}$' (*strong last*). Thus for any formula A, $\boldsymbol{\bullet}A$ is false, when interpreted at the beginning of time, while $\bullet A$ is true at that point. In particular, $\bullet$ **false** can only be satisfied when interpreted at the beginning of time.[3]

Models for PTL consist of a sequence of *states*, representing moments in time, i.e.,

$$\sigma = s_0, s_1, s_2, s_3, \ldots$$

Here, each state, s_i, contains those propositions satisfied in the ith moment in time. As formulae in PTL are interpreted at a particular moment, the satisfaction of a formula f is denoted by

$$(\sigma, i) \models f$$

---
[3] Although this is not necesary it avoids having to deal with negated past-time formulae like $\neg \bullet$ **false**.

where σ is the model and i is the state index at which the temporal statement is to be interpreted. For any well-formed formula f, model σ and state index i, then either $(\sigma, i) \models f$ or $(\sigma, i) \not\models f$. For example, a proposition symbol, 'p', is satisfied in model σ and at state index i if, and only if, p is one of the propositions in state s_i, i.e.,

$$(\sigma, i) \models p \quad \text{iff} \quad p \in s_i.$$

The full syntax and semantics of PTL will not be presented here, but can be found in [Fis91].

2.2 A Normal Form for PTL

Formulae in PTL can be transformed to a normal form, Separated Normal Form (SNF), which is the basis of the resolution method used in this paper. SNF was introduced first in [Fis91] and has been extended both in [FN92] and [Fis92]. While the translation from an arbitrary temporal formula to SNF will not be described here, we note that such a transformation preserves satisfiability and so any contradiction generated from the formula in SNF implies a contradiction in the original formula. Formulae in SNF are of the general form

$$\Box \bigwedge_i R_i$$

where each R_i is known as a *rule* and must be one of the following forms.

- $\mathbf{false} \Rightarrow \bigvee_{b=1}^{r} l_b$ (an *initial* $\Box$-rule)

- $\bigcirc \bigwedge_{a=1}^{g} k_a \Rightarrow \bigvee_{b=1}^{r} l_b$ (a *global* $\Box$-rule)

- $\mathbf{false} \Rightarrow \Diamond l$ (an *initial* $\Diamond$-rule)

- $\bigcirc \bigwedge_{a=1}^{g} k_a \Rightarrow \Diamond l$ (a *global* $\Diamond$-rule)

Here k_a, l_b, and l are literals. The outer '$\Box$' operator, that surrounds the conjunction of rules is usually omitted.

We take this opportunity to note a variant on SNF called merged-SNF (SNF$_m$) [Fis91], for combining pairs of rules, that will be used later in this paper. Given a set of rules in SNF, the relevant set of SNF$_m$ rules may be generated by repeatedly applying the following rule.

$$\frac{\bigcirc A \Rightarrow F \qquad \bigcirc B \Rightarrow G}{\bigcirc (A \wedge B) \Rightarrow F \wedge G}$$

The right hand side of this SNF rule generated may have to be further translated into Disjunctive Normal Form (DNF), if either F or G are disjunctive, to maintain the SNF rule structure. Thus, SNF$_m$ represents all possible conjunctive combinations of SNF rules.

3 Overview of the Resolution Procedure

We here present a review of the temporal resolution method and an overview of its implementation. This provides the background necessary to understand the implementation techniques presented in this paper. For further details of the temporal resolution method the reader is referred to the work presented in [Fis91, Fis92]. The algorithms implemented, together with their correctness are presented in §4 and §5, respectively.

The clausal temporal resolution method consists of repeated applications of both 'non-temporal' and 'temporal' resolution steps on sets of formulae in SNF, together with various simplification steps.

3.1 The Non-Temporal Resolution Procedure

'Non-temporal' resolution consists of the application of standard classical resolution rule to formulae representing constraints at a particular moment in time, together with simplification rules for transferring contradictions within states to constraints on previous states.

The non-temporal resolution rule is a form of classical resolution applied between $\Box$–rules, representing constraints applying to the same moment in time. Pairs of initial $\Box$–rules, or global $\Box$–rules, may be resolved using the following (non-temporal resolution) rule where $\mathcal{L}_1$ and $\mathcal{L}_2$ are both last-time formulae (with the same last-time operator).

$$\frac{\begin{array}{c}\mathcal{L}_1 \Rightarrow A \vee r \\ \mathcal{L}_2 \Rightarrow B \vee \neg r\end{array}}{(\mathcal{L}_1 \wedge \mathcal{L}_2) \Rightarrow A \vee B}$$

Once a contradiction within a state is found using non-temporal resolution, the following rule can be used to generate extra constraints.

$$\frac{\bigcirc P \Rightarrow \text{false}}{\bullet \text{true} \Rightarrow \neg P}$$

This rule states that if by satisfying P in the last moment in time a contradiction is produced then P must never be satisfied in *any* moment in time. The new constraint therefore represents $\Box \neg P$ (though it must first be translated into SNF before being added to the rule-set).

The non-temporal resolution process terminates when either no new resolvents are derived, or **false** is derived in the form of one of the following rules.

- ● true $\Rightarrow$ false
- ● false $\Rightarrow$ false
- ○ true $\Rightarrow$ false

3.2 Temporal Resolution

The temporal resolution process consists of (possibly multiple) applications of the temporal resolution rule resolving together formulae with the $\Box$ and $\Diamond$ operators. However, as described in §1, the interactions between the '$\bigcirc$' and '$\Box$' operators in PTL ensure that the definition of such a rule is non-trivial. Further, as the translation to SNF restricts the rules to be of a certain form, resolution will be between a $\Diamond$-rule and a *set* of rules that together imply an invariance which will contradict the $\Diamond$-rule. Thus, given a set of rules in SNF, then for every rule of the form $\mathcal{L}Q \Rightarrow \Diamond \neg p$ where $\mathcal{L}$ may be either of the last-time operators, temporal resolution may be applied between this $\Diamond$-rule and a set of global $\Box$-rules, which taken together force p always to be satisfied. Here if Q is satisfied then the rule is satisfiable unless the set of global $\Box$-rules force p to always be **true**. In this case, we must ensure that there is no situation where both Q and any of these rules are satisfied at the same moment. Similarly, once Q has been satisfied we must also ensure that none of the global $\Box$-rules are satisfied *unless* $\neg p$ is satisfied. To resolve with the above rule then, a set of rules must be identified so that, the set of rules, when combined, have the effect of $S \Rightarrow \Box p$. Here, the S may only be of the form $\bullet A$ where A is in DNF. So the general resolution rule, written as an inference rule, becomes

$$\frac{\bullet A \Rightarrow \Box p \\ \mathcal{L}Q \Rightarrow \Diamond \neg p}{\bullet \mathbf{true} \Rightarrow \neg A \vee \neg Q \\ \mathcal{L}Q \Rightarrow (\neg A) \, \mathcal{W} (\neg p)}$$

where the first resolvent shows that both A and Q cannot be satisfied together, and the second that once Q has occurred then $\neg p$ must occur (i.e. the eventuality must be satisfied) before A can be satisfied.

The full temporal resolution rule is given by the following

$$\frac{\begin{array}{c}\bullet A_0 \Rightarrow F_0 \\ \ldots \quad \ldots \\ \bullet A_n \Rightarrow F_n \\ \mathcal{L}Q \Rightarrow \Diamond \neg p\end{array}}{\begin{array}{c}\bullet \mathbf{true} \Rightarrow \neg Q \vee \bigwedge_{i=0}^{n} \neg A_i \\ \mathcal{L}Q \Rightarrow (\bigwedge_{i=0}^{n} \neg A_i) \, \mathcal{W} \neg p\end{array}} \quad \text{with side conditions} \quad \left\{\begin{array}{c}\text{for all } 0 \leq i \leq n \vdash F_i \Rightarrow p \\ \text{and } \vdash F_i \Rightarrow \bigvee_{j=0}^{n} A_j\end{array}\right\}$$

where the side conditions ensure that each $\Box$-rule makes p true and the right hand side of each $\Box$-rule ensures that the left hand side of one of the $\Box$-rules will be satisfied.[4] So if any of the A_i are satisfied then p will be *always* be satisfied, i.e.,

$$\bullet \bigvee_{k=0}^{n} A_k \Rightarrow \Box p.$$

---

[4] In its original presentation [Fis91] there was a third side condition but this was found to be unnecessary [Pei94].

Such a set of rules are known as a *loop* in p.

Thus, the first of the new resolvents ensures that none of the rules which form a loop in p can occur at the same time as a Q occurs, so none of the left-hand sides of the rules which ensure a loop in p, (i.e., A_i for $i = 0, 1, \ldots, n$) can be satisfied at the same time as Q. The second new resolvent ensures that once Q has been satisfied, meaning that the eventuality $\Diamond \neg p$ must be satisfied, none of the conditions for entering a loop is allowed to occur, i.e., none of the A_i (for $i = 0, 1, \ldots, n$) can be satisfied until the eventuality ($\neg p$) has been satisfied. For more details of the resolution method see [Fis91].

3.3 Implementing Temporal Resolution: An Overview

To apply the temporal resolution rule, an appropriate set of $\square$–rules must be detected for resolution with one of the $\Diamond$–rules. As such sets may not be immediately apparent from inspection of the rule-set, graph-theoretic techniques are used to identify them.

Representing Rules as Graphs. In order to apply various graph-based algorithms, we must first represent the SNF rules as a graph. We use global $\square$–rules in SNF to represent the edges in this graph. For example a rule $\bigcirc a \Rightarrow b$ could be represented as shown in Fig. 1a. Any rule containing a disjunct on its right-hand side, for example

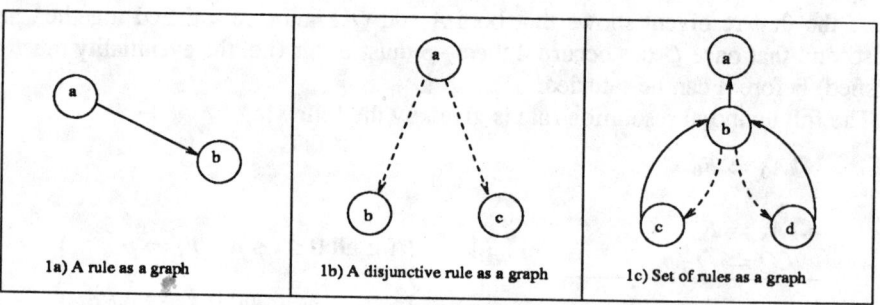

Fig. 1. Some rules represented as graphs

$\bigcirc a \Rightarrow b \vee c$, can be represented as in Fig. 1b using special 'disjunctive' arcs. In fact, for our purposes we need not distinguish between the two types of arcs as the edges leading out of a node will only relate to one rule (in SNF_m), however, we do so for clarity. The set of rules $\bigcirc b \Rightarrow c \vee d$, $\bigcirc c \Rightarrow b$, $\bigcirc d \Rightarrow b$, $\bigcirc b \Rightarrow a$ can be represented as shown in Fig. 1c.

Although there are other ways that sets of SNF rules could correspond to graph structures (e.g., see [Fis91]), the above representation is particularly suitable for our needs.

Overview of the Algorithm. The aim of our algorithm is to detect loops by constructing as little of the graph as possible. Assuming we wish to apply temporal resolution to $\mathcal{L}Q \Rightarrow \Diamond \neg p$, then we would like to construct a portion of the graph representing $S \Rightarrow \Box p$, i.e., a subgraph where p is generated at each node as the graph is built. If this subgraph forms a strongly-connected component, we have a loop in p, that represents $S \Rightarrow \Box p$, and we can apply the temporal resolution rule. In fact other subgraphs may be sufficient to ensure we have a loop, as long as each edge from a node leads back into the subgraph. So, the basic idea underlying the algorithm is to construct the graph piecemeal until either a suitable subgraph representing a loop is found, or until we have tried all possible rules and combinations of rules to construct the "next" node from each node in the graph and no loop has been found (in which case no temporal resolution was possible). Note, however, that the only edges that we use to construct the graph are ones which will give p at each node; all other possible edges are ignored.

For example if we are trying to resolve with $\mathcal{L}Q \Rightarrow \Diamond \neg p$, given a set of global $\Box$-rules R of which $\{\, \bullet b \Rightarrow c \vee d,\ \bullet c \Rightarrow b,\ \bullet d \Rightarrow b,\ \bullet b \Rightarrow a,\ \bullet a \Rightarrow p,\ \bullet b \Rightarrow p,\ \bullet c \Rightarrow p,\ \bullet d \Rightarrow p \,\}$ is a subset, the portion of the graph representing these rules is given in Fig. 1c. The subgraph formed from the nodes b, c and d represent a loop in p as once we generate one of these nodes then we will always be led back to a node in this subgraph and p is always generated.

Alternative Algorithms. An initial naive approach was to take all possible combinations of rules, represent them as an AND/OR graph and search for the strongly connected components in this graph where p held at every node [Fis91]. Our approach is an improvement on this basic algorithm in that we only construct as little of the graph structure as necessary, combining as few SNF rules as possible. Further, in contrast to the naive approach, we only use rules or combinations of rules that will ensure that p is produced at every node.

4 Implementing Temporal Resolution: Detailed Algorithms

The algorithms for loop detection, the different search methods, and selection and combination of rules into SNF_m are given in this section, followed by several examples.

4.1 Definitions of Terms

In the description of the algorithm and its usage the following terms will be used. Given a set of rules R, a graph of R, $G_R = (N, E)$ is the set of nodes, N, and the set of edges E. A *node*, n_i, is a set of literals l_k, i.e. $n_i = (l_1, l_2, \ldots, l_m)$. The set of edges E is the distributed union of *sets of disjunctive edges for rules*, i.e.

$$E = \bigcup_{i=1}^{n} D_i.$$

A *set of disjunctive edges for a rule*, D_i are directed edges in the graph formed from the rule

$$\bullet A \Rightarrow \bigvee_{j=1}^{n} B_j.$$

so that $D_i = \{(A, B_1), \ldots, (A, B_n)\}$. A *path* is a structured list of nodes

$$[n_1, n_2, \ldots, n_m, [n_{m11}, n_{m12} \ldots], [n_{m21}, n_{m22} \ldots], \ldots, [n_{mr1}, n_{mr2} \ldots]]$$

such that $(n_i, n_{i+1}) \in E$ for $i = 1, 2, \ldots, m-1$, $(n_{mxi}, n_{mxi+1}) \in E$ for $x = 1, 2 \ldots, r$, $i = 1, 2, \ldots$, and $(n_m, n_{mx1}) \in E$ for $x = 1, 2, \ldots, r$ where each sublist may be structured in the same way. A *loop* is a subgraph $G_l = (N_l, E_l)$ of G such that

1. for every edge $e_j = (n_j, n_k)$ such that $e_j \in E_l$, then $n_j \in N_l$ and $n_k \in N_l$; and
2. for every n_j in N_l there exists an edge $e_j = (n_j, n_k)$ leading out of this node and $e_j \in E_l$ and if $e_j \in D_r$ then $D_r \subseteq E_l$, where D_r is a set of disjunctive edges for a rule.

A *self-contained loop* is one where there is only one element in each set of disjunctive edges for a rule. A *partial loop* is a subgraph of the loop where $D_r \not\subseteq E_l$ for some r. Typically, we will have followed one of the disjuncts, but will have others remaining to be explored, and will have found a looping path back to one of the nodes in the partial loop.

4.2 The Loop Search Algorithm

The graph is searched for loops as it is constructed. It is constructed in a depth-first manner using the SNF rules i.e. when there is a choice of rules to use in order to expand the graph, only one is followed. If we find a situation where no rules may be applied to create a new node, then backtracking is invoked and a different choice is made at an earlier choice point.

The algorithm used (given that we want to resolve with $\mathcal{L}Q \Rightarrow \Diamond \neg p$) is given below. The forwards and backwards search algorithms mentioned in the steps 4 and 6 are described in §4.3.

1. Search for all the rules of the form $\bullet R_i \Rightarrow p$
2. Use the literal (or conjunction of literals) R_i as the initial nodes in the graph.
3. Set the current node, n_1, equal to the next initial node if one is available, and path equal to $[n_1]$, otherwise the algorithm terminates having found no new loops.
4. Perform a backwards search from n_1 until either
   (a) no loop has been detected from any of the successors to n_1 - repeat step 3;
   (b) a self-contained loop has been found - the algorithm terminates with this loop;
   (c) a partial loop has been found - remove any nodes that do not form part of the loop (the prefix to the loop) and continue processing with step 5.
5. Set n_1, the current node equal to a new disjunct if one is available, otherwise the algorithm terminates with a loop.
6. Perform a forward search from n_1 until either

(a) no loop has been detected from any of the successors to n_1, backtrack to where the disjunctive rule was used (i.e. if previously performing a backwards search to within step 4, or if previously performing a forwards search to within step 6) and continue processing;
(b) a self-contained loop has been found - the algorithm terminates with a loop;
(c) a partial loop has been found - continue processing with step 5.

For a rule-set with more than one eventuality the algorithm is carried out using each $\Diamond$–rule in turn until a new loop is detected or there are no more $\Diamond$–rules to process.

4.3 Backward and Forward Search Algorithms

If the initial nodes in our graph are R_j, for $j = 1, \ldots, m$ a new node n_{i+1} may be generated using the backwards or forwards depth-first search algorithms, where applicable, from the current node n_i (we abuse our notation slightly and are assuming that n_i represents a literal or conjunction of literals) if there exists a global $\Box$–rule

$$\mathbf{o} \bigwedge_{a=1}^{g} k_a \Rightarrow \bigvee_{b=1}^{r} l_b$$

and the following conditions are satisfied.

1. For backwards search there exists a b such that $1 \le b \le r$, $l_b \Rightarrow n_i$, and for all b, $l_b \Rightarrow \bigvee_{j=1}^{m} R_j$, and $\bigwedge_{a=1}^{g} k_a \Rightarrow \bigvee_{j=1}^{m} R_j$ then n_{i+1} will be $(k_1, k_2, \ldots, k_g)$. One of the disjuncts from the right hand side of the global $\Box$–rule implies the current node, the other disjuncts and the new node, n_{i+1} imply one of the initial nodes, thus ensuring p in the previous node and the new node will be the conjunction of literals from the left hand side of the rule.

2. For forwards search $n_i \Rightarrow \bigwedge_{a=1}^{g} k_a$ and for all b such that $1 \le b \le r$, $l_b \Rightarrow \bigvee_{j=1}^{m} R_j$ then n_{i+1} will be l_b for some b such that $1 \le b \le r$. The current node implies the literal or conjunction of literals on the left hand side of the rule, all the disjuncts on the right hand side of the global $\Box$–rule imply one of the initial nodes and the next node, n_{i+1}, will be one of the disjuncts.

When the global $\Box$–rule used for backwards or forwards search has $r > 1$ we store the disjuncts that have not been matched to the current node during backwards search, or used as the new node in forwards search, in a list for future processing.

4.4 Combinations of Rules

If no successor to the current node can be found, rules are combined using the SNF_m transformation in an attempt to generate a successor. To avoid the expense of combining all the rules, only the rules which may generate an appropriate successor are extracted from the rule-set for combination purposes. Initial rules can also be used for combination purposes. Any literals p (assuming the eventuality we are resolving with is $\Diamond \neg p$) generated through combinations on the right hand side of a combined rule, not being explicitly noted, as it is assumed that p holds at each node in the graph.

4.5 Addition of Rules

Before each cycle of temporal resolution we need to add the (valid) rule $\bigcirc \mathbf{true} \Rightarrow l \vee \neg l$ for any pair of complementary literals, l and $\neg l$, occurring on the left hand sides of any rules. This is to enable us to be able to detect loops in sets of clauses such as $\bullet \mathbf{false} \Rightarrow \Diamond \neg p$, $\bigcirc \neg l \Rightarrow p$, $\bigcirc l \Rightarrow p$.

4.6 Examples

The following examples, together with their explanations, show how the algorithm is used to detect loops from several rule-sets. It will be assumed that the original formulae have already been translated into SNF and, with each of these examples, the graph structure constructed during the search will be given.

Example 1: A Disjunctive Loop

1. $\bigcirc b \Rightarrow c \vee d$
2. $\bigcirc c \Rightarrow b$
3. $\bigcirc d \Rightarrow b$
4. $\bigcirc b \Rightarrow a$
5. $\bigcirc a \Rightarrow l$
6. $\bigcirc b \Rightarrow l$
7. $\bigcirc c \Rightarrow l$
8. $\bigcirc d \Rightarrow l$
9. $\bullet \mathbf{false} \Rightarrow \Diamond \neg l$

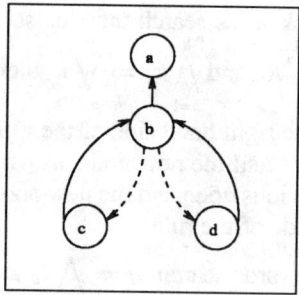

1. The $\Diamond$-rule is rule 9, and so we search for loops in l.
2. Initial nodes are (a), (b), (c), and (d), from rules 5–8. Without loss of generality the current node is set to (a) and path to $[(a)]$.
3. Backwards search matches a with the right hand side of rule 4. The new current node is b, and the path is $[(b), (a)]$.
4. Backwards search matches (b) with the right hand side of rule 2. The new current node becomes (c), and the path becomes $[(c), (b), (a)]$.
5. Backwards search matches (c) with the right hand side of the disjunctive rule 1. The new current node becomes (b), and (d) is stored waiting to be solved. The path is $[(b), [(c), (b), (a)], [(d)]]$. As (b) has already been visited we have found a partial loop, and need to begin the search from remaining disjuncts. The node (a) does not form part of this partial loop and is removed from the path. The path becomes $[(b), [(c), (b)], [(d)]]$.

6. The remaining disjunct d is solved using forward search. Forward search matches d with the left hand side of rule 3. The current node becomes (b), the sub-path for the disjunct becomes $[(d),(b)]$. As (b) has already been searched through before, we have found another partial loop, so the node (d) in the previous path is replaced by this sub-path to give a new path $[(b),[(c),(b)],[(d),(b)]]$.
7. There are no remaining disjuncts left to solve so the algorithm terminates.

If either of the other three initial nodes had been used to start the search, or we had selected rule 3 instead of rule 2 in step 4 above, the algorithm would still have found the same loop.

Example 2: A Loop Using Combinations of Rules

1.   ◯$a \Rightarrow l$
2.   ◯$b \Rightarrow l$
3.   ◯$c \Rightarrow l$
4. ◯$(a \wedge b) \Rightarrow a$
5. ◯$(b \wedge c) \Rightarrow b$
6. ◯$(a \wedge c) \Rightarrow c$
7. ● $\text{false} \Rightarrow \Diamond \neg l$

1. From rule 7, we look for a loop in l.
2. Initial nodes are (a), (b), and (c), from rules 1, 2, and 3. The current node is set to be (a) and the path is $[(a)]$.
3. Backwards search from a matches a with the right hand side of rule 4. The new current node is set to (a,b), and the path is $[(a,b),(a)]$.
4. Backwards search is carried out attempting to match the right hand side of a rule with node (a,b). As no rule matches, rules 4 and 5 are extracted as they will give us an a, or a b, on the right hand side, and are combined using SNF_m to give a new rule ◯$(a \wedge b \wedge c) \Rightarrow a \wedge b$. The right hand side of this new rule now matches the current node, so the new current node is set to (a,b,c) and the path becomes $[(a,b,c),(a,b),(a)]$.
5. Again no rule in SNF matches the current node (a,b,c). Rules 4, 5, and 6 are extracted as they will generate the required nodes on the right hand side to match with the current node. Combining these rules into SNF_m gives a new rule ◯$(a \wedge b \wedge c) \Rightarrow (a \wedge b \wedge c)$. The current node matches with the right hand side of this rule, the new current node is set to be the left hand side of the rule (a,b,c), and the path becomes $[(a,b,c),(a,b,c),(a,b),(a)]$. As we have already searched through node (a,b,c), we have found a loop. The nodes (a,b) and (a) are removed as they are not part of the loop, leaving the path as $[(a,b,c),(a,b,c)]$, and the algorithm terminates as there are no disjuncts to process.

If either of the other two initial nodes (b) and (c) had been had been used to start the search from, the algorithm would still have found the same loop.

4.7 Integration of the Temporal Resolution Step into a Theorem Prover

The implementation of the temporal resolution step has been integrated with other subprograms performing the translation of formulae into SNF and non temporal resolution to produce a complete resolution theorem prover for PTL. The resolution theorem prover has been implemented on Sun Workstations using SICSTUS PROLOG [CW91].

5 Formal Analysis of the Algorithm

In this section, we provide outline proofs exhibiting the correctness of the algorithm. Full versions of the proofs are available, but are omitted due to space restrictions.

5.1 Soundness

If the algorithm terminates detecting a loop for the literal p then the set of rules R which give rise to this loop imply the invariance of p.

Proof. Assume that a loop in the literal p has been found. Let the rules, R, in SNF_m which represent this loop be

$$\left\{ \begin{array}{c} \bigcirc A_0 \Rightarrow B_0 \\ \ldots \quad \ldots \\ \bigcirc A_n \Rightarrow B_n \end{array} \right\}$$

The matching of the left or right hand sides of rules to the current node, only stopping when we have reached a node we have visited before and when there are no remaining disjuncts means that we will always be brought back to a rule in the set. The restriction of only allowing the expansion of nodes using rules that will generate a p means that p is ensured at each node. So once one of the A_i rules is triggered, no edge can lead out of the loop and p will always be generated i.e.

$$\bigcirc \bigvee_{i=1}^{n} A_i \Rightarrow \Box p$$

So R does imply a $\Box$–formula.

5.2 Completeness

If there is a set of rules R of the form $\bigcirc A_i \Rightarrow B_i$ in SNF_m for $i = 0, \ldots, n$ such that

$$\bigcirc \bigvee_{i=1}^{n} A_i \Rightarrow \Box p$$

then, by taking suitable choices the algorithm will find the loop corresponding to this set of rules.

Proof. The proof refers to the algorithm described in §4.2, the numbers in brackets referring to the relevant part of the algorithm. Let the rules in SNF be

$$\chi = \left\{ \begin{array}{c} \bullet X_0 \Rightarrow Y_0 \\ \ldots \quad \ldots \\ \bullet X_p \Rightarrow Y_p \end{array} \right\} \quad \text{and} \quad \psi = \left\{ \begin{array}{c} \bullet Z_0 \Rightarrow W_0 \\ \ldots \quad \ldots \\ \bullet Z_q \Rightarrow W_q \end{array} \right\}$$

where the members of χ are the rules that have been combined to make the $\bullet A_i \Rightarrow B_i$ rules and the members of ψ are the remaining SNF rules.

Initially the algorithm searches for rules of the form $\bullet R_i \Rightarrow p$ (step 1 of the algorithm). Some of these are going to be in the set χ in order to generate p at every node of the loop. The search for a node always commences from one of these rules (step 2 and 3 of the algorithm) so at some point we will begin the search for the loop from a rule of the form $\bullet R_i \Rightarrow p$ which is in the set χ.

Assume we've picked the rule $\bullet A_i \Rightarrow B_i$ in SNF_m, then to generate the next node we need to choose a rule of the form $\bullet C \Rightarrow D$ in SNF_m. There are two options:

1. $\bullet C \Rightarrow D$ is made from combining at least one rule from ψ with zero or more of the rules from χ.
2. $\bullet C \Rightarrow D$ is made by combining one or more rules from χ (and none of the rules from ψ) and is equivalent to one of the $\bullet A_i \Rightarrow B_i$ rules.

By guiding the search always to take the second option, rules selected will all be equivalent to one of the $\bullet A_i \Rightarrow B_i$ rules and will be made from combining one or more of the rules from χ until a $\bullet A_i \Rightarrow B_i$ rule generated is already in the set. If we have no remaining disjuncts we are done and a loop has been detected (step 4(b) of the algorithm).

However, if we have used a disjunctive rule (step 4(c) of the algorithm), then we must search from each remaining disjunct in turn. Again by guiding the search always to take combinations of rules from χ to produce one of the $\bullet A_i \Rightarrow B_i$ rules until a the $\bullet A_i \Rightarrow B_i$ rule generated is already in the set, a partial loop will have been detected (step 6(c) of the algorithm). Further partial loops are found from the remaining disjuncts until there are no disjuncts remaining (step 5 of the algorithm). As we know there is a loop, then by combining one or more rules from χ it is always possible to detect it.

6 Conclusions and Further Work

6.1 Related Work

A variety of resolution proof systems have been developed for modal and temporal logics. Methods for modal resolution may be found in [Min90, Ohl88, EdC89] for example. However the problems considered in this report, in detecting the rules which together imply a $\square$–rule, with which to apply the resolution rule, due to the interaction between the $\square$ and $\bigcirc$ operators do not occur as generally modal logics do not use the $\bigcirc$ operator.

Resolution methods for temporal logics can be found in [Aba87, AM90, CdC84, Ven86] for example. The systems described in [CdC84, Ven86] are both for propositional linear temporal logic and, as we do, both convert the formula into a normal form before the application of any resolution rules, whereas the resolution method outlined in [Aba87, AM90] for propositional and first order temporal logic is, unlike ours, non-clausal. None of the systems, include any past-time temporal operators while the implementation outlined in [Aba87, AM90] is limited to propositional temporal logic for operators $\bigcirc$, $\square$ and $\diamond$.

6.2 Future Work

The use of resolution strategies to guide the refutation in classical logics, have been fully described in [CL73, WOLB84, Lov78]. The application of these strategies could certainly be implemented for the non-temporal resolution and their development investigated for use to direct the application of the temporal resolution rule.

So far the theorem prover has been developed for PTL only, but extensions for first order logic are being investigated. Although *full* First-Order Temporal Logic is undecidable [Aba87], it is conjectured that there are useful subsets of First-Order Temporal Logic to which this form of temporal resolution can be successfully applied.

The need to combine rules in SNF together into SNF_m, in certain cases, to enable the expansion of a node, is the clearly the costly aspect of this algorithm. Although the need to combine rules into SNF_m means the complexity of the algorithm is therefore exponential, a detailed study of the complexity still needs to be carried out.

The temporal resolution method may also be used for verification of programs or systems represented as state-transitions, as SNF is particularly useful for describing these. The development of a *temporal* logic programming language may also be possible using this method, just as the logic programming language PROLOG has been developed from resolution in classical logics. In fact, the normal form required for the resolution procedure, SNF, is derived from that developed for use in METATEM [BFG+89], which is itself an executable temporal logic.

6.3 Conclusions

The work described in this paper has shown how graph-theoretic techniques can be applied to the implementation of a temporal resolution step, which has been incorporated into a resolution theorem prover. This theorem prover has been used to determine the unsatisfiability of temporal formulae. The method is more widely applicable than the other resolution methods, some of which have been mentioned in §6.1, as it deals with past-time, future-time and a wider range of temporal operators directly and has been automated successfully.

A comparison of the resolution theorem prover and DP, [Gou84], an automated decision procedure for temporal logics based on the tableau method has been carried out. For the results of the comparison and further details of the full theorem-prover, see [Dix92].

References

[Aba87]   M. Abadi. *Temporal-Logic Theorem Proving*. PhD thesis, Department of Computer Science, Stanford University, March 1987.

[AM90]   M. Abadi and Z. Manna. Nonclausal Deduction in First-Order Temporal Logic. *ACM Journal*, 37(2):279–317, April 1990.

[BFG+89]   H. Barringer, M. Fisher, D. Gabbay, G. Gough, and R. Owens. METATEM: A Framework for Programming in Temporal Logic. In *Proceedings of REX Workshop on Stepwise Refinement of Distributed Systems: Models, Formalisms, Correctness*, Mook, Netherlands, June 1989.

[CdC84]   A. Cavali and L. Fariñas del Cerro. A Decision Method for Linear Temporal Logic. In R. E. Shostak, editor, *Proceedings of the 7th International Conference on Automated Deduction*, pages 113–127. LNCS 170, 1984.

[CL73]   C-L. Chang and R. Lee. *Symbolic Logic and Mechanical Theorem Proving*. Academic Press, 1973.

[CW91]   M. Carlsson and J. Widen. *SICStus Prolog User's Manual*. Swedish Institute of Computer Science, Kista, Sweden, September 1991.

[Dix92]   C. Dixon. A graph-based approach to resolution in temporal logic. Master's thesis, Department of Computer Science, University of Manchester, Oxford Road, Manchester, December 1992.

[EdC89]   P. Enjalbert and L. Fariñas del Cerro. Modal Resolution in Clausal Form. *Theoretical Computer Science*, 65:1–33, 1989.

[Fis91]   M. Fisher. A Resolution Method for Temporal Logic. In *Proceedings of the Twelfth International Joint Conference on Artificial Intelligence (IJCAI)*, Sydney, Australia, August 1991. Morgan Kaufman.

[Fis92]   M. Fisher. A Normal Form for First-Order Temporal Formulae. In *Proceedings of Eleventh International Conference on Automated Deduction (CADE)*, Saratoga Springs, New York, June 1992.

[FN92]   M. Fisher and P. Noël. Transformation and Synthesis in METATEM – Part I: Propositional METATEM. Technical Report UMCS-92-2-1, Department of Computer Science, University of Manchester, Oxford Road, Manchester M13 9PL, U.K., February 1992.

[Gou84]   G. D. Gough. Decision Procedures for Temporal Logic. Master's thesis, Department of Computer Science, University of Manchester, October 1984.

[Lov78]   D. Loveland. *Automated Theorem Proving: a Logical Basis*. North-Holland, Inc., 1978.

[Min90]   G. Mints. Gentzen-Type Systems and Resolution Rules, Part I: Propositional Logic. *Lecture Notes in Computer Science*, 417:198–231, 1990.

[Ohl88]   H-J. Ohlbach. A Resolution Calculus for Modal Logics. *Lecture Notes in Computer Science*, 310:500–516, May 1988.

[Pei94]   Martin Peim. Propositional Temporal Resolution Over Labelled Transition Systems. (Unpublished Technical Note), 1994.

[SV89]   S. Safra and M. Y. Vardi. On ω-Automata and Temporal Logic. In *STOC*, pages 127–137, Seattle, Washington, May 1989. ACM.

[Ven86]   G. Venkatesh. A Decision Method for Temporal Logic based on Resolution. *Lecture Notes in Computer Science*, 206:272–289, 1986.

[VW86]   M. Vardi and P. Wolper. Automata-theoretic Techniques for Modal Logics of Programs. *Journal of Computer and System Sciences*, 32(2):183–219, April 1986.

[Wol83]   P. Wolper. Temporal Logic Can Be More Expressive. *Information and Control*, 56, 1983.

[WOLB84]   L. Wos, R. Overbeek, E. Lusk, and J. Boyle. *Automated Reasoning – Introduction and Applications*. Prentice-Hall, Englewood Cliffs, New Jersey, 1984.

Annotation-Based Deduction in Temporal Logic [‡]

Hugh McGuire[*], Zohar Manna[*], and Richard Waldinger[**]

[*] Computer Science Department, Stanford University, Stanford, California 94305, U.S.A.; `mcguire@cs.stanford.edu` , `manna@cs.stanford.edu` .
[**] Artificial Intelligence Center, SRI International, Menlo Park, California 94025, U.S.A.; `waldinger@ai.sri.com` .

Abstract. This paper presents a deductive system for predicate temporal logic with induction.

Representing temporal operators by first-order expressions enables temporal deduction to use the already developed techniques of first-order deduction. But when translating from temporal logic to first-order logic is done indiscriminately, the ensuing quantifications and comparisons of time expressions encumber formulas, hindering deduction. So in the deductive system presented here, translation occurs more carefully, via *reification* rules. These rules paraphrase selected temporal formulas as nontemporal first-order formulas with *time annotations*. This time reification process suppresses quantifications (the process is analogous to quantifier skolemization) and uses addition instead of complicated combinations of comparisons. Some ordering conditions on arithmetic expressions can arise, but such are handled automatically by a special-purpose unification algorithm plus a decision procedure for Presburger arithmetic.

This deductive system is relatively complete.

Contents

1 Introduction
2 The Language: Annotated Temporal Logic
3 Deduction I: Fundamentals
4 Deduction II: Time Reification
5 Deduction III: Nonclausal Resolution
6 Deduction IV: Induction
7 Examples
8 Properties of this deductive system
9 Conclusion

[‡] This research was supported in part by the National Science Foundation under grant CCR-92-23226, by the Advanced Research Projects Agency under contract NAG2-703 and grant NAG2-892, and by the United States Air Force Office of Scientific Research under contract F49620-93-1-0139.

1 Introduction

Temporal logic has been found valuable for the specification of concurrent and reactive software and hardware systems (see, e.g., [MP91]). To use such specifications, however, it is necessary to have techniques for reasoning about temporal-logic formulas.

For example, model-checking or -exploring ('semantic tableau') systems such as [CES86, P86, KM93] are effective for temporal formulas which are essentially propositional. But so far, they have been inapplicable to general predicate temporal formulas. The deductive system of [AM90] extends nonclausal resolution for predicate logic [MW80] to handle temporal operators directly, but that system requires a cut rule which gratuitously introduces new formulas into proofs, and it imposes possibly distracting restrictions on its rules.

An alternative to reasoning directly with temporal operators is to translate temporal formulas into nontemporal first-order predicate logic, using quantifications and comparisons of explicit time parameters to express the temporal operators. Then, one can apply the accumulated technology for reasoning in first-order logic. Such an approach is used in [W89] and [O88] for modal logics other than temporal logic. But performing this translation indiscriminately may confuse otherwise simple modal-logic proofs, in which the modal operators are intuitively meaningful; proofs requiring induction are particularly susceptible to this disservice. (See also [O93] re efficiency.)

This paper presents a deductive system which employs such translation, but with some finesse. Temporal operators are translated gradually and selectively, and the resulting expressions representing time are simple: time is identified with the natural numbers, so temporal operators can be represented by addition (and subtraction). Then, reasoning about terms which represent time is done automatically by unification (as with "t" in the following simple example) plus a decision procedure for Presburger arithmetic derived from [S79].

An example of such a proof, of the formula $(\Box p \to p)$, is:

	Assertions	Goals	Explanations
1		$0 : (\Box p \to p)$	Given
2	$0 : \boxed{\Box p}^{\,-}$		1, split "$\to$"
3		$0 : \boxed{p}^{\,+}$	1, split "$\to$"
4	$t : \boxed{p}^{\,-}$		2, reify "$\Box^{-}$"
5		true	3&4, resolution, unifier $\{t \leftarrow 0\}$

This structure formalizes the goal-directed style of proofs that humans prefer to construct and find easiest to understand.

The deductive system here also provides induction for temporal operators, and rules such as rewriting, for convenience. This system subsumes the prior one of [AM90]; i.e., any valid formula which can be proven in that system can also be proven here.

2 The Language: Annotated Temporal Logic (ATL)

The basic language here is the Linear Temporal Logic of [MP91]; the complete language of ATL adds time annotations to formulas. For example, in the ATL formula $[3 : \Box p]$, the numeral "3" is the time annotation. This formula is satisfied if its basic subformula $\Box p$ is satisfied at time 3.

2.1 Syntax

The *basic* language here has function symbols, variable symbols, predicate symbols, the equality symbol "=", the six boolean operators "*true*", "*false*", "¬", "∧", "∨", and "→", the quantification operators "∀" and "∃", and temporal operators. The temporal operators (which apply to formulas) comprise "□", "◇", "○", and "$\mathcal{U}$" — named "henceforth", "eventually", "next", and "until", respectively — as well as their 'mirror' operators which refer to the past: "⊟", "⟡", "⊖", and "$\mathcal{S}$" — named "heretofore", "once", "previously", and "since", respectively. The operators "$\mathcal{U}$" and "$\mathcal{S}$" are binary; the other temporal operators are unary. An additional temporal operator is "⊙" — named "next-value" — which applies to terms (and is unary). Lastly, re basic symbols, some zero-arity function symbols and all zero-arity predicate symbols (i.e., *propositions*) are *dynamic*, i.e., 'flexible'. (The significance of dynamic symbols is semantic; see below.)

In addition to the preceding basic symbols, the complete language of ATL requires *time symbol*s: the natural numerals ("0", "1", etc.), the addition function symbol "+", the predecessor function symbol "*pred*", other uninterpreted function symbols (e.g., "h_1", "h_2"), variable symbols (e.g., "t_1", "t_2", "δ_1", "δ_2"), and comparison symbols "<" and "≤" (in addition to "="). Time symbols are not dynamic.

Terms and *incomplete* (unannotated) formulas are constructed from basic symbols and time symbols as is customary, applying function symbols, predicate symbols, and operators to appropriate arguments. A *time term* is either a time numeral, a time variable, or the application of a time function symbol to arguments which do not contain "⊙" or any dynamic symbols.

An elementary complete annotated formula of ATL has the form "$[\tau : \varphi]$", where τ is a time term — in this context called a *time annotation* — and φ is an incomplete formula. An elementary complete *time formula* is an unannotated comparison of two time terms. Further complete formulas are constructed from these elementary ones using nontemporal boolean operators and quantifiers as is customary (except that quantifiers are not allowed to bind variables in time annotations).

2.2 Semantics

A *model* is a structure with a *domain* — i.e., a nonempty set of *objects* — and a sequence of *states* indexed by the natural numbers $\mathbb{N}$. For an index $n \in \mathbb{N}$, state n of model σ is denoted by the expression "$\langle \sigma, n \rangle$". A model assigns concrete objects, functions, and relations (over the domain) to the language's nondynamic variable symbols, function symbols, and predicate symbols, respectively; a state does the same for the dynamic symbols. A model interprets the time numerals ("0" etc.) as the natural numbers $\mathbb{N}$, "+" as addition, "*pred*" as predecessor,[1] and so on (for the time symbols). For example, a time function symbol (e.g., "h") is interpreted as a function whose range is $\mathbb{N}$.

Interpretation of terms and unannotated formulas is as customary, with functions, operators, quantifiers, etc. An expression which contains no temporal operators or dynamic symbols is interpreted via the global assignments of a model; otherwise, it is interpreted by a state. Some examples for temporal operators are:

$\langle \sigma, n \rangle \models \bigcirc \varphi$ iff $\langle \sigma, n+1 \rangle \models \varphi$.

$\langle \sigma, n \rangle \models \ominus \varphi$ iff $0 < n$ and $\langle \sigma, n-1 \rangle \models \varphi$.

$\langle \sigma, n \rangle \models (\varphi \,\mathcal{U}\, \psi)$ iff   there exists a $k \geq n$ such that $\langle \sigma, k \rangle \models \psi$, and, for every i such that $n \leq i < k$, $\langle \sigma, i \rangle \models \varphi$.

$\langle \sigma, n \rangle \models (\varphi \,\mathcal{S}\, \psi)$ iff   there exists a $k \leq n$ such that $\langle \sigma, k \rangle \models \psi$, and, for every i such that $k < i \leq n$, $\langle \sigma, i \rangle \models \varphi$.

The value of the term $\odot \alpha$ in the nth state of model σ is the value of the term α in the $(n+1)$st state of σ.

A model σ interprets an annotated formula $[\tau : \varphi]$ as follows: Let $n \in \mathbb{N}$ be σ's interpretation of the time annotation τ. Then, $\sigma \models [\tau : \varphi]$ iff $\langle \sigma, n \rangle \models \varphi$. Interpretation of further complete formulas is then as customary.

A formula is *valid* if it is satisfied by all models.

A *theory* is specified by a set A of formulas which are called the *axioms* of the theory. A formula φ is *valid within the theory* specified by the set A if for every model σ that satisfies all the axioms A, $\sigma \models \varphi$.

3 Deduction I: Fundamentals

Our deductive framework, using deductive tableaux, is that of [MW93].

3.1 Notation: Deductive Tableaux

A *deductive tableau* is a table with four columns and any positive number of rows. The leftmost column simply contains row numbers; the middle two columns, titled "Assertions" and "Goals", contain complete ATL formulas (only one per row); and the rightmost column, titled "Explanations", contains texts

---
[1] A model's interpretation of "*pred*(0)" is arbitrary.

which explain how rows were derived. Each assertion or goal formula may have one (or more) of its subformulas highlighted with a box (boxes). For an example of a deductive tableau, see the Introduction.

The intuitive interpretation of a deductive tableau[2] is that the column of goals contains a formula being proved valid, followed by reductions of it to more tractable goals, and the column of assertions contains formulas which can be assumed true, to be used in the proof. A box highlights a subformula in a row when the subformula is particularly significant to a deduction rule which is applied to that row in the proof; this subformula is called a *target* of the rule.

Associated with every subformula in a deductive tableau is a characteristic called *polarity*, which is determined by the parity of the number of negations — whether explicit, or implicit via "$\rightarrow$" — within whose scope the subformula lies; one additional negation is implicit for assertions, reflecting their status of 'implying' the goals. Even parity of these negations occasions *positive* polarity, and odd parity occasions *negative* polarity. A superscript of either "+" or "−" on a formula indicates the formula's polarity.

3.2 Outline of the Deductive Scheme

An ATL proof that a basic formula φ is valid begins with the following deductive tableau:

	Assertions	Goals	Explanations
1		$0 : \varphi$	Given

Then, rows are added via rules. The proof succeeds when *true* is derived as a reduced goal formula.

Quasi-deductive primitive operations, acting within rows instead of adding new ones, include renaming variables and automatically simplifying expressions, for example changing $\neg\neg\psi$ to ψ and $0+\tau$ to τ.

If the validity being proved is relative to a theory, then any axiom of the theory can be added to the deductive tableau as an assertion. Previously proved lemmas can be added similarly.

Such axioms and lemmas for the natural numbers can be used for time.

3.3 The Rewriting Rule

This rule generates a new row from a prior one by replacing a target subformula with an equivalent (sub)formula. Examples of rewriting schemas are:

$$\neg(\varphi_1 \wedge \varphi_2) \Leftrightarrow (\neg\varphi_1 \vee \neg\varphi_2) \quad \text{and} \quad (\forall \nu_1)(\forall \nu_2)\varphi \Leftrightarrow (\forall \nu_2)(\forall \nu_1)\varphi \ .$$

Nontemporal rewritings are listed in [MW93]. Temporal ones are as follows. First:

$$\Box\varphi \Leftrightarrow \varphi \wedge \bigcirc\Box\varphi$$
$$\Diamond\varphi \Leftrightarrow \varphi \vee \bigcirc\Diamond\varphi$$
$$(\varphi_1 \, \mathcal{U} \, \varphi_2) \Leftrightarrow \varphi_2 \vee [\varphi_1 \wedge \bigcirc(\varphi_1 \, \mathcal{U} \, \varphi_2)]$$

---
[2] [MW93] gives formal semantics for deductive tableaux.

When used from left to right, these rewritings are called *expansions*.

Next, there are rewritings which specify that the temporal operator "$\bigcirc$" distributes over or commutes with nontemporal boolean operators ("$\neg$", "$\wedge$", "$\vee$", and "$\rightarrow$") and quantifiers. For example: $\bigcirc(\varphi_1 \wedge \varphi_2) \Leftrightarrow (\bigcirc\varphi_1 \wedge \bigcirc\varphi_2)$.
There are some similar rewritings involving the other temporal operators:

$$\begin{array}{ll}
\Box(\varphi_1 \wedge \varphi_2) \Leftrightarrow \Box\varphi_1 \wedge \Box\varphi_2 & (\varphi_1 \wedge \varphi_2)\,\mathcal{U}\,\varphi_3 \Leftrightarrow (\varphi_1\,\mathcal{U}\,\varphi_3) \wedge (\varphi_2\,\mathcal{U}\,\varphi_3) \\
\Diamond(\varphi_1 \vee \varphi_2) \Leftrightarrow \Diamond\varphi_1 \vee \Diamond\varphi_2 & \varphi_1\,\mathcal{U}\,(\varphi_2 \vee \varphi_3) \Leftrightarrow (\varphi_1\,\mathcal{U}\,\varphi_2) \vee (\varphi_1\,\mathcal{U}\,\varphi_3) \\
\Box(\forall\nu)\varphi \Leftrightarrow (\forall\nu)\Box\varphi & (\forall\nu)\varphi\langle\nu\rangle\,\mathcal{U}\,\psi \Leftrightarrow (\forall\nu)(\varphi\langle\nu\rangle\,\mathcal{U}\,\psi) \\
\Diamond(\exists\nu)\varphi \Leftrightarrow (\exists\nu)\Diamond\varphi & \psi\,\mathcal{U}\,(\exists\nu)\varphi\langle\nu\rangle \Leftrightarrow (\exists\nu)(\psi\,\mathcal{U}\,\varphi\langle\nu\rangle)
\end{array}$$

(In each of the quantification cases for $\mathcal{U}$, the formula ψ must not contain any free occurrence of the variable ν.)

There are similar rewritings for the operators "$\boxminus$", "$\diamondsuit$", "$\ominus$", and "$\mathcal{S}$".

3.4 The Time Distribution Rule

Time annotations distribute over or commute with nontemporal boolean operators ("$\neg$", "$\wedge$", "$\vee$", and "$\rightarrow$") and quantifiers. For example, the formula $[3 : (\Box p \rightarrow p)]$ is equivalent to the formula $([3 : \Box p] \rightarrow [3 : p])$. The following rewritings provide this *time distribution*:

$$\begin{array}{ll}
[\tau : (\neg\varphi)] \Rightarrow \neg[\tau : \varphi] & [\tau : (\varphi_1 \rightarrow \varphi_2)] \Rightarrow [\tau : \varphi_1] \rightarrow [\tau : \varphi_2] \\
[\tau : (\varphi_1 \wedge \varphi_2)] \Rightarrow [\tau : \varphi_1] \wedge [\tau : \varphi_2] & [\tau : (\forall\nu)\varphi] \Rightarrow (\forall\nu)[\tau : \varphi] \\
[\tau : (\varphi_1 \vee \varphi_2)] \Rightarrow [\tau : \varphi_1] \vee [\tau : \varphi_2] & [\tau : (\exists\nu)\varphi] \Rightarrow (\exists\nu)[\tau : \varphi]
\end{array}$$

(In each of the quantification cases, the time annotation τ must not contain ν.)

Time distribution can be (automatically) combined with the normal operations of other deduction rules, as desired. (For clear examples of such combination, see applications of the splitting rule — e.g., in the sample deductive tableau in the Introduction.)

3.5 The Duality Rule

Applied to a negated or simply *false* formula which is in one column (Assertions or Goals), this rule adds the formula's inverse to the other column.

3.6 The Splitting Rule

This rule disassembles a formula, easing further processing of its components.

Assertions	Goals
	$\varphi_1 \rightarrow \varphi_2$
φ_1	
	φ_2

Assertions	Goals
	$\varphi_1 \vee \varphi_2$
	φ_1
	φ_2

Assertions	Goals
$\varphi_1 \wedge \varphi_2$	
φ_1	
φ_2	

4 Deduction II: Time Reification

Quantifier elimination is presented here in some detail because it is crucial to time reification.

4.1 The Quantifier-Elimination Rule (Skolemization) [3]

A quantifier $(\mathcal{Q}\nu)$ has *universal force* if either its quantification operator $\mathcal{Q}$ is "$\forall$" and its polarity is positive, or if its quantification operator is "$\exists$" and its polarity is negative; otherwise, $(\mathcal{Q}\nu)$ has *existential force*.

The schema of the quantifier-elimination rule is: $[\tau : (\mathcal{Q}\nu)\varphi] \Rightarrow [\tau : \tilde{\varphi}]$,[4] where formula $\tilde{\varphi}$ is constructed from formula φ as follows:

- If the quantifier $(\mathcal{Q}\nu)$ being eliminated has universal force, then a function symbol "f_{new}" — new to the deductive tableau — is introduced. Additionally, let $\overline{\nu}$ be a list of the following variables of the entire row formula to which this rule is being applied: all the free variables, plus all the variables bound by quantifiers of existential force within whose scope the target subformula $[\tau : (\mathcal{Q}\nu)\varphi]$ lies. Then, the new subformula $\tilde{\varphi}$ is constructed from the original subformula φ by substituting the term "$f_{\text{new}}(\overline{\nu})$" for each occurrence of variable ν which was bound by the target quantifier $(\mathcal{Q}\nu)$.
- If the quantifier $(\mathcal{Q}\nu)$ being eliminated has existential force (and does not lie within the scope of any quantifier of universal force, in which case elimination of $(\mathcal{Q}\nu)$ would be disallowed), then the 'new' subformula $\tilde{\varphi}$ is simply identical to the original subformula φ; thus, $(\mathcal{Q}\nu)$ is simply removed.

(When eliminating $(\mathcal{Q}\nu)$ might change the quantifier binding of any variable via 'quantifier capturing', such a variable must be renamed.)

4.2 Introduction of the Time Reification Rule

Time reification removes a temporal operator from a formula via a process which, actually, is equivalent to the three-step operation of (1) translating the formula into nontemporal first-order predicate logic, (2) eliminating quantifiers introduced by step (1), and (3) translating back into temporal logic.

[AM90] gives a typical scheme for translating basic temporal formulas to semantically equivalent nontemporal formulas. An illustrative example is:

$$\bigcirc\bigcirc\bigcirc\Box p \quad \longmapsto \quad (\forall t)\bigl[(t \geq 3) \to p(t)\bigr] \ .$$

Reification, applied to the equivalent ATL formula, translates analogously for the "$\bigcirc$"-operators (yielding "3") but differently for the operator "$\Box$":

$$[0 : \bigcirc\bigcirc\bigcirc\Box p] \quad \stackrel{(1)}{\longmapsto} \quad (\forall \delta)\, p(3 + \delta) \ .$$

---

[3] Skolemization in other systems has somewhat reversed operations since the goal there is refutation while the goal here is validation.

[4] Or, for unannotated time formulas, simply: $(\mathcal{Q}\nu)\varphi \Rightarrow \tilde{\varphi}$.

(The choice of "δ" versus "t" indicates a time increment instead of an absolute time.) Assuming negative polarity, reification would continue as follows:

$$(\forall \delta)\, p(3+\delta) \overset{(2)}{\longmapsto} p(3+\delta) \overset{(3)}{\longmapsto} [3+\delta\,:\,p]\,.$$

Another example of ATL translation is:

$$[0\,:\,\bigcirc\bigcirc\bigcirc(p\,\mathcal{U}\,q)] \overset{(1)}{\longmapsto} (\exists \delta_1)\Big(q(3+\delta_1)\,\wedge\,(\forall\delta_2)\big[(\delta_2<\delta_1)\to p(3+\delta_2)\big]\Big)\,.$$

Reification of $\bigcirc$

As an illustrative example, translating the formula $[t\,:\,\bigcirc p]$ of ATL into nontemporal predicate logic yields $p(t+1)$. Translating back yields $[t+1\,:\,p]$. Generalizing (facilely), the schema for reification of the operator "$\bigcirc$" is:

$$[\tau\,:\,\bigcirc\varphi]\,\Rightarrow\,[\tau+1\,:\,\varphi]\,.$$

Reification of $\square$ and $\diamond$

An example for the operator "$\square$" is given above; the operator "$\diamond$" is treated analogously, with "$\exists$" instead of "$\forall$". The reification schemas are:

$$\big[\tau\,:\,(\square\varphi)^+\big]\,\Rightarrow\,\big[\tau+h_{\text{new}}(\overline{\nu})\,:\,\varphi\big] \qquad \big[\tau\,:\,(\square\varphi)^-\big]\,\Rightarrow\,\big[\tau+\delta_{\text{new}}\,:\,\varphi\big]$$

$$\big[\tau\,:\,(\diamond\varphi)^-\big]\,\Rightarrow\,\big[\tau+h_{\text{new}}(\overline{\nu})\,:\,\varphi\big] \qquad \big[\tau\,:\,(\diamond\varphi)^+\big]\,\Rightarrow\,\big[\tau+\delta_{\text{new}}\,:\,\varphi\big]$$

Since reification of these operators implicitly involves quantifiers, the details of the quantifier-elimination rule apply: In the two cases on the left, for "$\square^+$" and "$\diamond^-$", the meta-symbol "$\overline{\nu}$" denotes a list of variables as in skolemization; and in the other two cases, for "$\square^-$" and "$\diamond^+$", this rule is disallowed for any formula $[\tau\,:\,(\square\varphi)^-]$ or $[\tau\,:\,(\diamond\varphi)^+]$ that occurs within the scope of any quantifier of universal force. An example for "$\square^+$" is:

Asstns	Goals	Explanations
	$\big[t\,:\,\neg(d=f(x))\big]\,\wedge\,(\exists y)\Big[c\,:\,\boxed{\square(d=y)}^+\Big]$	
	$\big[t\,:\,\neg(d=f(x))\big]\,\wedge\,(\exists y)\big[c+h(t,x,y)\,:\,d=y\big]$	reify $\square^+$; "h" is new

Reification of $\odot$

An example of a formula containing the operator "$\odot$" is $[t\,:\,\odot d=a]$. The symbol d here should be dynamic; otherwise, the term $\odot d$ would be equivalent to d, so simplification would delete the operator "$\odot$" in the formula. Assume that the symbol a is not dynamic. Then translating this formula into nontemporal logic yields $d(t+1)=a$. Translating back yields $[t+1\,:\,d=a]$.

But if the symbol a were dynamic, that result would not be equivalent to the original formula. To see this, note that with a dynamic, translation of the formula $[t : \odot d = a]$ into nontemporal logic would yield $d(t+1) = a(t)$, while translation of $[t+1 : d = a]$ would yield $d(t+1) = a(t+1)$.

Considering these cases, the reification schema for the operator "$\odot$" is:
$$[\tau : \varphi\langle \odot\alpha_1, \ldots, \odot\alpha_n\rangle] \quad \Rightarrow \quad [\tau+1 : \varphi\langle \alpha_1, \ldots, \alpha_n\rangle] \;,$$
where (1) the target subformula $\varphi\langle\ldots\rangle$ must not contain any dynamic symbol outside of the specified terms $\odot\alpha_i$, and (2) none of the specified terms $\odot\alpha_i$ may occur within the scope of any other temporal operator.

Reification of $\ominus$, $\boxminus$, and $\diamondsuit\!\!\!\!-$

$$[\tau : \ominus\varphi] \quad \Rightarrow \quad (0 < \tau) \wedge [pred(\tau) : \varphi] \;.$$

Both $\left[\tau : (\boxminus\varphi)^+\right]$ and $\left[\tau : (\diamondsuit\!\!\!\!-\varphi)^-\right] \quad \Rightarrow \quad \left(h_{\text{new}}(\overline{\nu}) \leq \tau\right) \wedge \left[h_{\text{new}}(\overline{\nu}) : \varphi\right]\;.$

Both $\left[\tau : (\boxminus\varphi)^-\right]$ and $\left[\tau : (\diamondsuit\!\!\!\!-\varphi)^+\right] \quad \Rightarrow \quad (t_{\text{new}} \leq \tau) \wedge [t_{\text{new}} : \varphi]\;.$

The details for $\boxminus$ and $\diamondsuit\!\!\!\!-$ are the same as for the operators $\square$ and $\diamondsuit$.

Reification of $\mathcal{U}$ and $\mathcal{S}$ (Here, the subscript "$_n$" specifies newness.)

$\left[\tau : (\varphi\,\mathcal{U}\,\psi)^-\right] \Rightarrow \left[\tau + h_n(\overline{\nu}) : \psi\right] \wedge \left((\delta_n < h_n(\overline{\nu})) \to [\tau + \delta_n : \varphi]\right)$

$\left[\tau : (\varphi\,\mathcal{U}\,\psi)^+\right] \Rightarrow$

$\qquad [\tau + \delta_n : \psi] \wedge \left((h_n(\overline{\nu}, \delta_n) < \delta_n) \to [\tau + h_n(\overline{\nu}, \delta_n) : \varphi]\right)$

$\left[\tau : (\varphi\,\mathcal{S}\,\psi)^-\right] \Rightarrow$

$\qquad (h_n(\overline{\nu}) \leq \tau) \wedge [h_n(\overline{\nu}) : \psi] \wedge \left([(h_n(\overline{\nu}) < t_n) \wedge (t_n \leq \tau)] \to [t_n : \varphi]\right)$

$\left[\tau : (\varphi\,\mathcal{S}\,\psi)^+\right] \Rightarrow$

$\qquad (t_n \leq \tau) \wedge (t_n : \psi) \wedge \left([(t_n < h_n(\overline{\nu},t_n)) \wedge (h_n(\overline{\nu},t_n) \leq \tau)] \to [h_n(\overline{\nu},t_n) : \varphi]\right)$

5 Deduction III: Nonclausal Resolution

The schemas for nonclausal resolution are:[5]

Assertions	Goals
$\varphi\langle[\tau : \rho^+]\rangle$	
$\psi\langle[\tilde{\tau} : \tilde{\rho}^-]\rangle$	
$(\varphi\langle\text{true}\rangle \vee \psi\langle\text{false}\rangle) \circ \theta$	

Asstns	Goals
	$\varphi\langle[\tau : \rho^+]\rangle$
	$\psi\langle[\tilde{\tau} : \tilde{\rho}^-]\rangle$
	$(\varphi\langle\text{true}\rangle \wedge \psi\langle\text{false}\rangle) \circ \theta$

---

[5] As with the quantifier-elimination rule, target subformulas for the resolution rule can be unannotated time formulas.

Assertions	Goals
$\varphi\langle[\tau : \rho^+]\rangle$	
	$\psi\langle[\tilde{\tau} : \tilde{\rho}^-]\rangle$
	$\begin{pmatrix} \neg\varphi\langle true\rangle \\ \wedge \\ \psi\langle false\rangle \end{pmatrix} \circ \theta$

Assertions	Goals
	$\varphi\langle[\tau : \rho^+]\rangle$
$\psi\langle[\tilde{\tau} : \tilde{\rho}^-]\rangle$	
	$\begin{pmatrix} \varphi\langle true\rangle \\ \wedge \\ \neg\psi\langle false\rangle \end{pmatrix} \circ \theta$

In each case: (1) the formulas φ and ψ must not share any free variables; (2) neither target subformula $[\tau : \rho]$ or $[\tilde{\tau} : \tilde{\rho}]$ may contain any occurrence of any variable that is bound by a quantifier outside of the target; and (3) these target subformulas must unify via a most general unifier θ.

Additional restrictions apply if the unifier θ involves a replacement $\nu \leftarrow \alpha$ with term α containing a dynamic symbol. For then:

1. No free occurrence of variable ν in the formulas φ and ψ may lie within the scope of any temporal operator or inside any time annotation or time formula.

2. All the time annotations within whose scope(s) replaceable occurrences of ν lie must be unified (by θ).

These two restrictions ensure consistency of the time at which a dynamic symbol may be interpreted. Otherwise, such a symbol — contained in term α, placed in various contexts by the replacement $\nu \leftarrow \alpha$ — could refer to different objects, violating soundness.

When resolution fails only because the target time annotations τ and $\tilde{\tau}$ fail to unify, an option is that the equation $\tau = \tilde{\tau}$ can be conjoined with the desired result of resolution as something additional that needs to be proven. A Presburger-like decision procedure can handle time formulas such as this equation. For an example of this situation, see Section 7.2.

The Equality Rule

This rule involves resolution-based reasoning. Hence, like the resolution rule itself, this rule comprises four schemas because of the alternatives for columns. Only one such schema need be shown:

Assertions	Goals
	$\varphi\langle[\tau : \rho\langle\alpha\rangle]\rangle$
$\psi\langle[\tilde{\tau} : (\beta = \tilde{\alpha})^-]\rangle$	
	$\Big(\varphi\langle[\tau : \rho\langle\beta\rangle]\rangle \;\wedge\; \neg\psi\langle false\rangle\Big) \circ \theta$

Here, θ must unify term α with term $\tilde{\alpha}$ as well as time annotation τ with time annotation $\tilde{\tau}$. Further details are as for resolution.

6 Deduction IV: Induction

The logic may refer to a domain, such as the natural numbers or lists, for which principles of induction hold. Then, rules which implement these principles are available. An example, for the natural numbers, is:

Assertions	Goals
	$\tau : (\forall \nu)\varphi\langle\nu\rangle$
	$\tau : \big(\varphi\langle 0\rangle \wedge [\varphi\langle m\rangle \to \varphi\langle m+1\rangle]\big)$

where the initial row must not contain free variables and m must be new. Thus, a goal can be reduced to a base case plus an inductive step.

Induction also applies to temporal operators. For example:

Assertions	Goals
	$\tau : \Box\varphi$
	$[\tau : \varphi] \wedge \big([\tau + c_{\text{new}} : \varphi] \to [\tau + c_{\text{new}} + 1 : \varphi]\big)$

where the initial row must not contain free variables and c_{new} must be new. There are additional temporal induction schemas, e.g., for assertions of the form $[\tau : \Diamond\varphi]$ (see the extended version of this paper).

7 Examples

7.1 Validity of the Formula $\big([p \wedge \Box(p \to \bigcirc\Diamond p)] \to \Box\Diamond p\big)$

This example, which requires temporal induction, is used in [AM90].

	Assertions	Goals	Explans
1		$0 : \begin{pmatrix} [p \wedge \Box(p \to \bigcirc\Diamond p)] \\ \to \\ \Box\Diamond p \end{pmatrix}$	Given
2	$0 : \boxed{p}^{-}$		1, split
3	$0 : \boxed{\Box(p \to \bigcirc\Diamond p)}^{-}$		1, split
4		$0 : \boxed{\Box\Diamond p}$	1, split
5		$\big[0 : \boxed{\Diamond p}^{+}\big] \wedge \big([c_1 : \Diamond p] \to [c_1 + 1 : \Diamond p]\big)$	4, temporal induction (of $\Box$)

6		$[t_2 : \boxed{p}^+]$ $\wedge\ (\ldots \rightarrow \ldots)$	5, reify $\Diamond^+$
7		$[c_1 : \Diamond p]$ $\rightarrow$ $[c_1 + 1 : \Diamond p]$	6&2, resolution, unifier $\{t_2 \leftarrow 0\}$
8	$c_1 : \boxed{\Diamond p}$		7, split
9		$c_1 + 1 : \boxed{\Diamond p}^+$	7, split
10	$t_3 : (p \rightarrow \bigcirc\Diamond p)$		3, reify $\Box^-$
11	$[t_3 : p] \rightarrow [t_3 : \boxed{\bigcirc\Diamond p}]$		10, distribution
12	$[t_3 : p] \rightarrow [t_3 + 1 : \boxed{\Diamond p}^-]$		11, reify $\bigcirc$
13		$c_1 : \boxed{p}^+$	12&9, resolution, unifier $\{t_3 \leftarrow c_1\}$
14	$c_1 : (p \vee \bigcirc\Diamond p)$		8, expand $\Diamond$
15	$[c_1 : \boxed{p}^-] \vee [c_1 : \bigcirc\Diamond p]$		14, distribution
16	$c_1 : \boxed{\bigcirc\Diamond p}$		15&13, resolution; duality
17	$c_1 + 1 : \boxed{\Diamond p}^-$		16, reify $\bigcirc$
18		true	17&9, resolution

Highlights of this proof are:

- First, the splitting rule conveniently demarcates the given formula's premises and desired conclusion (goal #4). Time distribution is used implicitly.
- One rationalization of the decision to apply temporal induction to goal #4 is that when one attempts the simpler operation, namely reification (of the operator "$\Box$"), one gets stuck later in the proof.
- Goal #5, being the result of the induction rule, comprises two conjuncts: the base case and the inductive step. In this proof, the base case (the left conjunct) is proven first.
- Goal #7 is split for convenience.
- Assertion #8 is the inductive hypothesis, and goal #9 is the desired inductive conclusion. Experience with induction proofs leads one to attempt to reduce the latter to the former. Assertion #3 is used for this reduction.
- When goal #13 arises, assertion #8 can be used.
- The last two deduction-steps handle a case due to the disjunctive nature (consider assertion #14) of the inductive hypothesis.

7.2 Validity of the Formula $\left(\Box\Diamond((\neg p)\,\mathcal{S}\,q) \rightarrow \Box(p \rightarrow \Diamond q)\right)$

	Assertions	Goals	Explanations
1		$0: \begin{pmatrix} \Box^-\Diamond^-\left((\neg p)\,\mathcal{S}\,q\right) \\ \rightarrow \\ \Box^+(p \rightarrow \Diamond q) \end{pmatrix}$	Given
2	$t_1 + h_2(t_1) : \boxed{(\neg p)\,\mathcal{S}\,q}^{\,-}$		1, split; reify $\Box^-$ & $\Diamond^-$
3		$c_3 : p \rightarrow \Diamond^+ q$	1, split; reify $\Box^+$
4	$c_3 : \boxed{p}^{\,-}$		3, split
5		$c_3 + \delta_4 : q^+$	3, split; reify $\Diamond^+$
6	$h_5(t_1) \leq t_1 + h_2(t_1)$		2, reify $\mathcal{S}^-$; split
7	$h_5(t_1) : \boxed{q}^{\,-}$		2, reify $\mathcal{S}^-$; split
8	$\begin{array}{l}(h_5(t_1) < t_6) \wedge (t_6 \leq t_1 + h_2(t_1)) \\ \rightarrow \\ \left[t_6 : \neg \boxed{p}^{\,+}\right]\end{array}$		2, reify $\mathcal{S}^-$; split
9		$\begin{array}{c} h_5(t_1) < c_3 \\ \wedge \\ \boxed{c_3 \leq t_1 + h_2(t_1)}^{\,+} \end{array}$	8&4, resolution, $\{t_6 \leftarrow c_3\}$; and duality

At this point in the proof, one could resolve goal #5, containing q^+, with assertion #7, containing q^-; the system would automatically generate the goal $(h_5(t_1) = c_3 + \delta_4)$ and pass this new formula plus the deductive tableau's other time formulas — assertion #6 and goal #9 — to a Presburger-like decision procedure, altogether in the form '⟨assertions⟩ → ⟨goals⟩':

$$(\forall t_1)\left(h_5(t_1) \leq t_1 + h_2(t_1)\right) \rightarrow \begin{pmatrix} (\exists t_1)\left((h_5(t_1) < c_3) \wedge (c_3 \leq t_1 + h_2(t_1))\right) \\ \vee \\ (\exists t_1, \delta_4)\left(h_5(t_1) = c_3 + \delta_4\right) \end{pmatrix}$$

This formula is indeed valid (for the natural numbers), so these operations would satisfactorily finish the proof.

Alternatively, one may continue the proof 'manually' as follows:

	Assertions	Goals	Explanations
10	$\boxed{x \leq x + y}^{\,-}$		axiom for natural numbers
11		$\boxed{h_5(c_3) < c_3}^{\,+}$	10&9, resolution, $\{t_1 \leftarrow c_3,\ x \leftarrow c_3,\ y \leftarrow h_2(c_3)\}$

12	$\boxed{x < y}^{-} \lor (\exists z)(x = y + z)$		property of naturals
13	$\boxed{(\exists z)\bigl(h_5(c_3) = c_3 + z\bigr)}^{-}$		12&11, resolution, $\{x \leftarrow h_5(c_3),\ y \leftarrow c_3\}$; duality
14	$\boxed{h_5(c_3) = c_3 + c_7}^{-}$		13, eliminate $\exists^{-}$
15		$h_5(c_3) : \boxed{q}^{+}$	14&5, equality, $\{\delta_4 \leftarrow c_7\}$
16		true	15&7, resolution, $\{t_1 \leftarrow c_3\}$

7.3 Verification of List-processing

An assumption of list-processing code is that for any list l and at any time, if some dynamic symbol d is set equal to the list, and then (as time passes) d gets repeatedly truncated (while doing so is possible), then eventually d will equal the empty list NIL. Such a statement may be formalized as follows:

$$(\forall l)\square\Bigl(\bigl[(d = l) \land \square\bigl(\neg(d = NIL) \to (\odot d = tail(d))\bigr)\bigr] \to \Diamond(d = NIL)\Bigr).$$

This formula can be proven in ATL, using induction for lists, in thirty-three steps.

8 Properties of the ATL Deductive System

(Proofs of these properties are given in the extended version of this paper.)

8.1 Soundness

When all ATL formulas and operations are translated into nontemporal predicate logic, the soundness established in [MW93] applies; i.e., any formula which can be proven in ATL is indeed valid.

8.2 Relative Completeness

The ATL system can perform all the operations of the deductive system of [AM90]. Hence, as [AM90] identified a class of 'arithmetical' formulas (of predicate temporal logic) for which their deductive system is complete, the ATL system is also complete for that class.

8.3 Decision Procedure for Propositional Temporal Logic

The ATL deductive system can be used for a decision procedure for formulas of propositional temporal logic by mimicking more standard 'semantic' decision-algorithms such as the graph-based one of [KM93]. (This decision procedure using ATL also resembles the construction in [AM90]'s proof of completeness of their deductive system for propositional temporal formulas.)

9 Conclusion

The deductive system described here allows established techniques for reasoning in predicate logic with induction to be extended to temporal logic. This system has several features which enable it to avoid complications of removal of temporal operators, and it invokes special unification and decision procedures to reduce the burden on the user. It enables proofs of temporal formulas to be constructed with more formality, ease, and clarity than was previously possible.

References

[AM90] Abadi, M., and Manna, Z.: "Nonclausal Deduction in First-Order Temporal Logic," in *Journal of the Association for Computing Machinery (JACM)*, Volume 37 (1990), Number 2 (April), pp. 279–317.

[CES86] Clarke, E., Emerson, E., and Sistla, A.: "Automatic Verification of Finite-State Concurrent Systems Using Temporal Logic Specifications," in *ACM Transactions on Programming Languages and Systems*, Volume 8 (1986), Number 2 (April), pp. 244–263.

[KM93] Kesten, Y., Manna, Z., McGuire, H., and Pnueli, A.: "A Decision Algorithm for Full Propositional Temporal Logic," in Courcoubetis, C. (editor): *Computer Aided Verification (5th International Conference, CAV '93)* (LNCS #697), pp. 97–109. Springer-Verlag, Berlin, 1993.

[MP91] Manna, Z., and Pnueli, A.: *The Temporal Logic of Reactive and Concurrent Systems: Specification.* Springer-Verlag, New York, 1991.

[MW80] Manna, Z., and Waldinger, R.: "A Deductive Approach to Program Synthesis," in *ACM Transactions on Programming Languages and Systems*, Volume 2 (1980), pp. 90–121.

[MW93] Manna, Z., and Waldinger, R.: *The Deductive Foundations of Computer Programming.* Addison-Wesley, Reading, Massachusetts, 1993.

[O88]   Ohlbach, H.: "A Resolution Calculus for Modal Logics," in Lusk, E., and Overbeek, R. (editors): *9th International Conference on Automated Deduction (Proceedings)* (LNCS #310), pp. 500–516. Springer-Verlag, Berlin, 1988.

[O93]   Ohlbach, H.: "Translation Methods for Non-Classical Logics — An Overview," in *Automated Deduction in Nonstandard Logics* (Technical Report #FS-93-01), pp. 113–125. AAAI Press, Menlo Park, California, 1993.

[P86]   Plaisted, D.: "A Decision Procedure for Combinations of Propositional Temporal Logic and Other Specialized Theories", in *Journal of Automated Reasoning*, Volume 2 (1986), pp. 171–190.

[S79]   Shostak, R.: "A practical decision procedure for arithmetic with function symbols," in *JACM*, Volume 26 (1979), Number 2 (April), pp. 351–360.

[W89]   Wallen, L.: *Automated Proof Search in Nonclassical Logics.* The MIT Press, Cambridge, Massachusetts, 1989.

An Overview of Temporal and Modal Logic Programming

Mehmet A. Orgun[1] and Wanli Ma[2]

[1] Department of Computing, Macquarie University, Sydney, NSW 2109, Australia
[2] Computer Sciences Laboratory, RSISE, The Australian National University, Canberra, ACT 0200, Australia

Abstract. This paper presents an overview of the development of the field of temporal and modal logic programming. We review temporal and modal logic programming languages under three headings: (1) languages based on interval logic, (2) languages based on temporal logic, and (3) languages based on (multi)modal logics. The overview includes most of the major results developed, and points out some of the similarities, and the differences, between languages and systems based on diverse temporal and modal logics. The paper concludes with a brief summary and discussion.

1 Introduction

In logic programming, a program is a set of Horn clauses representing our knowledge and assumptions about some problem. The semantics of logic programs as developed by van Emden and Kowalski [96] is based on the notion of the least (minimum) Herbrand model and its fixed-point characterization. As logic programming has been applied to a growing number of problem domains, some of its limitations have also started to surface, especially in those problem domains requiring the notion of dynamic change. On the one hand, we would like to be able to broaden applications of logic programming to, say, temporal reasoning, deductive databases, knowledge representation, and dataflow computation. On the other hand, we would like to remain true to the original goals of logic programming outlined by Kowalski [61]: Logic programs should specify only *what* is to be computed; not how it is to be computed.

In order to overcome some of the limitations of logic programming, many non-logical constructs in the form of annotations, extra system predicates, and infinitary objects have been introduced; for instance, see [91] for a comprehensive survey on concurrent logic languages. These languages are an important contribution to the field of programming languages and their implementations, but the problem with them is that extended "logic" programs are no longer logic. The declarative meaning of an extended program cannot be reasoned about from its logical reading; we must know the underlying execution mechanism of a particular extension to understand what a program really *does* [50, 53, 100]. One argument for these extensions is that expressiveness and efficiency of logic programming systems have priority over declarative features of logic programs such as the minimum model semantics and completeness. For instance, logic programming with infinite terms does not have a minimum model semantics and therefore the intended meaning of an infinitary logic program is not characterized by the set of logical consequences of the program [97, 64]. And this is a major problem: what an infinitary program does is not what it says declaratively. But then how can we attack problems from different domains and still keep the declarative features of logic programming if the expressiveness of our tools is so limited?

In our opinion, the solution lies not in extending logic programming with non-logical tools, but in employing more powerful logical tools. In particular, if we want to model the notion of change in time in a given application (such as that in temporal reasoning) and the dynamic properties of certain problems such as simulation, why not use temporal logic in the first place? Temporal logics have already been used in program specification and verification [65, 63], temporal reasoning [86, 32], and temporal databases [34, 95, 47]. Or if we want to model knowledge and belief, why not use modal and/or epistemic logics? These logics have been extensively studied in philosophy and mathematics and applied in many problem domains (including those in artificial intelligence) with success [8, 57, 92, 93]. Therefore we advocate extending logic programming with temporal and modal logics (or with other non-classical logics) whenever appropriate.

Recently, several researchers have proposed extending logic programming with temporal logic, modal logic and other forms of intensional logic. There are a number of modal and temporal logic programming languages: Tempura [70, 52] and Tokio [9] are based on interval logic; THLP [99, 100], Chronolog [83, 74], Templog [3, 4] and Temporal Prolog [44] are based on temporal logic; there is also another Temporal Prolog [54] based on reified temporal logic; Brzoska [26] proposed temporal logic programming based on metric temporal logic; Molog [39] is based on user-elected modal logics; Modal Prolog [87] is based on modal logic; InTense [68] is a multi-dimensional language with temporal and spatial dimensions. There are also some multi-modal approaches to temporal and modal logic programming [7, 36]. Some other non-classical extensions of logic programming include multiple-valued logic programming schemes of Blair et al [24] and of Fitting [43]. However, these two approaches deal with non-classical semantics for logic programming.

The declarative and operational semantics of some of these languages have already been worked out, and some languages have already been implemented. For instance, Baudinet [16, 17, 18] shows the completeness of the proof procedure of Templog and provides the declarative semantics of Templog programs. Orgun and Wadge [74, 75, 76] develop the model-theoretic semantics of Chronolog and describe a general framework to deal with several intensional programming systems. It is shown in [78] that Chronolog admits a sound and complete resolution-type proof procedure. Balbiani et al [13] provide a declarative and operational semantics for a class of Molog programs. Gabbay [44] showed the soundness of a computation procedure for Temporal Prolog, based on branching time. Fitting [43] employs topological bilattices to treat the semantics of multiple-valued logic programming. For other languages such as InTense [67, 68] and Tokio [9, 60], some kind of extended operational semantics are usually provided. Progress on implementing modal extensions of Prolog has been reported in [15].

In short, there are a variety of logic programming languages based on diverse temporal and modal logics. The aim of this paper is to present a timely overview of the development of the field of temporal and modal logic programming. Some related languages and systems are not covered, for example, non-classical extensions based on intuitionistic logic, linear logic, multiple-valued logic, paraconsistent logic, fuzzy logic, etc, because our focus is on languages based on extensions of the classical logic such as temporal and modal logics, not on "alternatives" to the classical logic. In the sequel, we start by providing background on temporal and modal logics. Readers who are familiar with temporal and modal logics can skip this section. We then review temporal and modal logic programming languages under three headings: (1) languages based on interval logic, (2) languages based on temporal logic, and (3) languages based on (multi)modal logics. This classification is naturally provided by the logics these languages are based upon. The paper concludes with a brief summary and discussion.

2 Temporal and Modal Logics

Modal logic [56, 29] is the study of context-dependent properties such as necessity and possibility. In modal logic, the meaning of expressions depends on an implicit context, abstracted away from the object language. Temporal logic [82, 28] can be regarded as an instance of modal logic where the collection of contexts models a collection of moments in time. Therefore the following discussion on modal logic also applies to temporal logic. A modal logic is equipped with modal operators through which elements from different contexts can be combined. The underlying language of modal logic is obtained from a first-order language by extending it with formation rules for modal operators.

The collection of contexts is also called the *universe* or the set of *possible worlds*, and denoted by $\mathcal{U}$. The set of possible worlds $\mathcal{U}$ is not a disorganized collection. For instance, in temporal logic, $\mathcal{U}$ can be regarded as a linearly ordered set. In Kripke-style semantics, there is one ordering relation over $\mathcal{U}$, called an *accessibility* relation, associated with each modal operator [51, 82]. If $\triangledown$ is a unary modal operator, the ordering relation R associated with $\triangledown$ is a set of pairs of possible worlds from $\mathcal{U}$; in other words, $R \subseteq \mathcal{U} \times \mathcal{U}$. For any $< w, v > \in R$, it means that v is *accessible* from w.

Let ML denote the underlying modal language of a modal logic. A (modal) interpretation of ML basically assigns meanings to all elements of ML at all possible worlds in $\mathcal{U}$. A modal interpretation can also be viewed as a collection of first-order interpretations (Tarskian structures) one for each possible world in $\mathcal{U}$. Here the denotations of variables and function symbols are *extensional* (a.k.a. *rigid*), that is, independent of the elements of $\mathcal{U}$. This is not generally so in modal logic; but it is quite satisfactory for application to logic programming:

in logic programming, we deal with Herbrand interpretations over the Herbrand universe which consists of uninterpreted terms.

Let $P(X)$ denote the set of all subsets of the set X and $[X \to Y]$ the set of functions from X to Y. Then the formal definition of a modal interpretation can be given as follows.

Definition 1. A modal interpretation I of a modal language ML comprises a non-empty set $\mathbf{D}$, called the domain of the interpretation, over which the variables range, and for each variable, an element of $\mathbf{D}$; for each n-ary function symbol, an element of $[\mathbf{D}^n \to \mathbf{D}]$; and for each n-ary predicate symbol, an element of $[\mathcal{U} \to P(\mathbf{D}^n)]$.

All formulas of ML are *intensional*, that is their meanings may vary depending on the elements of $\mathcal{U}$. The fact that a formula A is true at world w in some modal interpretation I will be denoted as $\models_{I,w} A$. The definition of the satisfaction relation $\models$ in terms of modal interpretations is given as follows (bar the semantics of modal operators). Let $I(E)$ denote the value in $\mathbf{D}$ that I gives an ML term E.

Definition 2. The semantics of elements of ML are given inductively by the following, where I is a modal interpretation of ML, $w \in \mathcal{U}$, and A and B are formulas of ML.

(a) If v is a variable, then $I(v) \in \mathbf{D}$. If $f(e_0, \ldots, e_{n-1})$ is a term, then
$I(f(e_0, \ldots, e_{n-1})) = I(f)(I(e_0), \ldots, I(e_{n-1})) \in \mathbf{D}$.
(b) For any n-ary predicate p and terms $e_0, \ldots, e_{n-1}$, $\models_{I,w} p(e_0, \ldots, e_{n-1})$
iff $< I(e_0), \ldots, I(e_{n-1}) > \in I(p)(w)$.
(c) $\models_{I,w} \neg A$ iff $\not\models_{I,w} A$.
(d) $\models_{I,w} A \wedge B$ iff $\models_{I,w} A$ and $\models_{I,w} B$.
(e) $\models_{I,w} (\forall x)A$ iff $\models_{I[d/x],w} A$ for all $d \in \mathbf{D}$ where the interpretation $I[d/x]$ is just like I except that the variable x is assigned the value d in it.

Furthermore, $\models_I A$ means that A is true in I at all worlds, that is, I is a *model* of A, and $\models A$ means that A is true in any interpretation of ML. We regard the above definition as a framework for modal logics, which is enjoyed by most of the temporal and modal languages discussed in the sequel.

The definition is, however, incomplete. We must define the semantics of modal operators available in the language. For instance, consider two classical modal operators: $\Box$ (*necessary*) and $\Diamond$ (*possible*). In Kripke-style semantics for modal logic, the meanings of $\Box$ and $\Diamond$ are determined by an accessibility relation R. One R is enough, because $\Diamond$ can be defined using $\Box$ and $\neg$; and $\Box$ using $\Diamond$ and $\neg$, depending on which of the two modal operators is chosen as a primitive operator.

Informally, $\Box A$ is true at a world w if and only if A is true at all worlds accessible from w; and $\Diamond A$ is true at w if and only if A is true at some world accessible from w. More formally, given a modal interpretation I and $w \in \mathcal{U}$,

$\models_{I,w} \Box A$ iff $\models_{I,v} A$ for all v where $< w, v > \in R$, and

$\models_{I,w} \Diamond A$ iff $\models_{I,v} A$ for some v where $<w,v> \in R$.

Note that $\Diamond A$ and $\neg \Box \neg A$ are logically equivalent; in other words, $\Diamond$ is the dual of $\Box$. If R is an equivalence relation, this gives a Kripke-style semantics for the modal logic S5 [56].

The traditional Kripke approach is, however, too restrictive, because it limits us to a dual pair of modal operators. We could extend it with extra modalities in the obvious way, by allowing a family of dual pairs, each with its own accessibility relation. This is better but still not truly general because, as Scott [90] and others have pointed out, there are many natural modal operators that cannot be defined in terms of an accessibility relation alone.

There are more general approaches to the semantics of modal logic, including "neighborhood" semantics which is usually attributed to Scott [90] and Montague [69]. For a detailed exposition of more general approaches and their relative strengths, we refer the reader to the literature, for example, see [27] and [103]. Neighborhood semantics is used in [77] to provide a unifying theoretical framework for intensional (temporal and modal) logic programming languages. In most of the other reported works, the standard Kripke-style semantics is usually employed.

As mentioned above, temporal logic [82, 28] is regarded as the *modal* approach to time. In temporal logic, there are usually two sets of modalities, one referring to the past, P (*sometime in the past*) and H (*always in the past*); and the other referring to the future, F (*sometime in the future*) and G (*always in the future*). The semantics of formulas with temporal operators (for example, PA) are defined as above using accessibility relations. The accessibility relations associated with these temporal or tense operators exhibit the properties of a time-line: (usually) discrete, linear, branching, unbounded (in either direction), dense, and so on.

3 Interval Logic Programming

Interval logic is a form of temporal logic in which the semantics of formulas are defined using temporal interpretations and *intervals*, that is, pairs of moments in time which represent a duration of time. There are two temporal languages based on interval logic, namely, Tempura [70, 52] and Tokio [9]. The execution of a program in Tempura is a reduction or transformation process. In other words, Tempura is a temporal logic programming language in a broad sense; it is not based on the "logic programming" paradigm (resolution and unification). The execution of a program in Tokio is also a reduction process, but one which is combined with resolution and unification. The very nature of the execution mechanisms of these two languages and the properties of their underlying logic called ITL [70] set them apart from the other temporal languages. Therefore we discuss them under a separate heading of interval logic programming.

3.1 Tempura

Originally proposed by B. C. Moszkowski [70], Tempura is a programming language based on discrete-time Interval Temporal Logic (ITL). The fundamental temporal operators in ITL are □ (*always*) and ◇ (*sometimes*). To partition a time interval, the *chop* or *sequential* operator, "; (*semicolon*)", is introduced. If we write $w_1; w_2$, it means that there is an interval σ which can be divided into two consecutive subintervals, say σ_1 and σ_2, and $w_1; w_2$ is satisfied on σ if and only if w_1 is satisfied on σ_1 and w_2 on σ_2. In contrast to the *sequential* operator, Tempura uses *conjunction*, ∧, for concurrency. The conjunction is written as **and** in Tempura programs. For instance, $w_1 \wedge w_2$ means that both w_1 and w_2 are satisfied concurrently in the interval σ. As the intervals in Tempura are discrete, the next time operator ○ is also used to change the time to the next subinterval (the current interval without the first state).

The roots of Tempura are in the functional programming, imperative programming, and logic programming paradigms. As a descendant of imperative programming languages, it has a "destructive" assignment statement. Unlike the assignment statement in conventional imperative programming languages, Tempura's assignment only affects those variables that are explicitly mentioned; the values of other variables, which are not mentioned, will never be concerned. This is a special aspect of Tempura, which distinguishes it from the other temporal logic programming languages.

The programs of Tempura are a conjunction of executable ITL formulas, which are composed by the Tempura's operators. Disjunction and negation are not allowed for the sake of efficiency. To simplify writing Tempura programs, new and expressive operators are derived from the fundamental operators. Some of them are [52, page 97]:

$$
\begin{array}{lll}
\textbf{empty} \equiv_{def} \neg \bigcirc true & & \text{(the zero-length interval)} \\
\textbf{skip} \equiv_{def} \bigcirc \textbf{empty} & & \text{(the unit-length interval)} \\
\textbf{keep } w \equiv_{def} \Box (\neg \textbf{empty} \rightarrow w) & & \text{(on all but the last state)} \\
\text{A \textbf{ gets } B} \equiv_{def} \textbf{keep } ((\bigcirc A) = B) & & \text{(unit delay)} \\
\textbf{halt } C \equiv_{def} \Box (C \equiv \textbf{empty}) & & \text{(interval termination)} \\
\textbf{fin } w \equiv_{def} \Box(\textbf{empty} \rightarrow w) & & \text{(on the last state)} \\
A \leftarrow B \equiv_{def} \exists x:(x=B \wedge \textbf{fin } A=x) & & \text{(assignment)}
\end{array}
$$

Let us give some examples to show the meanings of the operators defined above. Suppose A, B, and C are temporal variables, $< seq >$ represents an interval with states seq. The values of A, B, and C over the states s, t, and u are given in Figure 1. Note that the value of a temporal variable is a sequence varying along the time axis, and that an interval with a single state is of length 0. Then we have:

$< u > \models \textbf{empty}$
$< st > \models \textbf{skip}$
$< ststsu > \models \textbf{fin}(B = 0)$
$< ststsu > \models B \leftarrow A$

st. var.	s	t	s	t	s	u
A	0	0	0	0	0	1
B	1	0	1	0	1	0
C	1	1	1	1	1	1

Fig. 1. The values of three variables over three states

Hale [52] gives a number of applications of Tempura, including motion representation, data transmission, and hardware design. The following program is an example of a Tempura program, which solves the *"Towers of Hanoi"* problem [52, page 100]:

```
/* Tempura solution to the "Towers of Hanoi" problem */
define hanoi(n) = exists L, C, R : {
  L = [0..n-1] and
  C = [ ] and
  R = [ ] and
  move_r(n, L, C, R) and
  always display(L, C, R)
}.
/* Move n rings from peg A to peg B */
define move_r(n, A, B, C) = {
  if n=0 then empty
  else {
    move_r(n-1, A, C, B);
    move_step(A, B, C);
    move_r(n-1, C, B, A)
  }
}.
/* Move the topmost ring from peg A to peg B */
define move_step(A, B, C) = {
  skip and
  A <- tail(A) and
  B <- append([head(A)], B) and
  C <- C
}.
```

The execution of a Tempura program is a transformation (or reduction) process. The program is repeatedly reduced according to the operations on time intervals until no interval can be divided, that is, the current subinterval contains a single state. For example, consider the following program which calculates the value of 2^m:

(M=4) ∧ (N=1) ∧ halt(M=0) ∧ (M gets M-1) ∧ (N gets 2*N).

It will be reduced to:

$$(M=4 \wedge N=1)$$
$$\wedge \bigcirc \ (M=3 \wedge N=2)$$
$$\wedge \bigcirc \bigcirc \ (M=2 \wedge N=4)$$
$$\wedge \bigcirc \bigcirc \bigcirc \ (M=1 \wedge N=8)$$
$$\wedge \bigcirc \bigcirc \bigcirc \bigcirc \ (M=0 \wedge N=16 \wedge \text{empty}).$$

The final formula is regarded as a state-description that satisfies the original formula. Therefore, the value sequence of M is 4, 3, 2, 1, 0 and that of N is 1, 2, 4, 8, 16, over an interval of length 5.

Moszkowski [70] described an interpreter for Tempura with the details of the way in which the temporal constructs are implemented. Some applications of Tempura including algorithm description and hardware specification are also given.

3.2 Tokio

Tokio was proposed by Aoyagi, Fujita, and Moto-oka [9] for the description of computer hardware. It is based on the first-order local Interval Temporal Logic (ITL), influenced by Tempura. It is also a superset of Prolog [35]. Tokio is based on discrete linear-time.

The temporal operators in Tokio are:

- *concurrency*(,): The clause $P :- Q, R$ means that Q and R are executed at the beginning of a time interval concurrently.
- *chop*(&&): This operator divides a time interval into two subintervals. The clause $P :- Q \ \&\& \ R$ means that Q will be executed at the first subinterval and R will be executed at the second subinterval.
- *next*(@): The clause $P :- @Q$ means that Q will be executed at the time interval after the current time interval.
- *always*(#): The clause $P :- \#Q$ means that Q will be executed at all subintervals which make up the P's interval.
- *sometime*(<>): The clause $P :- <> Q$ means that the execution of goal Q will be at some time in the interval in which P is executed.
- *keep*: The clause $P :- \text{keep}(Q)$ means that Q will be executed at every subinterval of P's except the final one.
- *final*(fin): The clause $P :- \text{fin}(Q)$, in contrast to $\text{keep}(Q)$, means that Q will only be executed at the final subinterval of P's.

As in Tempura [70], variables in a Tokio program may have different values at different time intervals. In other words, the value of a variable varies with time. This makes the unification in Tokio more complicated. There are two kinds of unification in Tokio: One is concerned with unifying two Tokio variables, that is, unifying the entire sequences of values for the two variables. The second one is concerned with unifying the values of Tokio variables at specific moments in time through the use of special unification primitives. For example, over a given interval, X <- Y means that the value of Y at the first state of the interval is unified with the value of X at the last state of the interval.

Intervals can be manipulated using certain builtin operators, such as `length`, `empty`, and `notEmpty`. For instance, the `length` operator is used to determine the length of an interval.

The execution of a Tokio program is a mixture of resolution and transformation (or reduction). There were two Tokio interpreters, one written in Prolog, and the other in C [60].

To the best of our knowledge, there have been no attempts at developing either the declarative or the operational semantics of Tokio programs. In order to give a formal semantics to Tokio, one would need to combine the semantics of ITL with a semantics of Prolog that explicitly represents the execution mechanism (e.g., that of Baudinet [20]).

Below are some examples of Tokio sentences and the results of their execution:

1. `length(2), @write(0) && length(2), #write(1)`.

   Time   0 1 2 3 4
   Result – 0 1 1 1

2. `length(5), <>write(1)`.

   Time   0 1 2 3 4 5
   Result – 1 1 1 1 –

3. `length(2), keep(write(0)) && length(3), #write(1)`.

   Time   0 1 2 3 4 5
   Result 0 0 1 1 1 –

For an application of Tokio to hardware specification, we refer the reader to Masahiro et al [66].

4 Temporal Logic Programming

This section discusses logic programming languages and systems, based on linear- and branching-time temporal logics. The languages such as Templog [3, 4], Chronolog [74, 83], Gabbay's Temporal Prolog [44], and Sakuragawa's Temporal Prolog [88] directly extend logic programming with temporal operators. One common feature of these languages is that they all use temporal versions of resolution-based proof procedures. Some other languages such as Hrycej's Temporal Prolog [54] and Starlog [33] model time-dependent properties using reified temporal logics and additional time-parameters. MTL [26] uses a method which translates into the CLP-scheme. Starlog uses a connection-graph theorem prover.

4.1 Templog

Templog was originally proposed by Abadi and Manna [3, 4]. It uses a discrete linear-time axis with an unbounded future based on the set of natural numbers. As a programming language, Templog imposes some restrictions on its syntax for efficiency reasons. Templog adopts the syntax of Prolog, except that it has three temporal operators: $\bigcirc$ (*the next moment in time*), $\square$ (*from now on*), and $\diamond$ (*sometime in the future*). With these temporal operators, some new concepts appear:

- *next-atom*: an atom, which has the same syntax as in Prolog, with a prefix of next operators, for example, $\bigcirc^k A$ where A is an atom and $\bigcirc^k$ is a k-folded application of the next-time operator (for $k \geq 0$). A formula composed by next-atoms is called a *next-formula*.
- *initial clause*: $\forall x_1, \ldots, \forall x_n (A_1 \wedge \ldots \wedge A_n \to B)$ or $B \leftarrow A_1, \ldots, A_n$ where $A_1, \ldots, A_n$, and B are *next-formulas*.
- *permanent clause*: $\forall x_1, \ldots, \forall x_n \square (A_1 \wedge \ldots \wedge A_n \to B)$ or $B \Leftarrow A_1, \ldots, A_n$ where $A_1, \ldots, A_n$, and B are *next-formulas*.

A Templog program is a conjunction of initial and permanent clauses. A goal is a conjunction of next-formulas.

The following Templog program simulates the states of a computer's cpu, which is adapted from a Chronolog program given in Orgun and Wadge [76]. Suppose that the definition of the predicate `job_queue` is defined elsewhere.

```
cpu(idle,0) ←
○cpu(idle,0) ⇐ cpu(S,0), job_queue([ ])
○cpu(X,N) ⇐ cpu(S,0), job_queue([[X,N]|R])
○cpu(S,N) ⇐ cpu(S,s(N))
```

The binary predicate `cpu/2` represents the state of a cpu. Its first parameter shows the cpu state (either `idle` or the current job's name) and the second parameter indicates how many more units of time the current job will run. The cpu will be available at the next moment if either its status is `idle` or the execution of the current job has ended.

To enhance its specification ability, Templog also permits the $\diamond$ operator to appear in the body of a clause, and $\square$ to appear in the head of an initial clause. For instance, the following are Templog's program clauses:

```
□ employee(X) ← employee(X)
reachable(X,Y) ⇐ ◇(at(X), ◇ at(Y))
```

The first clause says that an employee who is currently employed will always be employed in the future. The second clause says that position Y is reachable from position X if there eventually exists a case that we are at position X and eventually get to position Y.

The operational semantics of Templog is given in terms of a resolution-type proof procedure, called TSLD-resolution, which is based on a restricted form of a non-clausal temporal deduction system [2]. It has a number of rules for dealing with temporal operators. A Templog goal $\leftarrow G$ will be satisfied by an infinite sequence of related goals: $\leftarrow G, \leftarrow \bigcirc G, \leftarrow \bigcirc^2 G, \leftarrow \bigcirc^3 G$, and so on. Consider the following program which generates the sequence of Fibonacci numbers:

```
fib(0) ←
○fib(1) ←
○² fib(X) ⇐ fib(Y), ○fib(Z), X is Y+Z
```

and the goal $\leftarrow$ fib(X). At the first instant of time, fib(X) unifies with the first clause of the program, and yields the answer substitution $\{X \leftarrow 0\}$. And then, the next goal, $\bigcirc$fib(X) unifies with the the second clause of the program, and so the answer substitution is $\{X \leftarrow 1\}$. After these, a goal of the form $\bigcirc^k$fib(X) unifies with the third clause, and produce new goals of $\bigcirc^{k-2}$fib(Y) and $\bigcirc^{k-1}$fib(Z). Hence we obtain a sequence of answer substitutions for X from the original goal:

$$\{X \leftarrow 0\}, \{X \leftarrow 1\}, \{X \leftarrow 1\}, \{X \leftarrow 2\}, \{X \leftarrow 3\}, \ldots$$

representing the infinite sequence of Fibonacci numbers: $< 0, 1, 1, 2, 3, 5, \ldots >$.

Baudinet [16, 17, 18] showed that TSLD-resolution is a sound and complete proof procedure for Templog, and also developed the declarative semantics of Templog programs as a temporal extension of van Emden-Kowalski semantics [96]. It is shown that Templog enjoys the minimum model semantics based on temporal Herbrand models, and its fixed-point characterization. It is also shown that the expressiveness of Templog queries, in the propositional case, corresponds to a fragment of μTL of Vardi [98] allowing only least fixed-points to be applied to positive formulas. Note that Templog is in fact equivalent in expressive power to a fragment of itself, called TL1, in which the only temporal operator is the next-time operator $\bigcirc$ [18, section 5].

Brzoska [25] showed that Templog can be considered as an instance of the CLP scheme of Jaffar and Lassez [58] over a suitable algebra $\mathcal{A}$. Templog programs are translated into classical logic programs and Templog goals into classical goals through a meaning preserving transformation, Π. Translated programs contain additional function and predicate symbols and a temporal context is added to all the predicate symbols in a Templog program. The algebra $\mathcal{A}$ consists of the free term algebra of a Templog program plus the algebra of natural numbers with functions +1 (successor) and + (addition). It is shown by Brzoska that for any given Templog program P and goal G, $P \models G$ if and only if $\Pi(G) \models_\Pi \Pi(P)$. In Π-structures, the meaning of the additional symbols are fixed: they are interpreted over the algebra $(\mathcal{N}, 0, +1, +, =_\mathcal{N})$ where $\mathcal{N}$ is the set of natural numbers. TSLD-derivations are simulated using $(\Pi, \mathcal{A})$ derivations by introducing extra equational constraints on temporal contexts. Then the constraints are solved using the constraint-solving mechanism of the CLP-scheme. An alternative approach to the declarative and fixed-point semantics for Templog programs, as opposed to the direct approach of Baudinet [16, 17], is also given via translation. The meaning-preserving translation of Templog programs into CLP-programs yields a new proof procedure for the derivation of Templog goals.

These results suggest that a similar translation approach is possible for other languages such as Chronolog [74] and Temporal Prolog [44]. It would be interesting to investigate the sufficient conditions under which a temporal or modal logic programming language can be considered to be an instance of the CLP-scheme.

Chomicki and Imieliński [31] proposed an extension of Datalog which has the same expressive power as (the function-free subset of) TL1. Their language (called $Datalog_{1S}$) is not directly based on temporal logic: all the predicates are extended with an extra parameter for time. The time parameter can be constructed using a specific unary function symbol denoting the *successor* function, and can be viewed as interpreted over the natural numbers. The effects of temporal operators of TL1 are simulated by manipulating the extra time parameters in predicates. It is mentioned in [19] that $Datalog_{1S}$ can be considered as a syntactical variant of (function-free) Templog, because Templog programs can be translated into TL1 programs. It is also mentioned in [19] that there is a translation from $Datalog_{1S}$ into Templog. These results suggest that a variant of SLD-resolution based on a two-sorted logic can also be used as a proof procedure for (translated) Templog programs.

4.2 Chronolog

Wadge [99, 100] proposed a tensed extension of Horn logic programming (called THLP), which was later developed into Chronolog [74, 76, 83]. The design of Chronolog was influenced by the dataflow language Lucid [101]. Hence its original target application was modelling non-terminating dataflow computations. Chronolog has two temporal operators (borrowed from Lucid): **first** refers to the initial moment in time, and **next** to the next moment in time. Like Templog, Chronolog uses the set of natural numbers as the collection of moments in time.

Chronolog adopts a C-Prolog-like syntax [35]. It accepts all the syntax of a first-order language with two new extensions: if A is a formula, so are **first** A and **next** A. The temporal operators can only apply to formulas, not to terms of the language.

Any formula, say A, in Chronolog has a sequence of values along time axis. The formula **first** A means the initial or first value of A, and the **next** A means the value of A at the next instant of time. In other words, **next** has the same semantics as the next-time operator of Templog. The accessibility relations associated with both **first** and **next** are *functional*, and hence these two operators are self-dual. In other words, we have that **first** $A \leftrightarrow \neg \textbf{first} \neg A$ (and for **next**).

The following Chronolog program defines the predicate **fib** which at each time t is true of the $t + 1^{th}$ Fibonacci number:

```
first fib(0).
first next fib(1).
next next fib(N) <- next fib(X), fib(Y), N is X+Y.
```

The first clause says that at the first moment in time (time 0) the number 0, the first Fibonacci number, makes `fib(0)` true; the second clause gives the second Fibonacci number, 1. The third clause gives the general frame: the current Fibonacci number (at times greater than 1) is obtained as the sum of the previous two.

Thus, from the Fibonacci program, we can prove: `first fib(0)`, `first next fib(1)`, `first next`^{2} `fib(1)`, `first next`^{3} `fib(2)` and so on. As in Templog, a goal like `<- fib(X)` triggers an attempt to prove `fib(X)` at all moments in time. It will lead to a non-terminating computation.

Now we give another Chronolog program which has appeared in Templog; the simulation of a computer's cpu states [76].

```
first cpu(idle,0).
next cpu(idle,0) <- cpu(S,0), job_queue([]).
next cpu(X,N) <- cpu(S,0), job_queue([[X,N]|R]).
next cpu(S,N) <- cpu(S,s(N)).
```

The meaning of the program is the same as that in Templog. In Chronolog, all program clauses are read as assertions true at all moments in time; hence there is an implicit always operator □ applied to all program clauses. Using the terminology from Templog, all program clauses in Chronolog are permanent. Although there are no initial clauses a la Templog in Chronolog, program clauses in which all atoms have `first` applied to them can be regarded as initial clauses, for example, the first program clause above is an initial clause.

The declarative semantics of Chronolog programs are developed using temporal Herbrand interpretations [74, 76] as an extension of van Emden-Kowalski semantics for ordinary logic programs [96, 64]. It is shown that every Chronolog program has a unique minimum temporal Herbrand model, which is regarded as the canonical meaning of the program. A fixed-point characterization of the minimum model semantics is also given.

The implementation of Chronolog is based on its operational semantics, TiSLD-resolution (a Timely SLD-resolution), which is a temporal extension of the classical SLD-resolution. It is applied to a set of *canonical instances* of program clauses and goals in which all atoms are in the scope of `first` [78]. For example, consider a program specifying the simulation of a traffic light modeled by the time-varying `light` predicate:

```
first light(green).
next light(amber) <- light(green).
next light(red) <- light(amber).
next light(green) <- light(red).
```

It says that traffic light goes green, amber, red, green, amber, red, green, and so on. A TiSLD-refutation for a goal `<- first next light(Color)` is:

$G_0 =$ `<- first next light(Color)`,
$C_0 =$ `first next light(amber) <- first light(green)`,
$\theta_0 =$ {Color ← amber}.

$G_1 = $ <- first light(green)θ_0,
$C_1 = $ first light(green),
$\theta_1 = \{\}$.

C_0 is a canonical instance of the second program clause, and C_1 is the same as the first clause in the program. Thus we obtain that first next light(Color) is true of the program under the substitution $\theta_0 = \{\text{Color} \leftarrow \text{amber}\}$.

Chronolog($\mathcal{Z}$) is an extension of Chronolog with an unbounded past [78, 80], in which the set of integers $\mathcal{Z}$ is the collection of moments in time. The only extra operator in Chronolog($\mathcal{Z}$) is prev (*the previous moment in time*), which is the complete inverse of next. It is shown in [78] that TiSLD-resolution extended with rules for prev is a sound and complete proof procedure for Chronolog($\mathcal{Z}$), and hence for Chronolog.

When compared to Templog, Chronolog seems to lack expressive power because the operators $\square$ and $\diamond$ are not allowed in Chronolog. Recall that Templog has the same expressive power as one of its subsets, called TL1, in which the only allowed temporal operator is the next time operator $\bigcirc$ [18]. There is a simple transformation from TL1 into Chronolog: apply first to all initial clauses in Templog to preserve their meaning, and replace $\bigcirc$ by next and $\Leftarrow$ by <- in all clauses. The resulting set of clauses is a Chronolog program. It follows that Chronolog also has the same expressive power as Templog; but, a direct transformation from Chronolog to TL1 is not possible. Some clauses in Chronolog such as p <- first q cannot be directly transformed into TL1, because TL1 lacks first. These results are, however, more of a theoretical nature without any practical implications.

An early implementation of Chronolog, called μChronolog, can be found in Mitchell [67]; it is based on meta-interpretation in Prolog. Rolston [85] recently proposed an *eductive* implementation technique for ordinary logic programming (Prolog), which can also be used in implementing temporal languages such as Chronolog in a dataflow environment. Eduction [10, 42] is a standard demand-driven computation model for implementing intensional languages such as Lucid.

Another implementation based on a hybrid dataflow model is currently in progress [104]. The model is called CHEM (CHronolog Execution Model). A warehouse mechanism (such as a cache or a context-associative memory) is used to prevent repeated proving of the same goals so that the efficiency of the implementation can be improved. CHEM is a temporal extension of DIALOG [105] which is a parallel execution model for Prolog based on dataflow computation.

A non-deterministic extension of Chronolog is also proposed [76] and its semantics studied in detail in [79]. This extension is suitable for modeling certain resource-sharing problems such as that of the Dining Philosophers problem and non-deterministic dataflow computations. Rolston employed temporal logic programming to mitigate the frame problem in artificial intelligence [84]. Orgun [72] suggested that the function-free subset of Chronolog (called Temporal DATALOG) extended with temporal modules can form the basis of a temporal deductive database system.

4.3 Temporal Prolog (Gabbay)

Temporal Prolog proposed by Gabbay [44] is an extension of logic programming to allow modal and temporal connectives such as P (*sometime in the past*), F (*sometime in the future*), and $\Box$ (*always*). This language is very expressive as it also allows nested implications (and even negation as failure rule). But it also introduces a restriction: unlike in Templog, $\Box$ is not allowed in the heads of program clauses in Temporal Prolog. Thus the two languages are seemingly tailored for different kinds of applications.

The syntax of Temporal Prolog programs is defined as follows [44, definition 2.1]:

1. A *program* is a set of clauses.
2. A *clause* is either an ordinary clause or an always clause.
3. An *always clause* is $\Box A$ where A is an ordinary clause.
4. An *ordinary clause* is a head or an $A \to H$ where A is a body and H is a head.
5. A *head* is either an atomic formula or FA or PA where A is a conjunction of ordinary clauses.
6. A *body* is either an atomic formula, a conjunction of bodies, an FA or PA where A is a body.

The following are examples of program clauses:

$\Box(\Diamond(A(x) \wedge FB(y)) \to R(z))$,
$a \to F((b \to Pq) \wedge F(a \to Fb))$,

but $a \to \Box b$ is not.

Gabbay [44] outlined a computation procedure for Temporal Prolog. The soundness of the computation procedure is also shown, but it is not mentioned whether the procedure is complete or not. The computation rules are given for branching time temporal logic, and it is suggested that the procedure can be extended to consider linear-time temporal logic. The language is also extended with negation as failure, and the computation procedure is modified to handle negation in the propositional case. It is not clear whether a proof strategy on top of the computation procedure can be defined so that we can have a Prolog-like inference mechanism for Temporal Prolog.

As can be seen from the above definition, Temporal Prolog allows P and F to appear in the heads of program clauses. For instance, consider the following program:

$A(a)$
$A(x) \to FB(y)$

and the goal $\leftarrow F(B(a) \wedge B(b))$. From the second program clause, we can deduce either $B(a)$ or $B(b)$ but not both at the same time, therefore the goal fails. This is because the variable y is quantified over the temporal operator F, so the future point at which $B(y)$ is true depends on y, and may be different for different

y's. The computation procedure for Temporal Prolog requires that constants be substituted for all variables as soon as a computation rule is first applied, so that y would have a fixed value in the second clause. We do not have to know what value y is going to have when the rule is first applied, provided that y is replaced by a dummy constant. The actual value might be determined at later stages of a computation. For instance, given the goal $\leftarrow FB(z) \wedge A(z)$, we first use the second clause to deduce $FB(z)$ and replace the variables y in the second clause and z in the goal by a dummy constant, say $\bar{y}$, and then we can replace $\bar{y}$ by a using the first clause.

Temporal Prolog does not enjoy a minimum model semantics because of the use of operators such as F and P in the head of a program clause. Orgun and Wadge [77] showed that F and P are not *conjunctive*, and that the model intersection property does not hold for (intensional) logic programs when an operator which is not conjunctive is allowed to appear in the head of a program clause. The model intersection property is essential for the minimum model of a logic program to exist [77, 64]. For instance, Templog does have a minimum model semantics, because the use of $\Diamond$ is restricted to the body of a clause. Chronolog also has a minimum model semantics, because both `first` and `next` are conjunctive.

We conjecture that the meaning of the above Temporal Prolog program can be characterized with respect to the *minimal model* semantics. In any minimal model of the program, either $B(a)$ or $B(b)$ would be true at any given moment in time, but not both so that the minimality condition is satisfied. Hence either result would be justified with respect to a particular minimal model of the program. Of course, the computation procedure always tries to find the right one, that is, a minimal model in which the goal is true of the given program.

Gabbay [45] also proposed labeled deduction systems as the basis of a temporal logic programming machine. The language considered is basically Temporal Prolog enriched with extra temporal operators such as G (*always in the future*), H (*it has always been the case*), $\bigcirc$ (*the next moment in time*), and $\otimes$ (*the previous moment in time*). Also, time indicators are used to represent temporal data. For instance, a statement like "If $A(x)$ is true at t, then it will continue to be true" is represented as

$$t : A(x) \rightarrow GA(x).$$

Some restrictions are imposed on temporal program clauses to ensure tractability. For instance, Skolem functions are added to eliminate the existential connectives such as F and P. It is shown that labeled deduction systems can deal with three main flows of time: (1) general partial orders (including branching time), (2) linear orders, and (3) the integers or the natural numbers. Each flow of time requires the use of a different set of temporal rules. The soundness of the rules are also established [45].

Chen and Lin [30] studied the complexity of the satisfiability problem of temporal Horn clauses in the propositional case. They consider the future fragment of Temporal Prolog (in which the past temporal operators such as P are not allowed) and extend it with the next-time operator ○ of Templog. In particular they established two complexity results:

- The satisfiability problem for the future fragment of Temporal Prolog with only □ and ◇ is NP-complete.
- The satisfiability problem for Temporal Prolog with □, ◇ and ○ is PSPACE-complete.

They also claim that the second result also holds for Templog programs.

4.4 Temporal Prolog (Hrycej)

Hrycej [54] proposed an extension of Prolog (also called Temporal Prolog) capable of handling temporally referenced logical statements and temporal constraints. Temporal Prolog is not directly based on temporal logic: it uses Allen's temporal constraint model [8] for reasoning about time intervals and their relationships, but with some modifications for efficiency reasons [55, section 3].

As Temporal Prolog is built on top of Prolog, it introduces two additional clause types into Prolog:

1. *Temporal references* are used to assert that a certain statement holds exactly during a time interval, for example, P *in* T where P is a fact or a rule and T is an interval identifier (such as **morning**, **afternoon**, etc.) Temporal references are retrieved using the predicates P *dur* T or P *mkdur* T.
2. *Temporal constraints* axiomatize the relationships between time intervals. They are asserted by the predicate *constrain_rel*(T,S,R) where T and S are intervals and R is a relationship such as "<" (before), ">" (after), "m" (meets), "mi" (met by), "o" (overlaps), "di" (contains) and so on. Interval relations are retrieved by the predicate *get_rel*(T,S,R).

The following Temporal Prolog program is from Hrycej [54]:

```
is_to_speak(X) :- at_home(X).
is_to_speak(X) :- on_visit_at(X,Y), has_telephone(Y).

(is_to_speak(X) :- at_work(X)) in working_time.

at_home(tom) in morning.
on_visit_at(tom,john) in evening.
at_work(tom) in afternoon.

has_telephone(john).
```

The program contains both ordinary Prolog clauses and temporal references. A number of temporal constraints can be added to the program to express interval relationships, for example, **constrain_rel(afternoon,14,[di])** (*afternoon* contains time 14).

The axioms of the modified formalism are also implemented as Prolog rules using a number of additional predicates, one for each logical connective [55, section 5.2]. Temporal Prolog also requires the implementation of a temporal constraint solver (a slightly modified version of Allen's constraint solver [8]). The link between the constraint solver and Temporal Prolog is established through the *constraint-rel* predicate. The temporal constraint solver can be implemented either in Prolog, or in a procedural language with an interface to Prolog. The performance of a C-based implementation of the constraint solver is reported in Hrycej [55]: it outperforms a Prolog implementation by a thousand times.

For an application of Temporal Prolog to database queries and planning, we refer the reader to [54]. Hrycej [55] claimed that one of the potential application areas of Temporal Prolog is qualitative physics.

4.5 Temporal Prolog (Sakuragawa)

In [3, 4], there is a mention of another Temporal Prolog proposed by Sakuragawa. Although we do not have a copy of Sakuragawa's original paper [88], we include a brief review of this language based on a discussion in Abadi and Manna [3, 4] so that some common features of Temporal Prolog and Chronolog($\mathcal{Z}$) can be identified.

In Temporal Prolog, the meaning of predicates in the future depends on their meaning in the past; in other words, all the program clauses are *causal* as in Starlog [33] (see below). Just as in Chronolog, all program clauses in Temporal Prolog are permanent. Past temporal operators such as "$\bullet$" (*previous moment in time*) may occur in the body of a clause while future operators may occur in the head of a clause. Temporal Prolog is quite expressive. For instance, consider the following Temporal Prolog program:

$\Box$ *alarm* $\Leftarrow$ *dangerous(X)*.

It says that "If something is dangerous, alarm will always stay on." This program cannot be directly written in Templog, Gabbay's Temporal Prolog, or in Chronolog, because $\Box$ is not allowed in the heads of permanent clauses in Templog and in Gabbay's Temporal Prolog, and Chronolog lacks $\Box$. The following clauses in Chronolog can, however, be written to the same effect:

```
alarm <- dangerous(X).
next alarm <- alarm.
```

There is also a simple translation from these clauses into equivalent Templog clauses by replacing **next** by $\bigcirc$ and <- by $\Leftarrow$.

In [3, 4], it is mentioned that the suggested implementation method for Temporal Prolog is by translation into ordinary Prolog. Translation is performed after all program clauses are put into a normal form where the only allowed temporal operator is the previous time operator.

If we assume that Temporal Prolog is based on a temporal logic with an unbounded past and future, TiSLD-resolution of Chronolog($\mathcal{Z}$) can be directly applied to programs of Temporal Prolog in normal form. Therefore a restricted version of TiSLD-resolution, in which the only allowed temporal operator is **prev**, provides an operational semantics for Temporal Prolog. It can also form the basis of a direct implementation, but not necessarily an efficient one for the full language, based on unification and backtracking.

4.6 MTL

MTL (temporal logic programming with metric and past operators) proposed by Brzoska [26] has linear and unbounded (in both directions) time attributes. The set of integers $\mathcal{Z}$ is the collection of moments in time. The formulas of MTL are built with ordinary logical connectives and some temporal operators, which can only apply to formulas, not to terms. The basic temporal operators are:

1. $\Box_t$: *always within t time points*, $t \in \mathcal{Z} \cup \{-\infty, +\infty\}$;
2. $\Diamond_t$: *sometime within t time points*, $t \in \mathcal{Z} \cup \{-\infty, +\infty\}$;
3. $\circ$: *the next moment in time*;
4. $\bullet$: *the previous moment in time*.

There are also a number of derived temporal operators. Some of them are listed below:

$$\begin{array}{lll}
\Box A \equiv_{def} \Box_{+\infty} \Box_{-\infty} A & \text{(unrestricted always)} \\
\Diamond A \equiv_{def} \Diamond_{+\infty} \Diamond_{-\infty} A & \text{(unrestricted sometimes)} \\
\Box_+ A \equiv_{def} \Box_{+\infty} A & \text{(always in the future)} \\
\Box_- A \equiv_{def} \Box_{-\infty} A & \text{(always in the past)} \\
\Diamond_+ A \equiv_{def} \Diamond_{+\infty} A & \text{(sometime in the future)} \\
\Diamond_- A \equiv_{def} \Diamond_{-\infty} A & \text{(sometime in the past)}
\end{array}$$

A program of MTL is composed by a set of Prolog-like clauses with the temporal operators applied to their formulas [26, definition 2.3], and an MTL goal is a Prolog-like formula prefixed with some temporal operators (possibly none). As is the case in Templog, program clauses in MTL do not contain applications of $\Box$-operators in bodies, or $\Diamond$-operators in heads. MTL is, however, more expressive than either Templog or Chronolog.

It is shown that MTL can be considered as an instance of the CLP-scheme over a suitable algebra [26]. This work is in fact a continuation of the earlier results reported by Brzoska [25], but the translation function Π in MTL is different from that of [25]. It is based on the free-term algebra of an MTL program plus the algebra $(\mathcal{Z}, 0, +1, -1, +, =_{\mathcal{Z}}, \leq_{\mathcal{Z}})$ where $\mathcal{Z}$ is the set of integers. The function Π maps an MTL program P and a goal $\leftarrow G$ to the corresponding classical logic program $\Pi(P)$ and goal $\leftarrow \Pi(G)$. It is established that $P \models G$ if and only if $\Pi(P) \models_\Pi \Pi(G)$.

The operational semantics of MTL-programs is based on MTL-resolution, which is a restriction of the CLP-derivations over the considered algebra for tractability reasons. The soundness and completeness of MTL-resolution are also shown [26].

4.7 Starlog

Starlog [33] is a logic programming language which can handle some time-dependent problems. It is claimed that Starlog is a temporal programming language, but it is not directly based on temporal logic and it does not have any temporal operators. It simply adds an additional argument to every Prolog predicate. The first argument position of every predicate is reserved for this time-argument. For example, the statement "John is at school at time t" is expressed as `school(t,John)`. If we want to say that "John is at school from time 8.5 to 10.3" then:

```
school(t,John) <- t>=8.5, t<=10.3.
```

Note that Starlog uses the real number time axis and hence time-arguments can be real numbers. Program clauses are required to be *causal*, that is, the time values in the head of a clause must be greater than or equal to the time values in its body. Starlog does not use a standard resolution-type proof procedure for the execution of its programs, instead, it uses a variant of a connection-graph theorem prover and an intelligent arithmetic constraint solver.

5 Modal Logic Programming

This section discusses modal logic programming languages. Temporal logic languages (including interval languages) are tailored for modelling time-dependent and dynamic properties of certain problems such as simulation, temporal (historical) databases, dataflow computation, temporal reasoning and so on. On the other hand, modal logic languages target application areas involving the notions of knowledge, belief, and assumption. They also lift the requirement that the universe of possible worlds exhibit a certain intrinsic structure (such as a time-line in temporal logic), but at the same time they introduce some other requirements that the accessibility relations be reflexive, transitive, symmetric or antisymmetric, etc. Depending on the properties of the accessibility relations (or the corresponding set of axioms), we have a different modal logic.

We first discuss three modal logic programming languages (or schemes): Molog [39] (which has evolved into TIM [14, 15]), Modal Prolog [87], and Akama's proposal of modal logic programming [6]. In these works, some traditional modal logics such as B, D, S4, S5 and so on are considered, and proof procedures specialized for these modal logics are proposed. There are also multi-dimensional languages such as InTense [68, 67] and 3-Dimensional (3-D) logic programming [77] with extra spatial dimensions. These languages are based on a (multi)modal logic in which the accessibility relation associated with each modality is functional. Two other more standard approaches to (multi)modal logic programming based on translation [7, 36] are also summarized.

5.1 Molog

Molog is a modal system proposed by Fariñas del Cerro [39]. Molog extends Prolog to a modal logic programming language so that it can express concepts of belief, knowledge, or assumption. The user fixes a modal logic, and defines the rules to deal with modal operators; in this respect, Molog is a framework which can be "instantiated" with particular modal logics. Modal operators are grouped in two categories: universal operators (such as □), and existential operators (such as ◇). Standard Kripke-style semantics is employed for the semantics of modal operators.

The basic structure of Molog is a *modal Horn clause*, which is an ordinary Horn Clause governed by modal operators. Modal operators can only be applied to formulas. The general form of a modal Horn clause is as follows:

Modality (Modal atomic formula ← Conjunctive modal formula).

A sequence of modal operators is called a *Modality*, which can qualify a modal atomic formula, a conjunction of them, or a whole modal Horn clause. An example of a *Modal Horn Clause* is:

knows(paul)(come_back(X) ← comp(paul) it_rains & tired(X)).

It says that "Paul *knows* that if someone is tired and it is *compatible* with Paul's knowledge that it is raining then this person should come back." [21, page 762] A Molog program is a set of modal Horn clauses, and a goal is a conjunctive modal formula.

Molog uses a modal resolution method whose roots are in a deduction method for modal logics [38]. The general modal resolution rule (inference step) in Molog has the following form [39, page 40]:

$$\frac{M_1(E_1 \leftarrow B) \qquad \leftarrow M_2(E_2)}{\leftarrow M_3(E_3, B); \text{if } T(M_1 E_1, M_2 E_2) \text{ is verified.}}$$

where T possesses an associated procedure reflecting the corresponding modal proof theory (based on selected rules for modal operators) for a user-elected modal logic and M_1, M_2, and M_3 are modalities.

Molog requires the use of a different modal resolution method for different modal logics. Since the resolution rules can be created by the user (or selected from a library of resolution rules), Molog is very flexible, but it is not clear how the consistency (and the completeness) of the resulting resolution method can be guaranteed. As for standard modal logics such as S5, it is shown that an instance of Molog, based on S5, has a complete modal resolution method [39]. The following example of an S5-Molog program is taken from [39, page 42]:

knows(pierre) knows(jean) q
knows(pierre)(p ← knows(jean) q)

where "knows(a)" for any term a is regarded as a universal modal operator. Using the modal resolution method for S5, it can be shown that, for example, the goal "← knows(pierre) p" follows from the program.

Fariñas del Cerro [39] gives another deduction method for S5-Molog. It is based on a technique called *compilation*, by which all modal Horn clauses are first transformed into a simple form in which no modalities are applied to a whole modal Horn clause, and then a Prolog-like inference rule based on resolution and unification is used. The completeness of the deduction method by compilation is also shown.

The complexity of the satisfiability of modal Horn clauses, in the propositional case, is also studied [37]. A bottom-up algorithm is given for testing the satisfiability of sets of propositional modal Horn clauses. It is shown that, for some modal logics such as S5, the algorithm works in polynomial time, while for other modal logics such as K, Q, T, and S4 the worst case complexity can be considerably higher.

Balbiani et al [13] discussed the declarative semantics of an instance of Molog based on modal logic Q (in which all accessibility relations associated with modal operators are serial). There, it is shown that the interpretations of a Q-Molog program can be represented by a tree and each Q-Molog program has a minimum Kripke model, which is the limit (the least fixed-point) of a certain transformation over trees. A modal SLD-resolution method is also defined, and its completeness is shown. In order to ensure that modal resolution rules are closed, Skolem techniques are employed to eliminate the occurrences of the existential modal operator $\Diamond$. The results are also extended to instances of Molog based on modal logics T and S4. Orgun and Wadge [77] provided a general model-theory for intensional logic languages which can also be applied to these instances of Molog. They showed, using certain semantic properties of modal operators, that Molog programs considered in [13] enjoy the minimum model semantics, and its fixed-point characterization.

At first, there were two main assumptions in the design of Molog [14]:

- to keep the fundamental logic programming mechanisms: backward chaining, depth-first strategy, backtracking, and unification;
- to parameterize the inference step: it is the user who specifies how to compute a new goal from a given one, in logic programming style.

A further extension makes it possible for a user to select clauses which can not exactly unify with the current goal, but just resemble it in some way. This gave birth to a more general modal logic programming system—TIM (*Toulouse Inference Machine* [14]). As its predecessor, TIM is a general framework for modal logic programming. It can produce a concrete logic programming language, such as Prolog, Templog, modular Prolog, and an epistemic logic programming language etc., by a certain specification: the user specifies the inference rules for modal operators, and then provides a mechanism for clause selection. It is not mentioned whether it can produce languages such as Gabbay's Temporal Prolog in which P and F are allowed in the head of a clause.

In [14, 15], it is reported that a prototype implementation of TIM was written in LISP, and a compiled version in C, following the same principles as WAM (Warren's Abstract Machine [102, 5]). The underlying abstract machine of TIM implementations is called TARSKI (*Toulouse Abstract Reasoning System for Knowledge Inference*). These implementations of TIM led to a distributed implementation on a network of workstations linked by Internet sockets. In the distributed implementation, which is written in Ada, each workstation runs a TARSKI machine; the early performance of a top-down configuration of the network is also reported [15].

Owing to its parameterized inference mechanism, TIM can produce working implementations of modal logic programming languages (such as Q- or S5-Molog) with minimal effort.

5.2 Modal Prolog

Modal Prolog [87] extends Prolog to give it the abilities to express modularity, hierarchy, and/or structure. A modal logic program consists of two parts, a possible world description and a relationship description upon the possible worlds. Possible worlds descriptions of Modal Prolog resemble program clauses with time indicators (labels) in Gabbay's labeled deduction systems [45].

1. *The description of a possible world*: This is like a Prolog program with modal operators applied to atoms. The modal operators are $\Box$ (necessity) and $\Diamond$ (possibility). Modal operators can not be used in the head of a clause. Every world is identified by a unique name, and program clauses for a possible world form the axioms of that world. There are only finitely many possible worlds. Therefore the use of $\Box$ in the body of a clause (similar to the use of the universal quantifier $\forall$) does not cause any problems in Modal Prolog.

2. *Relationship description*: This gives the accessibility relations between the possible worlds and their attributes such as reflexivity, symmetry, transitivity etc. A relation atom is of the form $re(w_1, w_2)$, where w_1 and w_2 are the names of possible worlds. A relation clause is just like a Prolog clause, except that it describes the attributes of the accessibility relations. The following example describes an accessibility relation re which is reflexive, symmetric, and transitive:

    $re(X, X);$
    $re(X, Y) \leftarrow re(Y, X);$
    $re(X, Z) \leftarrow re(X, Y), re(Y, Z);$

The following Modal Prolog program is taken from [87, page 87]:

```
relation of
   re(w1,w2); re(w1,w3);
fo.
world w1 of
   p(X) ← □q(X);
   r(X) ← ◇s(X);
fo.
world w2 of
   q(a); q(b); s(a);
fo.
world w3 of
```

```
    q(b); q(c); s(b);
fo.
```

With the description of the relationship of the possible worlds, we can prove that in world w1, $\Box$q(b), $\Diamond$q(a), $\Diamond$q(c), $\Diamond$s(a), $\Diamond$s(b), p(b), r(a), and r(b) are all true.

Every possible world description corresponds to a module of the program. The interactions among modules are described in the relationship definition, and values from different worlds are combined using modal operators. Thus modal logic programming introduces a hierarchic structure into logic programming. We think that a similar approach can also be applied to "distributed logic programming [81]" where parts of a given program are stored at different sites in a distributed environment.

Modal Prolog enjoys a *possible-world model* and its operational semantics is based on resolution and unification. The soundness and completeness of the proof procedure of Modal Prolog are shown [87]. A simple meta-interpreter for the language, written in Prolog, is also given.

5.3 Modal Logic Programming (Akama)

Akama [6] proposed an extension of Prolog based on modal logic S5. A standard Kripke-style semantics is employed for modal operators. The basic building blocks of modal logic programs are *modal definite clauses*, which consist of a conjunction of modal literals with only one positive literal. A modal literal is either of the form ∇A or of the form $\neg(\nabla A)$ where ∇ is a modality (a sequence of modal operators $\Box$ and $\Diamond$) and A is an atomic formula. A modal logic program is a set of modal definite clauses. As a proof procedure for modal logic programs, an extension of the standard SLD-resolution with two additional inference rules for modal operators is defined, and its completeness is established.

A generalization of the modal logic programming to (multi)modal logic programming is also considered, but no realistic implementation of (multi)modal logic programming is suggested. Modal definite clauses considered in this work are analogous to those obtained by the compilation technique for S5-Molog [39], but, since there are no restrictions on the use of modal operators in Akama's work [6], the usefulness of the modal SLD-resolution in defining a satisfactory interpreter for the language is questionable. If restrictions are imposed (as in Templog or in Molog), we can conclude that Akama's proposal of modal logic programming is subsumed by Molog.

5.4 InTense

The multi-dimensional logic language InTense was invented by Mitchell and Faustini [68, 67]. It was inspired by both Chronolog and Lucid [101]. Lucid is a functional programming language, in fact, an intensional extension of ISWIM [11, 101]. In Lucid, as in Tempura, variables represent a sequence of values over the time dimension.

InTense has a large repertoire of Lucid-like intensional operators such as asa (*as soon as*), wvr (*whenever*), and upon, in addition to the temporal operators of Chronolog. Formulas in an InTense program vary not only along the time axis, but also along the spatial axis. The language can, in theory, accommodate a (possibly infinite) number of temporal and spatial dimensions; thus a possible world in InTense is a point in a time-space hyperfield, namely, $\mathcal{Z}^\omega$.

InTense is also a Prolog-like programming language. An InTense program consists of a set of intensional Horn clauses, which are first-order Horn clauses with intensional operators applied to formulas. All program clauses are interpreted as assertions true at *all* possible worlds (i.e., points in $\mathcal{Z}^\omega$). In other words, they are all permanent as in Chronolog.

Intensional operators are prev, first, and next on the time axis and prior, initial, and rest on the space axis. Temporal operators prev, first, and next are just like the operators of Chronolog($\mathcal{Z}$). The operator initial refers to the original point in space (point 0). Spatial operators prior and rest are analogous to prev and next, except that they operate on the space axis. There is also a dimension indicator attached to these operators. If it is omitted, intensional operators are for the default dimensions (that is, the first time and space dimensions). Note that the accessibility relations associated with the operators initial, prior, rest, first, prev, and next are functional; hence all the operators are self-dual.

The following InTense program is taken from Mitchell [67, page 32] (it uses the Sieve of Eratosthenes to produce the sequence of prime numbers). Note that InTense retains the extra-logical and non-logical features of Prolog, for instance, it has the cut "!", and assert and retract.

```
initial ints(2) :- !.
rest ints(X) :- ints(Y), X is Y+1.

first sieve(X) :- ints(X), !.
next sieve(X) :- initial sieve(Y),newsmallest(X),not(0 is X mod Y).

newsmallest(X) :- sieve(X).
newsmallest(X) :- rest newsmallest(X).

prime(X) :- initial sieve(X).
```

As shown in Figure 2, the sieve predicate represents the sequence of prime numbers along the time dimension at point 0 in space. The program makes use of the first time and space dimensions only.

There are two reported InTense interpreters [67]. The first one is written in Prolog and translates InTense programs into Prolog programs. The other is written in C, which is based on Warren's Abstract Machine WAM [102] with intensional extension. In the C implementation, up to four temporal dimensions and four spatial dimensions are supported. It is not clear how the Lucid-like operators (such as wvr and asa) can be given an operational semantics and proof

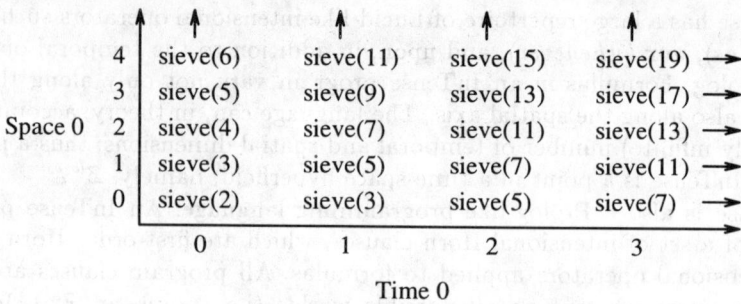

Fig. 2. The sieve/1 relation in two dimensions

rules can be defined for them, and yet they are implemented. The correctness issues of the implementation are not addressed. Note that, in InTense, Chronolog programs can be run unchanged.

Another approach similar to InTense is 3-D Logic Programming Language of Orgun and Wadge [77], which extends Chronolog with 2 spatial dimensions. There are some modal operators on the extra spatial dimensions, namely, side, edge, north, south, west, and east. The operators side and edge are analogous to initial for the first two spatial dimensions in InTense. Hence the 3-D language can be considered as a pure subset of InTense restricted to 3-D. It is shown that the 3-D logic programming enjoys the minimum model semantics [77], but no completeness results are offered for the language. Another multi-dimensional language which is in spirit similar to "pure" InTense is considered by Orgun and Du [73]. They propose a resolution-based proof procedure for the language and outline a spreadshet interface which can be used for both query formulation and the display of results.

5.5 Multi-Modal Approaches

A Meta-logical Foundation (Akama, 1989): Akama [7] proposed a foundation for modal logic programming. It is based on the simulation of a model (or proof) theory of modal logic in the Horn clause logic programming, using meta-programming techniques. It is a departure from the previous work by Akama [6] in which a modal resolution method was defined.

Modal programs are translated into programs of a two-sorted first-order logic by introducing world paths as extra parameters to function and predicate symbols. A meta-interpreter is given for the execution of translated programs. The translation method does not require the axiomatization of the accessibility relations associated with modal operators such as □ and ◇ in modal logic programs, rather it directly encodes the accessibility relations into the unification algorithm. This encoding technique is borrowed from Ohlbach's resolution calculus for modal logics [71]. It is claimed that the encoding technique improves the efficiency of the implementations (meta-interpreters) of modal logic programming, but it requires special unification algorithms for different types of modal logics. Considered modal logics are T, S4, S5, B, D, D4 and DB.

Although it is shown that a modal Herbrand property holds for translated modal formulas, the declarative semantics of modal logic programs are not given. A version of the minimum model semantics does not seem to hold for the class of modal logic programs considered; in fact some of the examples given in Akama [7] contain the use of negation in the heads of program clauses.

Multimodal Logic Programming (Debart et al, 1992): Debart et al [36] extended the automated theorem proving method in modal logic of Aufray and Enjalbert [12] to multimodal logic. The resulting method is called Σ-E-resolution. Then a restricted version of Σ-E-resolution is applied to modal logic programming. Modal logics considered are KD, KT, KD4, KT4, and KF. When restricted to a single modality, this approach to modal theorem proving is, in spirit, similar to Akama's [7].

For each modality i (chosen from the above mentioned modal logics), there is a pair of modal operators, $\Diamond_i$ and $\Box_i$, and $\Diamond_i$ is the dual of $\Box_i$. Standard accessibility relations are used to define the semantics of modal operators, and it is required that the relations be serial. (Multi)modal formulas are first translated into formulas of order-sorted equational theories, preserving satisfiability, and then Σ-E-resolution is used to show satisfiability of the translated formulas. The translation of modal formulas involves the addition of as many extra arguments (world paths) to function and predicate symbols as the total number of modalities. Translated formulas are called *path formulas*. Because of the extra world path arguments, a special unification algorithm is developed. The properties of modal operators are encoded in the unification algorithm as in Akama's work [7]; hence there is no need to axiomatize the accessibility relations associated with modal operators in translated programs.

When path formulas are restricted to Horn clauses, we have a logic programming language called Pathlog. Debart et al [36] showed that Σ-E-resolution with the restriction to Horn clauses is a sound and complete proof procedure for Pathlog. This result provides a basis for multimodal logic programming for the case where multimodal logic programs can be translated into the Horn subset of path formulas. In particular, it is shown that for any given Templog program [3, 4], there is a corresponding Pathlog program; hence it is suggested that Pathlog can be the target language for "compiling" Templog programs. Whether such a compilation technique would provide an efficient interpreter for Templog or not is not discussed.

6 Discussion

An overview of the progress to date in temporal and modal logic programming has been presented. The three classes of interval logic languages, temporal logic languages, and (multi)modal logic languages are *logical*, at least at this stage. The classes are naturally identified by the logics these languages are based upon.

In our opinion, considering that temporal and modal logic programming is a relatively new field of research which gained momentum in the mid-eighties, the depth and variety of the reported work have been quite extensive. The origins of most of the languages are in modal and temporal theorem proving, and logic programming (Prolog), but languages such as Tempura [70] and Tokio [9] have certain features also found in imperative languages. Figure 3 summarizes all the reviewed languages with a few exceptions (such as Sakuragawa's Temporal Prolog [88] and multi-dimensional logic programming [73]).

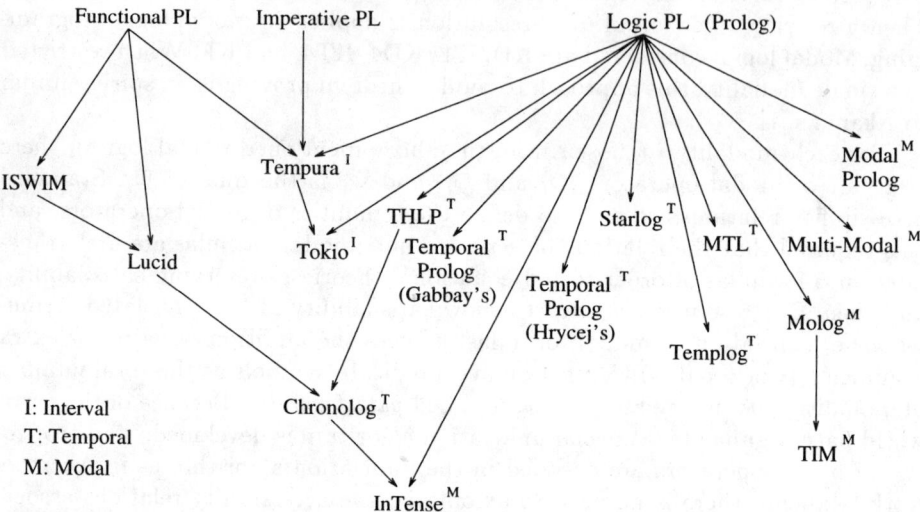

Fig. 3. Temporal and modal logic programming languages

Results on the completeness of temporal and modal logic programming are very encouraging, given that first-order temporal and modal logics are inherently incomplete [1, 94]. We now know that there are complete subsets of temporal and modal logics which can be used as the basis of logic programming. In standard logic programming, the declarative and operational semantics of logic programs coincide [96, 64]. The completeness results (such as those of Templog [16, 17, 18], Chronolog [74, 78], Molog [13], MTL [26]) guarantee that the declarative and operational semantics of temporal and modal logic programs of those languages also coincide. As a consequence, we can use the tools and techniques from temporal and modal logics to *reason about* programs and study their properties, but this is not yet a well-explored area. Similar completeness results are also established for other non-classical languages; we refer the reader to the literature for more details on languages based on linear, paraconsistent, intuitionistic, defeasible logics, and so on, and their applications [23, 89, 41, 49, 59].

There are also some general semantic frameworks for temporal and modal logic programming. Orgun and Wadge [77] provide a unified model-theory for intensional (temporal and modal) logic programming languages, based on Scott-Montague neighborhood semantics for intensional operators. It explains the reasons behind the restriction on the uses of □ and ◇ in Templog and Molog. ◇ is not a *conjunctive operator*, and conjunctivity is related to the model intersection property. If ◇ is used in the head of a clause, the minimum model semantics no longer applies. For example, Gabbay's Temporal Prolog does not enjoy the minimum model semantics, because it allows the use of non-conjunctive P and F in the head of a clause. □ is not a *continuous* operator, and continuity is related to computability. If □ is used on the body of a clause, the fixed-point semantics no longer applies. Kriaučiukas [62] also proposed a semantics framework for non-classical logic programming based on Kripke-style semantics for modal logic, and investigated the connections between logic programming and dynamic logic. Another general framework is provided by Blair et al [24, 22] for non-classical extensions of logic programming.

We believe that the utility of temporal and modal logic programming has been demonstrated in the works reviewed in this paper; we are, however, yet to see their use in more realistic applications. One of the key features of temporal and modal logic programming is that it provides an abstraction, which we call *context abstraction*. Context abstraction plays an important rôle in many applications involving the notion of dynamic change, time, belief, and knowledge. Through context abstraction, temporal and modal logic programming can be used to describe dynamic and context-dependent properties of certain problems in a natural and problem-oriented way. It also introduces a form of parallelism, that is, *context parallelism*: goals at different contexts (possible worlds) can be executed in parallel. This form of parallelism is in addition to the standard AND- and OR-parallelism existing in logic programs.

Other related issues such as meta-logics, meta-interpretation, translation vs direct implementation of temporal and modal logics are also extensively discussed in the literature; for example, see [48, 44, 46, 36, 40]. Many implementations of temporal and modal languages are built on top of Prolog, either with some modifications on the unification algorithm, or with some extra rules for temporal and modal operators. We are in favour of direct implementations of temporal and modal logic programming so that we can, for instance, exploit context-parallelism, but translation and also meta-interpretation provide a good basis for rapid-prototyping some experimental systems. Meta-interpretation can also be used to investigate the language features which may or may not be feasible and/or practical for inclusion in production languages. In this respect, TIM [14, 15] is a very promising system which has the ability to produce many concrete non-classical languages with minimal effort.

We include a timeless quotation from Scott [90, page 143] which we believe is also valid for temporal and modal logic programming:

> "One often hears that modal (or some other) logic is pointless because
> it can be translated into some simpler language in a first-order way.
> Take no notice of such arguments. There is no weight to the claim that

the original system must therefore be replaced by the new one. What is essential is to single out important concepts and to investigate their properties."

In the near future, as the implemented systems mature and efficient parallel implementations emerge, we think we are going to see more and more languages applied to a broader range of applications with success.

Acknowledgements

Wanli Ma has been partially supported by an Overseas Postgraduate Research Award (OPRA) from the Australian Government. Thanks are due to Lee Flax, Jan Hext and Kang Zhang for their useful comments and suggestions on an earlier draft of this paper.

References

1. M. Abadi. The power of temporal proofs. *Theoretical Computer Science*, 65(1989):35–83, 1989.
2. M. Abadi and Z. Manna. Nonclausal temporal deduction. In R. Parikh, editor, *Proc. of Conference on Logics of Programs*, volume 193 of *LNCS*, pages 1–15. Springer-Verlag, 1985.
3. M. Abadi and Z. Manna. Temporal logic programming. In *Proceedings of the 1987 Symposium on Logic Programming*, pages 4–16, San Fransisco, Calif, 1987. IEEE Computer Society Press.
4. M. Abadi and Z. Manna. Temporal logic programming. *Journal of Symbolic Computation*, 8:277–295, 1989.
5. Hassan Ait-Kaci. *The WAM: A (Real) Tutorial*. Paris Research Laboratory, Digital Equipment Corporation, Paris, France, 1990.
6. Seiki Akama. A proposal of modal logic programming (extended abstract). In *Proc. of the 6th Canadian Conference on Artificial Intelligence*, pages 99–102, École Polytechnique de Montréal, Montréal, Québec, Canada, May 1986. Presses de l'Université du Québec.
7. Seiki Akama. A meta-logical foundation of modal logic programming. 1-20-1, Higashi-Yurigaoka, Asao-ku, Kawasaki-shi, 215, Japan, December 1989.
8. J. F. Allen. Maintaining knowledge about temporal intervals. *Communications of the ACM*, 26:832–843, November 1983.
9. T. Aoyagi, M. Fujita, and T. Moto-oka. Temporal logic programming language Tokio. In E. Wada, editor, *Logic Programming'85*, volume 221 of *LNCS*, pages 138–147. Springer-Verlag, 1986.
10. E. A. Ashcroft, A. A. Faustini, and B. Huey. Eduction: A model of parallel computation and the programming language Lucid. In *Proc. of Phoenix Conference on Computers and Communications*, pages 9–15. IEEE Computer Society Press, 1985.
11. E. A. Ashcroft and W. W. Wadge. Lucid – a formal system for writing and proving programs. *SIAM Journal on Computing*, 5:336–54, September 1976.
12. Y. Aufray and P. Enjalbert. Modal theorem proving: an equational viewpoint. To appear in *Journal of Logic and Computation*, 1992.

13. Philippe Balbiani, Luis Fariñas del Cerro, and Andreas Herzig. Declarative semantics for modal logic programs. In *Proceedings of the 1988 International Conference on Fifth Generation Computer Systems*, pages 507–514. ICOT, 1988.
14. Philippe Balbiani, Andreas Herzig, and Mamede Lima-Marques. TIM: The Toulouse inference machine for non-classical logic programming. In *PDK'91: International Workshop on Processing Declarative Knowledge*, volume 567 of *LNAI*, pages 365–382. Springer-Verlag, 1991.
15. Philippe Balbiani, Andreas Herzig, and Mamede Lima-Marques. Implementing Prolog extensions: a parallel inference machine. In *Proc. of the 1992 International Conference on Fifth Generation Computer Systems*, pages 833–842. ICOT, 1992.
16. M. Baudinet. On the semantics of temporal logic programming. Technical Report STAN–CS–88–1203, Computer Science Department, Stanford University, Stanford, Calif, June 1988.
17. M. Baudinet. Temporal logic programming is complete and expressive. In *Conference Record of the Sixteenth ACM Symposium on Principles of Programming Languages*, pages 267–280, Austin, Texas, January 1989. The Association for Computing Machinery.
18. M. Baudinet. A simple proof of the completeness of temporal logic programming. In L. Fariñas del Cerro and M. Penttonen, editors, *Intensional Logics for Programming*, pages 51–83. Oxford University Press, 1992.
19. M. Baudinet, J. Chomicki, and P. Wolper. Temporal deductive databases. In A. Tansel and et al, editors, *Temporal Databases: Theory, Design, and Implementation*. Benjamin/Cummings Publishing Company, Redwood City, CA, 1993.
20. Marianne Baudinet. Proving termination properties of Prolog programs: a semantic approach. *Journal of Logic Programming*, 14(1&2):1–30, October 1992.
21. Pierre Bieber, Luis Fariñas del Cerro, and Andreas Herzig. MOLOG: A Modal Prolog. In E. Lusk and R. Overbeek, editors, *Proceedings of the 9th International Conference on Automated Deduction*, pages 762–763. Springer-Verlag, 1988.
22. H. A. Blair, A. L. Brown, and V. S. Subrahmanian. Monotone logic programming. In L. Fariñas del Cerro and M. Penttonen, editors, *Intensional Logics for Programming*, pages 1–22. Oxford University Press, 1992.
23. H. A. Blair and V. S. Subrahmanian. Paraconsistent logic programming. *Theoretical Computer Science*, 68:135–154, 1989.
24. H.A. Blair et al. A logic programming semantics scheme, Part I. Technical Report LPRG–TR–88–8, Logic Programming Research Group, Syracuse University, 1988.
25. Christoph Brzoska. Temporal logic programming and its relation to constraint logic programming. In V. Saraswat and K. Ueda, editors, *Proceedings of the 1991 International Logic Programming Symposium*, pages 661–677, San Diego, Calif, October 28-31 1991.
26. Christoph Brzoska. Temporal logic programming with metric and past operators. Universität Karlsruhe, P.O.Box 6980, D-7500 Karlsruhe, Germany, January 1992.
27. R. Bull and K. Segerberg. Basic modal logic. In D. M. Gabbay and F. Guethner, editors, *Handbook of Philosophical Logic, Vol. II*, pages 1–88. D. Reidel Publishing Company, 1984.
28. J. P. Burgess. Basic tense logic. In D. M. Gabbay and F. Guethner, editors, *Handbook of Philosophical Logic, Vol. II*, pages 89–134. D. Reidel Publishing Company, 1984.
29. B. F. Chellas. *Modal Logic: An Introduction*. Cambridge University Press, 1980.
30. Cheng-Chia Chen and I-Peng Lin. The computational complexity of satisfiability of temporal Horn formulas in propositional linear-time logic. *Information Processing Letters*, 45:131–136, March 1993.

31. Jan Chomicki and Tomasz Imieliński. Temporal deductive databases and infinite objects. In *Proceedings of the Seventh ACM SIGACT-SIGMOD-SIGART Symposium on Principles of Database Systems*, pages 61–73. The Association for Computing Machinery, 1988.
32. Eugène Chouraqui. Formal expression of time in a knowledge base. In P. Smets, A. Mamdani, D. Dubois, and H. Prade, editors, *Non-Standard Logics for Automated Reasoning*, pages 81–103. Academic Press, 1988.
33. John G. Cleary and Vinit N. Kaushik. Updates in a temporal logic programming language. Technical report, Department of Computer Science, University of Calgary, Calgary, Alberta, Canada, 1991.
34. James Clifford and David S. Warren. Formal semantics for time in databases. *ACM Transactions on Database Systems*, 8(2):214–254, June 1983.
35. W. Clocksin and C. Mellish. *Programming in Prolog*. Springer-Verlag, 1981.
36. Françoise Debart, Patrice Enjalbert, and Madeleine Lescot. Multimodal logic programming using equational and order-sorted logic. *Theoretical Computer Science*, 105(1992):141–166, 1992.
37. L. Fariñas del Cerro and M. Penttonen. The complexity of the satisfiability of modal Horn clauses. *Journal of Logic Programming*, 4:1–10, March 1987.
38. Luis Fariñas del Cerro. A simple deduction method for modal logic. *Information Processing Letters*, 14(2), 1982.
39. Luis Fariñas del Cerro. MOLOG: A system that extends PROLOG with modal logic. *New Generation Computing*, 4:35–50, 1986.
40. Luis Fariñas del Cerro and Andreas Herzig. Metaprogramming through intensional deduction: some examples. In *Meta-Programming in Logic: Proc. of the Third International Workshop, META-92, Uppsala, Sweden*, pages 11–25. Springer-Verlag, June 1992.
41. Luis Fariñas del Cerro and Martti Penttonen, editors. *Intensional Logics for Programming*. Oxford University Press, 1992. ISBN 019-853775-1.
42. A. A. Faustini and W. W. Wadge. An eductive interpreter for pLucid. Technical Report TR-006-86, Department of Computer Science and Engineering, Arizona State University, 1986.
43. Melvin Fitting. Logic programming on a topological bilattice. *Fundamenta Informaticae*, XI:209–18, 1988.
44. D. M. Gabbay. Modal and temporal logic programming. In A. Galton, editor, *Temporal Logics and Their Applications*, pages 197–237. Academic Press, 1987.
45. D. M. Gabbay. A temporal logic programming machine [modal and temporal logic programming, Part 2]. Department of Computing, Imperial College, November 1989.
46. D. M. Gabbay. Metalevel features in the object level: modal and temporal logic programming III. In L. Fariñas del Cerro and M. Penttonen, editors, *Intensional Logics for Programming*, pages 85–123. Oxford University Press, 1992.
47. Dov Gabbay and Peter McBrien. Temporal logic & historical databases. In *Proceedings of the 17th Very Large Data Bases Conference*, pages 423–430, Barcelona, Spain, September 1991. Morgan Kauffman, Los Altos, Calif.
48. A. Galton. Temporal logic and computer science: an overwiev. In A. Galton, editor, *Temporal Logics and Their Applications*, pages 1–52. Academic Press, 1987.
49. A. Galton, editor. *Temporal Logics and Their Applications*. Academic Press, 1987.

50. R. Gerth, M. Codish, Y. Lichtenstein, and E. Y. Shapiro. Fully abstract denotational semantics for Flat Concurrent Prolog. Technical Report CS88-03, Department of Applied Math. and Computer Science, The Weizmann Institute of Science, Rehovot, Israel, April 1988.
51. R. Goldblatt. *Logics of Time and Computation*. CSLI – Center for the Study of Language and Information, Stanford University, 1987. Lecture Notes no:7.
52. R. Hale. Temporal logic programming. In A. Galton, editor, *Temporal Logics and Their Applications*, pages 91–119. Academic Press, 1987.
53. C. Hewitt and Gul Agha. Guarded Horn clause languages: Are they deductive and logical? In *Proceedings of the 1988 International Conference on Fifth Generation Computer Systems*, pages 650–657. ICOT, 1988.
54. Tomas Hrycej. Temporal Prolog. In *Proc. of the European Conference on Artificial Intelligence*, pages 296–301, Munich, Germany, 1988.
55. Tomas Hrycej. A temporal extension of Prolog. *Journal of Logic Programming*, 15(1&2):113–145, January 1993.
56. G. E. Hughes and M. J. Creswell. *An Introduction to Modal Logic*. Methuen and Co Ltd, London, 1968.
57. P. Jackson, H. Reichgelt, and F. van Harmelen, editors. *Logic-Based Knowledge Representation*. MIT Press, 1989.
58. J. Jaffar and J-L. Lassez. Constraint logic programming. In *Conference Record of the Fourteenth ACM Symposium on Principles of Programming Languages*, pages 111–119, Munich, Germany, 1987. ACM Press.
59. Max I. Kanovich. Linear logic as a logic of computations. In *Proc. of the 7th Annual IEEE Symposium on Logic in Computer Science*, pages 200–210, Santa Cruz, Calif, June 1992. IEEE Computer Society Press.
60. S. Kono, T. Aoyagi, M. Fujita, and H. Tanaka. Implementation of temporal logic programming language Tokio. In E. Wada, editor, *Logic Programming'85*, volume 221 of *LNCS*, pages 138–147. Springer-Verlag, 1986.
61. R. A. Kowalski. Predicate logic as programming language. In *Proceedings of IFIP'74*, pages 569–574, Amsterdam, 1974. North–Holland.
62. Valentinas Kriaučiukas. Non-classical models for logic programs. In *PDK'91: International Workshop on Processing Declarative Knowledge*, volume 567 of *LNAI*, pages 179–190. Springer-Verlag, 1991.
63. Fred Kroger. *Temporal Logic of Programs*. Springer-Verlag, Berlin Heidelberg, 1987.
64. J. W. Lloyd. *Foundations of Logic Programming*. Springer-Verlag, 1984.
65. Z. Manna and A. Pnueli. Verification of concurrent programs: the temporal framework. In Boyer and Moore, editors, *Correctness Problem in Computer Science*, pages 215–273. Academic Press, 1981.
66. Fujita Masahiro et al. Using the temporal logic programming language Tokio for algorithm description and automatic CMOS gate array synthesis. In E. Wada, editor, *Logic Programming'85*, volume 221 of *LNCS*, pages 246–255. Springer-Verlag, 1986.
67. W. H. Mitchell. Intensional Horn clause logic as a programming language – it's use and implementation. Master's thesis, Department of Computer Science and Engineering, Arizona State University, Tempe, Arizona, 1988.
68. W. H. Mitchell and A. A. Faustini. The intensional logic language InTense. In *Proceedings of the 1989 International Symposium on Lucid and Intensional Programming*, Arizona State University, May 8 1989.

69. R. Montague. *Formal Philosophy, Selected Papers of Richard Montague*. Yale University Press, 1974. edited by Richmond Thomason.
70. B. Moszkowski. *Executing Temporal Logic Programs*. Cambridge University Press, 1986.
71. Hans Jürgen Ohlbach. A resolution calculus for modal logics. In E. Lusk and R. Overbeek, editors, *Proceedings of the 9th International Conference on Automated Deduction*, pages 500–516. Springer-Verlag, 1988.
72. M. A. Orgun. On temporal deductive databases. Technical Report 93-140C, Department of Computing, Macquarie University, Sydney, NSW 2109, Australia, December 1993.
73. M. A. Orgun and W. Du. Multi-dimensional logic programming. In *Proceedings of ICCI'94: The Sixth International Conference on Computing and Information*, Trent University, Peterborough, Ontario, Canada, May 26–28 1994. To appear.
74. M. A. Orgun and W. W. Wadge. Chronolog: A temporal logic programming language and its formal semantics. Department of Computer Science, University of Victoria, Victoria, B.C., Canada, January 1988.
75. M. A. Orgun and W. W. Wadge. A theoretical basis for intensional logic programming. In *Proceedings of the 1988 International Symposium on Lucid and Intensional Programming*, pages 33–49, Sidney, B.C., Canada, April 7-8 1988.
76. M. A. Orgun and W. W. Wadge. Theory and practice of temporal logic programming. In L. Fariñas del Cerro and M. Penttonen, editors, *Intensional Logics for Programming*, pages 23–50. Oxford University Press, 1992.
77. M. A. Orgun and W. W. Wadge. Towards a unified theory of intensional logic programming. *Journal of Logic Programming*, 13(4):413–440, August 1992.
78. M. A. Orgun and W. W. Wadge. Chronolog admits a complete proof procedure. In *Proceedings of the Sixth International Symposium on Lucid and Intensional Programming*, pages 120–135, Université Laval, Québec City, Québec, Canada, April 26–27 1993.
79. M. A. Orgun and W. W. Wadge. Extending temporal logic programming with choice predicates non-determinism. To appear in *Journal of Logic and Computation*, 1994.
80. M. A. Orgun, W. W. Wadge, and W. Du. Chronolog(Z): Linear-time logic programming. In *Proceedings of ICCI'93: The Fifth International Conference on Computing and Information*, Laurentian University, Sudbury, Ontario, Canada, May 27–29 1993. IEEE Computer Society Press.
81. R. Ramanujam. Semantics of distributed logic programs. *Theoretical Computer Science*, 68:203–220, 1989.
82. N. Rescher and A. Urquhart. *Temporal Logic*. Springer-Verlag, 1971.
83. D. W. Rolston. Chronolog: A pure tense-logic-based infinite-object programming language. Department of Computer Science and Engineering, Arizona State University, Tempe, Arizona, August 1986.
84. D. W. Rolston. Toward a tense-logic-based mitigation of the frame problem. In F. M. Brown, editor, *Proceedings of the 1987 Workshop on the Frame Problem in AI*, Lawrence, Kansas, April 1987. Morgan Kaufmann, Los Altos, Calif.
85. D. W. Rolston. *Parallel Logic Programming Using an Intensional Model of Computation*. PhD thesis, Department of Computer Science and Engineering, Arizona State University, Tempe, Arizona, 1992.
86. F. Sadri. Three approaches to temporal reasoning. In A. Galton, editor, *Temporal Logics and Their Applications*, pages 121–168. Academic Press, 1987.

87. Y. Sakakibara. Programming in modal logic: An extension of PROLOG based on modal logic. In E. Wada, editor, *Logic Programming'86*, volume 264 of *LNCS*, pages 81–91. Springer-Verlag, 1987.
88. Takahashi Sakuragawa. Temporal Prolog. In *Proc. of RIMS Conference on Software Science and Engineering*. Springer-Verlag, 1987.
89. P. Schroeder-Heister, editor. *International Workshop on Extensions of Logic Programming*, volume 475 of *LNAI*. Springer-Verlag, 1991.
90. D. Scott. Advice on modal logic. In K. Lambert, editor, *Philosophical Problems in Logic*, pages 143–173. D.Reidel Publishing Company, 1970.
91. E. Shapiro. The family of concurrent logic programming languages. In Friedrich L. Bauer, editor, *Logic, Algebra, and Computation*, pages 359–485. Springer-Verlag, Berlin Heidelberg, 1991.
92. Yoav Shoham. *Reasoning About Change*. MIT Press, 1988.
93. P. Smets, A. Mamdani, D. Dubois, and H. Prade, editors. *Non-Standard Logics for Automated Reasoning*. Academic Press, 1988.
94. A. Szalas. Concerning the semantic consequence relation in first-order temporal logic. *Theoretical Computer Science*, 47:329–334, 1986.
95. Alexander Tuzhilin and James Clifford. A temporal relational algebra as a basis for temporal relational completeness. In D. McLeod, R. Sacks-Davis, and H. Schek, editors, *Proceedings of the 16th International Conference on Very Large Data Bases*, pages 13–23, Brisbane, Australia, August 13–16 1990. Morgan Kaufmann Publishers Inc., Los Altos, Calif.
96. M.H. van Emden and R.A. Kowalski. The semantics of predicate logic as a programming language. *Journal of the Association for Computing Machinery*, 23:733–42, 1976.
97. M.H. van Emden and M.A. Nait Abdallah. Top-down semantics of fair computations of logic programs. Technical Report CS–84–27, Department of Computer Science, University of Waterloo, October 1984.
98. M. Y. Vardi. A temporal fixpoint calculus. In *Conference Record of the Sixteenth ACM Symposium on Principles of Programming Languages*, pages 250–259, San Diego, Calif, January 1988. ACM Press.
99. W. W. Wadge. Tense logic programming: a respectable alternative. Department of Computer Science, University of Victoria, Victoria, B.C., Canada, 1985.
100. W. W. Wadge. Tense logic programming: a respectable alternative. In *Proceedings of the 1988 International Symposium on Lucid and Intensional Programming*, pages 26–32, Sidney, B.C., Canada, April 7-8 1988.
101. W. W. Wadge and E. A. Ashcroft. *Lucid, the Dataflow Programming Language*. Academic Press, 1985.
102. David H. D. Warren. An abstract Prolog instruction set. Technical report, SRI International, Menlo Park, Calif, October 1983.
103. R. Wojcicki. *Theory of Logical Calculi*. Kluwer Academic Publishers, 1988.
104. K. Zhang and M. A. Orgun. Parallel execution of temporal logic programs using dataflow computation. In *Proceedings of ICCI'94: The Sixth International Conference on Computing and Information*, Trent University, Peterborough, Ontario, Canada, May 26–28 1994. To appear.
105. Kang Zhang and R. Thomas. DIALOG – A dataflow model for parallel execution of logic programs. *Future Generation Computer Systems*, 6(4):373–388, September 1991.

A Survey of Concurrent METATEM
— The Language and its Applications

Michael Fisher

Department of Computing
Manchester Metropolitan University
Manchester M1 5GD
United Kingdom

M.Fisher@mmu.ac.uk

Abstract. In this paper we present a survey of work relating to the Concurrent METATEM programming language. In addition to a description of the basic Concurrent METATEM system, which incorporates the direct execution of temporal formulae, a variety of extensions that have either been implemented or proposed are outlined. Although still in the development stage, there appear to be many areas where such a language could be applied. We present a variety of sample applications, highlighting the particular features of Concurrent METATEM that we believe will make it appropriate for use in these areas.

1 Introduction

Concurrent METATEM is a language based upon the direct execution of temporal formulae [15]. It consists of two distinct aspects: an execution mechanism for temporal formulae in a particular form; and an operational model that treats single executable temporal logic programs as asynchronously executing objects in a concurrent object-based system. The motivation for the development of this language has been provided by many areas. For example, being based upon executable logic, the language can be used as part of the specification and prototyping of reactive systems. Also, as it uses *temporal*, rather than classical, logic the language provides a high-level programming notation in which the dynamic attributes of individual components can be concisely represented. Finally, it incorporates a novel model of concurrent computation which has a range of applications in distributed systems.

The logic used as a basis for Concurrent METATEM is a discrete, linear temporal logic. This presents a simple view of time and, in doing so, provides the systems designer with a direct analogy between the models for the logic and the discrete, linear execution sequences with which he or she is familiar. This, together with the fact that a restricted form of this temporal logic is executed, ensures that the temporal aspects of Concurrent METATEM are manageable, both for the programmer and for the implementation of the system.

Each object executes its own set of temporal formulae and, in doing so, (effectively) generates an infinite sequence of states. As each object executes asynchronously, it constructs a separate execution sequence. Further, within Concurrent METATEM, a mechanism is provided for communication between separate objects. This simply consists of a partition of each object's propositions into those controlled by the object and those controlled by the environment. To fit in with this logical view of communication, whilst also providing a flexible and powerful message-passing mechanism, *broadcast* message-passing is used to pass information between objects.

This relatively straightforward combination of executable temporal logic, a concurrent object-model and broadcast message-passing forms the basis of Concurrent METATEM. Together, these features provide an coherent and consistent programming model within which a variety of reactive systems can be represented and implemented.

This review is structured as follows. Firstly, in §2, we will present the core elements of the language, including basic temporal execution, together with its operational and communication model. In §3, we will outline various extensions to the core language that have either been proposed or implemented. Thus, together, these two sections will provide a survey of the language itself. In §4, we provide a range of sample applications of Concurrent METATEM, utilising both the core features of the language and its extensions. Although, in some cases, these are only *potential* applications, we will argue that the properties that Concurrent METATEM exhibits make it suitable for application in these areas. Thus, the range of examples in this section provide a survey of the current and potential applications of the language. Finally, in §5, we present concluding remarks, including brief overviews of related work, current status of the implementation, and possible future work.

2 Concurrent METATEM

In this section, we will provide an introduction to the Concurrent METATEM system, which consists of objects, whose behaviour is implemented using executable temporal logic, communicating via broadcast message-passing. Concurrent METATEM itself was originally developed as an extension of the sequential execution of temporal logic programs provided by METATEM, an executable temporal logic described in [3, 10]. The rules that are executed are based upon the normal form developed for temporal theorem-proving [14], while the concurrent operational model was outlined in [9] and developed to its current state in [15].

2.1 An Overview of the Approach

While it is possible to program objects in Concurrent METATEM using a small range of temporal operators (just the *last-time* and *sometime in the future* operators), a large range of temporal operators are available in the interests of convenience. Most of these operators are eliminated during the transformation from the rules input by the programmer to rules that are actually executed. These transformations follow those used in producing a normal form for temporal formulae [14]. The *transformed* rules are then executed directly, providing the dynamic behaviour of the individual object. The components of this core language represent the basic descriptive elements of our system:

- logical properties of individual states, through the use of classical logic to represent declarative description of a state;

- properties of state transformation steps, through the use of the *last-time* operator in conjunction with constraints on the present state;
- global properties of temporal sequences, through the use of multiple state transformation steps together with the *sometime in the future* operator.

An important aspect of the language is the mixture, within the execution of these temporal formulae, of both declarative and imperative aspects. For example, we might provide a *declarative* description of a particular state through a formula such as $p \wedge q$, while also providing an *imperative* rule describing *how* to generate the current state from the last one, such as $\bigcirc\!\!\!\bullet\, p \Rightarrow q$ (here, '$\bigcirc\!\!\!\bullet$' is the *last-time* operator).

2.2 Temporal Logic

Temporal logic can be seen as classical logic extended with various modalities representing temporal aspects of logical formulae. The temporal logic we use is based on a discrete, linear model and, thus, time is modelled as an infinite sequence of discrete states, with an identified starting point, called 'the beginning of time'. Classical formulae are used to represent constraints within individual states, while temporal formulae represent constraints *between* states. As formulae are interpreted at particular states in this sequence, operators which refer to both the past and future are required.

The logic we use, called First-Order METATEM Logic (FML) is a simple first-order temporal logic, based on discrete, linear models with finite past and infinite future. Below we give an outline of its syntax and semantics (for a more detailed presentation, see [3, 14]).

Syntax As in classical logics, the *terms* the language are constructed from a set of *constant symbols* ($\mathcal{L}_c$), a set of *variable symbols* ($\mathcal{L}_v$) and a set of *function symbols* ($\mathcal{L}_f$). From these elements, the set of terms, $\mathcal{L}_t$, can be generated. The other symbols used in FML are as follows.

- A set, $\mathcal{L}_p$, of *predicate symbols*, contains elements usually represented by strings of lower-case alphabetic characters.
- Classical connectives, $\neg$, $\vee$, $\wedge$, $\Rightarrow$, and $\Leftrightarrow$.
- Future-time temporal operators, including unary operators $\bigcirc$, $\Diamond$ and $\square$, and binary operators $\mathcal{U}$ and $\mathcal{W}$.
- Past-time temporal operators, including unary operators $\bigcirc\!\!\!\bullet$, $\bullet$, $\blacklozenge$ and $\blacksquare$, and binary operators $\mathcal{S}$ and $\mathcal{Z}$.
- Quantifiers, $\forall$ and $\exists$.
- '(' and ')' which are, as usual, used to avoid ambiguity.

The set of well-formed formulae of FML (WFF$_f$) is defined as follows.

1. If $t_1, \ldots, t_n$ are in $\mathcal{L}_t$, and p is a predicate of arity n, then $p(t_1, \ldots, t_n)$ is in WFF$_f$.
2. if A and B are in WFF$_f$, then the following are in WFF$_f$

$\neg A$	$A \vee B$	$A \wedge B$	$A \Rightarrow B$	(A)
$\Diamond A$	$\square A$	$A\,\mathcal{U}\,B$	$A\,\mathcal{W}\,B$	$\bigcirc A$
$\blacklozenge A$	$\blacksquare A$	$A\,\mathcal{S}\,B$	$A\,\mathcal{Z}\,B$	$\bigcirc\!\!\!\bullet\, A$	$\bullet A$

3. If A is in WFF$_f$ and v is in $\mathcal{L}_v$, then $\exists v. A$ and $\forall v. A$ are both in WFF$_f$.

Sub-classifications of WFF$_f$ are defined as follows. A *literal* is defined as either a predicate symbol applied to an appropriate term, or the negation of such a predicate. A *State-formula* is either a literal or a non-temporal combination of other state-formulae. *Future-time formulae* contain only classical and future-time temporal operators, while *past-time formulae* contain only classical and past-time temporal operators. *Strict* versions of both these categorisations can be formed by removing from them formulae that refer to the present.

Semantics The basic models of FML are discrete, linear structures with finite past and infinite future. To this structure a domain, $\mathcal{D}$, and mappings from elements of the language to denotations are added. Thus the full model structure for FML is

$$\mathcal{M} = \langle \sigma, \mathcal{D}, \pi_c, \pi_f, \pi_p \rangle$$

where

- σ, a sequence of states $s_0, s_1, s_2, s_3, \ldots$,
- $\mathcal{D}$ is the object-level domain,
- π_c is a map from $\mathcal{L}_c$ to $\mathcal{D}$,
- π_f is a map from $\mathcal{L}_f$ to $\mathcal{D}^n \to \mathcal{D}$, where n is the arity of f, and,
- π_p is a map from $\mathbf{N} \times \mathcal{L}_p$ to $\mathcal{D}^n \to \{T, F\}$.

Thus, for a particular state s, and a particular predicate p of arity n, $\pi_p(s, p)$ represents a map from n-tuples of elements of $\mathcal{D}$ to T or F. Note that the *constant domain* assumption is used, i.e. that $\mathcal{D}$ is constant for every state, and that both constant and function symbols have fixed interpretations.

The semantics of FML is given with respect to a model, $\mathcal{M}$, a state, s_i, at which the formula is to be interpreted, and a variable assignment, V. From the model and the variable assignment, we are able to generate a *term assignment*, $\tau_{v\pi}$, which maps every term to its appropriate element of the domain, $\mathcal{D}$.

We first consider the semantics of atomic predicates:

$$\langle \mathcal{M}, s_i, V \rangle \models p(x_1, \ldots, x_n) \quad \text{iff} \quad \pi_p(i, p)(\tau_{v\pi}(x_1), \ldots, \tau_{v\pi}(x_n)) = T.$$

The semantics of the standard propositional connectives is as in classical logic, e.g.,

$$\langle \mathcal{M}, s_i, V \rangle \models \varphi \vee \psi \quad \text{iff} \quad \langle \mathcal{M}, s_i, V \rangle \models \varphi \text{ or } \langle \mathcal{M}, s_i, V \rangle \models \psi$$

The semantics of the unary future-time temporal operators is defined as follows

$$\langle \mathcal{M}, s_i, V \rangle \models \bigcirc \varphi \quad \text{iff} \quad \langle \mathcal{M}, s_{i+1}, V \rangle \models \varphi$$
$$\langle \mathcal{M}, s_i, V \rangle \models \Diamond \varphi \quad \text{iff there exists a } j \geq i \text{ such that } \langle \mathcal{M}, s_j, V \rangle \models \varphi$$
$$\langle \mathcal{M}, s_i, V \rangle \models \Box \varphi \quad \text{iff for all } j \geq i \text{ then } \langle \mathcal{M}, s_j, V \rangle \models \varphi.$$

The informal semantics of these operators is as follows: $\bigcirc \varphi$ means that φ must be satisfied in the *next* state; $\Diamond \varphi$ means that φ must be satisfied at *some* state in the future; $\Box \varphi$ means that φ must be satisfied at *all* states in the future.

The two binary future-time temporal operators that we use are interpreted as follows

$\langle \mathcal{M}, s_i, V \rangle \models \varphi \mathcal{U} \psi$ iff there exists a $k \geq i$ such that $\langle \mathcal{M}, s_k, V \rangle \models \psi$
and for all $i \leq j < k$ then $\langle \mathcal{M}, s_j, V \rangle \models \varphi$
$\langle \mathcal{M}, s_i, V \rangle \models \varphi \mathcal{W} \psi$ iff for all $j \geq i$ then $\langle \mathcal{M}, s_j, V \rangle \models \varphi$
or $\langle \mathcal{M}, s_i, V \rangle \models \varphi \mathcal{U} \psi$

If past-time temporal formulae are interpreted at a particular state, s_i, then states with indices less than i are 'in the past' of the state s_i. The semantics of unary past-time operators is given as follows:

$\langle \mathcal{M}, s_i, V \rangle \models \blacklozenge \varphi$ iff $i = 0$ or $\langle \mathcal{M}, s_{i-1}, V \rangle \models \varphi$
$\langle \mathcal{M}, s_i, V \rangle \models \lozenge \varphi$ iff $i > 0$ and $\langle \mathcal{M}, s_{i-1}, V \rangle \models \varphi$
$\langle \mathcal{M}, s_i, V \rangle \models \blacklozenge \varphi$ iff there exists j such that $0 \leq j < i$ and $\langle \mathcal{M}, s_j, V \rangle \models \varphi$
$\langle \mathcal{M}, s_i, V \rangle \models \blacksquare \varphi$ iff for all j such that $0 \leq j < i$ then $\langle \mathcal{M}, s_i, V \rangle \models \varphi$

Note that, in contrast to the future-time operators, the $\blacklozenge$ and $\blacksquare$ operators are interpreted as being *strict*, i.e. the current index is not included in the definition. Also, as there is a unique start state, termed the *beginning of time*, two different last-time operators are used. The difference between '$\lozenge$' and '$\blacklozenge$' is that for any formula φ, $\lozenge \varphi$ is false at the beginning of time, while $\blacklozenge \varphi$ is true at the beginning of time. In particular, $\blacklozenge$**false** is only true when interpreted at the beginning of time; otherwise it is false. Note that, as the formula $\blacklozenge$**false** appears so regularly in Concurrent METATEM, we often abbreviate it with the nullary operator '**start**', thus making obvious its association with the beginning of time (and the beginning of execution).

The semantics for the binary past-time operators $\mathcal{S}$ and $\mathcal{Z}$ relates to that for $\mathcal{U}$ and $\mathcal{W}$ just as the unary past-time operators relate to the unary future-time operators. Finally, the semantics of quantifiers is defined as follows.

$\langle \mathcal{M}, s, V \rangle \models \forall x. \varphi$ iff for all $d \in \mathcal{D}. \langle \mathcal{M}, s, V \dagger [x \mapsto d] \rangle \models \varphi$
$\langle \mathcal{M}, s, V \rangle \models \exists x. \varphi$ iff there exists $d \in \mathcal{D}.$ such that $\langle \mathcal{M}, s, V \dagger [x \mapsto d] \rangle \models \varphi$

As the interpretation consists of a triple, comprising model, state, and assignment components, a well-formed formula, φ, is *satisfied* in a particular model, $\mathcal{M}$, at the beginning of time, s_0, under a particular variable assignment, V, if $\langle \mathcal{M}, s_0, V \rangle \models \varphi$.

2.3 Executable Rules

Next, we define the subset of FML that can be used in a Concurrent METATEM program. The description of a Concurrent METATEM object is a set of *rules*, represented by

$$\Box \bigwedge_i R_i$$

where each R_i is, in turn, of the form

'past and present formula' **implies** 'present or future formula'

Taking quantification into account, the general form of these rules becomes

$$\Box \bigwedge_{i=1}^{n} \forall \bar{X}_i.\ P_i(\bar{X}_i) \Rightarrow F_i(\bar{X}_i)$$

where '$\bar{X}_i$' represents a vector of variables, $X_{i_1}, X_{i_2}, \ldots, X_{i_m}$.

Each rule is further restricted to be one of the following.

$$\forall \bar{X}. \quad [\mathbf{start} \wedge \bigwedge_{b=1}^{h} l_b(\bar{X})] \Rightarrow \bigvee_{j=1}^{r} m_j(\bar{X}) \quad \text{(an } \textit{initial } \Box\text{-rule)}$$

$$\forall \bar{X}. \ [(\mathbf{O} \bigwedge_{a=1}^{g} k_a(\bar{X})) \wedge \bigwedge_{b=1}^{h} l_b(\bar{X})] \Rightarrow \bigvee_{j=1}^{r} m_j(\bar{X}) \quad \text{(a } \textit{global } \Box\text{-rule)}$$

$$\forall \bar{X}. \quad [\mathbf{start} \wedge \bigwedge_{b=1}^{h} l_b(\bar{X})] \Rightarrow \Diamond l(\bar{X}) \quad \text{(an } \textit{initial } \Diamond\text{-rule)}$$

$$\forall \bar{X}. \ [(\mathbf{O} \bigwedge_{a=1}^{g} k_a(\bar{X})) \wedge \bigwedge_{b=1}^{h} l_b(\bar{X})] \Rightarrow \Diamond l(\bar{X}) \quad \text{(a } \textit{global } \Diamond\text{-rule)}$$

where each k_a, l_b, m_j or l is a literal. Note that the left-hand side of each initial rule is a constraint only on the *first* state, while the left-hand side of each global rule represents a constraint upon a state together with its predecessor. The right-hand side of each $\Box$-rule is simply a disjunction of literals referring to the current state, while the right-hand side of each $\Diamond$-rule is a single eventuality (i.e., '$\Diamond$' applied to a literal).

Note also that, although arbitrary FML formulae can be transformed into a set of rules of the form

$$\forall \bar{X}. \quad [(\forall \bar{Y}.\ \mathbf{start} \wedge \bigwedge_{b=1}^{h} l_b(\bar{X}, \bar{Y})) \Rightarrow \exists \bar{Z}. \bigvee_{j=1}^{r} m_j(\bar{X}, \bar{Z})]$$

$$\forall \bar{X}. \ [(\forall \bar{Y}.\ (\mathbf{O} \bigwedge_{a=1}^{g} k_a(\bar{X}, \bar{Y})) \wedge \bigwedge_{b=1}^{h} l_b(\bar{X}, \bar{Y})) \Rightarrow \exists \bar{Z}. \bigvee_{j=1}^{r} m_j(\bar{X}, \bar{Z})]$$

$$\forall \bar{X}. \quad [(\forall \bar{Y}.\ \mathbf{start} \wedge \bigwedge_{b=1}^{h} l_b(\bar{X}, \bar{Y})) \Rightarrow \exists \bar{Z}.\ \Diamond l(\bar{X}, \bar{Z})]$$

$$\forall \bar{X}. \ [(\forall \bar{Y}.\ (\mathbf{O} \bigwedge_{a=1}^{g} k_a(\bar{X}, \bar{Y})) \wedge \bigwedge_{b=1}^{h} l_b(\bar{X}, \bar{Y})) \Rightarrow \exists \bar{Z}.\ \Diamond l(\bar{X}, \bar{Z})]$$

where each k_a, l_b, m_j or l is a literal, we here choose to execute only a subset of this normal form [14].

2.4 Execution within Objects

We now describe how a set of rules for a given object in Concurrent METATEM is executed in order to provide its basic behaviour. Concurrent METATEM uses a set of 'rules' of the above form to represent an object's internal definition. Due to the outer '$\Box$' operator present in this rule form, the rules are applied at every moment in time (i.e., at every step of the execution) during the construction of a model for the formula.

As an example of a simple set of rules for a single object, consider the following. (Note that these rules are not meant to form a 'meaningful' program – they are only given for illustrative purposes.)

$$\begin{aligned} \textbf{start} &\Rightarrow \texttt{popped(a)} \\ \bullet\texttt{pop(X)} &\Rightarrow \Diamond\texttt{popped(X)} \\ \bullet\texttt{push(Y)} &\Rightarrow \texttt{stack-full()} \vee \texttt{popped(Y)} \end{aligned}$$

Note that the 'X' and 'Y' here represent universally quantified variables. Looking at these program rules, we see that popped(a) is satisfied at the beginning of time and whenever pop(X) is satisfied in the previous moment in time, a commitment to eventually satisfy popped(X) is given. Similarly, whenever push(Y) is satisfied in the previous moment in time, then either stack-full() or popped(Y) must be satisfied.

The temporal language which forms the basis of execution within particular objects in Concurrent METATEM also provides two orthogonal mechanisms for representing choice. These are

- static indeterminacy, through the classical operator '$\vee$', and,
- temporal indeterminacy, through '$\Diamond$', the temporal operator representing *sometime in the future*.

A logical description, containing the '$\vee$' operator, of the properties of a given state represents a choice about the exact nature of the state. Although other constraints upon the state might restrict this choice, the potential for a completely non-deterministic choice is present. A formula such as $\Diamond a$ states that the proposition a must be satisfied *at some time in the future*. Thus, it represents a form of temporal indeterminacy. However, we do not model this as a truly non-deterministic choice. Rather, given a constraint, such as $\Diamond a$, the execution mechanism attempts to satisfy a as soon as possible (taking in to account any other temporal constraints). Thus, a formula such as

$$\Diamond a \wedge \Diamond b \wedge \Diamond c$$

when executed, would ensure that a, b and c are all satisfied as soon as possible. If necessary, we can add further temporal formulae representing extra ordering within these constraints, for example to ensure that c can not be satisfied until both a and b have been satisfied.

Once the object has commenced execution, it continually follows a cycle of checking which rules have antecedents satisfied by the previous state, conjoining together the consequents of these rules, rewriting this conjunction into a disjunctive form and choosing one of these disjuncts to execute. From this disjunct a state is constructed, with predicates remaining false unless otherwise constrained. The computation then moves forward to the next state where this cycle begins again. If a contradiction is found, it may be possible to backtrack to a previous choice (but see §2.5 for restrictions). This choice is constrained by the currently outstanding eventualities, i.e., formulae of the form $\Diamond a$ that have not yet been satisfied. As many as possible of these formulae are satisfied in the state constructed, starting with oldest outstanding eventuality.

Note that this basic execution mechanism is similar to that provided for (sequential) METATEM [3, 10].

We now describe the general operational model for Concurrent METATEM objects executing in the above manner. This incorporates the asynchronous execution of individual objects, dynamic attributes of object interfaces, and the communication mechanism between objects.

2.5 Concurrent Object-Based Operational Model

The computational model used in Concurrent METATEM combines the two notions of *objects* and *concurrency*. Objects are here considered to be self contained entities, encapsulating both data and behaviour, and communicating via message-passing. In particular, Concurrent METATEM has the following fundamental properties.

1. The basic mechanism for communication between objects is *broadcast* message-passing.
2. Objects are not message driven — they begin executing from the moment they are created and continue even while no messages are received.
3. Each object contains a set of temporal rules representing its behaviour.
4. Objects execute asynchronously.

So, rather than seeing computation as objects sending mail messages to each other, and thus invoking some activity (as in Actor systems [2]), computation in a collection of Concurrent METATEM objects can be visualised as independent entities *listening* to messages broadcast from other objects.

In addition to this basic framework, the operational model exhibits several important features which are described in more detail below.

Communication Mechanism Broadcast communication is used, not only as it is a flexible and powerful communication mechanism, but also as it has a logical meaning within our system, namely that of passing valuations for predicates throughout the system.

In order to implement communication as an integral part of execution, we categorise the basic predicates used in the language, with several categories of predicate corresponding to messages to and from the object, as follows.

- *Environment* predicates, which represent incoming messages.
  An environment predicate can be made true if, and only if, the corresponding message has just been received. Thus, a formula containing an environment predicate, such as 'push(Y)', is only true if a message of the form 'push(b)' has just been received (for some argument 'b', which unifies with 'Y').

- *Component* predicates, which represent messages broadcast from the object.
  When a component predicate is made true, it has the (side-)effect of broadcasting the corresponding message to the environment. For example, if the formula 'popped(e)' is made true, where popped is a component predicate, then the message 'popped(e)' is broadcast.
- *Internal* predicates, which have no external effect.
  These predicates are used as part of formulae participating in the internal computation of the object and, as such, do not correspond either to message-sending or message reception.
  This category of predicates may include various *primitive* operations.

Object Interfaces Networks of Concurrent METATEM objects communicate via broadcasting messages and individual objects only act upon certain identified messages. Thus, an object must be able to filter out messages that it wishes to recognise, ignoring all others. The definition of which messages an object recognises, together with a definition of the messages that an object may itself produce, is provided by the *interface definition* for that particular object.

The interface definition for an object, for example 'stack', is defined in the following way

```
stack(pop,push)[popped,stack-full].
```

Here, {pop, push} is the set of messages the object recognises, while the object itself is able to produce the messages {popped, stack-full}. Note that these sets of messages need not be disjoint – an object may broadcast messages that the object itself recognises. In this case, messages sent by an object to itself are recognised immediately. Note also that many distinct objects may broadcast and recognise the same messages.

Backtracking In general, if an object's execution mechanism is based on the execution of logical statements, then a computation may involve backtracking. Objects may backtrack, with the proviso that an object may not backtrack past the broadcasting of a message. Consequently, in broadcasting a message to its environment, an object effectively *commits* the execution to that particular path. Thus, the basic operation of an object can be thought of as a period of internal execution, possibly involving backtracking, followed by appropriate broadcasts to its environment. The analogy with a collections of humans is of a period of thinking, followed by some (broadcasted) action, e.g. speech. Backtracking can occur during thinking, but once an action has been carried out, it cannot be undone.

3 Extensions of the Basic System

Above, we have described the principles behind the basic Concurrent METATEM system. In this section, we will outline various extensions that have ither been implemented, or are actively under development.

3.1 Autonomous Objects

There are a variety of extensions to Concurrent METATEM that allow objects to have control, to some extent, over their own execution. In particular, these extensions allow objects to control their interface with their environment.

Dynamic Interfaces The interface definition of an object defines the initial set of messages that are recognised by that object. However, the object may dynamically change the set of messages that it recognises. In particular, an object can either start 'listening' for a new type of message, or start 'ignoring' previously recognised message types. For example, given an original object interface such as

stack(pop,push) [popped,stack-full]

the object may dynamically choose to stop recognising 'pop' messages, for example by executing 'ignore(pop)'. This effectively gives the object the new interface

stack(push) [popped,stack-full].

Dynamic Message Queues Each object in the system has, associated with it, a *message queue* representing the messages that the object has recognised, but has yet to process. The number of messages that an object reads from its message queue during an execution step is initially defined by the object's interface. In principle, the execution of an object is based on the set of messages received by the object since the last execution step it completed. Thus, Concurrent METATEM objects process sets of messages, rather than enforcing some linearisation on the order of arrival of messages.

We are developing extensions to enable the object to dynamically modify its own behaviour regarding the manipulation of its message queue. The default behaviour of an object is for it to, at every execution step, read a sequence of messages from its input queue up to either the end of the queue, or the repetition of a message. For example, if the message queue is

pop(a), push(c), pop(d), pop(e), push(c), pop(f), pop(a), ...

then in one execution step the set of messages read in is

{ pop(a), push(c), pop(d), pop(e) }

and the message queue remaining is

push(c), pop(f), pop(a), ...

This behaviour can be modified dynamically so that, for example, objects read only one message at a time from their input queue, or read up to the second occurrence of either of several messages. Complex varieties of message queue manipulation may be defined, depending on the properties of the Concurrent METATEM objects. Such an extension is being developed through the use of extended primitive predicates (an alternative approach is through the use of meta-level features [4]).

3.2 Synchronisation Mechanisms

There are two approaches to the synchronisation of asynchronously executing objects in Concurrent METATEM, outlined as follows.

1. An object asks for something, continues processing, but does something when the answer arrives.

    Thus, a request is made and execution continues, but when a reply is received from the environment, some suitable action is taken.

    $$\ldots \Rightarrow \texttt{ask}$$
    $$\bigcirc \;\; \texttt{answer} \Rightarrow \texttt{do\_it}$$

    Here ask is a component predicate and answer is an environment predicate.

2. An object asks for something, *waits* for an answer, and does something when the answer arrives.

    To suspend execution whilst waiting for a synchronisation message, an extra synchronisation rule is added to (1) above, giving:

    $$\ldots \Rightarrow \texttt{ask}$$
    $$\bigcirc \;\; \texttt{answer} \Rightarrow \texttt{do\_it}$$
    $$\bigcirc \;\; \texttt{ask} \Rightarrow \texttt{answer}$$

    If an ask message has been sent, then the only way to satisfy this last rule is to ensure that answer is received in the next state. Thus, the object cannot execute further until the required message arrives, and consequently it is suspended.

    Note that, until the appropriate synchronisation message arrives, execution of the process suspends and no further incoming messages are processed (though they are recorded).

3.3 Point-to-point message-passing

Although broadcast message passing has been defined as the primitive mechanism for communication, point-to-point message-passing can be implemented on top of this. Such a mechanism can either be provided as part of the implementation of Concurrent METATEM, and made available through extra primitives, or can be provided by utilising meta-level features within each Concurrent METATEM object (see [4] for an outline of these basic features). To provide point-to-point message-passing using meta-level features, every Concurrent METATEM object (that wishes to take part in such a scheme) must include a 'meta-rule' of the form

$$\texttt{send(me,X)} \Rightarrow \texttt{X}$$

within its definition. Here 'me' is the name of that particular object. Thus, whenever an object, obj1, wishes to send the message 'p(a)' to another object, obj2, the following message must be broadcast by obj1:

$$\texttt{send(obj2,p(a))}.$$

If such a scheme is enforced in all objects, then point-to-point message passing is available throughout the system.

An alternative approach is to add an extra *destination* argument to each message. Thus, to send the message 'p(a)' to object 'obj2', we broadcast

$$p(obj2,a)$$

and ensure that the rules in obj2 itself deal with this appropriately.

3.4 Synchronous Concurrent METATEM

A variation on the basic execution scheme for Concurrent METATEM is the development of *Synchronous* Concurrent METATEM. In this system, each object executes in step and messages sent from any object reach all other objects by the start of the next step. This simplification removes the need for synchronisation using environment predicates, but also reduces the flexibility of the approach (see [5] for an outline of this approach).

3.5 Groups

Finally, objects are also members of *groups*. Each object may be a member of several groups. When an object sends a message, that message is, by default, broadcast to all the members of its group(s), but to no other objects. Alternatively an object can select to broadcast only to certain groups (of which it is a member). This mechanism allows the development of complex structuring within the object space and provides the potential for innovative applications, such as the use of groups to represent physical properties of objects. For example, if we assume that any two objects in the same group can 'see' each other, then movement broadcast from one object can be detected by the other object. Similarly, if an object moves 'out of sight', it moves out of the group and thus the objects that remain in the group 'lose sight' of it. Examples, such as this (see also §4.3), show some of the power of the group concept.

4 Examples

In this section we provide a range of examples exhibiting some of the applications of Concurrent METATEM. Several of these examples represent abstractions of particular application areas and, within each example, we will attempt to identify the properties of Concurrent METATEM that make it suitable for use in that particular area. In addition to those provided here, other 'standard' examples, such as the dining philosophers and producer/consumer problems can be defined using Concurrent METATEM.

4.1 Snow White and The Seven Dwarves — A tale of 8 objects

We will first give a 'toy' example. This not only provides a simple and appealing introduction to Concurrent METATEM, but also exhibits some of the features used later in more complex systems. This example, taken from [15], is a descendent of the 'resource controller' example used in earlier papers on METATEM and, as such, is related to a variety of resource allocation systems. However, the individual objects can be specified so that they show a form of 'intelligent' behaviour, and so this example system is also related to applications in Distributed AI [12]. First, a brief outline of the properties of the leading characters in this example will be given.

The Scenario Snow White has a bag of sweets. All the dwarves want sweets, though some want them more than others. If a dwarf asks Snow White for a sweet, she will give him one, but maybe not straight away. Snow White is only able to give away one sweet at a time.

Snow White and the dwarves are going to be represented as a set of objects in Concurrent METATEM. Each dwarf has a particular strategy that it uses in asking for sweets, which is described below.

1. eager initially asks for a sweet and, from then on, whenever he receives a sweet, asks for another.
2. mimic asks for a sweet whenever he sees eager asking for one.
3. jealous asks for a sweet whenever he sees eager receiving one.
4. insistent asks for a sweet as often as he can.
5. courteous asks for a sweet only when eager, mimic, jealous and insistent have all *asked* for one.
6. generous asks for a sweet only when eager, mimic, jealous, insistent and courteous have all *received* one.
7. shy only asks for a sweet when he sees no one else asking.
8. snow-white can only allocate one sweet at a time. She keeps a list of outstanding requests and attempts to satisfy the oldest one first.
If a new request is received, and it does not occur in the list, it is added to the end. If it does already occur in the list, it is ignored. Thus, if a dwarf asks for a sweet n times, he will eventually receive at most n, and at least 1, sweets.

This example may seem trivial, but it represents a set of objects exhibiting different behaviours, where an individual object's internal rules can consist of both safety and liveness constraints, and where complex interaction can occur between autonomous objects.

The Program The Concurrent METATEM program for the scenario described above consists of the definitions of 8 objects, given below. To give a better idea of the meaning of the temporal formulae representing the internals of these objects, a brief description will be given with each object's definition. Requests to Snow White are given in the form of an ask() message with the name of the requesting dwarf as an argument. Snow White gives a sweet to a particular dwarf by sending a give() message with the name of the dwarf as an argument. Finally, upper-case alphabetic characters, such as X and Y represent universally quantified variables.

1.   eager(give)[ask] :
                start $\Rightarrow$ ask(eager)
           ◯give(eager) $\Rightarrow$ ask(eager)

Initially, eager asks for a sweet and, whenever he has just received a sweet, he asks again.

2. mimic(ask)[ask]:
$$\bullet \text{ask(eager)} \Rightarrow \text{ask(mimic)}$$
If eager has just asked for a sweet then mimic asks for one.

3. jealous(give)[ask]:
$$\bullet \text{give(eager)} \Rightarrow \text{ask(jealous)}$$
If eager has just received a sweet then jealous asks for one.

4. insistent[ask]:
$$\text{start} \Rightarrow \square \text{ask(insistent)}$$
From the beginning of time insistent asks for a sweet as often as he can.

5. courteous(ask)[ask]:
$$\begin{bmatrix} (\neg \text{ask(courteous)}) \, \mathcal{S} \, \text{ask(eager)} \, \wedge \\ (\neg \text{ask(courteous)}) \, \mathcal{S} \, \text{ask(mimic)} \, \wedge \\ (\neg \text{ask(courteous)}) \, \mathcal{S} \, \text{ask(jealous)} \, \wedge \\ (\neg \text{ask(courteous)}) \, \mathcal{S} \, \text{ask(insistent)} \end{bmatrix} \Rightarrow \text{ask(courteous)}$$
If courteous has not asked for a sweet since eager asked for one, has not asked for a sweet since mimic asked for one, has not asked for a sweet since jealous asked for one, and, has not asked for a sweet since insistent asked for one, then he will ask for a sweet.

6. generous(give)[ask]:
$$\begin{bmatrix} (\neg \text{ask(generous)}) \, \mathcal{S} \, \text{give(eager)} \, \wedge \\ (\neg \text{ask(generous)}) \, \mathcal{S} \, \text{give(mimic)} \, \wedge \\ (\neg \text{ask(generous)}) \, \mathcal{S} \, \text{give(jealous)} \, \wedge \\ (\neg \text{ask(generous)}) \, \mathcal{S} \, \text{give(insistent)} \, \wedge \\ (\neg \text{ask(generous)}) \, \mathcal{S} \, \text{give(courteous)} \end{bmatrix} \Rightarrow \text{ask(generous)}$$
If generous has not asked for a sweet since eager received one, has not asked for a sweet since mimic received one, has not asked for a sweet since jealous received one, has not asked for a sweet since insistent received one, and, has not asked for a sweet since courteous received one, then he will ask for a sweet!

7. shy(ask)[ask]:
$$\text{start} \Rightarrow \Diamond \text{ask(shy)}$$
$$\bullet \text{ask(X)} \Rightarrow \neg \text{ask(shy)}$$
$$\bullet \text{ask(shy)} \Rightarrow \Diamond \text{ask(shy)}$$
shy initially wants to ask for a sweet but is prevented from doing so whenever he sees some other dwarf asking for one. Thus, he only succeeds in asking for one when he sees no one else asking and, as soon as he has asked for a sweet, he wants to try to ask again!

8. snow-white(ask)[give]:
$$\bullet \text{ask(X)} \Rightarrow \Diamond \text{give(X)}$$
$$\text{give(X)} \wedge \text{give(Y)} \Rightarrow X=Y$$
If snow-white has just received a request from a dwarf, a sweet will be sent to that dwarf eventually. The second rule ensures that sweets can not be sent to two dwarves at the same time by stating that if both give(X) and give(Y) are to be broadcast, then X must be equal to Y.

Note that, in this example, several of the dwarves were only able to behave as required because they could observe all the ask() and give() messages that were broadcast. The dwarves can thus be programmed to have strategies that are dependent on the behaviour of other dwarves. Also, the power of executable temporal logic is exploited in the definition several objects, particularly those using the '$\Diamond$' operator to represent multiple goals.

Though this example is fairly simple, it does give some idea of how complex interacting systems can be developed using Concurrent METATEM. It also shows how useful the model is, particularly when 'intelligent' objects (agents) are considered.

We also note that, as the objects behaviour is represented explicitly, and in a logical way, the verification of properties of the system is possible. For example, given the objects' definitions, we are able to prove that every dwarf, except 'shy' will eventually receive a sweet. (For further work on the verification of properties of such systems, see [13].)

4.2 Cooperative and Competitive Behaviour

We now extend the type of example given above to incorporate the notion of some other resource that the dwarves can attempt to exchange with Snow White for sweets, e.g. money. Note that this example is taken from [11].

The purpose of this extension of the scenario is to show how not only can the behaviours of single objects be developed, but also how more complex social structures can be represented. In particular, we outline how both cooperation and competition can be represented in Concurrent METATEM.

Bidding Initially, we will simply change the ask predicate so that it takes an extra argument representing the amount the dwarf is willing to pay for a sweet. This enables dwarves to 'bid' for a sweet, rather than just asking for one. For example, dwarf1 below asks for a sweet, bidding '2'.

$$\text{dwarf1()[ask]} :$$
$$\textbf{start} \Rightarrow \text{ask(dwarf1,2)}$$

We can further modify a dwarf's behaviour so that it does not bid more than it can afford by introducing some record of the amount of money that the dwarf has at any one time. Thus, the main rule defining the 'bidding' behaviour of a dwarf might become something like

$$\bullet[\text{money(N)} \land \text{N} \geq 2] \Rightarrow \text{ask(dwarf1,2)}$$

Note that the behaviour of Snow White might also change so that all the bids are recorded then a decision over which bid to accept is made based upon the bids received. Once a decision is made, give is again broadcast, but this time having an extra argument showing the amount paid for the sweet. For example, if Snow White accepts the bid of '2' from dwarf1, then give(dwarf1,2) is broadcast.

Finally, a dwarf whose bid has been accepted, in this case dwarf1, must remember to record the change in finances:

$$\bullet[\text{money(N)} \land \text{give(dwarf1,C)}] \Rightarrow \text{money(N-C)}$$

Renewable Resources Dwarves who keep buying sweets will eventually run out of money. Thus, we may want to add the concept of the renewal of resources, i.e., being paid. This can either happen at a regular period defined within each dwarf's rules, e.g.

$$\textbf{start} \Rightarrow \texttt{money(100)} \land \texttt{paid}$$
$$\bigcirc[\texttt{money(N)} \land \bigcirc\bigcirc\bigcirc\bigcirc\texttt{paid}) \Rightarrow \texttt{money(N+100)} \land \texttt{paid}$$

or the dwarf can replenish its resources when it receives a particular message from its environment, e.g.

dwarf1(go)[ask] :
$$\textbf{start} \Rightarrow \texttt{money(100)}$$
$$\bigcirc(\texttt{money(N)} \land \texttt{go}) \Rightarrow \texttt{money(N+100)}$$

Competitive Bidding As the bids that individual dwarves make are broadcast, other dwarves can observe the bidding activity and can revise their bids accordingly. We saw earlier that the 'mimic' dwarf asks for a sweet when it sees the 'eager' dwarf asking for one. Similarly, dwarf2 might watch for any bids by dwarf1 and then bid more, e.g.,

$$\bigcirc[\texttt{ask(dwarf1,B)} \land \texttt{myhigh(M)} \land \texttt{B>M}] \Rightarrow \texttt{ask(dwarf2,B+1)} \land \texttt{myhigh(B+1)}$$

Although we will not give further detailed examples in this vein, it is clear that a range of complex behaviours based upon observing others' bids can be defined.

Borrowing Money Above we showed how individual dwarves might compete with each other for Snow White's sweets. Now, we will consider how dwarves might *co-operate* in order to get sweets from Snow White. In particular, we consider the scenario where one dwarf on its own does not have enough money to buy a sweet, and thus requires a loan from other dwarves. In order to borrow money from other dwarves to enable a single dwarf to buy a sweet, the dwarf can broadcast a request for a certain amount. For example, if the dwarf (dwarf3 in this case) knows that the highest amount bid for a sweet so far is X and he only has Y, then he can ask to borrow X-Y, possibly as follows.

dwarf3(lend)[borrow, ask] :
$$\bigcirc[\texttt{highest(X)} \land \texttt{money(Y)} \land \texttt{X>Y}] \Rightarrow \texttt{borrow(dwarf3,(X-Y)+1)}$$

Now, if another dwarf, say dwarf4, offers to lend a certain amount, say Z, to dwarf3, then another rule recording the loan must be added to dwarf3's rule-set:

$$\bigcirc[\texttt{lend(dwarf4,dwarf3,Z)} \land \texttt{money(Y)}] \Rightarrow \texttt{money(Y+Z)} \land \texttt{owe(dwarf4,Z)}$$

Lending Behaviour Dwarves might have various strategies of lending and borrowing money. For example, perhaps a dwarf won't lend any more money to any dwarf who still owes money. Further, a dwarf might be less likely to lend money to any dwarf who has never offered to help his previous requests.

Again, a variety of strategies for lending and borrowing in this way can be coded in Concurrent METATEM. Rather than giving further examples of this type, we next consider the use of groups in the development of structured systems of interacting objects.

4.3 Societies of Dwarves

As described earlier, as well as the notion of autonomous objects, Concurrent METATEM also provides a larger structuring mechanism through the 'groups' extension. This restricts the extent of an object's communications and thus provides an extra mechanism for the development of strategies for organisations. Rather than giving detailed examples, we will outline how the group mechanism could be used in Concurrent METATEM to develop further cooperation, competition and interaction amongst objects.

Again, we will consider a scenario similar to Snow White and the Seven Dwarves described above, but will assume the existence of a large number of dwarves, and possibly several Snow White's! We will outline several examples of how the grouping of these objects can be used to represent more complex or refined behaviour.

Collective Bidding If we again have a situation where dwarves bid for sweets, then we can organise cooperation within groups so that the group as a whole puts together a bid for a sweet. If successful, the group must also decide who to distribute the sweet to. Thus, a number of groups might be cooperating internally to generate bids, but competing (with other groups) to have their bid accepted.

Forming Subgroups Within a given group, various subgroups can be formed. For example, if several members of a group are unhappy with another member's behaviour, they might be able to create a new subgroup within the old grouping which excludes the unwanted object. Note that members of the subgroup can hear the outer group's communications, while members of the outer one cannot hear the inner group's communications. Although we have described this as a retributive act, such dynamic restructuring is natural as groups increase in size.

As we have seen above, by using a combination of individual object strategies and of grouping objects together, we are able to form simple societies. In particular, we can represent societies where individuals cooperate with their fellow group members, but where the groups themselves compete for some global resource. Although our examples have been based upon objects competing and cooperating in order to get a certain resource, many other types of multi-agent system can be developed in Concurrent METATEM. Finally, it is important to note that there is no explicit global control or global plan in these examples. Individual objects perform local interactions with each other and their environment.

4.4 Distributed Problem Solving

The next example we will look at is taken from [13], and defines simple, abstract distributed problem solving systems. We will assume that individual problem-solving objects can be implemented in Concurrent METATEM and will look at how such objects can be organised to form useful problem-solving architectures.

Hierarchical Problem Solving Once we have defined individual problem solvers, for example simple planner objects, we can construct distributed problem solving systems. The simplest approach is for one object to assign sub-plans to individual problem-solvers. This can be achieved by defining an appropriate manager object which knows about a variety of other problem-solving objects, can split a plan up and can assign a particular sub-problem to a specific object. This approach would involve the use of point-to-point message passing as the manager only wishes to pass each task on to one specific object.

Group Problem Solving While the above approach is often used in real problem-solving systems, a more dynamic and flexible architecture can be defined by utilising the facility, in Concurrent METATEM, for having multiple objects recognising the same messages. In particular, we can, in a way similar to the Contract Net approach [22], broadcast sub-problems to be solved to a set of objects. Each object might attempt to solve the problem in its own way, but the top-level object again waits until at least one solution has been returned from the set of objects. Note that the manager object in this case need not know exactly what other problem-solving objects exist.

We might define such a Concurrent METATEM system below.

```
manager(solution1)[problem1,solved1]:
    start ⇒ ◇problem1;
    ●solution1 ⇒ solved1.

solvera(problem2)[solution2]:
    ●problem2 ⇒ solution2.

solverb(problem1)[solution2]:
    ●problem1 ⇒ ◇solution1.

solverc(problem1)[solution1]:
    ●problem1 ⇒ ◇solution1.
```

Here, solvera can solve a different problem from the one manager poses, while solverb can solve the desired problem, but doesn't announce the fact (as solution1 is not a component proposition for solverb); solverc can solve the problem posed by manager, and will *eventually* reply with the solution.

Cooperative Problem-Solving We now look at a refinement of the above system, where solverc has been removed and replaced by by two objects which together can solve problem1, but can not manage this individually. These objects, called solverd and solvere are defined below.

```
solverd(problem1,solution1.2)[solution1]:
    (●solution1.2 ∧ ◆problem1) ⇒ ◇solution1.

solvere(problem1)[solution1.2]:
    ●problem1 ⇒ ◇solution1.2.
```

Thus, when `solverd` receives the problem it cannot do anything until it has heard from `solvere`. When `solvere` receives the problem, it broadcasts the fact that it can solve part of the problem (i.e., it broadcasts `solution1.2`). When `solverd` sees this, it knows it can solve the other part of the problem and broadcasts the whole solution.

More complex problem-solving architectures can be developed in a similar way.

4.5 Process Control

Next, we will describe an example system presented in [8] which models simple rail networks and the movement of trains within them. The essential properties of this example are that each station is represented by a Concurrent METATEM object, and each object knows the identity of the next stations on its line(s). We then define a general protocol in terms of permissions for trains to enter stations and the general goal of each station to move trains on. Finally, we add initial conditions such as the initial placement of trains within the system. As the set of Concurrent METATEM objects execute, they communicate with each other in order to organise the movement of trains around the system.

The Scenario First we will give a brief outline of the problem, then we show how the elements can be abstractly modelled and prototyped using Concurrent METATEM. Here, we model, *abstractly*, the behaviour of several networks of stations within a railway system. We do not intend to describe *all* the details, but simply present executable temporal logic programs that characterise the general behaviour of the system. Consequently, the model we use will not correspond directly to a real-life transport system. It represents an abstraction of such a system exhibiting several features fundamental to the behaviour of transport systems. We will look at various configurations of lines and stations, and will use the following simplifying restrictions.

- Each station can be occupied by at most one train.
- Each train is assigned to a particular line, and can only ever travel on that line.
- Trains can only travel in one direction on a particular line, and each line has a direction associated with it.
- Networks consists only of stations, connected to each other.

Sample Network We will now show how a particular rail configuration can be represented in Concurrent METATEM by constraints within objects. This example, while being simple, represents an abstraction of the typical properties of many networks. We describe the network using the following predicates:

`station(S)` — the names of the stations.
`line(T, L)` — the line L of each train T.
`has(S, T)` — a predicate storing the name of the train T currently at station S.
`next(S1, L, S2)` — a predicate which stores, for each station S1, and line L, the next station S2.

Whereas, in general, predicates can change value as the execution progresses, we constrain the station and next predicates as being constant throughout time.

The particular network given here consists of two lines, one moving clockwise and one moving anti-clockwise, as shown in Figure 1. Thus, the lines join at station 'D',

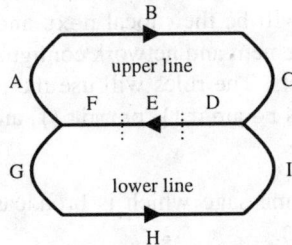

Fig. 1. A double ring network

split at station 'F', and use same track at station 'E'. So, if 'upper' is the line running through A–B–C–D–E–F–A and 'lower' is the line running through G–H–I–D–E–F–G, then the definitions of next are as follows

```
next(A,upper,B),  next(D,upper,E),  next(D,lower,E),  next(G,lower,H),
next(B,upper,C),  next(E,upper,F),  next(E,lower,F),  next(H,lower,I),
next(C,upper,D),  next(F,upper,A),  next(F,lower,G),  next(I,lower,D)
```

The initial configuration uses two trains on each line: trains '1' and '2' run only on the upper line, while trains '3' and '4' run only on the lower line. The initial placement of these four trains is train '1' at station 'C', train '2' at station 'E', train '3' at station 'F', and train '4' at station 'H'.

Station 'D' differs from the other stations in this example in that it has *two* entry points — station 'C' on the upper line and station 'I' on the lower line. This 'merging' station requires more sophisticated control rules than the single-entry stations. The initial configuration for the double ring is:

```
       has(train1,C),    has(train3,F),
       has(train2,E),    has(train4,G)
```

The Program We will now show how the rail configurations described above together with the movement of trains between stations can be modelled in Concurrent METATEM. Each station will be represented as a separate Concurrent METATEM object and we will introduce a communication protocol which will enable stations to negotiate the passing of trains between stations. Thus, rather than having a global model of the system, the state of each station will be represented locally, with the station's has and next constraints being private to that station. Thus, for example, the Concurrent METATEM object representing station 'F' in the double ring network described above would contain the following initial constraints

$$\text{station}(F), \quad \text{has}(\text{train3})$$

together with the following network details

$$\text{next}(\text{upper}, A), \quad \text{next}(\text{lower}, G)$$

Each separate station will use the same set of Concurrent METATEM rules. The only differences between stations will be their local next and has constraints. These formulae representing initial placement and network configuration will thus be partitioned amongst the appropriate objects. The rules will use the predicates defined earlier, together with the extra predicates request(), permit(), and moved(), whose effect can be described as follows.

request(S, T) — a request message which is broadcast asking for permission for train T to move to station S.

permit(S, T) — a message which is broadcast permitting train T to move to station S.

moved(S, T) — a message which is broadcast as train T moves to station S.

Thus, in order for a train to move, the station it occupies must request the 'next' station for permission to move the train on. A station can only give permission for a train to move to it if the station is currently empty and no permission has been given to another train. The interface definition for each station object is simply

$$\text{station}(\text{request}, \text{permit}, \text{moved})[\text{request}, \text{permit}, \text{moved}].$$

showing that each station object both recognises and broadcasts instances of request, permit and moved messages. The internal definition of each station is represented by a set of 9 rules, described below. Note that, as usual, all upper-case letters represent universally quantified variables. Also, the variables T1, T2 and T3 represent trains, while S1 and S2 represent stations.

1. $\bigcirc[\text{has}(T1) \wedge \text{line}(T1, L) \wedge \text{next}(L, S1) \wedge \neg\text{moved}(S1, T1)] \Rightarrow \text{has}(T1)$
   If a station had a train, and the train has not moved, then the station still has that train. Note that this can be seen as a simple frame condition.

2. $\left[\begin{array}{c} \bigcirc[\text{has}(T1) \wedge \text{line}(T1, L) \wedge \text{next}(L, S1)] \\ \wedge \\ (\neg\text{request}(S1, T1))\mathcal{Z}(\neg\text{has}(T1)) \end{array} \right] \Rightarrow \text{request}(S1, T1)$

   If a station had a train, and a request to move the train on has not been made during this time, then the station will make such a request to the next station on the train's line.

3. $\bigcirc[\text{has}(T1) \wedge \text{line}(T1, L) \wedge \text{next}(L, S1) \wedge \text{permit}(S1, T1)] \Rightarrow \text{moved}(S1, T1)$
   If a station had a train, and it has just received permission to move the train on to its next station, then the train can be moved to its next station.

4. $\bigcirc[\text{station}(S2) \wedge \text{moved}(S2, T1)] \Rightarrow \text{has}(T1)$
   If a train has just moved to a station, then that station has it.

5. ⬤ [station(S2) ∧ moved(S2,T1)] ⟹ ¬permit(S2,T2)
   If a train has just moved to a station, then that station cannot give permission to any other trains to move to that station.
6. ⬤ [station(S2) ∧ request(S2,T1)] ⟹ ◇permit(S2,T1)
   If a station receives a request for a train to be moved to the station, then eventually it will give permission for this move.
7. [station(S2) ∧ permit(S2,T2) ∧ permit(S2,T3)] ⟹ T2 = T3
   A station can only ever give permission to one move at a time.
8. ⬤ [station(S2) ∧ has(T1)] ⟹ ¬permit(S2,T2)
   If a station has a train, then it can not give permission to any train wanting to move to the station.
9. ⬤ station(S2) ∧ (¬moved(S2,T1)) $\mathcal{S}$ permit(S2,T1) ⟹ ¬permit(S2,T2)
   If a station has given permission for a move, but the move has not yet occurred, then the station can not give permission for any move (until the outstanding move has been completed).

Again, though the above examples are simple abstractions of railway networks, they exhibit features found in much larger, more complex rail systems. We also note that the temporal rules within each object encapsulate both the protocol for transferring trains and the goal of moving them along the line. As each station is represented by an object, it has complete control over both the station's state and the permissions granted to other stations to send trains forward.

4.6 Fault Tolerant Systems

We next consider an application which utilises the combination of broadcast message-passing and group structuring found in Concurrent METATEM. Many real-life fault-tolerant systems, such as Distributed Operating Systems [6], utilise the power of broadcast message-passing to provide duplication of important resources. One common approach is that of 'process pairs' [7]. In this section, we outline the use of process pairs for fault-tolerance, exhibiting the prower of the model of communication that we use. It will hopefully be obvious to the reader that such systems can also be developed in Concurrent METATEM.

Consider an 'essential' process in a distributed operating system, for example a network server of some kind. This can be represented as the *primary* process in Figure 2. Here, this process, if represented in Concurrent METATEM, might have an interface of the form

$$\text{primary}(r1, \ldots, rn)[a1, \ldots, am]$$

where r1 to rn are the requests that the process recognises, while a1 to am are the replies that it might send. The *secondary* object listens for exactly the same messages — this is possible since all requests are broadcast. Further, the secondary object *simulates* the behaviour of the primary object, with the exception that it does not broadcast any messages of its own. Thus the secondary object would have the following interface.

$$\text{secondary}(r1, \ldots, rn)[\ ]$$

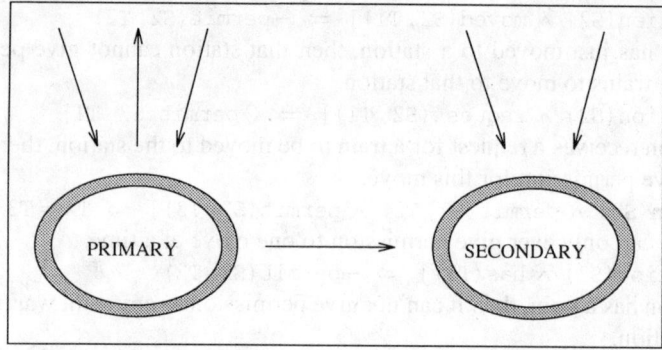

Fig. 2. A Process and its 'shadow object'

Now, the secondary object also notes the primary object's activity. If the primary object has not responded to a request for a specific amount of time, the secondary object assumes that the primary object's processor has 'crashed' and takes over the duties associated with the service. This requires changing the object's interface so that it now broadcasts replies. At this stage, the secondary object has become the primary one and it spawns a new copy of itself to act as the new secondary one. This final configuration is shown in Figure 3. In this way, fault-tolerance can be added 'seamlessly' to certain

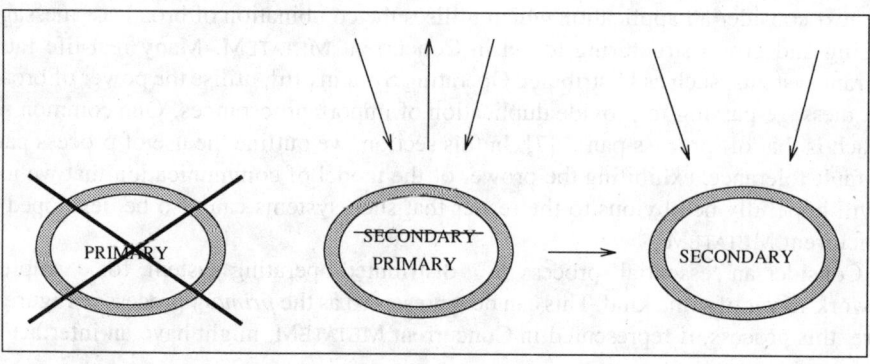

Fig. 3. Secondary object takes over.

types of distributed systems. Note that, to implement this in Concurrent METATEM we require both the ability to dynamically change an object's interface and the ability to create new objects. It should be clear from this example, that it is the use of *broadcast* message-passing that is the main reason for the utility of such an approach.

4.7 Heterogeneous Systems

Finally, we note that the set of rules in each Concurrent METATEM object is effectively a specification of the behaviour of that object (under a particular operational interpretation). Thus, we can also view Concurrent METATEM as a framework for specifying heterogeneous systems, with each object being implemented in several possible ways. As long as each object satisfies its specification and observes the protocols regarding communication, then the *real* implementation of the object can be anything: a Cobol program, a Prolog program, a 'real' item of hardware, even a human.

5 Concluding Remarks

Concurrent METATEM not only provides a novel model for the simulation and implementation of reactive systems, but also incorporates executable temporal logic to implement individual objects. Consequently, such objects have explicit logical semantics and the behaviour of certain types of system is easily coded as temporal logic rules. Concurrent METATEM has potential applications in a wide range of areas, for example simulation and programming in concurrent and distributed systems, the development of distributed algorithms, distributed process control, distributed learning/problem solving, and multi-agent AI. An advantage of this approach, at least when representing certain systems from Distributed AI, is that the model follows the way in which humans communicate and cooperate.

5.1 Implementation

An implementation of (propositional) Concurrent METATEM has been developed. This program provides a platform on which simple experiments into both synchronous and asynchronous systems can be carried out. Experience with this system is directing the implementation of full (first-order) Concurrent METATEM. Future implementation work will include the development of efficient techniques for recognising and compiling both point-to-point message-passing and groups via multi-cast message-passing.

5.2 Related Work

In [16], Gehani describes Broadcasting Sequential Processes (BSP) which is also based on the asynchronous broadcasting of messages. As in Concurrent METATEM, objects may screen out certain messages, but not only is the identity of the sender always incorporated into a message, but also objects cannot manipulate their message queue as is intended in Concurrent METATEM. One, more fundamental, difference between BSP and our approach is that objects in BSP are message driven. The Linda model [17] has some similarities with this approach in that the shared data structures represented in the Linda tuple space can be seen as providing a broadcast mechanism for data. However, our computational model fixes much more than just the basic communication and distribution system. As mentioned earlier, the Concurrent METATEM model has some similarities with the Actor model, the main differences being the ability to act in a non message-driven way, the ability to process sets of incoming messages, and the ability to synchronise with other objects. Maruichi et. al. [18] use a model of computation similar to Concurrent METATEM for their investigation in DAI systems, while several distributed operating systems also use the notion of process groups (what we call 'groups') in order to group processes (objects) together [6].

Various executable temporal logics have been developed, for example [19, 1], but few have incorporated the notions of concurrency and we know of none that are based upon a computational model similar to the one described here. However, one comparable approach is Shoham's work on *Agent Oriented-Programming (AOP)* [21]. Here, individual agents are represented within a multi-modal logic, which is more concerned with the *beliefs*, *intentions* and actions of agents that we are here. Further, both the logical basis and the model of computation use in AOP are different to that considered here.

5.3 Future work

Given an object's interface definition, the internal computation method can be defined in any way that is consistent with the interface. The possibility of using a mixture of languages for the each object's internal computation will be investigated. Within the Concurrent METATEM model, both point-to-point message-passing and synchronous processes can be developed. The current Concurrent METATEM interpreter already includes the possibility of executing objects synchronously, but more support needs to be added to ensure that, if point-to-point message-passing is defined using broadcast message-passing, it remains efficient.

We also intend to look at the incorporation of dynamic object creation into the Concurrent METATEM system. On the formal side, we are developing a specification and development framework for Concurrent METATEM systems and looking at giving an algebraic semantics for Concurrent METATEM, for example based upon [20].

5.4 Acknowledgements

The author would like to thank Michael Wooldridge, Richard Owens, Howard Barringer and Marcello Finger who have collaborated on several of the papers summarised here, and Mark Reynolds who provided useful suggestions relating to the language.

References

1. M. Abadi and Z. Manna. Temporal Logic Programming. *Journal of Symbolic Computation*, 8: 277–295, 1989.
2. G. Agha. *Actors - A Model for Concurrent Computation in Distributed Systems*. MIT Press, 1986.
3. H. Barringer, M. Fisher, D. Gabbay, G. Gough, and R. Owens. METATEM: A Framework for Programming in Temporal Logic. In *Proceedings of REX Workshop on Stepwise Refinement of Distributed Systems: Models, Formalisms, Correctness*, Mook, Netherlands, June 1989. (Published in LNCS volume 430, Springer Verlag).

4. H. Barringer, M. Fisher, D. Gabbay, and A. Hunter. Meta-Reasoning in Executable Temporal Logic. In *Proceedings of the International Conference on Principles of Knowledge Representation and Reasoning (KR)*, Cambridge, Massachusetts, April 1991.
5. H. Barringer and D. Gabbay. Executing temporal logic: Review and prospects (Extended Abstract). In *Proceedings of Concurrency '88*, 1988.
6. K. Birman. The Process Group Approach to Reliable Distributed Computing. Techanical Report TR91-1216, Department of Computer Science, Cornell University, July 1991.
7. A. Borg, J. Baumbach, and S. Glazer. A Message System Supporting Fault Tolerance. In *Proceedings of the Ninth ACM Symposium on Operating System Principles*, New Hampshire, October 1983. ACM. (In ACM Operating Systems Review, vol. 17, no. 5).
8. M. Finger, M. Fisher, and R. Owens. METATEM at Work: Modelling Reactive Systems Using Executable Temporal Logic. In *Sixth International Conference on Industrial and Engineering Applications of Artificial Intelligence and Expert Systems (IEA/AIE-93)*, Edinburgh, U.K., June 1993. Gordon and Breach Publishers.
9. M. Fisher and H. Barringer. Concurrent METATEM Processes — A Language for Distributed AI. In *Proceedings of the European Simulation Multiconference*, Copenhagen, June 1991.
10. M. Fisher and R. Owens. From the Past to the Future: Executing Temporal Logic Programs. In *Proceedings of Logic Programming and Automated Reasoning (LPAR)*, St. Petersberg, Russia, July 1992. (Published in LNCS volume 624, Springer Verlag).
11. M. Fisher and M. Wooldridge. A Logical Approach to the Representation of Societies of Agents. In *Proceedings of Second International Workshop on Simulating Societies (SimSoc)*, Certosa di Pontignano, Siena, Italy, July 1993.
12. M. Fisher and M. Wooldridge. Executable Temporal Logic for Distributed A.I. In *Twelfth International Workshop on Distributed A.I.*, Hidden Valley Resort, Pennsylvania, May 1993.
13. M. Fisher and M. Wooldridge. Specifying and Verifying Distributed Intelligent Systems. In *Portuguese Conference on Artificial Intelligence (EPIA)*. Springer-Verlag, October 1993.
14. M. Fisher. A Normal Form for First-Order Temporal Formulae. In *Proceedings of Eleventh International Conference on Automated Deduction (CADE)*, Saratoga Springs, New York, June 1992. (Published in LNCS volume 607, Springer Verlag).
15. M. Fisher. Concurrent METATEM — A Language for Modeling Reactive Systems. In *Parallel Architectures and Languages, Europe (PARLE)*, Munich, Germany, June 1993.
16. N. Gehani. Broadcasting Sequential Processes. *IEEE Transactions on Software Engineering*, 10(4):343–351, July 1984.
17. D. Gelernter, N. Carriero, S. Chandran, and S. Chang. Parallel programming in Linda. In *International Conference on Parallel Processing*, August 1985.
18. T. Maruichi, M. Ichikawa, and M. Tokoro. Modelling Autonomous Agents and their Groups. In *Decentralized AI 2 – Proceedings of the 2nd European Workshop on Modelling Autonomous Agents and Multi-Agent Worlds (MAAMAW)*. Elsevier/North Holland, 1991.
19. B. Moszkowski. *Executing Temporal Logic Programs*. Cambridge University Press, Cambridge, U.K., 1986.
20. K. V. S. Prasad. A Calculus of Broadcasting Systems. In *Proceedings of the International Joint Conference on Theory and Practice of Software Development (TAPSOFT)*, pages 338–358, Brighton, U.K., April 1991. (Published in LNCS volume 493, Springer Verlag).
21. Y. Shoham. Agent Oriented Programming. Technical Report STAN–CS–1335–90, Department of Computer Science, Stanford University, California, USA, 1990.
22. R. G. Smith and R. Davis. Frameworks for Cooperation in Distributed Problem Solving. *IEEE Transactions on Systems, Man and Cybernetics*, 11(1):61–70, 1981.

Temporal Query Languages: a Survey*

Jan Chomicki**

Kansas State University

Category: Survey
Area: Temporal Databases

Abstract. We define formal notions of *temporal domain* and *temporal database*, and use them to survey a wide spectrum of temporal query languages. We distinguish between an abstract temporal database and its concrete representations, and accordingly between abstract and concrete temporal query languages. We also address the issue of incomplete temporal information.

1 Introduction

A *temporal database* is a repository of temporal information. A *temporal query language* is any query language for temporal databases. In this paper we propose a formal notion of temporal database and use this notion in surveying a wide spectrum of temporal query languages.

The need to store temporal information arises in many computer applications. Consider, for example, records of various kinds: financial, personnel, medical, or judicial.

There has been a lot of research in temporal databases in the past 15 years, as evidenced by a recent book [TCG+93]. We think that the field of temporal databases is mature enough to warrant a unifying mathematical framework. Our proposed framework attempts to integrate the research on temporal databases with database theory and logic research. The framework is also relevant for the research on knowledge representation in the area of Artificial Intelligence. We hope that it will serve as a foundation for further, more systematic advances in the field of temporal databases.

The well-known ANSI/SPARC architecture [Ull88] distinguishes three different levels in a database: physical, conceptual, and external. In the context of temporal databases, we propose to split the conceptual level even further and distinguish between an *abstract* temporal database and a *concrete* one. Intuitively, an abstract temporal database captures the formal, representation-independent

---

* Based on the invited tutorial presented at the 12th ACM SIGACT-SIGMOD-SIGART Symposium on Principles of Database Systems, May 1993, Washington, D.C.
** Address: Computing and Information Sciences, Kansas State University, Manhattan, KS 66506, USA, E-mail: chomicki@cis.ksu.edu

meaning of a temporal database, while a concrete one provides a specific, finite representation for it in terms of a specific temporal data model. Accordingly, we study *abstract* query languages whose semantics is defined for abstract temporal databases and *concrete* query languages defined for specific concrete representations. Among the languages of the first group are: first-order logic, temporal logic, relational algebra, and a number of deductive languages; among those of the second – several languages described in [TCG+93]. We study the following issues in particular: formal semantics, expressiveness, query processing and its computational complexity, and representation-independence (for concrete query languages). We also address the implications for query languages of allowing *incomplete* information in temporal databases. We list many open problems.

The present survey is different from other existing surveys [BCW93, MS91, Sno92], mainly because of the emphasis on a single unifying formal framework applicable to a wide spectrum of languages. [BCW93] discusses the technical issues involved in representing infinite temporal databases, while the scope of [MS91, Sno92] is limited to concrete temporal query languages. None of the cited surveys deals with the issue of incomplete temporal information. We do not claim that the present survey is exhaustive. So many different temporal data models and query languages have been proposed that it is not possible to describe them in detail in a single paper. However, we have attempted to extract the most important features of those models and languages, and present them in a single unified framework. We believe that this framework is also applicable to the approaches that are not explicitly covered here.

The plan of the paper is as follows. In section 2 we introduce a formal framework for temporal databases. In particular we specify the notions of *temporal domain*, *abstract* temporal database, and *concrete* temporal database. In section 3 we introduce a number of general properties of interest of temporal query languages. In section 4 we discuss *abstract* temporal query languages, and in section 5 a number of *concrete* temporal query languages. In section 6 we discuss incomplete temporal information, and in section 7 we summarize related work in Artificial Intelligence. In section 8 we draw conclusions.

2 Temporal databases

Within a formal framework for temporal databases the following issues need to be addressed:

- *choice of temporal domains:* points vs. intervals; linear vs. branching, dense vs. discrete, bounded vs. unbounded time.
- definitions of *abstract* and *concrete* temporal database.
- *query languages:* formal semantics, expressiveness, implementation.

In this section we address the first two issues. Temporal query languages are discussed in sections 3, 4, and 5.

2.1 Temporal domains

Temporal ontology. Here basically two options exist: *points* vs. *intervals*. In the database context the point-based view is predominant and we concentrate on it here. We use the term *instant* for a time point. However, most of the AI research (for an up-to-date survey see [MP93a]) takes the interval-based view. In the point-based view intervals are obtained as pairs of points. In the interval-based view it is common to have designators for interval endpoints. Thus usually moving between both views is easy in the first-order case [3].

Mathematical structure. As a rule some kind of mathematical structure is imposed on time instants. The most common is *order*: partial or total (linear). The orders studied in temporal databases are usually linear. However, exceptions exist: a linear transitive order that is not irreflexive or asymmetric can be used to model *cyclic time*[4] and a partial order that satisfies *left-linearity* ("no branching to the left") – to model *branching time* [Eme90]. In this survey we consider only linear orders.

A temporal domain, like any other domain in databases, is not defined axiomatically but rather as a specific structure. The basic linearly-ordered sets in this context are: $\mathbf{N} = (N, <)$ (natural numbers), $\mathbf{Z} = (Z, <)$ (integers), $\mathbf{Q} = (Q, <)$ (rationals), and $\mathbf{R} = (R, <)$ (reals). Often the signatures of the above structures are expanded to contain additional symbols: a unary *successor* function symbol s, a binary function symbol $+$ for *addition* (such symbols enable us to talk about *succession* or *relative distance* of time points), or a binary function symbol $\equiv_k$ for *congruence* modulo k (to talk about *periodicity*). We assume that equality is always available in the temporal domain. Sometimes the temporal domains considered have universes that are finite or bounded subsets of one of the above sets.

In most temporal database applications, it is commonly assumed that the time is discrete and isomorphic to natural numbers [Sno92]. However, continuous time has turned out to be extremely valuable in mathematics and physics. Our view is that dense and continuous time provides a useful abstraction. For example, if we prefer to think of time as discrete but with instants that are very close, we may often be dealing with very large sets of instants and this in turn may be difficult to implement efficiently. Query evaluation is also often easier in $\mathbf{Q}$ or $\mathbf{R}$ than in $\mathbf{N}$ or $\mathbf{Z}$ [KKR90, Rev90], mainly because of the absence of a notion of successor in the former. The emerging consensus [DSS94] is that many different temporal domains should be supported in a truly general temporal DBMS.

Time granularity. All of the temporal domains mentioned so far are "flat". To handle multiple time granularities, e.g., days vs. weeks, it is necessary to consider multiple interrelated temporal domains. An instant in a "higher-level" domain

---
[3] The propositional case is different but it does not concern us here.
[4] Linear-time cyclicity can also be handled using *ultimately periodic sets*: see the discussion later in this section.

corresponds to a contiguous set of instants in another, "lower-level" domain. We do not consider multiple time granularities in this survey, as they have been treated recently in [WJS93].

2.2 Abstract temporal databases

We propose here a formal notion of an *abstract* temporal database which captures the representation-independent meaning of a temporal database. An abstract temporal database may be viewed in several different but equivalent ways. The model-theoretic view, which is the most basic, treats an abstract temporal database as a first-order structure. The snapshot view treats it as a function that maps every instant to a set of tuples. Finally, the timestamp view treats it as a mapping associating a set of instants with every tuple.

For simplicity, we assume first a single temporal dimension, a single temporal domain $\mathcal{T}$, and a single data domain $\mathcal{U}$ (the latter domain contains standard database constants). We show then how to lift these restrictions. We will work in the context of the relational data model but the definition can be generalized to other data models formulated within a first-order framework, e.g., certain object-oriented models described by Beeri [Bee90]. Moreover, we assume that the database schema is fixed and consists of a fixed set of relations.

Model-theoretic view. It is well known that a relational database may be viewed as a finite structure $D = (\mathcal{U}, P_1, \ldots, P_k)$ for the first-order language L_D containing relation symbols $p_1, \ldots, p_k$ for all the relations $P_1, \ldots, P_k$ in the database and constant symbols for all the elements of $\mathcal{U}$. A corresponding *abstract temporal database* (called a *temporal structure*) is a structure $D^{\mathcal{T}} = (\mathcal{U}, \mathcal{T}, P_1^{\mathcal{T}}, \ldots, P_k^{\mathcal{T}})$ for the two-sorted first-order language $L_D^{\mathcal{T}}$ containing a new temporal relation symbol $p_i^{\mathcal{T}}$ for every temporal relation $P_i^{\mathcal{T}}$ and constant symbols for at least all the elements of $\mathcal{U}$ (and possibly some elements of $\mathcal{T}$ as well). The arity of $p_i^{\mathcal{T}}$ is the arity of p_i plus 1. The extra argument of $p_i^{\mathcal{T}}$ (last by convention) is called *temporal*, other arguments are called *data*. The intuition behind the definition is as follows: $D^{\mathcal{T}} \models p^{\mathcal{T}}(x_1, \ldots, x_n, t)$ iff $p(x_1, \ldots, x_n)$ holds at instant t. The language $L_D^{\mathcal{T}}$ contains also the relation and function symbols, e.g., "<" or "+", from the signature of the temporal domain $\mathcal{T}$ (those symbols are not interpreted in $D^{\mathcal{T}}$ but have a fixed meaning).

Note that this definition is parameterized by the temporal domain $\mathcal{T}$ and no assumptions about $\mathcal{T}$ are made. We are going to say that $D^{\mathcal{T}}$ is *finite* if it consists of finite relations. Thus, the above definition allows finite and infinite databases. In fact, in some applications, e.g., dealing with infinite periodic data, it is more natural to consider infinite databases [CI88, KSW90, BCW93].

Snapshot view. In the snapshot view, an abstract temporal database (called a *snapshot* database) is a function that associates with each instant $t \in \mathcal{T}$ and relation symbol p from L_D the set od tuples $(x_1, \ldots, x_n)$ such that $p(x_1, \ldots, x_n)$ holds at t. A snapshot database may be thought of as consisting of *snapshot relations*, one for every relation symbol. Each such relation has arity two and is a non-1NF relation (the second attribute values are sets of tuples).

Timestamp view. In the timestamp view, an abstract temporal database (called a *timestamp* database) is a function that associates with each tuple $(x_1, \ldots, x_n)$ (consisting of elements of the data domain $\mathcal{U}$) and relation symbol p from L_D the set of instants when $p(x_1, \ldots, x_n)$ holds. A timestamp database may be thought of as consisting of *timestamp relations*, one for every relation symbol. A timestamp relation has a data attribute for every data argument of the corresponding relation symbol, and a timestamp attribute. Each timestamp relation is also a non-1NF relation (the timestamp attribute values are sets of instants).

It should be clear now that all the three views presented above have the same expressive power. Below, we show in Table 1 a European history database viewed as a timestamp relation. In Table 2, we present a fragment of the corresponding snapshot relation and in Table 3 a relation that is a part of the corresponding temporal structure.

Multiple temporal dimensions. Multiple temporal dimensions are necessary to model *intervals* (as pairs of points) or multiple kinds of time, e.g., *valid time* (the time when a fact holds) and *transaction time* (the time when a fact is recorded in the database) [SA86]. In the case of multiple temporal dimensions with temporal domains $\mathcal{T}_1, \ldots, \mathcal{T}_m$ (not necessarily distinct), the notion of a temporal database needs to be appropriately generalized in an obvious way. In the model-theoretic view, instead of $D^{\mathcal{T}}$ we need to consider the structure denoted by $D^{\mathcal{T}_1, \ldots, \mathcal{T}_m}$. There will be now more than one temporal argument per relation symbol. The different interpretations of multidimensional temporal databases can be captured by additional axioms serving as integrity constraints. For example, a constraint may state that the beginning of an interval always precedes its end. The snapshot and timestamp views can be modified along the same lines as the model-theoretic view.

There does not seem to be any natural limit on the number of temporal dimensions in a temporal database. Recently, it has been argued by Clifford and Isakowitz [CI94] that in addition to *valid time* and *transaction time* another kind of time, *reference time*, is necessary. If each of those kinds of time is represented by intervals, we get six-dimensional time! However, adding time dimensions results in considerably more complicated query languages that are also harder to implement, and has a negative influence on the computational complexity of query processing. We will illustrate these points later in the paper.

Properties of abstract temporal databases. We claim that any of the above definitions of an abstract temporal database provides a representation-independent meaning for any database defined within any of the extant temporal data models that are based on the relational model [TCG+93]. Any model-specific database (called hereafter a *concrete* temporal database) is then just a *representation* of an abstract temporal database. Two concrete temporal databases are equivalent if they represent the same abstract temporal database. The set of abstract temporal databases representable within a temporal data model precisely captures the *expressiveness* of the model. Thus different temporal data models can be formally compared in terms of expressiveness. (This notion of expressiveness is called *data expressiveness* [BNW91, BCW93] and should be distinguished from *query expressiveness* discussed in section 3.) Moreover, as we show later in this section the notion of abstract temporal database makes possible the study of integration and interoperability of different temporal data models within a single, precise framework. Finally, the semantics of many query languages and integrity constraints can be defined directly on abstract temporal databases, without any concern for the way they are represented. We discuss this topic in section 4. In particular, integrity constraints can be just first-order axioms if the model-theoretic view is adopted.

2.3 Concrete temporal databases

The notion of abstract temporal database is not sufficient to deal with temporal database applications. There is a fundamental reason for it: an abstract temporal database may be infinite, while only finite objects can be explicitly represented in computer storage. Moreover, many extant temporal database models provide already specific (and often incompatible) notions of temporal database. Therefore, it is necessary to consider *concrete* temporal databases that are specific finite representations of abstract temporal databases. For simplicity we discuss initially only the case of one temporal dimension.

Concrete timestamp databases. Out of the three different views of an abstract temporal database, the timestamp view leads to the most natural and useful notion of a concrete database. Timestamps can be infinite sets but such sets can often be implicitly represented using *timestamp formulas:* first-order formulas with one free variable in the language of the temporal domain. Example: $0 < t < 5 \lor t > 10$. An example concrete timestamp database is shown in table 4.

We may ask what subsets of the temporal domain can serve as timestamps, i.e., can be defined by timestamp formulas. For example, if the temporal domain is $(N, <)$, then the timestamps are all (and only) finite or co-finite subsets of N. A more interesting case is that of Presburger arithmetic $(N, 0, +, <)$ where the timestamps are all ultimately-periodic subsets of N. For example, the set of natural numbers with period 7 beginning with 0 ("Sundays"), is described by the Presburger formula

$$\exists y.\ t = y + y + y + y + y + y + y$$

Another equivalent formulation is $t \equiv_7 0$ where "$\equiv_k$" means congruent modulo k. (This is an example of a *congruence formula*.) Consequently, if we allow Presburger or congruence formulas as timestamps, then infinite ultimately periodic temporal databases can be finitely represented [KSW90]. (A temporal database D^T is *ultimately periodic* if there is an instant t_0 and a natural number d such that for all $p_i, \bar{x}$ and $t > t_0$, $D^T \models p_i(\bar{x}, t) \Leftrightarrow D^T \models p_i(\bar{x}, t + d)$.)

Infinite periodic sets are useful because they can define *calendars*, e.g., the set of all business days or the set of all Sundays. Even finite periodic sets are better represented as infinite sets plus constraints guaranteeing finiteness. For example, consider the set of all Sundays in a given year.

The main technical tool in the theory of timestamp databases is *quantifier elimination*[End72]. A theory is said to admit quantifier elimination if for every formula in the language of this theory an equivalent quantifier-free formula can be effectively constructed. Additionally, we require that quantifier-free formulas thus obtained can be effectively tested for satisfaction. This additional requirement is especially important as we want to be able to effectively test whether a specific instant belongs to a timestamp defined by a timestamp formula. Fortunately, the theories of all the temporal domains that we proposed earlier in this section admit quantifier elimination.

Example 1. Consider the formula:

$$\phi = \exists t_1 .\, t < t_1 < 6.$$

In $(N, <)$, quantifier elimination yields $\phi' = t < 5$. In $(Q, <)$, quantifier elimination yields $\phi'' = t < 6$.

Quantifier elimination plays two distinct roles in timestamp databases. First, it makes possible limiting the attention to quantifier-free timestamp formulas. Second, it is a basic tool of query evaluation (see section 4).

In fact, timestamp formulas can be further simplified. Consider a tuple w in a timestamp relation, whose quantifier-free timestamp formula is ϕ. The formula ϕ can be transformed into a disjunctive normal form $\phi_1 \vee \cdots \vee \phi_m$. Subsequently, negation can be eliminated from every disjunct because we are dealing with linear orders, giving $\phi' = \phi'_1 \vee \cdots \vee \phi'_m$. As a result every disjunct ϕ'_j, $1 \leq j \leq m$, becomes a conjunction of atomic formulas (called *constraints*). Now we can replace the tuple w by m tuples $w_1, \ldots, w_m$. Each new tuple w_j, $1 \leq j \leq m$, has data components identical to w and has ϕ'_j as the timestamp formula. Thus, disjunction and negation can be eliminated from timestamp formulas. A finite timestamp relation in this form is a *generalized relation* in the sense of Kanellakis, Kuper and Revesz [KKR90].

If we want to admit more than one time dimension, e.g., to represent intervals, then we need to have timestamps which are formulas with *two or more* free variables. Example: $t_1 = 0 \wedge t_2 = 5$. This should be distinguished from ranges of instants which are timestamps defined by single-variable formulas, e.g., $0 \leq t \wedge t \leq 5$. Multi-dimensional timestamp formulas can describe, for example, the set of all subintervals of an interval: $0 \leq t_1 < t_2 \leq 5$.

Most existing temporal data models [TCG+93] choose (a variant of) the timestamp representation. However, it is usually assumed that timestamps are finite (or at least bounded) sets. Using timestamp formulas to represent infinite sets makes it possible to skirt this limitation. It is an open question whether timestamps could be defined in a richer language than the first-order theory of the time domain.

Other concrete temporal databases. For snapshot databases or temporal structures there is no natural notion of implicit finite representation. Therefore, if they are to be used as concrete databases, one has to enforce the restriction that they are finite and describe only a *finite* subset of the time domain. Also, as should be obvious from the examples above, both snapshot databases and temporal structures are quite wasteful in terms of space usage in typical applications.

Logic programs. An abstract temporal database, being an infinite structure, can often be finitely represented using a *logic program* consisting of a set of deductive (Horn) rules and a finite database. The abstract temporal database corresponding to such a program is its least Herbrand model. For example, "Sundays" can be represented by the following program:

$$sunday(0).$$
$$sunday(s^7(T)) :- sunday(T).$$

Chomicki and Imielinski [CI88] introduced this kind of representation for abstract temporal databases over the time domain $(N, 0, s)$ where s is the unary successor symbol. The syntax of logic programs is restricted by requiring that the successor symbol s is the only function symbol and can occur only in one argument of relations. The resulting language, which is an extension of *Datalog* (the language of function-free logic programs), is called $Datalog_{1S}$. ($Datalog_{1S}$ is discussed as a query language over temporal databases in section 4.) An equally expressive logic programming language, Templog, whose syntax is based on temporal logic was proposed by Abadi and Manna [AM89] and Baudinet [Bau89]. There is a simple syntactic translation between $Datalog_{1S}$ and Templog. It is interesting to note that the data expressiveness of these approaches is identical to that of the timestamp representation with Presburger timestamps: both express exactly ultimately periodic temporal databases [BCW93].

2.4 Interoperability

Our framework for temporal databases makes it possible to formulate the issue of temporal database interoperability in a very natural way. (This part represents ongoing work.)

Suppose that we have two temporal data models P_1 and P_2. Assume first that they use the same underlying temporal domains. The *meaning* of P_1 (resp. P_2) is defined as a *total* mapping ϕ_1 (resp. ϕ_2) from concrete temporal databases defined under P_1 (resp. P_2) to abstract temporal databases. The inverse mappings ϕ_1^{-1} and ϕ_2^{-1} may be *partial* because not necessarily every abstract temporal database is representable in the given data model (for example the model can be capable of representing only finite temporal databases). Now the database $d_1 = \phi_1^{-1} \circ \phi_2(d_2)$, when it is defined, represents the concrete database under P_1 corresponding to a concrete database d_2 under P_2. The database d_1 can then be queried using the query languages defined for P_1, providing thus the access to the database d_2.

In this way interoperability is given a formal basis which also provides a general direction for the implementation. The data expressiveness of a data model sets exact limits on the interoperability. E.g., if ϕ_1^{-1} is not defined on $\phi_2(d_2)$, then there is no hope to use d_2 under P_1.

Notice that in the case when the semantics of a query language are defined directly on abstract temporal databases, there are no inverse mappings to deal with and the whole process is considerably simplified.

If the heterogenous databases have different underlying temporal domains, then additionally one has to construct a coercion between the domains. Such a coercion is often very natural, for example between **Z** and **Q**.

3 Properties of query languages

The semantics of *abstract* query languages are defined with respect to abstract temporal databases, while that of *concrete* query languages – with respect to concrete temporal databases.

In the next two sections we survey the following properties of abstract and concrete temporal query languages:

- declarative semantics,
- representation-independence,
- expressiveness,
- data complexity of query evaluation, and
- efficient implementation.

We explain these notions below.

A semantics for a query language is *declarative* if it assigns a meaning to a query without referring to the way how the query is evaluated. Preferably, such semantics should be *logical* and provide a precise notion of a *model* of a query.

Representation-independence means that the answer to a query should be the same for every two concrete databases representing the same abstract temporal database (this property was identified by Gadia [Gad93]). It should be clear that every abstract query language is representation-independent if it is correctly implemented. However, for concrete query languages representation-independence needs to be separately proved.

Query expressiveness was defined by Chandra and Harel [CH80] as follows. Considering only yes-no queries, two queries are said to be *equivalent* if they return the same answer for every database. Now a query language L_1 is *at least as expressive* as another query language L_2 if for every query formulated in L_2, there is an equivalent query in L_1 (L_1 is *more expressive* than L_2 if additionally there is a query in L_1 for which there is no equivalent query in L_2). This notion of expressiveness is called *query expressiveness* [BNW91] and is formally defined to be the class of sets of abstract temporal databases for which the queries in the language evaluate to true. Query expressiveness can be considered not only for abstract temporal databases but also for concrete ones. This notion has been very important in the theory of query languages for relational databases [Cha88a, AV92].

Data complexity of query evaluation was defined by Chandra and Harel [CH82] and Vardi [Var82] to mean the computational complexity of the set of finite databases for which a given, fixed query evaluates to true. One can also study *combined complexity* where the query is also a part of the input. However, data complexity seems to measure better the computational effort necessary for evaluating queries formulated in a given language. This notion has also been extensively used in database theory [Cha88a].

4 Abstract query languages

4.1 Relational calculus

The first-order language $L_D^{T_1,\ldots,T_m}$ of an abstract temporal database (assuming the model-theoretic view) can be used as a query language, as suggested by Kabanza, Stevenne and Wolper [KSW90], and Tuzhilin and Clifford [TC90]. It is commonly known as the *domain relational calculus*. Its semantics is the standard Tarskian semantics [End72] which is obviously declarative. The answer to a first-order query is the set of valuations that make the query formula true in the given database.

Example 2. Consider the query *"list all countries that lost and regained independence"*. This query can be formulated in first-order logic as follows (I is a shorthand for *Independent*):

$$\exists t_1, t_2, t, s_1, s_2. \forall s. \ I(x, s_1, t_1) \land I(x, s_2, t_2) \land \neg I(x, s, t) \land t_1 < t < t_2.$$

Example 3. The functional dependency that *"an employee can have only one salary at a given time"* is simply expressed as:

$$\forall x, s, t, s', t'. \ Emp(x, s, t) \land Emp(x, s', t') \land t = t' \Rightarrow s = s'.$$

Another type of constraint may limit the range of the time instants that appear in a temporal relation. For example, suppose the constraint is *"a transaction can be done only on business days"* (i.e., Monday through Friday) and the time point 0 corresponds to a Sunday. This constraint can be expressed as:

$$\forall x, t. \ Transaction(x, t) \Rightarrow (t \not\equiv_7 0 \land t \not\equiv_7 6).$$

Jensen and Snodgrass [JS92] use first-order logic to formulate a *taxonomy* of temporal databases. The resulting formulas can be viewed as *constraint dependencies* [BCW94] that generalize the traditional dependencies [Ull88].

Under the snapshot view, relational calculus can be treated in two different ways. If the language $L_D^{T_1,\ldots,T_m}$ is to be used, then it is necessary to redefine the standard semantics of first-order logic in the context of the snapshot view (which is straightforward). However, if the non-1NF relational presentation is used, as in table 2, then relational calculus requires second-order constructs for dealing with sets. Such a language has been proposed by Wang, Jajodia, and Subrahmanian [WJS93]. This issue has been extensively researched in the context of data models with non-1NF relations or complex values ([ADA93] contains a overview of this area). Similar issues arise under the timestamp view. The occurrence of infinite sets presents an interesting twist, not present in the up-to-date research in that area.

First-order logic can also be used as a concrete query language. This is straightforward for finite temporal structures. For finite timestamp databases, there are essentially two possibilities to implement query evaluation:

1. translate the query to relational algebra (for example using the algorithm of van Gelder and Topor [VGT91]) and use generalized versions of algebraic operations that work on timestamp databases, or
2. evaluate the query in closed form (i.e., the result should also be a temporal relation) using quantifier elimination procedures for the temporal domain and the data domain.

For the first approach to work, it seems necessary to introduce new database relations for the atomic formulas of the temporal domain. This is perfectly valid because such relations, although infinite, are finitely representable using timestamp formulas. For example the order relation $<$ can be represented using a binary relation Less consisting of a single tuple with no data arguments and the timestamp formula $t_1 < t_2$. The implementation of relational algebra operators for timestamp databases is discussed later in this section. Quantifier elimination for the temporal domain is still necessary there to implement projection (note the difference with the second approach above where quantifier elimination for both the temporal and the data domains needs to be used).

Example 4. The query *"list all countries that lost and regained independence"* can be formulated in relational algebra (with renaming) as:

$$\pi_X(P - \pi_{X,T}(I(X,S,T))).$$

where

$$P = \pi_{X,T}(I(X,S_1,T_1) \bowtie Less(T_1,T) \bowtie I(X,S_2,T_2) \bowtie Less(T,T2)).$$

For the second approach to work, the theory of the data domain has to admit quantifier elimination. Kanellakis, Kuper and Revesz [KKR90] proposed an elegant quantifier elimination algorithm for infinite domains with inequality constraints that would be applicable here (we do not assume any mathematical structure of the data domain). But still more work needs to be done in order to see whether quantifier elimination can be implemented with an efficiency approaching that of standard relational database operations.

We discuss now the issue of *data complexity* of first-order logic queries over concrete temporal databases with different temporal domains.

Kanellakis, Kuper and Revesz [KKR90] analyzed the computational complexity of various *constraint query languages*. Their results apply to finite timestamp databases with timestamp formulas that are conjunctions of constraints (atomic formulas). Any fixed number of temporal dimensions over a single temporal domain T is allowed. They characterized the data complexity of processing of first-order logic queries as being in:

- LOGSPACE if T is a countably infinite dense linear order, e.g., $T = \mathbf{Q}$,
- NC if T is real arithmetic, i.e., $T = (R, 0, +, *, <)$.

Kabanza, Stevenne and Wolper [KSW90] considered timestamp formulas that are constraints in Presburger arithmetic, i.e., $T = (N, 0, +, <)$. They showed that data complexity of first-order queries is in PTIME. Their results also apply to any fixed number of temporal dimensions.

4.2 Relational algebra

The semantics of relational algebra is defined set-theoretically for arbitrary, not necessarily finite, relations, so it fits well with the model-theoretic view of abstract temporal databases.

Under the snapshot view, relational algebra operations can only be used *pointwise*, i.e., on the snapshots corresponding to the *same* time instants in different relations. One needs to consider also their generalized versions that simultaneously apply to *all* instants. (This is analogous to the *filter* or *apply_to_all* operation proposed first in the area of functional programming and applied to databases with complex values in [Bee90].) Also, additional operations that relate different instants are necessary. Such an approach has been pursued by Orgun and Müller [Org93] in the context of the temporal domain $\mathbf{N}$. To deal with the timestamp view, relational algebra should also be appropriately extended. We are not aware of any general proposal along those lines.

In practice relational algebra is considered to be a procedural language applied to finite relations. As shown above, in the context of finite timestamp databases relational algebra is still a natural counterpart of first-order logic. This has been explored by Kabanza, Stevenne and Wolper [KSW90] for the temporal domain $T = (N, 0, +, <)$. They've shown that fixed-schema complexity of all the algebraic operations is in PTIME. (This result should be interpreted with some caution: for databases with m temporal dimensions, the degree of the polynomial may be as high as m^2.) A similar method could be applied to $(R, 0, +, <)$.

Developing efficient implementation methods and query optimization techniques that appropriately generalize those developed for relational databases is very much an open problem. Some recent work in constraint databases by Kanellakis et al. [KRVV93] and Brodsky, Jaffar, and Maher [BJM93] may be applicable here. Also, the possibility of generalizing the methods of query processing developed for finite temporal databases (Leung and Muntz [LM93] and Segev [Seg93]) should be explored.

For timestamp databases the hardest operation to implement is the projection of a temporal argument because it involves quantifier elimination. A specialized quantifier elimination procedure for each temporal domain needs to be supplied. We are not aware of any work dealing with the optimization of the projection operation in this context.

4.3 Temporal logic

Here we consider first databases with one temporal dimension. Instead of the first-order language L_D^T, we can use a temporal extension of L_D, denoted by $tl(L_D)$. This language contains binary temporal connectives **since** and **until** with the following meaning:

- A **since** B is true at instant i iff for some j, $j < i$, B is true at instant j, and for every k, $j < k \leq i$, A is true at instant k.
- A **until** B is true at instant i iff for some j, $j > i$, B is true at instant j and for every k, $i \leq k < j$, A is true at instant k.

We will use "$\blacklozenge A$" (*sometime in the past A*) as a shorthand for "*true* **since** A" and "$\lozenge A$" (*sometime in the future A*) as a shorthand for "*true* **until** A".

Example 5. The query *"list all countries that lost and regained independence"* from example 2 can be formulated in temporal logic as follows:

$$\exists s_1, s_2.\ \lozenge\,(\,\blacklozenge\, I(x, s_1) \wedge\ \lozenge\, I(x, s_2) \wedge \forall s.\ \neg I(x, s)).$$

Temporal logic is of interest because of its ubiquity in different areas of computer science. In databases temporal logic has been used as a language for querying *finite* snapshot databases (Tuzhilin and Clifford [TC90], Gabbay and McBrien [Gab89, GM91]), and for formulating temporal integrity constraints (Lipeck and Saake [LS87], Chomicki [Cho92], Sistla and Wolfson [SW94]).

One possible implementation method for temporal logic queries is to translate them to first-order logic queries and use the implementation for first-order logic queries described earlier. It may also be possible to develop query evaluation and optimization methods that are specific to temporal logic.

We believe that the central research issue here is *query expressiveness* of temporal logic. Is $tl(L_D)$ equally expressive as L_D^T (assuming the signature of the temporal domain T contains only the order relation symbol $<$)? This seems to be one of the most interesting open research problems in the area of temporal databases. There are some relevant early results due mainly to H. Kamp. He showed that propositional temporal logic is equally expressive as monadic first-order logic [Kam68] over Dedekind-complete temporal domains (e.g., **N**, **Z**, and **R**). Later, he showed that $tl(L_D)$ is strictly less expressive than L_D^T for arbitrary temporal databases [Kam71]. His construction, however, used in an essential way infinite temporal databases and was applicable only to dense temporal domains. We conjecture that the separation between first-order and temporal logics holds even for finite databases and discrete domains.

Another important result is that of Gabbay, Pnueli, Shelah and Stavi [GPSS80] who proved that over **N** the connective **until** suffices in the propositional case; **since** can be eliminated. However, their proof uses Kamp's result [Kam68], so the same question for first-order temporal logic remains open.

If there is more than one temporal dimension in a temporal database, then the above **since/until** temporal logic is no longer applicable. The choice of a more powerful temporal logic is determined by the semantics of the temporal dimensions. If they are meant to represent intervals, then some kind of *interval* temporal logic is called for (Halpern and Shoham [HS91], Venema [Ven90], van Benthem [vB91]). If they are meant to represent multiple kinds of time, then some form of *multi-dimensional* temporal logic should be used (Finger [Fin92], Montanari and Pernici [MP93b]).

4.4 Inductive query languages

The temporal query languages discussed so far are all *first-order*. However, there are many natural queries that are not first-order but inductive or even second-order.

Example 6. The following is an example of an inductive temporal query that may very well be posed by an epidemiologist:

> Find all the persons at risk where "being at risk" is defined in the following way: a person is *at risk* at a given time if she has been earlier infected or she has been in contact with someone already at risk.

Notice that having been in contact with someone who only became at risk later should not result in classifying the person as at risk. It should be clear that the above query is not first-order because of the inherent recursion.

Inductive temporal queries can be formulated in a number of *logic programming languages*. Those languages extend *Datalog*, the language of function-free logic programs, in various ways. We consider the following:

- *Datalog*$^{<z}$ [Rev90, Rev93]: *Datalog* with integer order constraints,
- *Datalog*$^{<Q}$ [KKR90]: *Datalog* with rational order constraints,
- *Datalog*$_{1S}$ [CI88, CI93]: *Datalog* with a unary successor symbol in one argument,
- *Datalog* with a unary successor symbol and linear arithmetic constraints [BNW91].

The formulation in *Datalog*$^{<z}$ (or *Datalog*$^{<Q}$) of the query from example 6 is very natural:

$$atRisk(X, T) \leftarrow infected(X, T'), T' < T.$$
$$atRisk(X, T) \leftarrow contacted(X, Y, T'), atRisk(Y, T'), T' < T.$$

The formulation in *Datalog*$_{1S}$ is less transparent because the order $<$ relation symbol is not directly available.

The above deductive languages can be extended with stratified negation [ABW88] in clause bodies. Such an extension of *Datalog*$_{1S}$ will be denoted *Stratified Datalog*$_{1S}\neg$. The addition of negation makes the deductive languages capable of expressing also all first-order logic queries over abstract temporal databases with one temporal dimension, in particular the query from example 2. The semantics of logic programs without negation is given by their least Herbrand models [vEK76]. In the presence of stratified negation, the semantics is given by perfect models [Prz88].

We discuss now the issue of the *data complexity* of the deductive languages.

Kanellakis, Kuper and Revesz [KKR90, Rev90, Rev93] obtained closed-form evaluation and PTIME computability for $Datalog^{<Q}$ and $Datalog^{<z}$. They also considered the extension of $Datalog^{<Q}$ with negation but under a different, *inflationary* semantics [KP91], which also has closed-form evaluation and PTIME computability. Extending $Datalog^{<z}$ with stratified negation is problematic, as one can define then a relation coding the successor function symbol and obtain all Turing-computable functions [Rev90, Rev93]. Extending $Datalog^{<z}$ with congruence constraints preserves, however, closed-form evaluation and PTIME computability [TCR94].

Chomicki and Imielinski have shown that the data complexity of $Datalog_{1S}$ (*Stratified $Datalog_{1S}\neg$*) queries on finite temporal databases is PSPACE-complete [CI88, Cho90a] but there are subsets of $Datalog_{1S}$ for which queries can be evaluated in PTIME [Cho90b]. *Datalog* with a unary successor symbol and linear arithmetic constraints, proposed by Baudinet, Niézette and Wolper [BNW91], is a very expressive language and the termination of query evaluation can not be guaranteed.

Another way to obtain inductive queries is to extend logic query languages with *fixpoint operators*. In fixpoint query languages, queries are defined as least (or greatest) solutions of equations $X = \phi(X)$ where ϕ is a logic formula and X is a relation or a set.

Two best known fixpoint query languages are: least fixpoint queries on finite relational databases [CH80] and temporal fixpoint calculus (*propositional* temporal logic extended with fixpoint operators) [Var88, Gab89]. Both languages add extra expressive power to the underlying logic languages. For example, the first language makes possible the expression of transitive closure queries which are not expressible in first-order logic. The second language is capable of defining a property EVEN true of every *even-numbered* state (in the case of $T = (N, <)$ a temporal database may be viewed as a sequence of states), which is not expressible in propositional temporal logic [Wol83]. In the context of temporal databases, it is natural to consider the fixpoint extensions of both first-order logic and temporal logic. This is a topic of current and future research.

We have conjectured earlier that temporal logic is strictly less expressive than first-order logic. Does a similar separation hold for the fixpoint extensions of both languages? Moreover, it is worthwhile to study the relationship between deductive, fixpoint and *second-order* languages. Baudinet [Bau94] has shown that propositional temporal fixpoint calculus and Monadic Stratified $Datalog_{1S}\neg$ are equally expressive and that there is a similar correspondence between the positive fixpoint calculus and Monadic $Datalog_{1S}$. Propositional temporal fixpoint calculus and *second-order* monadic logic (with quantification over sets of time points) are known to be equally expressive [Tho81].

Tuzhilin and Clifford [TC90] and Gabbay and McBrien [GM91] have proposed to extend relational algebra with recursively-defined operators. These operators are essentially limited propositional temporal fixpoint operators. They are, however, restricted to finite temporal databases. Moreover, Tuzhilin and Clifford's algebra does not even achieve the full power of propositional fixpoints, as it is unable to express EVEN. Their version of the relational algebra is equally expressive as the **since/until** temporal logic discussed earlier. Gabbay and McBrien additionally proposed to introduce a special *time* relation that keeps track of the flow of time. Using this relation and arithmetic selection conditions, EVEN can be expressed. The exact expressive power and the data complexity of the resulting query language have not been studied, however.

5 Concrete Query Languages

In this section we discuss a number of temporal query languages whose semantics is defined with respect to concrete temporal databases. We then summarize the main limitations of those languages.

5.1 TQuel

TQuel, proposed by Snodgrass [Sno87], is a well-known temporal query language derived from Quel. It supports a single temporal domain which is discrete, infinite and multi-level (there is a way to refer to a specific day, hour etc.), and two temporal dimensions: valid time and transaction time. For simplicity, however, we will consider valid time only.

The data model of TQuel is a variant of the timestamp representation. Specifically, a timestamp is an interval (a, b) where a is the start and b is the end of the interval. Associating the interval (a, b) with a fact $p(\bar{x})$ (or, equivalently with a tuple $\bar{x}$ in the relation corresponding to p) means that $p(\bar{x})$ holds for every t, $a \leq t \leq b$. If $b = \infty$, then $(a, b) = \{t : a \leq t\}$. The intervals are required to be maximal, so the timestamps of identical facts are coalesced if they denote overlapping intervals. This requirement imposes quite a burden on database *update* procedures (an insertion might trigger a coalescing operation) but is essential for the expressiveness of the language: see example 7.

It is important to see that the data model of TQuel is point-based, not interval-based. Intervals serve only as a representational device. The truth values of facts are associated with points, not intervals. For example, it is impossible to represent in the database the situation where a fact is true of an interval but not of some of its subintervals.

TQuel databases are finite. The semantics of TQuel queries is given by a translation to Quel. This translation is problematic if the token ∞ appears in the database, as pointed out recently by Clifford and Isakowitz [CI94]. TQuel presently has no declarative, logical semantics, but such semantics is in principle possible.

Example 7. The query *"list all countries that lost and regained independence"* from example 2 can be written in TQuel as follows:

```
range of O1 is IndependentCountries
range of O2 is IndependentCountries
retrieve (O1.Name)
valid at begin of O1
where O1.Name=O2.Name
when (end of O1) precede (begin of O2)
```

The negation is not necessary here as it is assumed that the timestamps represent maximal intervals that can not be coalesced.

The expressiveness of TQuel depends on the repertoire of interval constructors (like `precede` in the example above). With a sufficiently rich set of those constructors [All83] sequences of TQuel queries can simulate temporal logic queries. TQuel is essentially first-order so it is incapable of expressing inductive temporal queries. The data complexity of TQuel queries has not been analyzed but is clearly polynomial.

The syntax and semantics of TQuel become quite cumbersome when transaction time is also considered. There is no support for more than two time dimensions.

5.2 HRDM

HRDM (Historical Relational Data Model) is one of the most influential temporal data models. It was proposed by Clifford and Croker [CC87, CC93], based on the earlier work of Clifford, Tansel and Warren [CT85, CW83]. A similar model was used by Gadia [Gad88].

HRDM supports a single, discrete, and infinite temporal domain, and a single time dimension. HRDM relations are finite. The treatment of relation attributes is not uniform: some of them are designated as parts of the *key*. Non-key attributes can take values that are *functions* whose domains are finite subsets of the temporal domain (thus HRDM provides a limited form of *non-1NF* relations). Specific ways of efficiently storing such functions are also proposed. For example, if a function is constant, storing one value together with the domain of the function is sufficient. Stepwise constant functions are handled similarly.

HRDM has an algebra obtained from the relational algebra by redefining most of the relational operators in the context of the HRDM data model. The semantics of the HRDM relational algebra operations are defined set-theoretically. As noted by Clifford, Croker and Tuzhilin [CCT94] the algebra has a rather limited expressive power because it can not express queries that relate database states at different time instants. Consequently, it can not express the query from example 2. Because of its essentially first-order character, the algebra can not also express inductive temporal queries. The data complexity of the algebra has not been analyzed but is clearly polynomial.

5.3 Backlogs

A rather different temporal data model, based on *backlog relations*, was proposed by Jensen and Mark [JM93]. The model supports a single, discrete and infinite temporal domain, and two temporal dimensions: valid and transaction-time (which are, however, not independent as in TQuel). Backlog relations store not raw data but rather requests to change the data. Therefore only *transaction time* instants are stored. To find out whether a specific fact was true at a given *valid* time instant, the appropriate backlog relation has to be scanned to make sure that the appropriate tuple was inserted and not subsequently deleted (modification is handled similarly). Backlog relations are finite and can represent only finite abstract temporal databases. An example backlog relation representing a part of the European history database introduced in section 2 is shown in table 5.

The backlog model can support any relational query language, for example, relational algebra or calculus. In particular, the query *"list all countries that lost and regained independence"* from example 2 can be written in the algebra (with renaming) as:

$$\sigma_{O_1=Insert \land O_2=Delete \land O_3=Insert \land T_1 < T < T_2}(R)$$

where

$$R = B(Id_1, O_1, T_1, X, S_1) \bowtie B(Id_2, O_2, T, X, S_2) \bowtie B(Id_3, O_3, T_2, X, S_3))$$

and B is the backlog relation whose first attribute contains the consecutive number of the update operation, second – the name of the operation itself, third – transaction time, fourth and fifth – the relevant data (country, capital). This is illustrated in Table 5.

The data complexity of processing relational algebra or calculus queries to backlog relations is clearly polynomial. Jensen and Mark [JM93] describe many techniques for the incremental evaluation of queries in this model.

5.4 Limitations of temporal data models and query languages

There are many more different temporal data models and query languages. Some of them are, like HRDM, not fully relational by allowing non-1NF relations or limited forms of object identity [CCT94]. The recent book [TCG+93] presents at least 12 different temporal data models that are extensions of the relational model. A specific temporal data model is obtained, as above, by choosing a single fixed temporal domain, adopting a specific framework for concrete temporal databases (usually of the timestamp variety), and defining one or more query languages that are applicable only to this particular framework. If a timestamp framework is selected, then it is usually assumed that the timestamps are finite or bounded sets. Consequently, the abstract temporal databases represented have to be finite. Moreover, the signatures of the time domain contain only the order relation symbol. A serious limitation of the expressiveness of the temporal query languages discussed in the book [TCG+93] (except for the languages discussed in one of its chapters [BCW93]) is their inability to express inductive temporal queries.

Virtually all temporal data models are mutually incompatible. This situation seriously hinders further systematic progress in the area of temporal databases, as remarked by Jensen, Soo and Snodgrass [JSS93]. Their solution, the provision of a single unifying temporal data model to which other models could be mapped, is not sufficient. They still fall short of defining a *representation-independent* abstract semantics for temporal databases. In fact, their model uses another notion of concrete temporal database, admittedly simpler and more general than others. This model is still limited as it deals with a single temporal domain and timestamps that are finite sets. We also think that the model is unnecessarily complicated. It violates Ockham's razor[5] by introducing a new type of entity - the "bitemporal element". The objectives of Jensen, Soo and Snodgrass can be achieved in a much simpler way by adopting the framework described here in section 2.

Current work on the interoperability of temporal databases, e.g., Wang, Jajodia and Subrahmanian [WJS93], addresses similar concerns as the present paper. The snapshot and timestamp views are identified without, however, referring to the underlying model-theoretic view. Only finite snapshot databases and a single, fixed temporal domain are considered. On the other hand, it should be pointed out that the referenced work [WJS93] deals with the issue of multiple time granularities that we do not address here.

By treating a temporal database as a special kind of first-order structure the transfer of results and techniques from mathematical logic and database theory becomes possible. Database theory in particular supplies the concepts of *query expressiveness* and *data complexity* that are necessary for the analysis of the mathematical properties of query languages. We think that there is no single notion of *completeness* of query languages for temporal databases. Clifford, Croker, and Tuzhilin [CCT94] proposed first-order logic (the language L_D^T) as a "temporally-complete" query language. But in this language inductive temporal queries are not expressible. Moreover, the claim of Clifford, Croker, and Tuzhilin's that temporal logic is equally expressive as first-order logic has not been substantiated yet. There is strong evidence to the contrary [Kam71]. We conjecture that their claim does not hold even for finite databases. In the context of relational databases it has been shown by Chandra, Harel, Abiteboul, and Vianu [CH80, CH82, AV92] that there is no single complete query language but rather many different classes of query languages related to one another in intricate ways. We believe the situation in temporal databases is similar.

6 Incomplete temporal information

Temporal information may often be incompletely specified. For example, only a partial ordering of events may be given. One way of dealing with such problems

---

[5] "A rule stating that entities should not be multiplied needlessly, which is interpreted to mean that the simplest of two or more competing theories is preferable or that an explanation for unknown phenomena should first be attempted in terms of what is already known." *Webster's II New Riverside University Dictionary.*

is to give up linearity and well-known mathematical structures like **N** and **R**, and develop a temporal logic based on events [vB91]. Another: to generalize the notion of a temporal database. We describe the latter solution as it has been predominant in the database area.

To deal with incomplete temporal information, several authors proposed to apply the classic framework of Imielinski and Lipski [IL84] to *timestamp* databases. Imielinski and Lipski modeled incomplete information in the context of the relational data model using *marked nulls* – placeholders standing for *some* value in the domain. The same null value may appear in different columns and different rows of a table. Moreover, every row has a (quantifier-free) local condition associated with it that contains some nulls of this row. Finally, the entire table has a (quantifier-free) global condition relating nulls in different rows. The semantics of such a table is the *set* of relations obtained by substituting domain values for the occurrences of nulls in such a way that all the local conditions and the global condition are satisfied.

It should be clear now that there is a close correspondence between local conditions in tables and quantifier-free timestamp formulas. Koubarakis [Kou93, Kou94, Kou] pursued this correspondence in full generality in the context of the temporal domains **Q** and **Z**. In his approach, there may be one or two temporal dimensions. Timestamp formulas which already contain variables (e.g., $5 < t$) may additionally contain nulls (e.g., $5 < c \wedge c = t$ where c is a null). We call such timestamp formulas *indefinite*. Indefinite timestamp formulas define indefinite timestamps which are just sets of timestamps. (One should note, however, that eliminating quantifiers from timestamp formulas is no longer possible because of the presence of implicit existential quantifiers corresponding to null values.)

Example 8. Suppose we do not know the exact time of the split of Czechoslovakia into two different countries but we know that it was before 1993. This incomplete knowledge can be represented as the table in Table 6 with a global condition $c < 1993$.

As far as query languages for incomplete temporal databases are concerned, *abstract* query languages discussed in section 4 generalize easily. The semantics of an incomplete temporal database is given by a set X of abstract temporal databases, thus an answer to a query q is now an incomplete relation representing the set of relations corresponding to the answers to q obtained separately for the individual members of X. In addition, modal queries can be asked: which facts are *certain* (true in every member of X), and which are *possible* (true in some member of X)?

The implementation of the abstract query languages for incomplete timestamp databases raises new problems because of the presence of global conditions and timestamp formulas with nulls. Also, modal queries need to be supported. Koubarakis [Kou93, Kou94] studied relational calculus and algebra for incomplete timestamp databases. He has shown [Kou] than the complexity of query processing for such databases is no worse than that for incomplete relational databases [Gra91].

Van der Meyden [vdM92] obtained the corresponding lower bounds in a much more restricted framework. Only single values, null or non-null, can constitute timestamps. (Consequently, only finite abstract temporal databases can be represented in this framework.) Nulls can be related through a global conjunctive condition. There may be arbitrarily many temporal dimensions but not surprisingly the complexity of evaluating queries crucially depends on this number. Van der Meyden showed that the data complexity of obtaining *certain* answers to first-order queries was:

- in PTIME for one-dimensional time (for disjunctive queries the original proof is non-constructive, although it can be made constructive [Ron van der Meyden, personal communication]),
- co-NP-complete for n-dimensional time ($n \geq 2$).

He had also many very interesting results about the combined complexity of evaluating queries that we can not discuss here because of a lack of space. Van der Meyden's results apply to $\mathbf{Z}$, $\mathbf{Q}$, as well as finite orders.

Dyreson and Snodgrass [DS93] and Gadia [Gad93] pursued the problem of incomplete temporal information for *concrete* query languages. They didn't base their approaches on the work of Imielinski and Lipski but to some degree their approaches can be recast using the framework of Koubarakis'.

In the context of the discrete temporal domain $\mathbf{Z}$, Dyreson and Snodgrass postulated to represent an indefinite point by a *possible interval*, which could be viewed as an indefinite timestamp formula (e.g., $t = c \wedge 0 < c < 5$ meaning "*some* instant between 0 and 5"), and an indefinite interval by a pair of possible intervals. Null values appearing in timestamp formulas associated with different tuples can not be compared, even for equality. Thus the incomplete temporal relation from example 8 can not be represented in their framework. On the other hand, a quadratic query evaluation algorithm becomes possible for queries expressed in a (slightly restricted) subset of TQuel [Curtis Dyreson, personal communication]. Dyreson and Snodgrass also proposed to incorporate several kinds of *probabilistic* information in incomplete temporal databases and described a possible implementation.

In the context of the discrete temporal domain $\mathbf{N}$, Gadia proposed to represent an indefinite timestamp by a pair (*lower bound*, *upper bound*) where both bounds are finite unions of intervals. This can be easily represented using an indefinite timestamp formula. For this approach, precise expressiveness and complexity bounds on query processing have still to be determined.

Chaudhuri [Cha88b] was historically the first to deal with the problem of incomplete temporal information. His approach, however, does not completely fit in the framework presented above. In his approach a temporal database records information about events (points or intervals) and their relationships. Characteristically, the relationships are just attribute values, e.g., *before* is a constant that may appear in a database. This makes it possible to formulate queries asking for all relationships between two events in a purely first-order framework. The set of allowed relationships is user-defined but the derivable relationships between events should be specified in a special, essentially Horn, clausal form. This makes query evaluation efficient (polynomial). However, the derivation of disjunctions is not possible, which makes the aproach incomplete for disjunctive queries. Also, Chaudhuri discussed only the evaluation of atomic queries. Some of Chaudhuri's work can be recast using incomplete temporal databases. Point events can be viewed as null values and their ordering captured by a global condition.

7 Related work in AI

We briefly summarize here the differences between the database and the AI perspectives on the representation and processing of temporal information.

First, most AI approaches are restricted to the propositional case (the non-temporal part is propositional). Second, they take an interval-based view of time that originates in a very influential paper by Allen [All83]. The approach of Allen's was to associate every proposition with an interval in which it holds. Allen proposed an algebra of intervals based on 13 basic kinds of interval relationships, e.g. **precedes**, and operations on those relationships, e.g. composition. Moreover, two intervals could be related through a set of relationships which represented a disjunction. Thus a rich array of disjunctive information could be represented. Unfortunately, determining satisfaction of a set of such relationships turned out to be NP-complete [VK86]. Thus more restricted algebras were proposed. One of them, the point algebra of Vilain and Kautz [VK86] is representable in the framework of Koubarakis described in section 6. Later works [DMP91, KL91, Mei91] introduced constructs for representing distance information. Again, this can be represented in a first-order framework, assuming the signature of the time domain contains successor or addition.

The third difference is in the query languages supported and query evaluation. Allowed queries typically involve establishing a relationship between points or intervals. The representation languages are propositional, thus unable to represent a relational database. Consequently, first-order queries involving quantifiers are not supported. Queries are evaluated using constraint satisfaction algorithms. It is an open question whether the above approaches can be generalized to first-order logic.

8 Conclusions

In this survey we have proposed a single and uniform formal framework for studying temporal databases and temporal query languages. We have applied this framework to a wide spectrum of query languages. We have also identified a number of important mathematical properties of query languages. We believe that the framework can serve as a foundation for further, more systematic, advances in the area of temporal databases.

Acknowledgments

We are very grateful to Marianne Baudinet, Surajit Chaudhuri, Curtis Dyreson, Manolis Koubarakis, Peter Revesz, Rick Snodgrass, and Pierre Wolper for sharing with us their insights in and knowledge of temporal databases. Thanks also go to David Toman for a careful reading of the manuscript.

References

[ABW88] K.R. Apt, H.A. Blair, and A. Walker. Towards a Theory of Declarative Knowledge. In J. Minker, editor, *Foundations of Deductive Databases and Logic Programming*, pages 89–148. Morgan Kaufmann, 1988.

[ADA93] P. Atzeni and V. De Antonellis. *Relational Database Theory*. Benjamin/Cummings, 1993.

[All83] J.F. Allen. Maintaining Knowledge about Temporal Intervals. *Communications of the ACM*, 26(11):832–843, November 1983.

[AM89] M. Abadi and Z. Manna. Temporal Logic Programming. *Journal of Symbolic Computation*, 8(3), September 1989.

[AV92] S. Abiteboul and V. Vianu. Expressive Power of Query Languages. In *Theoretical Studies in Computer Science*. Academic Press, 1992.

[Bau89] M. Baudinet. Temporal Logic Programming is Complete and Expressive. In *ACM SIGACT-SIGPLAN Symposium on Principles of Programming Languages*, 1989.

[Bau94] M. Baudinet. On the Expressiveness of Temporal Logic Programming. *Information and Computation*, 1994. To appear.

[BCW93] M. Baudinet, J. Chomicki, and P. Wolper. Temporal Deductive Databases. In A. Tansel, J. Clifford, S. Gadia, S. Jajodia, A. Segev, and R. Snodgrass, editors, *Temporal Databases: Theory, Design, and Implementation*, pages 294–320. Benjamin/Cummings, 1993.

[BCW94] M. Baudinet, J. Chomicki, and P. Wolper. Constraint-Generating Dependencies. In *Workshop on Principles and Practice of Constraint Programming*, Orcas Island, Washington, May 1994.

[Bee90] C. Beeri. A formal approach to object-oriented databases. *Data and Knowledge Engineering*, 5:353–382, 1990.

[BJM93] A. Brodsky, J. Jaffar, and M.J. Maher. Towards Practical Constraint Databases. In *International Conference on Very Large Data Bases*, 1993.

[BNW91] M. Baudinet, M. Niézette, and P. Wolper. On the Representation of Infinite Temporal Data and Queries. In *ACM SIGACT-SIGMOD-SIGART Symposium on Principles of Database Systems*, 1991.

[CC87] J. Clifford and A. Croker. The Historical Relational Data Model (HRDM) and Algebra Based on Lifespans. In *IEEE International Conference on Data Engineering*, 1987.

[CC93] J. Clifford and A. Croker. The Historical Relational Data Model (HRDM) Revisited. In A. Tansel, J. Clifford, S. Gadia, S. Jajodia, A. Segev, and R. Snodgrass, editors, *Temporal Databases: Theory, Design, and Implementation*, pages 6–27. Benjamin/Cummings, 1993.

[CCT94] J. Clifford, A. Croker, and A. Tuzhilin. On Completeness of Historical Relational Query Languages. *ACM Transactions on Database Systems*, 19(1):64–116, March 1994.

[CH80]   A.K. Chandra and D. Harel. Computable Queries for Relational Databases. *Journal of Computer and System Sciences*, 21:156–178, 1980.

[CH82]   A.K. Chandra and D. Harel. Structure and Complexity of Relational Queries. *Journal of Computer and System Sciences*, 25:99–128, 1982.

[Cha88a]   A.K Chandra. Theory of Database Queries. In *ACM SIGACT-SIGMOD-SIGART Symposium on Principles of Database Systems*, pages 1–9, 1988.

[Cha88b]   S. Chaudhuri. Temporal Relationships in Databases. In *International Conference on Very Large Data Bases*, 1988.

[Cho90a]   J. Chomicki. *Functional Deductive Databases: Query Processing in the Presence of Limited Function Symbols*. PhD thesis, Rutgers University, New Brunswick, New Jersey, January 1990. Also Laboratory for Computer Science Research Technical Report LCSR-TR-142.

[Cho90b]   J. Chomicki. Polynomial-Time Computable Queries in Temporal Deductive Databases. In *ACM SIGACT-SIGMOD-SIGART Symposium on Principles of Database Systems*, Nashville, Tennessee, April 1990.

[Cho92]   J. Chomicki. History-less Checking of Dynamic Integrity Constraints. In *IEEE International Conference on Data Engineering*, Phoenix, Arizona, February 1992.

[CI88]   J. Chomicki and T. Imieliński. Temporal Deductive Databases and Infinite Objects. In *ACM SIGACT-SIGMOD-SIGART Symposium on Principles of Database Systems*, Austin, Texas, March 1988.

[CI93]   J. Chomicki and T. Imieliński. Finite Representation of Infinite Query Answers. *ACM Transactions on Database Systems*, 18(2):181–223, June 1993.

[CI94]   J. Clifford and T. Isakowitz. On the Semantics of (Bi) Temporal Variable Databases. In *International Conference on Extending Database Technology*, Cambridge, UK, March 1994.

[CT85]   J. Clifford and A.U. Tansel. On an Algebra for Historical Relational Databases: Two Views. In *ACM SIGMOD International Conference on Management of Data*, 1985.

[CW83]   J. Clifford and D.S. Warren. Formal Semantics for Time in Databases. *ACM Transactions on Database Systems*, 8(2):214–254, June 1983.

[DMP91]   R. Dechter, I. Meiri, and J. Pearl. Temporal Constraint Networks. *Artificial Intelligence*, 49:61–95, 1991.

[DS93]   C. E. Dyreson and R.T. Snodgrass. Historical Indeterminacy. In *IEEE International Conference on Data Engineering*, 1993.

[DSS94]   C.E. Dyreson, M.D. Soo, and R.T. Snodgrass. The TSQL2 Data Model for Time. A TSQL2 Commentary, March 1994.

[Eme90]   E.A. Emerson. Temporal and Modal Logic. In Jan van Leeuwen, editor, *Handbook of Theoretical Computer Science*, volume B, chapter 16, pages 995–1072. Elsevier/MIT Press, 1990.

[End72]   H.B. Enderton. *A Mathematical Introduction to Logic*. Academic Press, 1972.

[Fin92]   M. Finger. Handling Database Updates in Two-Dimensional Temporal Logic. *Journal of Applied Non-Classical Logic*, 1992.

[Gab89]   D. Gabbay. The Declarative Past and Imperative Future: Executable Temporal Logic for Interactive Systems. In B. Banieqbal, B. Barringer, and A. Pnueli, editors, *Temporal Logic in Specification*, volume 398, pages 409–448. Springer-Verlag, LNCS 398, 1989.

[Gad88]   S.K. Gadia. A Homogenous Relational Model and Query Languages for Temporal Databases. *ACM Transactions on Database Systems*, 13(4):418–448, December 1988.

[Gad93]   S.K. Gadia. Temporal Databases: A Prelude to Parametric Data. In A. Tansel, J. Clifford, S. Gadia, S. Jajodia, A. Segev, and R. Snodgrass, editors, *Temporal Databases: Theory, Design, and Implementation*, pages 28–66. Benjamin/Cummings, 1993.

[GM91]   D. Gabbay and P. McBrien. Temporal Logic and Historical Databases. In *International Conference on Very Large Data Bases*, 1991.

[GPSS80]   D. Gabbay, A. Pnueli, S. Shelah, and S. Stavi. On the Temporal Analysis of Fairness. In *ACM SIGACT-SIGPLAN Symposium on Principles of Programming Languages*, 1980.

[Gra91]   G. Grahne. *The Problem of Incomplete Information in Relational Databases*. Springer-Verlag, LNCS 554, 1991.

[HS91]   J.Y. Halpern and Y. Shoham. A Propositional Modal Logic of Time Intervals. *Journal of the ACM*, 38(4):935–962, October 1991.

[IL84]   T. Imielinski and W. Lipski. Incomplete Information in Relational Databases. *Journal of the ACM*, 31(4):761–791, 1984.

[JM93]   C.S. Jensen and L. Mark. Differential Query Processing in Transaction-Time Databases. In A. Tansel, J. Clifford, S. Gadia, S. Jajodia, A. Segev, and R. Snodgrass, editors, *Temporal Databases: Theory, Design, and Implementation*, pages 457–496. Benjamin/Cummings, 1993.

[JS92]   C.S. Jensen and R.T. Snodgrass. Temporal Specialization. In *IEEE International Conference on Data Engineering*, pages 594–603, 1992.

[JSS93]   C.S. Jensen, M.D. Soo, and R.T. Snodgrass. Unification of Temporal Data Models. In *IEEE International Conference on Data Engineering*, 1993.

[Kam68]   J.A.W. Kamp. *Tense Logic and the Theory of Linear Order*. PhD thesis, University of California, Los Angeles, 1968.

[Kam71]   H. Kamp. Formal properties of 'now'. *Theoria*, 37:227–273, 1971.

[KKR90]   P.C. Kanellakis, G.M. Kuper, and P.Z. Revesz. Constraint Query Languages. In *ACM SIGACT-SIGMOD-SIGART Symposium on Principles of Database Systems*, pages 299–313, Nashville, Tennessee, April 1990. To appear in Journal of Computer and System Sciences.

[KL91]   H. Kautz and P. Ladkin. Integrating Metric and Qualitative Temporal Reasoning. In *National Conference on Artificial Intelligence*, 1991.

[Kou]   M. Koubarakis. The Complexity of Query Evaluation in Indefinite Temporal Constraint Databases. Manuscript.

[Kou93]   M. Koubarakis. Representation and Querying in Temporal Databases: the Power of Temporal Constraints. In *IEEE International Conference on Data Engineering*, 1993.

[Kou94]   M. Koubarakis. Foundations of Indefinite Constraint Databases. In *Workshop on Principles and Practice of Constraint Programming*, Orcas Island, Washington, May 1994.

[KP91]   P.G. Kolaitis and C.H. Papadimitriou. Why not Negation by Fixpoint? *Journal of Computer and System Sciences*, 43:125–144, 1991.

[KRVV93]   P.C. Kanellakis, S. Ramaswamy, D.E. Vengroff, and J.S. Vitter. Indexing for Data Models with Constraints and Classes. In *ACM SIGACT-SIGMOD-SIGART Symposium on Principles of Database Systems*, 1993.

[KSW90] F. Kabanza, J-M. Stevenne, and P. Wolper. Handling Infinite Temporal Data. In *ACM SIGACT-SIGMOD-SIGART Symposium on Principles of Database Systems*, pages 392–403, Nashville, Tennessee, April 1990.

[LM93] T.Y. Cliff Leung and R.R. Muntz. Stream Processing: Temporal Query Processing and Optimization. In A. Tansel, J. Clifford, S. Gadia, S. Jajodia, A. Segev, and R. Snodgrass, editors, *Temporal Databases: Theory, Design, and Implementation*, pages 329–355. Benjamin/Cummings, 1993.

[LS87] U.W. Lipeck and G. Saake. Monitoring Dynamic Integrity Constraints Based on Temporal Logic. *Information Systems*, 12(3):255–269, 1987.

[Mei91] I. Meiri. Combining Qualitative and Quantitative Constraints in Temporal Reasoning. In *National Conference on Artificial Intelligence*, 1991.

[MP93a] A. Montanari and B. Pernici. Temporal Reasoning. In A. Tansel, J. Clifford, S. Gadia, S. Jajodia, A. Segev, and R. Snodgrass, editors, *Temporal Databases: Theory, Design, and Implementation*, pages 534–562. Benjamin/Cummings, 1993.

[MP93b] A. Montanari and B. Pernici. Towards a Temporal Logic Reconstruction of Temporal Databases. In *Proc. International Workshop on an Infrastructure for Temporal Databases*, Arlington, Texas, June 1993.

[MS91] L.E. Jr. McKenzie and R.T. Snodgrass. Evaluation of Relational Algebras Incorporating the Time Dimension in Databases. *ACM Computing Surveys*, 23(4):501–543, December 1991.

[Org93] M.A. Orgun. A Temporal Algebra Based on an Abstract Model. In M.E. Orlowska and M. Papazoglou, editors, *Advances in Database Research: Proceedings of the 4th Australian Database Conference*, pages 301–316. World Scientific, 1993.

[Prz88] T. C. Przymusinski. On the Declarative Semantics of Deductive Databases and Logic Programs. In J. Minker, editor, *Foundations of Deductive Databases and Logic Programming*, pages 193–216. Morgan Kaufmann, 1988.

[Rev90] P.Z. Revesz. A Closed Form for Datalog Queries with Integer Order. In *International Conference on Database Theory*, pages 187–201. Springer-Verlag, LNCS 470, 1990. To appear in Theoretical Computer Science.

[Rev93] P.Z. Revesz. A Closed-Form Evaluation for Datalog Queries with Integer (Gap)-Order Constraints. *Theoretical Computer Science*, 116:117–149, 1993.

[SA86] R. Snodgrass and I. Ahn. Temporal Databases. *IEEE Computer*, 19(9), 1986.

[Seg93] A. Segev. Join Processing and Optimization in Temporal Relational Databases. In A. Tansel, J. Clifford, S. Gadia, S. Jajodia, A. Segev, and R. Snodgrass, editors, *Temporal Databases: Theory, Design, and Implementation*, pages 356–387. Benjamin/Cummings, 1993.

[Sno87] R. Snodgrass. The Temporal Query Language TQuel. *ACM Transactions on Database Systems*, 12(2):247–298, June 1987.

[Sno92] R.T. Snodgrass. Temporal Databases. In A.U. Frank, I. Campari, and U. Formentini, editors, *Theories and Methods of Spatio-Temporal Reasoning in Geographic Space*, pages 22–64. Springer-Verlag, LNCS 639, 1992.

[SW94] A.P. Sistla and O. Wolfson. Temporal Triggers in Active Databases. *IEEE Transactions on Knowledge and Data Engineering*, 1994. To appear.

[TC90] A. Tuzhilin and J. Clifford. A Temporal Relational Algebra as a Basis for Temporal Relational Completeness. In *International Conference on Very Large Data Bases*, 1990.

[TCG+93] A. Tansel, J. Clifford, S. Gadia, S. Jajodia, A. Segev, and R. Snodgrass, editors. *Temporal Databases: Theory, Design, and Implementation*. Benjamin/Cummings, 1993.

[TCR94] D. Toman, J. Chomicki, and D.S. Rogers. Datalog with Integer Periodicity Constraints. Technical Report TR-CS-94-1, Computing and Information Sciences, Kansas State University, 1994. Submitted for publication.

[Tho81] W. Thomas. A Combinatorial Approach to the Theory of ω-automata. *Information and Control*, 48(3):261–283, 1981.

[Ull88] J.D. Ullman. *Principles of Database and Knowledge-Base Systems*, volume 1. Computer Science Press, 1988.

[Var82] M.Y. Vardi. The Complexity of Relational Query Languages. In *ACM SIGACT Symposium on Theory of Computing*, pages 137–146, 1982.

[Var88] M. Y. Vardi. A Temporal Fixpoint Calculus. In *ACM SIGACT-SIGPLAN Symposium on Principles of Programming Languages*, 1988.

[vB91] J.F.A.K. van Benthem. *The Logic of Time*. D.Reidel, 2nd edition, 1991.

[vdM92] R. van der Meyden. The Complexity of Querying Indefinite Data about Linearly Ordered Domains. In *ACM SIGACT-SIGMOD-SIGART Symposium on Principles of Database Systems*, 1992.

[vEK76] M.H. van Emden and R.A. Kowalski. The Semantics of Predicate Logic as a Programming Language. *Journal of the ACM*, 23(4):733–742, 1976.

[Ven90] Y. Venema. Expressiveness and Completeness of an Interval Tense Logic. *Notre Dame Journal of Formal Logic*, 31(4):529–547, 1990.

[VGT91] A. Van Gelder and R.W. Topor. Safety and Translation of Relational Calculus Queries. *ACM Transactions on Database Systems*, 16(2):235–278, June 1991.

[VK86] M. Vilain and H. Kautz. Constraint Propagation Algorithms for Temporal Reasoning. In *National Conference on Artificial Intelligence*, 1986.

[WJS93] X.S. Wang, S. Jajodia, and V.S. Subrahmanian. Temporal Modules: An Approach Toward Federated Temporal Databases. In *ACM SIGMOD International Conference on Management of Data*, pages 227–236, 1993.

[Wol83] P. Wolper. Temporal Logic Can Be More Expressive. *Information and Control*, 56:72–99, 1983.

Country	Capital	Years of Independence
Czech Kingdom	Prague	$\{1198,\ldots,1620\}$
Czechoslovakia	Prague	$\{1918,\ldots,1938\} \cup \{1945,\ldots,1992\}$
Czech Republic	Prague	$\{1993,\ldots\}$
Slovakia	Bratislava	$\{1939,\ldots,1945\} \cup \{1993,\ldots\}$
Poland	Gniezno	$\{1025,\ldots,1039\}$
Poland	Cracow	$\{1040,\ldots,1595\}$
Poland	Warsaw	$\{1596,\ldots,1794\} \cup \{1918,\ldots,1938\} \cup \{1945,\ldots\}$

Table 1. Timestamp view

Year	Independent Countries
...	...
1917	{}
1918	{(Czechoslovakia,Prague),(Poland,Warsaw)}
...	...
1939	{(Slovakia,Bratislava)}
1945	{(Czechoslovakia,Prague),(Poland,Warsaw)}
...	...
1993	{(Czech Republic,Prague),(Slovakia,Bratislava), (Poland,Warsaw)}
...	...

Table 2. Snapshot view

Country	Capital	Year of Independence
...	...	...
Czechoslovakia	Prague	1918
Poland	Warsaw	1918
...	...	...
Slovakia	Bratislava	1939
...	...	...
Czechoslovakia	Prague	1945
Poland	Warsaw	1945
...	...	...
Czech Republic	Prague	1993
Slovakia	Bratislava	1993
Poland	Warsaw	1993
...	...	...

Table 3. Model-theoretic view

Country	Capital	Years of Independence
Czech Kingdom	Prague	$1198 \leq t < 1621$
Czechoslovakia	Prague	$1918 \leq t < 1939 \vee 1945 \leq t \leq 1992$
Czech Republic	Prague	$1992 < t$
Slovakia	Bratislava	$1939 \leq t < 1945 \vee 1992 < t$
Poland	Gniezno	$1025 \leq t < 1040$
Poland	Cracow	$1040 \leq t < 1596$
Poland	Warsaw	$1596 \leq t < 1795 \vee 1918 \leq t < 1939 \vee 1945 \leq t$

Table 4. Timestamp view with timestamp formulas

Id	Op	Time	Country	Capital
1	Insert	1198	Czech Kingdom	Prague
2	Delete	1621	Czech Kingdom	Prague
3	Insert	1918	Czechoslovakia	Prague
4	Delete	1939	Czechoslovakia	Prague
5	Insert	1939	Slovakia	Bratislava
6	Delete	1945	Slovakia	Bratislava
7	Insert	1945	Czechoslovakia	Prague
8	Delete	1993	Czechoslovakia	Prague
9	Insert	1993	Czech Republic	Prague
10	Insert	1993	Slovakia	Bratislava
...				

Table 5. Backlog representation

Country	Capital	Years of Independence
Czech Kingdom	Prague	$1198 \leq t < 1621$
Czechoslovakia	Prague	$1918 \leq t < 1939 \vee 1945 \leq t \leq c$
Czech Republic	Prague	$c < t$
Slovakia	Bratislava	$1939 \leq t < 1945 \vee c < t$
Poland	Gniezno	$1025 \leq t < 1040$
Poland	Cracow	$1040 \leq t < 1596$
Poland	Warsaw	$1596 \leq t < 1795 \vee 1918 \leq t < 1939 \vee 1945 \leq t$

Global condition
$c < 1993$

Table 6. Incomplete Information

Improving Temporal Logic Tableaux using Integer Constraints

Reiner Hähnle and Ortrun Ibens

University of Karlsruhe
Institute for Logic, Complexity and Deduction Systems
76128 Karlsruhe, Germany
{reiner,ortrun}@ira.uka.de

1 Introduction

In this position paper we present some ideas that aim to improve analytic tableau for temporal logics with the ultimate goal of reviving the interest in using them for temporal logic satisfiability checking. Although tableau formulations for several propositional temporal logics exist [9], these are not used much in practice, because the tableau size becomes intractable already for quite small formulas. Moreover, checking tableau closure is complicated and expensive to implement. For practical purposes, usually automata theoretic approaches are preferred [3]. It might, however, still be interesting to have competitive tableau formulations as will be pointed out in Section 4.

The ideas for the research reported here were stimulated by [6, 7] where an extension of analytic tableau with linear constraints led to an efficient (mixed) integer programming formulation of a tableau system for finitely and infinitely-valued propositional logics. The usefulness of linear constraints for many-valued logics comes from the fact that the set of truth values can be identified with (a subset of) the integers, hence many-valued connectives can be characterized using integer constraints. If we consider, say, a discrete linear temporal logic without past time operators, then the *always*, *sometimes*, etc. operators realize certain properties of the natural numbers when Kripke worlds are identified with numbers and the linear order on the worlds with the successor function. It is tempting, in our opinion, to try using linear constraints over world variables in order to give more concise tableau systems for such logics than it is possible without the use of constraints. In this note we give one possible formulation of a constraint tableau system for one of the simplest propositional temporal logics, namely for discrete linear temporal logic with (future) always and sometimes.

2 Basic Definitions

Let $\mathcal{L} = \langle L_0, \neg, \vee, \wedge, \Box, \Diamond \rangle$ be a propositional language of *linear temporal logic* (LTL) with negation, disjunction, conjunction, always, and sometimes defined as usual over a non-empty set of propositional variables L_0.

Call $\mathcal{S} = \{k + x_1 + \cdots + x_n | k, n \in \mathbb{N}_0, x_1, \ldots, x_n \in S_0\}$ the set of *signs* over a non-empty set S_0 of world variables disjoint with L_0. The language of *signed linear temporal logic* (S-LTL) is $\mathcal{L}^* = \{s : \phi | s \in \mathcal{S}, \phi \in \mathcal{L}\}$.

An S-LTL-$Interpretation$ v maps each propositional variable $p \in L_0$ to the set of time points p holds at and each world variable $s_0 \in S_0$ to a time point. The extension to formulas $f \in \mathcal{L}$ and to signs $s \in \mathcal{S}$ is immediate.

Let $s : \phi \in \mathcal{L}^*$, then $s : \phi$ is *satisfiable* if there is an S-LTL interpretation v such that $v(s) \in v(\phi)$. The extension to sets of S-LTL-formulas is immediate.

$\mathcal{C} = \{s_1 \leq_{\mathbb{N}_0} s_2 | s_1, s_2 \in \mathcal{S}\}$ is the set of S-LTL *constraints*. Usually we omit the subscript of $\leq$. We make use of the standard abbreviations $<, =, \neq$ etc.

3 Constraint Tableau Calculus

We assume the reader is familiar with tableau calculi for propositional logics, and in particular with uniform notation for propositional modal formulas as defined in [5]. Just recall that type α formulas are conjunctive, type β are disjunctive, type ν are universal or 'always', and type π are existential or 'sometimes'.

In contrast to classical propositional tableaux, nodes consist of *sets of signed formulas* rather than single formulas. Moreover, we must accomodate linear constraints. Hence, a *tableau node* K is a pair $\langle \Phi, C \rangle \in (2^{\mathcal{L}^*} \times 2^{\mathcal{C}})$. A node $K = \langle \Phi, C \rangle$ is satisfiable if Φ has a model v which as well solves C.

In Table 1 we give correspondences between premises and conclusions of tableau rules for all types of compound formulas. The superscripts '... $^\nu$' in the ν rules are to be understood as markers indicating that no logical rule is to be applied anymore to the formula thus superscribed. This is a crucial difference to conventional temporal tableau systems, and we will see how completeness is restored by different means. There are four types of tableau rules which will be explained in the following.

Logical tableau rules can be specified as in Table 2. Note that taking the disjoint union in the premise implies that the main formula of the premise does not occur anymore in the conclusion. The letter k in the π rules is a new variable from S_0 that occurs not on the current branch. Informally, the conclusion of a π rule says that there is some point in the future when ϕ comes true.

Table 1. Correspondence between premises and conclusions of temporal constraint tableau rules.

α	α_1	α_2	ν	ν_1	ν_2
$s : \phi \wedge \psi$	$s : \phi$	$s : \psi$	$s : \Box \phi$	$s : \phi$	$(s : \Box \phi)^\nu$
$s : \neg(\phi \vee \psi)$	$s : \neg \phi$	$s : \neg \psi$	$s : \neg \Diamond \phi$	$s : \neg \phi$	$(s : \neg \Diamond \phi)^\nu$
$s : \neg\neg\phi$	$s : \phi$	$s : \phi$			

β	β_1	β_2	π	$\pi_1(k)$	$c_1(k)$
$s : \phi \vee \psi$	$s : \phi$	$s : \psi$	$s : \Diamond \phi$	$s+k : \phi$	$k \geq 0$
$s : \neg(\phi \wedge \psi)$	$s : \neg \phi$	$s : \neg \psi$	$s : \neg \Box \phi$	$s+k : \neg \phi$	$k \geq 0$

The rôle of the linear constraints is to record the information on possible models for each branch in a more concise form than in conventional tableaux.

Table 2. Temporal tableau rules for compound formulas.

$$\frac{\langle\{\alpha\}\dot\cup\Phi,C\rangle}{\langle\{\alpha_1,\alpha_2\}\cup\Phi,C\rangle} \qquad \frac{\langle\{\beta\}\dot\cup\Phi,C\rangle}{\langle\{\beta_1\}\cup\Phi,C\rangle|\langle\{\beta_2\}\cup\Phi,C\rangle}$$

$$\frac{\langle\{\nu\}\dot\cup\Phi,C\rangle}{\langle\{\nu_1,\nu_2\}\cup\Phi,C\rangle} \qquad \frac{\langle\{\pi\}\dot\cup\Phi,C\rangle}{\langle\{\pi_1(k)\}\cup\Phi,\{c_1(k)\cup C\}\rangle}$$

The justification is that every class of counter models can in fact be characterized by a set of linear inequations with integral solutions. One link between formulas and constraints is constituted by the π rules as we have seen, another link is obtained from the observation that, obviously, a formula and its complement cannot be true simultaneously at the same time point. This concept is formalized in the following *conflict rule*.

$$\frac{\langle\{s:\phi,t:\neg\phi\}\cup\Phi,C\rangle}{\langle\{s:\phi,t:\neg\phi\}\cup\Phi,C\cup\{s\ne t\}\rangle}$$

As a matter of fact, there are other possible constraints one could generate, for instance, $s<t$ from $s:\phi$ and $t:\square\neg\phi$ etc., but these are being subsumed by the *synchronization rule* below which we need for completeness anyway.

In conventional temporal tableau systems such as in [9] two more ingredients are needed for which we have no substitute so far. First, for completeness one must have the information that an always-formula does hold in all subsequent states. Usually this is achieved by reintroducing always-formulas in the conclusion of a rule and thus allowing them to be applied more than once. As a consequence, in order to obtain an effective tableau system, nodes N all whose formulas have occurred already in a previous node M are 'looped back' to this node M such that temporal tableaux really are directed graphs with a unique root. We intend to combine both rules, the 'restart' rule on always-formulas and the 'loop back' rule, into a single rule called *synchronization rule*. The rationale behind it is to 'synchronize' always-formulas so that they are available in the 'present' state. This has a similar effect as the restart rule, but because in the constraint part of a node we can accumulate information from different time points, a finite number of applications of this rule is sufficient.

In slight abuse of notation let now ν denote the unsigned part of an always-formula, and ν_1 the unsigned part of its correspondent according to Table 1; let $\lambda\in L_0\cup\{\neg p|p\in L_0\}$, i.e. λ is a literal. Then we can write down the synchronization rule as follows:

$$\frac{\langle\{s:\lambda,t:\nu\}\cup\Phi,C\rangle}{\langle\{s:\lambda,t:\nu\}\cup\Phi,C\cup\{s\le t\}\rangle \;\; | \;\; \langle\{s:\lambda,t:\nu,(s:\nu_1)^{\mathrm{syn}}\}\cup\Phi,C\cup\{s>t\}\rangle}$$

The *syn* marker prevents indefinite application of the synchronization rule, because the synchronization rule can only be applied to formulas without a *syn* marker.

But there are examples where the synchronization rule must be applied to a formula which itself is a synchronization result (see Fig. 1). So we need a rule which activates formulas with a *syn* marker, called *activation rule*. The idea is to activate a literal $(s : \lambda)^{\text{syn}} \in \Phi$ if a literal $s_0 : \lambda \in \Phi$ together with an S-LTL-interpretation v which maps $v(s) = v(s_0)$ cannot be found, and to activate an always formula $(s : \nu)^{\text{syn}} \in \Phi$ if an always formula $s_0 : \nu \in \Phi$ together with an S-LTL-interpretation v which maps $v(s) \geq v(s_0)$ cannot be found. There is an effective test for the applicability of the activation rule, but the details must be omitted due to lack of space. Let $(s : \phi)^{\text{syn}}$ be a formula to be activated. Then the activation rule is simply the following:

$$\frac{\langle \{(s : \phi)^{\text{syn}}\} \cup \Phi, C \rangle}{\langle \{s : \phi\} \cup \Phi, C \rangle}$$

A node $K = \langle \Phi, C \rangle$ in our tableau is *contradictory* if C is unsolvable. A tableau is *closed* if every branch contains a contradictory node. A formula $\phi \in \mathcal{L}$ is valid iff the tableau with root $\langle \{s : \neg\phi\}, \{s \geq 0\} \rangle$, $s \in S_0$, is closed. In Fig. 1 we give a tableau proof of $\neg\Diamond\Box p \vee \neg\Box\Diamond\neg p$.

$$\langle \{s : \Diamond\Box p, s : \Box\Diamond\neg p\}, \{s \geq 0\} \rangle$$
$$|$$
$$\langle \{s+k : \Box p, s : \Diamond\neg p, (s : \Box\Diamond\neg p)^{\nu}\}, \{k \geq 0, \ldots\} \rangle$$
$$|$$
$$\langle \{s+k : p, (s+k : \Box p)^{\nu}, s+l : \neg p, (s : \Box\Diamond\neg p)^{\nu}\}, \{l \geq 0, \ldots\} \rangle$$

$\langle \{s+l : \neg p, (s+k : \Box p)^{\nu}, \ldots\},$ $\langle \{(s+l : p)^{\text{syn}}, \ldots\},$
$\{l \leq k, \ldots\} \rangle$ $\{l > k, \ldots\} \rangle$

$\langle \{s+k : p,$ $\langle \{(s+k : \Diamond\neg p)^{\text{syn}}, \ldots\},$ $\langle \{(s+l : p)^{\text{syn}},$
$(s : \Box\Diamond\neg p)^{\nu}, \ldots\},$ $\{k > 0, \ldots\} \rangle$ $s+l : \neg p, \ldots\},$
$\{k \leq 0, \ldots\} \rangle$ $\{l \neq l, \ldots\} \rangle$
$|$ $|$
$\langle \{s+k : p,$ $\langle \{(s+k+m : \neg p)^{\text{syn}}, \ldots\},$
$s+l : \neg p, \ldots\},$ $\{m \geq 0, \ldots\} \rangle$
$\{k \neq l, \ldots\} \rangle$ $|$
   $\langle \{s+k+m : \neg p, \ldots\}, \{\ldots\} \rangle$
$\langle \{s+k+m : \neg p, (s+k : \Box p)^{\nu} \ldots\},$ $\langle \{(s+k+m : p)^{\text{syn}}, \ldots\},$
$\{m \leq 0, \ldots\} \rangle$ $\{m > 0, \ldots\} \rangle$
$|$ $|$
$\langle \{s+k+m : \neg p, s+k : p, \ldots\},$ $\langle \{(s+k+m : p)^{\text{syn}}, s+k+m : \neg p, \ldots\},$
$\{m \neq 0, \ldots\} \rangle$ $\{k+m \neq k+m, \ldots\} \rangle$

Fig. 1. We omitted the first steps which are purely propositional. The next two steps are straightforward ν and π rule applications. Next follow two applications of the synchronisation rule that split the proof into three branches. The first and the third of these can be closed after a conflict rule application. The second branch can be closed after the activation rule and again the synchronization rule are applied. (Due to space restrictions only such formulas and constraints are displayed that are newly generated or that are needed in that step. The formulas between which a synchronization takes place appear as the first two formulas in the left leaf which is generated by the synchronization rule.)

4 Conclusion

It is fairly easy, if somewhat tedious, to prove that our rules preserve satisfiability. Completeness, however, is a much tougher question, and the proof is quite involved. This is one of the reasons why the present note is merely a position paper. The other reason is that so far we did not show that the idea of incorporating linear constraints into temporal logic tableaux is in some sense a real improvement over the current state of the art. On the other hand, we would like to point out at least a few advantages that might be gained from it:

- Better performance if compared to conventional temporal tableaux. Admittedly, this is not obvious from the present appearance of the rules, but there is ample room for optimisation, for example, in the synchronisation rules.
- Implementing real-time logics like the ones suggested in [1].
- Extend the use of linear constraints to a translation of temporal logic satisfiability into integer programming (as it has been done in [6] for many-valued logic). This would allow easy amalgamation with other non-standard features like multiple truth values (needed, for instance, for hardware verification at the switch level [8]) or non-monotonicity [2] in a homogenous framework.
- Perspective of a constraint-based approach to non-classical deduction if integrated with ideas from [4].
- Lifting to restricted first-order versions.

References

1. R. Alur and T. A. Henzinger. Real-time logics: Complexity and expressiveness. In *Proc. 5th LICS*, pages 390–401. IEEE Press, 1990.
2. C. Bell, A. Nerode, R. Ng, and V. Subrahmanian. Mixed integer programming methods for computing nonmonotonic deductive databases. *JACM, to appear*, 1994.
3. J. Burch, E. Clarke, K. McMillan, and D. Dill. Sequential circuit verification using symbolic model checking. In *Proc. 27th Design Automation Conference (DAC 90)*, pages 46–51, 1990.
4. M. D'Agostino and D. M. Gabbay. Labelled refutation sustems: A case study. In *Proc. Second Workshop on Theorem Proving with Tableaux and Related Methods, Marseille*. Technical Report, MPI Saarbrücken, 1993.
5. M. C. Fitting. *Proof Methods for Modal and Intutionistic Logics*. Reidel, Dordrecht, 1983.
6. R. Hähnle. A new translation from deduction into integer programming. In J. Calmet and J. A. Campbell, editors, *Proc. Int. Conf. on Artificial Intelligence and Symbolic Mathematical Computing AISMC-1, Karlsruhe, Germany*, pages 262–275. Springer, LNCS 737, 1992.
7. R. Hähnle. *Automated Theorem Proving in Multiple-Valued Logics*, Oxford University Press, 1993.
8. R. Hähnle and W. Kernig. Verification of switch level designs with many-valued logic. In A. Voronkov, editor, *Proc. LPAR'93, St. Petersburg*, pages 158–169. Springer, LNAI 698, 1993.
9. P. Wolper. Temporal logic can be more expressive. *Information and Control*, 56:72–99, 1983.

A System for Automated Deduction in Graphical Interval Logic

P. M. Melliar-Smith, L. E. Moser, Y. S. Ramakrishna, G. Kutty, L. K. Dillon

Department of Electrical and Computer Engineering
Department of Computer Science
University of California, Santa Barbara, CA 93106

The Graphical Interval Logic (GIL) automated deduction system, developed at the University of California, Santa Barbara, supports specification and verification of concurrent and real-time systems. Graphical Interval Logic is a linear-time temporal logic, the key construct of which is the interval. An interval defines a context within which properties are asserted to hold and bounds the scope of temporal operators nested within it. Real-time constraints can be placed on the duration of an interval.

The GIL automated deduction system has evolved out of our previous work on the EHDM specification and verification system and on the design verification of the SIFT aircraft flight control computer. The automated deduction system for GIL includes
- A syntax-directed editor for constructing and editing graphical specifications and proofs on a workstation display
- An automated theorem prover for checking the validity of proofs in the logic and for producing counterexamples to invalid proofs
- A proof management and database system for tracking proof dependencies and for storing and retrieving graphical formulas.

The Graphical User Interface and Syntax-Directed Editor

The graphical user interface to the syntax-directed editor is shown in Figure 1 as it appears on a workstation display. The editor provides high-level editing operations based on the abstract syntax of Graphical Interval Logic formulas, and displays the formulas graphically to the user.

The user selects buttons, menus and formulas on the display with the mouse. The buttons on the upper left of the display (New, Del, Cut, Paste, etc) provide editing operations that create a new formula, delete a selected subformula, store a selected subformula in a buffer and replace a subformula with that stored subformula. The buttons on the lower left (Text, etc) allow the user to select appropriate Graphical Interval Logic constructs to apply to the currently highlighted subformula. The pull-down menus at the top of the display (File, Edit, Misc) contain commands for storing and retrieving formulas, for changing the layout of formulas, and for invoking the theorem prover. Scroll bars allow the user to view large formulas.

The editor sizes formulas to fit the available context length. If a formula does not fit into the allotted space, the user can resize the context length or the search arrows to allow the formula to be drawn correctly; all subformulas of the formula are automatically resized to scale. The user can also align corresponding points in the formulas that comprise a proof. Appropriate alignment shows how states in different

---

This research was supported in part by NSF/ARPA grant CCR-9014382.

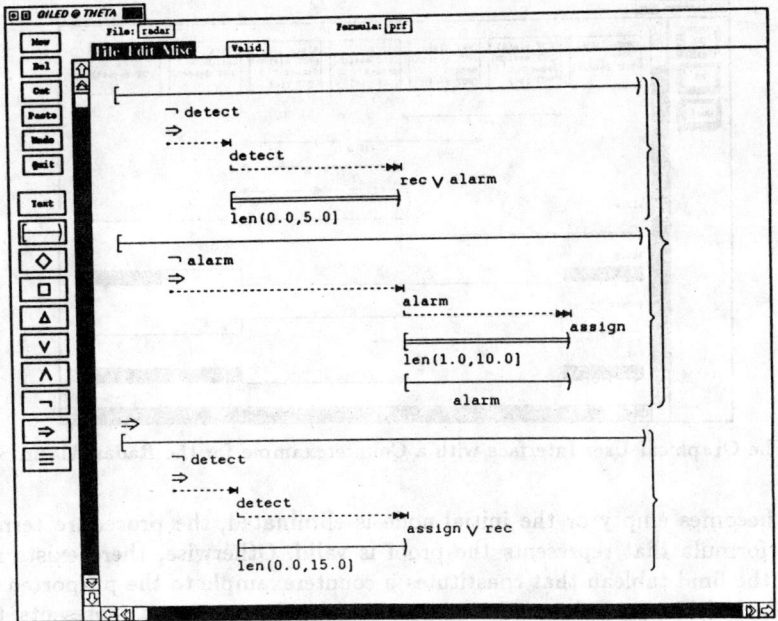

Fig. 1. The Graphical User Interface with a Valid Proof for a Radar Alarm System.

formulas are ordered relative to one another, how intervals are positioned relative to each other, and how durations of intervals are related to satisfy real-time constraints. Alignment facilitates the construction of proofs and the debugging of fallacious proofs.

The Automated Theorem Prover

To prove a theorem T, the human selects a subset of the specifications, lemmas and theorems as the premises $S_1, S_2, \ldots, S_n$ of the proof. The intelligence and understanding of the human are necessary to select a set of premises sufficient to establish the theorem but small enough to keep the proof time reasonable; intermediate lemmas and theorems may be required. The editor displays the proof as the formula $S_1 \wedge S_2 \wedge \ldots \wedge S_n \Rightarrow T$ in its graphical form. The graphical representation is converted into a Lisp S-expression, negated, and then submitted to the decision procedure for refutation.

The Graphical Interval Logic decision procedure is implemented as a tableau-theoretic method with mechanisms to handle nested interval constructs. For real-time constraints, a semantic notion of time durations is encoded into the tableau. Because the underlying time domain is dense, this construction is non-trivial; it uses concepts based on the timed automata and region graphs of Alur and Dill. The procedure begins by constructing an untimed tableau and performing the standard eventuality-based pruning of this tableau. Using the duration formulas in the nodes of the remaining tableau, it then constructs a timed tableau by adding timing constraints to the edges of the original untimed tableau. Timing consistency of this tableau is checked using Dill's algorithm. This step may eliminate some possible traces from the original tableau and, thus, a further round of eventuality-based pruning is necessary. If at any stage the

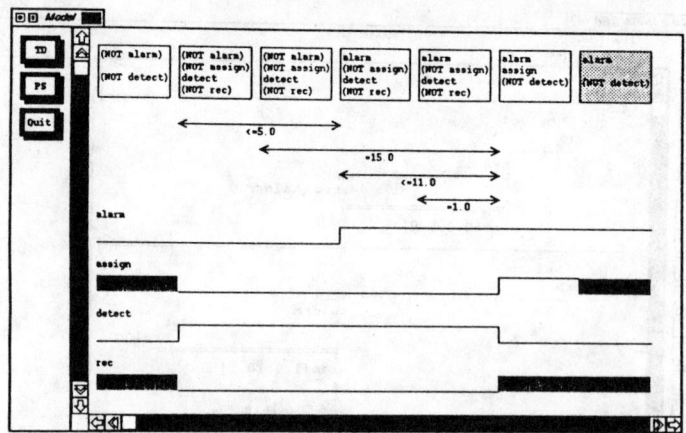

Fig. 2. The Graphical User Interface with a Counterexample for the Radar Alarm System.

tableau becomes empty or the initial node is eliminated, the procedure terminates and the formula that represents the proof is valid. Otherwise, there exists a trace through the final tableau that constitutes a counterexample to the purported proof.

If the decision procedure determines that the formula that represents the attempted proof is invalid, the user can request a counterexample. The counterexample, extracted from the final tableau, is diplayed as a sequence of states with their predicate valuations and also as a timing diagram, shown in Figure 2.

The complexity of the decision procedure for the logic without real time is $2^{O(n^{2k})}$, and for the real-time logic is $2^{O(n^{2k} \cdot 2k \cdot \lg n + n^{2k} + t \cdot \lg t)}$, where n is the number of logical connectives in the formula, k is the depth of interval nesting, and t is the binary encoding of the largest duration constant in the formula.

The Proof Management and Database System

Graphical Interval Logic formulas are stored on disk in a simple database consisting of Unix files. Several formulas can be stored in the same file by giving a unique name to each of them. The editor can be invoked to display the names of the formulas in a file and also to add a formula to a file or to delete a formula from it.

For each proof that succeeds, the proof manager records the success of the proof, the premises of the proof, and the time at which the proof was performed. The user can, thus, determine if a proof for the formula already exists and can list the premises of an existing proof. To ensure that a proof is up-to-date, the proof manager checks that neither the theorem nor any of the premises has been modified since the time of the proof. It also detects circularities in a proof and ensures that the proof dependency graph is acyclic.

Availability of the System

The GIL automated deduction system is implemented in Lucid Common Lisp on SUN Sparcstations with 32 Mbytes of main memory. The tools, documentation, and related papers can be accessed from directory /pub/GIL at nu.ece.ucsb.edu by anonymous ftp.

$\mathcal{SCDBR}$: A Reasoner for Specifications in the Situation Calculus of Database Updates.[*]

Leopoldo E. Bertossi[**] and J. Cristian Ferretti

Pontificia Universidad Catolica de Chile, Escuela de Ingenieria, Departamento de Ciencia de la Computacion, Casilla 306, Santiago 22, Chile. e-mail: {bertossi,cfs}@ing.puc.cl

In [2], Reiter presents a formalism for database updates specification. He uses the "situation calculus", that is, many-sorted first-order languages for representing knowledge about dynamically changing worlds that evolve through different states when actions are executed. In these worlds, properties of their objects depend on the current state of the world. These properties are called fluents and correspond in our context to database tables.

The specification of a dynamically changing world, like a database, typically contains the following information: (1) state independent knowledge about the objects of the world, (2) knowledge about the state of the world at the initial situation S_0, (3) preconditions for performing the different actions; a binary predicate $Poss$ is introduced to say that the execution of an action is possible at a state, (4) the immediate effects of actions, when they are possible, in terms of the fluents whose truth values are known to be changed by the execution of the action. The knowledge contained in items (1) and (2) is called the initial database $\mathcal{D}_0$.

EXAMPLE: Let us consider a library database. For reasons of space we outline the preliminary specification. In this case, language has the following ingredients: Fluents (Tables): $Classified(isbn, id, s)$, $Stock(isbn, quantity, s)$, $Unclassified(isbn, copies, s)$, $LostBook(id, s)$, $SoldOut(isbn, s)$. Update Actions (Atomic Transactions): $deleteBook(id)$, $classifyBook(isbn, id)$, $order(isbn, copies)$.
Preconditions for action $classifyBook$: $Poss(classifyBook(isbn, id), s) \equiv$
  $(\exists\, copies)\, Unclassified(isbn, copies, s) \wedge \neg(\exists\, isbn')\, Classified(isbn', id, s)$
Two effect laws:
$Poss(classifyBook(isbn, id), s) \wedge Unclassified(isbn, copies, s) \rightarrow$
  $\neg Unclassified(isbn, copies, do(classifyBook(isbn, id), s))$
$Poss(classifyBook(isbn, id), s) \rightarrow Classified(isbn, id, do(classifyBook(isbn, id), s))$
One (initial) integrity constraint:
  $Classified(isbn, id, S_0) \wedge Classified(isbn', id, S_0) \rightarrow isbn = isbn'$. ∎

An specification that contains only this kind of information does not solve the frame problem by itself. In [1], Reiter presents a simple first-order solution to the frame problem for specifications given in the situation calculus. The restriction

---

[*] Category: System Description.
[**] For many and different reasons we are grateful to Ray Reiter, Tania Bedrax, Pablo Saez, Rich Scherl, Aris Zachintinos and Javier Pinto. This project has been partially financed by DIUC, Fundacion DICTUC and FONDECYT (#1930554) Grants.

of Reiter's general solution to actions with deterministic effects is usually satisfied by database transactions. Reiter solves the frame problem, starting from a preliminary specification in the lines of the previous example, by a syntactic transformation of the given axioms, and the final specification turns out to be entirely first-order. In [2], Reiter applies his solution to the specification of database updates. The final specification contains (maybe properly contains), the following axioms: (1) for each fluent R, one "successor state axiom" of the form $(\forall a \forall s).Poss(a,s) \supset (R(do(a,s)) \equiv \Phi_R(a,s))$, (2) for each action name A, a precondition axiom of the form $Poss(A(\bar{x}),s) \leftrightarrow \pi_A(\bar{x},s)$, (3) unique names axioms for actions and states, (4) the initial database $\mathcal{D}_0$. The successor state axiom for a fluent R gives all the conditions under which the fluent becomes (or remains) true in a successor state resulting from executing an action at a predecessor state.

Our system $\mathcal{SCDBR}$ automatically derives a Reiter specification starting from a preliminary specification as described above. The system is written in PROLOG. This language is used as a programming language. The specifications handled by the system are fully first-order.

The user must specify the language first. In particular, types of objects must be declared. Formulas are internally represented as PROLOG ground terms. For example, the formula $\forall a \exists s(a = order(isbn,5) \land Stock(isbn,5,do(a,s)))$ is represented by the PROLOG ground term `forall(a): exists(s):(a eq order(isbn,5) & stock(isbn,5,do(a,s)))`. Here, the basic terms `stock(isbn,5,do(a,s))`, `forall(a)`, etc., are composed by application of the PROLOG operators `<=>,=>,&,v,:,neg,eq`, that are defined in the system. The direct effect laws are given by means of an incidence table that stores the fluents that are affected by each action. With this information at hand, the system is able to build the successor state axioms for each fluent. The PROLOG call : `ssa(fluent,X).`, instantiates X to the successor state axiom for the fluent. For example, the result from calling ssa in $\mathcal{SCDBR}$ for fluent *Classified* is:

```
| ?- ssa(classified, SSAF).
SSAF = forall(a): precond(a,s) => (classified(isbn,id,do(a,s)) <=>
       a eq classify_book(isbn,id) v (classified(isbn,id,s) &
       neg (a eq delete_book(id) & classified(isbn,id,s)))).
```

This answer is equivalent to the following successor state axiom for fluent *Classified*:

$(\forall a,s).(Poss(a,s) \supset (Classified(isbn,id,do(a,s)) \equiv a = classifyBook(isbn,id) \lor$

$Classified(isbn,id,s) \land \neg a = deleteBook(id))))$.

The head of the defining clause for ssa looks like the following:

```
ssa( F, forall(A):precond(A,S) => ( FPD <=> PHI ) :-   ...
```

The construction of the successor state axioms using this clause relies heavily on unification. The syntactical form of the successor state axiom is known, and this is reflected in the second parameter n the definition of ssa. The values for FPD (a PROLOG variable for the fluent in the successor state) and PHI are obtained executing the clauses in the body. FPD is easily constructed using the declarations for the fluent F and the specification language variables. PHI is built collecting pairs (*action, fluent precondition*) that affects F from the incidence table, and calling a clause that, starting from these pairs, builds PHI. In this process we also take advantage of its known syntax.

$\mathcal{SCDBR}$ is also able to perform other reasoning tasks like: (1) answering historical queries, (2) answering queries about the legality of sequences of transactions of the form $do(\alpha_1, do(\alpha_2, do(\alpha_3, \ldots) \ldots)))$, (3) "regressing" queries posed to a virtually updated database to a query about the initial database state, (4) doing "planning". For these last three tasks, the system processes the "list" of transactions leading to the updated database, taking into account the axioms that specify the successor states of tables in the database and preconditions on actions. This is based on a regression operator $\mathcal{R}$ proposed by Reiter in [1, 2]. For example, the regression of the formula *classified*$(isbn, id, do(order(isbn, id), s0))$ provided by the system is the following:

```
| ?- reg( classified(isbn,id,do(order(isbn,id),s0)), R ).
R = order(isbn,id) eq classify_book(isbn,id) v
    classified(isbn,id,s0) &
      neg @(order(isbn,id) eq delete_book(id) &
    classified(isbn,id,s0)).
```

We are working on extensions to the system in order to solve the following tasks: (1) updates of/through views (defined fluents), (2) interface to an Automated Theorem Prover, which is to be used mainly for proving statements with respect to initial database $\mathcal{D}_0$ (the regression operator reduces evaluation of queries posed to virtually updated databases to evaluation with respect to the initial database), (3) management of static and dynamic integrity constraints. In particular, mechanical modification of the initial specification so that the updated databases satisfy the given constraint, and mechanical proofs of integrity constraints from the specification. This last task requires inductive proofs, (4) Physical update of the database from a virtual update (i.e. from the initial database, the successor state axioms and a given sequence of transactions).

References

1. R. Reiter. *The frame problem in the situation calculus: a simple solution (sometimes) and a completeness result for goal regression.* In *Artificial Intelligence and Mathematical Theory of Computation: Papers in Honor of John McCarthy*, V. Lifschitz (ed.). Academic Press, 1991, pp. 359-380.
2. R. Reiter. *On Specifying Database Updates.* Technical Report KRR-TR-92-3, Department of Computer Science, University of Toronto, 1992

Authors Index

Avau, I. 165
Badaloni, S. 101
Baeten, J.C.M. 30
Barringer, H. 415
Van Belleghem, K. 301
Berati, M. 101
Bergstra, J.A. 30
Bernholtz, O. 210
Bertossi, L.E. 543
Blackburn, P 225
Böhlen, M. 283
Bol, R.N. 30
Bonner, A.J. 67
Chomicki, J. 506
Denecker, M. 301
Di Maio, M.C. 265
Dillon, L.K. 195, 540
Dixon, C. 415
Doherty, P. 82
Felder, M. 365
Ferretti, J.C. 543
Fiadeiro, J.L. 48
Fisher, M. 317, 415, 480
Gardent, C 225
Goranko, V. 133
Grumberg, O. 180, 210
Hähnle, R. 535
Hansen, O.E. 1
Hwang, C.H. 238
Ibens, O. 535
Katz, S. 17
Kifer, M. 67
Kurshan, R.P. 180
Kutty, G. 195, 540
Kwiatkowska, M. 398
Lewi, J. 165
Løvengreen, H.H. 1
Łukaszewicz, W. 82
Ma, W. 445
Maibaum, T. 48
McGuire, H. 430
Manna, Z. 430
Marti, R. 283
Melliar-Smith, P.M. 195, 540
Méry, D. 382
Mokkedem, A. 382
Morzenti, A. 365
Moser, L.E. 195, 540
Orgun, M.A. 445
Peled, D. 398
Penczek, W. 398
Porto, A. 349
Poté, A. 165
Ramakrishna, Y.S. 195, 540
Reynolds, M. 117
Ribeiro, C. 349
de Rijke, M. 225
De Schreye, D. 301
Schubert, L.K. 238
Sørensen, M.U. 1
Venema, Y. 149
Vergauwen, B. 165
Waldinger, R. 430
Wooldridge, M. 317
Xu, M. 332
Zanardo, A. 265

Lecture Notes in Artificial Intelligence (LNAI)

Vol. 622: F. Schmalhofer, G. Strube, Th. Wetter (Eds.), Contemporary Knowledge Engineering and Cognition. Proceedings, 1991. XII, 258 pages. 1992.

Vol. 624: A. Voronkov (Ed.), Logic Programming and Automated Reasoning. Proceedings, 1992. XIV, 509 pages. 1992.

Vol. 627: J. Pustejovsky, S. Bergler (Eds.), Lexical Semantics and Knowledge Representation. Proceedings, 1991. XII, 381 pages. 1992.

Vol. 633: D. Pearce, G. Wagner (Eds.), Logics in AI. Proceedings. VIII, 410 pages. 1992.

Vol. 636: G. Comyn, N. E. Fuchs, M. J. Ratcliffe (Eds.), Logic Programming in Action. Proceedings, 1992. X, 324 pages. 1992.

Vol. 638: A. F. Rocha, Neural Nets. A Theory for Brains and Machines. XV, 393 pages. 1992.

Vol. 642: K. P. Jantke (Ed.), Analogical and Inductive Inference. Proceedings, 1992. VIII, 319 pages. 1992.

Vol. 659: G. Brewka, K. P. Jantke, P. H. Schmitt (Eds.), Nonmonotonic and Inductive Logic. Proceedings, 1991. VIII, 332 pages. 1993.

Vol. 660: E. Lamma, P. Mello (Eds.), Extensions of Logic Programming. Proceedings, 1992. VIII, 417 pages. 1993.

Vol. 667: P. B. Brazdil (Ed.), Machine Learning: ECML – 93. Proceedings, 1993. XII, 471 pages. 1993.

Vol. 671: H. J. Ohlbach (Ed.), GWAI-92: Advances in Artificial Intelligence. Proceedings, 1992. XI, 397 pages. 1993.

Vol. 679: C. Fermüller, A. Leitsch, T. Tammet, N. Zamov, Resolution Methods for the Decision Problem. VIII, 205 pages. 1993.

Vol. 681: H. Wansing, The Logic of Information Structures. IX, 163 pages. 1993.

Vol. 689: J. Komorowski, Z. W. Raś (Eds.), Methodologies for Intelligent Systems. Proceedings, 1993. XI, 653 pages. 1993.

Vol. 695: E. P. Klement, W. Slany (Eds.), Fuzzy Logic in Artificial Intelligence. Proceedings, 1993. VIII, 192 pages. 1993.

Vol. 698: A. Voronkov (Ed.), Logic Programming and Automated Reasoning. Proceedings, 1993. XIII, 386 pages. 1993.

Vol. 699: G.W. Mineau, B. Moulin, J.F. Sowa (Eds.), Conceptual Graphs for Knowledge Representation. Proceedings, 1993. IX, 451 pages. 1993.

Vol. 723: N. Aussenac, G. Boy, B. Gaines, M. Linster, J.-G. Ganascia, Y. Kodratoff (Eds.), Knowledge Acquisition for Knowledge-Based Systems. Proceedings, 1993. XIII, 446 pages. 1993.

Vol. 727: M. Filgueiras, L. Damas (Eds.), Progress in Artificial Intelligence. Proceedings, 1993. X, 362 pages. 1993.

Vol. 728: P. Torasso (Ed.), Advances in Artificial Intelligence. Proceedings, 1993. XI, 336 pages. 1993.

Vol. 743: S. Doshita, K. Furukawa, K. P. Jantke, T. Nishida (Eds.), Algorithmic Learning Theory. Proceedings, 1992. X, 260 pages. 1993.

Vol. 744: K. P. Jantke, T. Yokomori, S. Kobayashi, E. Tomita (Eds.), Algorithmic Learning Theory. Proceedings, 1993. XI, 423 pages. 1993.

Vol. 745: V. Roberto (Ed.), Intelligent Perceptual Systems. VIII, 378 pages. 1993.

Vol. 746: A. S. Tanguiane, Artificial Perception and Music Recognition. XV, 210 pages. 1993.

Vol. 754: H. D. Pfeiffer, T. E. Nagle (Eds.), Conceptual Structures: Theory and Implementation. Proceedings, 1992. IX, 327 pages. 1993.

Vol. 764: G. Wagner, Vivid Logic. XII, 148 pages. 1994.

Vol. 766: P. R. Van Loocke, The Dynamics of Concepts. XI, 340 pages. 1994.

Vol. 770: P. Haddawy, Representing Plans Under Uncertainty. X, 129 pages. 1994.

Vol. 784: F. Bergadano, L. De Raedt (Eds.), Machine Learning: ECML-94. Proceedings, 1994. XI, 439 pages. 1994.

Vol. 795: W. A. Hunt, Jr., FM8501: A Verified Microprocessor. XIII, 333 pages. 1994.

Vol. 798: R. Dyckhoff (Ed.), Extensions of Logic Programming. Proceedings, 1993. VIII, 360 pages. 1994.

Vol. 799: M. P. Singh, Multiagent Systems: Intentions, Know-How, and Communications. XXIII, 168 pages. 1994.

Vol. 804: D. Hernández, Qualitative Representation of Spatial Knowledge. IX, 202 pages. 1994.

Vol. 808: M. Masuch, L. Pólos (Eds.), Knowledge Representation and Reasoning Under Uncertainty. VII, 237 pages. 1994.

Vol. 810: G. Lakemeyer, B. Nebel (Eds.), Foundations of Knowledge Representation and Reasoning. VIII, 355 pages. 1994.

Vol. 814: A. Bundy (Ed.), Automated Deduction — CADE-12. Proceedings, 1994. XVI, 848 pages. 1994.

Vol. 822: F. Pfenning (Ed.), Logic Programming and Automated Reasoning. Proceedings, 1994. X, 345 pages. 1994.

Vol. 827: D. M. Gabbay, H. J. Ohlbach (Eds.), Temporal Logic. Proceedings, 1994. XI, 546 pages. 1994.

Lecture Notes in Computer Science

Vol. 795: W. A. Hunt, Jr., FM8501: A Verified Microprocessor. XIII, 333 pages. 1994. (Subseries LNAI).

Vol. 796: W. Gentzsch, U. Harms (Eds.), High-Performance Computing and Networking. Proceedings, 1994, Vol. I. XXI, 453 pages. 1994.

Vol. 797: W. Gentzsch, U. Harms (Eds.), High-Performance Computing and Networking. Proceedings, 1994, Vol. II. XXII, 519 pages. 1994.

Vol. 798: R. Dyckhoff (Ed.), Extensions of Logic Programming. Proceedings, 1993. VIII, 360 pages. 1994. (Subseries LNAI).

Vol. 799: M. P. Singh, Multiagent Systems: Intentions, Know-How, and Communications. XXIII, 168 pages. 1994. (Subseries LNAI).

Vol. 800: J.-O. Eklundh (Ed.), Computer Vision – ECCV '94. Proceedings 1994, Vol. I. XVIII, 603 pages. 1994.

Vol. 801: J.-O. Eklundh (Ed.), Computer Vision – ECCV '94. Proceedings 1994, Vol. II. XV, 485 pages. 1994.

Vol. 802: S. Brookes, M. Main, A. Melton, M. Mislove, D. Schmidt (Eds.), Mathematical Foundations of Programming Semantics. Proceedings, 1993. IX, 647 pages. 1994.

Vol. 803: J. W. de Bakker, W.-P. de Roever, G. Rozenberg (Eds.), A Decade of Concurrency. Proceedings, 1993. VII, 683 pages. 1994.

Vol. 804: D. Hernández, Qualitative Representation of Spatial Knowledge. IX, 202 pages. 1994. (Subseries LNAI).

Vol. 805: M. Cosnard, A. Ferreira, J. Peters (Eds.), Parallel and Distributed Computing. Proceedings, 1994. X, 280 pages. 1994.

Vol. 806: H. Barendregt, T. Nipkow (Eds.), Types for Proofs and Programs. VIII, 383 pages. 1994.

Vol. 807: M. Crochemore, D. Gusfield (Eds.), Combinatorial Pattern Matching. Proceedings, 1994. VIII, 326 pages. 1994.

Vol. 808: M. Masuch, L. Pólos (Eds.), Knowledge Representation and Reasoning Under Uncertainty. VII, 237 pages. 1994. (Subseries LNAI).

Vol. 809: R. Anderson (Ed.), Fast Software Encryption. Proceedings, 1993. IX, 223 pages. 1994.

Vol. 810: G. Lakemeyer, B. Nebel (Eds.), Foundations of Knowledge Representation and Reasoning. VIII, 355 pages. 1994. (Subseries LNAI).

Vol. 811: G. Wijers, S. Brinkkemper, T. Wasserman (Eds.), Advanced Information Systems Engineering. Proceedings, 1994. XI, 420 pages. 1994.

Vol. 812: J. Karhumäki, H. Maurer, G. Rozenberg (Eds.), Results and Trends in Theoretical Computer Science. Proceedings, 1994. X, 445 pages. 1994.

Vol. 813: A. Nerode, Y. N. Matiyasevich (Eds.), Logical Foundations of Computer Science. Proceedings, 1994. IX, 392 pages. 1994.

Vol. 814: A. Bundy (Eds.), Automated Deduction—CADE-12. Proceedings, 1994. XVI, 848 pages. 1994. (Subseries LNAI).

Vol. 815: R. Valette (Eds.), Application and Theory of Petri Nets 1994. Proceedings. IX, 587 pages. 1994.

Vol. 816: J. Heering, K. Meinke, B. Möller, T. Nipkow (Eds.), Higher-Order Algebra, Logic, and Term Rewriting. Proceedings, 1993. VII, 344 pages. 1994.

Vol. 817: C. Halatsis, D. Maritsas, G. Philokyprou, S. Theodoridis (Eds.), PARLE '94. Parallel Architectures and Languages Europe. Proceedings, 1994. XV, 837 pages. 1994.

Vol. 818: D. L. Dill (Ed.), Computer Aided Verification. Proceedings, 1994. IX, 480 pages. 1994.

Vol. 819: W. Litwin, T. Risch (Eds.), Applications of Databases. Proceedings, 1994. XII, 471 pages. 1994.

Vol. 820: S. Abiteboul, E. Shamir (Eds.), Automata, Languages and Programming. Proceedings, 1994. XIII, 644 pages. 1994.

Vol. 821: M. Tokoro, R. Pareschi (Eds.), Object-Oriented Programming. Proceedings, 1994. XI, 535 pages. 1994.

Vol. 822: F. Pfenning (Ed.), Logic Programming and Automated Reasoning. Proceedings, 1994. X, 345 pages. 1994. (Subseries LNAI).

Vol. 823: R. A. Elmasri, V. Kouramajian, B. Thalheim (Eds.), Entity-Relationship Approach — ER '93. Proceedings, 1993. X, 531 pages. 1994.

Vol. 824: E. M. Schmidt, S. Skyum (Eds.), Algorithm Theory - SWAT '94. Proceedings. IX, 383 pages. 1994.

Vol. 825: J. L. Mundy, A. Zisserman, D. Forsyth (Eds.), Applications of Invariance in Computer Vision. Proceedings, 1993. IX, 510 pages.

Vol. 826: D. S. Bowers (Ed.), Directions in Databases. Proceedings, 1994. X, 234 pages. 1994.

Vol. 827: D. M. Gabbay, H. J. Ohlbach (Eds.), Temporal Logic. Proceedings, 1994. XI, 546 pages. 1994. (Subseries LNAI).

Vol. 828: L. C. Paulson, Isabelle. XVII, 321 pages. 1994.

Vol. 829: A. Chmora, S. B. Wicker (Eds.), Error Control, Cryptology, and Speech Compression. Proceedings, 1993. VIII, 121 pages. 1994.

Vol. 831: V. Bouchitté, M. Morvan (Eds.), Orders, Algorithms, and Applications. Proceedings, 1994. IX, 204 pages. 1994.